Compact Modeling

Gennady Gildenblat
Editor

Compact Modeling

Principles, Techniques and Applications

Springer

Editor
Dr. Gennady Gildenblat
Motorola Professor of Electrical Engineering
Arizona State University
University Drive and Mill Avenue
Tempe, AZ 85287-9309
USA
Gennady.Gildenblat@asu.edu

ISBN 978-90-481-8613-6 e-ISBN 978-90-481-8614-3
DOI 10.1007/978-90-481-8614-3
Springer Dordrecht Heidelberg London New York

Library of Congress Control Number: 2010929773

Cover design: eStudio Calamar

Printed on acid-free paper

Springer is part of Springer Science+Business Media (www.springer.com)

Preface

Models of circuit elements which are sufficiently simple to be incorporated in circuit simulators and are sufficiently accurate to make the outcome useful to circuit designers are called compact. The conflicting objectives of model simplicity and accuracy make the compact modeling field an exciting and challenging research area for device physicists, electronic engineers and applied mathematicians. Continued down-scaling of semiconductor devices has made it necessary to incorporate new physical phenomena, while extended applications have led to the inclusion of the secondary and ternary effects in order to achieve the required model accuracy. In addition several rigid requirements in terms of model continuity and qualitative behavior ("benchmarks") have been imposed over the years. At the same time, the increased size of the integrated circuits, that can now be subjected to the full SPICE analysis, disallowed proportional increase in the model execution time. Hence considerable effort went into compact model reformulation in such a way that dramatically increased accuracy and model sophistication are accomplished without prohibitive decrease in the computational efficiency.

The models of MOS transistors underwent revolutionary change in the last few years and are now based on new principles. The recent models of diodes, passive elements, noise sources and bipolar transistors were developed along the more traditional lines. Following this evolutionary development they became highly sophisticated and much more capable to reflect the increased demands of the advanced integrated circuit technology. The latter depends on the compact models for the shortening of the design cycle and eliminating the elements of overdesign which is often undesirable in today's competitive environment. At the same time, statistical modeling of semiconductor devices received new significance following the dramatic reduction of the device dimensions and of the power supply voltage. Finally, despite the complexity of the fabrication process, the multi-gate MOS transistors are now seriously considered for the purpose of controlling the small geometry effects. To evaluate the potential impact of these devices on the IC design, compact models of these devices, preferably based on the same principles as the models of traditional MOSFETs, are becoming important.

The last comprehensive description of the compact models of various semiconductor devices reflects the state of the art in late 1980s [2]. While remaining an influential and valuable source of information, it can no longer serve the needs of the compact modeling community in the 21st century. In this volume we present contemporary compact models of both active and passive semiconductor devices and discuss general modeling techniques. Given the limited size of the book, and the wide choice of compact models available today, the selection of the topics represented an interesting problem. In the end, it was decided to include compact models which are both heavily used in the industry and at the same time are theoretically significant. For experimental devices like multiple-gate transistors where the optimal choice of the compact models is not yet entirely clear, we have presented two particularly promising approaches. We have also included chapters on the MOSFET noise theory, benchmarking of MOSFET compact models, modeling of the power MOSFET, and overview of the bipolar modeling field. The book concludes with two chapters describing the variability modeling including some recent developments in the field. Once again, since the field of variability modeling is far from maturity, we present several alternative approaches to the subject.

The present volume is comprehensive but is in no way encyclopedic. It does not include several popular and experimental models that have already been described in considerable detail in a book form [1, 5, 6, 8, 11, 13, 15]. Another important subject that is not covered, is the application of behavioral languages in the compact modeling field. Many of the models discussed in this book (e.g. PSP, PSP-SOI, MOSVAR, R3) were developed using verilog-A language which not only significantly reduces the development time, but eliminates numerous potential sources of errors and simplifies model implementation in circuit simulators [10, 12, 14]. Furthermore, we concentrate on the most recent developments in the compact modeling. The reader interested in the evolution of the field can find extensive information in [2, 3, 7, 9] while [4] contains a useful tutorial on compact MOSFET models.

We hope that this volume will be useful to the engineers actively involved in the design of integrated circuits, developing compact models and for the graduate students of the corresponding disciplines.

Arizona State University, Tempe, AZ, USA Gennady Gildenblat

References

1. Aaen, P.H., Plá, J.A., Wood, J.: Modeling and Characterization of RF and Microwave Power FETs. Cambridge University Press, Cambridge (2007)
2. Antognetti, P., Massobrio, G., Massobrio, G.: Semiconductor Device Modeling with SPICE. McGraw-Hill, New York (1993)
3. Arora, N.D.: MOSFET Models for VLSI Circuit Simulation: Theory and Practice. Springer, New York (1993)
4. Bhattacharyya, A.B.: Compact MOSFET Models for VLSI Design. Wiley, New York (2009)
5. Cheng, Y., Hu, C.: MOSFET Modeling and BSIM3 User's Guide. Kluwer Academic, Boston (1999)
6. Enz, C., Vittoz, E.: Charge-Based MOS Transistor Modeling. Wiley, New York (2006)
7. Foty, D.: MOSFET Modeling with SPICE: Principles and Practice. Prentice-Hall, Upper Saddle River (1997)

8. Galup-Montoro, C., Schneider, M.: MOSFET Modeling for Circuit Analysis and Design. World Scientific, Singapore (2007)
9. Graaff, H., Klaassen, F.: Compact Transistor Modelling for Circuit Design. Springer, Berlin (1990)
10. Lemaitre, L., McAndrew, C., Hamm, S.: ADMS: Automated Device Model Synthesize, pp. 27–30 (2002)
11. Liu, W.: MOSFET Models for SPICE Simulation Including BSIM3v3 and BSIM4. Wiley, New York (2001)
12. McAndrew, C.C.: The real barrier to standardized compact models. In: Proceedings of 8th International Conference on Mixed Design of Integrated Circuits and Systems (MIXDES), pp. 53–58 (2001)
13. Miura-Mattausch, M., Mattausc, H.J., Ezaki, T.: The Physics and Modeling of MOSFETs: Surface-Potential Model HiSIM. World Scientific, Singapore (2008)
14. Troyanovsky, B., O'Halloran, P., Mierzwinski, M.: Compact modeling in Verilog-A. In: Grabinski, W., Nauwelaers, B., Schreurs, D. (eds.) Transistor Level Modeling for Analog/RF IC Design, pp. 271–291. Springer, Berlin (2001)
15. Ytterdal, T., Cheng, Y., Fjeldly, T.: Device Modeling for Analog and RF CMOS Circuit Design. Wiley, New York (2003)

Contents

Part II Compact Models of Bipolar Junction Transistors

Part I
Compact Models of MOS Transistors

Chapter 1
Surface-Potential-Based Compact Model of Bulk MOSFET

Gennady Gildenblat, Weimin Wu, Xin Li, Ronald van Langevelde, Andries J. Scholten, Geert D.J. Smit, and Dirk B.M. Klaassen

Abstract We review surface-potential-based approach to compact modeling of bulk MOS transistors and provide introduction to the widely used PSP model jointly developed by the Arizona State University and NXP Semiconductors. The emphasis is on the interplay between the mathematical structure of the compact model and its capabilities for the circuit design applications.

1.1 Introduction

There is presently a widely held consensus in the industry that the surface-potential-based models represent the best approach to the modeling of bulk and SOI MOS devices. This paradigm change came after careful and deliberate evaluation of different approaches to compact modeling [18, 81] and was enabled by finding innovative solutions for several long-standing problems of compact modeling such as the computationally efficient evaluation of the surface potential and mathematical techniques for incorporating small-geometry effects within the context of the surface-potential-based approach. The motivation for switching from the threshold-voltage-based to surface-potential-based approach comes from the desire to increase the physical content of the compact model and subsequently make it more suitable for modeling advanced MOS devices including low-V_{dd}, analog and RF applications where traditional compact models are not compatible with the circuit design

G. Gildenblat (✉) · W. Wu · X. Li
Ira A. Fulton School of Engineering, Department of Electrical Engineering, Arizona State University, Tempe, AZ 85287, USA
e-mail: gildenblat@asu.edu

R. van Langevelde
Philips Research Europe, 5656 AE Eindhoven, The Netherlands

A.J. Scholten · G.D.J. Smit · D.B.M. Klaassen
NXP-TSMC Research Center, 5656 AE Eindhoven, The Netherlands

G. Gildenblat (ed.), *Compact Modeling*,
DOI 10.1007/978-90-481-8614-3_1, © Springer Science+Business Media B.V. 2010

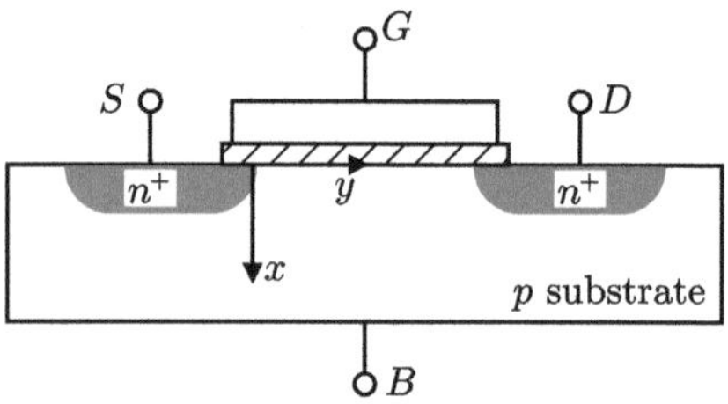

Fig. 1.1 A MOSFET cross-section

requirements in terms of qualitative behavior, $C(V)$ characteristics [36] or model symmetry [3, 31, 67]. The PSP model jointly developed by Arizona State University and NXP is the most advanced MOSFET model incorporating the surface-potential-based approach and has been verified and used in circuit design for the technology nodes from 250 to 32 nm. The description of the early versions of PSP can be found in [68] and [21] in a somewhat condensed form. In this chapter, we provide a more detailed description of the theoretical foundation of PSP, including extension of the symmetric linearization method for all regions of operation and some recent results, such as the origin of the surface potential equation [22], new circuit applications, and modeling of the non-uniformly doped devices. This allows us to examine and illustrate the main features of the surface-potential-based model formulation using PSP as a case study. In what follows, we concentrate on the quasi-static PSP model. The non-quasi-static version is described in [30, 37, 52, 79] and is also surface-potential-based.

1.2 Surface Potential Equation

Surface potential, ψ_s, i.e. the potential at the Si/SiO$_2$ interface is an implicit function of the terminal voltages that is usually obtained by solving the surface potential equation (SPE) occasionally known also as the input voltage equation. It is derived under several simplifying assumptions essential in the compact model formulation. The first simplification is the Shockley's gradual channel approximation (GCA) that assumes (cf. Fig. 1.1)

$$\left|\frac{\partial^2 \psi}{\partial y^2}\right| \ll \left|\frac{\partial^2 \psi}{\partial x^2}\right| \tag{1.1}$$

where ψ denotes the electrostatic potential. With this simplification the Poisson equation becomes

$$\frac{\partial^2 \psi}{\partial x^2} = -\frac{\rho}{\varepsilon_s} \tag{1.2}$$

where ρ denotes the charge density and ε_s is the dielectric permittivity of silicon. Denoting the electron and hole concentrations as n and p respectively,

$$\rho = q(p - n - N_a^-) \tag{1.3}$$

where N_a^- is the concentration of ionized acceptors and we consider an n-channel MOS device. Since the hole current component is negligible, so is the gradient of the hole imref F_p and the hole concentration is given by the Boltzmann relation $p = p_b \exp(-\beta\psi)$ where $\beta = 1/\phi_t$ and ϕ_t denotes the thermal potential. This form assumes that the reference point for the potential is in the neutral bulk region where the majority and minority carrier concentrations are p_b and n_b respectively. For electrons it is necessary to take into account the imref gradient so that

$$n = n_b \exp[\beta(\psi - \phi_n)] \tag{1.4}$$

where the normalized imref splitting (a.k.a."channel voltage")

$$\phi_n = (1/q)(F_p - F_n). \tag{1.5}$$

For most applications it is sufficient to assume the complete ionization of the channel dopants. Then N_a^- coincides with the total acceptor concentration $N_a = p_b - n_b$ (for the uniformly doped channel). Hence

$$\rho = q\left[p_b\left(e^{-\beta\psi} - 1\right) - n_b\left(ke^{\beta\psi} - 1\right)\right] \tag{1.6}$$

where

$$k = \exp\left(-\beta\phi_n\right). \tag{1.7}$$

From (1.2) and the boundary condition $\partial\psi/\partial x = 0$ for $\psi = 0$ it follows that

$$E_s^2 = -\frac{2}{\varepsilon_s}\int_0^{\psi_s} \rho\, d\psi \tag{1.8}$$

where the surface electric field $E_s = -(d\psi/dx)_{x=0}$. Continuity of the normal component of the displacement vector at the Si/SiO$_2$ interface provides SPE in the form

$$(V_{gb} - V_{fb} - \psi_s)^2 = \gamma^2\phi_t h \tag{1.9}$$

where V_{fb} is the flat-band voltage,

$$\gamma = \sqrt{2q\varepsilon_s p_b}/C_{ox} \tag{1.10}$$

denotes the body factor, unit area oxide capacitance $C_{ox} = \varepsilon_{ox}/t_{ox}$, t_{ox} is the oxide thickness and ε_{ox} is the oxide permittivity. The dimensionless variable h represents the normalized square of the surface electric field:

$$h = \frac{\varepsilon_s E_s^2}{2q\phi_t p_b} = -\frac{1}{q\phi_t p_b}\int_0^{\psi_s} \rho\, d\psi. \tag{1.11}$$

With the exception of the special case of an MOS capacitor where $F_n = F_p$ and $k = 1$, this integral with the charge density (1.6) cannot be obtained in a closed form and further approximations are inevitable in a compact model. The simplest

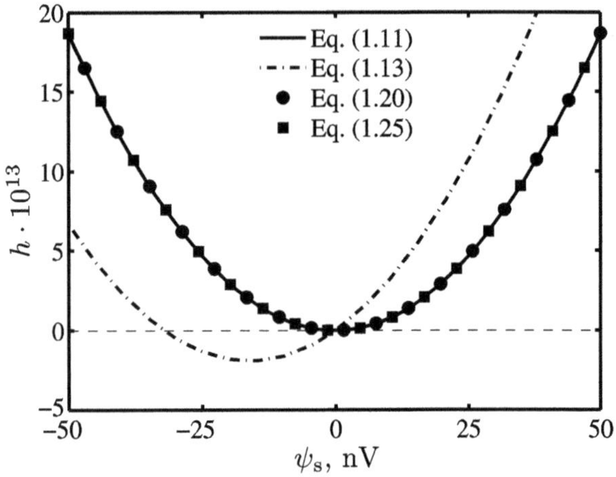

Fig. 1.2 Normalized square of the electrical field as a function of surface potential; $N_a = 2.2 \cdot 10^{17}$ cm^{-3}, $t_{ox} = 2$ nm, and $V_{bs} = 0.47$ V

approximation is obtained by neglecting the position dependence of the electron imref in which case $k = k_0$ where

$$k_0 = \exp(-\beta\phi_{n0}) \tag{1.12}$$

is determined by the imref splitting at the interface. Then $h = h_{PS}$ where [44]

$$h_{PS} = e^{-u} + u - 1 + \frac{n_b}{p_b} k_0 \left(e^u - \frac{u}{k_0} - 1 \right) \tag{1.13}$$

and $u = \beta\psi_s$ is the normalized surface potential. This result reproduced in classic textbooks (e.g. [61]), is exact for MOS capacitors and works well for MOSFETs except for a narrow region near the flat-band (i.e. near $\psi_s = 0$) where h_{PS} becomes negative [40, 83]. This property of (1.13) is incompatible with SPE (1.9) or physical interpretation of h (as the normalized square of the surface electric field) based on (1.11) is illustrated in Fig. 1.2 and can be explained as follows. For $u \to 0$ expansion

$$h_{PS} = (n_b/p_b)(k_0 - 1)u + O(u^2) \tag{1.14}$$

contains a linear term and may become negative if u and $k_0 - 1$ have different signs. As pointed out in [83], the physical origin of this difficulty lies in the assumption $F_n = F_{n0}$, i.e. $\partial\phi_n/\partial x = 0$ made in the derivation of (1.13). In reality, imref splitting ϕ_n is a strong function of both x and y [23, 61] with the standard approximation $\phi_n = \phi_n(y)$ introducing spurious behavior near the flat-band. Despite the fact that the region where $h_{PS} < 0$ is extremely narrow, it may result in the occasional non-convergence of circuit simulations.

We now proceed without assuming $\partial\phi_n/\partial x = 0$. From (1.11)

$$h(u) = e^{-u} + u - 1 + \frac{n_b}{p_b} \left[\int_0^u k(e^w - 1)\,dw + \int_0^u (k - 1)\,dw \right]. \tag{1.15}$$

Since the function $e^w - 1$ does not change sign on the interval $(0, u)$ (or $(u, 0)$ if $u < 0$), with reference to the first mean value theorem of the integral calculus one has

$$\int_0^u k(e^w - 1)\,dw = k(\xi)\int_0^u (e^w - 1)\,dw; \quad \xi \in (0, u) \tag{1.16}$$

so that [22]

$$h(u) = e^{-u} + u - 1 + \frac{n_b}{p_b}\left[k(\xi)\left(e^u - u - 1\right) + (k_{av} - 1)u\right] \tag{1.17}$$

where

$$k_{av} = \frac{1}{u}\int_0^u k\,dw. \tag{1.18}$$

Comparison with (1.13) shows that h_{PS} is obtained from (1.17) using approximation $k(\xi) = k_{av} = k_0$ which is an immediate consequence of neglecting $\phi_n(x)$ dependence. To improve this result we look for the best approximation of the type $k(\xi) = k_1$ and $k_{av} = k_2$ where coefficients k_1 and k_2 still do not depend on u but are allowed to have different values. Requiring $h(u) = h_{PS}(u)$ in strong inversion region where $h_{PS}(u)$ is physically sound we have $k_1 = k_0$. The second condition

$$h(0) = 0 \quad \text{and} \quad h(u) > 0; \quad u \neq 0 \tag{1.19}$$

implies $k_2 = 1$ whence

$$h = e^{-u} + u - 1 + \frac{n_b}{p_b}k_0\left(e^u - u - 1\right). \tag{1.20}$$

The analysis leading to this result is a variation of that in [22, 40]. It has been recently extended to account for the impurity freeze-out that is essential in low-temperature CMOS applications [22].

The results of the proposed modification of the SPE are shown in Fig. 1.2 illustrating condition (1.19) satisfied by (1.20). Note that apart from the narrow region near $\psi_s = 0$ the difference between (1.13) and (1.20) is small and consequently is invisible to the model user except for eliminating the occasional simulation crash.

Since the SPE (1.9) does not allow for an exact solution, it is solved using either iterative Newton-Raphson procedure [4, 41, 47] or analytical approximations for the surface potential [10, 19–21, 68, 70]. As shown in Fig. 1.3, the approximation used in [10, 84, 85] is particularly accurate leading to the absolute error under 1 nV. As shown in Fig. 1.4, this approximation is sufficiently accurate to evaluate $\Delta\psi$ including the subthreshold region where

$$\Delta\psi \propto \exp(-\beta\psi_{ss})\left[1 - \exp(-\beta V_{ds})\right] \tag{1.21}$$

is exceedingly small [26].

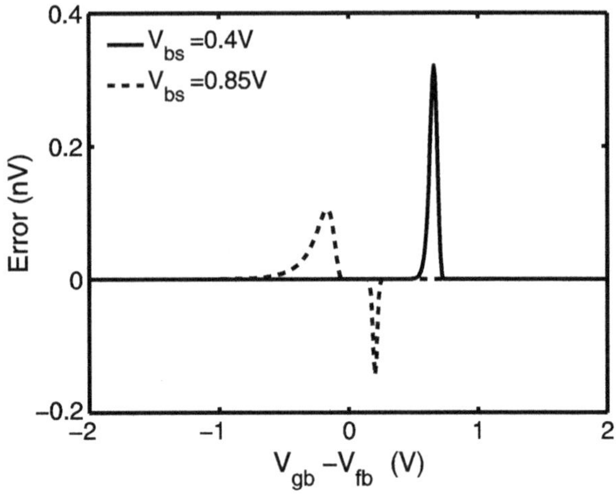

Fig. 1.3 The error in surface potential calculated from analytical approximation relative to the numerical solution for different forward bias; $N_a = 5 \cdot 10^{17}$ cm^{-3}, $t_{ox} = 2$ nm and $T = 300$ K

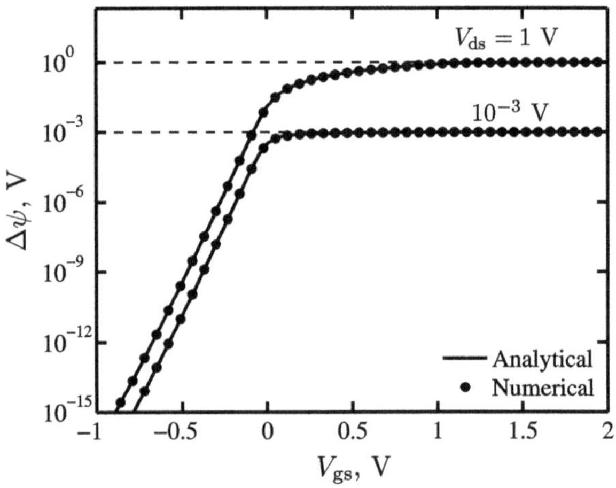

Fig. 1.4 $\Delta\psi$ as a function of the gate bias for various drain-source voltages; $t_{ox} = 2$ nm, $N_a = 10^{17}$ cm^{-3}, $V_{sb} = 0$ V and $T = 300$ K. *Circles* represent the numerical solution of (1.20), while *solid lines* are obtained using analytical approximations for ψ_{ss} and ψ_{sd} [21, 84]

SPE (1.9) with h given by (1.20) implies that derivative

$$\left(\frac{\partial \psi_s}{\partial V_{gb}}\right)_{V_{gb}=V_{fb}} = \frac{1}{1+\gamma\sqrt{(\beta/2)(1+\Delta)}} \tag{1.22}$$

where (ϕ_b denotes the "bulk potential")

$$\Delta = \exp[-\beta(2\phi_b + \phi_n)] \tag{1.23}$$

has different values at the source ($\phi_n = V_{sb}$) and drain ($\phi_n = V_{db}$) ends of the channel. Hence

$$\left(\frac{\partial \Delta\psi}{\partial V_{gb}}\right)_{V_{gb}=V_{fb}} \neq 0 \tag{1.24}$$

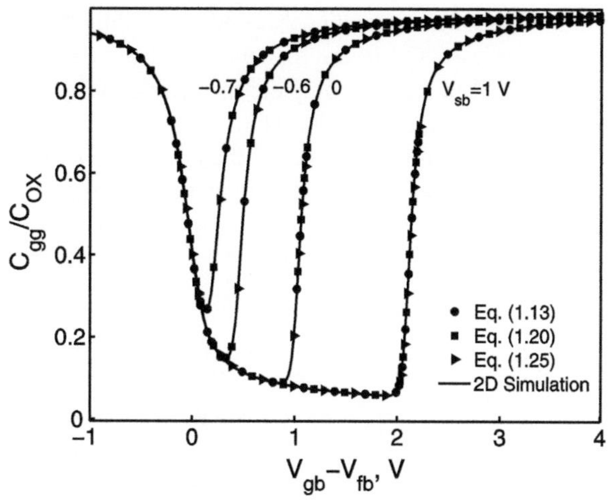

Fig. 1.5 Normalized capacitances computed using (1.13), (1.20) and (1.25). After [83]

where $\Delta\psi = \psi_{sd} - \psi_{ss}$ is the surface potential variation across the channel. This makes it impossible to simplify the model by setting $\Delta\psi = 0$ in accumulation region where $\Delta\psi$ is negligible.

To achieve this simplification in the PSP model h is selected in the form [84]

$$h = e^{-u} + u - 1 + \frac{n_b}{p_b} k_0 \left(e^u - u - 1 - \frac{u^2}{u^2 + 1} \right). \tag{1.25}$$

With this modification, condition (1.19) remains in place but the derivative

$$\left(\frac{\partial \psi_s}{\partial V_{gb}} \right)_{V_{gb}=V_{fb}} = \frac{1}{1 + \gamma\sqrt{\beta/2}} \tag{1.26}$$

does not depend on Δ and remains the same at both ends of the channel enabling approximation $\Delta\psi = 0$ for $V_{gb} < V_{fb}$. As shown in Fig. 1.5 the difference between (1.20) and (1.25) does not affect the device output characteristics and is a matter of convenience.

1.3 Symmetric Linearization Method

All compact surface-potential-based MOSFET models are based on the charge-sheet approximation [5, 64] justified by comparison with the Pao-Sah double integration formula [44]. Since the expressions for current and especially for the terminal charges in Brews' model are quite complicated [40, 77] they need to be simplified before implementation in the compact models [20, 21, 74]. Symmetric linearization method (SLM) developed in [11, 77] (also mentioned in [73]) represents a systematic way to address this problem. The original work in [11, 77] and its subsequent exposition [21, 68] deal with a simplified formulation applicable when the surface potential satisfies inequality $\psi_s > 3\phi_t$ excluding flat-band condition and accumulation region essential in the formulation of the complete model. In this section, we present SLM in a more complete form which is actually used in the formulation of the SP and PSP models [20, 21].

For this purpose, we introduce the inversion, q_i and bulk, q_b charges per unit area normalized to unit area oxide capacitance. Charge neutrality implies

$$q_i = -(V_{gb} - V_{fb} - \psi_s) - q_b \tag{1.27}$$

where we used the physical signs for q_i and q_b. In its usual form the charge-sheet approximation assumes that depletion theory expression

$$q_b = -\gamma\sqrt{\psi_s - \phi_t}; \quad \psi_s > 3\phi_t \tag{1.28}$$

remains valid even after the formation of the inversion layer [5, 64]. To remove the $\psi_s > 3\phi_t$ limitation, the expression for q_b can be modified as follows [81]:

$$q_b = -\gamma\sqrt{\psi_s - \phi_t[1 - \exp(-u)]} \cdot \mathrm{sgn}(\psi_s) \tag{1.29}$$

where the term $\phi_t \exp(-u)$ accounts for the majority carrier contribution and assures that the argument of the square root in (1.29) is non-negative.

In what follows, it is convenient to introduce functions

$$P(u) = e^{-u} + u - 1 \tag{1.30}$$

$$D(u) = \Delta(\phi_n)[e^u - u - 1 - \chi(u)] \tag{1.31}$$

where

$$\Delta(\phi_n) = \exp[-\beta(2\phi_b + \phi_n)] \tag{1.32}$$

and normalized body factor $G = \gamma\sqrt{\beta}$. Then SPE (1.9) becomes

$$[\beta(V_{gb} - V_{fb}) - u]^2 = G^2[P(u) + D(u)]. \tag{1.33}$$

Then $q_b = -\phi_t G\sqrt{P}$ and from (1.28) and (1.33)

$$q_i = -\frac{\phi_t G D(u)}{\sqrt{P(u) + D(u)} + \sqrt{P(u)}}. \tag{1.34}$$

To develop SLM we also introduce the average surface potential

$$\psi_m = \frac{1}{2}(\psi_{ss} + \psi_{sd}) \tag{1.35}$$

and call the point in the channel where $\psi_s = \psi_m$ "the surface potential midpoint". The values of u, q_i, D and P at the surface potential midpoint are denoted as u_m, q_{im}, D_m and P_m respectively.

The main approximation of the symmetric linearization method is [11]

$$q_i = q_{im} + \alpha \cdot s \tag{1.36}$$

where

$$s = \psi_s - \psi_m \tag{1.37}$$

and

$$\alpha = \left(\frac{dq_i}{d\psi_s}\right)_{\psi_s=\psi_m}. \tag{1.38}$$

From (1.27) and (1.29)

$$\alpha = 1 - \left(\frac{dq_b}{d\psi_s}\right)_{\psi_s=\psi_m} = 1 + \frac{G(1-e^{-u_m})}{2\sqrt{P_m}}. \tag{1.39}$$

To use expression (1.34) for q_{im} it is first necessary to obtain D_m which can be done by manipulating SPE (1.33) at the source and drain ends of the channel where the values of imref splitting ϕ_n are known. This yields

$$D_m = \frac{D_s + D_d}{2} + \frac{P_s + P_d}{2} - P_m - \frac{\varphi^2}{4G^2} \tag{1.40}$$

where $D_s = D(\beta\psi_{ss})$, $D_d = D(\beta\psi_{sd})$, P_s and P_d are similarly defined and $\varphi = \beta\Delta\psi$. Using (1.30)

$$D_m = \frac{D_s + D_d}{2} + e^{-u_m}\left(\cosh\frac{\varphi}{2} - 1\right) - \frac{\varphi^2}{4G^2}. \tag{1.41}$$

Since either the second term is negligible or $\varphi \ll 1$ (in subthreshold) it is safe to make approximation

$$\cosh\frac{\varphi}{2} \approx 1 + \frac{\varphi^2}{8} \tag{1.42}$$

so that finally

$$D_m = \frac{D_s + D_d}{2} + \left(e^{-u_m} - \frac{2}{G^2}\right)\frac{\varphi^2}{8} \tag{1.43}$$

which is the form used in SP and PSP models.

Combining (1.36) with the standard expression of the charge-sheet model [5, 64]

$$I_d = -\mu W C_{ox}\left(q_i\frac{d\psi_s}{dy} - \phi_t\frac{dq_i}{dy}\right) \tag{1.44}$$

one finds

$$I_d = -\mu W C_{ox}\,(q_{im} + \alpha \cdot s - \alpha\phi_t)\,\frac{ds}{dy}. \tag{1.45}$$

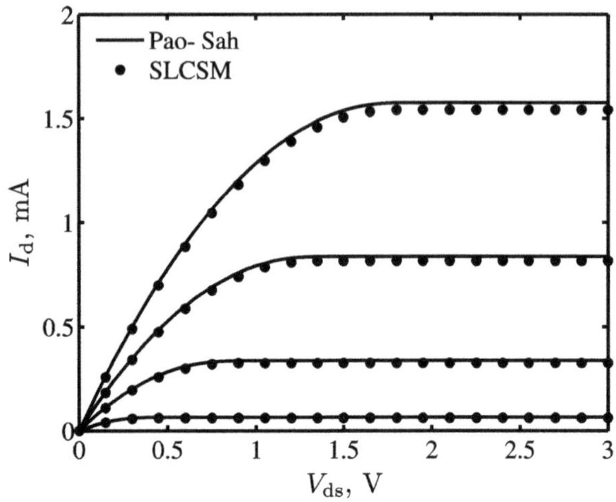

Fig. 1.6 Comparison of $I(V)$ characteristics for Pao-Sah and SLCSM models; $N_a = 2 \cdot 10^{17}$ cm^{-3}, $t_{ox} = 2$ nm and $W/L = 10/10$ μm

Here μ denotes the effective channel mobility while W and L are the width and length of the device, respectively. Separating variables and integrating yields expression for the drain current in the form

$$I_d = \mu \frac{W}{L} C_{ox} \left(-q_{im} + \alpha \phi_t\right) \Delta\psi \tag{1.46}$$

where $\Delta\psi = \psi_{sd} - \psi_{ss}$ is the surface potential variation along the channel. The accuracy of this result is demonstrated by comparison with the Pao-Sah model shown in Fig. 1.6 where abbreviation SLCSM stands for symmetrically-linearized charge-sheet-model.

Since in the rectangular channel device I_d is position-independent, (1.45) can be regarded as a differential equation for the surface potential (or, equivalently, s) as a function of position, y. From (1.45) and (1.46)

$$\frac{dy}{ds} = \frac{L(H - s)}{H\Delta\psi} \tag{1.47}$$

where $H = H_0$ and

$$H_0 = -\frac{q_{im}}{\alpha} + \phi_t. \tag{1.48}$$

After integration

$$y = y_m + \frac{Ls}{H\Delta\psi}\left(H - \frac{s}{2}\right) \tag{1.49}$$

where y_m denotes the coordinate of the "surface potential midpoint" where $\psi_s = \psi_m$. To find y_m note that $s = -\Delta\psi/2$ for $y = 0$ whence

$$y_m = \frac{L}{2}\left(1 + \frac{\Delta\psi}{4H}\right). \tag{1.50}$$

Unlike the case of the complete CSM [77] not only $y(\psi_s)$ but also $\psi_s(y)$ dependence becomes available in a closed form. Indeed, from (1.49) it follows that

$$\psi_s = \psi_m + H\left[1 - \sqrt{1 - (2\Delta\psi/HL)(y - y_m)}\right]. \tag{1.51}$$

The accuracy of this result has been demonstrated in [77].

To insure charge conservation, modern MOSFET models are charge-based with transcapacitances

$$C_{ij} = (2\delta_{ij} - 1)\frac{\partial Q_i}{\partial V_j} \tag{1.52}$$

provided directly by the circuit simulators. In (1.52) $i, j \in \{g, s, d, b\}$ and δ_{ij} denotes the "Kronecker" delta. The physical meaning of the terminal charges

$$Q_g = WC_{ox}\int_0^L (V_{gb} - V_{fb} - \psi_s)\,dy \tag{1.53}$$

and

$$Q_b = WC_{ox}\int_0^L q_b\,dy \tag{1.54}$$

is clear, however, the partition of the inversion charge $Q_i = -Q_g - Q_b$ into the source Q_s and drain $Q_d = Q_i - Q_s$ components requires a few comments. Commonly used Ward-Dutton partition [80]

$$Q_d = WC_{ox}\int_0^L (y/L)q_i\,dy \tag{1.55}$$

is appropriate for uniformly doped channels but not in general [1, 14]. Since in advanced MOSFETs lateral channel non-uniformity is essential (e.g. HALO doping [6]) Ward-Dutton partition should be regarded as an approximation. Numerical studies have shown that while the error introduced by (1.55) is noticeable, it does not at present justify switching back to capacitance-based modeling for circuit simulations.

Symmetric linearization method allows one to obtain simple closed form expressions for the terminal charges. For example, with reference to (1.47) and (1.49)

$$\frac{Q_d}{WLC_{ox}} = \frac{1}{H\Delta\psi}\int_{-\Delta\psi/2}^{\Delta\psi/2}\left[\frac{y_m}{L} + \frac{s(H - \frac{s}{2})}{H\Delta\psi}\right](q_{im} + \alpha s)(H - s)\,ds. \tag{1.56}$$

After integration

$$\frac{Q_d}{WLC_{ox}} = q_{im} + \frac{\alpha\Delta\psi}{12}\left(1 - \frac{\Delta\psi}{2H} - \frac{\Delta\psi^2}{20H^2}\right) \tag{1.57}$$

where the physical signs have been retained for both q_{im} and Q_d. Similarly,

$$\frac{Q_g}{WLC_{ox}} = V_{oxm} + \frac{\Delta\psi^2}{12H} \tag{1.58}$$

and

$$\frac{Q_i}{WLC_{ox}} = q_{im} + \frac{\alpha\Delta\psi^2}{12H} \tag{1.59}$$

where $V_{oxm} = V_{gb} - V_{fb} - \psi_m$ denotes the oxide voltage at the surface potential midpoint. In the SP and PSP models these expressions are used for $V_{gb} > V_{fb}$. As explained above for $V_{gb} \leq V_{fb}$ it is safe to set $\Delta\psi = 0$ so that $Q_g = WLC_{ox}V_{oxm}$ while $Q_s = Q_d = 0$ and $Q_b = -Q_g$. Since I_d is also negligible for $V_{gb} \leq V_{fb}$, this eliminates the need to evaluate the linearization coefficient α in the accumulation region.

As has been done with the expression (1.46) for the drain current, the accuracy of the symmetric linearization method application to the terminal charges has been verified by comparison with Pao-Sah model. Typical results are shown in Fig. 1.7. In addition for $\psi_{ss} > 3\phi_t$ one can also compare the results of the symmetric linearization method with conventional charge-sheet model. This has been done in [11, 68, 77].

To simplify the exposition of the symmetric linearization method we have neglected two important phenomena affecting the MOSFET performance: the quantum corrections and the effect of the polysilicon depletion layer. Both of these effects are included in PSP essentially as discussed in [17] and [74]. The presentation of symmetric linearization method without the assumption of the complete ionization of acceptors can be found in [22].

1.4 The Effective Channel Mobility

The effective channel mobility μ in the MOSFET channel is reduced relative to the bulk mobility as a result of the changed electron-phonon interaction, surface roughness scattering and increased Coulomb scattering. In the engineering models of the transistors these effects are usually combined via Matthiessen's rule to obtain semi-empirical expressions. The model used in PSP is as follows

$$\mu = \frac{\mathbf{UO} \cdot \mu_x}{1 + (\mathbf{MUE} \cdot E_{eff})^{\mathbf{THEMU}} + \mathbf{CS} \cdot [q_{bm}/(q_{bm} + q_{im})^2] + G_R} \tag{1.60}$$

where **UO** is the model parameter roughly corresponding to the low-field mobility (at this point we only discuss the effects of the vertical field), parameters **MUE** and **THEMU** describe effective mobility reduction by the effective vertical field E_{eff}, **CS** is the model parameter introducing the Coulomb scattering following [27] and G_R accounts for the series resistance. PSP also allows for the introduction of the source and drain series resistances directly (using extra nodes that are collapsed by

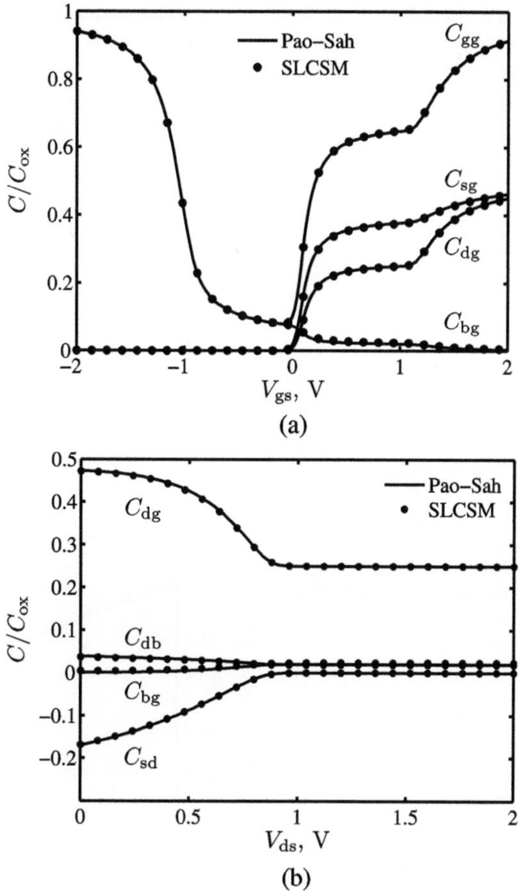

Fig. 1.7 Comparison of $C(V)$ characteristics for Pao-Sah and SLCSM models; $N_a = 2 \cdot 10^{17}$ cm^{-3}, $t_{ox} = 2$ nm, $V_{sb} = 0$, in **(a)** $V_{ds} = 1$ V and in **(b)** $V_{gs} = 1$ V

default) in which case $G_R = 0$. If G_R is used source and drain series resistances R_s are assumed to be equal and

$$G_R = \mathbf{UO} \cdot (W/L) \cdot q_{im} \cdot R_s. \tag{1.61}$$

The effective lateral field is usually defined as

$$E_{eff} = \frac{q_{bm} + \eta_\mu \cdot q_{im}}{\varepsilon_{Si}} \tag{1.62}$$

with $\eta_\mu = 1/2$ for the n-channel [49] and $\eta_\mu = 1/3$ for the p-channel [2] devices. Since, generally speaking, η varies with technology, PSP allows one to adjust η using parameter **FETA** with the default value of 1:

$$\eta_\mu = \begin{cases} \mathbf{FETA}/2 & \text{for NMOS,} \\ \mathbf{FETA}/3 & \text{for PMOS.} \end{cases} \tag{1.63}$$

Fig. 1.8 Linear transconductance for different values of mobility model parameters; $V_{ds} = 50$ mV

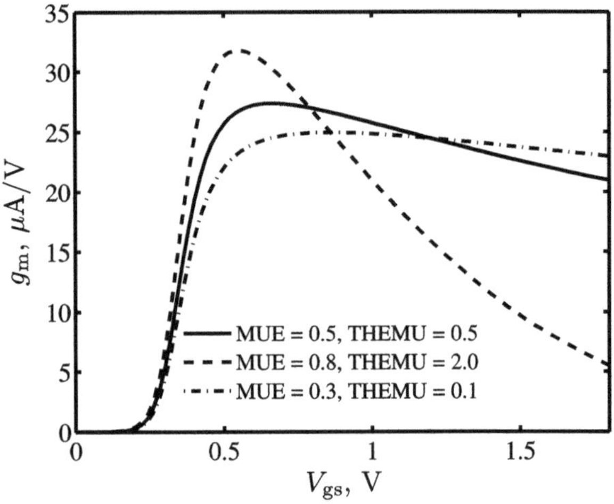

Fig. 1.9 Normalized channel mobility as a function of electrical field; $V_{ds} = 50$ mV

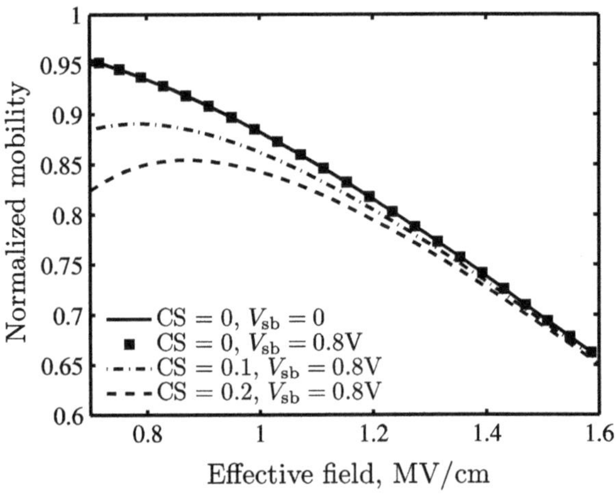

Strictly speaking, the effective channel mobility varies across the MOSFET channel as a result of the position dependence of E_{eff}, q_i, q_b and some other factors (e.g. doping level). While in principle it is possible to include such dependence explicitly, the resulting model complication is not warranted in the compact model. For this reason all variables affecting the effective channel mobility are evaluated at the surface potential midpoint to assure the Gummel symmetry of the model.

Typical results are shown in Fig. 1.8. By changing parameters **MUE** and **THEMU** one can control both the magnitude and the functional dependence of the mobility degradation by the vertical field. In the absence of the Coulomb scattering PSP mobility model is "universal". The introduction of the Coulomb scattering further reduces the effective mobility and introduces the non-universality effects, especially close to threshold and in the subthreshold region (cf. Fig. 1.9). Additional non-universality effects are described by the function μ_x ($\mu_x = 1$ for the default pa-

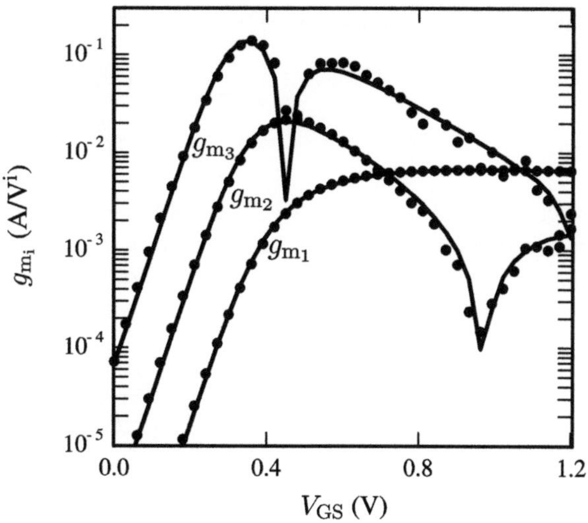

Fig. 1.10 High-order transconductances g_{m_i} versus V_{gs} for $V_{sb} = 0$ V, $T = 25°C$ and $W/L = 10/0.12$ μm; n-channel MOSFET; $i = 1$ (*lower curve*), 2 (*middle curve*) and 3 (*upper curve*). After [21]

rameter set) which also empirically accounts for the channel doping non-uniformity in the direction normal to the Si/SiO_2 interface. The ability of the PSP mobility model to accurately fit experimental data is illustrated in Fig. 1.10 where the higher order transconductances are defined as $g_{m_i} = \partial^i I_d/\partial V_g^i$. The PSP model also accurately reproduces the important g_m/I_d ratio further discussed in [34] and in Chap. 3 of this volume. We note also that in PSP the effect of the pocket implants ("halo doping") on the effective channel mobility is included through the geometry dependence of the average inverse doping concentration in the channel [53]. This, in turn, makes the effective mobility a function of the device geometry and provides an additional flexibility for fitting the experimental data [46].

1.5 Velocity Saturation

The symmetric linearization method was developed in Sect. 1.3 without velocity saturation effects in order to facilitate its verification by comparison with the Pao-Sah model. The actual drift velocity-field relation used in the core PSP model includes velocity saturation effects produced by electron scattering by the optical phonons and acoustic phonons with the wavelengths small relative to the lattice constant [59]. The core PSP model includes non-linear velocity-field dependence for electrons in the form

$$v_d = \frac{\mu E_y}{\sqrt{1 + (\mu E_y/v_{sat})^2}} \tag{1.64}$$

where v_d denotes the drift velocity, v_{sat} is saturation velocity and it is convenient to define the lateral field as $E_y = d\psi_s/dy$. For holes PSP uses an adaptation of the

Scharfetter-Gummel model [51]

$$v_d = \frac{\mu E_y}{\sqrt{1 + \frac{(\mu E_y / v_c)^2}{\mathcal{G} + \mu E_y / v_c}}} \tag{1.65}$$

where v_c corresponds to the velocity of the longitudinal acoustic phonons and $\mathcal{G}$ is parameter set to 1 in PSP (it was found to be of minor influence in fitting MOS $I(V)$ characteristics). In order to arrive at the formulation similar for both n-channel and p-channel devices, expression (1.65) is simplified by setting

$$v_d = \frac{\mu E_y}{\sqrt{1 + \frac{(\mu E_y / v_c)^2}{\mathcal{G} + \mu E_{av} / v_c}}} \tag{1.66}$$

where $E_{av} = \Delta\psi / L$ is the average longitudinal field in the MOSFET channel. With this change the velocity-field relation for holes can be written in the form (1.64) provided that v_{sat} is changed into

$$v_{sat,p} = v_c \sqrt{1 + (\mu \Delta\psi / L v_c)}. \tag{1.67}$$

Thus all equations subsequently derived for n-channel transistors remain valid for p-channel devices after v_{sat} is changed into $v_{sat,p}$.

With drift velocity given by (1.64) the differential equation (1.44) of the charge-sheet model becomes modified as follows

$$I_d = -\frac{\mu W C_{ox} (q_i \frac{d\psi_s}{dy} - \phi_t \frac{dq_i}{dy})}{\sqrt{1 + (\mu E_y / v_{sat})^2}}. \tag{1.68}$$

While exact integration is possible, the results are far too complex to be useful in the development of the compact MOSFET model. Instead we follow [74] to arrive at approximation

$$I_d = -\mu W C_{ox} \left(q_i^* - \frac{I_d^2}{2W^2 v_{sat}^2 q_i^*} \right) \frac{d\psi_s}{dy} \tag{1.69}$$

where

$$q_i^* = q_i - \phi_t \frac{dq_i}{d\psi_s} \tag{1.70}$$

and the second term in the last expression accounts for the diffusion current. After integration

$$I_d = -\mu \frac{W}{L} C_{ox} \left(q_1 - \frac{I_d^2}{2W^2 v_{sat}^2 q_2} \right) \Delta\psi \tag{1.71}$$

where

$$q_1 = \frac{1}{\Delta\psi}\int_{\psi_{ss}}^{\psi_{sd}} q_i^* d\psi_s, \tag{1.72}$$

$$q_2 = \frac{\Delta\psi}{\int_{\psi_{ss}}^{\psi_{sd}} d\psi_s / q_i^*}. \tag{1.73}$$

Symmetric linearization approximation (1.36) yields

$$q_i^* = q_{im} + \alpha s - \alpha\phi_t \tag{1.74}$$

whence by (1.72) $q_1 = q_{im}^*$ where

$$q_{im}^* = q_{im} - \alpha\phi_t. \tag{1.75}$$

Solving quadratic (1.71) with respect to I_d we find the drain current expression used in the core PSP model [75]

$$I_d = -\mu\frac{W}{L}C_{ox}\frac{q_{im}^*\Delta\psi}{\mathcal{G}_{vsat}} \tag{1.76}$$

where factor

$$\mathcal{G}_{vsat} = \frac{1}{2}\left[1 + \sqrt{1 + 2(\theta_{sat}\Delta\psi)^2}\right] \tag{1.77}$$

describes the reduction of the drain current associated with the velocity saturation effect and

$$\theta_{sat} = \frac{\mu\Delta\psi}{v_{sat}L}\sqrt{\frac{q_1}{q_2}}. \tag{1.78}$$

While the integral for q_2 is readily evaluated in the symmetric linearization approximation, it was found that better results are achieved when θ_{sat} is turned into a model parameter **THESAT** [46].

To obtain expressions for the terminal charges it is necessary to evaluate the position dependence of the surface potential. Combining (1.69) and (1.74) we find

$$\frac{dy}{d\psi_s} = \frac{L}{\Delta\psi}\cdot\frac{q_{im}^* - \alpha' s - \dfrac{I_d^2}{2W^2 v_{sat}^2 q_{im}^*}}{q_{im}^* - \dfrac{I_d^2}{2W^2 v_{sat}^2 q_{im}^*}} \tag{1.79}$$

where

$$\alpha' = \alpha\left[1 + \frac{I_d^2}{2(W v_{sat} q_{im}^*)^2}\right]. \tag{1.80}$$

After some algebra we recover (1.47) and (1.49) but with

$$H = -\frac{q_{im}^{*}}{\alpha' \mathcal{G}_{vsat}}. \tag{1.81}$$

In particular, expressions (1.57)–(1.59) of the symmetric linearization method remain valid in the presence of the velocity saturation effect provided H is changed from (1.48) to (1.81). This conclusion is not dependent on a particular form of the velocity-field relation. As shown in [20], expressions (1.57)–(1.59) are still applicable with a totally different parametrization of the drift velocity as long as H is modified accordingly.

Direct incorporation of the non-linear velocity-field relation into the charge-sheet model (or any model based on the GCA) is well-known to produce unphysical region of negative output conductivity. The physical origin of this difficulty is the failure of the GCA in the so-called pinch-off region of the MOSFET channel. In PSP and some earlier models [45] this problem is solved by softly clamping the drain bias using "effective drain-source voltage" V_{dse}. The asymptotic upper limit V_{dsat} for V_{dse} is selected so that $(\partial I_d/\partial V_{ds})_{V_{ds}=V_{dsat}} = 0$ eliminating the negative output conductance problem. Specific form of the $V_{dse}(V_{ds})$ dependence used in PSP is

$$V_{dse} = \frac{V_{ds}}{[1+(V_{ds}/V_{dsat})^{a_x}]^{1/a_x}} \tag{1.82}$$

where a_x is a local model parameter scaled according to expression

$$a_x = \frac{\mathbf{AXO}}{1+\mathbf{AXL}/L} \tag{1.83}$$

where **AXO** and **AXL** are global PSP parameters. This scaling of a_x allows model users to account for the more gradual transition from the triode to saturation region in short-channel devices. The minimum value of a_x allowed in PSP is 2. This assures the existence of the third derivative d^3I_d/dV_{ds}^3 for $V_{ds} = 0$. If higher order derivatives are required at $V_{ds} = 0$ then parameter extraction procedure is becoming more restrictive to assure the higher values of a_x for all device dimensions [33, 34].

The PSP model also includes several semi-empirical expressions for the channel-length modulation effects [13] and the degradation of the output conductance in long-channel devices with halo doping [9, 42, 48]. Such detailed description of the saturation region is necessary in order to achieve not only a good fit of MOSFET output characteristics but also of high-order output conductances $g_{ds_i} = \partial^i I_d/\partial V_{ds}^i$ essential in analog and RF applications [69]. Typical results shown in Fig. 1.11 indicate a good agreement with the experimental data.

The effect of the velocity saturation on the MOSFET transcapacitances is illustrated in Fig. 1.12. Qualitative behavior of transcapacitances remains essentially the same as for long-channel devices.

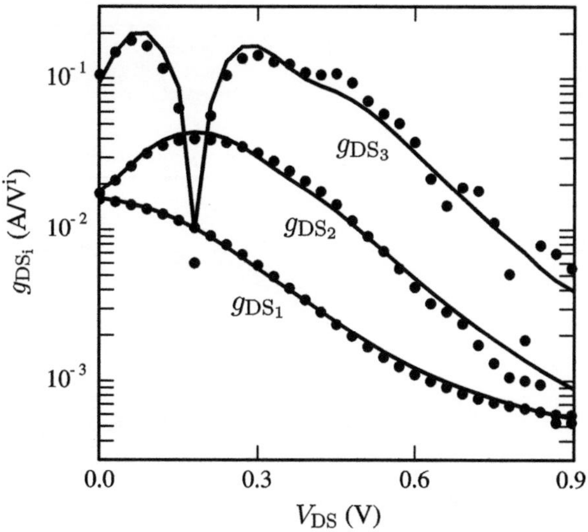

Fig. 1.11 High-order output conductances g_{ds_i} versus V_{ds} for $V_{sb} = 0$ V, $T = 25$°C and $W/L = 10/0.12$ μm; n-channel MOSFET; $i = 1$ (*lower curve*), 2 (*middle curve*) and 3 (*upper curve*). After [21]

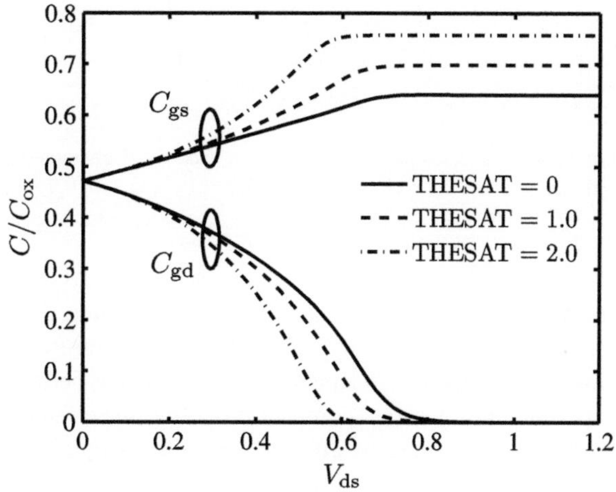

Fig. 1.12 The effect of the velocity saturation on the MOSFET transcapacitances; $V_{gs} = 1$ V

1.6 Lateral Doping Non-uniformity

Halo (or pocket) implants are often used to suppress short-channel effects in scaled MOS transistors. Consequently, advanced MOSFETS exhibit a strong lateral (in the direction along the current flow) doping non-uniformity indicated in Fig. 1.13. This non-uniformity has several consequences for the device characteristics and has to be accounted for in the state-of-the-art MOSFET models. The first effect is the qualitative change in the potential distribution along the channel that manifests itself as a stronger than expected DIBL effect in long-channel transistors. In PSP this is included in the channel length modulation effects leading to an accurate description of the output conductance [21]. The second effect is the apparent dependence of the effective channel mobility on the device geometry. Its physical origin can be

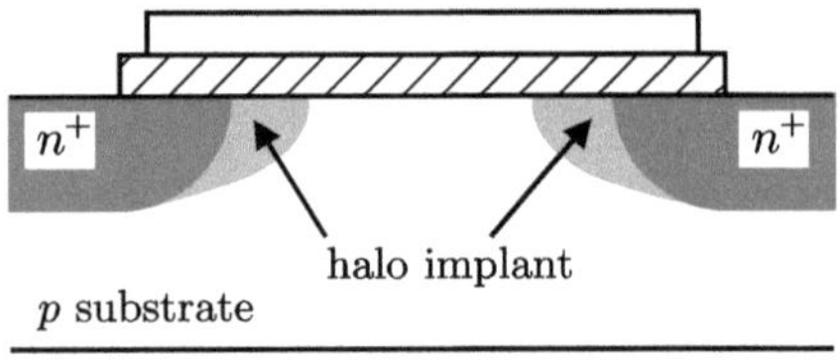

Fig. 1.13 Illustration of the cross section of a halo doped n-channel MOSFET

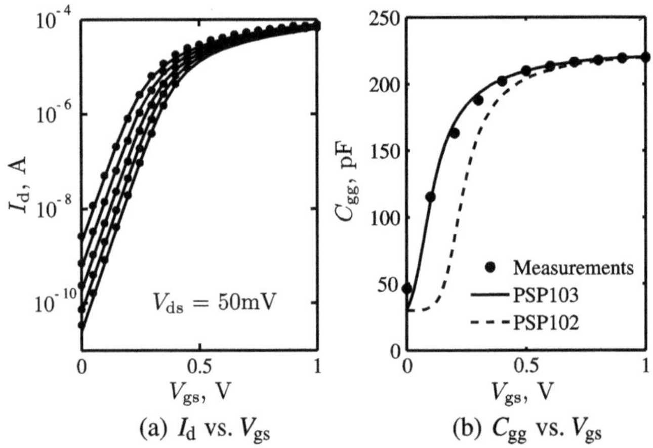

Fig. 1.14 The discrepancy in fitting I–V and C–V characteristics produced by lateral non-uniform doping. For I–V characteristics, $V_{sb} = 0, 0.25, 0.5, 0.75, 1$ V

traced to the fact that mobility decreases as the average channel doping is increased so that in shorter devices the tails of the doping distribution near the gate edges directly affect the mobility. A simple model assuming that these tails are exponential with the parameters independent of the device geometry leads to an accurate and experimentally verified compact model of this effect. Another assumption inherent in this model is that the effective mobility is inversely proportional to the average channel doping [21, 46, 53].

A much more drastic manifestation of the lateral channel non-uniformity is the difficulty in achieving a simultaneous fit of the $I(V)$ and $C(V)$ characteristics for advanced technology nodes. A typical parameter extraction procedure for almost any compact model starts with fitting $C(V)$ characteristics and in the process extracting the effective doping concentration and the oxide thickness. The next step which is the fitting of the device $I(V)$ characteristics requires readjustment of the doping level and as a consequence the fitting of the $C(V)$ characteristics is compromised. As shown in Fig. 1.14 the discrepancy can be as much as 200 mV. What happens physically is that differently doped regions can be regarded as connected in series for the evaluation of the drain current and in parallel as far as $C(V)$ characteristics are concerned. This means that different (and geometry dependent) impurity levels (and hence surface potentials) are needed to accurately reproduce both $C(V)$ and $I(V)$ characteristics [42, 48].

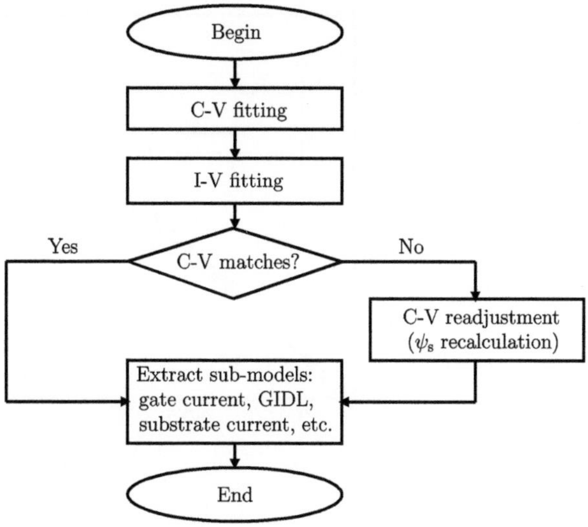

Fig. 1.15 Parameter extraction procedure which allows the decoupling of the surface potential calculations for currents and charges

When this is done a near perfect fit is achieved for both (cf. Fig. 1.14) at the price of an increased computational complexity. Hence this option now available in PSP103 and subsequent versions should be used sparingly. A modified parameter extraction procedure is illustrated in Fig. 1.15 indicating that $C(V)$ readjustment is an optional part of the parameter extraction procedure usually not required except for the most recent technology nodes. In order to provide physical consistency and to further simplify the fitting process the bulk potential can be also selected independently for $C(V)$ and $I(V)$ characteristics.

1.7 Punch-Through Effect and Vertical Doping Non-uniformity

The vertical doping non-uniformity usually comes in the form of the retrograde or the inverse retrograde profiles. These can be used, for example, in order to achieve the desired trade-off between the device characteristics for the high-speed and low-power applications [6]. From the modeling point of view the inverse retrograde (or "high-low") doping profile brings about the reduction of the back-bias sensitivity for high back biases. This means that the threshold voltage dependence on the back bias dependence no longer follows the classic result [50, 62]

$$V_{th} = V_{th0} + \gamma \cdot z \tag{1.84}$$

where

$$z = \sqrt{V_{sb} + 2\phi_b} - \sqrt{2\phi_b}. \tag{1.85}$$

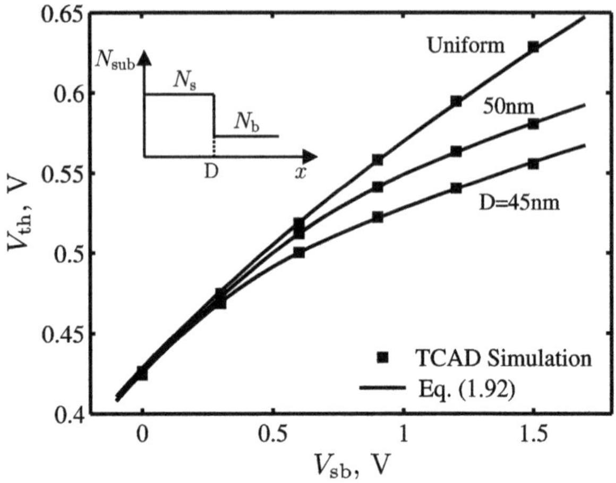

Fig. 1.16 The effect of the vertical doping non-uniformity (NUD) on the $V_{th}(V_{sb})$ dependence. For non-uniform cases $N_s = 8 \cdot 10^{17}$ cm^{-3} and $N_b = 8 \cdot 10^{16}$ cm^{-3}; for uniform case $N_s = N_b = 8 \cdot 10^{17}$ cm^{-3}. In all cases $t_{ox} = 2.2$ nm and $L = 10$ µm. The inset shows the step doping profile; x is the distance from the Si/SiO$_2$ interface

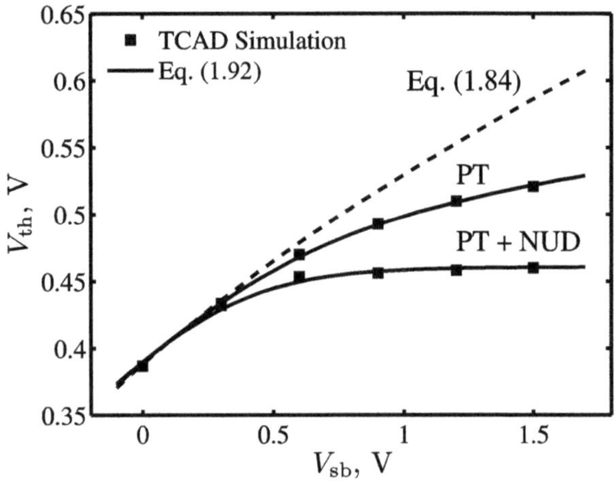

Fig. 1.17 The effect of the vertical non-uniformity and PT on the $V_{th}(V_{sb})$ dependence. For PT simulation $N_s = N_b = 8 \cdot 10^{17}$ cm^{-3}; for PT and vertical non-uniformity $N_s = 8 \cdot 10^{17}$ cm^{-3} and $N_b = 8 \cdot 10^{16}$ cm^{-3}; for (1.84) $N_{sub} = 8 \cdot 10^{17}$ cm^{-3}. In all cases $t_{ox} = 2.2$ nm and $L = 80$ nm

Note that although V_{th} is not used in the PSP model formulation it is convenient to use it in the qualitative discussion of the back bias effect. Another effect that is often responsible for the reduction and sometimes collapse of the back bias sensitivity is the punch-through (PT) phenomenon [63]. The results of the TCAD simulations shown in Fig. 1.16 for the step profile (see the inset in Fig. 1.16) and Fig. 1.17 for the PT effect indicate that both effects produce quantitatively similar $V_{th}(V_{sb})$ characteristics and can be modeled simultaneously. The main difference is that if PT dominates then only short-channel devices exhibit the behavior shown in Fig. 1.17. In order to introduce both effects in the compact model we redefine the body factor as $\gamma = dV_{th}/dz$ and plot it as a function of z in Fig. 1.18. Thus defined γ gradually decreases from γ_1 obtained from (1.10) with $N_{sub} = N_s$ to γ_2 corresponding to

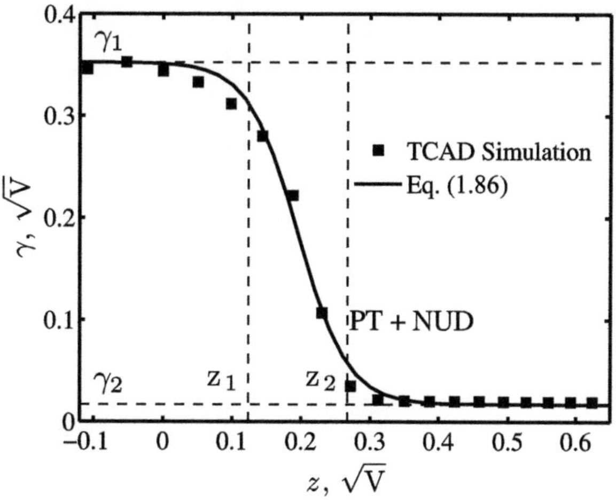

Fig. 1.18 Bias dependence of the effective body factor $\gamma = dV_{th}/du$ corresponding to the lower (PT + NUD) curve in Fig. 1.17. The parameters used in (1.86) are $\gamma_1 = 0.35$, $\gamma_2 = 0.017$, $u_1 = 0.124$ and $u_2 = 0.267$

$N_{sub} = N_b$. This $\gamma(z)$ dependence can be approximated as

$$\gamma = \gamma_1 - \frac{1}{2}(\gamma_1 - \gamma_2)[1 + \tanh(v)] \tag{1.86}$$

where

$$v = \frac{2(z - z_1)}{z_2 - uz_1} - 1 \tag{1.87}$$

and z_1 and z_2 are constant factors related to the model parameters V_{sb1} and V_{sb2} which are selected as the corresponding values of V_{sb}:

$$V_{sbi} = \left(z_i + \sqrt{2\phi_b}\right)^2 - 2\phi_b; \quad i = 1, 2. \tag{1.88}$$

The implementation of the $\gamma(z)$ dependence in the compact model can be simplified if instead of modifying γ we keep $\gamma = \gamma_1$ and modify z relative to (1.85). To assure the identical transistor characteristics in both cases we require

$$\gamma dz = \gamma_1 d\tilde{z} \tag{1.89}$$

where $\tilde{z}$ denotes the modified value of z corresponding to the modified back bias

$$\tilde{V}_{sb} = \left(\tilde{z} + \sqrt{2\phi_b}\right)^2 - 2\phi_b. \tag{1.90}$$

After integration

$$\tilde{z} = z - \frac{1}{4} \cdot (1 - \gamma_2/\gamma_1) \cdot (z_2 - z_1) \ln\left(1 + e^{2v}\right) + C \tag{1.91}$$

where the integration constant C is set to zero so that $\tilde{V}_{sb} \approx V_{sb}$ for $V_{sb} < V_{sb1}$ (cf. Fig. 1.19). Subsequently, the expression of V_{th} for the devices affected by PT or

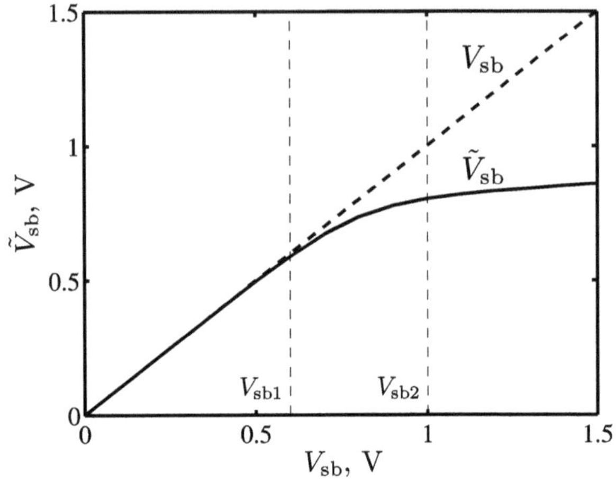

Fig. 1.19 Transformation of the back bias. $V_{sb1} = 0.6$ V, $V_{sb2} = 1.0$ V and $\gamma_2/\gamma_1 = 0.4$

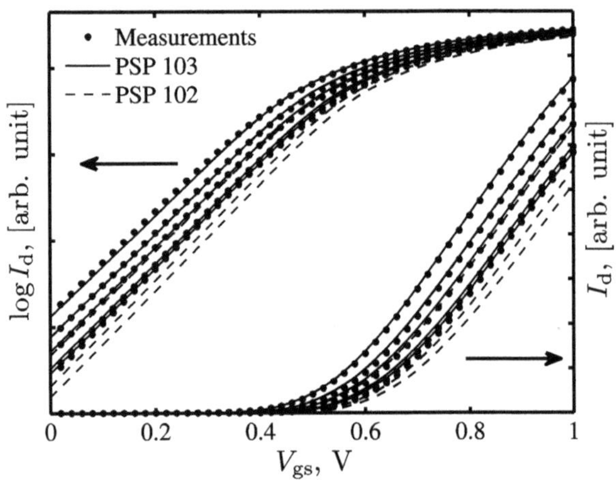

Fig. 1.20 Collapse of the back bias sensitivity for large V_{sb}. PSP 103 includes NUD and PT effects; W = 10 µm, $L = 0.06$ µm, $V_{sb} = 0$, 0.3, 0.6, 1.2, 1.44 V and $V_{ds} = 50$ mV

vertical non-uniformity becomes

$$V_{th} = V_{th0} + \gamma_1 \cdot \tilde{z}\,. \tag{1.92}$$

The surface potential is now computed from (1.9) but with V_{sb} changed into $\tilde{V}_{sb}$ so that V_{gb} is transformed into $\tilde{V}_{gb} = V_{gs} + \tilde{V}_{sb}$. Hence the modified imrefs at the source and drain ends of the channel are given by $\tilde{V}_{sb}$ and $\tilde{V}_{sb} + V_{ds}$, respectively. Apart from this modification the structure of the SPE remains unchanged and so are the analytical or numerical techniques for the evaluation of the surface potentials.

Verification of this approach to modeling the effects of the vertical non-uniformity is presented in Fig. 1.20. The collapse of the back bias sensitivity for large V_{sb} is easily captured by PSP103 and the subsequent versions of the PSP model.

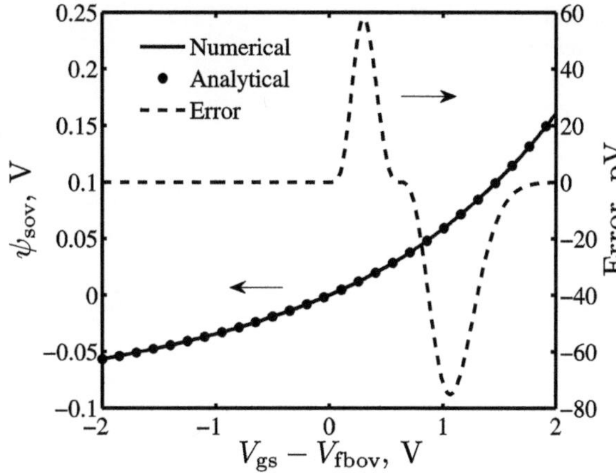

Fig. 1.21 Analytical approximation of the surface potential for the overlap region; $t_{ox} = 3$ nm, and $N_{ov} = 10^{20}$ cm^{-3}

1.8 The Extrinsic Model

The performance of the state of the art MOSFETs is to a large extent determined by the secondary effects such as overlap capacitances, impact ionization and tunneling currents, GIDL and related effects. The modeling of these effects in PSP is also surface-potential-based in order to increase the physical content of the model and to provide a more accurate and flexible description conforming to various benchmark tests [33].

1.8.1 Overlap Region Charges

The contribution of the source/drain overlap regions to the terminal charges is given by expressions

$$Q_{sov} = \mathbf{CGOV} \cdot (V_{gs} - V_{fbov} - \psi_{sov}) \tag{1.93}$$

and

$$Q_{dov} = \mathbf{CGOV} \cdot (V_{gs} - V_{fbov} - \psi_{dov}) \tag{1.94}$$

respectively. Here **CGOV** is the model parameter, $V_{fbov} \approx 0$ is the overlap region flat-band voltage and ψ_{sov} and ψ_{dov} are the surface potentials in the overlap regions. Since the inversion of the source and drain overlap regions does not occur, ψ_{sov} and ψ_{dov} can be computed from (1.9) with minority carrier contribution neglected., i.e. with (1.25) changed into

$$h(u) = e^{-u} + u - 1. \tag{1.95}$$

This simplifies not only the SPE but also its analytical approximation [76] which is illustrated in Fig. 1.21.

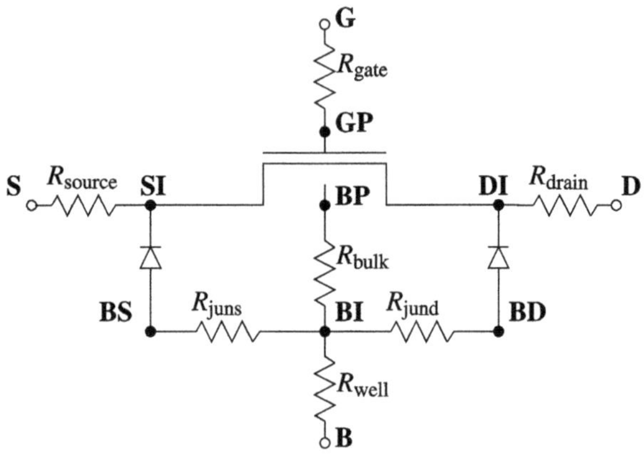

Fig. 1.22 Network of parasitic resistances in PSP

In order to provide an additional flexibility during the model parameter extraction, PSP allows for the difference in the oxide thickness in the active (**TOX**) and the overlap (**TOXOV**) regions. Hence

$$\mathbf{CGOV} = \frac{\varepsilon_{ox} \cdot W_{E,CV} \cdot \mathbf{LOV}}{\mathbf{TOXOV}} \tag{1.96}$$

where **LOV** denotes the length of the overlap region and $W_{E,CV}$ is the effective channel width for the evaluation of the terminal charges.

1.8.2 Parasitic Resistances

Starting with PSP103 one can include series resistances for both quasi-static and non-quasi-static PSP models. The corresponding equivalent circuit is shown in Fig. 1.22. If the external series resistances are used then G_R term in (1.60) is set to zero. The model allows for the use of the asymmetric source and drain series resistances

$$R_{source} = \mathbf{NRS} \cdot \mathbf{RSH} \tag{1.97}$$

and

$$R_{drain} = \mathbf{NRD} \cdot \mathbf{RSHD} \tag{1.98}$$

where **NRS** (instance parameter) is the number of squares of the source diffusion and **RSH** (model parameter) denotes the corresponding sheet resistance; parameters **NRD** and **RSHD** for the drain diffusion are defined similarly.

The gate resistance R_{gate} is important for RF applications including noise modeling. In the PSP model R_{gate} includes the distributed resistance of the silicide layer and the silicide-to-polysilicon contact resistance [35]:

$$R_{gate} = \frac{\rho_{si}}{3 \cdot N_{gc}^2 \cdot N_f^2} \cdot \frac{W}{L} + \frac{\rho_{con} + \rho_{poly}}{W \cdot L} \tag{1.99}$$

where ρ_{si} is the sheet resistance of the silicide, N_{gc} denotes the number of gate contacts, N_f is the number of fingers, $W = N_f W_f$ is the total gate width, ρ_{con} is specific contact resistance between the silicide and polysilicon and ρ_{poly} describes the vertical resistance of the polysilicon. It follows that the second component of the gate resistance describing the contact resistance cannot be reduced by increasing the number of fingers of the device.

The four resistances R_{juns}, R_{jund}, R_{bulk} and R_{jund} describing the substrate are optional in the model (they are essential for some RF applications) and at present are non-scalable model parameters. The junction model JUNCAP2 [55] is described in a companion chapter of this volume. It provides an accurate description of all components of the junction current as a function of bias and temperature. By using JUNCAP2 in conjunction with JUNCAP2 Express this is achieved in a manner compatible with computational efficiency required from the general-purpose compact model [57, 58].

1.8.3 Impact Ionization Current

Impact ionization (or weak avalanche) current I_{avl} is modeled in PSP essentially following the standard approach in [10, 16, 38, 78] but with an essential modification introduced in [25] in order to obtain the proper asymptotic behavior in the subthreshold region:

$$I_{avl} = \mathbf{A1} \cdot I_d \cdot \theta(w) \cdot w \cdot \exp(-a_2^*/w) \tag{1.100}$$

where θ denotes the unit-step function, $w = V_{ds} - \mathbf{A3} \cdot \Delta\psi$, and

$$a_2^* = \mathbf{A2}\left[1 + \mathbf{A4}\left(\sqrt{V_{sb}^* + \phi_b} - \sqrt{\phi_b}\right)\right] \tag{1.101}$$

and coefficients **A1**–**A4** are local model parameters which (except for **A2**) are scalable with the device geometry. Parameters **A1** and **A2** control the magnitude of I_{avl}, **A2** and **A3** determine the shape of the $I_{avl}(V_{ds})$ characteristics and **A4** introduces the back-bias effect on I_{avl} (beyond that already present in I_d).

In the strong inversion (1.101) leads to a standard expression [10, 16, 38, 78] while in subthreshold $\Delta\psi \ll V_{ds}$ and one recovers the correct asymptotic behavior

$$I_{avl} \propto \exp\left(-a_2^*/V_{ds}\right). \tag{1.102}$$

This parametrization of I_{avl} allows for an accurate reproduction of the experimental device characteristics [25].

1.8.4 Gate Tunneling Current

The accurate modeling of the gate tunneling is essential in advanced MOS devices with reduced oxide thickness [7, 82]. A simple one-electron theory for the tunneling

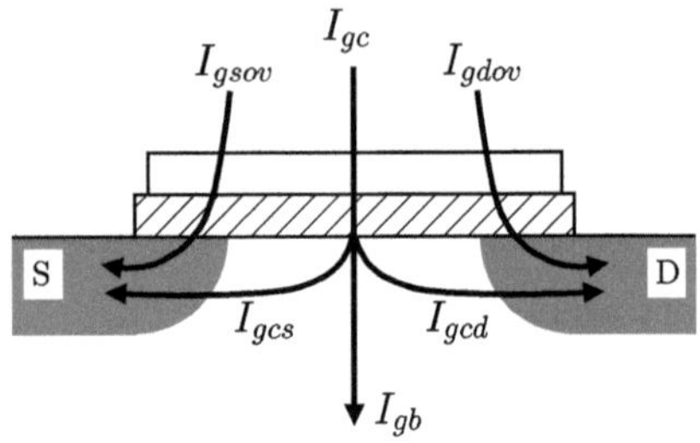

Fig. 1.23 Components of the gate tunneling current

current results in the Tsu-Esaki equation [65] that can only be evaluated numerically [8]. For example, the channel component of the tunneling current density is given by expression

$$J_g(y) = \frac{q^2 m^* \phi_t}{2\pi^2 \hbar^3} \int_0^\infty D(E, y) F_s(E, y)\, dE \tag{1.103}$$

where m^* denotes the effective electron mass, $D(E, y)$ is the transmission coefficient, and F_s is the supply function reflecting the difference in the occupation numbers on the two sides of the tunneling barriers [65]. The inclusion of F_s assures that the current density vanishes when there is no imref splitting across the oxide. Similar expressions apply to the components of the tunneling current associated with the source/drain overlap region, except that the position dependence of the current density can usually be neglected in these regions.The channel component of the gate tunneling current is obtained by integrating along the channel

$$I_{gc} = W \int_0^L J_g(y)\, dy \tag{1.104}$$

while the partitioning of I_{gc} between the source, I_{gcs}, and the drain, I_{gcd} components (cf. Fig. 1.23 is accomplished using expressions [60, 72] similar to the Ward-Dutton result (1.55):

$$I_{gcd} = W \int_0^L J_g(y)(y/L)\, dy \tag{1.105}$$

and $I_{gcs} = I_{gc} - I_{gcd}$.

With J_g given by (1.103), closed form expressions for I_{gcs} and I_{gcd} cannot be obtained. Instead, in PSP we use "monoenergetic approximation" to (1.103) by assuming that all electrons tunnel from the same energy level close to the conduction band edge (cf. Fig. 1.24). The precise value of this energy is adjusted using model parameters [46]. With this approximation, analytical expressions for all components of the tunneling current can be obtained in a relatively simple form [24] sufficient to accurately reproduce the experimental device characteristics shown in Fig. 1.25.

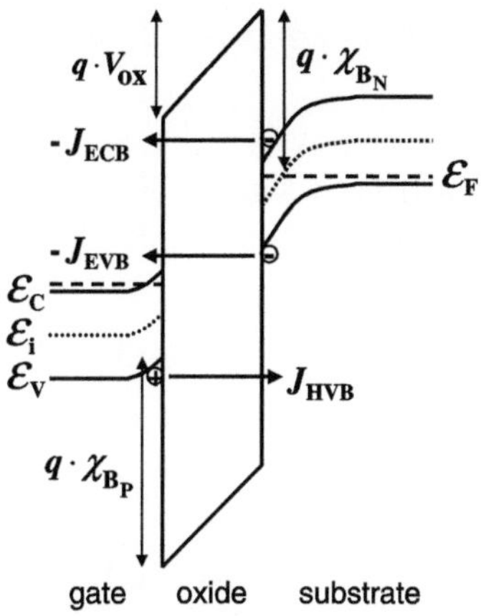

Fig. 1.24 Energy-band diagram of an n-MOS in inversion

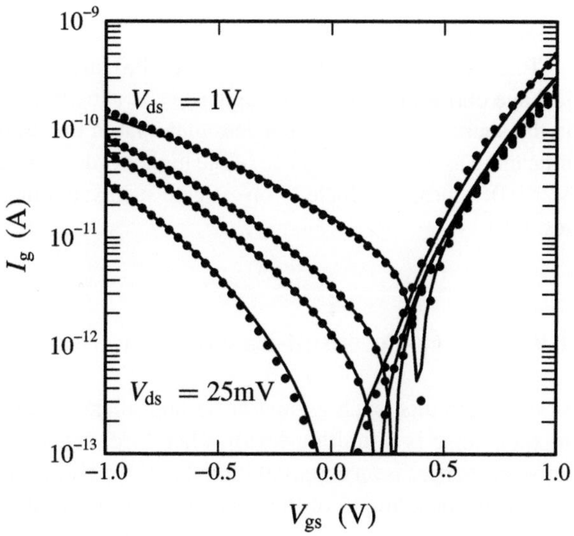

Fig. 1.25 Gate current I_g as a function of V_{gs} at $V_{sb} = 0$ for different V_{ds}. *Symbols* represent measurement and *lines* are the simulated results using PSP. After [21]

We note in passing that integration along the channel is enabled by the symmetric linearization method. For example the integral (1.105) is evaluated as

$$I_{gcd} = \frac{W}{L} \int_{-\Delta\psi/2}^{\Delta\psi/2} J_g(y(s))y(s)\frac{dy}{ds}\,ds \tag{1.106}$$

where $y(s)$ and dy/ds are given by (1.49) and (1.47), respectively.

Further details are presented in [24, 46]. For the bulk MOSFETs it is usually sufficient to include only the electron conduction band tunneling (ECB) and hole valence band tunneling (HVB) [8]. The electron valence band tunneling becomes important in SOI devices and can be modeled in a similar manner [86].

1.8.5 Gate-Induced Drain Leakage Current

Gate-induced drain leakage current (GIDL) is important in the off-state of a MOSFET and is associated with the high transversal field for high negative V_{gd} [12]. A simple model of GIDL can be developed following the approach in [32, 74]. The result is

$$I_{GIDL} = A_{GIDL} \cdot V_{db} \cdot V_{ox}^2 \cdot \exp\left(-B_{GIDL}/V_{tov}\right) \tag{1.107}$$

where A_{GIDL} and B_{GIDL} are local model parameters (physically scaled with the device geometry) and the oxide transversal overlap voltage V_{tov}

$$V_{tov} = \sqrt{V_{ov}^2 + (C_{GIDL} \cdot V_{db})^2} \tag{1.108}$$

includes a model parameter C_{GIDL} adjusting the effect of the V_{db} on the magnitude of the GIDL. The inclusion of the V_{db} multiplier in (1.107) ensures that I_{GIDL} changes sign as V_{db} passes through zero.

To assure the symmetry of the model, PSP also includes the Gate-induced source leakage current (GISL). Although I_{GISL} is usually small since the transversal field in the source overlap region is less than that in the drain region, its inclusion is significant to assure the existence of the higher order derivatives of the drain current for $V_{ds} = 0$. A proper vehicle to investigate such subtleties is the McAndrew symmetry test [39].

1.9 Surface-Potential-Based Noise Model

In many circuits, such as low-noise amplifiers, voltage-controlled oscillators, and mixers, noise is a limiting factor. Therefore, accurate simulation of the MOSFET noise properties is an essential ingredient of a compact model. PSP provides a comprehensive description of the noise behavior of modern MOSFETs, taking into account flicker noise (see Sect. 1.9.1), thermal and induced gate noise (see Sect. 1.9.2), as well as the noise related to gate tunneling, impact ionization, and junction leakage currents (see Sect. 1.9.3).

1.9.1 Flicker Noise

At low frequencies, flicker noise, or $1/f$-noise, is the dominant source of noise. Different mechanisms to explain this noise have been proposed, but the bulk of the evidence points towards an explanation in terms of carrier trapping in the oxide. In the widely accepted 'unified flicker noise model' [28, 29], the trapping of a charge carrier in the oxide leads to both a number fluctuation in the MOSFET inversion layer as well as a (correlated) mobility fluctuation. In the noise spectrum, a single oxide

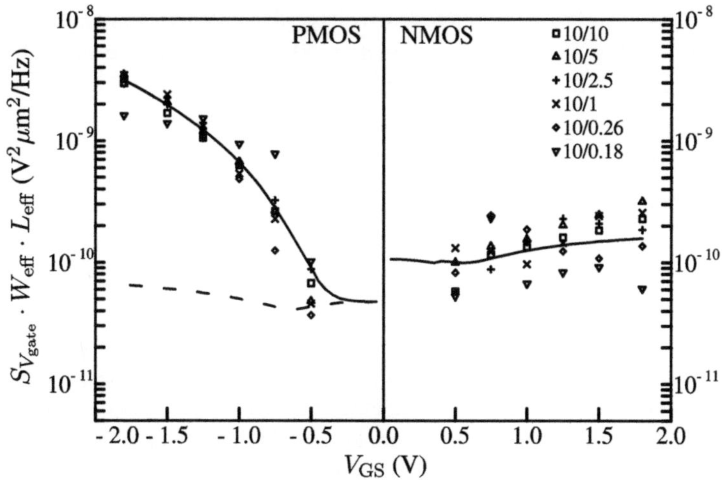

Fig. 1.26 Input-referred $1/f$ noise in 0.18 μm technology, multiplied by the effective device area, plotted versus gate-source voltage for several geometries, and for n- and p-channels. The unified $1/f$ noise model is used here to fit the data. For NMOS, only number fluctuations were taken into account in the fit (*solid line*). For PMOS, the *solid line* corresponds to the best fit with both number and mobility fluctuations, and the *dashed line* shows the number fluctuation component only. For clarity, only the model curves for the 10/10 devices are shown. Curves for the other geometries are very similar. After [54]

trap leads to a so-called Lorentzian noise spectrum (f^0 behavior at low frequencies turning over to f^{-2} behavior at higher frequencies). For an ensemble of oxide traps with a varying distance to the Si/SiO$_2$ interface, the individual Lorentzians add up to a $1/f$-like spectrum. Indeed, this is what is observed in the low-frequency noise spectra of MOSFETs with relatively large area. The noise spectra of small-area MOSFETs (which have a limited number of oxide traps), on the other hand, show clear signatures of the individual Lorentzians, and also show a huge variability, as expected from this theory [15]. Their ensemble-averaged behavior, however, is still $1/f$-like [54], justifying the description of the low-frequency noise of small-area devices along the same lines as their larger-area counterparts.

In PSP, the theory developed in [28, 29] has been reformulated in the surface-potential based framework. This naturally leads to a smooth transition from the subthreshold region to the strong inversion region. PSP includes several model parameters to fit experimental data: the local parameter **EF** tunes the spectral dependence $S_{fl} \propto 1/f^{\mathbf{EF}}$ of the flicker noise, where the default value $\mathbf{EF} = 1$ corresponds to pure $1/f$-noise; the local parameter **NFA** is used to model the number fluctuation effect; finally the local parameters **NFB** and **NFC**, are used to model the (correlated) mobility fluctuation effect.

In Fig. 1.26, we show an example where the unified noise model is used to fit flicker noise data from an 0.18 μm CMOS technology. For NMOS, the input-referred flicker noise displays only a limited dependence on the gate-source voltage. In this case, the number fluctuation component (corresponding to the parameter **NFA**) suffices to fit the data. For PMOS the gate-voltage dependence is much stronger, and correlated mobility fluctuations (parameters **NFB** and **NFC**) are needed to get a good description of the experimental data.

1.9.2 Thermal Noise

Above the so-called knee frequency (which is in the 100 MHz–1 GHz range in deep submicron MOSFETs) flicker noise is reduced so much that drain current thermal noise becomes the dominant source of noise. Thermal noise is caused by the random thermal (or Brownian) motion of the charge carriers. It manifests itself not only in the drain current, but, due to the capacitive coupling between channel and gate, also in the gate current noise. The channel thermal noise emanating from the gate is proportional to f^2 and is called 'induced gate noise'. Because drain current thermal noise and induced gate noise have the same physical origin, they are (partly) correlated. Note that induced gate noise also exists in the case of flicker noise ('induced flicker noise', linearly proportional to f), but that effect is generally considered to be negligible [54].

PSP provides a *predictive* thermal noise model that gives a consistent description of drain current thermal noise, induced gate noise, and their correlation. This is important for, e.g., the correct prediction of the minimum noise figure. The noise model has two important ingredients, i.e., (i) the local noise source, and (ii) the transfer of this local noise to the device terminals. For the local noise sources in the channel, the Nyquist law is used. In the calculation of the transfer to the drain and gate terminals, velocity saturation is accounted for [43, 71]. Finally, the total drain current noise, gate current noise, and their correlation are calculated by integrating all the contributions form source to drain, under the assumption that all local noise sources in the channel fluctuate independently. The noise model has been verified on various technology nodes, see Refs. [54, 56, 71].

The PSP thermal noise model only has one parameter, **FNT**. Its serves to switch the thermal noise on (**FNT** $= 1$) and off (**FNT** $= 0$) for diagnostical purposes. In Fig. 1.27 a circuit-design example is shown where the PSP noise model successfully predicts the measured noise of an LNA [58].

1.9.3 Other Noise Sources

Apart from flicker noise and thermal noise, PSP models several other noise mechanisms. First, the parasitic resistances, treated in Sect. 1.8.2, produce thermal noise. Moreover, the gate tunneling current (Sect. 1.8.4) and junction leakage current have full shot noise. Finally, the avalanche noise of the impact ionization current (Sect. 1.8.3) is modelled after [66]. Note that shot noise observed in the drain current when the MOSFET is biased in the subthreshold region, is *not* separately modelled.

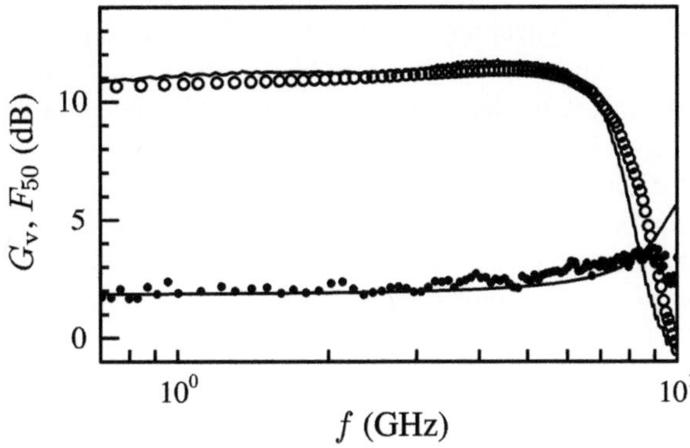

Fig. 1.27 Voltage gain G_v (*open circles*) and 50-Ω noise figure F_{50} (*solid dots*) of a low-noise amplifier, designed in a 65 nm CMOS technology, as a function of the frequency. *Symbols* represent measurements and *lines* represent simulations with the PSP model

It follows naturally from the model of thermal noise presented in Sect. 1.9.2 [54]. Specifically, provided that V_{ds} is significantly larger than ϕ_t, the spectral density of the channel thermal noise in subthreshold becomes $2qI_d$.

1.10 Conclusions

The fundamental problem facing any developer of a compact model is the inherent conflict between the requirements of the model accuracy and computational efficiency. More recently, the model development environment has become even more challenging by first insisting on the increase in the physical content of the models in order to enhance its statistical applications and then imposing several new requirements (known as "benchmarks") on the qualitative model behavior [31, 34]. As a matter of principle, these new objectives cannot be met by simply adding new features to the existing compact models. The development of the surface potential based approach achieves both of the new objectives by aligning the MOSFET compact model with the intrinsic device physics. In particular, many of the benchmarks simply state that the MOSFET model should behave qualitatively in the same manner as the long-channel physical MOSFET models which are, naturally, surface potential based [5, 44]. So using these models as a starting point for the development of the compact models almost automatically guarantees the correct qualitative behavior of the model simultaneously with an increased physical content. More challenging is the requirement of the computational efficiency. Indeed, the brute-force formulation of the surface-potential-based models is prohibitively complex even in the case of the long-channel devices once the terminal charges are included [39]. Hence, the new mathematical techniques are critical for the shift from the threshold-voltage-based to the surface-potential-based approach. In case of the PSP model the unifying theme is the symmetric linearization method that is consistently used to develop all model components: drain current, terminal charges, gate tunneling current, noise model, etc. As demonstrated in [57], this allows one to formulate a surface potential based model which computational efficiency on par with that of the mature threshold-voltage based models. Furthermore, development of the analytical approximation for the surface potential eliminated iterative loops and potential convergence problems which are occasionally encountered in less sophisticated implementations of the surface potential based models. In the last few years it has also become necessary to account for the doping profile non-uniformity in a very elaborate matter still consistent with the compact nature of the model. The experience of the SP, MM11 and, finally, PSP model development makes it likely that further advances in the area of compact MOSFET models will require progress in two related directions: including new physical phenomena in advanced MOSFETs and development of the mathematical techniques required to implement them in a compact manner.

Acknowledgments This work is supported in part by the Semiconductor Research Corporation and by the Compact Model Council (CMC). The detailed evaluation by the CMC members of several versions of the PSP model is deeply appreciated. The authors are much indebted to Dr. C.C. McAndrew for the numerous discussions of the material presented in this chapter. We are grateful to Dr. J. Watts for the test data shown in Fig. 1.14 and to G. Dessai for reading the manuscript and numerous useful comments.

References

1. Aarts, A.C.T., van der Hout, R., Paasschens, J.C.J., Scholten, A.J., Willemsen, M.B., Klaassen, D.B.M.: New fundamental insights into capacitance modeling of laterally nonuniform MOS devices. IEEE Trans. Electron Devices **53**(2), 270–278 (2006)
2. Arora, N.D., Gildenblat, G.S.: A semi-empirical model of the MOSFET inversion layer mobility for low-temperature operation. IEEE Trans. Electron Devices **34**, 89–93 (1987)
3. Bendix, P., Rakers, P., Wagh, P., Lemaitre, L., Grabinski, W., McAndrew, C.C., Gu, X., Gildenblat, G.: RF distortion analysis with compact MOSFET models. In: Proc. IEEE Custom Integr. Circuits Conf., pp. 9–12, 3–6 October 2004
4. Boothroyd, A.R., Tarasewicz, S.W., Slaby, C.: MISNAN—a physically based continuous MOSFET model for CAD applications. IEEE Trans. Electron Devices **10**, 1512–1529 (1991)
5. Brews, J.R.: A charge-sheet model of the MOSFET. Solid-State Electron. **21**, 345–355 (1978)
6. Brews, J.R.: Threshold shifts due to nonuniform doping profiles in surface channel MOSFETs. IEEE Trans. Electron Devices **ED-26**, 1696 (1979)
7. Choi, C.-H., Nam, K.-Y., Yu, Z., Dutton, R.W.: Impact of gate direct tunneling current on circuit performance: A simulation study. IEEE Trans. Electron Devices **48**, 2823–2829 (2001)
8. Cai, J., Sah, C.T.: Gate tunneling currents in ultrathin oxide metal–oxide–silicon transistors. J. Appl. Phys. **89**(4), 2272–2285 (2001)
9. Cao, K.M., Lee, W.C., Liu, W., Jin, X., Su, P., Fung, S.K.H., An, J.X., Yu, B., Hu, C.: BSIM4 gate leakage model including source-drain partition. In: IEDM Tech. Dig., pp. 815–818 (2000)
10. Chen, T.L., Gildenblat, G.: Analytical approximation for the MOSFET surface potential. Solid-State Electron. **45**(2), 335–339 (2001)
11. Chen, T.L., Gildenblat, G.: Symmetric bulk charge linearization of charge-sheet MOSFET model. Electron. Lett. **37**(12), 791–793 (2001)
12. Chen, J., Chan, T.Y., Ko, P.K., Hu, C.: Subbreakdown drain leakage current in MOSFET. IEEE Electron Device Lett. **8**, 515–517 (1987)
13. El-Mansy, Y.A., Boothroyd, A.R.: A simple two-dimensional model for IGFET operation in the saturation region. IEEE Trans. Electron Devices **24**, 254–262 (1977)
14. Enz, C.: An MOS transistor model for RF IC design valid in all regions of operation. IEEE Trans. Microw. Theory Tech. **50**, 342–359 (2002)

15. Ghibaudo, G., Roux-dit Buisson, O.: Low-frequency fluctuations in scaled down silicon CMOS devices status and trends. In: Proc. Eur. Solid-State Device Res. Conf., vol. 94, pp. 693–700 (1994)
16. Gildenblat, G., Chen, T.L.: Overview of an advanced surface-potential-based MOSFET model. In: Int. Conf. on Modeling and Simul. Microsyst., pp. 657–661 (2002)
17. Gildenblat, G., Chen, T.L., Bendix, P.: Analytical approximation for the perturbation of MOSFET surface potential by polysilicon depletion layer. Electron. Lett. **35**, 1999 (1974)
18. Gildenblat, G., Cai, X., Chen, T.L., Gu, X., Wang, H.: Reemergence of the surface-potential-based compact models. In: IEDM Tech. Dig., pp. 863–866 (2003)
19. Gildenblat, G., Chen, T.L., Gu, X., Wang, H., Cai, X.: SP: an advanced surface-potential-based model. In: Proc. IEEE Custom Integr. Circuits Conf., pp. 233–240 (2003)
20. Gildenblat, G., Wang, H., Chen, T.L., Cai, X.: SP: an advanced surface-potential-based compact MOSFET model. IEEE J. Solid-State Circuits **39**, 1394–1406 (2004)
21. Gildenblat, G., Li, X., Wu, W., Wang, H., Jha, A., van Langevelde, R., Smit, G.D.J., Scholten, A.J., Klaassen, D.B.M.: PSP: an advanced surface-potential-based MOSFET model for circuit simulation. IEEE Trans. Electron Devices **53**(9), 1979–1993 (2006)
22. Gildenblat, G., Zhu, Z., McAndrew, C.C.: Surface potential equation for bulk MOSFET. Solid-State Electron. **53**(1), 11–13 (2009)
23. Grove, A.: Physics and Technology of Semiconductor Devices. Wiley, New York (1967)
24. Gu, X., Chen, T.L., Gildenblat, G., Workman, G.O., Veeraraghavan, S., Shapira, S., Stiles, K.: A surface potential-based compact model of n-MOSFET gate-tunneling current. IEEE Trans. Electron Devices **51**(1), 127–135 (2004)
25. Gu, X., Gildenblat, G., Workman, G., Veeraraghavan, S., Shapira, S., Stiles, K.: A surface-potential-based extrinsic compact MOSFET model. In: Tech. Proc. Nanotechnol. Conf., pp. 364–367 (2003)
26. Gu, X., Wang, H., Chen, T.L., Gildenblat, G.: Substrate current in surface-potential-based compact MOFET models. In: Tech. Proc. Nanotechnol. Conf., pp. 310–313 (2003)
27. Huang, C., Arora, N.: Characterization and modeling of the n- and p-channel MOSFET's inversion-layer mobility in the range 25–125°C. Solid-State Electron. **37**, 97–103 (1994)
28. Hung, K.K., Ko, P.K., Hu, C., Cheng, Y.C.: A physics-based MOSFET noise model for circuit simulators. IEEE Trans. Electron Devices **37**(5), 1323–1333 (1990)
29. Hung, K.K., Ko, P.K., Hu, C., Cheng, Y.C.: A unified model for the flicker noise in metal-oxide-semiconductor field-effect transistors. IEEE Trans. Electron Devices **37**(3), 654–665 (1990)
30. Hwang, S., Yoon, T., Kwon, D., Yu, Y., Kim, K.: A physics-based, SPICE-compatible non-quasi-static MOS transient model based on the collocation method. Jpn. J. Appl. Phys. **37**(2A), 119–121 (1998)
31. Joardar, K., Gullapalli, K.K., McAndrew, C.C., Burnham, M.E., Wild, A.: An improved MOSFET model for circuit simulation. IEEE Trans. Electron Devices **45**(1), 134–148 (1998)
32. Kane, E.: Zener tunneling in semiconductors. J. Phys. Chem. Solids **12**(2), 181–188 (1959)
33. Li, X., Wu, W., Jha, A., Gildenblat, G., Langevelde, R.V., Smit, G.D.J., Scholten, A.J., Klaassen, D.B.M., McAndrew, C.C., Watts, J., Olsen, M., Coram, G., Chaudhry, S., Victory, J.: Benchmarking PSP compact model of MOS transistors. In: IEEE Int. Conf. on Microelectron. Test Structures, pp. 259–264 (2007)
34. Li, X., Wu, W., Jha, A., Gildenblat, G., van Langevelde, R., Smit, G.D.J., Scholten, A.J., Klaassen, D.B.M., McAndrew, C.C., Watts, J., Olsen, C.M., Coram, G.J., Chaudhry, S., Victory, J.: Benchmark tests for MOSFET compact models with application to the PSP model. IEEE Trans. Electron Devices **56**(2), 243–251 (2009)
35. Litwin, A.: Overlooked interfacial silicide-polysilicon gate resistance in MOS transistors. IEEE Trans. Electron Devices **48**(9), 2179–2181 (2001)
36. Liu, W., Bowen, C., Chang, M.C.: A CAD-compatible non-quasi-static MOSFET model. In: IEDM Tech. Dig., pp. 151–154 (1996)
37. Mancini, P., Turchetti, C., Masetti, G.: A non-quasi-static analysis of the transient behavior of the long-channel MOST valid in all regions of operation. IEEE Trans. Electron Devices **ED-34**, 325–334 (1987)

38. McAndrew, C.C.: Practical modeling for circuit simulation. IEEE J. Solid-State Circuits **33**(3), 439–448 (1998)
39. McAndrew, C.C.: Validation of MOSFET model source-drain symmetry. IEEE Trans. Electron Devices **53**(9), 2202–2206 (2006)
40. McAndrew, C.C., Victory, J.J.: Accuracy of approximations in MOSFET charge models. IEEE Trans. Electron Devices **49**(1), 72–81 (2002)
41. Miura-Mattausch, M., Sadachika, N., Navarro, D., Suzuki, G., Takeda, Y., Miyake, M., Warabino, T., Mizukane, Y., Inagaki, R., Ezaki, T., Mattausch, H.J., Ohguro, T., Iizuka, T., Taguchi, M., Kumashiro, S., Miyamoto, S.: Hisim2: Advanced MOSFET model valid for RF circuit simulation. IEEE Trans. Electron Devices **53**(9), 1994–2007 (2006)
42. Mudanai, S., Shih, W.K., Rios, R., Xi, X., Rhew, J.H., Kuhn, K., Packan, P.: Analytical modeling of output conductance in long-channel halo-doped MOSFETs. IEEE Trans. Electron Devices **53**(9), 2091–2097 (2006)
43. Paasschens, J.C.J., Scholten, A.J., van Langevelde, R.: Generalizations of the Klaassen-Prins equation for calculating the noise of semiconductor device. IEEE Trans. Electron Devices **52**(11), 2463–2472 (2005)
44. Pao, H.C., Sah, C.T.: Effects of diffusion current on characteristics of metal-oxide (insulator)-semiconductor transistors. Solid-State Electron. **9**, 927–937 (1966)
45. Park, H.J., Ko, P.K., Hu, C.: A charge sheet capacitance model of short channel MOSFETs for SPICE. IEEE Trans. Comput.-Aided Des. Integr. Circuits Syst. **10**(3), 376–389 (1991)
46. PSP Group: PSP Manual Version 102.2 (2007). http://pspmodel.asu.edu/downloads/psp1022_summary.pdf
47. Rios, R., Mudanai, S., Shih, W.K., Packan, P.: An efficient surface potential solution algorithm for compact MOSFET models. In: IEDM Tech. Dig., pp. 755–758 (2004)
48. Rios, R., Shih, W.K., Shah, A., Mudanai, S., Packan, P., Sandford, T., Mistry, K.: A three-transistor threshold voltage model for halo processes. In: IEDM Tech. Dig., pp. 113–116 (2002)
49. Sabnis, A.G., Clemens, J.T.: Characterization of the electron mobility in the inverted <100> Si surface. In: IEDM Tech. Dig., vol. 25, pp. 18–21 (1979)
50. Sah, C.T.: Fundamentals of Solid-State Electronics. World Scientific, Singapore (1991)
51. Scharfetter, D.L., Gummel, H.K.: Large-signal analysis of a silicon read diode oscillator. IEEE Trans. Electron Devices **16**, 64–77 (1969)
52. Scholten, A.J., Tiemeijer, L.F., Vreede, P.W.H.D., Klaassen, D.B.M.: A large signal non-quasi-static MOS model for RF circuit simulation. In: IEDM Tech. Dig., pp. 5–8 (1999)
53. Scholten, A., Duffy, R., van Langevelde, R., Klaassen, D.: Compact modelling of pocket-implanted MOSFETs. In: Proc. Eur. Solid-State Device Res. Conf., pp. 311–314 (2001)
54. Scholten, A.J., Tiemeijer, L.F., van Langevelde, R., Havens, R.J., Zegers-van Duijnhoven, A.T.A., Venezia, V.C.: Noise modeling for RF CMOS circuit simulation. IEEE Trans. Electron Devices **50**(3), 618–632 (2003)
55. Scholten, A.J., Smit, G.D.J., Durand, M., van Langevelde, R., Klaassen, D.B.M.: The physical background of JUNCAP2. IEEE Trans. Electron Devices **53**(9), 2098–2107 (2006)
56. Scholten, A.J., van Langevelde, R., Tiemeijer, L.F., Klaassen, D.B.M.: Compact modeling of noise in CMOS. In: Proc. IEEE Custom Integr. Circuits Conf., pp. 711–716 (2006)
57. Scholten, A.J., Smit, G.D.J., De Vries, B.A., Tiemeijer, L.F., Croon, J.A., Klaassen, D.B.M., van Langevelde, R., Li, X., Wu, W., Gildenblat, G.: The new CMC standard compact MOS model PSP: advantages for RF applications (invited). In: IEEE Radio Freq. Integr. Circuits Symp., pp. 247–250 (2008)
58. Scholten, A.J., Smit, G.D.J., De Vries, B.A., Tiemeijer, L.F., Croon, J.A., Klaassen, D.B.M., van Langevelde, R., Li, X., Wu, W., Gildenblat, G.: The new CMC standard compact MOS model PSP: advantages for RF applications. IEEE J. Solid-State Circuits **44**(5), 1415–1424 (2009)
59. Seeger, K.: Semiconductor Physics: An Introduction, 9th edn. Springer, Berlin (2004)
60. Shih, W.K., Rios, R., Packan, P., Mistry, K., Abbott, T.: A general partition scheme for gate leakage current suitable for MOSFET compact models. In: IEDM Tech. Dig., pp. 293–296 (2001)

61. Sze, S.M., Ng, K.K.: Physics of Semiconductor Devices, 3rd edn. Wiley, New York (2006)
62. Taur, Y., Ning, T.: Fundamentals of Modern VLSI Devices. Cambridge University Press, Cambridge (1998)
63. Taylor, G.W.: The effects of two-dimensional charge sharing on the above-threshold characteristics of short-channel IGFETS. Solid-State Electron. **22**(8), 701–717 (1979)
64. Tsividis, Y.: Operation and Modeling of the MOS Transistor, 2nd edn. McGraw-Hill, New York (1999)
65. Tsu, R., Esaki, L.: Tunneling in a finite superlattice. Appl. Phys. Lett. **22**, 562–564 (1973)
66. van der Ziel, A., Chenette, E.R.: Noise in solid state devices. IEEE Trans. Electron Devices **46**, 313 (1978)
67. van Langevelde, R.: A compact MOSFET model for distortion analysis in analog circuit design. Ph.D. thesis (1998)
68. van Langevelde, R., Gildenblat, G.: PSP: An Advanced Surface-Potential-Based MOSFET Model. Springer, Dordrecht (2006), Chap. 2, pp. 22–69
69. van Langevelde, R., Klaassen, F.M.: Accurate drain conductance modeling for distortion analysis in MOSFETs. In: IEDM Tech. Dig., pp. 313–316 (1997)
70. van Langevelde, R., Klaassen, F.M.: An explicit surface-potential-based MOSFET model for circuit simulation. Solid-State Electron. **44**, 409–418 (2000)
71. van Langevelde, R., Paasschens, J.C.J., Scholten, A.J., Havens, R.J., Tiemeijer, L.F., Klaassen, D.B.M.: New compact model for induced gate current noise. In: IEDM Tech. Dig., pp. 867–870 (2003)
72. van Langevelde, R., Scholten, A.J., Duffy, R., Cubaynes, F.N., Knitel, M.J., Klaassen, D.B.M.: Gate current modeling: Δl extraction and impact on RF performance. In: IEDM Tech. Dig., pp. 289–292 (2001)
73. van Langevelde, R., Scholten, A.J., Havens, R.J., Tiemeijer, L.F., Klaassen, D.B.M.: Advanced compact MOS modelling. In: Proc. Eur. Solid-State Device Res. Conf., pp. 81–88 (2001)
74. van Langevelde, R., Scholten, A.J., Klaassen, D.B.M.: Physical background of MOS Model 11. Nat. Lab. Unclassified Report, NL-TN 2003/00239 (2003). http://www.nxp.com/models/mos_models/model11/.
75. van Langevelde, R., Scholten, A.J., Klaassen, D.B.M.: Recent enhancements of MOS model 11. In: Tech. Proc. Nanotechnol. Conf., pp. 60–65 (2004)
76. Victory, J., Yan, Z., Gildenblat, G., McAndrew, C.C., Zheng, J.: A physically based, scalable MOS varactor model and extraction methodology for RF applications. IEEE Trans. Electron Devices **52**(7), 1343–1353 (2005)
77. Wang, H., Chen, T.L., Gildenblat, G.: Quasi-static and non-quasi-static compact MOSFET models based on symmetrically linearization of the bulk and inversion charges. IEEE Trans. Electron Devices **50**(11), 2262–2272 (2003)
78. Wang, H., Gildenblat, G.: Scattering matrix based compact MOSFET model. In: IEDM Tech. Dig., pp. 125–128 (2002)
79. Wang, H., Li, X., Wu, W., Gildenblat, G., van Langevelde, R., Smit, G.D.J., Scholten, A.J., Klaassen, D.B.M.: A unified non-quasi-static MOSFET model for large-signal and small-signal simulations. IEEE Trans. Electron Devices **53**(9), 2035–2043 (2006)
80. Ward, D.E., Dutton, R.W.: A charge-oriented model for MOS transistor capacitances. IEEE J. Solid-State Circuits **13**, 703–708 (1978)
81. Watts, J., McAndrew, C.C., Enz, C., Galup-Montoro, C., Gildenblat, G., Hu, C., van Langevelde, R., Miura-Mattausch, M., Rios, R., Sah, C.T.: Advanced compact models for MOSFETs. In: Tech. Proc. Workshop on Compact Modeling, pp. 3–12 (2005)
82. Wright, P.J., Saraswat, K.C.: Thickness limitations of SiO_2 gate dielectrics for MOS ULSI. IEEE Trans. Electron Devices **37**, 1990 (1884–1892)
83. Wu, W., Chen, T.L., Gildenblat, G., McAndrew, C.C.: Physics-based mathematical conditioning of the MOSFET surface potential equation. IEEE Trans. Electron Devices **51**(7), 1196–1199 (2004)
84. Wu, W., Li, X., Wang, H., Gildenblat, G., Workman, G.O., Veeraraghavan, S., McAndrew, C.C.: SP-SOI: A third generation surface potential based compact SOI MOSFET model. In: Proc. IEEE Custom Integr. Circuits Conf., pp. 819–822 (2005)

85. Wu, W., Li, X., Gildenblat, G., Workman, G., Veeraraghavan, S., McAndrew, C.C., van Langevelde, R., Smit, G.D.J., Scholten, A.J., Klaassen, D.B.M., Watts, J.: PSP-SOI: A surface potential based compact model of partially depleted SOI MOSFETs (invited). In: Proc. IEEE Custom Integr. Circuits Conf., pp. 41–48 (2007)
86. Wu, W., Li, X., Gildenblat, G., Workman, G.O., Veeraraghavan, S., McAndrew, C.C., van Langevelde, R., Smit, G.D.J., Scholten, A.J., Klaassen, D.B.M.: A compact model for valence-band electron tunneling current in partially depleted SOI MOSFETs. IEEE Trans. Electron Devices **54**(2), 316–322 (2007)

Chapter 2
PSP-SOI: A Surface-Potential-Based Compact Model of SOI MOSFETs

Weimin Wu, Wei Yao, and Gennady Gildenblat

Abstract Surface-potential-based models, which represent the mainstream approach to compact modeling of bulk MOSFETs, are now in the process of being applied to SOI devices. In this chapter we discuss two advanced SOI models—PSP-SOI-PD for partially depleted devices and PSP-SOI-DD including the dynamic depletion effects. Both models are based on the popular PSP model of bulk MOSFETs. The theoretical foundation of all PSP-family models is the symmetric linearization method that allows one to raise the physical contents of the compact model without prohibitive increase in its computational complexity. In addition to the physics-based structure of the new models inherited from bulk PSP, they account for phenomena specific to SOI devices (e.g. floating body, and valence band tunneling current) and include a detailed description of parasitic effects. We discuss both the theoretical developments and verification of the model against test data and TCAD simulations with particular emphasis on the interplay between the model structure and its simulation capabilities.

2.1 Introduction

Compact models of SOI devices provide a bridge between the manufacturing process and circuit design. They are required to accurately reproduce the device characteristics responsible for placing SOI technology in the mainstream for low-power high performance ULSI applications [12, 46]. These include reduced junction capacitance, elimination of body effect in stacked devices (e.g. NMOS transistors in NAND gates), dynamic threshold voltage shift brought about by the floating body effect (FBE) and the corresponding increase of the ON/OFF ratio which is beneficial for the low power CMOS SOI applications. At the same time particular attention

W. Wu (✉) · W. Yao · G. Gildenblat
School of Electrical, Computer, and Energy Engineering, Arizona State University, Tempe, AZ 85287, USA
e-mail: Weimin.Wu@asu.edu

G. Gildenblat (ed.), *Compact Modeling*,
DOI 10.1007/978-90-481-8614-3_2,

should be paid to the reproduction of numerous parasitic elements (e.g. body contact resistance) and secondary effects (e.g. self-heating) which are not present or less important in bulk CMOS technology. More recently growing RF applications of SOI technology placed even more stringent requirements on the qualitative behavior of compact SOI models: it is now common to request that advanced SOI compact models have the capability to model intermodulation effects. This requires preserving the model symmetry with respect to the source-drain interchange in order to obtain non-singular $I(V)$ and $C(V)$ characteristics at zero drain bias. In other words, it is now expected that SOI compact models provide all the advanced features of the latest bulk MOSFET models. With this in mind it is natural to take one of these models (PSP) as a starting point in developing compact models of SOI. From a more general point of view, we are witnessing the paradigm change in the development of SOI compact models. With a few years delay relative to the bulk CMOS technology, threshold-voltage based models of SOI MOSFETs are gradually being replaced by the more advanced surface-potential-based models.

The surface-potential-based approach to compact models is not new—for bulk devices it goes all the way back to the classic work of Pao and Sah [41] and its efficient approximation in the charge-sheet model (CSM) [3]. The feasibility of this approach for modeling SOI MOSFETs has been demonstrated in [43, 49, 69]. In particular, complete surface-potential model of partially-depleted SOI devices including all secondary effects and charge model has been developed and verified in [69] starting with the SP model [20] which together with MM11 [60] is one of the predecessors of the industry standard PSP. The new element of the present situation is that for bulk devices the transition to the surface-potential-based models is completed and hence it is time to move from experimental models and feasibility studies to the industrial strength surface potential based models of SOI transistors.

Two such models based on the PSP model [19] and the experiences gained in the process of development of the SP-SOI [69] are discussed in the present chapter. We start with the partially depleted SOI MOSFET model PSP-SOI-PD. Since the model inherits drain current and terminal charges formulation from the bulk PSP model, we concentrate on the effects specific for the SOI devices. In Sect. 2.2 we discuss the modeling of the FBE including impact ionization Junction diode and parasitic bipolar transistor. We also describe the compact model of the electron tunneling from the valence band (EVB) which is usually negligible in bulk MOSFETs but is often required to catch the fine details of the SOI MOSFET transfer characteristics. Both experimental data and circuit implications of the EVB current are considered in some detail. Following the discussion of the self-heating effect in Sect. 2.3, non-linear body resistance model in Sect. 2.4 and noise sources are in Sect. 2.5 we proceed with the parameter extraction procedure and model verification against the experimental data in Sect. 2.6.

The second part of the chapter describes a more general "dynamic depletion" model, PSP-SOI-DD, which can model the transition between the partial depletion and the full depletion regimes of SOI MOSFETs. Superficially, it may appear that the development of the dynamic depletion model obviates the need for a separate model of the partially depleted SOI devices. However, the surface potential formu-

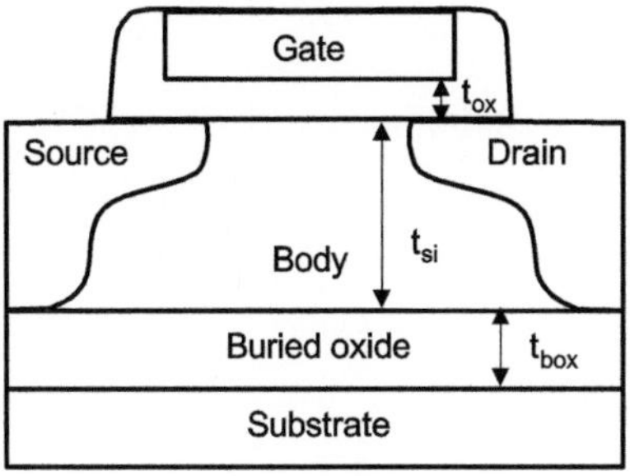

Fig. 2.1 Cross-sectional view of a PD-SOI MOSFET; t_{ox} is the gate oxide thickness, t_{si} is the channel region silicon thickness, and t_{box} is the buried oxide thickness

lation of the DD model is inevitably more complex and requires the solution of coupled surface potential equations at the two (or even three) interfaces. This is justified if there is a real possibility of the DD behavior in the simulated circuits. However, for the PD technology, using separate SOI model of partially depleted devices is more computationally efficient and should be preferred.

Theoretical background of the PSP-SOI-DD model is developed in Sect. 2.7 by developing well-conditioned system of surface potential equations and obtaining a particular version of the symmetric linearization method which underlies all PSP family models and has been recently extended to multiple-gate devices [13, 14]. We also provide a detailed discussion of how the electrostatics of the SOI-DD transistors can be simplified without noticeable effects for the output characteristics. Since PSP-SOI-DD inherits from both PSP and PSP-SOI-PD the discussion of the SOI-specific effects in Sect. 2.2 remains directly applicable to PSP-SOI-DD and is not repeated. Section 2.8 presents model verification by TCAD computations followed by conclusions in Sect. 2.9.

2.2 PD-SOI Floating Body Effect Modeling

PD-SOI with floating body is often the most desirable configuration in SOI technology. In this case, the active channel of a PD-SOI MOSFET is electrically isolated from its underneath substrate by the insulating buried oxide (Fig. 2.1). Unlike the conventional bulk MOSFET, the individual SOI MOSFETs have different body potentials which are determined by various physical mechanisms and needs to be modeled accurately for circuit simulations. The threshold voltage is a function of the body potential and affects the device characteristics. Thus, an accurate characterization and modeling of these mechanisms is essential in determining the body potential and capturing the floating body effect in PD-SOI MOSFETs.

The floating body effect in a PD-SOI MOSFET is manifested by the "kinks" in output characteristics at high drain biases. These are usually caused by the forward bias of the junction produced by the impact ionization current in the body. With scaled ultra-thin gate oxide, the direct gate tunneling current also injects charge carriers into the body and causes "kinks" which can be observed in the transfer characteristics [11, 29, 74].

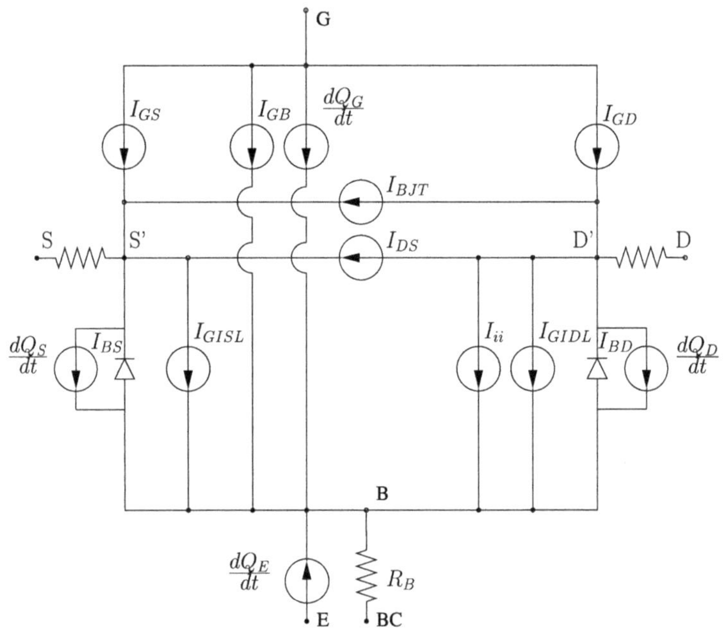

Fig. 2.2 Circuit representation of PSP-SOI. I_{DS}—intrinsic drain current, I_{BJT}—parasitic bipolar current, I_{ii}—impact ionization current, I_{GIDL}/I_{GISL}—gate induced drain (source) leakage. I_{BS}/I_{BD} is source/drain junction current. $I_{GS}/I_{GD}/I_{GB}$ is gate to source/drain/body tunneling current. Q_S, Q_D, Q_G, Q_E are source, drain, gate and back-gate charges. For body-contacted SOI, an extra node (BC) is provided to control the internal body node (B) voltage through a body resistance R_B

All known physical mechanisms that affect the body potential are included in PSP-SOI and are shown in the circuit representation in Fig. 2.2. For the floating body configuration, there are four external nodes: source (S), drain (D), gate (G) and substrate/back-gate (E). In the DC regime, the body bias relative to the source, V_{BS}, is established by the equilibrium between the injection of majority carriers (holes in nMOS) into the body and the removal of majority carriers out of the body. The injection primarily includes: (i) impact ionization near the channel-drain side caused by the high lateral electric field, (ii) reversed biased drain-body junction leakage, (iii) direct tunneling between the gate and the body, (iv) gate-induced drain/source leakage (GIDL/GISL). The majority carriers are removed mainly by (i) forward-biased body-source junction when the body potential is high enough, and (ii) thermal recombination in the junctions. In the transient or AC regime, capacitive coupling between the body and external nodes (source, drain, gate and substrate) also influences the body potential.

For the body-contacted configuration, an external body contact node (BC) connects to the internal body node (B) through a resistive path (body resistance R_B). In this configuration, the balance of currents from the external body contact node, capacitive coupling (displacement currents), and above-mentioned DC paths determines the body potential.

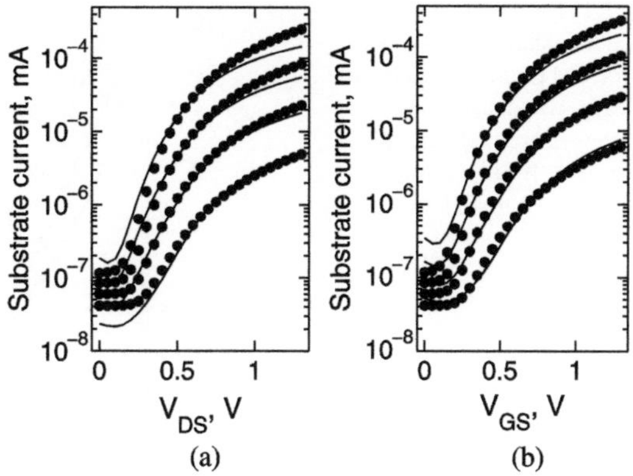

Fig. 2.3 PSP-SOI simulated (*lines*) and measured (*symbols*) substrate current of a body-contacted PD-SOI at 25°C. (**a**) $V_{GS} = 1.0, 1.1, 1.2, 1.3$ V; (**b**) $V_{DS} = 0.8, 0.9, 1.0, 1.3$ V. $W/L = 3$ μm/ 0.065 μm. After [72]

2.2.1 Impact Ionization

PSP-SOI uses the same impact ionization model as PSP [20, 23]. It includes accurate descriptions of the subthreshold region and the effect of body bias V_{BS}:

$$I_{ii} = a_1(V_{DS} - a_3\Delta\psi)\exp\left(-\frac{a_2^*}{V_{DS} - a_3\Delta\psi}\right) I_{DS}, \tag{2.1}$$

$$a_2^* = a_2\,(T_{KD}/T_{KR})^{\kappa_{a2}}\left[1 + a_4\left(\sqrt{2\phi_B - V_{BS}} - \sqrt{2\phi_B}\right)\right]. \tag{2.2}$$

Here a_1, a_2, a_3 and a_4 are model parameters, $\Delta\psi = \psi_{sd} - \psi_{ss}$, ψ_{ss} and ψ_{sd} are the surface potentials at the source and drain ends of the channel, respectively, $\phi_B = \phi_t \ln(N_A/n_i)$ is the Fermi potential, N_A is the substrate doping density and n_i is the intrinsic carrier density of silicon. In SOI MOSFETs, the impact ionization current can be enhanced due to self-heating which increases the channel current [54]. This temperature dependence of I_{ii} is accounted for by introducing the temperature dependence parameter κ_{a2} of impact ionization exponent a_2^*. T_{KD}, T_{KR} are the device and reference temperatures, respectively. Self-heating and its effect on device characteristics are further discussed in Sect. 2.3.

Figure 2.3 shows the model fits the substrate current of a short-channel SOI device very well, including the low drain and low gate bias regions. This is important to capture the experimentally observed gradual turn-on of the "kink" effect that may occur at a drain bias below the impact ionization threshold owing to various energy gain mechanisms [1, 17].

2.2.2 Junction Diode

The characteristics of junction diodes formed by the body and the source or drain are essential to the electrical behaviors of SOI MOSFETs. In PD-SOI technologies,

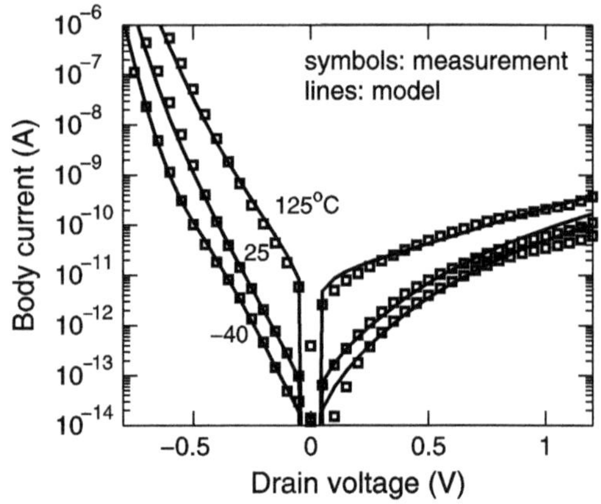

Fig. 2.4 Model fit of junction leakage current. The source and drain terminals of the body-contacted n-channel SOI device are tied together. $V_{GS} = 0$ V. $W/L = 3$ μm/ 0.065 μm. After [72]

the junction leakage characteristics are highly non-ideal, due to the high doping concentration in the HALO region. Thus, the band-to-band tunneling (BTB) and trap-assisted tunneling (TAT) components must be included to model the junction current of PD-SOI MOSFETs in addition to the ideal drift and diffusion current, Shockley-Read-Hall recombination and generation current:

$$I_{B,S/D} = I_{ideal} + I_{SRH} + I_{TAT} + I_{BBT}. \tag{2.3}$$

To achieve the accurate modeling of the junction current, PSP-SOI includes JUNCAP2 [45] diode model. The junction current is very sensitive to the temperature variations. Accordingly, the body potentials and device performance (drain current) will be affected as well. JUNCAP2 includes temperature dependence of all physical diode leakage components, making it ideal for modeling the PD-SOI MOSFETs. The resulting temperature dependence of the body current is shown to be in a good agreement with experimental data. Typical results are illustrated in Fig. 2.4 for three different temperatures.

2.2.3 Parasitic Bipolar Current

The major issue in the PD-SOI technology is the parasitic bipolar effect primarily resulting from the floating body under the gate, which acts as the base of the parasitic bipolar transistor (shown in Fig. 2.1). The source and drain act as the emitter and collector, respectively. The base current is supplied by the impact ionization current and can be amplified by the forward biased emitter-base junction (i.e. source-body).

In circuit operation, the parasitic bipolar effect primarily manifests itself during transient switching if there is a large voltage across the body-to-source junction of the parasitic bipolar transistor. This current can cause extra power consumption, degrade noise margins in static CMOS configurations, or lead to logical state errors in some dynamic circuits [33]. Hence, it needs to be included in PD-SOI compact models.

In PSP-SOI, parasitic bipolar current is modeled by adding a parasitic BJT current element described by a simplified version of the Gummel-Poon model [18, 25]:

$$I_{BJT} = \alpha_{BJT}(I_S/q_B)\left[\exp(\beta V_{BS}) - \exp(\beta V_{BD})\right], \tag{2.4}$$

where the bipolar transport factor $\alpha_{BJT} = \mathrm{sech}(L/L_n)$, L_n is the minority carrier diffusion length and I_S is the saturation current. The normalized base charge q_B given by

$$q_B = \frac{q_1}{2} + \sqrt{\left(\frac{q_1}{2}\right)^2 + q_2}, \tag{2.5}$$

where

$$q_1 = 1 + (V_{BS} + V_{BD})/V_A, \tag{2.6}$$

and

$$q_2 = (4I_S/I_K)\left[\exp(\beta V_{BS}) + \exp(\beta V_{BD}) - 2\right]. \tag{2.7}$$

Here V_A is the Early voltage of the parasitic bipolar transistor and I_K is the knee current. The recombination current in the quasi-neutral body are also included. The source junction contribution is

$$I_{JS,rec} = (1 - \alpha_{BJT})(I_S/f)\left[\exp(\beta V_{BS}) - 1\right], \tag{2.8}$$

where f accounts for the high-level injection effect

$$f = \frac{1}{2} + \sqrt{\frac{1}{2} + \frac{I_S}{I_K}\exp(\beta V_{BS})}. \tag{2.9}$$

A similar expression is used for the drain junction contribution $I_{JD,rec}$. Thus, the total junction leakage current consists of the recombination-generation current in the junction depletion regions, hole and electron diffusion currents and the recombination current in the neutral body region. For completeness, the diffusion capacitances are included as well by introducing transit time coefficient τ_t

$$Q_{JS,diff} = \tau_t \cdot I_{JS,rec}. \tag{2.10}$$

Figure 2.5 shows Gummel plot measured on a body-contacted SOI nMOSFET, which can be used to extract parasitic bipolar current model parameters. A maximum bipolar gain of 10 is observed for this device. At large forward V_{BS}, the bipolar gain becomes smaller due to high-level injection and series resistance.

Figure 2.6 shows a simulation of pass-gate logic with the PSP-SOI model. Initially, the control signal "C" and the input signal "IN" are "High" (V_{DD}). With both the nMOS and pMOS turned on, the drain node and internal body potentials settle to V_{DD}. If the source node (IN) is pulled down after switching the control signal "C" to "Low", a large body-to-source voltage is created . This turns on the parasitic npn BJT and causes a transient bipolar current I_{BJT} to flow. Once the body is discharged, this current disappears.

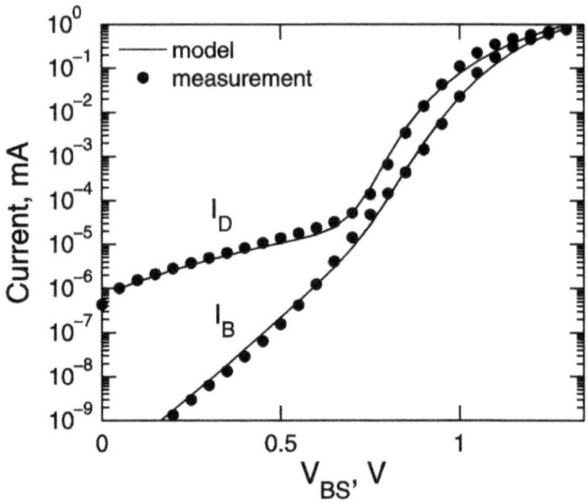

Fig. 2.5 Gummel plot of parasitic BJT in a body-contacted SOI (nMOS). The gate and source are grounded and $V_{DB} = 0$ while sweeping V_{BS}. $W/L = 3\ \mu\text{m}/0.055\ \mu\text{m}$. After [72]

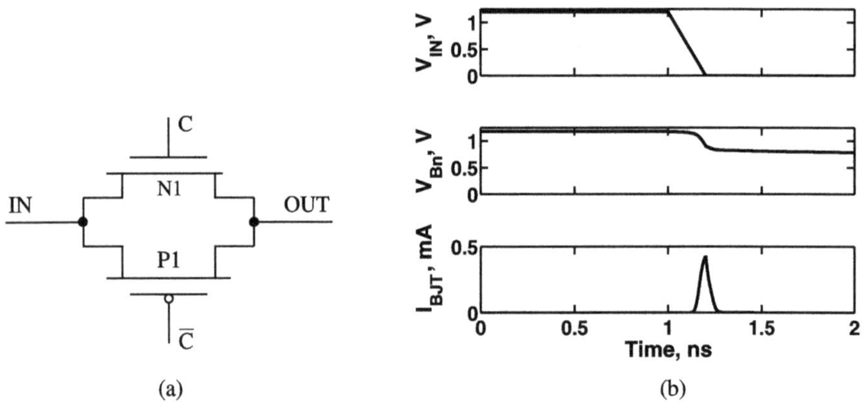

Fig. 2.6 (**a**) Circuit diagram of a basic pass-gate logic. (**b**) Waveforms of input signal V_{IN}, body potential V_{Bn} and parasitic bipolar current I_{BJT} in nMOS. The model parameters are extracted from typical 65 nm PD-SOI technology; $W/L = 30\ \mu\text{m}/0.065\ \mu\text{m}$. After [72]

2.2.4 Gate-to-Body Tunneling Current

As the gate oxide thickness t_{ox} scales to 1.0 nm, the oxide tunneling current increases dramatically. Gate-to-body tunneling includes several components: ECB (electron tunneling from conduction band), HVB (hole tunneling from valence band) and EVB (electron tunneling from valence band) [4]. In bulk MOSFETs EVB tunneling generates the substrate current, which is much less than the gate-to-channel tunneling current (ECB or HVB) and therefore can be neglected. However, in SOI MOSFETs the EVB tunneling current charges and discharges the body, and consequently affects the threshold voltage V_T by altering the body potential. The impact of gate-to-body tunneling current on PD-SOI CMOS circuits has been investigated in [11, 29].

In PSP-SOI, the EVB model is developed from a surface potential based approach. In the Tsu-Esaki formulation [59], the tunneling current density has the form

$$J_{EVB} = \frac{4\pi q m^*}{h^3} \int_0^{qV_{ox}-E_g} D(E_x)(qV_{ox} - E_g - E_x)\,dE_x, \tag{2.11}$$

where q is the magnitude of electron charge, m^* is the effective electron mass in the valence band of Silicon, h is the Planck's constant, E_g is the energy gap, and V_{ox} is the voltage across the oxide which is position dependent. $D(E_x)$ is the tunneling transmission coefficient which for simplicity is evaluated in a WKB approximation:

$$D(E_x) = \exp\left\{-2\int_0^{t_{ox}} \sqrt{\frac{2m_{ox}^*}{\hbar^2}\left[E_C(x) + E_x\right]}\,dx\right\}, \tag{2.12}$$

where integration is over the component of the electron's energy E_x in the direction normal to the potential barrier, $E_C(x)$ is the conduction band energy in the oxide and m_{ox}^* is the effective mass of electrons in SiO_2. To obtain an explicit and simple expression, we assume that the EVB tunneling current is mainly from electrons having energy $E_x = 0$ in the valence band (mono-energetic approximation already used for ECB in [24]). The total EVB tunneling current is obtained by integrating the current density along the channel

$$I_{EVB} = W\int_0^L J_{EVB}\,dy. \tag{2.13}$$

As Fig. 2.7 shows, the drain current I_{DS} is increased due to higher body potential induced by the EVB tunneling. The model accurately reproduces the "linear kink effect" induced by the EVB tunneling current [15, 37], as observed in the transconductance g_m, particularly at low V_{DS}.

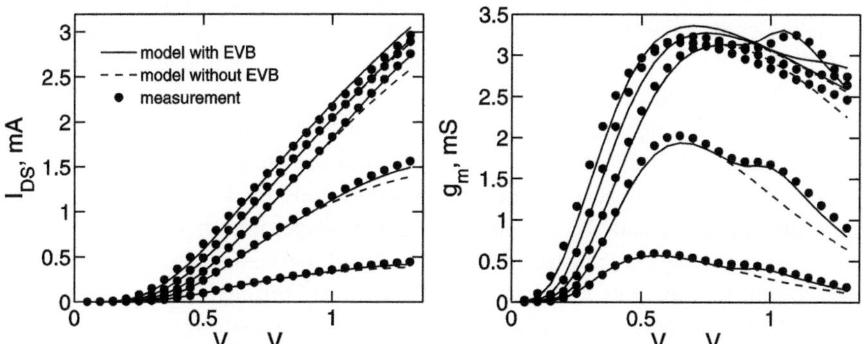

(a) Drain current (b) Transconductance

Fig. 2.7 Impact of gate-to-body tunneling current (EVB) on the DC transfer characteristics of floating body SOI MOSFETs. $V_{DS} = 0.05, 0.2, 0.6, 1.0\ 1.3$ V, $W/L = 3\ \mu m/0.13\ \mu m$. After [72]

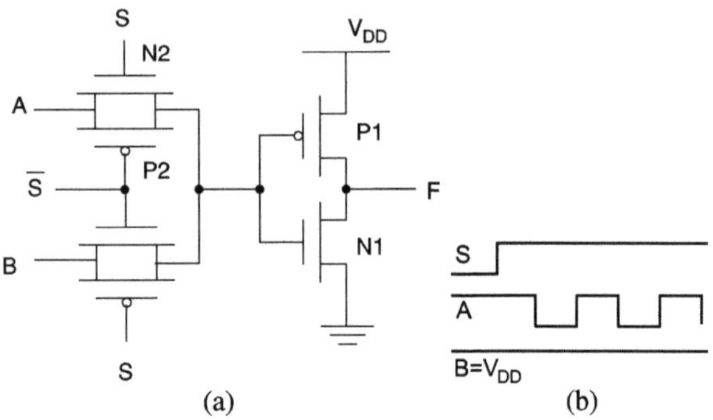

Fig. 2.8 (**a**) Circuit diagram of a transmission-gate multiplexer. (**b**) Input signal waveforms used in simulations. Signal "A" has a period of 2 ns and slew time of 20 ps. After [72]

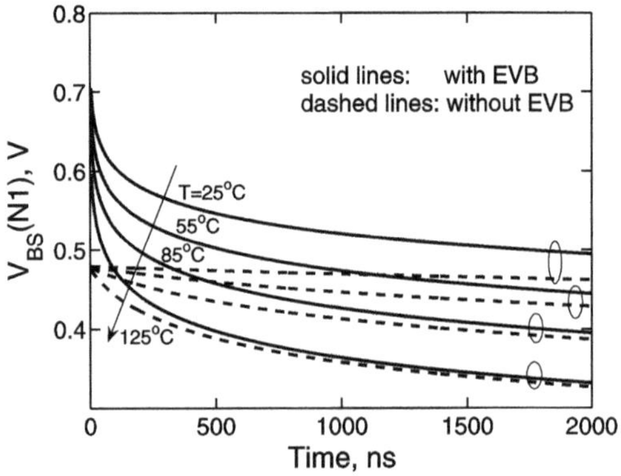

Fig. 2.9 Body potential of nMOS (N1) before the input falling transition. After [72]

Figure 2.8 shows a transmission-gate multiplexer implementing the Boolean function $F = \overline{A \cdot S + B \cdot \bar{S}}$. If initially the inputs "A", "B" are "High" ($V_{A,B} = V_{DD}$) and control signal "S" is "Low", the input of the inverter is settled at "High". Under this scenario the pre-switch body potential of nMOS (N1) is determined by the EVB tunneling current and two forward biased junction currents. The body potential of pMOS (P1) is determined by the balance of back-to-back junction leakage currents. The EVB tunneling current has little impact on the body potential of P1. Consequently, the initial input-fall delay t_{pLH} is larger and the input-rise delay t_{pHL} is smaller because N1 is "stronger" (a lower V_T) in the presence of EVB tunneling current. In the transient steady state, the body potentials are determined by both the capacitive coupling and DC currents (junction leakage, impact ionization current and EVB tunneling), as illustrated in Figs. 2.9 and 2.10. Also, the nMOS (N2) of the top transmission-gate becomes "slightly stronger" during input rise transitions due to the EVB tunneling current. This makes the input rise delay even faster.

As temperature increases, the junction leakage current become larger and the amount of body potential increase caused by the EVB tunneling, which is less temperature sensitive, becomes smaller. The impact of EVB tunneling current on circuit delay times becomes less significant. This is also illustrated in Figs. 2.9 and 2.10.

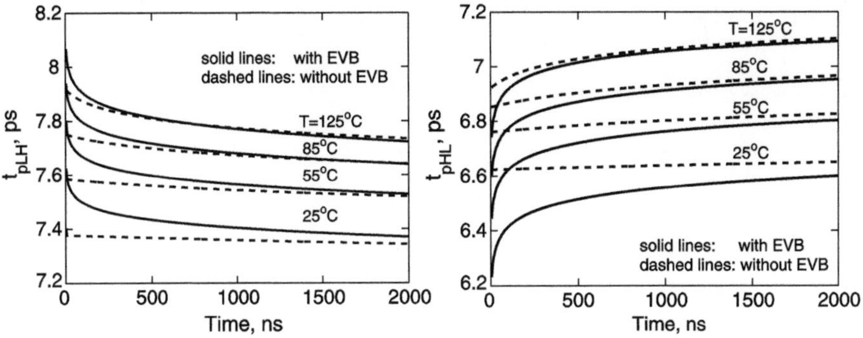

(a) Input-fall delay. (b) Input-rise delay.

Fig. 2.10 Impact of EVB tunneling current on the delay times of a transmission-gate multiplexer with initial "High" condition. Simulations are done at several ambient temperatures. The model parameter sets used in simulations are representing typical 65 nm PD-SOI technology. After [72]

2.2.5 Gate-Induced Drain Leakage Current

In both bulk and SOI MOSFETs, high electrical field may be induced in the gate-to-drain (source) overlap regions when the MOSFET is in off-state (gate bias $V_{GS} \leq 0$). This causes significant leakage current flow between the drain (source) and the body [7]. In addition, for floating body PD-SOI MOSFETs, this current can raise the body potential if the device is biased into accumulation region and may affect the hysteresis behavior [8, 10]. In PSP-SOI, this leakage current is modeled by the same expressions as in bulk MOSFETs [19, 42]

$$I_{GIDL} = A_{GIDL} V_{DB} V_{tov} V_{ov} \exp\left(-B_{GIDL}/V_{tov}\right), \tag{2.14}$$

$$V_{tov} = \sqrt{V_{ov}^2 + (C_{GIDL} \cdot V_{DB})^2}, \tag{2.15}$$

where $A_{GIDL} \propto W \cdot L_{OV}$, W is the channel width and L_{OV} is the overlap length of the gate-to-drain (source) overlap region. V_{ov} is the voltage across the overlap region. V_{tov} is proportional to the maximum electrical field at the Si/SiO_2 interface in the overlap region. B_{GIDL} and C_{GIDL} are model parameters.

2.3 Self-Heating Effect

It is well known that SOI MOSETs suffer from the self-heating effect (SHE) because the buried oxide is an efficient thermal insulator. The Joule heat generated in the channel region can not be transferred outside of the local device quickly, consequently raising the chip temperature. The elevated chip temperature lowers the device performance (by lowering the drain current and transconductances). The self-heating effect in SOI devices and circuits has been extensively studied [51, 56].

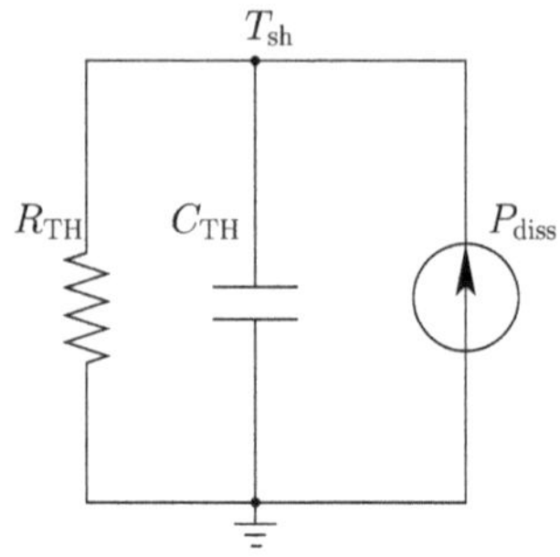

Fig. 2.11 Auxiliary self-heating network. $P_{diss} = I_{DS} \times V_{DS}$. The thermal node T_{sh} is accessible by the user to monitor the device temperature rise in SPICE simulations

In PSP-SOI, the self-heating effect is modeled by a standard auxiliary R_{th}-C_{th} subcircuit [53, 62] (cf. Fig. 2.11). The nodal voltage on T_{sh} is interpreted as the increased local device temperature while R_{th} and C_{th} are thermal resistance and capacitance, respectively. Multifinger SOI devices are also considered in the model of self-heating effect. For example, R_{th} is given by [63]

$$R_{th} = \frac{\mathrm{RTHW}}{\mathrm{NF} \cdot (\mathrm{W} + \mathrm{WTH0})}, \tag{2.16}$$

where RTHW is the normalized thermal resistance, WTH0 is the width offset, and NF is the number of fingers. This is important for low power RF applications where multifinger devices are commonly used.

To accurately model the impact of local device temperature rise on the device characteristics, the temperature dependence of key model parameters are also included. In PSP and PSP-SOI, the temperature dependence of flat-band voltage V_{FB} is accounted for by

$$V_{FB} = V_{FB0} + \kappa_{V_{FB}} (T_{KD} - T_{KR}) . \tag{2.17}$$

Here V_{FB0} is the flat-band voltage at the reference temperature, and $\kappa_{V_{FB}}$ is the temperature coefficient of V_{FB}. The temperature dependence of mobility, carrier saturation velocity and series resistance are modeled by the following empirical equation

$$P = P_0 (T_{KR}/T_{KD})^{\kappa_P} , \tag{2.18}$$

where P is the corresponding model parameter, P_0 is the value of P at the reference temperature, and κ_P is its temperature coefficient.

Figure 2.12 shows the measured and simulated static output characteristics of a floating body SOI device. Simulations for high gate and drain bias without self-heating predict a larger drain current I_{DS} than when self-heating is enabled. The rise of device temperature is about 60–100 K.

In most high performance logic circuit applications, self-heating is negligible since the power consumption per device under switching condition is low [39]. Furthermore, the thermal time constant (0.1–1 μs) is much larger than the switching clock rate and self-heating is therefore effective suppressed. The same simulation

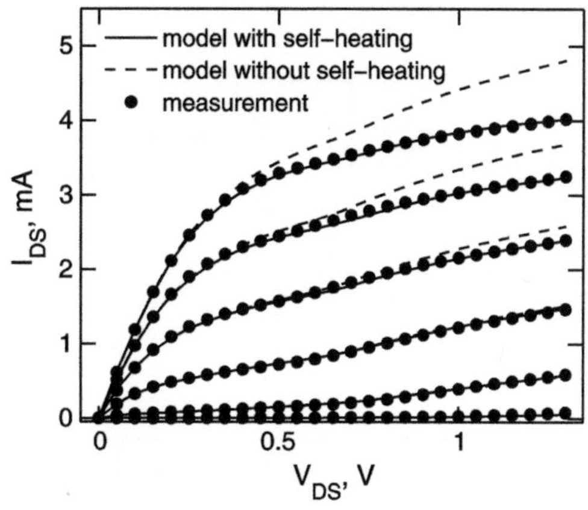

Fig. 2.12 Measured and modeled output characteristics of a floating body SOI nMOSFET. $W/L = 3\ \mu m/0.065\ \mu m$. After [72]

results can be obtained with self-heating disabled in order to improve the computational efficiency of the model.

However, self-heating effect should be taken into account while extracting model parameters. The data are usually taken from DC measurements where self-heating effect is present. The first step in PD-SOI model calibration to hardware data is to determine the thermal resistance and capacitance. They are usually extracted from the AC conductance measurement [28, 39]:

$$G_{DS} = G_{DS0} - \frac{I_{DS0} + G_{DS0} V_{DS}}{V_{DS} - (G_{th} + j\omega C_{th})(\frac{\partial I_{DS}}{\partial T})^{-1}}, \tag{2.19}$$

where G_{DS0} is the intrinsic drain conductance when self-heating is suppressed at high frequency. G_{th} is the thermal conductance. $\frac{\partial I_{DS}}{\partial T}$ can be measured experimentally.

2.4 Body Contact Model

In some critical circuits, especially in high voltage I/O and analog applications, where the instabilities of the threshold voltage are not acceptable, body contacts are used to control the body potential and suppress the floating body effect. A common configuration of the body-contacted (BC) devices is the T-gate structure shown in Fig. 2.13. The body contact is connected to the internal body of the intrinsic device through the region directly under the T-shaped extrinsic gate.

The body contact introduces extra capacitances associated with the extrinsic gate which are modeled by empirical expressions in PSP-SOI. The body resistance, which provides a path for the body currents flowing out the device, depends strongly on the doping profile and channel silicon film thickness. A simple bias-independent model often used to estimate the body resistance is given by

$$R_B = R_{bsh}(W/L), \tag{2.20}$$

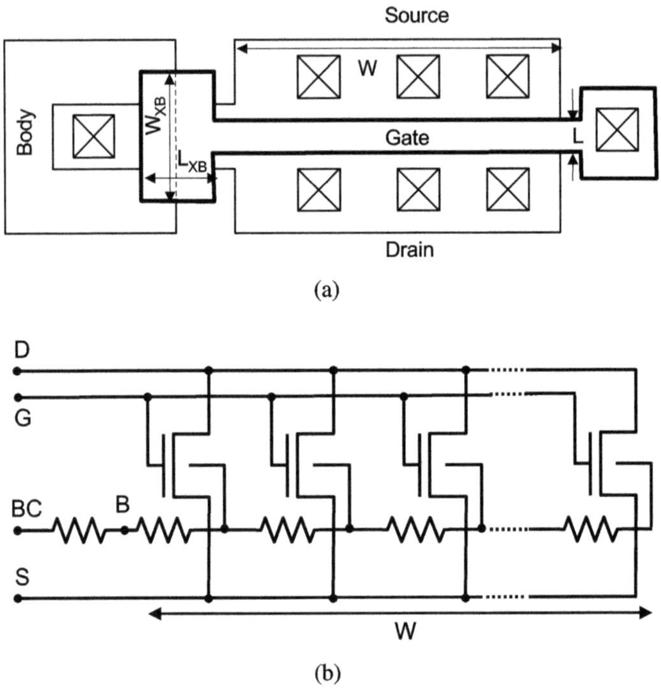

Fig. 2.13 (**a**) Typical structure of a T-gate SOI device. For an H-gate SOI device, another body contact is patterned at the other end of the gate. (**b**) Schematic representation of the T-gate SOI subcircuit taking into account the distributed nature of body resistance

where R_{bsh} is the body sheet resistance and W and L are the width and length of the device. In practical applications the body resistance is highly bias-dependent. In some cases, the silicon film can become fully depleted and the resistance of the body region becomes so high that the body terminal is effectively disconnected from the internal body of the device.

In PSP-SOI, a bias-dependent body resistance model is provided to capture the variation of R_B with the terminal voltages:

$$R_B = W^2/(\mu_B Q_{nbr}), \tag{2.21}$$

where μ_B is the mobility of majority carriers in the body (holes in nMOSFETs), and Q_{nbr} is the absolute value of the mobile charge in the quasi-neutral body region. It can be expressed as

$$Q_{nbr} = qN_{EFF}t_{si}WL - Q_B. \tag{2.22}$$

Here N_{EFF} is the effective channel doping including the effect of halo implants, and t_{si} is the channel silicon film thickness. The total bulk charge Q_B includes the (front) gate induced bulk charge Q_B^f, the junction depletion charges $Q_{j,S/D}$ and the back-gate induced bulk charge Q_E, as illustrated in Fig. 2.14.

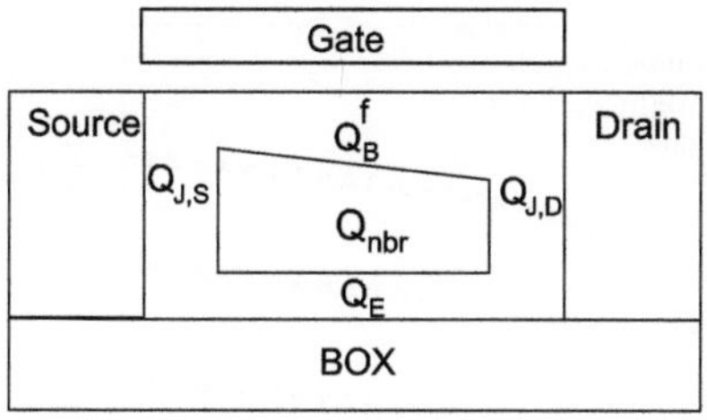

Fig. 2.14 Illustration of the calculation of mobile charge Q_{nbr} in the neutral body region

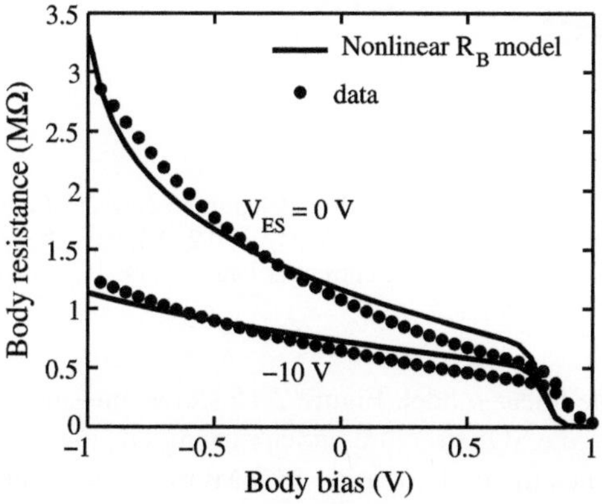

Fig. 2.15 Measured and modeled body resistance of a body-contacted SOI MOSFET (H-gate). $V_{GS} = -0.3$ V, $V_{DS} = 0$ V, back-gate bias $V_{ES} = 0$, -10 V; $W/L = 3$ μm/ 0.065 μm. The sudden drop of body resistance near $V_{BS} = 0.8$ V is caused by the leakage current of highly forward biased junction). After [71]

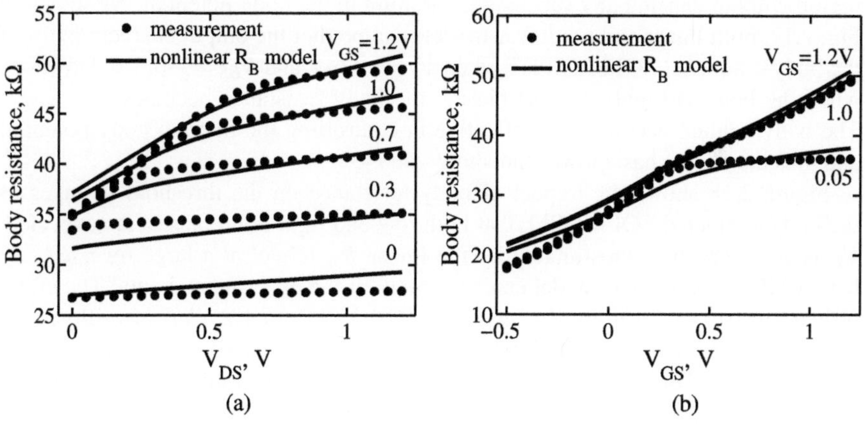

Fig. 2.16 Measured and modeled body resistance of an H-gate SOI structure under varying DC bias condition; $W/L = 4$ μm/2μm. After [70]

Figure 2.15 shows the fitting results of body resistance measured on a body-contacted SOI MOSFET with the nonlinear body resistance model. It clearly shows that the body resistance varies significantly with the external body bias V_{BS} even when the device is off. If the linear, bias-independent model is used, the body resistance is commonly set between the minimum and maximum measured values. In the simulations presented in this work, we set $R_B = 1.09$ MΩ (value at $V_{BS} = 0$ V)

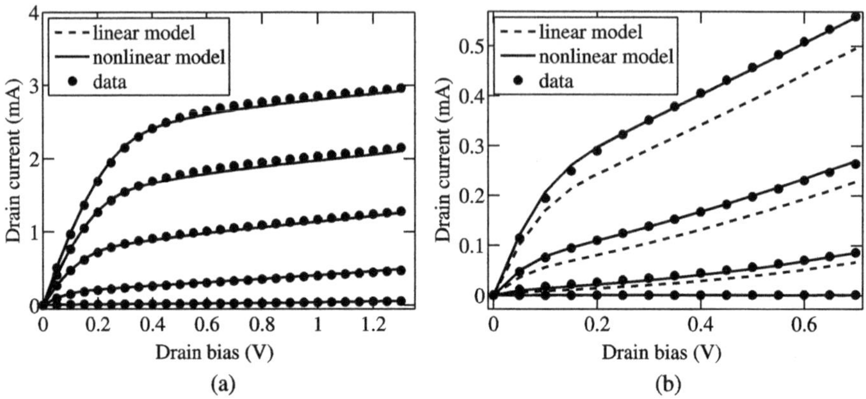

Fig. 2.17 (**a**) Measured and modeled drain current of a body-contacted SOI MOSFET at large forward body bias. $V_{BS} = 0.6$ V, $V_{GS} = 0.2, 0.4, 0.6, 0.8, 1.0$ V; $W/L = 3\ \mu\text{m}/0.065\ \mu\text{m}$. (**b**) Measured and modeled drain current at large reverse body bias. $V_{BS} = -0.6$ V, $V_{GS} = 0.2, 0.4, 0.5, 0.6$ V. After [71]

for the linear model. Figure 2.16 shows measured and modeled body resistance of an H-gate SOI structure under varying DC bias conditions.

When the body of a PD-SOI MOSFET is highly forward biased, the body resistance becomes small enough to bleed off all the DC currents injected into the body region without causing any substantial variation in the body potential. As shown in Fig. 2.17, both linear and nonlinear models can predict the output characteristics of the body-contacted SOI MOSFET when $V_{BS} = 0.6$ V and $V_{BS} = -0.6$ V. However, when the body is highly reverse biased, the body resistance becomes very large. The body contact becomes less effective in controlling the internal body potential and thus the device has a lower threshold voltage .

Figure 2.18 shows the impact of body resistance on the threshold voltages of the body-contacted SOI MOSFET at both low and high drain biases. As expected, V_T is significantly overestimated by the linear R_B model at a large reverse body bias, while the nonlinear model can accurately fit the experimental data. The effect of body resistance on Idlin (I_{DS} at $V_{DS} = 0.05$ V, $V_{GS} = 1.2$ V) is also illustrated in Fig. 2.19. Simulation results indicate that Idlin is increased if the nonlinear R_B model is used. This is consistent with the corresponding results for V_T shown in Fig. 2.18.

In several sensitive applications (e.g. DTMOS), the distributed nature of the body resistance needs to be included in SOI compact models [52, 66]. In PSP-SOI, the internal body node (B) is accessible to the model user so that the distribution effect can be accounted for by partitioning a wide device into several segments along the width direction with proper handling of the narrow-width effects [cf. Fig. 2.14(b)]. For each segment, the width $W_i = W/N_i$, where N_i is an integer number and $\sum_i 1/N_i = 1$ [52]. In this chapter, the effect of body resistance on device characteristics is studied using distributed R_B model.

In principle, segmentation can pose a significant penalty in terms of simulation time. Practically, a lumped body resistance model with the effective resistance

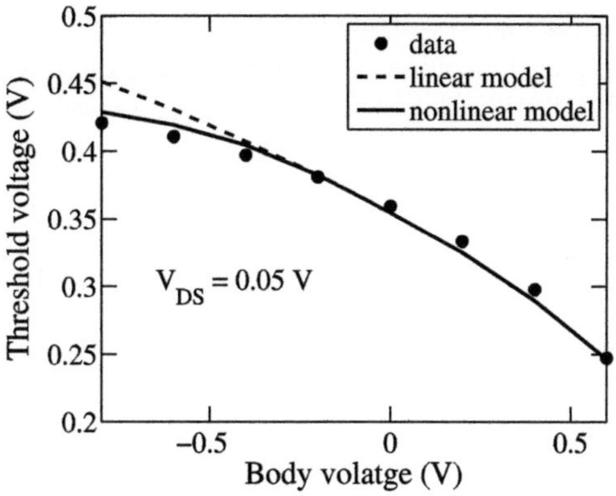

Fig. 2.18 Measured and modeled threshold voltage at low drain bias ($V_{DS} = 0.05$ V). After [71]

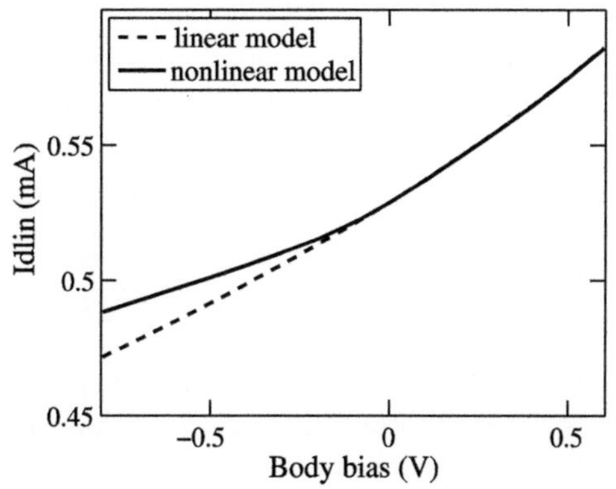

Fig. 2.19 Simulated Idlin (drain current at $V_{GS} = 1.2$ V and $V_{DS} = 0.05$ V). Model parameters for PSP-SOI are extracted from IBM 65 nm PD-SOI technology data. After [71]

$R_{B,\mathit{eff}} = R_B/3$ for T-gate SOI devices or $R_{B,\mathit{eff}} = R_B/12$ for H-gate SOI devices is often adequate [22, 66]. This lumped model works well when hysteresis is negligible, and is even accurate to within 10% in predicting delays when the device is hysteretic [66].

2.5 Noise Modeling

SOI technology has become a viable option for RF applications and RF systems-on-chip. Consequently, accurate noise descriptions in compact models of SOI MOSFETs become essential.

The two main noise sources in MOSFETs are the low frequency noise (also called $1/f$ noise) and the thermal noise. In PSP-SOI, these noise sources, together with the channel induced gate noise are modeled physically following the description developed for the bulk PSP model including velocity saturation effects [19, 44]. In particular, the shot noise in the subthreshold region is recovered automatically from the channel thermal noise using surface potential based formulation. Several other noise sources common to both bulk and SOI MOSFETs are included, such as shot noises associated with gate-tunneling current, junction leakage current, GIDL/GISL and impact ionization. PSP-SOI includes three additional noise sources relative to the bulk PSP model: shot noise associated with the EVB tunneling current with the spectral density [47]

$$S_{I_{EVB}}(\mathrm{G},\mathrm{B}) = 2qI_{EVB}, \tag{2.23}$$

and the shot noise associated with the parasitic BJT [67]

$$S_{I_{BJT}}(\mathrm{D},\mathrm{S}) = 2qI_{BJT}. \tag{2.24}$$

For body-contacted SOI device, the thermal noise generated in the body resistance [16] is

$$S_{V,R_B}(\mathrm{B},\mathrm{BC}) = 4k_BTR_B, \tag{2.25}$$

where the body resistance R_B is given by (2.21).

For floating body PD-SOI MOSFET, experimental data indicate the presence of excess Lorentzian-like noise overshoot in the low-frequency range [27, 57, 67]. The frequency dependence of the excess noise spectral density has been found to be

$$S_f = \frac{S_0}{1+(f/f_c)^2}, \tag{2.26}$$

where S_0 corresponds to the low-frequency plateau and f_c is the corner frequency, which is determined by the small signal impedance of the body node and hence strongly depends on the drain bias through impact ionization current.

The excess noise in the drain current is caused by the floating body effect which amplifies the shot noises associated with the body-source junction current [27], impact ionization current, gate-to-body tunneling current, and for body-contacted SOI, the thermal noise of the body resistance. Since the parasitic body currents and the associated shot noises are physically modeled in PSP-SOI, the excess noise at low frequency with the spectral density (2.26) comes out automatically. The simulated equivalent drain output noise is shown in Fig. 2.20, demonstrating the qualitative agreement of simulation results with experimental observations [27].

2.6 PD-SOI MOSFET Model Verification

PSP-SOI has been verified against several PD-SOI processes, including 90 nm and 65 nm nodes from both Freescale and IBM. The first step in the model verification process is the treatment of self-heating effect for which two methods have been proposed [22]. The common approach is to characterize the thermal resistance through the electrical resistance measurement of the polysilicon gate [51]. The subsequent model fitting and calibration involve the tuning of both temperature-independent and temperature-dependent model parameters simultaneously (cf. "method 1" in

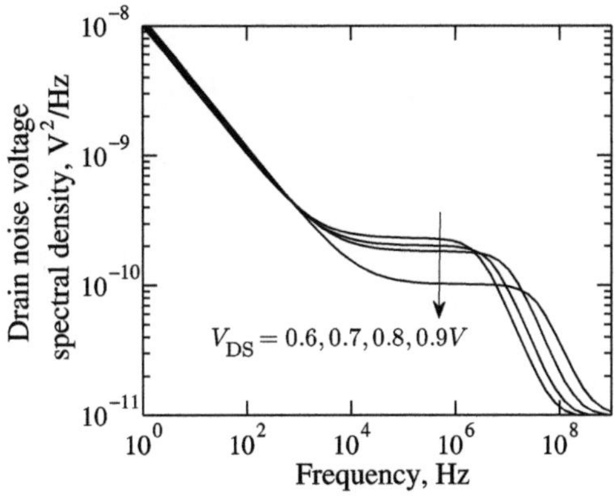

Fig. 2.20 Simulated equivalent drain output noise spectral density; $V_{GS} = 0.8$ V. $W/L = 3$ μm/0.065 μm. Same set of model parameters as in Figs. 2.22 and 2.23 are used in simulation. After [72]

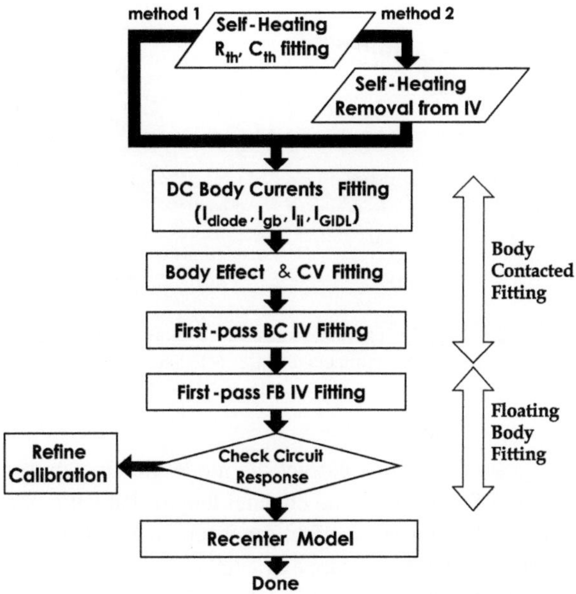

Fig. 2.21 PD-SOI model parameter extraction flow. After [22]

Fig. 2.21). The second approach starts from generating self-heating free $I(V)$ data from static measurements [9, 39] (cf. "method 2" in Fig. 2.21). The self-heating removal is applied to various body currents, such as impact ionization current. In method 2, the temperature-independent and temperature-dependent model parameters are tuned separately which requires fewer iterations relative to method 1.

To fit the floating body device characteristics, parasitic capacitance and current components should be fitted first on body-contacted devices. The reason is that impact ionization current, junction current, gate-to-body tunneling, and GIDL/GISL, etc. are all important in determining the floating body effect but can not be measured directly on floating body devices.

Figures 2.22 and 2.23 show typical model fitting results on an n-channel floating body SOI MOSFET with channel length $L = 55$ nm. In the parameter extrac-

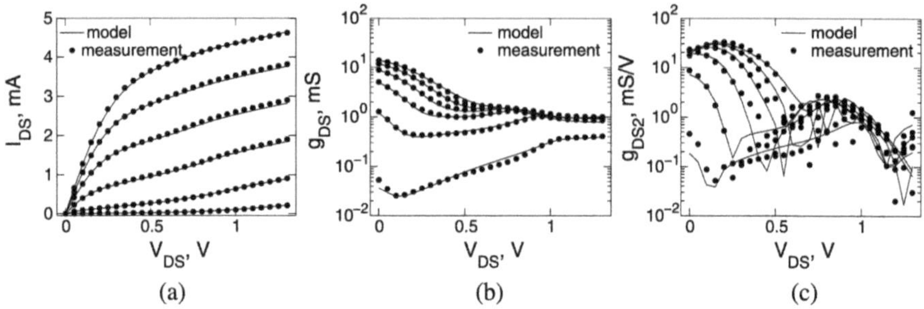

Fig. 2.22 Output characteristics of a short channel floating body nMOSFET. (**a**) Drain current; (**b**) Conductance g_{DS}; (**c**) Second-order conductance $g_{DS2} = \partial g_{DS}/\partial V_{DS}$. $W/L = 3\ \mu m/0.055\ \mu m$. $V_{GS} = 0.2, 0.4, 0.6, 0.8, 1.0, 1.2$ V. After [72]

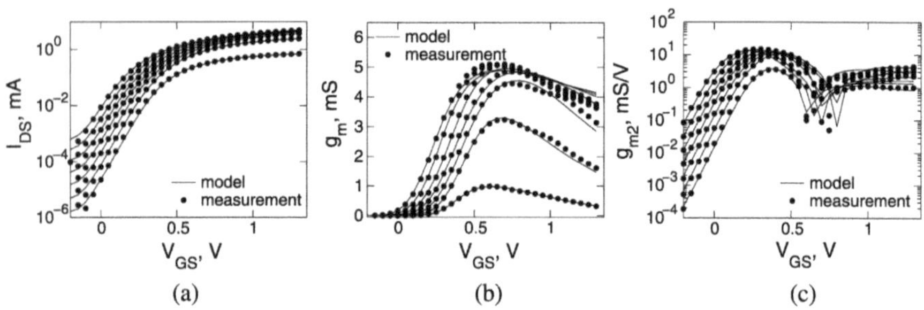

Fig. 2.23 Transfer characteristics of a short channel floating body nMOSFET. (**a**) Drain current; (**b**) Transconductance g_m; (**c**) Second-order transconductance $g_{m2} = \partial g_m/\partial V_{GS}$. $W/L = 3\ \mu m/0.055\ \mu m$. $V_{DS} = 0.05, 0.2, 0.4, 0.6, 0.8, 1.0, 1.2$ V. After [72]

tion routine, we first extract the model parameters on a body-contacted n-channel MOSFET with the same channel length. Parasitic current components which control the floating body effect (as discussed in Sect. 2.2), such as impact ionization, junction leakage, etc., are extracted from separate dc measurements. As we can see, not only the drain current is accurately reproduced by the PSP-SOI model, the conductance [Fig. 2.22(b)] and transconductance [Fig. 2.23(b)] are also reproduced by the model. This demonstrates that PSP-SOI is suitable for both digital and analog applications.

In order to simulate the signal distortion associated with the nonlinearities of PD-SOI MOSFETs, the compact model should reproduce the higher-order conductances and transconductances. Typical results are shown in Figs. 2.22(c) and 2.23(c).

2.7 Modeling of Dynamically Depleted SOI MOSFETs

An SOI MOSFET switching between the PD and FD modes for different terminal voltages is said to be operating in the dynamic depletion mode [49, 55]. Accurate

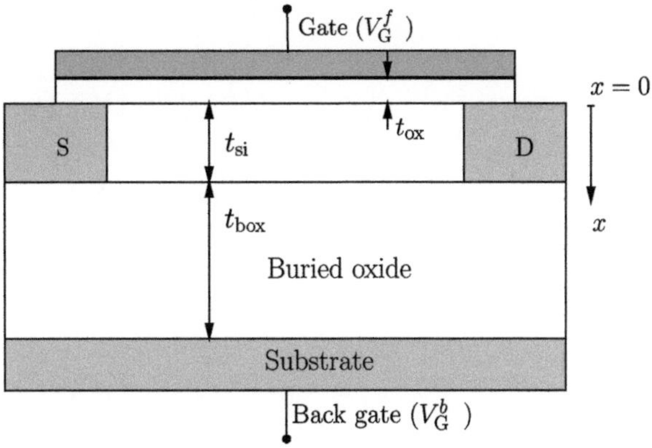

Fig. 2.24 Cross-section of a DD-SOI MOSFET structure. V_G^f is the (front) gate voltage, V_G^b is the back gate (substrate) voltage

modeling of the DD effect can be based on the well-known principles, but the challenge is to make the formulation compatible with the needs of compact modeling without any significant loss of the accuracy essential for the SOI circuit simulations.

There are several approaches to develop compact models of dynamically depleted SOI devices. In the traditional threshold-voltage-based formulation this has been accomplished using the concept of the body-to-source built-in potential lowering to provide a link between the PD and FD modes [55, 63]. The surface-potential-based approach, which has emerged as the mainstream of the compact modeling for both bulk [19, 20] and PD-SOI [38, 69, 72] devices, has been already introduced for the DD-SOI in [2, 26, 30, 49, 75]. It starts from the first integral of Poisson's equation in the channel region. The surface potential equation (SPE) is then obtained by including boundary conditions at the two Si/SiO_2 interfaces. Together with the coupling equation between the front and back surfaces, the front and back surface potentials (denoted as ψ_s^f and ψ_s^b) are calculated and then used to formulate the drain current and the terminal charges models.

PSP-SOI-DD is based on a mathematically well-conditioned SPE and an approximate coupling equation which explicitly includes the back gate effect. To obtain explicit expressions for both the drain current and terminal charges, a new symmetric linearization method is developed specifically for DD-SOI. This allows us to formulate the model within the context of the PSP and PSP-SOI-PD models but with DD effects included. The secondary effects can be then adopted directly from PSP-SOI-PD as in [19, 70, 72].

2.7.1 Surface Potential and Coupling Equations

The surface potential equation for dynamically depleted SOI MOSFETs can be obtained from Possion's equation considering the boundary conditions at the front and back interfaces [40] (cf. Fig. 2.24):

$$\left(V_G^f - V_{FB}^f - \psi_s^f\right)^2 - \left(\frac{t_{ox}}{t_{box}}\right)^2 \left(V_G^b - V_{FB}^b - \psi_s^b\right)^2 \\ = G(\beta\psi_s^f, \Delta_n) - G(\beta\psi_s^b, \Delta_n). \tag{2.27}$$

In PSP-SOI-DD, the form of G is somewhat different than in [40] in order to take advantage of the experience gained in the development of the bulk PSP model:

$$G(x, \Delta_n) = \gamma^2 \phi_t \left[e^{-x} + x - 1 + \Delta_n \left(e^x - x - 1 - \frac{x^2}{x^2+2}\right)\right], \tag{2.28}$$

where ψ_s^f and ψ_s^b are the front and back surface potentials, and V_{FB}^f and V_{FB}^b are the front and back gate flat-band voltages, respectively. γ is the body factor, $\Delta_n = \exp[-\beta(2\phi_B + V_{CB})]$ where V_{CB} is imref splitting. Function $G(x, \Delta_n)$ has been conditioned to prevent the possibility of being negative ($G < 0$) and allows us to set $\Delta\psi_s^f = 0$ in accumulation without producing spurious $C(V)$ characteristics [21, 35, 68].

The rigorous form of the coupling equation required to solve the surface potentials (ψ_s^f and ψ_s^b) from (2.27) is given by [34, 40]

$$t_{si} = \int_{\psi_s^b}^{\psi_s^f} d\phi / E_x, \tag{2.29}$$

where

$$E_x = \mathrm{sgn}(\phi) \cdot \sqrt{\left(E_s^f\right)^2 - (C_{ox}/\varepsilon_s)^2 \left[G(\beta\psi_s^f, \Delta_n) - G(\beta\phi, \Delta_n)\right]} \tag{2.30}$$

is the electric field component along the x direction and E_s^f is the electric field at the front interface. Since the integral in (2.29) requires numerical evaluation, this formulation is not conducive to the compact model development. It is used primarily to verify the accuracy of the approximate closed form coupling equation developed below.

The simplest approximate coupling equation is obtained by neglecting the inversion charge and assuming that Si film is fully depleted [32]:

$$\psi_s^b = \psi_s^f - \psi_c + r_c \left(V_G^b - V_{FB}^b - \psi_s^b\right), \tag{2.31}$$

where $r_c = C_{box}/C_b$, $C_{box} = \varepsilon_{ox}/t_{box}$, $C_b = \varepsilon_s/t_{si}$, and

$$\psi_c = \frac{qN_A t_{si}^2}{2\varepsilon_s}. \tag{2.32}$$

From (2.31)

$$\psi_s^b = \frac{1}{1+r_c}\left(\psi_s^f - \psi_c\right) + \frac{r_c}{1+r_c}\left(V_G^b - V_{FB}^b\right). \tag{2.33}$$

The second term describes the back gate effect.

The coupling equation (2.33) is applicable only in the FD mode of operation. In the PD mode, ψ_s^b is decoupled from ψ_s^f and its value ψ_{s0}^b is determined from

$$\left(\frac{t_{ox}}{t_{box}}\right)^2 \left(V_G^b - V_{FB}^b - \psi_{s0}^b\right)^2 = G(\beta\psi_{s0}^b, 0). \tag{2.34}$$

To explicitly describe the transition between the PD and FD modes, smoothing function is employed:

$$\psi_s^b = \psi_{s0}^b + \frac{\phi_t}{1+r_c}\ln\left\{\frac{1+\exp[\beta(\psi_s^f - \psi_c^*)]}{1+\exp(-\beta\psi_c^*)}\right\}, \tag{2.35}$$

where

$$\psi_c^* = \psi_c - r_c\left(V_G^b - V_{FB}^b\right) + (1+r_c)\psi_{s0}^b. \tag{2.36}$$

For $\beta(\psi_s^f - \psi_s^b) \gg 1$ we recover the FD mode described by (2.33), while for $\psi_s^f < \psi_c^* - 3\phi_t$ the device operates in the PD mode and $\psi_s^b \approx \psi_{s0}^b$. The inclusion of the small factor $\exp(-\beta\psi_c^*)$ assures that $\psi_s^b = \psi_{s0}^b$ for $\psi_s^f = 0$, as expected from the physical considerations.

Equation (2.35) is conceptually similar to that used in [49] but does not assume $r_c = 0$ and uses a more complete form of ψ_c^* including the back gate effect.

2.7.2 Symmetrically Linearized Charge-Sheet Model for DD-SOI

Compact surface-potential-based models of bulk and SOI MOSFETs are based on charge-sheet approximation [3]. Following [40], the front channel inversion charge density (normalized to WLC_{ox}) is approximated by

$$q_i = -(V_G^f - V_{FB}^f - \psi_s^f) + \sqrt{\left(V_G^f - V_{FB}^f - \psi_s^f\right)^2 - \gamma^2\phi_t\Delta_n\left(e^{\beta\psi_s^f} - \beta\psi_s^f - 1\right)}. \tag{2.37}$$

Once surface potentials and inversion charge are available, the intrinsic drain current can be computed from the drift and diffusion equation [3, 58]

$$I_{DS} = -W\mu_n C_{ox}\left(q_i\frac{\partial\psi_s^f}{\partial y} - \phi_t\frac{\partial q_i}{\partial y}\right), \tag{2.38}$$

y is the position along the channel. After integration

$$I_{DS} = -\mu_n \left(\frac{W}{L}\right) C_{ox} \left[\int_{\psi_{ss}^f}^{\psi_{sd}^f} q_i d\psi_s^f + \phi_t (q_{is} - q_{id})\right] \tag{2.39}$$

where ψ_{ss}^f, ψ_{sd}^f and q_{is} and q_{id} are the values of ψ_s^f and q_i at the source and drain ends of the channel, respectively.

In SP [6, 20] and PSP [19] models, the drain current and the intrinsic terminal charges are formulated using symmetric linearization method. The inversion charge (per unit area) is approximated by its first order Taylor expansion at the potential middle point ψ_m:

$$q_i = q_{im} - \alpha_m u, \tag{2.40}$$

where $q_{im} = q_i|_{\psi_s=\psi_m}$ is the inversion charge at the surface potential middle point, $u = \psi_s - \psi_m$, and the linearization coefficient

$$\alpha_m = -\left.\frac{\partial q_i}{\partial \psi_s}\right|_{\psi_s=\psi_m}. \tag{2.41}$$

For an SOI MOSFET operating in the PD mode, (2.40) is the same as for a bulk MOSFET. When MOSFET enters the FD mode, the channel silicon film is fully depleted and the bulk charge density $-qN_A t_{si}$ becomes position-independent and also no longer depends on ψ_s^f. Once it happens [34]

$$q_i = -\left[V_G^f - V_{FB}^f - \psi_s^f - \frac{qN_A t_{si}}{C_{ox}} - \frac{C_b}{C_{ox}}\left(\psi_s^f - \psi_s^b\right)\right] \tag{2.42}$$

becomes an approximately linear function of ψ_s^f. This results in unit value of the linearization coefficient because in FD mode, $\alpha_m^{DD} \approx 1 + C_b/C_{ox} \approx 1$. Hence the original symmetric linearization method [19, 20] can not be applied directly to the modeling of dynamically depleted SOI MOSFETs.

Instead, the following adaptation of symmetric linearization method is used:

$$q_i = q_{im} - \alpha_m^{DD} u_f, \tag{2.43}$$

where $u_f = \psi_s^f - \psi_m^f$ and $\psi_m^f = (\psi_{ss}^f + \psi_{sd}^f)/2$. The linearization coefficient is approximated as

$$\alpha_m^{DD} = \frac{q_{is} - q_{id}}{\Delta\psi_f}, \tag{2.44}$$

where $\Delta\psi_f = \psi_{sd}^f - \psi_{ss}^f$.

A similar form for the linearization coefficient has been already used in the recent core compact model of a double-gate MOSFET [13, 14]. In [14] its accuracy was also verified for a bulk MOSFET case. In the final analysis, the accuracy of (2.43) and (2.44) is established by comparison with numerical computations presented below.

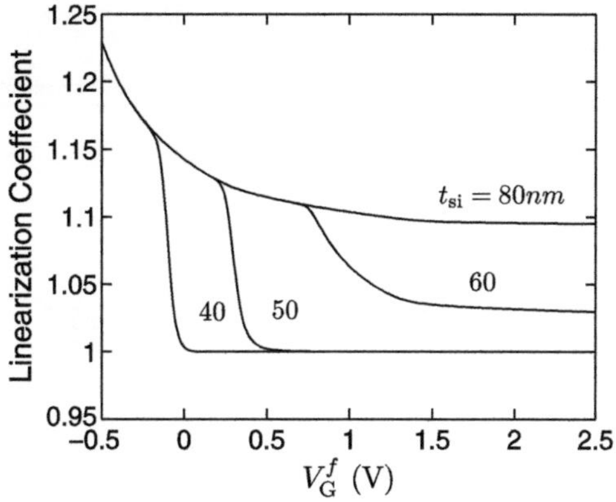

Fig. 2.25 Linearization coefficient α_m^{DD} as a function of the front gate bias for different silicon film thicknesses. $t_{ox} = 2$ nm, $t_{box} = 200$ nm, $N_A = 5 \times 10^{17}$ cm^{-3}. $V_{FB}^f = -0.9$ V, $V_G^b = 0$ V, $V_{FB}^b = 0$ V and $V_{DS} = 1$ V. After [73]

The expression for the drain current follows from (2.39) and (2.43) and has the same form as in [19, 20]

$$I_{DS} = \mu(W/L)C_{ox}\left(q_{im} + \alpha_m^{DD}\phi_t\right)\Delta\psi_f \tag{2.45}$$

except that now q_{im} and α_m^{DD} are different functions of the terminal voltages.

Figure 2.25 shows α_m^{DD} calculated from (2.44) as a function of front gate voltage for several different silicon film thicknesses. For $t_{si} = 80$ nm, the device is always partially depleted, thus, the dependence of α_m^{DD} on the front gate bias is similar to that in a bulk MOSFET. For $t_{si} = 40$, 50 nm, we can see the smooth transition of α_m^{DD} between the PD and FD modes of operation. For $t_{si} = 60$ nm, the silicon film is always partially depleted at the source end, but it can be fully depleted at the drain end, depending on the front gate bias. For this case, we can see that α_m^{DD} lies between its values in PD and FD modes. It follows that without using any smoothing functions expression (2.44) provides the expected limits of the linearization coefficient in the PD and FD modes as well as the smooth transition between the two limiting cases.

The derivation of the compact expressions for the terminal charges requires position dependence of the front surface potential which follows from (2.38) and (2.45):

$$y = y_m + \frac{L}{\Delta\psi_f} \cdot \left[\psi_{ss}^f - \psi_{sm}^f - \frac{(\psi_{ss}^f - \psi_{sm}^f)^2}{2H}\right], \tag{2.46}$$

where

$$y_m = \frac{L}{2} \cdot \left(1 + \frac{\Delta\psi_f}{4H}\right) \tag{2.47}$$

is the coordinate of the front gate surface potential middle point and

$$H = \phi_t - q_{im}/\alpha_m^{DD}. \tag{2.48}$$

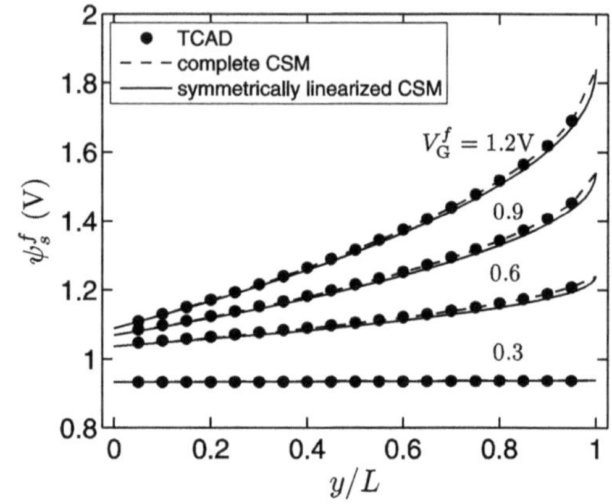

Fig. 2.26 Front surface potential along the channel for the complete and symmetrically linearized charge-sheet model. $t_{ox} = 2$ nm, $N_A = 5 \times 10^{17}$ cm^{-3}, $t_{si} = 40$ nm, $t_{box} = 200$ nm, $V_{FB}^f = -0.8$ V, $V_{FB}^b = 0$ V, $V_{DS} = 1.2$ V. After [73]

Solving for $\psi_s^f(y)$ yields

$$\psi_s^f = \psi_{sm}^f + H\left[1 - \sqrt{1 - \left(\frac{2\Delta\psi_f}{HL}\right)(y - y_m)}\,\right]. \tag{2.49}$$

Comparison of (2.49) with $\psi_s^f(y)$ dependence corresponding to the complete CSM is shown in Fig. 2.26. This further confirms the accuracy of the proposed version of the symmetric linearization method.

With $\psi_s^f(y)$ given by form (2.49) the total gate charge (all charges are normalized to WLC_{ox})

$$Q_G = (1/L)\int_0^L \left(V_G^f - V_{FB}^f - \psi_s^f\right) dy \tag{2.50}$$

can be evaluated in a closed form:

$$Q_G = V_G^f - V_{FB}^f - \psi_{sm}^f + \frac{\Delta\psi_f^2}{12H}. \tag{2.51}$$

The source and drain terminal charges are obtained in the Ward-Dutton partition [65]. We have

$$Q_D = (1/L)\int_0^L (y/L) q_i \, dy, \tag{2.52}$$

or, explicitly

$$Q_D = \frac{q_{im}}{2} + \frac{\alpha_m^{DD}\Delta\psi_f}{12}\left(1 - \frac{\Delta\psi_f}{2H} - \frac{\Delta\psi_f{}^2}{20H^2}\right). \tag{2.53}$$

The source charge $Q_S = Q_I - Q_D$ where

$$Q_I = q_{im} + \frac{\alpha_m^{DD}\Delta\psi_f{}^2}{12H}. \tag{2.54}$$

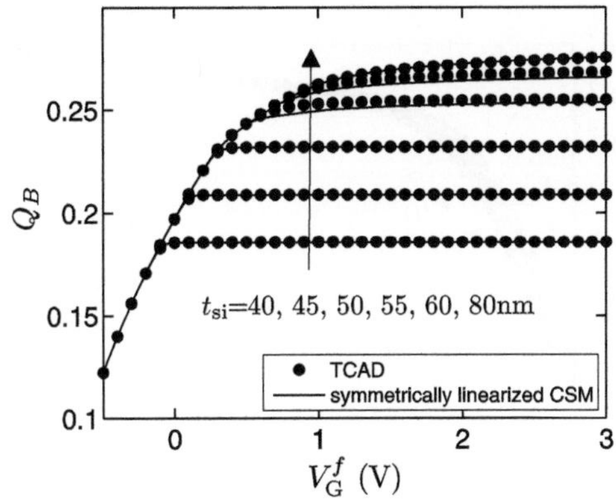

Fig. 2.27 Normalized bulk charge Q_B as a function of the front gate bias. In the simulation, $t_{ox} = 2$ nm, $t_{box} = 200$ nm, $N_A = 5 \times 10^{17}$ cm^{-3}. $V_{FB}^f = -0.9$ V, $V_G^b = 0$, $V_{FB}^b = 0$, $V_{DS} = 0.5$ V. After [73]

Finally, the bulk charge is obtained from the neutrality condition $Q_B = Q_G - Q_I$.

Of the various charges, Q_B is most sensitive to the operating mode of the SOI MOSFET. For this reason it is chosen to investigate the accuracy of the symmetrically linearized CSM relative to the TCAD simulations. The results shown in Fig. 2.27 indicate that introducing symmetric linearization has little or no effect on the CSM output. In particular, the qualitative change in the $Q_B(V_G^f)$ dependence associated with the transition to the FD mode of operation is accurately reproduced.

It is worth noting that for bulk MOSFET analytical expressions for terminal charges are complex but, nevertheless, available in a closed form [36, 64] so that the use of symmetric linearization is a matter of efficiency. For DD-SOI analytical evaluation of the terminal charges using complete CSM is impossible. Hence symmetric linearization is the enabling factor in the formulation of compact model for the terminal charges.

Apart from the different bias dependence of α_m^{DD} and H, the final forms of the expressions (2.51), (2.53) and (2.54) is the same for bulk, PD, DD and multi-gate transistors [13, 19, 20]. The accuracy is about the same in all cases.

2.8 DD-SOI Model Verification and Discussion

In the previous section we considered the core PSP-SOI-DD model and accuracy of symmetric linearization method relative to the complete CSM formulation. A more detailed model verification is accomplished by comparing the $I(V)$ and $C(V)$ characteristics with the results of TCAD simulations. The complete PSP-SOI-DD model including small-geometry effects is implemented in circuit simulators using Verilog-A approach [31]. The front and back surface potentials are solved using analytical approximation conceptually similar to the PSP bulk and PSP-SOI-PD models. Secondary effects, such as quantum mechanical correction, polysilicon depletion, velocity saturation, etc. are included as well. The model retains the floating body simulation capability which is important to model the device characteristics in the PD mode.

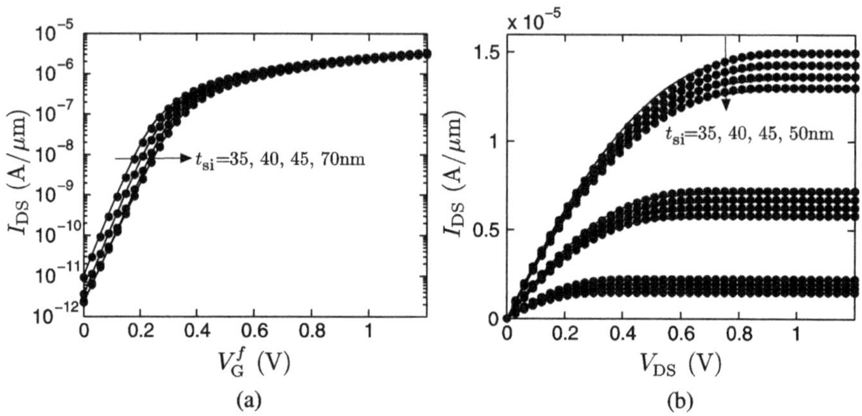

Fig. 2.28 (**a**) Transfer characteristics of SOI MOSFETs with different silicon film thicknesses. $V_{DS} = 0.1$ V and $V_G^b = 0$; (**b**) Output characteristics of SOI MOSFETs with different silicon film thicknesses. The front gate bias $V_G^f = 0.6$, 0.9 and 1.2 V and the back gate bias $V_G^b = 0$; The *symbols* represent the TCAD simulation data and the lines stand for the PSP-SOI-DD model. After [73]

Figure 2.28(a) shows the transfer characteristics of SOI MOSFETs with different silicon film thicknesses. The compact model accurately reproduces the TCAD simulation data. The device which is fully depleted for $t_{si} = 35$ nm becomes partially depleted for $t_{si} = 70$ nm. The transition between PD and FD modes is clearly indicated by the subthreshold slope change as a function of silicon film thickness. In FD mode, the SOI MOSFET has nearly ideal subthreshold slope due to the very small capacitance of the buried oxide layer. In PD mode, the subthreshold slope increases as a consequence of the relatively large capacitance of the depletion layer.

Figure 2.28(b) shows the output characteristics of SOI MOSFETs with different silicon film thickness. The output current increases as t_{si} is reduced. This follows from the fact that an FD SOI device with smaller t_{si} has a lower threshold voltage with other device parameters being the same.

Figure 2.29(a) shows the simulated and modeled transfer characteristics of a long-channel SOI MOSFET. As V_G^b changes from −4 V to 4 V, the subthreshold slope is decreasing to its ideal value (60 mV/dec) as the device enters the FD mode. Figure 2.29(b) shows the output characteristics of an SOI MOSFET with $t_{si} = 40$ nm for different gate and substrate biases. At large positive substrate bias ($V_G^b = 5$ V) the device is fully depleted and shows minimum floating body effect ("kink"). However, for $V_G^b = -5$ V, the device is partially depleted and exhibits pronounced floating body effect.

As all modern compact models, PSP-SOI-DD is charge based in order to preserve charge conservation [5]. However, it is a common practice to evaluate the model accuracy in terms of $C(V)$ characteristics which are more sensitive than terminal charges to any problems in model formulation. Figure 2.30 compares the simulated gate transcapacitances obtained from TCAD simulation and the compact model. For $t_{si} = 40$ nm, the device becomes fully depleted around $V_G^f \simeq 0$ V; without channel

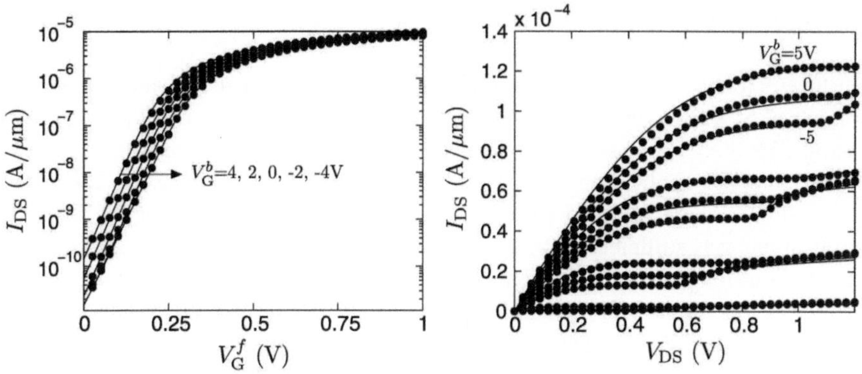

(a) (b)

Fig. 2.29 (**a**) Transfer characteristics of an SOI MOSFET with $t_{si} = 40$ nm at different back gate biases. $V_{DS} = 0.1$ V; (**b**) Output characteristics of an SOI MOSFET with $t_{si} = 40$ nm at different back gate biases. The front gate bias $V_G^f = 0.3$, 0.6, 0.9 and 1.2 V, the back gate bias $V_G^b = -5$, 0, 5 V; The *symbols* represent the TCAD simulation data and the lines stand for the PSP-SOI-DD model. After [73]

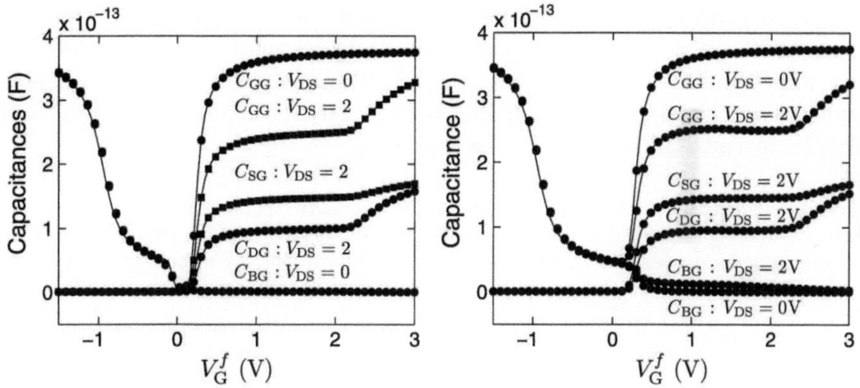

(a) $t_{si} = 40$ nm (b) $t_{si} = 70$ nm

Fig. 2.30 Capacitances for $t_{si} = 40$ and 70 nm. $N_A = 5 \times 10^{17}$ cm^{-3}, $V_{DS} = 0$ and 2 V. The *symbols* represent the TCAD data and the *lines* stand for the PSP-SOI-DD model results. After [73]

inversion, the gate capacitance becomes

$$C_{gg} \approx \frac{C_{ox} C_b}{C_{ox} + C_b} \approx C_b \tag{2.55}$$

assuming $C_{ox} \gg C_b$. This behavior has been already experimentally observed and captured in a threshold-voltage based DD-SOI model [48]. Finally, as shown in Fig. 2.30(b) PSP-SOI-DD also accurately describes the transcapacitances of a partially depleted device with a thicker channel film thickness ($t_{si} = 70$ nm).

2.9 Conclusions

Surface-potential-based approach provides a foundation for the advanced compact models of SOI transistors compatible with the compact models of bulk MOSFETs. Using PSP model as a starting point we present two fully developed SOI models for PD and DD applications which satisfy all requirements imposed by the growing analog-mixed signal and RF applications of SOI technology. For the SOI-DD applications it is sufficient to solve two rather than three coupled surface potential equations as long as $t_{box} \gg t_{ox}$. This enables analytical approximations for the front end surface potential and obviates the need for the iterative solution. Efficient formulation of the SOI-DD electrostatics allows one to capture the details of the transistor behavior including the smooth transition between the PD and FD modes on all $C(V)$ and $I(V)$ characteristics and the back bias effect. Symmetric linearization method can be extended to both SOI-PD and SOI-DD models to achieve a unified formulation of the drain current and terminal charges in all PSP-family models (SP [20], PSP [19], PSP-SOI-PD [72], PSP-SOI-DD [73], MOSVAR [61], PSP-DGFET [13, 14, 50]). For the core model the details of the device physics enter through the bias and geometry dependence of the function H and linearization coefficient. Surface potential-based formulation of the effects specific for SOI devices such as the bias dependence of the body resistance and of the EVB tunneling current has been developed as well. When combined with the advanced junction model [45], and thermal node for the description of the self-heating effects, new formulation provides accurate description of SOI devices already verified against experimental data for several SOI technology nodes.

Acknowledgments The development of PSP-SOI is partially supported by Semiconductor Research Corporation. The authors are grateful to C.C. McAndrew, J. Watts, and G.O. Workman for insightful discussions, and H. Barnaby and G. Dessai for reading the manuscript and valuable comments.

References

1. Anil, K.G., Mahapatra, S., Eisele, I.: Role of inversion layer quantization on sub-bandgap impact ionization in deep-sub-micron n-channel MOSFETs. In: IEDM Tech. Dig., pp. 675–678 (2000)
2. Bolouki, S., Maddah, M., Afzali-Kusha, A., El Nokali, M.: A unified IV model for PD/FD SOI MOSFETs with a compact model for floating body effects. Solid-State Electron. **47**(11), 1909–1915 (2003)
3. Brews, J.R.: A charge-sheet model of the MOSFET. Solid-State Electron. **21**, 345–355 (1978)
4. Cai, J., Sah, C.T.: Gate tunneling currents in ultrathin oxide metal–oxide–silicon transistors. J. Appl. Phys. **89**(4), 2272–2285 (2001)
5. Chatterjee, P.K., Leiss, J.E., Taylor, G.W.: A dynamic average model for the body effect in ion implanted short channel ($L = 1$ μm) MOSFET's. IEEE Trans. Electron Devices **28**(5), 606–607 (1981)
6. Chen, T.L., Gildenblat, G.: Symmetric bulk charge linearization of charge-sheet MOSFET model. Electron. Lett. **37**(12), 791–793 (2001)

7. Chen, J., Chan, T.Y., Ko, P.K., Hu, C.: Subbreakdown drain leakage current in MOSFET. IEEE Electron Device Lett. **8**, 515–517 (1987)
8. Chen, Q., Suryagandh, S., Goo, J.S., An, J.X., Thuruthiyil, C., Icel, A.B.: Impact of gate induced drain leakage and impact ionization currents on hysteresis modeling of PD SOI circuits. In: Tech. Proc. Workshop on Compact Modeling, pp. 570–573 (2007)
9. Chen, Q., Wu, Z.Y., Su, R.Y.K., Goo, J.S., Thuruthiyil, C., Radwin, M., Subba, N., Suryagandh, S., Ly, T., Wason, V., An, J.X., Icel, A.B.: Extraction of self-heating free I-V curves including the substrate current of PD SOI MOSFETs. In: IEEE Int. Conf. on Microelectron. Test Structures, pp. 272–275 (2007)
10. Chuang, C.T., Joshi, R.V., Puri, R., Kim, K.: Design considerations of scaled sub-0.1 μm PD/SOI CMOS circuits. In: Proc. Int. Symp. on Quality Electron. Des., pp. 153–158 (2003)
11. Chuang, C.T., Puri, R.: Effects of gate-to-body tunneling current on PD/SOI CMOS latches. In: Proc. Int. Conf. Simul. of Semicond. Processes and Devices, pp. 291–294 (2003)
12. Colinge, J.P.: Silicon-On-Insulator Technology: Materials to VLSI, 3rd edn. Springer, Berlin (2004)
13. Dessai, G., Dey, A., Gildenblat, G., Smit, G.D.J.: Symmetric linearization method for double-gate and surrounding-gate MOSFET models. Solid-State Electron. **53**(5), 548–556 (2009)
14. Dessai, G., Wu, W., Gildenblat, G.: Compact charge model for independent-gate asymmetric DGFET. IEEE Trans. Electron Devices (submitted)
15. Dieudonne, F., Jomaah, J., Balestra, F.: Gate-induced floating body effect excess noise in partially depleted SOI MOSFETs. IEEE Electron Device Lett. **23**(12), 737–739 (2002)
16. Faccio, F., Anghinolfi, F., Heijne, E.H.M., Jarron, P., Cristoloveanu, S.: Noise contribution of the body resistance in partially-depleted SOI MOSFETs. IEEE Trans. Electron Devices **45**(5), 1033–1038 (1998)
17. Fischetti, M.V., Sano, N., Laux, S.E., Natori, K.: Full-band Monte Carlo simulation of high-energy transport and impact ionization of electrons and holes in Ge, Si, and GaAs. In: Proc. Int. Conf. Simul. of Semicond. Processes and Devices, pp. 43–44 (1996)
18. Getreu, I.E.: Modeling the Bipolar Transistor. Tektronix, Beaverton (1976)
19. Gildenblat, G., Li, X., Wu, W., Wang, H., Jha, A., van Langevelde, R., Smit, G.D.J., Scholten, A.J., Klaassen, D.B.M.: PSP: an advanced surface-potential-based MOSFET model for circuit simulation. IEEE Trans. Electron Devices **53**(9), 1979–1993 (2006)
20. Gildenblat, G., Wang, H., Chen, T.L., Cai, X.: SP: an advanced surface-potential-based compact MOSFET model. IEEE J. Solid-State Circuits **39**(9), 1394–1406 (2004)
21. Gildenblat, G., Zhu, Z., McAndrew, C.C.: Surface potential equation for bulk MOSFET. Solid-State Electron. **53**(1), 11–13 (2009)
22. Goo, J.S., Williams, R.Q., Workman, G.O., Chen, Q., Lee, S., Nowak, E.J.: Compact modeling and simulation of PD-SOI MOSFETs: current status and challenges. In: Proc. IEEE Custom Integr. Circuits Conf., pp. 265–272 (2008)
23. Gu, X., Wang, H., Chen, T.L., Gildenblat, G.: Substrate current in surface-potential-based compact MOFET models. In: Tech. Proc. Nanotechnol. Conf., pp. 310–313 (2003)
24. Gu, X., Chen, T.L., Gildenblat, G., Workman, G.O., Veeraraghavan, S., Shapira, S., Stiles, K.: A surface potential-based compact model of n-MOSFET gate-tunneling current. IEEE Trans. Electron Devices **51**(1), 127–135 (2004)
25. Gummel, H.K., Poon, H.C.: An integral charge control model of bipolar transistors. Bell Syst. Tech. J. **49**(5), 827–852 (1970)
26. Jang, S.L., Huang, B.R., Ju, J.J.: A unified analytical fully depleted and partially depleted SOI MOSFET model. IEEE Trans. Electron Devices **46**(9), 1872–1876 (1999)
27. Jin, W., Chan, P.C.H., Fung, S.K.H., Ko, P.K.: Shot-noise-induced excess low-frequency noise in floating-body partially depleted SOI MOSFET's. IEEE Trans. Electron Devices **46**(6), 1180–1185 (1999)
28. Jin, W., Fung, S.K.H., Liu, W., Chan, P.C.H., Hu, C.: Self-heating characterization for SOI MOSFET based on AC output conductance. In: IEDM Tech. Dig., pp. 175–178 (1999)
29. Joshi, R.V., Chuang, C.T., Fung, S.K.H., Assaderaghi, F., Sherony, M., Yang, I., Shahidi, G.: Effects of gate-to-body tunneling current on PD/SOI CMOS SRAM. In: Symp. on VLSI Technol. Dig. of Tech. Papers, pp. 75–76 (2001)

30. Kusu, S., Ishimura, K., Ohyama, K., Miyoshi, T., Hori, D., Sadachika, N., Murakami, T., Ando, M., Mattausch, H.J., Miura-Mattausch, M., Baba, S., Ida, J.: Consistent dynamic depletion model of SOI-MOSFETs for device/circuit optimization. In: Proc. IEEE Int. SOI Conf., pp. 59–60 (2008)
31. Lemaitre, L., McAndrew, C.C., Hamm, S.: ADMS: automated device model synthesize. In: Proc. IEEE Custom Integr. Circuits Conf., pp. 27–30 (2002)
32. Lim, H.K., Fossum, J.G.: Threshold voltage of thin-film silicon-on-insulator (SOI) MOSFET's. IEEE Trans. Electron Devices **30**(10), 1244–1251 (1983)
33. Lu, P.F., Chuang, C.T., Ji, J., Wagner, L.F., Hsieh, C.M., Kuang, J.B., Hsu, L.L.C., Pelella, M.M., Chu, S.F.S. Jr., Anderson, C.J.: Floating-body effects in partially depleted SOI CMOS circuits. IEEE J. Solid-State Circuits **32**(8), 1241–1253 (1997)
34. Mallikarjun, C., Bhat, K.: Numerical and charge sheet models for thin-film SOI MOSFETs. IEEE Trans. Electron Devices **37**(9), 2039–2051 (1990)
35. McAndrew, C.C.: Practical modeling for circuit simulation. IEEE J. Solid-State Circuits **33**(3), 439–448 (1998)
36. McAndrew, C.C., Victory, J.J.: Accuracy of approximations in MOSFET charge models. IEEE Trans. Electron Devices **49**(1), 72–81 (2002)
37. Mercha, A., Rafi, J.M., Simoen, E., Augendre, E., Claeys, C.: "Linear kink effect" induced by electron valence band tunneling in ultrathin gate oxide bulk and SOI MOSFETs. IEEE Trans. Electron Devices **50**(7), 1675–1682 (2003)
38. Murakami, T., Ando, M., Sadachika, N., Yoshida, T., Miura-Mattausch, M.: Modeling of floating-body effect in silicon-on-insulator metal-oxide-silicon field-effect transistor with complete surface-potential-based description. Jpn. J. Appl. Phys. **47**(4), 2556–2559
39. Nakayama, H., Su, P., Hu, C., Nakamura, H., Komatsu, H., Takeshita, K., Komatsu, Y.: Methodology of self-heating free parameter extraction and circuit simulation for SOI CMOS. In: Proc. IEEE Custom Integr. Circuits Conf., pp. 381–384 (2001)
40. Ortiz-Conde, A., Garcia Sanchez, F.J., Schmidt, P.E., Sa-Neto, A.: The nonequilibrium inversion layer charge of the thin-film SOI MOSFET. IEEE Trans. Electron Devices **36**(9), 1651–1656 (1989)
41. Pao, H.C., Sah, C.T.: Effects of diffusion current on characteristics of metal-oxide (insulator)-semiconductor transistors. Solid-State Electron. **9**, 927–937 (1966)
42. PSP group: PSP Manual Version 102.2 (2007). http://pspmodel.asu.edu/downloads/psp1022_summary.pdf
43. Sadachika, N., Kitamaru, D., Uetsuji, Y., Navarro, D., Yusoff, M.M., Ezaki, T., Mattausch, H.J., Miura-Mattausch, M.: Completely surface-potential-based compact model of the fully depleted SOI-MOSFET including short-channel effects. IEEE Trans. Electron Devices **53**(9), 2017–2024 (2006)
44. Scholten, A.J., Tiemeijer, L.F., van Langevelde, R., Havens, R.J., Zegers-van Duijnhoven, A.T.A., Venezia, V.C.: Noise modeling for RF CMOS circuit simulation. IEEE Trans. Electron Devices **50**(3), 618–632 (2003)
45. Scholten, A.J., Smit, G.D.J., Durand, M., van Langevelde, R., Klaassen, D.B.M.: The physical background of JUNCAP2. IEEE Trans. Electron Devices **53**(9), 2098–2107 (2006)
46. Shahidi, G.G., Ajmera, A., Assaderaghi, F., Bolam, R.J., Hovel, H., Leobandung, E., Rausch, W., Sadana, D., Schepis, D., Wagner, L.F., Wissel, L., Wu, K., Davari, B.: Device and circuit design issues in SOI technology. In: Proc. IEEE Custom Integr. Circuits Conf., pp. 339–346 (1999)
47. Simoen, E., Mercha, A., Claeys, C., Lukyanchikova, N.B., Garhar, N.: Electron valence band tunnelling induced excess Lorentzian noise in fully depleted SOI transistors. In: Proc. Eur. Solid-State Device Res. Conf., pp. 279–282 (2003)
48. Sinitsky, D., Fung, S., Tang, S., Su, P., Chan, M., Ko, P., Hu, C.: A dynamic depletion SOI MOSFET model for SPICE. In: Symp. on VLSI Technol. Dig. of Tech. Papers, pp. 114–115 (1998)
49. Sleight, J., Rios, R.: A continuous compact MOSFET model for fully- and partially-depleted SOI devices. IEEE Trans. Electron Devices **45**(4), 821–825 (1998)

50. Smit, G.D.J., Scholten, A.J., Serra, N., Pijper, R.M.T., van Langevelde, R., Mercha, A., Gildenblat, G., Klaassen, D.B.M.: PSP-based compact FinFET model describing dc and RF measurements. In: IEDM Tech. Dig., pp. 1–4 (2006)
51. Su, L.T., Chung, J.E., Antoniadis, A.D., Goodson, K.E., Flik, M.I.: Measurement and modeling of self-heating in SOI nMOSFET's. IEEE Trans. Electron Devices **41**(1), 69–75 (1994)
52. Su, P., Fung, S.K.H., Assaderaghi, F., Hu, C.: A body-contact SOI MOSFET model for circuit simulation. In: Proc. IEEE Custom Integr. Circuits Conf., pp. 50–51 (1999)
53. Su, P., Fung, S.K.H., Tang, S., Assaderaghi, F., Hu, C.: BSIMPD: a partial-depletion SOI MOSFET model for deep-submicron CMOS designs. In: Proc. IEEE Custom Integr. Circuits Conf., pp. 197–200 (2000)
54. Su, P., Goto, K., Sugii, T., Hu, C.: Enhanced substrate current in SOI MOSFETs. IEEE Electron Device Lett. **23**(5), 282–284 (2002)
55. Su, P., Fung, S.K.H., Wyatt, P.W., Wan, H., Chan, M., Niknejad, A.M., Hu, C.: A unified model for partial-depletion and full-depletion SOI circuit designs: using BSIMPD as a foundation. In: Proc. IEEE Custom Integr. Circuits Conf., pp. 241–244 (2003)
56. Tenbroek, B.M., Lee, M.S.L., Redman-White, W., Bunyan, R.J.T., Uren, M.J.: Impact of self-heating and thermal coupling on analog circuits in SOI CMOS. IEEE J. Solid-State Circuits **33**(7), 1037–1046 (1998)
57. Tseng, Y.C., Huang, W.M., Mendicino, M., Monk, D.J., Welch, P.J., Woo, J.C.S.: Comprehensive study on low-frequency noise characteristics in surface channel SOI CMOSFETs and device design optimization for RF ICs. IEEE Trans. Electron Devices **48**(7), 1428–1437 (2001)
58. Tsividis, Y.: Operation and Modeling of the MOS Transistor, 2nd edn. McGraw-Hill, New York (1999)
59. Tsu, R., Esaki, L.: Tunneling in a finite superlattice. Appl. Phys. Lett. **22**(11), 562–564 (1973)
60. van Langevelde, R., Scholten, A.J., Klaassen, D.B.M.: MOS Model 11, Level 1102 (2004). http://www.nxp.com/models/mos_models/
61. Victory, J., Zhu, Z., Zhou, Q., Wu, W., Gildenblat, G., Yan, Z., Cordovez, J., McAndrew, C.C., Anderson, F., Paassches, J.C.J., van Langevelde, R., Kolev, P., Cherne, R., Yao, C.: PSP-based scalable MOS varactor model. In: Proc. IEEE Custom Integr. Circuits Conf., pp. 495–502 (2007)
62. Vogelsong, R., Brzezinski, C.: Simulation of thermal effects in electrical systems. In: IEEE Appl. Power Electron. Conf. and Expos. (APEC), pp. 353–356 (1989)
63. Wan, H., Xi, X., Niknejad, A., Hu, C.: BSIMSOI4.0 MOSFET Model. University of California, Berkeley, CA (2005)
64. Wang, H., Chen, T.L., Gildenblat, G.: Quasi-static and non-quasi-static compact MOSFET models based on symmetric linearization of the bulk and inversion charges. IEEE Trans. Electron Devices **50**(11), 2262–2272 (2003)
65. Ward, D.E., Dutton, R.W.: A charge-oriented model for MOS transistor capacitances. IEEE J. Solid-State Circuits **13**, 703–708 (1978)
66. Workman, G.O., Fossum, J.G.: A comparative analysis of the dynamic behavior of BTG/SOI MOSFETs and circuits with distributed body resistance. IEEE Trans. Electron Devices **45**(10), 2138–2145 (1998)
67. Workman, G.O., Fossum, J.G.: Physical noise modeling of SOI MOSFETs with analysis of the Lorentzian component in the low-frequency noise spectrum. IEEE Trans. Electron Devices **47**(6), 1192–1201 (2000)
68. Wu, W., Chen, T.L., Gildenblat, G., McAndrew, C.C.: Physics-based mathematical conditioning of the MOSFET surface potential equation. IEEE Trans. Electron Devices **51**(7), 1196–1199 (2004)
69. Wu, W., Li, X., Wang, H., Gildenblat, G., Workman, G.O., Veeraraghavan, S., McAndrew, C.C.: SP-SOI: a third generation surface potential based compact SOI MOSFET model. In: Proc. IEEE Custom Integr. Circuits Conf., pp. 819–822 (2005)
70. Wu, W., Li, X., Gildenblat, G., Workman, G., Veeraraghavan, S., McAndrew, C.C., van Langevelde, R., Smit, G.D.J., Scholten, A.J., Klaassen, D.B.M., Watts, J.: PSP-SOI: a surface potential based compact model of partially depleted SOI MOSFETs (invited). In: Proc. IEEE Custom Integr. Circuits Conf., pp. 41–48 (2007)

71. Wu, W., Li, X., Gildenblat, G., Workman, G.O., Veeraraghavan, S., Watts, J.: A nonlinear body resistance model for accurate PD/SOI technology characterization. In: Proc. IEEE Int. SOI Conf., pp. 151–152 (2008)
72. Wu, W., Li, X., Gildenblat, G., Workman, G.O., Veeraraghavan, S., McAndrew, C.C., van Langevelde, R., Smit, G.D.J., Scholten, A.J., Klaassen, D.B.M., Watts, J.: PSP-SOI: An advanced surface potential based compact model of partially depleted SOI MOSFETs for circuit simulations. Solid-State Electron. **53**, 18–29 (2009)
73. Wu, W., Yao, W., Gildenblat, G.: Surface-potential-based compact modeling of dynamically depleted SOI MOSFETs. Solid-State Electron (2010). doi:10.1016/j.sse.2009.12.040
74. Yang, J.W., Fossum, J.G., Workman, G.O., Huang, C.L.: A physical model for gate-to-body tunneling current and its effects on floating-body PD/SOI CMOS devices and circuits. Solid-State Electron. **48**(2), 259–270 (2004)
75. Yu, Y., Kim, S., Hwang, S., Ahn, D.: All-analytic surface potential model for SOI MOSFETs. IEE Proc. Circuits Device Syst. **152**(2), 183–188 (2005)

Chapter 3
Benchmark Tests for MOSFET Compact Models

**Xin Li, Weimin Wu, Gennady Gildenblat,
Colin C. McAndrew, and Andries J. Scholten**

Abstract It has long been recognized that, apart from computational efficiency and accuracy of fitting experimental data, compact MOS transistor models should exhibit qualitatively correct physical behavior for drain current, terminal charges, noise, and all derivatives. Physics-based models may automatically embody the correct physical behavior for long-channel devices, but compact models of scaled transistors inevitably involve approximations that can introduce unphysical qualitative characteristics. Over time, several "benchmark tests" were developed to ensure that transistor characteristics predicted by a compact model satisfy the needs of the circuit designers, especially for analog and mixed-signal design. As the importance of RF CMOS circuits increases, the requirements for qualitatively correct physical behavior of compact MOSFET models are becoming more stringent (for example, it is now common to require the existence of fifth order derivatives) and several new benchmark tests, targeted for RF design needs, were developed. This chapter describes both traditional and newly developed MOSFET model benchmark tests and applies them to the PSP model.

3.1 Introduction

Compact MOSFET models should provide both an accurate reproduction of fine details of transistor behavior (e.g. higher order transconductances) and be computa-

X. Li (✉) · W. Wu · G. Gildenblat
Ira A. Fulton School of Engineering, Department of Electrical Engineering, Arizona State University, Tempe, AZ 85287, USA
e-mail: xinli@asu.edu

C.C. McAndrew
Freescale Semiconductor, Tempe, AZ 85284, USA

A.J. Scholten
NXP-TSMC Research Center, 5656 AE Eindhoven, The Netherlands

G. Gildenblat (ed.), *Compact Modeling*,
DOI 10.1007/978-90-481-8614-3_3, © Springer Science+Business Media B.V. 2010

tionally efficient. The latter requires the use of empirical relations and approximations to minimize the number of computationally expensive transcendental function evaluations. While fully physics-based models of idealized MOSFETs (e.g. the Pao-Sah model [21]) automatically exhibit correct qualitative behavior in all regions of operation, this is not always the case for compact models, which can exhibit unphysical behavior [1, 11, 16, 34]. In digital applications this may not be a problem, but for analog and mixed-signal design correct qualitative behavior can be critical to accurate prediction of circuit performance.

The first pleas to improve MOSFET modeling accuracy for analog design, with examples of where there were significant qualitative errors in available MOSFET models, were voiced more than 25 years ago [31, 33] but were mostly ignored. The warning bells about troubles with MOSFET models, and the need for circuit designers to be aware of model limitations, were rung again more than 15 years ago [34], and although this alarm was heard a little more widely it still did not trigger fundamental improvements in mainstream MOSFET models.

Two recent developments significantly improved the qualitative behavior of advanced MOSFET models. First, the transition to the surface potential based formulation [6, 7, 36, 37, 44] automatically assures that the core model is fully physical and only small-geometry effects need to be checked for physical behavior. When combined with the improved surface potential equation [8, 38, 45], the surface potential based approach gives a single set of equations for all regions of operation. This eliminates singularities at the boundaries between regions. Second, following the principle of "asymptotic correctness," which in essence states that model developers should verify physically correct behavior of their approximations in extreme or limiting cases [17], helps ensure secondary effects are included in a consistent and correct manner. Both developments go a long way toward assuring desired model behavior, but in the end a model still needs to be subjected to benchmark tests to verify that it exhibits qualitatively and physically correct behavior for all design applications.

Over the years numerous benchmark tests were developed for, and applied to, MOSFET models. However, the increase in importance of RF CMOS applications has imposed more stringent requirements on qualitative model behavior, for example symmetry with respect to source-drain interchange, class $C^{(3)}$ (or even $C^{(5)}$) continuity, and asymptotic correctness of noise and NQS (non-quasi-static) models over frequency [1, 11, 14, 15, 26]. This has led to the development of new benchmarks, for example of symmetry in the presence of impact ionization and gate tunneling currents [19]. This means that the submodels for these effects are now subject to the same stringent requirements as the core model. In the final analysis, the value of all benchmark tests is how they improve IC design. With reduced margin for error in modern applications, we need to be assured that any feature of device behavior predicted by a compact model is physical and is not an artifact of a model deficiency.

In this chapter we provide a comprehensive description of both classic and more recent benchmark tests for dc, ac, and RF characteristics. While most of the supporting examples are based on the PSP model [7], the material is generic and can be applied to any compact MOSFET model.

3.2 Benchmark Tests

3.2.1 Weak and Moderate Inversion Regions

With continued scaling of the power supply voltage and increased use of CMOS technologies for analog and RF applications, the moderate inversion region (between the weak and strong inversion regions) becomes more important. Historically, this region is modeled unphysically by threshold-voltage based models because they use purely mathematical techniques to connect the weak and strong inversion regions, which are dominated by diffusion and drift currents, respectively. In addition, such models generally have an ideal exponential dependence of diffusion current on V_{gs}, which is physically incorrect. Therefore, several tests were designed to check for physical behavior in weak and moderate inversion (collectively called the subthreshold region of operation). Two important traditional benchmark tests are the slope-ratio and tree-top tests, which evaluate the subthreshold $I_d(V_{ds})$ behavior and the g_m/I_d ratio behavior of a model, respectively [20]. These tests essentially require a model to be qualitatively similar to the Pao-Sah [21] or charge sheet models [3, 32].

3.2.1.1 Slope Ratio Test

The slope ratio test checks whether a MOSFET model correctly distinguishes the qualitative difference in $I_{ds}(V_{ds})$ behavior between weak and strong inversion operation. The slope ratio is defined as

$$S_R = \frac{(I_2 + I_1)(V_{db2} - V_{db1})}{(I_2 - I_1)(V_{db2} + V_{db1})} \tag{3.1}$$

where I_1 and I_2 are the drain currents corresponding to $V_{db} = V_{db1}$ and $V_{db} = V_{db2}$, respectively, chosen in the range where I_d varies with V_{ds} in weak inversion. S_R should reduce smoothly and monotonically from about 1.3 to 1 going from weak to strong inversion.

Figure 3.1 shows that PSP passes the slope ratio test, using the common values $V_{db1} = 0.01$ V and $V_{db2} = 0.02$ V.

A problem with the slope ratio (3.1) is that its limiting value in weak inversion S_{RO} depends on temperature (cf. Fig. 3.1). Indeed, the subthreshold current can be expressed as [21]

$$I_d = \mu \frac{W}{L} C_{ox} \cdot \frac{\gamma \phi_t^2}{2\sqrt{\phi_{ss}}} e^{\beta(\phi_{ss} - 2\phi_B - V_{sb})} (1 - e^{-\beta V_{ds}}) \tag{3.2}$$

where μ is the effective channel mobility, C_{ox} the oxide capacitance per unit area, γ the body factor, ϕ_{ss} the surface potential at the source end of the channel, ϕ_B the

Fig. 3.1 Traditional slope ratio test results for PSP using default parameters with $W/L = 10/10$ μm for various temperatures, $V_{db1} = 10$ mV, $V_{db2} = 20$ mV and $V_{sb} = 0$

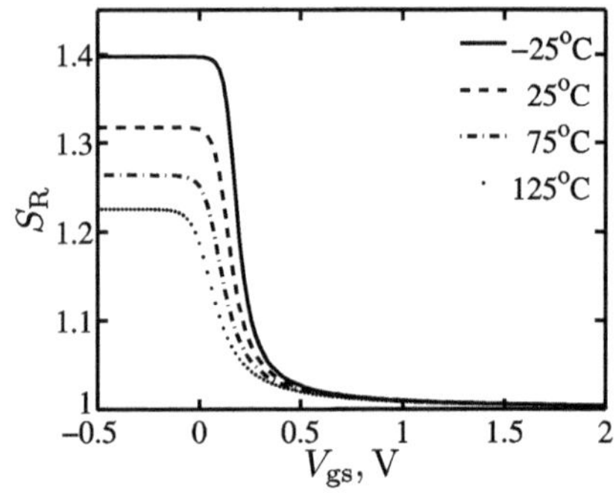

"Fermi potential," $\phi_t = k_B T/q$ the "thermal potential," and $\beta = 1/\phi_t$. Then

$$S_{RO} = \frac{e^{\beta V} - \cosh(\beta \Delta V)}{\sinh(\beta \Delta V)} \cdot \frac{\Delta V}{V} \tag{3.3}$$

where

$$V = \frac{V_{db1} + V_{db2}}{2} \tag{3.4}$$

and

$$\Delta V = \frac{V_{db2} - V_{db1}}{2}. \tag{3.5}$$

If V_{db1} and V_{db2} are temperature independent then βV and $\beta \Delta V$, and therefore S_{RO}, depend on temperature, producing the results shown in Fig. 3.1. To make the test results independent of temperature we select [15]

$$V_{db1} = \frac{\phi_t}{2} - \Delta V \tag{3.6}$$

and

$$V_{db2} = \frac{\phi_t}{2} + \Delta V \tag{3.7}$$

where $\Delta V \ll \phi_t$. Then $V_{db1} + V_{db2} = \phi_t$ and from (3.3)

$$S_{RO} = 2(\sqrt{e} - 1) \approx 1.297 \tag{3.8}$$

regardless of the temperature.

The symmetric linearization method developed for PSP [5, 41] allows it to physically and correctly model MOSFET behavior in subthreshold operation (cf. Fig. 1.4 in Chap. 1). Typical results for S_{RO} for PSP are shown in Fig. 3.2 for $\Delta V = 1$ mV, and Fig. 3.3 shows the sensitivity of S_{RO} to the value of ΔV.

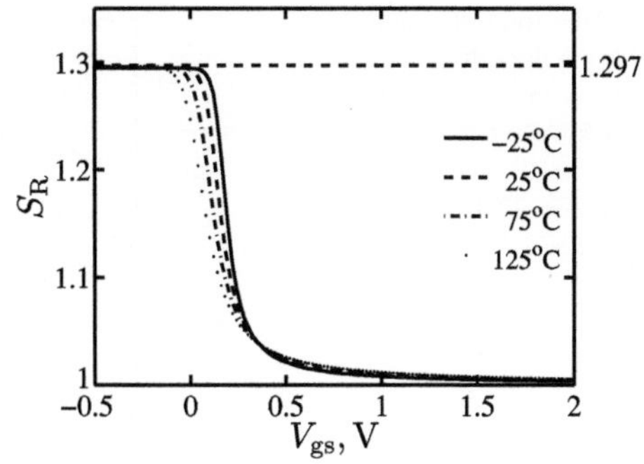

Fig. 3.2 Modified temperature independent slope ratio test results for PSP with default parameters, $\Delta V = 1$ mV

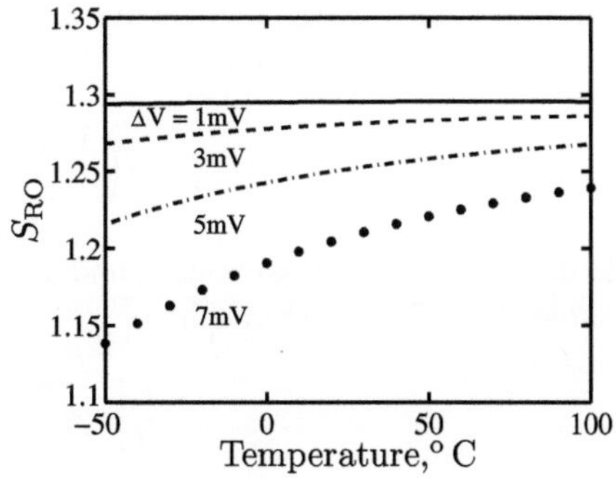

Fig. 3.3 Sensitivity of S_{RO} to ΔV for PSP

3.2.1.2 Tree-Top Test

The tree-top test checks the physical correctness of modeling the ratio g_m/I_d from weak to strong inversion, which is important for analog design. In weak inversion, from (3.2),

$$\frac{g_m}{I_d} = \left(\beta - \frac{1}{2\phi_{ss}}\right)\frac{\partial \phi_{ss}}{\partial V_{gs}} \tag{3.9}$$

or, equivalently,

$$\frac{g_m}{I_d} = \left(\beta - \frac{1}{2\phi_{ss}}\right)\left(1 - \frac{C_{gg}}{C_{ox}}\right) \tag{3.10}$$

where

$$C_{gg} = \frac{\partial Q_g}{\partial V_g}. \tag{3.11}$$

The increase of g_m/I_d going from weak to moderate inversion follows the corresponding decrease of C_{gg}. As shown in Fig. 3.4 the ratio never actually reaches its possible maximum value (for $\phi_{ss} \gg \phi_t$) of β since $C_{gg} > 0$.

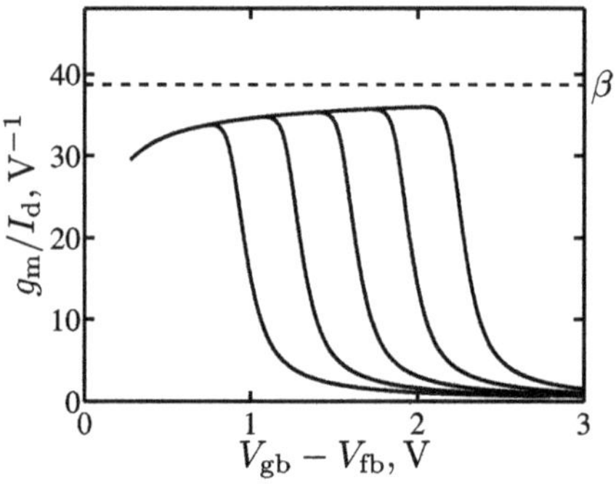

Fig. 3.4 Tree-top test results for PSP using default parameters with $V_{ds} = 0.1$ V, V_{bs} varies from -1 to 0.2 V in 0.3 V steps, $W/L =$ 10/10 μm, $T = 27$°C. The *horizontal dashed line* represents $\beta = 38.7$ V^{-1}

In threshold voltage-based models the weak inversion region is often described using the semi-empirical relation

$$I_d \propto \exp\left(\frac{V_{gb} - V_t}{n\phi_t}\right) \tag{3.12}$$

where V_t denotes the threshold voltage and n is a fixed constant used to control the subthreshold slope. In this approach

$$\frac{g_m}{I_d} = \frac{\beta}{n} \tag{3.13}$$

is constant in the weak inversion region and the effect of the variation of C_{gg} with V_{gb} on the bias dependence of g_m/I_d is not modeled.

3.2.1.3 Weak Inversion Conductance Test

This test, like the slope-ratio test, verifies that a model captures the $I_d(V_{ds})$ dependence predicted by (3.2), but is formulated in terms of output conductance which is more relevant to both model parameter extraction and design applications. From (3.2)

$$\frac{\partial}{\partial V_d}\left[g_{ds} \cdot \exp\left(\beta V_{ds}\right)\right] = 0 \tag{3.14}$$

which implies that $g_{ds} \cdot \exp(\beta V_{ds})$ vs. V_{ds} should be flat. This is observed in weak inversion, while increasing the gate bias beyond the point where (3.2) is a good approximation leads to a finite slope (see Fig. 3.5). There is an interesting connection between this characteristic behavior and the Gummel symmetry test (GST) to be discussed in Sect. 3.2.3.2. When velocity saturation is directly included into the charge sheet MOSFET model it gives an unphysical negative g_{ds} in saturation. This problem is usually solved by using a smooth limiting function to transition from V_{ds} in non-saturation to an effective drain bias V_{dse} in saturation [7]. To pass the GST, the $V_{dse}(V_{ds})$ function must be odd [18] and has been suggested to require a unity slope at $V_{ds} = 0$ [11]. The latter condition is in fact not required but is useful to

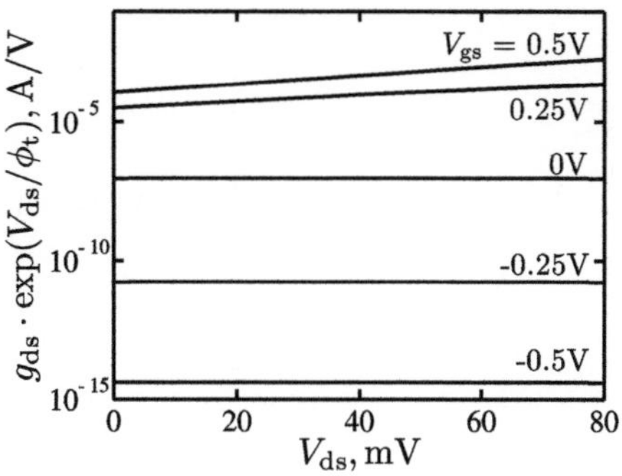

Fig. 3.5 PSP results for the subthreshold region, V_{gs} varies from −0.5 to 0.5 V in 0.25 V steps. Default parameters are used

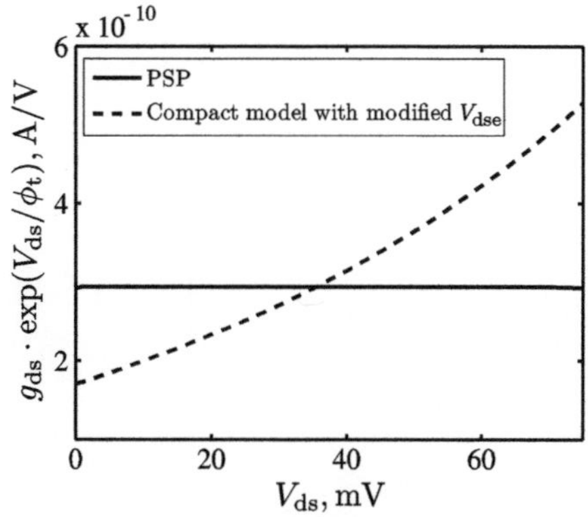

Fig. 3.6 Comparison of PSP and another compact model with modified V_{dse} for which dV_{dse}/dV_{ds} differs from 1 appreciably for $V_{ds} < 3\phi_t$, $V_{gs} = -0.1$ V. After [15]

ensure that (3.14) is satisfied. More precisely, a model should have $dV_{dse}/dV_{ds} = 1$ for V_{ds} less than about $3\phi_t$. If this condition is violated then so is (3.14), as Fig. 3.6 shows.

3.2.2 Capacitances

3.2.2.1 Transcapacitances

MOSFET (trans-)capacitances C_{ij} $(i, j = d, g, s$, or $b)$ are defined as

$$C_{ij} = (2 \cdot \delta_{ij} - 1) \cdot \frac{\partial Q_i}{\partial V_j} \tag{3.15}$$

where Q_i and V_j are terminal charge and voltage respectively, and δ_{ij} is the Kronecker delta. For a four terminal MOSFET, there are 16 different capacitances. Of these, 9 are independent and the rest can be expressed as linear combinations of these [32]. Transcapacitances as functions of terminal biases, with fine bias step size, are traditionally investigated to detect possible singular or other unphysical behavior of the terminal charges as functions of the terminal voltages. Figures 3.7

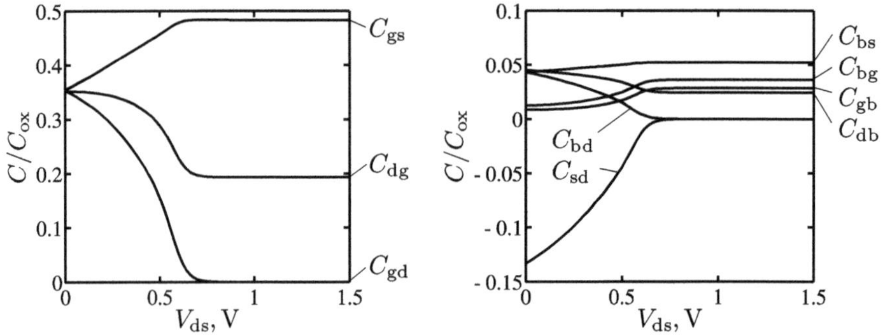

Fig. 3.7 Various transcapacitances for PSP with $W/L = 10/10$ μm, $V_{gs} = 1$ V. Default model parameters are used

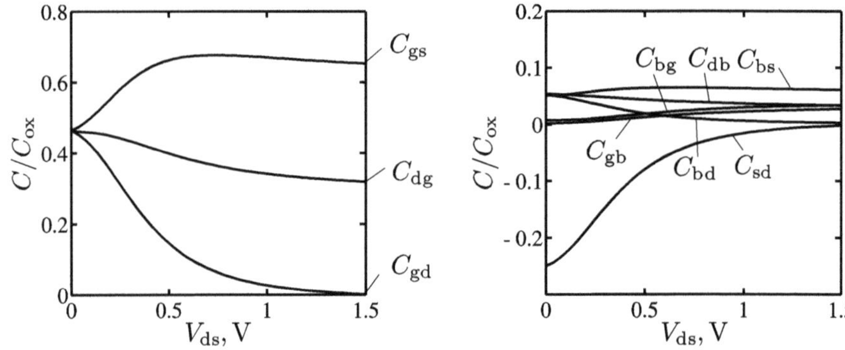

Fig. 3.8 Various transcapacitances for PSP with $W/L = 10/0.2$ μm, $V_{gs} = 1$ V. A short channel model parameter set is used. Note that the negative sign of C_{sd} is physical

and 3.8 show typical results that verify the continuous and physical behavior of the transcapacitances of PSP for both long and short channel devices, including the correct sign for C_{bd} and C_{gb}. Many models exhibit capacitances with the wrong sign, but in PSP the symmetric linearization technique [7] provides the desired behavior seen in Figs. 3.7 and 3.8.

3.2.2.2 Reciprocity at $V_{ds} = 0$

One requirement of MOSFET capacitances is that they should be reciprocal at zero drain-source bias:

$$C_{ij} = C_{ji} \quad \text{for } V_{ds} = 0. \tag{3.16}$$

In general MOSFET capacitances are non-reciprocal, i.e., $C_{ij} \neq C_{ji}$. However, reciprocity is recovered for $V_{ds} = 0$. This special case is of concern since it is encountered during parameter extraction. More important, satisfying the condition (3.16)

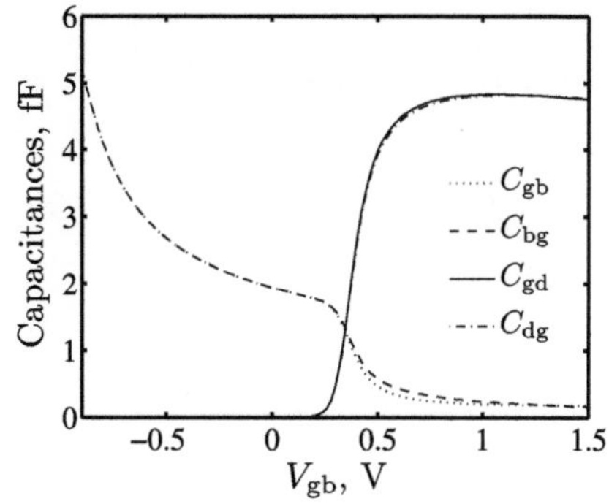

Fig. 3.9 C_{gd}, C_{dg}, C_{gb} and C_{bg} for PSP with $W/L = 10/0.08$ μm at $V_{ds} = 0$. Default parameters are used. After [15]

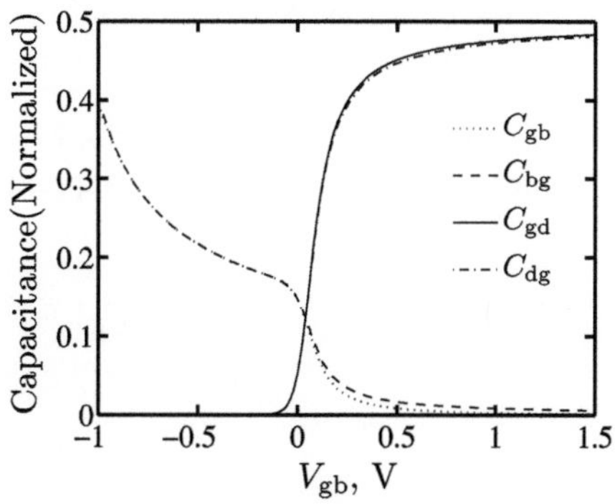

Fig. 3.10 C_{gd}, C_{dg}, C_{gb} and C_{bg} for the charge sheet model at $V_{ds} = 0$; $t_{ox} = 4$ nm, $N_a = 3 \times 10^{23}$ m^{-3}. After [15]

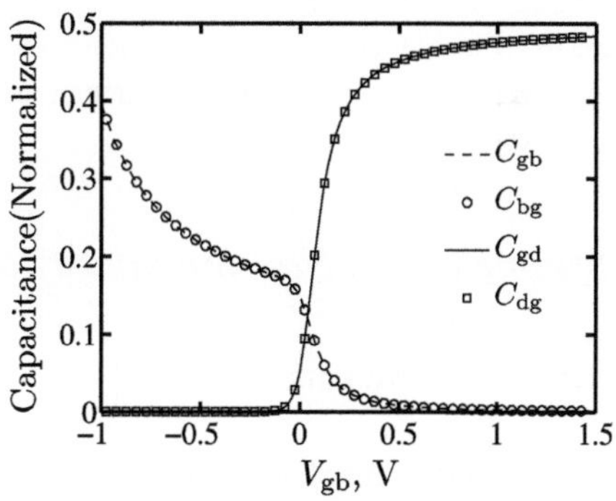

Fig. 3.11 C_{gd}, C_{dg}, C_{gb} and C_{bg} for the Pao-Sah model at $V_{ds} = 0$; $t_{ox} = 4$ nm, $N_a = 3 \times 10^{23}$ m^{-3}. After [15]

validates the physical nature of a MOSFET model. Models can be tested for reciprocity by examining C_{ij} and C_{ji} vs. V_{gs}, at $V_{ds} = 0$; as Fig. 3.9 shows the reciprocity condition is approximately satisfied in PSP with accuracy sufficient for engineering applications. The charge sheet model (Fig. 3.10) [3] shows the same approximate reciprocity for $V_{ds} = 0$. If the charge sheet approximation is not introduced and the complete Pao-Sah model [21] is used then the reciprocity at $V_{ds} = 0$ becomes exact (cf. Fig. 3.11). Thus the small deviation from reciprocity at $V_{ds} = 0$ in PSP is caused by the charge sheet approximation [37].

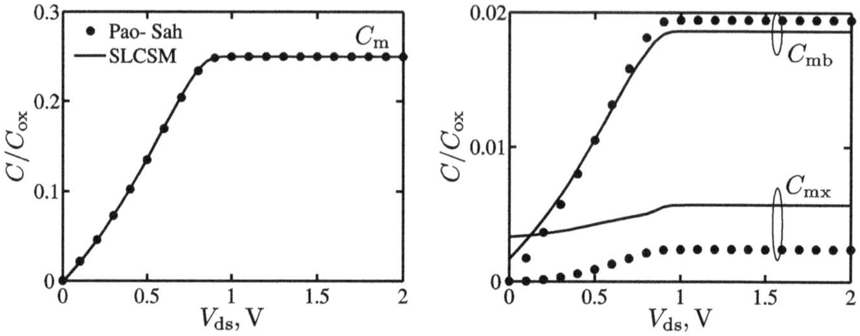

Fig. 3.12 C_m, C_{mb} and C_{mx} for the symmetrically linearized charge sheet model and the Pao-Sah model at $V_{gs} = 1$ V, $t_{ox} = 2$ nm, $N_a = 2 \times 10^{23}$ m^{-3}

For $V_{ds} > 0$, in particular in saturation, MOSFET transcapacitances are not reciprocal regardless of the charge sheet approximation. Figure 3.12 shows this for the quantities

$$C_m = C_{dg} - C_{gd}, \tag{3.17}$$

$$C_{mb} = C_{db} - C_{bd} \tag{3.18}$$

and

$$C_{mx} = C_{bg} - C_{gb} \tag{3.19}$$

which are used to emphasize the non-reciprocity (they also appear naturally in the small-signal equivalent circuit of the intrinsic MOSFET [32]).

3.2.3 Symmetry and Non-Singularity at Zero Drain-Source Bias

3.2.3.1 General Comments

Both the physical structure and electrical behavior of MOS transistors are symmetric with respect to source-drain interchange. During the last decade it was discovered that certain circuits, most importantly passive mixers and pass gates that switch though $V_{ds} = 0$, cannot be simulated even qualitatively correctly without a model that preserves this symmetry [1, 11]. The main reasons for modeling problems at $V_{ds} = 0$ are: the use of a source-referenced threshold voltage; a singular velocity-field relation; and asymmetric limiting of the terminal voltages (such limiting is ubiquitous in all practical compact models).

The situation is rather subtle: regardless of whether a model formulation is symmetric or asymmetric (we use these terms to mean with respect to source and drain terminals), most model implementations interchange the definitions of source and drain for $V_{ds} < 0$. If a model formulation is symmetric, this "terminal flip" is inconsequential. However, a terminal flip for an asymmetric model formulation results in a model that is symmetric but singular at $V_{ds} = 0$. More specifically, second and higher order derivatives may cease to exist at $V_{ds} = 0$ making it impossible to perform a Taylor series expansion of the drain current around $V_{ds} = 0$. Consequently, simulation of intermodulation effects, which requires the possibility of such an expansion, becomes impossible. In this section we study this phenomenon in detail and present tests for model symmetry.

As befits a subtle concept, the symmetry issue has given rise to several misconceptions. For example, it is often stated that a model has a higher order derivative discontinuity (jump) at $V_{ds} = 0$. This is wrong: according to a classic Darboux theorem [2] derivatives of a function cannot have discontinuity of the first kind (and discontinuities of the second kind so far were not encountered in MOSFET compact models). What actually happens is that derivatives of certain orders do not exist at $V_{ds} = 0$. Another piece of modeling folklore is that the symmetry violation is brought about by the reference point used for the potential definition ("Do you use source or body as a reference?"). This is not true either, since equations of semiconductor physics are gauge-invariant and one can use the potential at any point inside or outside the device as a reference—if it is done correctly [12, 32]. For example, the problem with threshold-voltage based models described in [1, 11] is not that the source is used as a potential reference, but that threshold voltage is traditionally defined as the gate-to-source bias corresponding to the onset of strong inversion onset near the source—not near the drain. Finally, with all the attention that has been given to the symmetry issue it may seem that this is the only reason for singular behavior at $V_{ds} = 0$. However, a perfectly symmetric core model can still give rise to singular behavior if it adopts the popular velocity-field relation [30]

$$v_d = \frac{\mu E_y}{1 + \mu |E_y| / v_{sat}} \tag{3.20}$$

where v_d denotes the drift velocity, μ the low field mobility, E_y the lateral field, and v_{sat} is the saturation velocity. Then the second order derivative $d^2 v_d / d E_y^2$ does not exist for $E_y = 0$ or, equivalently, for $V_{ds} = 0$ [1].

3.2.3.2 Gummel Symmetry Test (GST)

The first symmetry test deals with symmetry of the core model (not including the gate and substrate currents). It is based on the observation that, as explained above, regardless of the internal symmetry of a model for the test arrangement shown in Fig. 3.13 (V_b denotes the body bias, $V_x = V_{ds}/2$, and $V_{bo} = (V_{db} + V_{sb})/2$ [20]) the drain current is an odd function of V_x:

$$I_d(V_x) = -I_d(-V_x). \tag{3.21}$$

The test consists of investigating the behavior of the derivatives of I_d with respect to V_x around $V_x = 0$ to detect a singularity. Typical results for PSP are presented

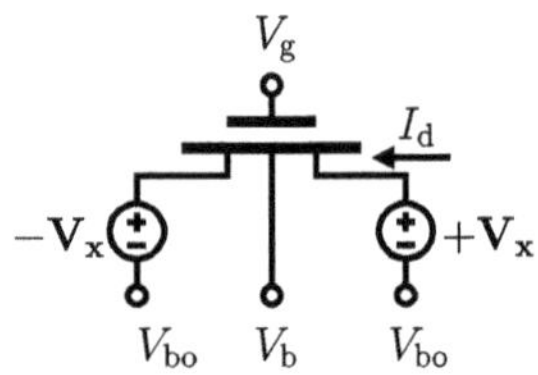

Fig. 3.13 Biasing scheme for Gummel symmetry test

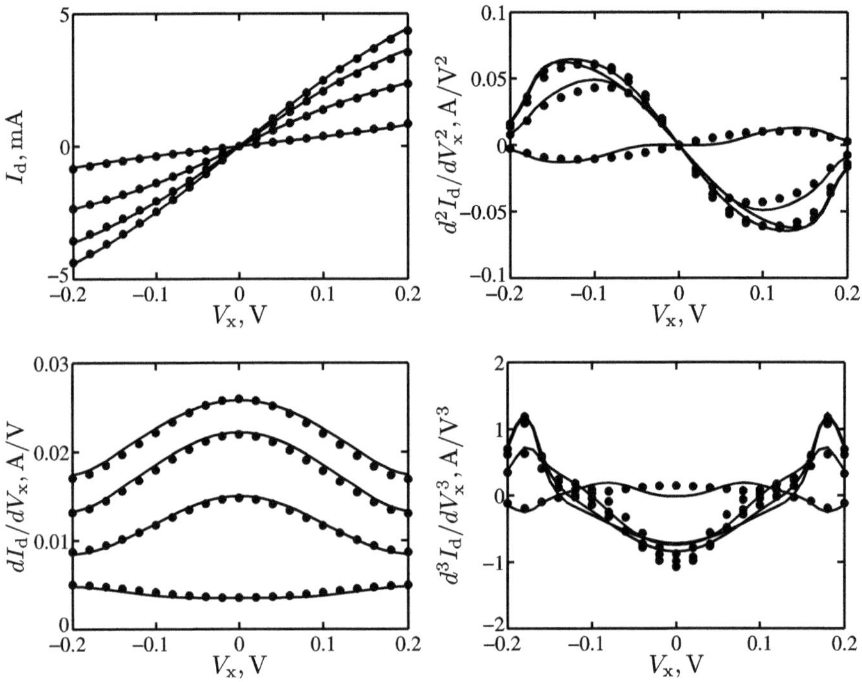

Fig. 3.14 Drain current and derivatives for the Jazz Semiconductor 0.18 μm technology, n-channel 10/0.18 μm MOSFET, V_g varies from 0.6 to 1.8 V in steps of 0.4 V. Symbols represent experimental data while solid lines represent the PSP model, $V_b = 0$ V, $V_{bo} = 0$ V. After [14]

as solid lines in Fig. 3.14, indicating that PSP has no singularity and passes this test. This is because PSP has a nonsingular velocity saturation model [1, 7], an odd limiting function for the transition to saturation [7, 18], and a symmetric linearization procedure for the inversion and bulk charges as functions of the surface potential [7]. The results of Fig. 3.14 show not only that PSP is nonsingular but also that it accurately reproduces experimental data (symbols). Note that this is the first experimental investigation of higher order derivatives for a MOS transistor around $V_{ds} = 0$ [14, 15] and that until recently it was impossible to model even the second derivative for $V_x = 0$ with industry standard models.

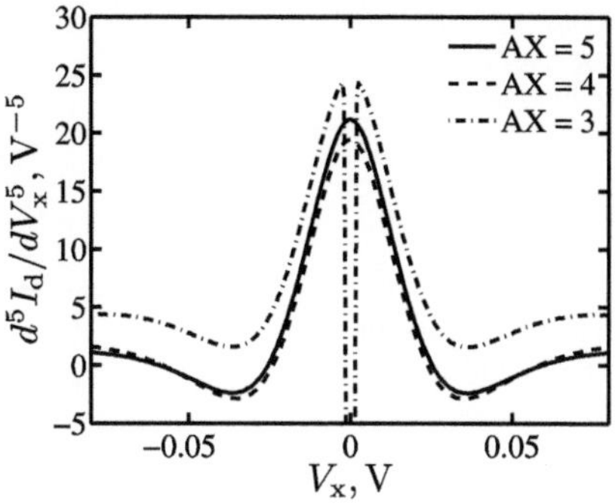

Fig. 3.15 5th order derivatives for different **AX** (using the GST biasing scheme). For **AX** = 3, condition (3.23) is violated

3.2.3.3 Higher Order Derivatives at $V_{ds} = 0$

To model harmonic distortion in RF circuits operating at $V_{ds} = 0$ MOSFET models need to be not just class $C^{(3)}$ (derivatives continuous to third order) but, at least, class $C^{(5)}$. The nature of the limiting function

$$V_{dse} = \frac{V_{ds}}{\left[1 + \left(\frac{|V_{ds}|}{V_{dsat}}\right)^{\mathbf{AX}}\right]^{\frac{1}{\mathbf{AX}}}} \tag{3.22}$$

used in PSP ensures class $C^{(3)}$ behavior (the model parameter **AX** is restricted to be greater than 2), but in order for the n-th derivative to exist at $V_{ds} = 0$ it is necessary that

$$\mathbf{AX} > n - 1. \tag{3.23}$$

An exception to this requirement is when **AX** is an even integer, in which case the derivatives of all order exist.

The local parameter **AX** scales with device effective channel length L as

$$\frac{\mathbf{AXO}}{1 + \mathbf{AXL}/L} \tag{3.24}$$

where **AXL** is a global (scaling) parameter. For short-channel devices this limits the range of **AXL** making it more difficult to fit g_{ds} as a function of drain bias. Typical GST results for $n = 5$ for different values of **AX** are shown in Fig. 3.15. The lack of existence of higher order derivatives at $V_{ds} = 0$ is not specific to PSP; in standard threshold-voltage based models the second derivative does not exist [11].

3.2.3.4 Harmonic Balance Simulation Test

Distortion analysis is important for evaluation of RF circuit performance. Traditional MOSFET models can produce unphysical harmonic balance simulation results for the third and higher order harmonics due to their singular behavior at

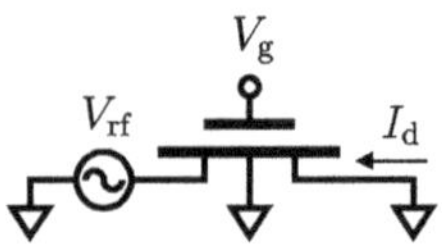

Fig. 3.16 Biasing scheme for a single tone harmonic balance simulation

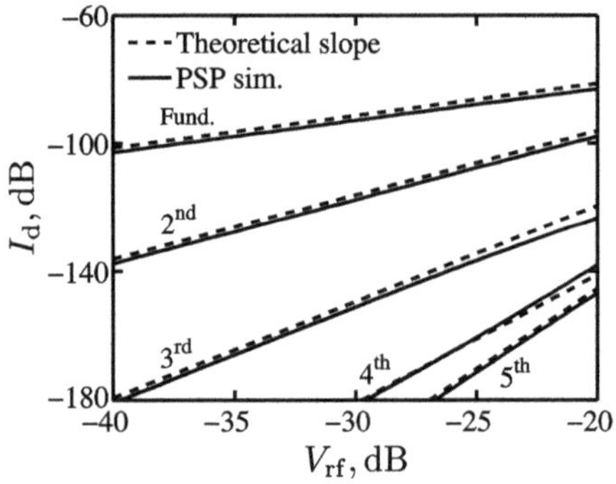

Fig. 3.17 Single tone harmonic balance simulation results for PSP comparing with theoretical slopes, $V_g = 0.4$ V, $W/L = 10/0.4$ μm, fundamental frequency is 100 MHz. Default parameters are used for PSP

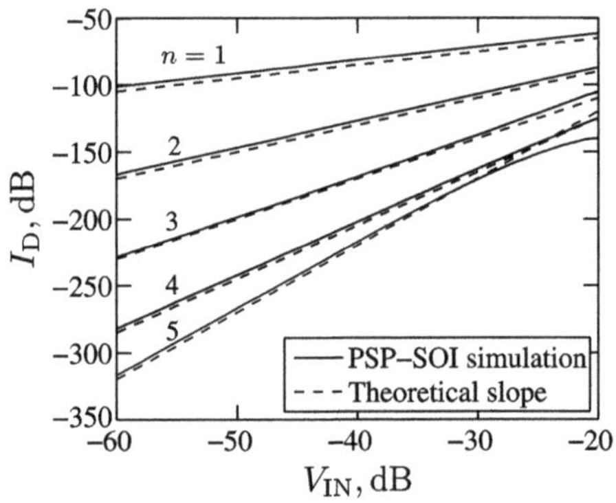

Fig. 3.18 Single tone harmonic balance simulation results for PSP-SOI. $V_g = 0.8$ V; $W/L = 3/0.055$ μm; fundamental frequency is 100 MHz. After [47]

$V_{ds} = 0$ [1, 23]. Theory [13] indicates that the second harmonic should be proportional to the square of the input signal level, the third harmonic should be proportional to the cube of the input signal level, and so on. This test consists of biasing a MOSFET as shown in Fig. 3.16 and plotting the harmonic content of the drain current as a function of the amplitude of the single tone excitation V_{rf}. The slope of the nth harmonic should be n times the slope of the fundamental frequency.

Figure 3.17 shows harmonic balance simulation results for PSP; it produces correct slopes for up to the fifth order harmonic. The PSP-SOI model [47], which is formulated within the framework of the PSP model, inherits its symmetry nature and also produces correct results when subjected to the same test, see Fig. 3.18.

The origin of the unphysical 2 dB/dB slope of the third harmonic characteristic for singular transistor models has been investigated in [23] for a simple MESFET model. Here we extend the method of [23] to a more general case and factor in details of how MOSFET models are implemented in circuit simulators.

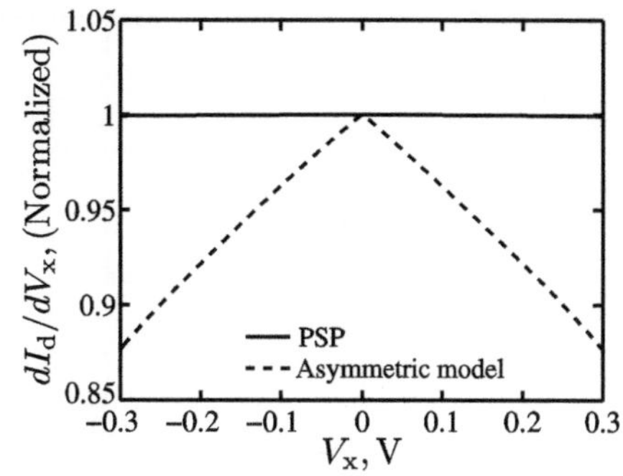

Fig. 3.19 Illustration of (3.27) and (3.31). For PSP $b = 0$, for an asymmetric model $b = -b' \neq 0$

As discussed previously, MOSFET models are generally implemented with a source and drain terminal flip for $V_{ds} < 0$, so the condition

$$I_d = f(V_{sb}, V_{db}, V_{gb}) = -f(V_{db}, V_{sb}, V_{gb}) \tag{3.25}$$

is automatically imposed. In the case of the GST (cf. Fig. 3.13) V_g and V_{bo} are fixed and the "symmetry condition" takes the form (3.21) where

$$I_d(V_x) = f(-V_{bo} - V_x, -V_{bo} + V_x, V_{gb}). \tag{3.26}$$

For an asymmetric core model this terminal flip introduces a singularity at $V_{ds} = 0$, which commonly manifests as nonexistence of the second derivative $d^2 I_d/dV_x^2$ for $V_x = 0$ (cf. Fig. 3.19).

To investigate the consequences, consider first the symmetric single tone excitation corresponding to Fig. 3.13 with $V_{bo} = 0$ and $V_x = \frac{1}{2} V_o \cos(\omega t)$. In what follows $V = 2V_x = V_{ds}$.

Assuming that the $I_d(V)$ characteristic is such that the first three derivatives exist at $V = 0$,

$$I_d = aV + bV^2 + cV^3 + O(V^3) \tag{3.27}$$

where a, b and c are constants. For a symmetric model $b = 0$, therefore

$$I_d = I_1 \cos(\omega t) + I_3 \cos(3\omega t) + O(V_o^3), \tag{3.28}$$

$$I_1 = aV_o + \frac{3}{4} c V_o^3 \tag{3.29}$$

and

$$I_3 = \frac{cV_o^3}{4} \tag{3.30}$$

corresponding to the correct 3 dB/dB slope of the third harmonic (cf. Fig. 3.20).

For an asymmetric model, after the terminal flip, instead of (3.27) one has

$$I_d = \begin{cases} aV + bV^2 + cV^3 + O(V^3), & V \geq 0, \\ aV + b'V^2 + c'V^3 + O(V^3), & V < 0. \end{cases} \tag{3.31}$$

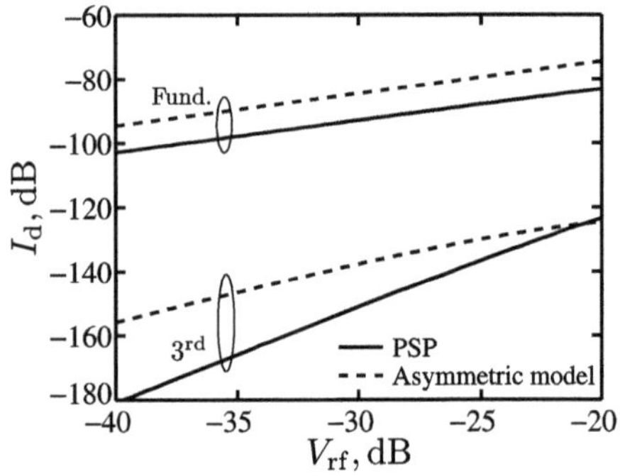

Fig. 3.20 Harmonic balance simulation results for symmetric excitations, $V_g = 0.4$ V, $W/L = 10/0.4$ μm, fundamental frequency is 100 MHz. Default parameters are used for PSP

Here as in [11, 23] the first derivative exists and is continuous,

$$a = I_d'(0), \tag{3.32}$$

$$b = \frac{1}{2} I_d''(0+), \qquad b' = \frac{1}{2} I_d''(0-) \tag{3.33}$$

and

$$c = \frac{1}{6} I_d'''(0+), \qquad c' = \frac{1}{6} I_d'''(0-). \tag{3.34}$$

By (3.21), $b' = -b$, $c' = c$.

In [23] the condition (3.25) was not imposed and it was assumed that $c = c' = 0$. Regardless of the singularity issue, $I_d(V)$ can be expanded as a Fourier series in the form (for the symmetric excitation in Fig. 3.13, $I_0 = I_2 = 0$)

$$I_d = \frac{I_0}{2} + \sum_{n=1}^{\infty} I_n \cos(n\omega t). \tag{3.35}$$

The amplitude of the third harmonic

$$I_3 = \frac{4}{T} \int_0^{\frac{T}{2}} I(t) \cos(3\omega t) dt \tag{3.36}$$

is given by [15]

$$I_3 = \frac{8bV_o^2}{15\pi} + \frac{cV_o^3}{4} + O(V_o^3). \tag{3.37}$$

Note that the second term in (3.37) is given by (3.30) while the first term is associated with the singularity described by (3.31) and first studied in [23]. For small V_o, $I_3 \propto V_o^2$ which explains the erroneous slope of 2 dB/dB for the third harmonic observed in harmonic balance simulations of asymmetric models (see Fig. 3.20). Note also that (3.37) shows there is a gradual transition from $I_3 \propto V_o^2$ to $I_3 \propto V_o^3$ as $|b|$ is reduced relative to $|c|V_o$. For symmetric models $b = 0$ exactly and the correct $I_3 \propto V_o^3$ behavior emerges.

For asymmetric harmonic balance excitation, see Fig. 3.16,

$$I_d = f(-V_{rf}, 0, V_g) \tag{3.38}$$

where f is the same function as in (3.25). Then

$$I_d = \begin{cases} a_{rf}V_{rf} + b_{rf}V_{rf}^2 + c_{rf}V_{rf}^3 + O(V_{rf}^3), & V_{rf} \geq 0, \\ a_{rf}V_{rf} + b'_{rf}V_{rf}^2 + c'_{rf}V_{rf}^3 + O(V_{rf}^3), & V_{rf} < 0. \end{cases} \tag{3.39}$$

Generally speaking, the coefficients a_{rf}, b_{rf} and c_{rf} are different from the coefficients a, b and c in (3.31). Furthermore, the symmetry property (3.25) still applies but does not require I_d to be an odd function of V_{rf}. In this case

$$I_3 = \frac{4(b_{rf} - b'_{rf})V_o^2}{15\pi} + \frac{(c_{rf} + c'_{rf})V_o^3}{8} + O(V_o^3). \tag{3.40}$$

For a symmetric model the condition $b = b' = 0$ can be reformulated in terms of f to show that $b_{rf} = b'_{rf} = 0$. Then $I_3 \propto V_o^3$ and it exhibits the correct 3 dB/dB slope. However, for an asymmetric model this is not the case and once again the erroneous 2 dB/dB slope appears, as described in [1, 23]. Note also that for asymmetric excitation the even harmonics do not vanish. For example,

$$I_2 = \frac{V_o^2(b_{rf} + b'_{rf})}{4} + O(V_o^3) \tag{3.41}$$

has the correct slope of 2 dB/dB regardless of the model symmetry.

3.2.3.5 Modified Symmetry Test (MST)

Modern MOSFET models include extrinsic components such as gate and substrate currents. The principle that the device should maintain symmetry when $V_{ds} = 0$ still holds in the presence of these extrinsic currents. If these are included, then the traditional Gummel symmetry test does not apply. From Fig. 3.21

$$I_d(V_x) = I_{ds}(V_x) - I_{gd}(V_g - V_x) - I_{bd}(V_b - V_x). \tag{3.42}$$

So when $V_x = 0$ the drain current $I_d(0) = -I_{gb}(V_g) - I_{bd}(V_b)$ may be different from zero. Hence, generally speaking, I_d is not an odd function of V_x. For this reason, a modified symmetry test has been developed in [19] where

$$I_x = \frac{I_d - I_s}{2} \tag{3.43}$$

is still an odd function of V_x in the presence of extrinsic currents and devolves to the GST when these are not present. For example, including gate current,

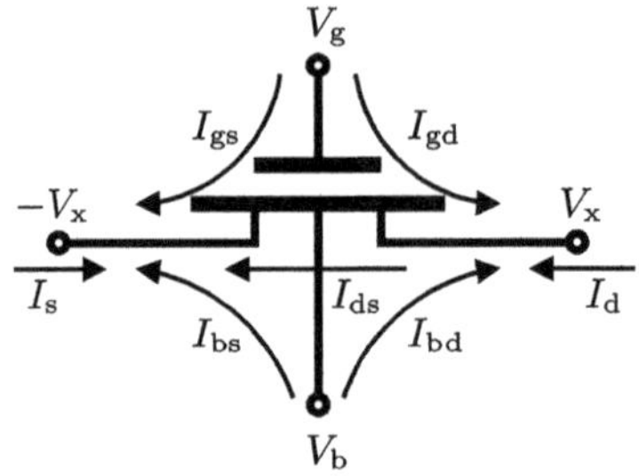

Fig. 3.21 Current balance for MST

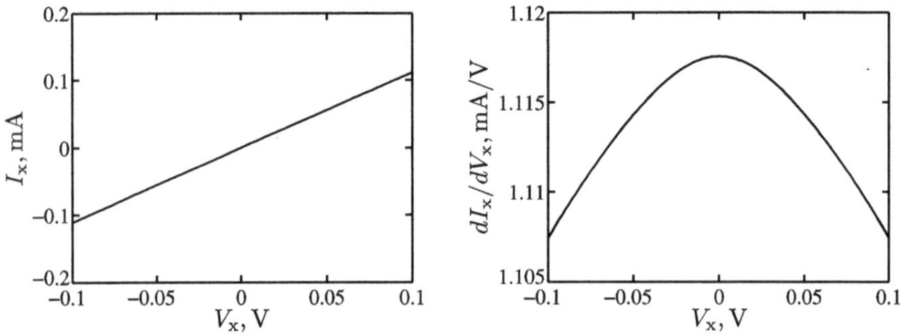

Fig. 3.22 MST for PSP with $V_g = 1$ V, $V_b = -0.2$ V and $V_{bo} = 0$ V (cf. Fig. 3.13). Default model parameters are used

$$I_x(V_x) = I_{ds}(V_x) + \frac{1}{2} \cdot \left[I_{gs}(V_g + V_x) - I_{gd}(V_g - V_x)\right] \tag{3.44}$$

$$= -I_{ds}(-V_x) - \frac{1}{2} \cdot \left[I_{gs}(V_g - V_x) - I_{gd}(V_g + V_x)\right] \tag{3.45}$$

$$= -I_x(-V_x). \tag{3.46}$$

A similar analysis holds if substrate current is included.

The same biasing scheme as for the GST (cf. Fig. 3.13) is used to test models symmetry in the presence of extrinsic currents. As for the GST, a model should not be singular at $V_x = 0$, see Fig. 3.22. Another example is shown in Fig. 3.23 for a floating body SOI MOSFET [47] where (unlike for the case of the bulk PSP model) valence-band electron (EVB) current is included [46].

In addition to dc symmetry, the modified symmetry tests [19] also includes a transcapacitance check. Figure 3.24 shows the biasing scheme for this test. I_{g+}, I_{d+} and I_{s+} are the gate, drain and source currents respectively, with in-phase source and drain ac excitations applied while I_{g-}, I_{d-} and I_{s-} are the corresponding currents with anti-phase source and drain ac excitations applied. These small signal elements form the quantities which can be used to test the symmetry of the terminal charge models. For example, δ_{cg}, defined as

$$\delta_{cg} = \frac{i_{g-}}{i_{g+}} = \frac{C_{gs} - C_{gd}}{C_{gs} + C_{gd}} \tag{3.47}$$

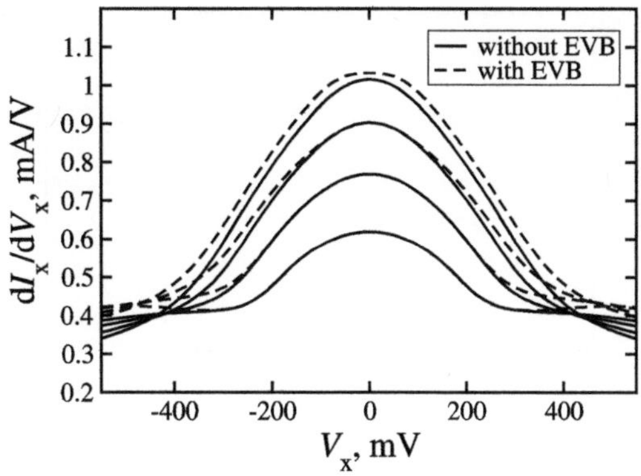

Fig. 3.23 Modified symmetry test for PSP-SOI. The model parameters are extracted from a 90 nm SOI technology. V_g varies from 0.7 to 1.0 V in 0.1 V steps; $W/L = 2/1$ μm. After [14]

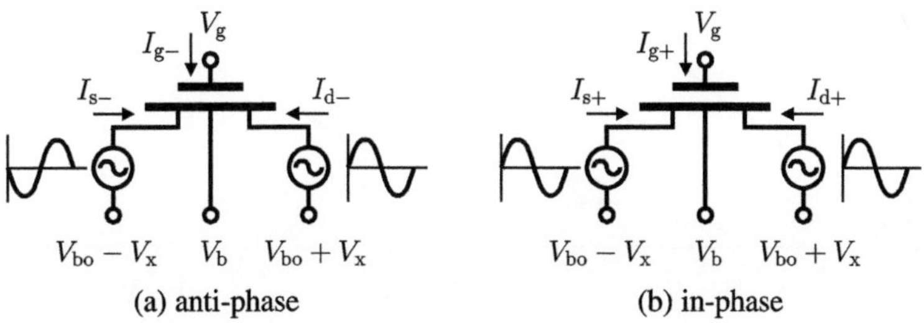

Fig. 3.24 Biasing scheme for modified ac symmetry test

where $i_{g-} = \mathrm{Im}(I_{g-})$ and $i_{g+} = \mathrm{Im}(I_{g+})$, should be an odd function of V_x and can be used to test the symmetry of the gate charge model. Similarly, symmetry of the source and drain charges can be tested by noting that

$$\delta_{csd} = \frac{(i_{s-} + i_{d-}) + (i_{s+} - i_{d+})}{(i_{s-} - i_{d-}) + (i_{s+} + i_{d+})} = \frac{C_{ss} - C_{dd}}{C_{ss} + C_{dd}} \tag{3.48}$$

(where $i_{s-} = \mathrm{Im}(I_{s-})$ and similarly for the other terms) should also be an odd function of V_x. These quantities (and an analogous form for the bulk charge, see [19]) should be nonsingular at $V_x = 0$ for properly symmetric MOSFET charge models. Figure 3.25 shows PSP results for these ac symmetry tests, verifying that the model passes and is properly symmetric.

3.2.4 Non-Quasi-Static (NQS) and Noise Model Tests

Modeling of NQS effects is important for some circuit design applications, in particular when relatively long-channel devices are used in RF circuits. Many MOSFET models include user selectable NQS options [9, 40, 42, 43]. NQS model benchmarking is in an embryonic state, although some circuits that are particularly difficult to model have been suggested to assess the qualitative performance of the

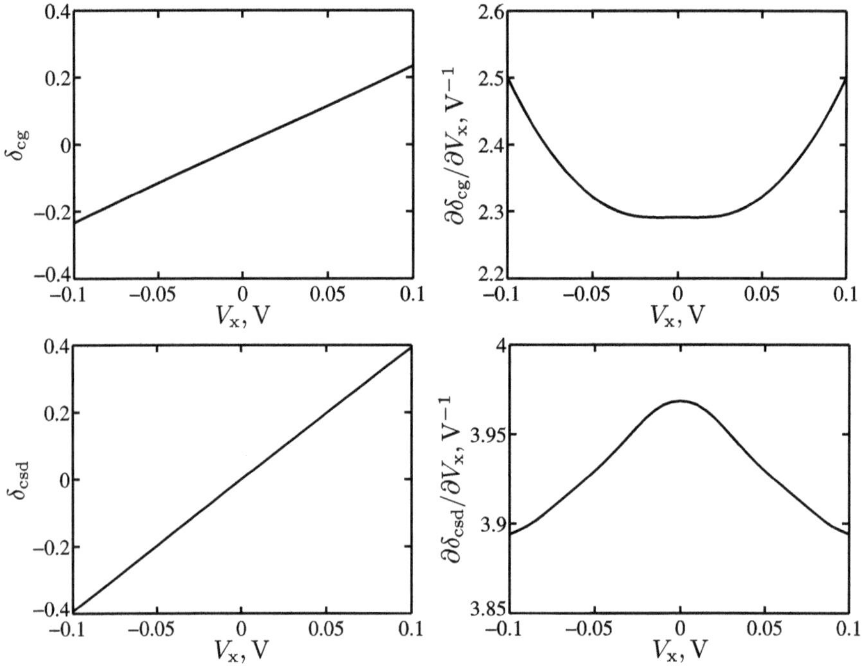

Fig. 3.25 Modified ac symmetry tests for PSP with $V_g = 1$ V, $V_b = -0.2$ V and $V_{bo} = 0$ V (cf. Fig. 3.24)

NQS models [16], and there is the obvious requirement that an NQS version of a model should asymptotically approach the quasi-static base model for low frequencies.

Here we describe two recent benchmark tests to verify that an NQS model gives theoretically expected results under quasi-static operation. For $V_{ds} = 0$ a long-channel MOSFET is essentially a transmission line whose properties are controlled by the gate bias; at low frequencies it is a voltage-controlled resistor with a distributed capacitance from channel to gate. With this in mind it is easy to show (see Appendix 1) that for long-channel devices [14]

$$R_{dg} \equiv \lim_{f\to 0} \mathrm{Re}\left(\frac{1}{y_{dg}}\right) = \frac{1}{6\cdot g_{ds0}}\cdot\left(1+\frac{C_s}{C_{ox}}\right) \tag{3.49}$$

where C_s denotes the capacitance of the space charge (i.e. bulk depletion) region per unit area and g_{ds0} is the channel conductance at $V_{ds} = 0$. A similar analysis gives [43]

$$R_{in} \equiv \lim_{f\to 0} \mathrm{Re}\left(\frac{1}{y_{gg}}\right) = \frac{1}{12\cdot g_{ds0}}. \tag{3.50}$$

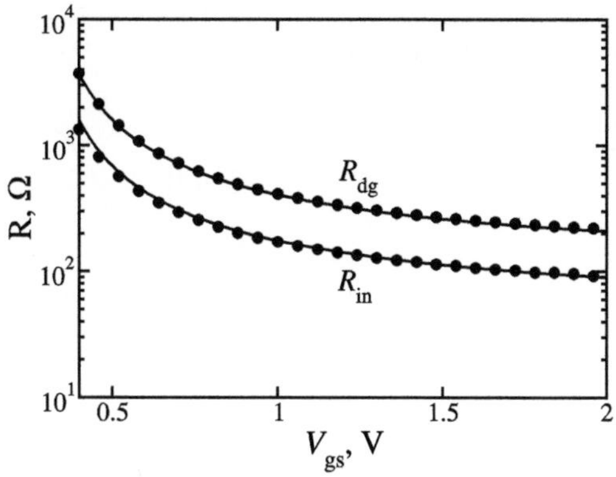

Fig. 3.26 R_{in} and R_{dg} as functions of gate-source bias. *Symbols* represent PSP simulations using a long-channel parameter set while *lines* refer to theoretical values of R_{in} and R_{dg}. After [14]

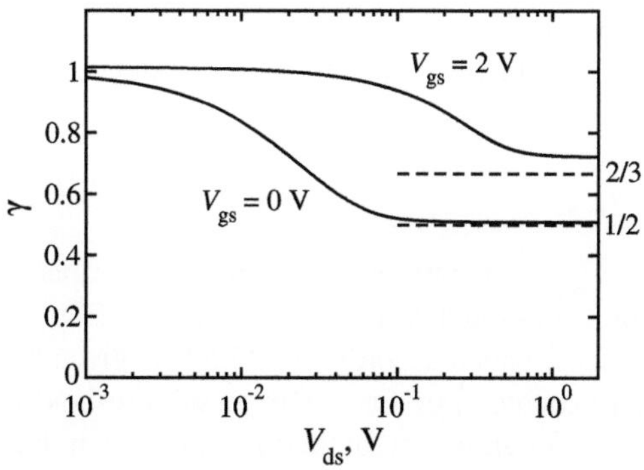

Fig. 3.27 Simulation results of white noise gamma factor as a function of drain-source voltage for PSP. After [14]

A physically correct model should satisfy both relations for all gate biases. Typical results for PSP (with NQS model activated by setting **SWNQS** = 9) are shown in Fig. 3.26.

Advanced compact MOSFET models should include all noise sources in a physical manner. In particular, thermal channel noise should automatically reduce to the Schottky shot noise formula in weak inversion. Further qualitative tests [14, 25] involve the so-called white-noise gamma factor

$$\gamma = S_{I_d}/(4k_B T g_{ds0}). \tag{3.51}$$

For $V_{ds} = 0$ V, the Nyquist law applies and $\gamma = 1$ should hold exactly irrespective of the level of inversion. At high V_{ds}, γ should approach $1/2$ in weak inversion (corresponding to "shot noise"), whereas in strong inversion saturation it should approximately approach the well-known value of $2/3$. This is no physical reason for γ to be exactly $2/3$ in this case. Figure 3.27 presents results that show that PSP complies with the requirements of this test.

The correlation coefficient between gate and drain thermal noise for a long channel device in saturation should be $c \approx 0.395j$ [35]. By symmetry, c should be equal to zero at $V_{ds} = 0$ [25] (see Appendix 2). Figure 3.28 shows that for PSP the imaginary part of c increases from 0 at $V_{ds} = 0$ to approximately 0.395 in saturation, as

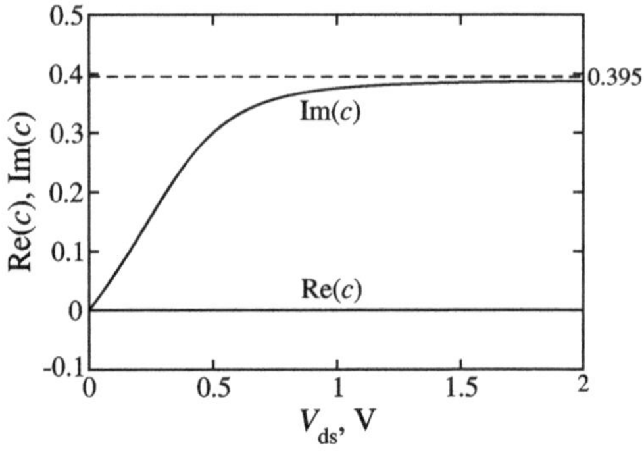

Fig. 3.28 Simulated real and imaginary parts of correlation coefficient versus drain-source voltage for PSP. After [14]

expected for this test (which should be done below the transition frequency f_T of the transistor).

Another benchmark for the noise model is based on the expression [7, 10, 22] for the spectral density of the induced gate noise at $V_{ds} = 0$:

$$S_{I_g} = 4k_B T (2\pi f C_{gc})^2 / (12 g_{ds0}) \tag{3.52}$$

where C_{gc} is the intrinsic channel-to-gate capacitance. The above equation can be interpreted as the thermal noise emanating from the real part of the MOSFET input impedance, which according to (3.50) is given by $1/(12 g_{ds0})$ at $V_{ds} = 0$]. Indeed, for zero source-drain bias, both in subthreshold and strong inversion, the PSP expressions for drain current thermal noise and induced gate noise reduce to $4k_B T g_{ds0}$ and (3.52), respectively. This is demonstrated in Fig. 3.29. Note that the zero source-drain bias condition applies to the situation when the MOSFET is used as a capacitor. In this case, drain current thermal noise has no contribution to the device noise, since source and drain are connected to each other, thereby short-circuiting the thermal noise source between source and drain. Induced gate noise, as given by (3.52), becomes the dominant source of noise. Noise from the gate resistance [24] is less important when a long-channel MOSFET is used as a capacitor.

3.2.5 Self-Heating Effect Test (SHE)

The self-heating effect in MOSFET devices and circuits has been extensively studied, especially for SOI devices where it significantly affects the parameter extraction procedure [27, 29]. The heat generated in the channel raises the local temperature due to the low thermal conductivity (two orders of magnitude smaller than that of Si) of the buried SiO_2. SHE modeling is typically done using a thermal network [28, 39]. The following benchmark test was designed to verify the correctness of the SHE model [4]. At fixed drain and gate biases, the thermal resistance is swept with self-heating enabled. The simulated channel temperature and drain current are denoted as T_{ch} and I_{ds}. Next, under the same bias but with self-heating disabled, the

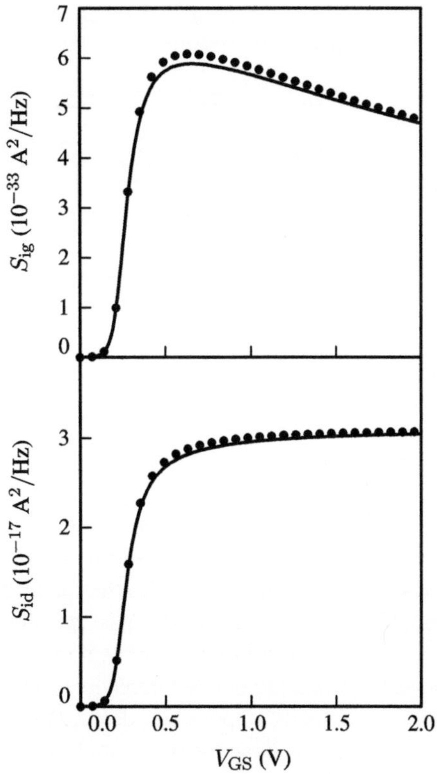

Fig. 3.29 Noise spectral densities of drain current thermal noise (*symbols*) and induced gate noise ($f = 1$ kHz, *symbols*), calculated with the PSP model for a $W/L = 10/0.1$ μm n-channel MOSFET using the default parameter set, with **QMC** $= 0$, $T = 290$ K, and $V_{ds} = 0$ V. The *solid lines* are the expected values for drain current noise and induced gate noise, i.e., $4k_B T g_{ds0}$ and (3.52), respectively

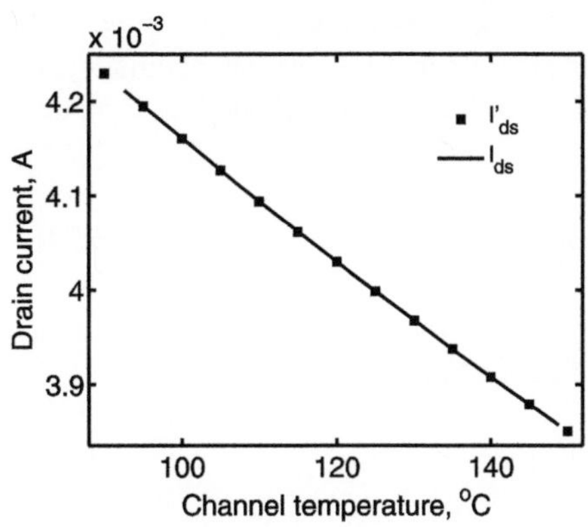

Fig. 3.30 Benchmark test results of self-heating implementation in PSP-SOI model. $W/L = 3/0.065$ μm, $V_{gs} = V_{ds} = 1.2$ V. After [47]

drain current I'_{ds} is computed at the ambient temperature $T_{amb} = T_{ch}$. If self-heating is implemented correctly, then $I_{ds} = I'_{ds}$. The procedure is illustrated for PSP-SOI in Fig. 3.30.

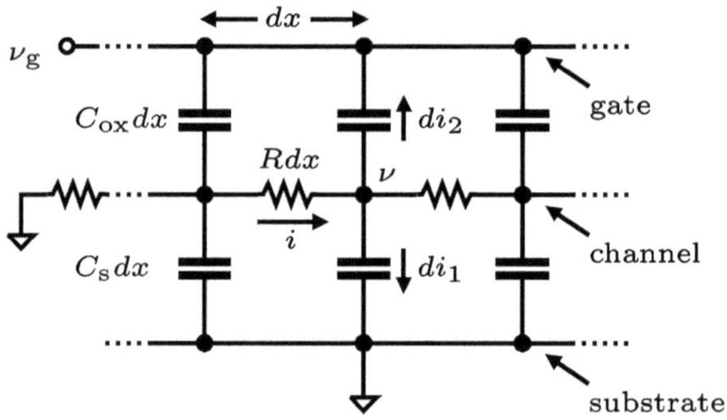

Fig. 3.31 Small-signal equivalent circuit for computing y_{dg} at $V_{ds} = 0$; $R = (g_{ds0} \cdot L)^{-1}$; $W = 1$

3.3 Conclusion

Despite a decades long indifference of the MOSFET modeling community to fundamental problems with, and unphysical characteristics in, MOSFET models, there has been significant improvement in the qualitative physical behavior exhibited by advanced MOSFET models in the past decade. A major driver for this improvement has been the on-going development of benchmark tests that verify whether or not a model exhibits the desired behavior: MOSFET models are so complex these days that it is virtually impossible, despite the best intentions, to build a "perfect" model. Benchmark tests are instrumental in verifying that a model has the correct behavior. In addition, models that pass the benchmark tests give designers a much better assurance that their simulation results will be accurate. All MOSFET models, including PSP, adopted or under development for advanced industrial applications should pass the benchmark tests presented in this chapter.

The development of the new benchmark tests for MOSFET models, especially in the areas of noise and NQS modeling, is an active and on-going area of research.

Appendix 1: Derivation of (3.49) and (3.50)

We derive (3.49) using a transmission line model representing MOSFET at $V_{ds} = 0$. With reference to Fig. 3.31

$$\frac{di_1}{dx} = j\omega C_s v \tag{3.53}$$

and

$$\frac{di_2}{dx} = j\omega C_{ox}(v - v_g). \tag{3.54}$$

Hence

$$\frac{di}{dx} = j\omega C_{ox} u \tag{3.55}$$

where

$$u = \left(1 + \frac{C_s}{C_{ox}}\right) v - v_g. \tag{3.56}$$

With this definition

$$\frac{du}{dx} = -ir \tag{3.57}$$

where

$$r = R\left(1 + \frac{C_s}{C_{ox}}\right). \tag{3.58}$$

From (3.55) and (3.57)

$$\frac{d^2u}{dx^2} + \frac{u}{\lambda^2} = 0 \tag{3.59}$$

where

$$\lambda = (j\omega r C_{ox})^{-1/2}. \tag{3.60}$$

The general solution of (3.59) is

$$u = a \cdot \sinh \frac{x}{\lambda} + b \cdot \cosh \frac{x}{\lambda}. \tag{3.61}$$

The boundary conditions used to compute

$$y_{dg} = \left. \frac{i(L)}{v_g} \right|_{v(0)=v(L)=0} \tag{3.62}$$

are $v(0) = v(L) = 0$ or, equivalently,

$$u(0) = u(L) = -v_g. \tag{3.63}$$

Hence

$$a = v_g \frac{\cosh \Theta - 1}{\sinh \Theta} \tag{3.64}$$

and

$$b = -v_g \tag{3.65}$$

where

$$\Theta = \frac{L}{\lambda} = L\sqrt{j\omega r C_{ox}}. \tag{3.66}$$

From (3.57) and (3.61)

$$i(L) = -\frac{1}{r}\left(\frac{du}{dx}\right)_{x=L} = -\frac{a \cdot \cosh \Theta + b \cdot \sinh \Theta}{\lambda r}. \tag{3.67}$$

With reference to (3.64) and (3.65)

$$i(L) = \frac{v_g}{\lambda r} \cdot \frac{\cosh \Theta - 1}{\sinh \Theta}. \tag{3.68}$$

It follows that

$$\frac{1}{y_{dg}} = \lambda r \frac{\sinh \Theta}{\cosh \Theta - 1} = \lambda r \coth \frac{\Theta}{2}. \tag{3.69}$$

For $f \to 0$, we have $\Theta \propto \sqrt{f} \to 0$ and

$$\coth \frac{\Theta}{2} = \frac{2}{\Theta} + \frac{\Theta}{6} + O(\Theta^2) \tag{3.70}$$

or

$$\frac{1}{y_{dg}} = \frac{2}{j\omega C_{ox} L} + \frac{rL}{6} + O(f). \tag{3.71}$$

The first term in (3.71) is imaginary and the second term is real. Therefore

$$\lim_{f \to 0} \mathrm{Re}\left(\frac{1}{y_{dg}}\right) = \frac{rL}{6} \tag{3.72}$$

which is equivalent to (3.49) after substituting r from (3.58).

To obtain (3.50), note that

$$y_{gg} = -\left.\frac{i_2}{i_g}\right|_{v(0)=v(L)=0}. \tag{3.73}$$

With reference to (3.54) and (3.56)

$$i_2 = \frac{j\omega C_{ox} v_g L}{1 + \frac{C_s}{C_{ox}}} \left(\frac{1}{v_g L} \int_0^L u dx - \frac{C_s}{C_{ox}}\right). \tag{3.74}$$

Using (3.61), (3.64) and (3.65) to compute the integral we found

$$y_{gg} = \frac{j\omega C_{ox} v_g L}{1 + \frac{C_s}{C_{ox}}} \left(\frac{\tanh \frac{\Theta}{2}}{\frac{\Theta}{2}} + \frac{C_s}{C_{ox}}\right). \tag{3.75}$$

For $f \to 0$ we have $\Theta \to 0$ and

$$\frac{\tanh \frac{\Theta}{2}}{\frac{\Theta}{2}} = 1 - \frac{\Theta^2}{12} + O(\Theta^4) \tag{3.76}$$

whence

$$\frac{1}{y_{gg}} = \frac{1}{j\omega C_{ox} v_g L}\left[1 + \frac{\Theta^2}{12(1 + \frac{C_s}{C_{ox}})} + O(\Theta^4)\right]. \tag{3.77}$$

Substituting (3.66) we have

$$\mathrm{Re}\left(\frac{1}{y_{gg}}\right) = \frac{rL}{12(1+\frac{C_s}{C_{ox}})} + O(f^2). \tag{3.78}$$

Now (3.50) follows once we note that

$$\frac{1}{g_{ds0}} = RL = \frac{rL}{(1+\frac{C_s}{C_{ox}})}. \tag{3.79}$$

Appendix 2: Correlation Coefficient Between Gate and Drain Thermal Noise at $V_{ds} = 0$

Here we consider the correlation coefficient c between gate and drain thermal noise. It is well-known that its theoretical value for a long channel device in saturation is $\sim 0.395j$ [35]. Its value of c at $V_{ds} = 0$ is less-well known but can be derived from a symmetry argument.

The correlation coefficient is defined as:

$$c = \frac{\langle i_g \cdot i_d^* \rangle}{\sqrt{\langle i_g \cdot i_g^* \rangle \cdot \langle i_d \cdot i_d^* \rangle}} \tag{3.80}$$

where i_d and i_g are the noise currents (in $\mathrm{A}/\sqrt{\mathrm{Hz}}$) to the drain and gate terminals, respectively.

Since the sum of noise currents through the four terminals is zero,

$$\langle i_g \cdot i_d^* \rangle = -\langle i_g \cdot i_g^* \rangle - \langle i_g \cdot i_s^* \rangle - \langle i_g \cdot i_b^* \rangle \tag{3.81}$$

where we have introduced i_s and i_b, analogous to i_d and i_g. Because of the source-drain symmetry, at $V_{ds} = 0$ (3.81) becomes

$$2 \cdot \langle i_g \cdot i_d^* \rangle = -\langle i_g \cdot i_g^* \rangle - \langle i_g \cdot i_b^* \rangle. \tag{3.82}$$

The noise current i_b corresponds to "induced bulk current noise", i.e., the counterpart of induced gate noise in the bulk terminal. Because the depletion capacitance is much smaller than the gate capacitance, the induced bulk current noise is much smaller than induced gate noise and is generally ignored. This is the case in PSP and other compact models. Neglecting i_b, (3.82) simplifies to

$$\langle i_g \cdot i_d^* \rangle = -\frac{1}{2} \cdot \langle i_g \cdot i_g^* \rangle. \tag{3.83}$$

Because $S_{id} = \langle i_d \cdot i_d^* \rangle$ and $S_{ig} = \langle i_g \cdot i_g^* \rangle$, Combining (3.80) and (3.83) yields

$$c = -\frac{1}{2} \cdot \sqrt{\frac{S_{ig}}{S_{id}}}. \tag{3.84}$$

Since at low frequencies $S_{ig} = O(f^2)$ and $S_{id} = O(1)$,

$$\lim_{f \to 0} c = 0. \tag{3.85}$$

At higher frequencies, as long as $f \ll f_T$, we have $S_{ig} \ll S_{id}$ and $c \approx 0$. These conclusions remain valid even if the induced bulk current noise is not neglected. The derivation, however, becomes more involved and is omitted here.

References

1. Bendix, P., Rakers, P., Wagh, P., Lemaitre, L., Grabinski, W., McAndrew, C.C., Gu, X., Gildenblat, G.: RF distortion analysis with compact MOSFET models. In: Proc. IEEE Custom Integr. Circuits Conf., pp. 9–12 (2004)
2. Bressoud, D.M.: A Radical Approach to Real Analysis. The Mathematical Association of America (2007)
3. Brews, J.R.: A charge-sheet model of the MOSFET. Solid-State Electron. **21**(2), 345–355 (1978)
4. Chen, Q., Wu, Z.Y., Su, R., Goo, J.S., Thuruthiyil, C., Radwin, M., Subba, N., Suryagandh, S., Ly, T., Wason, V., An, J., Icel, A.: Extraction of self-heating free I-V curves including the substrate current of PD SOI MOSFETs. In: Proc. IEEE Int. Conf. on Microelectron. Test Structures, pp. 272–275 (2007)
5. Chen, T.L., Gildenblat, G.: Symmetric bulk charge linearisation in charge-sheet MOSFET model. Electron. Lett. **37**(12), 791–793 (2001)
6. Gildenblat, G., Wang, H., Chen, T.L., Gu, X., Cai, X.: SP: an advanced surface-potential-based compact MOSFET model. IEEE J. Solid-State Circuits **39**(9), 1394–1406 (2004)
7. Gildenblat, G., Li, X., Wu, W., Wang, H., Jha, A., van Langevelde, R., Smit, G.D.J., Scholten, A.J., Klaassen, D.B.M.: PSP: An advanced surface-potential-based MOSFET model for circuit simulation. IEEE Trans. Electron Devices **53**(9), 1979–1993 (2006)
8. Gildenblat, G., Zhu, Z., McAndrew, C.C.: Surface potential equation for bulk MOSFET. Solid-State Electron. **53**(1), 11–13 (2009)
9. Jin, X., Ou, J.J., Chen, C.H., Liu, W., Deen, M.J., Gray, P.R., Hu, C.: An effective gate resistance model for CMOS RF and noise modeling. In: IEDM Tech. Dig., pp. 961–964 (1998)
10. Jindal, R.P.: Effect of induced gate noise at zero drain bias in field-effect transistors. IEEE Trans. Electron Devices **52**(3), 432–434 (2005)
11. Joardar, K., Gullapalli, K.K., McAndrew, C.C., Burnham, M.E., Wild, A.: An improved MOSFET model for circuit simulation. IEEE Trans. Electron Devices **45**(1), 134–148 (1998)
12. Klaassen, D.B.M., van Langevelde, R., Scholten, A.J.: Compact CMOS modelling for advanced analogue and RF applications. IEICE Trans. Electron. **E87-C**(6), 854–866 (2004)
13. Lee, T.H.: The Design of CMOS Radio-Frequency Integrated Circuits, 2nd edn. Cambridge University Press, Cambridge (2004)
14. Li, X., Wu, W., Jha, A., Gildenblat, G., van Langevelde, R., Smit, G.D.J., Scholten, A.J., Klaassen, D.B.M., McAndrew, C.C., Watts, J., Olsen, M., Coram, G., Chaudhry, S., Victory, J.: Benchmarking the PSP compact model for MOS transistors. In: Proc. IEEE Int. Conf. on Microelectron. Test Structures, pp. 259–264 (2007)
15. Li, X., Wu, W., Jha, A., Gildenblat, G., van Langevelde, R., Smit, G.D.J., Scholten, A.J., Klaassen, D.B.M., McAndrew, C.C., Watts, J., Olsen, C.M., Coram, G.J., Chaudhry, S., Victory, J.: Benchmark tests for MOSFET compact models with application to the PSP model. IEEE Trans. Electron Devices **56**(2), 243–251 (2009)
16. Liu, W.: MOSFET Models for SPICE Simulation: Including BSIM3V3 and BSIM. Wiley-Interscience, New York (2001)

17. McAndrew, C.C.: Practical modeling for circuit simulation. IEEE J. Solid-State Circuits **33**(3), 439–448 (1998)
18. McAndrew, C.C.: Useful numerical techniques for compact modeling. In: Proc. IEEE Int. Conf. on Microelectron. Test Structures, pp. 121–126 (2002)
19. McAndrew, C.C.: Validation of MOSFET model source–drain symmetry. IEEE Trans. Electron Devices **53**(9), 2202–2206 (2006)
20. McAndrew, C.C., Gummel, H.K., Singhal, K.: Benchmarks for compact MOSFET models. In: SEMATECH Compact Models Workshop (1995)
21. Pao, H.C., Sah, C.T.: Effects of diffusion current on characteristics of metal-oxide (insulator)-semiconductor transistors. Solid-State Electron. **9**(10), 927–937 (1966)
22. Paasschens, J.C.J., Scholten, A.J., van Langevelde, R.: Generalizations of the Klaassen-Prins equation for calculating the noise of semiconductor device. IEEE Trans. Electron Devices **52**(11), 2463–2472 (2005)
23. Scheinberg, N., Pinkhasov, A.: A computer simulation model for simulating distortion in FET resistors. IEEE Trans. Comput.-Aided Des. Integr. Circuits Syst. **19**(9), 981–989 (2000)
24. Scholten, A.J., Tiemeijer, L.F., van Langevelde, R., Havens, R.J., Zegers-van Duijnhoven, A.T.A., Venezia, V.C.: Noise modeling for RF CMOS circuit simulation. IEEE Trans. Electron Devices **50**, 618–632 (2003)
25. Scholten, A.J., van Langevelde, R., Tiemeijer, L.F., Klaassen, D.B.M.: Compact modeling of noise in CMOS. In: Proc. IEEE Custom Integr. Circuits Conf., pp. 711–716 (2006)
26. Scholten, A.J., Smit, G.D.J., De Vries, B.A., Tiemeijer, L.F., Croon, J.A., Klaassen, D.B.M., van Langevelde, R., Li, X., Wu, W., Gildenblat, G.: The new CMC standard compact MOS model PSP: Advantages for RF applications. IEEE J. Solid-State Circuits **44**(5), 1415–1424 (2009)
27. Su, L., Chung, J., Antoniadis, D., Goodson, K., Flik, M.: Measurement and modeling of self-heating in SOI nMOSFET's. IEEE Trans. Electron Devices **41**(1), 69–75 (1994)
28. Su, P., Fung, S., Tang, S., Assaderaghi, F., Hu, C.: BSIMPD: a partial-depletion SOI MOSFET model for deep-submicron CMOS designs. In: Proc. IEEE Custom Integr. Circuits Conf., pp. 197–200 (2000)
29. Tenbroek, B., Lee, M., Redman-White, W., Bunyan, R., Uren, M.: Impact of self-heating and thermal coupling on analog circuits in SOI CMOS. IEEE J. Solid-State Circuits **33**(7), 1037–1046 (1998)
30. Trofimenkoff, F.: Field-dependent mobility analysis of the field-effect transistor. Proc. IEEE **53**(11), 1765–1766 (1965)
31. Tsividis, Y.: Problems with precision modeling of analog MOS LSI. In: IEDM Tech. Dig., vol. 28, pp. 274–277 (1982)
32. Tsividis, Y.: Operation and Modeling of the MOS Transistor, 2nd edn. McGraw-Hill, New York (1999)
33. Tsividis, Y., Masetti, G.: Problems in precision modeling of the MOS transistor for analog applications. IEEE Trans. Comput.-Aided Des. Integr. Circuits Syst. **3**(1), 72–79 (1984)
34. Tsividis, Y.P., Suyama, K.: MOSFET modeling for analog circuit CAD: problems and prospects. IEEE J. Solid-State Circuits **29**(3), 210–216 (1994)
35. van der Ziel, A.: Noise in Solid State Devices and Circuits. Wiley-Interscience, New York (1986)
36. van Langevelde, R., Scholten, A.J., Havens, R.J., Tiemeijer, L.F., Klaassen, D.B.M.: Advanced compact MOS modelling. In: Proc. Eur. Solid-State Device Res. Conf., pp. 81–88 (2001)
37. van Langevelde, R., Scholten, A.J., Klaassen, D.B.M.: Physical background of MOS model 11. Nat.Lab. Unclassified Report, NL-TN 2003/00239 (2003). http://www.nxp.com/models/mos_models/model11/
38. Victory, J., Yan, Z., Gildenblat, G., McAndrew, C.C., Zheng, J.: A physically based, scalable MOS varactor model and extraction methodology for RF applications. IEEE Trans. Electron Devices **52**(7), 1343–1353 (2005)
39. Vogelsong, R., Brzezinski, C.: Simulation of thermal effects in electrical systems. In: Proc. IEEE Appl. Power Electron. Conf. and Expos. (APEC), pp. 353–356 (1989)

40. Wang, H., Gildenblat, G.: A robust large signal non-quasi-static MOSFET model for circuit simulation. In: Proc. IEEE Custom Integr. Circuits Conf., pp. 5–8 (2004)
41. Wang, H., Chen, T.L., Gildenblat, G.: Quasi-static and nonquasi-static compact MOSFET models based on symmetric linearization of the bulk and inversion charges. IEEE Trans. Electron Devices **50**(11), 2262–2272 (2003)
42. Wang, H., Li, X., Wu, W., Gildenblat, G., van Langevelde, R., Smit, G.D.J., Scholten, A.J., Klaassen, D.B.M.: Unified non-quasi-static MOSFET model for large-signal and small-signal simulations. In: Proc. IEEE Custom Integr. Circuits Conf., pp. 823–826 (2005)
43. Wang, H., Li, X., Wu, W., Gildenblat, G., van Langevelde, R., Smit, G.D.J., Scholten, A.J., Klaassen, D.B.M.: A unified nonquasi-static MOSFET model for large-signal and small-signal simulations. IEEE Trans. Electron Devices **53**(9), 2035–2043 (2006)
44. Watts, J., McAndrew, C.C., Enz, C., Galup-Montoro, C., Gildenblat, G., Hu, C., van Langevelde, R., Miura-Mattausch, M., Rios, R., Sah, C.T.: Advanced compact models for MOSFETs. In: Tech. Proc. Workshop on Compact Modeling, pp. 3–12 (2005)
45. Wu, W., Chen, T.L., Gildenblat, G., McAndrew, C.C.: Physics-based mathematical conditioning of the MOSFET surface potential equation. IEEE Trans. Electron Devices **51**(7), 1196–1199 (2004)
46. Wu, W., Li, X., Gildenblat, G., Workman, G.O., Veeraraghavan, S., McAndrew, C.C., van Langevelde, R., Smit, G.D.J., Scholten, A.J., Klaassen, D.B.M.: A compact model for valence-band electron tunneling current in partially depleted SOI MOSFETs. IEEE Trans. Electron Devices **54**(2), 316–322 (2007)
47. Wu, W., Li, X., Gildenblat, G., Workman, G.O., Veeraraghavan, S., McAndrew, C.C., van Langevelde, R., Smit, G.D.J., Scholten, A.J., Klaassen, D.B.M., Watts, J.: PSP-SOI: An advanced surface potential based compact model of partially depleted SOI MOSFETs for circuit simulations. Solid-State Electron. **53**(1), 18–29 (2009)

Chapter 4
High-Voltage MOSFET Modeling

E. Seebacher, K. Molnar, W. Posch, B. Senapati,
A. Steinmair, and W. Pflanzl

Abstract In many new applications like communication and automotive electronics the usage of integrated high voltage MOS transistors (LDMOS and DMOS) requires highly accurate compact models. In this chapter we present a deep look into special LDMOS transistor behavior and discuss state of the art sub-circuit modeling with BSIM/EKV core and JFET/Resistor approach. Parasitic diode and bipolar effects are discussed and modeling suggestions are presented. The EKV high voltage model developed by Swiss Federal Institute of Technology (EPFL) and the MM20 high voltage model introduced by NXP Research (formerly Philips Research) Laboratories is demonstrated in detail. The first CMC (Compact Modeling Council) standard high voltage MOSFET model HiSIM_HV developed by Hiroshima University is explained as well. Finally, characterization and measurement strategies for LDMOS modeling are described.

Keywords Compact modeling · LDMOS · LDMOS sub-circuit · EKV-HV · MM20 · HiSIM_HV · Parasitic modeling · Pulsed measurements · HV technology

---

E. Seebacher (✉) · K. Molnar · W. Posch · B. Senapati · A. Steinmair · W. Pflanzl
Austriamicrosystems AG, Tobelbaderstr. 30, 8141 Unterpremstaetten, Austria
e-mail: ehrenfried.seebacher@austriamicrosystems.com

K. Molnar
e-mail: kund.molnar@austriamicrosystems.com

W. Posch
e-mail: werner.posch@austriamicrosystems.com

B. Senapati
e-mail: biswanath.senapati@austriamicrosystems.com

A. Steinmair
e-mail: alexander.steinmair@austriamicrosystems.com

W. Pflanzl
e-mail: walter.pflanzl@austriamicrosystems.com

G. Gildenblat (ed.), *Compact Modeling*,
DOI 10.1007/978-90-481-8614-3_4,

4.1 Introduction

Compact modeling usually reflects the actual technology and product development requirements. High voltage MOSFET modeling has become especially important in recent years. Highly accurate SPICE models are needed in applications like Power Management, Sensor Interfaces and Actuators (MEMS), high-end Display Drivers, Line drivers (DSL, ADSL, SLIC), Printer Drivers, DC-Motor Drivers, FET Drivers, HPA standard products (LED drivers), DC-DC Converters, Switched Power Supplies, and Power Amplifiers. Used technologies are Bipolar-CMOS-DMOS (BCD) [19] and HV CMOS. Both use LDMOS and DMOS. HV CMOS [44] can achieve comparable performance to BCD at the lower process complexity of CMOS, with additional HV wells but without a buried layer. BCD uses process complexity to control parasitics, while LDMOS transistors in HV CMOS need extensive modeling of parasitic bipolar and diodes. The main differences between standard MOS and LDMOS transistor or power n-MOSFET are shown in Fig. 4.1 [35]. The basic n-channel MOSFET structure is shown in Fig. 4.1a. It is typically fabricated by diffusing or implanting phosphorus into a p-type silicon substrate to form the drain and source regions. For LDMOS as shown in Fig. 4.1b a lightly doped n-drift region is generated between the n^+ drain and the channel region to sustain a higher blocking voltage. The channel is usually formed below the silicon dioxide gate dielectric with a polysilicon or polycide gate electrode. The principle behavior of the intrinsic part of both transistors Figs. 4.1a and b is equal where the LDMOS structure is differentiated in the addition of the n-drift region. Beside the lateral structures, two commonly used vertical structures produced in BCD technology are the double diffused MOS (DMOS) and the UMOS shown in Figs. 4.2a and b. The n^+ source and p-body of the DMOS transistor is generated by a double diffused process through a common mask opening defined by the edge of the polysilicon gate. The channel region is defined by the difference of the lateral diffusion of the n^+ and p-body. The gate region of the UMOS has a U-form shaped by silicon reactive ion etching.

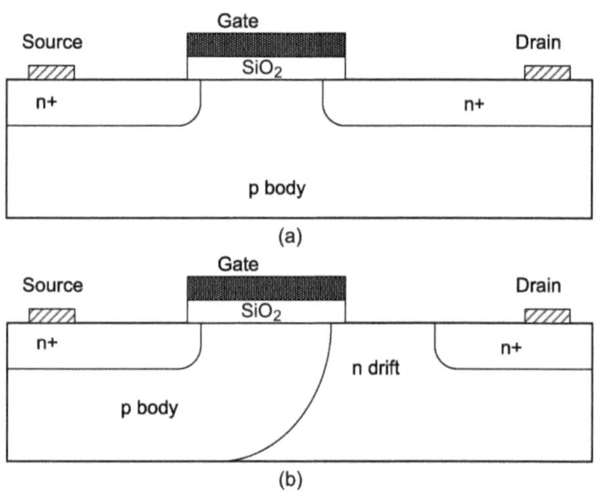

Fig. 4.1 **a** Plain n-MOSFET structure. **b** Power n-MOSFET structure with n-drift region added for higher voltages

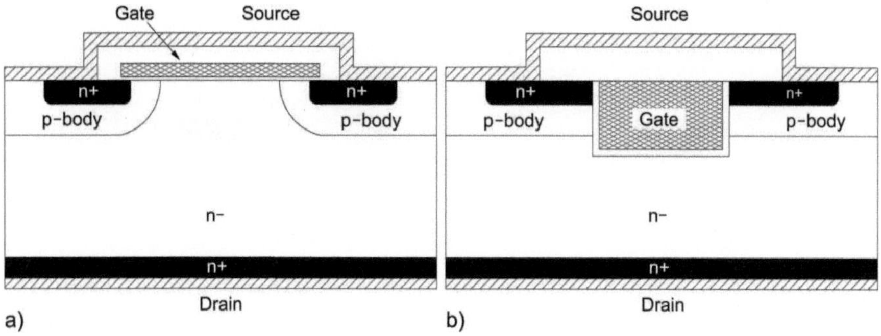

Fig. 4.2 Structure of **a** DMOS, **b** UMOS

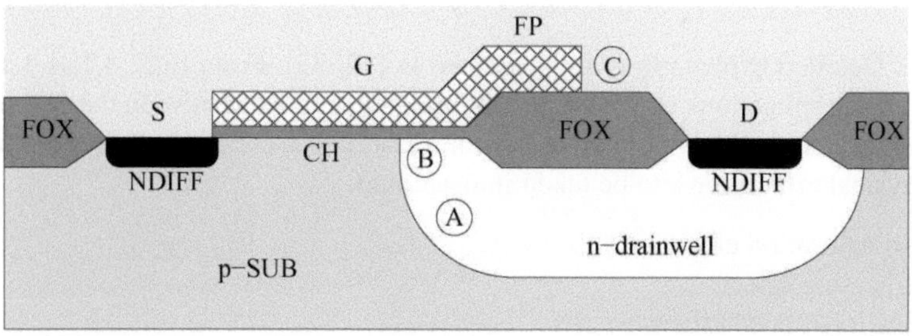

Fig. 4.3 Substrate based n-channel HV MOSFET

Grading effects in the p-body as well in the drain diffusion regions have to be taken into account for correct modeling.

Lateral LDMOS transistors fabricated in HV CMOS processes are shown in Fig. 4.3 as substrate based and in Fig. 4.4 as isolated devices. Increased junction-breakdown voltage [25] of the drain diffusion is achieved by using a deep drain well. Light well doping and large radii of the cylindrical and spherical junctions at the border increase the breakdown voltage. As small on-resistance and high break-down voltage are contrary effects, the optimization of the tradeoff between both quantities is of major interest.

The gate length is extended beyond the body-drain well junction, which increases the junction breakdown voltage. The gate acts as a field plate and bends the electric field in a way that the critical field strength occurs at increased drain source voltages, commonly known as RESURF effect [20]. As there are high voltages at the drain, the electric field at the end corner of the gate electrode becomes quite large due to the small radius. Therefore field oxide or shallow trench isolation is used in order to separate the critical gate region and the drain region.

As devices generally operate at high drain source voltages, reliability and hot carrier degradation are major concerns [18]. As mentioned above, the on-resistance is increased. Further more a so called quasi-saturation is found for short channel devices at high gate and drain bias. One reason is the Kirk-Effect [36] in the drift

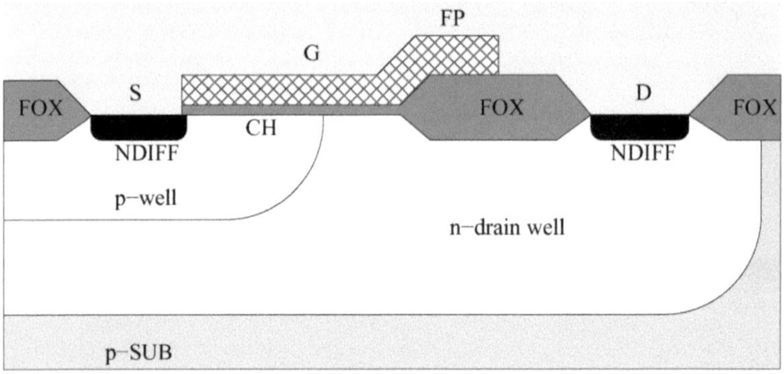

Fig. 4.4 Body isolated n-channel HV MOSFET

region. Detailed explanations are presented in [32, 33]. From Figs. 4.1–4.3 and the mentioned applications above we see the main differences between the HV transistor and the standard low voltage transistor. For reliable compact models the following physical effects have to be taken into account:

- quasi-saturation effect
- self-heating effect
- geometry-related effects
- graded channel effects
- bulk current
- impact ionization in the drift region
- high-side switch effect
- parasitic BJT effect

An additional challenge is the LDMOS Modeling for analog and RF circuit design. The integration of complex analog and RF functions such as RF power amplifiers in state of the art wireless circuits requires accurate compact models in the high frequency range. The compact model must account for all the additional LDMOS effects in DC, AC and RF large signal regimes [16]. The integration of these devices is done in RF CMOS, BiCMOS and BCD technologies.

Figures of Merit for LDMOS devices are mainly the on resistor R_{on}, the gate drain capacitance C_{gd} (Miller cap) and the breakdown voltage BV. For RF applications the parameters transit frequency FT and maximum oscillation frequency $FMAX$ are added to the list. These requirements have to be taken into account for the selected compact model and their capabilities.

4.2 HV LDMOS Modeling with Sub-Circuits

Foundries first offered HV LDMOS devices in the nineties and SPICE realizations which were able to consider the special HV LDMOS effects became important. The

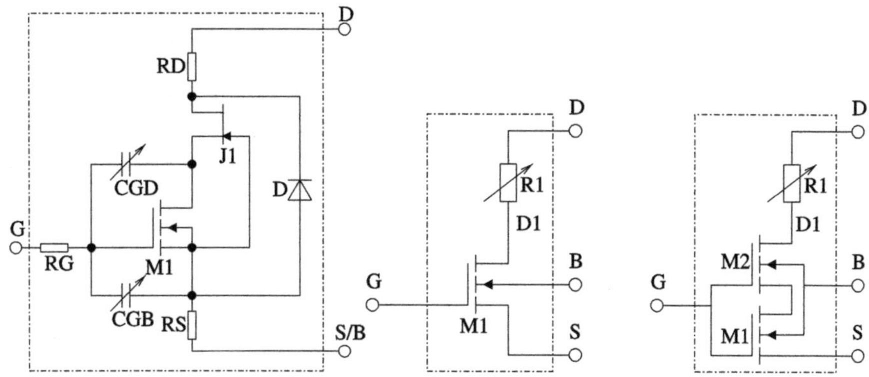

Fig. 4.5 *Left*: sub-circuit presented in [24]; *center*: sub-circuit using a controlled drain resistance; *right*: sub-circuit consisting of a controlled drain resistance and a second FET to consider the graded channel doping [26]

lack of HV compact models has driven the development of sub-circuits. A sub-circuit realization of a device is a network composed of different active and linear or nonlinear passive components. They are connected in a specific way such that the device characteristics are reproduced most accurately. Furthermore, sub-circuits allow high flexibility. They can be easily adapted by adding or removing components in the netlist. There is a well defined node interface between the sub-circuit box and the external circuit, where the sub-circuit appears to be a single device. The major disadvantage is an increased simulation time due to a larger effective number of circuit nodes and components. Furthermore, special attention needs to be paid to the convergency in the circuit simulator.

Numerous sub-circuit solutions were published during the last years (cf. Fig. 4.5). Sub-circuits for RF LDMOS modeling are introduced in [42] and [47]. Inductors and capacitors are included to model the RF behavior, additionally a JFET is used to improve the DC modeling. This JFET approach was applied in many circuits, e.g. [24] and [23], where improved capacitance modeling has been achieved by adding voltage-controlled gate-bulk and gate-drain capacitors or PFET devices to an NFET LDMOS structure. In [23] the threshold voltage of the JFET depends on the gate-source bias to circumvent a JFET pinch-off. Controlled drain resistances depending on the gate-source and the drain-source voltage are used in [26]. Additionally, two FETs are connected in series to describe the graded channel doping effect. Since the JFET approach in combination with an advanced core FET model, e.g. BSIM or EKV, was mainly used for modeling the quasi-saturation effect, [23, 24, 26, 42, 47], JFET parameter extraction procedures [43] are presented in the subsequent sections. The starting point is the drain resistor extension of a standard MOSFET compact model which in some cases can be successfully utilized, e.g. for PFET-LDMOS structures. These devices show a weaker quasi-saturation behavior compared to n-type devices due to a lower carrier mobility.

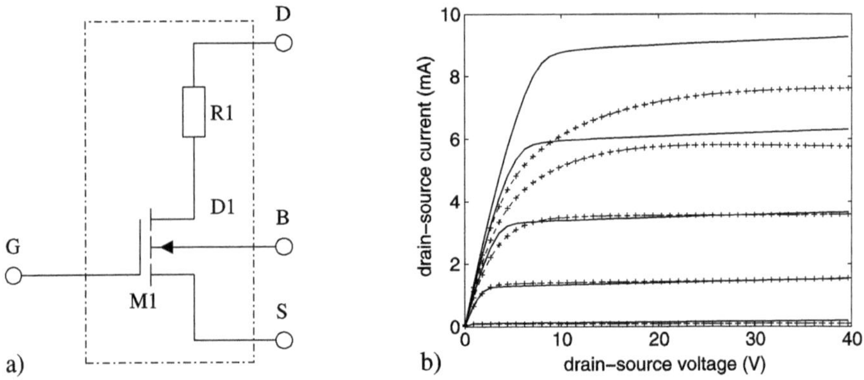

Fig. 4.6 **a** Sub-circuit consisting of a MOSFET device and a drain resistor. **b** Output characteristics I_{DS} vs. V_{DS} of a HV LDMOSFET modeled with the sub-circuit shown in (**a**). + measurement, – simulation

4.2.1 HV MOSFET Sub-Circuit Using a Drain Resistor

A basic approach for the HV MOSFET modeling is found by considering the structure of the device. The drift region is taken into account by a width dependent resistor positioned between the internal drain node $D1$ of the core MOSFET and the drain contact D of the HV device, as shown in Fig. 4.6a. The value of the drain resistor R_1 is defined by

$$R_1 = \frac{\mathbf{RHV}}{W + \mathbf{dWHV}} \tag{4.1}$$

where **RHV** corresponds to the sheet resistance of the drift region, W is equal to the channel width of the FET device and **dWHV** is a width offset parameter.

The linear region is well modeled, as illustrated in Fig. 4.6b. Since the depletion widths of the relevant drain-body and drain-channel junctions are small at low V_{DS} the resistivity of the drift region is hardly influenced. Nevertheless, an increased on-resistance has been observed at high drain-substrate bias for isolated HV FET devices, which is discussed in Sect. 4.5. The drain current in the saturation region cannot be modeled accurately by using a series resistor because I_{DS} is mainly determined by the FET acting as a current source.

4.2.2 HV MOSFET Sub-Circuit Using a JFET

A JFET is now used instead of a resistor (Fig. 4.7a) in order to increase the simulation accuracy in the quasi-saturation region. This intuitive approach came up because large depletion zones in the drift region were thought to be the reason for the special current limitation behavior. The gate of J_1 is connected to the source of the sub-circuit to describe the influence of V_{DS}. It turned out that the depletion

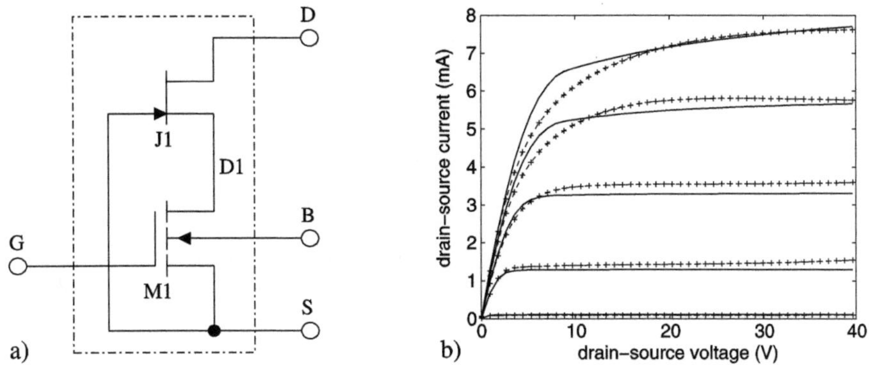

Fig. 4.7 **a** Sub-circuit consisting of a MOSFET device $M1$ and a JFET J_1. **b** Output characteristics I_{DS} vs. V_{DS} of a HV LDMOSFET modeled with the sub-circuit shown in (**a**). + measurement, – simulation

width of the drift-bulk junction hardly influences the quasi-saturation current and therefore the bulk potential is not used to control the JFET. Actually, this approach is a kind of feedback-loop which is a circuit approximation for modeling the drain current in the quasi-saturation region. In this case, three additional JFET (LEVEL1) parameters must be extracted properly. These are the threshold voltage **VT0**, the current gain factor **BETA** and a channel length modulation parameter **LAMDA**. By introducing the definition

$$V_{gst} = V_{gs} - \mathbf{VT0} \tag{4.2}$$

we can distinguish between three different regions of JFET operation, which are defined by V_{gst} and V_{ds}. Lower case letters for indices belong to JFET quantities, whereas upper case letters refer to the quantities of the entire sub-circuit.

- *linear region*, where $0 < V_{ds} < V_{gst}$:

$$I_{ds} = \mathbf{BETA}\ V_{ds} \cdot (2V_{gst} - V_{ds}) \cdot (1 + \mathbf{LAMDA} \cdot V_{ds}) \tag{4.3}$$

- *saturation region*, where $0 < V_{gst} < V_{ds}$:

$$I_{ds} = \mathbf{BETA} \cdot (V_{gst})^2 \cdot (1 + \mathbf{LAMDA} \cdot V_{ds}) \tag{4.4}$$

- sub-threshold region, where $V_{gst} < 0$: $I_{ds} = 0$

The first trial to optimize the core MOSFET model and JFET parameters simultaneously will result in unphysical parameter values due to strong correlations. It is expected to keep the same model accuracy in the R_{on} region as obtained for the simple circuit of Fig. 4.6a, since the HV FET is frequently used for switching. The idea is to find the correspondence between the JFET resistance and the formerly extracted R_1 for correct on-resistance modeling. The derivative of (4.3) for **LAMBDA** $= 0$ and for small V_{GS} and V_{DS} yields:

$$\mathbf{RHV} = \frac{1}{g_{ds,lin}} = \frac{1}{2 \cdot \mathbf{BETA} \cdot (-\mathbf{VT0})} \tag{4.5}$$

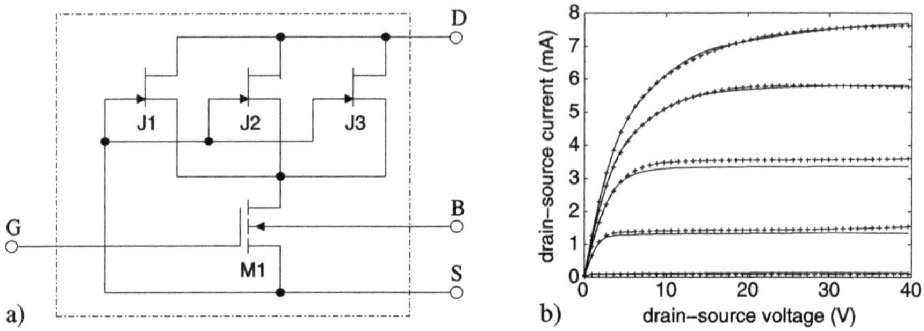

Fig. 4.8 **a** Sub-circuit consisting of a MOSFET device $M1$ and JFETs J_1, J_2, J_3. **b** Output characteristics I_{DS} vs. V_{DS} of a HV LDMOSFET modeled with the sub-circuit shown in (**a**). + measurement, – simulation

where $g_{ds,lin}$ is the output conductance of the JFET. IF **RHV** was determined properly and now **VT0** is adjusted during the device modeling, then the corresponding value of **BETA** is provided by (4.5). Therefore, the R_{on} region is modeled accurately, independent of the **VT0** parameter value. **VT0** and **LAMDA** are adjusted to model the quasi-saturation current behavior but the transition region from linear to saturation can only be partly covered by this approach. Furthermore, the g_{DS} of the whole sub-circuit depends strongly on **LAMDA** which makes an accurate g_{DS} modeling quite difficult.

4.2.3 HV MOSFET Sub-Circuit Using Three JFETs

In order to increase the quality of I_{DS} modeling in the quasi-saturation region two additional JFETs are added in parallel, as shown in Fig. 4.8a. Again, an advanced extraction strategy must be defined to reach at least the performance of the former circuit. Here the current is split into three branches

$$I_{DS} = I_{DS1} + I_{DS2} + I_{DS3} = m_1 \cdot I_{DS} + m_2 \cdot I_{DS} + m_3 \cdot I_{DS} \tag{4.6}$$

where

$$m_1 + m_2 + m_3 = 1 \quad 0 < m_i < 1. \tag{4.7}$$

In the same way as done for a single JFET, parameters **VT01**, **VT02** and **VT03** of each JFET can be adjusted to approximate the I_{DS} quasi-saturation characteristics. The relevant parameters $BETA1$, $BETA2$ and $BETA3$ are defined again by (4.5). To perform optimizations in a single step, weighting parameters m_1, m_2 can be adjusted to split the current in a defined way, whereas m_3 is defined by (4.7), giving $m_3 = 1 - m_1 - m_2$. In the final sub-circuit model SPICE parameters **BETA1** $= m_1 \cdot BETA1$, **BETA2** $= m_2 \cdot BETA2$ and **BETA3** $= (1 - m_1 - m_2) \cdot BETA3$ are implemented. As shown in Fig. 4.8b, this sub-circuit allows improved modeling of the quasi-saturation region by still preserving the performance of a well modeled R_{on} regime. It turned out that g_{DS} data shows unrealistic small values if V_{DS}

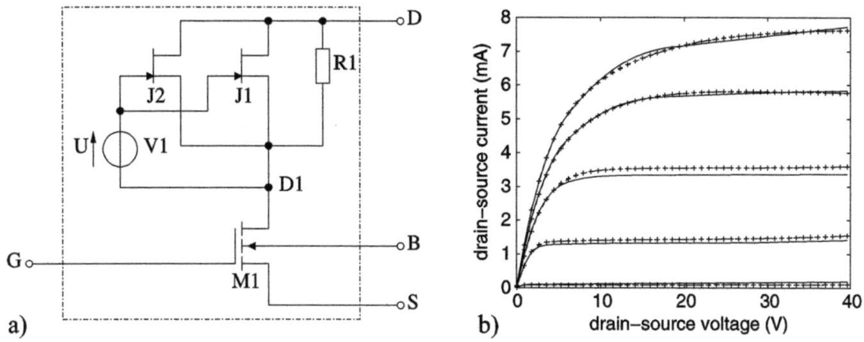

Fig. 4.9 **a** Sub-circuit consisting of a MOSFET device M_1, JFETs J_1, J_2, a resistor R_1 and a voltage controlled voltage source V_1. **b** Output characteristics I_{DS} vs. V_{DS} of a HV LDMOSFET modeled with the sub-circuit shown in (**a**). + measurement, − simulation

is larger than the highest absolute JFET threshold voltage parameter. Therefore, parameter **VT03** is generally chosen to be larger than $V_{DS,max}$, which led to the development of the next sub-circuit.

4.2.4 HV MOSFET Sub-Circuit Using JFETs, Resistors and Controlled Sources

In the sub-circuit (Fig. 4.9a) a JFET was replaced by a simple resistor. This modification can be done because of the high threshold voltage needed for the relevant JFET J_3 in the previous circuit. The resistance value is given by

$$R_1 = \frac{\mathbf{RHV}}{1 - m_1 - m_2}. \tag{4.8}$$

The voltage controlled voltage source V_1 is implemented to model the channel length dependency of the quasi-saturation current. V_1 mainly maps the voltage $V_{D1,S}$ to the gate of the JFETs. If unity gain is selected the JFET gates are virtually connected to the source node and the gate control remains the same as for the previous circuit. This controlled source approach allows an adjustment of the quasi-saturation current over a wide range, whereas the on-resistance is hardly influenced. In the R_{on} regime $V_{D1,S} = -V_{gs,J}$ is quite small and the JFETs operate as resistors. Adjustment of gain from $G = 0.5$ to $G = 5$ negligibly influences each JFET resistance, because gate voltages $V_{gs,J}$ are still small compared to the threshold voltages. For high $V_{D1,S}$ applied, the JFETs significantly influence the current behavior. The length dependency of the gain G is defined by

$$G = \mathbf{KVD0} + \frac{\mathbf{KVD}}{L}. \tag{4.9}$$

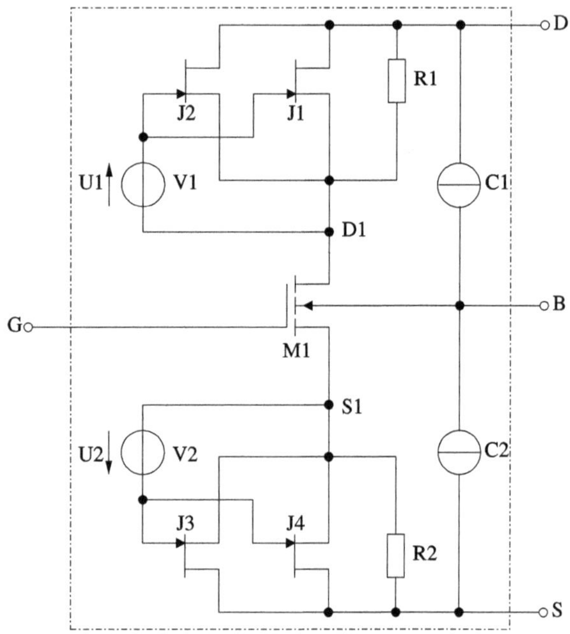

Fig. 4.10 Symmetrical HV MOSFET sub-circuit consisting of a MOSFET device M_1, JFETs J_1 to J_4, resistors R_1, R_2 and voltage controlled voltage sources U_1, U_2 as well as controlled current sources C_1, C_2 for bulk current modeling

If a proper gain value $G_{L_{min}}$ is found e.g. for the L_{min} device, then the quasi-saturation currents at increased channel lengths can be optimized by adjusting only the **KVD** parameter, whereas **KVD0** is implicitly changed to

$$\mathbf{KVD0} = G_{L_{min}} - \frac{\mathbf{KVD}}{L} \tag{4.10}$$

which guarantees the same value $G_{L_{min}}$ independent of the **KVD** parameter value. Finally, to introduce channel width W scaling, all **BETA** parameters are multiplied by the effective drift region width $(W + \mathbf{dWHV})$, whereas the resistor value R_1 must be divided by $(W + \mathbf{dWHV})$.

Summarizing the results, this sub-circuit approach provides a HV LDMOSFET device model which is scalable in channel length and width. The R_{on} region is well modeled and the quasi-saturation drain current behavior is piecewise approximated.

4.2.5 Symmetrical HV MOSFET Sub-Circuit with Bulk Current Modeling

A sub-circuit for symmetrical HV FET modeling is presented in Fig. 4.10. The device structure is fully symmetrical and each drift region terminal can operate either as source or as drain, which depends on the applied bias conditions. Therefore the asymmetric approach of the previously discussed sub-circuit is implemented at drain and source terminals. The sub-circuit devices are equal for both sides and the sign of the $V_{D1,S1}$ voltage determines if the relevant devices act as resistance at the source side or as JFET at the drain. Generally, the gate-source voltage $V_{gs,J}$ of the JFETs must be negative to control the JFET drain current behavior. $V_{gs,J}$ is positive for a symmetrical transistor biased in reverse direction. In order to guarantee a constant source resistor behavior, $V_{gs,J}$ is limited to maximum $V_{gs,J} = 0$ V. By using this restriction on both sides, the appropriate modes of operation are found for all sub-circuit devices.

An analogous approach is found for the controlled current sources which are implemented to model the bulk current. C_1 is activated for normal mode, whereas C_2 is intended to be active in reverse mode. It is not possible to use the intrinsic BSIM or EKV bulk current model of M_1 because the node potential of $D1$ does not provide the proper V_{DS} voltage for bulk current calculation. Therefore the internal bulk current model equations were adapted in order to use controlled sources for a SPICE implementation. In Table 4.1 a summary of the RES/JFET HV sub-circuit model features is demonstrated.

4.3 EKV High-Voltage MOSFET Model

The EKV-HV compact model for vertical and lateral DMOS has been developed by the Swiss Federal Institute of Technology (EPFL) [17]. It relies on the core of the charge based EKV 2.6 (**E**nz-**K**rummenacher-**V**ittoz) MOSFET model for the intrinsic MOSFET, while the drift region is modeled with a bias dependent resistance. The EKV MOSFET model [21] has the advantage of being physical and continuous in all regions of operation. The model is formulated on an inversion-charge description, where the voltage versus surface-potential equation is transformed into an approximate equation of node-charge versus node voltage drop. Similarly, the current versus surface potential equation can also be transformed into an approximate current versus node-charge equation. Thus, the node charges become the independent variables. The single expression formulation ensures continuity of first and higher-order derivatives with respect to any terminal voltage in the entire operating range. The following physical effects are included in the EKV MOSFET model:

- Basic geometrical and process related aspects as oxide thickness, junction depth, effective channel length and width
- Effects of doping profile, substrate effect
- Modeling of weak, moderate and strong inversion behavior
- Modeling of mobility effects due to vertical and lateral fields, velocity saturation
- Short-channel effects as channel-length modulation (CLM), source and drain charge-sharing (included for narrow channel widths), reverse short channel effect (RSCE)
- Modeling of substrate current due to impact ionization
- Quasi-static charge-based dynamic model
- Thermal and flicker noise modeling
- A first-order non-quasi-static model for the trans-admittances
- Short-distance geometry- and bias-dependent device matching

Table 4.1 Summary of HV MOSFET model features

PHYSICAL EFFECTS	RES/JFET HV SUB-CIRCUIT	EKV-HV	MM20	HiSIM_HV
TECHNOLOGY RELATED DEVICE EFFECTS				
symmetric/asymmetric device	+	−	−	+
quasi-saturation	+	+	+	+
R_{on}	+	+	+	+
mobility	+	+	+	+
carrier velocity saturation	+	+	+	+
channel length modulation	+	+	+	+
drain induced barrier lowering	+	−	+	+
impact ionization current	+[b]	+	+	+
poly-Si gate depletion	+	−	−	+
channel length scaling	+	−	−	+
channel width scaling	+	+	+	+
ASYMMETRIC MOS CAPACITANCES				
intrinsic capacitance	+	+	+	+
overlap capacitance	+	+	+	+
fringing capacitance	+	−	−	+
BULK DIODES				
diode current	+	−	−	+
diode capacitance	+	−	−	+
TEMPERATURE DEPENDENCE				
threshold voltage	+	+	+	+
mobility	+	+	+	+
quasi saturation	+	+	+	+
R_{on}	+	+	+	+
bulk current	+	−	+	+
self-heating	−[a]	+	+	+
NOISE				
SPICE2 noise model	+	−	−	−
Flicker noise model	+	−	+	+
short channel thermal noise model	−[a]	−	+	+
induced noise in gate	−[a]	−	+	+
induced noise in substrate	−[a]	−	−	+
RF MODELING				
gate resistance model	−[a]	−	−	+

Table 4.1 (Continued)

PHYSICAL EFFECTS	RES/JFET HV SUB-CIRCUIT	EKV-HV	MM20	HiSIM_HV
substrate resistance model	–[a]	–	–	+
multi finger transistors	–[a]	+	–	+
NON-QUASI STATIC (NQS)				
NQS	+	–	–	+

[a]BSIM3v3, depends on the features of the core FET model

[b]Intrinsic bulk current model not applicable

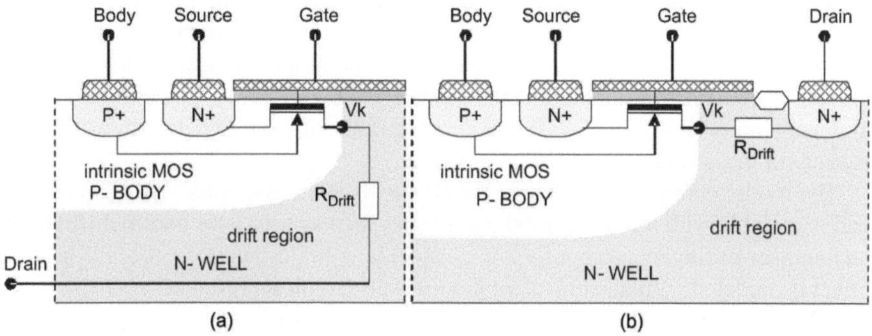

Fig. 4.11 Schematic representation of the n-channel high-voltage **a** VDMOS and **b** LDMOS device

4.3.1 EKV-HV DC Model

In this modeling approach, the HV transistor (Vertical and Lateral DMOS) is divided into an intrinsic MOSFET part and a drift region part at the channel/drift junction (boundary between the p-region and the extended drain n-region, as shown in Fig. 4.11). This location is called the key-point (K). Its potential V_K is used to define the intrinsic MOS drain-source voltage and the voltage drop on the drift region [11]. The variation of V_K with respect to gate voltage V_G and drain voltage V_D can be calculated with 2D device simulation in all regions of HV transistor operation. It has been demonstrated that the intrinsic drain voltage V_K always remains at low values for the entire bias operation and the intrinsic MOS part can be accurately modeled using the low voltage MOSFET model. The intrinsic MOSFET drain current is expressed as

$$I_{KS} = I_S.(i_f - i_r) \tag{4.11}$$

where I_S is the specific current, i_f and i_r are the normalized forward and reverse currents respectively, given as:

$$i_f = \left[\ln\left(1 + e^{\frac{V_P - V_S}{2}}\right)\right]^2, \tag{4.12}$$

$$i_r = \left[\ln\left(1 + e^{\frac{V_p - V_k}{2}}\right)\right]^2 \tag{4.13}$$

where V_p, V_s and V_k are the normalized pinch-off voltage, source voltage and intrinsic drain voltage as defined in the EKV model [21]. The modeling of drift region is carried out using a bias dependent resistance, i.e. the drift resistance is a function of the drain voltage V_D and the gate voltage V_G. The expression of drift resistance for a given drift length L_{DR}, width W and number of fingers N_F is obtained as [17]

$$R_{Drift} = \frac{R_{Drift0} \cdot L_{DR}}{W \cdot N_F} \left[\frac{1 + (\frac{V_D - V_K}{VSAT})^{\alpha vsat}}{1 + \theta_{Acc}.V_G} \right] \left[1 \pm (k_{rd} - 1) \frac{N_F - 1}{N_F + N_{CRIT}} \right] \times (1 - \alpha_T.\Delta T) \tag{4.14}$$

where R_{Drift0} is the drift resistance at low bias voltage, which is determined by the doping and geometry of the drift region, the velocity saturation effect is taken into account by $\alpha vsat$ and VSAT, θ_{Acc} is the gate bias modulation parameter introduced by the accumulation charge, N_{CRIT} and k_{rd} are empirical parameters for N_F scaling, α_T is the temperature coefficient and ΔT is the difference between device and ambient temperature. The "+" sign is used for drain-on-side devices while "−" sign is used for drain-all-around vertical devices.

The model correctly reproduces the effects of HV devices like quasi-saturation and is scalable with many physical parameters such as transistor width, drift length and number of fingers, except for e.g. channel length. Furthermore, the self-heating effect is modeled with an equivalent thermal sub-circuit representation [17].

4.3.2 EKV-HV Charge Model

The total gate charge is calculated as the sum of gate charge associated with the intrinsic MOS and the drift region. The intrinsic MOS gate charge Q_{GK} can be expressed as [17]:

$$Q_{GK} = Q_K + Q_S + Q_B \tag{4.15}$$

where Q_K, Q_S and Q_B are the charges related to intrinsic-drain (V_K), source and body nodes. These charges are obtained as a function of normalized forward current i_f and normalized reverse current i_r as used in the EKV model [21].

The normalized accumulation charge q_{Drift} of the drift region can be expressed with a simple approximation as

$$q_{Drift} = (V_G - V_{FB} - \psi_S) \cdot C_{ox} \tag{4.16}$$

where V_G, V_{FB} are the gate voltage and flat band voltage of the drift region, ψ_S is the surface potential and C_{ox} is the oxide capacitance per unit area above the drift region. The total accumulation charge Q_{Drift} of the drift region is calculated by integrating q_{Drift} (4.16) over drift length. The total gate charge including the drift region can be written as:

$$Q_G = Q_K + Q_S + Q_B + Q_{Drift}. \tag{4.17}$$

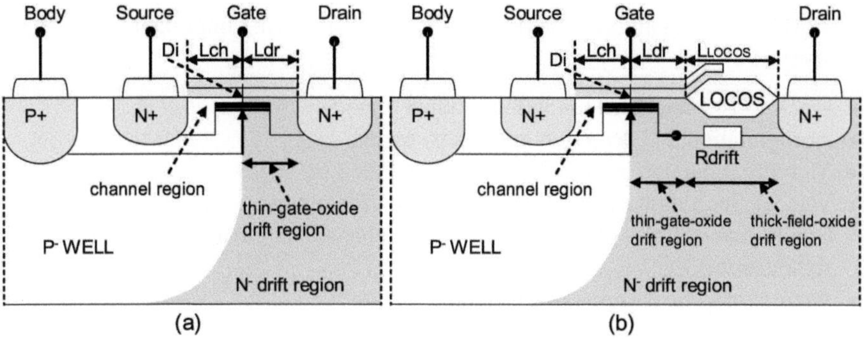

Fig. 4.12 Cross section of (**a**) a low-voltage LDMOS transistor for which the MM20 model is derived and (**b**) a high-voltage LDMOS transistor, for which the MM20 model is used in series with a constant drift resistor

The charge model is valid for a wide range of gate and drain voltage. Especially, the gate-to-source and gate-to-drain capacitance characteristics of an LDMOS are accurately predicted. The model provides excellent trade-off between speed, convergence and accuracy for circuit simulation. It is suitable for circuit design in any regime of operation of HV MOSFETs including self-heating and impact ionization effect. In Table 4.1 a summary of the EKV HV transistor model features is demonstrated.

4.4 MM20 High-Voltage MOSFET Model

MOS Model 20 level 2001 (MM20) is an asymmetrical, surface-potential-based LDMOS model, developed by NXP Research (formerly Philips Research) Laboratories. It is aimed to replace the macro model as the combination of MOS Model 9 (MM9) for the channel region in series with MOS Model 31 (MM31) [2] for the drift region under the gate oxide of various high-voltage MOSFET devices. Based on the calculation of the voltage at the transition node D_i between channel region and drift region (Fig. 4.12a), the MOS Model 20 has been developed especially to improve the convergency behavior during circuit simulation. The model combines the description of the MOSFET channel region behavior with that of the drift region under the gate oxide in HV transistors like Lateral Double-diffused MOS (LDMOS) devices or extended-drain MOSFETs.

A complete description of all transistor-action related quantities are provided. Nodal currents, nodal charges and noise-power spectral densities are formulated on surface-potential, resulting in equations valid over all operation regimes i.e. accumulation, depletion and inversion in both the channel and the drift region. The surface potential as a function of terminal voltages is obtained by the explicit expression as used for MOS Model 11 (MM11), [2]. In addition, several important physical effects, especially for the drift region have been included in the model. In Table 4.1 a summary of the MM20 HV MOSFET model features is demonstrated.

MM20 includes an accurate description of the following important LDMOS device effects:

- Weak and strong inversion in the channel region
- Accumulation and depletion in the drift region
- Mobility reduction in both the channel and the drift region
- Velocity saturation in both the channel and drift region
- Conductance effects of the channel region (channel length modulation, DIBL and static feedback)
- Weak avalanche currents in the channel region

4.4.1 MM20 DC Model

The DC model concept can be explained by the cross-section of the low-voltage LDMOS transistor shown in Fig. 4.12a. A graded channel is formed in a diffused p-well bulk (B) from the source-side under the gate (G). The internal drain D_i delimits the region where the graded channel turns into the lightly doped n-drift region (boundary between the graded channel of length L_{ch} and the lightly doped n-drift region of length L_{dr} and thickness t_{Si}). Due to the gate extension over the drift region, an accumulation layer forms in the drift region underneath the gate oxide. The key point of the MM20 compact model is the derivation of expressions for the current I_{ch} through the inversion channel as well as for the current I_{dr} through the drift region. These currents are calculated as a function of the known external drain, gate, source, and bulk voltages, V_D, V_G, V_S, and V_B, as well as of the unknown internal drain voltage V_{Di}. In comparison to a sub-circuit model, V_{Di} is expressed explicitly in terms of the external terminal voltages. The expression for V_{Di} is derived by equating I_{ch} to I_{dr}. Subsequently, V_{Di} is used to calculate the surface-potential. The final drain-to-source current I_{DS} is formulated in terms of the external terminal voltages. For the calculation of the internal-drain quasi-Fermi potential, the channel current I_{ch} is expressed as a function of surface-potential drop $\Delta\psi_s = V_{DiS} = V_{Di} - V_S$, [3, 4], which yields

$$I_{ch} = \frac{W\mu_{eff}^{ch}C_{ox}}{L_{ch}} \frac{(V_{invo} - \frac{1}{2}\xi V_{DiS} + \xi\phi_T)V_{DiS}}{1 + \theta_3 V_{DiS}}. \tag{4.18}$$

μ_{eff}^{ch} is the effective electron mobility in the channel region, $C_{ox} = \epsilon_{ox}/t_{ox}$ is the gate-oxide capacitance per unit area with thickness t_{ox} and permittivity ϵ_{ox}, $V_{invo} = -Q_{invo}/C_{ox}$ represents the inversion charge Q_{invo} per unit area at the source side, and ϕ_T is the thermal voltage. The effect of velocity saturation in the channel region is included in the term $\theta_3 = \mu_0^{ch}/(L_{ch}\vartheta_{sat})$, where μ_0^{ch} is the zero-field electron mobility in the channel region and ϑ_{sat} the saturated drift velocity of electrons. The variation of inversion charge with surface potential is accounted by the factor $\xi = 1 + \gamma_0/(2\sqrt{V_1 + \psi_0})$, where γ_0 is the body factor at the source, $V_1 = 1$ V and ψ_{so} is the surface potential at the source.

The channel current saturates before the onset of depletion in the drift region when the drift region length L_{dr} and the inversion channel length L_{ch} is in the same order of magnitude [4]. Therefore neglecting the diffusion current in the linear operating regime, the drift region current I_{dr} is given by

$$I_{dr} = \frac{1}{R_D}\left(1 - \lambda\frac{\sqrt{\theta_0 + V_{SB}} - \sqrt{\theta_0}}{\sqrt{\theta_0}}\right)V_{DDi} + \frac{W\mu_{acc}C_{ox}}{2L_D}\left(V_{accDi}^2 - V_{accD}^2\right) \tag{4.19}$$

where λ is a model parameter, μ_{acc} is the electron mobility in the accumulation layer, θ_0 is the built-in potential of the pn-junction between body and drain, V_{accDi} and V_{accD} represent the accumulation charge at the internal drain D_i and at the drain D, respectively. In the model the resistance R_D of the drift region is given by [4]

$$R_D = \frac{L_{dr}}{W\mu_{dr}(qN_Dt_{Si} - K_b\sqrt{\theta_0})} \tag{4.20}$$

where N_D is the concentration of donors in the drift region and K_b determines the effective thickness of the drift region, taking into account the depletion layer from the body.

For HV transistors, where the gate and the drain are separated by field oxide (LOCOS) or shallow trench oxide, MM20 can be extended to increase the model accuracy. To realize the very high voltage between source (S) and drain (D), the drift region is extended in length. It consists of two different sections: the first section is the gate-oxide drift region of length L_{dr}, and the second is the field-oxide or shallow trench oxide drift region of length L_{LOCOS}. In this case MOS Model 20 can be used in series with a separate model for the drift region under the field oxide or the shallow trench oxide. In this sub-circuit modeling approach, MM20 describes only the total region underneath the gate oxide. Since the gate voltage has negligible influence on the electrons in the drift region underneath the field oxide or shallow trench oxide, the current through this drift region is modeled with a constant resistance R_{drift}. The value of this resistance is given by [3]

$$R_{drift} = \frac{L_{LOCOS}}{W}R_{sheet} \tag{4.21}$$

where W is the width of the device and R_{sheet} is the sheet resistance of the drift region under field oxide or shallow trench oxide.

MOS Model 20 level 2001 only provides a model for the intrinsic HV MOSFET behavior of the region under the gate oxide of a high-voltage MOS device, as well as the gate-to-source and gate-to-drain overlap regions. Junction charges, junction leakage currents, interconnect capacitances and parasitic bipolar transistors should be covered by separate sub-circuit models. Furthermore, MM20 has both temperature scaling and geometry (only width) scaling rules included. Self-heating of the device can be incorporated externally via a thermal network. Due to complexity of the technology, MM20 is under further development. In the mean time a newer version of MOS20 level 2002 is available with more modeling features.

4.4.2 MM20 Charge Model

The nodal charge is determined for all different operation regimes in strong inversion, depletion and accumulation in both the drift region and the channel region. The gate charge Q_G of the whole device consists of the gate charge of the channel region and the gate charge of the drift region [4], according to

$$Q_G = Q_G^{ch} + Q_G^{dr} \tag{4.22}$$

where the gate charge of the channel and the drift region are given by

$$Q_G^{ch} = -W \int_0^{L_{ch}} (Q_{acc} + Q_{dep} + Q_{inv}) dx, \tag{4.23}$$

$$Q_G^{dr} = -W \int_{L_{ch}}^{L_{ch}+L_{dr}} (Q_{acc}^{dr} + Q_{dep}^{dr} + Q_{inv}^{dr}) dx. \tag{4.24}$$

Therefore, the nodal gate charge consists of the opposite of the total accumulation, depletion and inversion charges underneath the gate oxide.

The total bulk charge Q_B of the transistor is given by the sum of the bulk charge due to the channel region and that of the drift region, i.e. $Q_B = Q_B^{ch} + Q_B^{dr}$, where the bulk charge of the channel and the drift region [4] are given by

$$Q_B^{ch} = W \int_0^{L_{ch}} (Q_{acc} + Q_{dep}) dx, \tag{4.25}$$

$$Q_B^{dr} = W \int_{L_{ch}}^{L_{ch}+L_{dr}} Q_{inv}^{dr} dx. \tag{4.26}$$

Hence the bulk charge of the channel consists of the total accumulation and depletion charges underneath the gate oxide. Due to sufficiently negative gate voltages, holes enter the drift region from the p-well bulk, which gives rise to an inversion charge in the drift region. So the bulk charge of the drift consists only of the inversion charge underneath the thin gate oxide.

Since the LDMOS transistor is asymmetric, two limits are identified for the distribution of the charge underneath the gate oxide. The first limit is valid well-above threshold (i.e. for the gate voltage sufficiently large), and the drain charge is approximated by the Ward-Dutton charge partitioning scheme [38] (valid in case of a uniform MOSFET). The second limit is valid below threshold (i.e. for the gate voltage sufficiently small), and the drain charge is approximated by means that all accumulation charge of the drift region is attributed to the drain. Therefore, the drain charge Q_D is expressed in terms of the nodal charges of both the channel and the drift region [4], given by

$$Q_D = F_L Q_{Dinv}^{ch} + Q_{Dacc}^{dr} + F_L Q_{Sacc}^{dr} + Q_{Ddep}^{dr} \qquad V_{GS} \geq V_{TH}, \tag{4.27}$$

$$Q_D = F_L Q_{Dinv}^{ch} + Q_{Dacc}^{dr} + Q_{Sacc}^{dr} + Q_{Ddep}^{dr} \qquad V_{GS} \leq V_{TH}. \tag{4.28}$$

Here, $F_L = L_{ch}/(L_{ch} + L_{dr})$ and Q_{Dinv}^{ch}, Q_{Dacc}^{dr}, Q_{Sacc}^{dr} and Q_{Ddep}^{dr} are the different charges in the channel and the drift region in accumulation, depletion and inversion charge, respectively. In the model, the transition of the drain charge from (4.27) to (4.28) has been implemented in a smooth and continuous way.

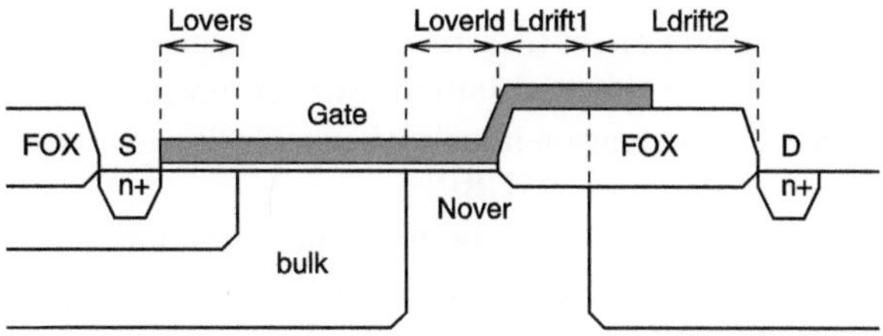

Fig. 4.13 Definition of an asymmetrical LDMOS structure (**COSYM** = 0)

4.5 HiSIM_HV High-Voltage MOSFET Model

In 2008 the CMC selected HiSIM_HV as the first standard high-voltage MOSFET model [1, 49]. HiSIM_HV has been developed as an extension of the advanced low-voltage MOSFET model HiSIM2 (Hiroshima-University STARC IGFET Model) [37]. The model concept is based on the drift-diffusion theory using charge sheet approximation of the inversion layer with zero thickness including gradual channel approximation. HiSIM2, unlike other surface potential based models which use simplified analytical equations, solves the Poisson equation iteratively. Less smoothing functions are needed due to smooth model equations resulting in less iteration steps. The applied iteration procedure ensures no run-time penalty. The surface potentials at the source (ϕ_{S0}), at the pinch-off point (ϕ_{SL}), at the channel/drain junction ($\phi_S(\Delta L)$), and at the drain contact ($\phi_{S0} + V_{ds}$) are determined, which are at the same time implicit functions of the applied terminal voltages referenced to the source node thus model-internal iteration procedures are required only for calculating ϕ_{S0} and ϕ_{SL}. The device physics based approach makes the model easily adaptable to special MOSFET devices like LDMOS.

HiSIM_HV solves the Poisson equation along the MOSFET channel iteratively including the resistance effect in the drift region. The complete surface potential based compact model offers a unified description of device characteristics for all bias regions. Model options can be selected by more than 20 model flags offering high flexibility. HiSIM_HV scales with the gate width, the gate length, the number of gate fingers (**NF**), the drift-region length and is capable of modeling symmetric and asymmetric HV devices. Figure 4.13 shows the asymmetrical LDMOS case introduced by setting the flag **COSYM** to zero. **COSYM** = 1 defines a HV MOS structure either with symmetrical or asymmetrical drift regions. For symmetrical HV MOS modeling the resistance model is applied to the source side as well thereby reducing the effective V_{gs}, V_{ds} and V_{bs} [39].

4.5.1 HiSIM_HV Model Features

Since the model is surface potential based intrinsic device parameters such as the gate-oxide thickness (**TOX**), the substrate doping concentration (**NSUBC**) and the flat-band voltage (**VFBC**) determine the device behavior. Threshold voltage shift effects like the short-channel (Vth reduction with decreasing gate length) and reverse-short-channel effect (description of vertical and lateral channel inhomogeneity effects) are incorporated into the surface potential iteration. The penetration of the pocket/halo extension into the channel is described by the model parameter **LP** with peak concentration **NSUBP**.

At low-field the mobility model describes three independent mechanisms: Coulomb, Phonon and Surface-roughness scattering:

$$\frac{1}{\mu_0} = \frac{1}{\mu_{CB}} + \frac{1}{\mu_{PH}} + \frac{1}{\mu_{SR}}. \tag{4.29}$$

At high-field the mobility becomes a function of the lateral electric field (E_y) and the maximum velocity saturation (V_{max}):

$$\mu = \frac{\mu_0}{[1 + (\frac{\mu_0 \cdot E_y}{V_{max}})^{\mathbf{BB}}]^{\frac{1}{\mathbf{BB}}}}. \tag{4.30}$$

The temperature dependence is automatically included into the surface potentials through the thermal voltage ($1/\beta$). Additionally, the band gap, the intrinsic carrier concentration, the carrier mobility, the carrier saturation velocity and resistance effects include temperature dependence. The self-heating effect is modeled by a thermal network by setting the flag **COSELFHEAT** and **RTH0** > 0. The temperature dependences of the self-heating and of the thermal dissipation are also taken into account. The following effects are also included:

- Depletion Effect of the Gate Poly-Si
- Quantum-Mechanical Effects
- Channel-Length Modulation
- Narrow-Channel Effects
- Shallow Trench Isolation (STI) effects
- Leakage Currents (gate, substrate and GIDL currents)
- Source/Bulk and Drain/Bulk Diode Models
- Noise Models ($1/f$, thermal noise, induced gate noise, coupling noise)
- Non-Quasi-Static (NQS) Model

In Table 4.1 a summary of the HiSIM_HV model features is demonstrated.

4.5.2 Resistance Modeling with HiSIM_HV

Up to version 1.2.0 empirical equations are used for drift resistance modeling. The following options can be selected by the **CORSRD** flag:
CORSRD $= 1$ (resistance with external nodes):

$$R_{drift} = (R_d + V_{ds} \cdot R_{DVD}) \left(1 + \mathbf{RDVG11} - \frac{\mathbf{RDVG11}}{\mathbf{RDVG12}} \cdot V_{gs}\right) \times (1 - V_{bs} \cdot \mathbf{RDVB}). \tag{4.31}$$

R_d and R_{DVD} are functions of the gate width, the gate length and the drift length.

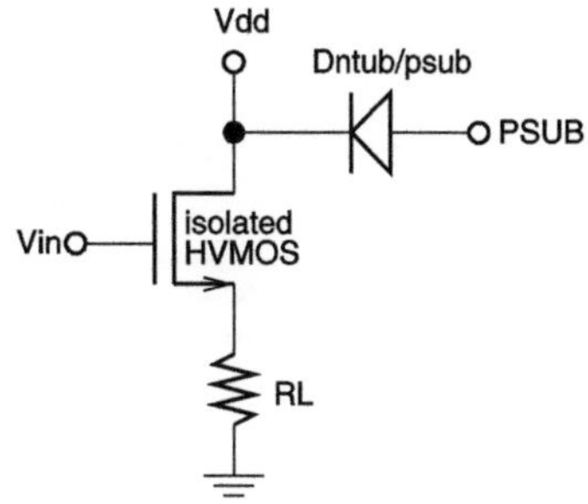

Fig. 4.14 HV device used on the high-side of a load

Analytical description of the resistance effect (**CORSRD** = 2):

$$I_{ds} = \frac{I_{ds0}}{1 + I_{ds0}\frac{R_d}{V_{ds}}} \tag{4.32}$$

where I_{ds0} is the drain current without resistance effect and

$$R_d = \frac{1}{W_{eff}}\left(R'_d \cdot V_{ds}^{\mathbf{RD21}} + V_{bs} \cdot V_{ds}^{\mathbf{RD22D}} \cdot \mathbf{RD22}\right). \tag{4.33}$$

R'_d is again a function of the gate width and the gate length. The third option is the combination of **CORSRD** = 1 and 2 (**CORSRD** = 3). In many cases where the quasi-saturation is very pronounced this option is needed. For future model versions a more physical drift-region model is planned. For RF applications the gate resistance and substrate resistance network implemented the same way as in BSIM4 can be used.

In the so called low-side switching application the MOSFET source is held to about the same potential as the wafer substrate while the drain can be at high potential relative to the substrate. When the device is used on the high-side of a load (Fig. 4.14) the source region can also be placed at high potential. The on-resistance (R_{on}) of the device is changing with the potential difference between substrate and source. This effect is known as high-side switch effect.

HiSIM_HV is the first compact model having this effect implemented. Figure 4.15 shows the cross-section of an isolated NMOS device. The depletion layer width at the drift side of the drift/substrate junction is calculated as:

$$W_{dep} = \sqrt{\frac{2\epsilon_{Si}[\mathbf{VBISUB} - (\mathbf{RDVDSUB} \cdot V_{ds} + \mathbf{RDVSUB} \cdot V_{sub,s})]}{q}} \times \sqrt{\frac{\mathbf{NSUBSUB}}{\mathbf{NOVER} \cdot (\mathbf{NSUBSUB} + \mathbf{NOVER})}} \tag{4.34}$$

where **NSUBSUB** and **NOVER** are the impurity concentrations and **VBISUB** is the built-in voltage of the junction. V_{sub} modulates the effective depth of the drift-region. Decreasing the potential on the substrate side will increase the width of the

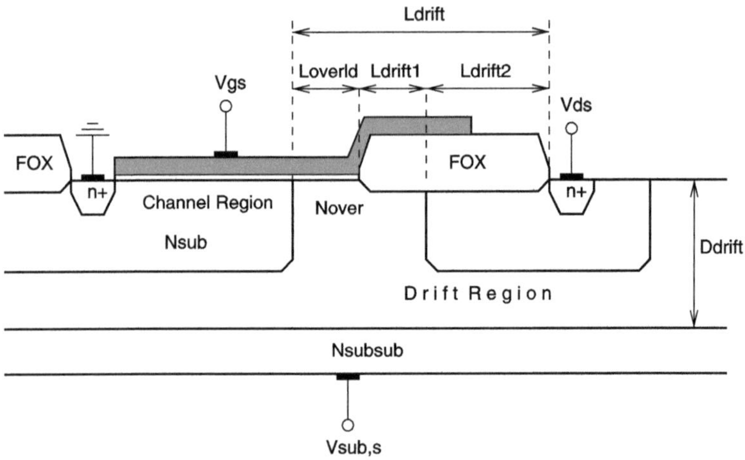

Fig. 4.15 Cross-section of an isolated HV device

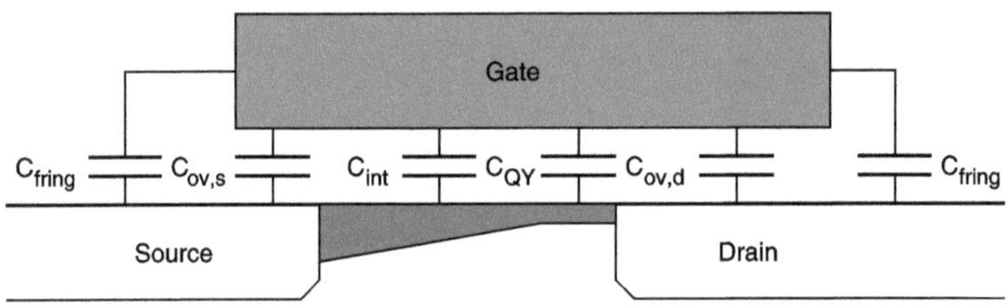

Fig. 4.16 Intrinsic and extrinsic capacitance components in HiSIM_HV

depletion layer thereby reducing the effective drift depth (**DDRIFT**) and increasing the drift resistance. Equation (4.31) is extended by a term including W_{dep}:

$$R_{drift} = (R_d + V_{ds} \cdot R_{DVD}) \left(1 + \mathbf{RDVG11} - \frac{\mathbf{RDVG11}}{\mathbf{RDVG12}} \cdot V_{gs}\right) (1 - V_{bs} \cdot \mathbf{RDVB}) \times \left(\frac{\mathbf{LDRIFT1} + \mathbf{LDRIFT2}}{\mathbf{DDRIFT} - W_{dep}}\right). \tag{4.35}$$

4.5.3 Capacitance Modeling with HiSIM_HV

The charges are explicit functions of the surface potential. From the charges the intrinsic capacitances are calculated by analytical solutions. Therefore no additional model parameters are required. At high-field condition the lateral electric field induces a capacitance C_{QY} parallel to the other MOS capacitance components as shown in Fig. 4.16.

In the overlap regions the bias dependent surface potentials are solved the same way as in the channel assuming negligible potential variation along the overlap region, describing the formation of the accumulation, depletion or inversion region. There are three options available for overlap capacitance calculations selected by the flags **COOVLP** (drain side), **COOVLPS** (source side) and the impurity concentrations in the overlap regions (**NOVER** and **NOVERS**): Constant overlap capacitance, surface-potential based model and simplified bias-dependent model. Setting **COOVLP** = 1 and **NOVER** larger than zero activates the surface-potential based model at the drain side. The overlap charge is determined in addition by the flat-band voltage (**VFBOVER**) and the overlap length (**LOVERLD**):

In accumulation and depletion:

$$Q_{over} = W_{eff} \cdot \mathbf{NF} \cdot \mathbf{LOVERLD} \cdot \left[\sqrt{\frac{2\epsilon_{Si} q \cdot \mathbf{NOVER}}{\beta}} \sqrt{\beta(\phi_S + V_{ds}) - 1} \right]. \tag{4.36}$$

In inversion:

$$Q_{over} = W_{eff} \cdot \mathbf{NF} \cdot \mathbf{LOVERLD} \cdot C_{OX}(V_{gs} - \mathbf{VFBOVER} - \phi_S). \tag{4.37}$$

For overlap capacitance calculations both the potential at the channel/drift junction and at the external drain can be considered. The ratio can be adjusted by the model parameter **CVDSOVER**:

$$C_{ov} = (1 - \mathbf{CVDSOVER}) \cdot C_{ov}(int) + \mathbf{CVDSOVER} \cdot C_{ov}(ext) \tag{4.38}$$

where $C_{ov}(int)$ is calculated using the potential at the channel/drift junction while $C_{ov}(ext)$ is calculated applying the external node potential.

4.6 Modeling of HV MOSFET Parasitics in HV CMOS Technology

The complexity of a HV LDMOS transistor structure introduces parasitic junctions which can significantly influence the device performance. Depending on the architecture of various integrated circuits the relevance of the different parasitic effects varies. Three main parasitic effects can be identified: leakage currents, forward biased bipolar transistors and capacitive effects.

Leakage currents flow through different wells down to the substrate. These currents are always present and are strongly temperature dependent, more precisely they increase with rising temperature. Low power applications are most sensitive to those currents, but due to their relevance for stand-by power consumption, leakage currents are of interest to all applications.

Possible forward bipolar action also provides a current path down to the substrate. In order to prevent this action especially switched devices must be treated with great care. Applications like DC-DC converters or H-bridges can be mentioned here.

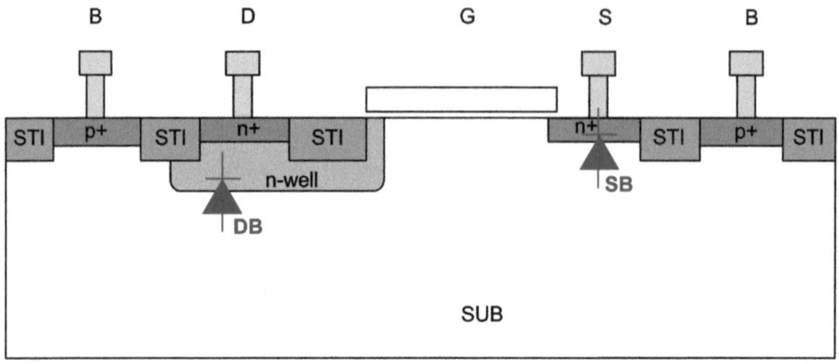

Fig. 4.17 Cross-section of a substrate based MOS transistor

There are two types of parasitic capacitances, the MOS overlap and the junction depletion capacitance. The first one (gate to drain and gate to source), can couple a signal from one terminal to the other causing large problems regarding signal distortion as well as speed losses. The second one (well to well or well to substrate), can result in speed reduction of a circuit.

The generally used design approach to simulate parasitic effects after parasitic extract can fail. A more advanced strategy in parasitic modeling is to provide a solution already during circuit design phase. Therefore, all parasitic effects should be covered by the device models. For covering of substrate currents, no possibility for implementation is given so far. The lack of compact models including parasitic effects implies the usage of sub-circuit solutions. Examples showing the distinct implementation are discussed below.

4.6.1 Substrate Based Devices

Non-isolated HV MOS devices are placed directly in substrate. Therefore they are also called substrate based MOS transistors. In the case of p-type substrate such devices can only be realized as n-MOS transistors.

The parasitic drain to bulk (DB) and source to bulk (SB) diodes are shown in the cross-section Fig. 4.17. The additional n-well under the drain diffusion indicates the difference in doping concentrations and size of the active pn-junction. Due to this asymmetry of the wells forming the drain and source area, the parasitic diodes will also differ in their behavior. This indicates different diode models connected to drain and source. Going a little more into detail and treating the well area of the device individually, we see that the current can be divided into two parts. The first flowing vertically through the well and the second through its sidewall. Figure 4.18 shows these two current components. The thickness of the space charge region again can be split into area and perimeter parts. For modeling of the current and capacitance behavior the level 1 diode model can be used which already provides an area and a perimeter part. Both can be modeled as functions of the MOS transistors geometry.

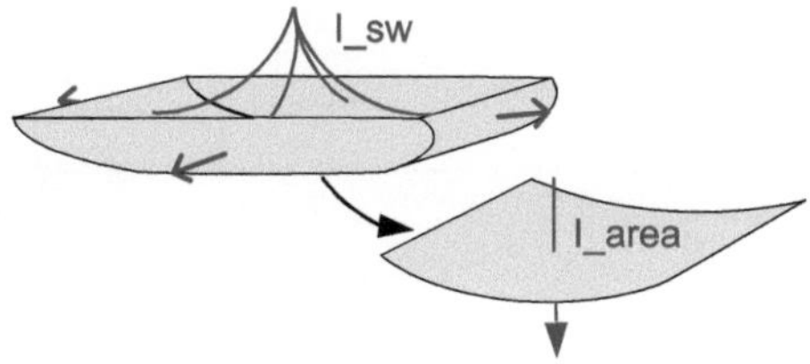

Fig. 4.18 The components of the diode currents: sidewall part (I_{sw}) and area part (I_{area})

It is obvious that the well scales with the width of the MOS device. For the diode current we get

$$I_{area} = J_S(T) \cdot \left(e^{\frac{V}{\mathbf{N} \cdot V_T}} - 1 \right) \cdot \text{AREA}, \tag{4.39}$$

$$I_{peri} = J_{SW}(T) \cdot \left(e^{\frac{V}{\mathbf{NS} \cdot V_T}} - 1 \right) \cdot \text{PERIMETER} \tag{4.40}$$

where J_S, J_{SW} are the saturation current densities for area and sidewall respectively, **N**, **NS** are the emission coefficients for area and sidewall and V_T is the thermal voltage. The temperature effect can be described by

$$J_S(T) = \mathbf{JS} \cdot \exp \left[\frac{\frac{E_{g0}}{V_{T_0}} - \frac{E_g}{V_T} + \mathbf{XTI} \cdot \ln(\frac{T}{T_0})}{\mathbf{N}} \right], \tag{4.41}$$

$$J_{SW}(T) = \mathbf{JSSW} \cdot \exp \left[\frac{\frac{E_{g0}}{V_{T_0}} - \frac{E_g}{V_T} + \mathbf{XTI} \cdot \ln(\frac{T}{T_0})}{\mathbf{NS}} \right] \tag{4.42}$$

where E_{g0} and E_g are the band gap at nominal temperature T_0 and device temperature T. **XTI** is the temperature exponent coefficient of the diode.

The capacitances are modeled as

$$C_{area} = \frac{\mathbf{CJ}}{(1 - \frac{V}{\mathbf{VJ}})^{\mathbf{MJ}}} \cdot \text{AREA}, \tag{4.43}$$

$$C_{peri} = \frac{\mathbf{CJSW}}{(1 - \frac{V}{\mathbf{VJSW}})^{\mathbf{MJSW}}} \cdot \text{PERIMETER} \tag{4.44}$$

where **CJ**, **CJSW** are the zero-bias junction capacitances for area and perimeter, **VJ**, **VJSW** are the junction potentials and **MJ**, **MJSW** are the grading coefficients. In the equations above bold capital names indicate process specific SPICE parameters. Improved modeling of the forward behavior can be provided by adding a series resistor, which is strongly layout dependent. It has to be stated that this forward behavior is absolutely unwanted and has to be avoided in any way. From design point of view the potential difference between source/drain and substrate has to be equal or exceed the forward voltage of the specific diode over the whole operation period.

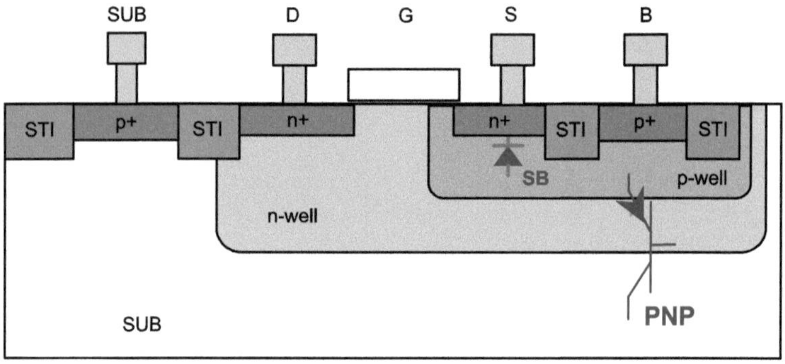

Fig. 4.19 Cross-section of an isolated n-MOS transistor

4.6.2 Isolated Devices

Owing to the additional well in the structure of isolated devices, which can be n-MOS as well as p-MOS devices, the parasitic structure down to substrate forms a bipolar transistor of type pnp (in p-type substrate). As an example, Fig. 4.19 shows an n-MOS transistor. The parasitic device is formed by the substrate (collector), drain (base) and bulk (emitter) of the MOS transistor. For a p-type device it is be substrate (collector), bulk (base) and drain (emitter). All the parasitic effects which hold for the substrate based devices also hold here for the distinct wells.

Only the bipolar effect needs to be considered in addition. The forward operation is again an undesired operating condition. But opposite to a single diode this effect can occur when the appropriate base connection (drain in n-MOS, bulk in p-MOS) is switched illegally. For example the bulk pin of an n-MOS transistor is below the forward voltage of the base to emitter diode of the parasitic bipolar transistor. To avoid this, the bulk of the MOSFET has to be connected to the highest potential in the case of p-MOSFET and to substrate in the case of n-MOSFET. For modeling of the bipolar transistor a Gummel-Poon model providing all the relevant effects can be used. The behavior of the parasitic bipolar device is strongly correlated to the MOSFET geometry. To find the correct scaling of specific bipolar model parameters is especially interesting. The total terminal current of the bipolar device consists of a vertical and a lateral component of the terminal current. The vertical component current depends on the area, which is the product of length and width of the NMOS transistor. For example the vertical component of the saturation current can be scaled as

$$\mathrm{IS}_{vert} = \mathbf{IS_{latw}} \cdot W + \mathbf{IS_{latl}} \cdot L. \tag{4.45}$$

For the variation of the length and width of the NMOS transistor, the lateral component current has to be split into a length dependent and a width dependent part to provide accurate gain modeling and the best fit can be found according to the following equation

$$\mathrm{IS}_{lat} = \mathbf{IS_{vert0}} + \mathbf{IS_{vertlw}} \cdot L \cdot W. \tag{4.46}$$

Bold type in the above formula indicates parameters that are used for scaling. In a PMOS transistor, a similar modeling approach can be used, but the scaling equations may differ depending on the device structure configuration.

4.7 Measurement Requirements for HV MOS Modeling

The basis for accurate SPICE modeling are precise and consistent characterization data of the device under test (DUT). Especially for HV devices and also for sub micron technologies, the power density increases dramatically [46], thus also thermal properties influence the electrical behavior. Thermal data in general is difficult to determine since either thermal power or thermal voltages/resistances are not directly measurable. The self-heating effect can be avoided (self heating free, SHF) if the DUT is switched on shortly enough without being heated up. In this case the long off time of the DUT and long integration time results in long measurement times. Low rise time of the whole measurement setup is another essential characteristic since the response of the DUT should be determined and not the properties of the measurement system itself [14].

4.7.1 DC Measurements for HV MOS Modeling

The transfer and output characteristics over a wide range of geometries are mandatory for the extraction of geometry scalable models. Additionally, Gummel plots and output characteristics of the parasitic bipolar transistors for different channel lengths are needed as well as the DC characteristics of the junctions. In order to extract parameters describing the temperature dependence, selected characteristic curves are recorded over a wide temperature range. Standard parameter analyzer systems [7, 29] are capable of handling voltages up to ± 200 V e.g. on the drain side.

4.7.2 AC Measurements for HV MOS Modeling

AC characterization can be divided in two categories of measurements: Capacitance-Voltage (CV) [6, 28] and S-parameter measurement. The CV characterization is a useful method in order to reveal the voltage dependence of the bulk diodes or other parasitic junctions and also to provide valuable information on the intrinsic/extrinsic MOS capacitance. Accurate CV measurement requires large structures resulting in a capacitance preferably in the pF range. In the case of diode characterization area and perimeter structures are widely used in order to extract model parameters. For MOSFET CV characterization several unit structures connected in parallel are

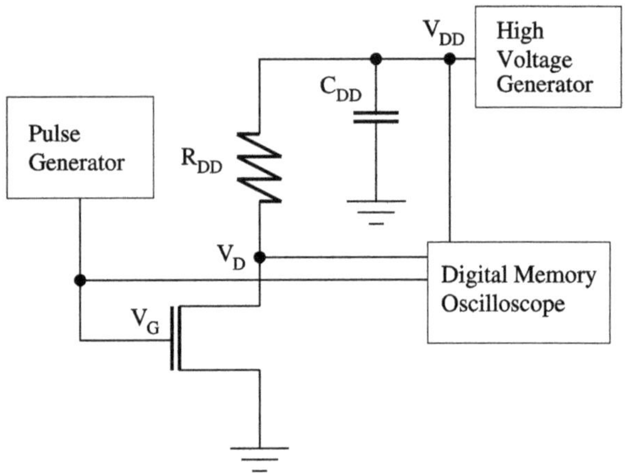

Fig. 4.20 General schematic of a pulsed *I–V* measurement system [12]

needed. The test signal frequency is usually in the range of 1 kHz to 1 MHz. Different types of connection schemes are possible: Some of the terminals are connected either to the "High" or to the "Low" terminals of the LCR meter at the same time other terminals are biased by an SMU of a parameter analyzer or occasionally kept floating. The investigation of the drain-source voltage dependence of the capacitance requires small structures otherwise the large DC currents would overload the LCR meter. Especially for this purpose the S-parameter characterization is more preferable [48].

For S-parameter measurements a vector signal analyzer (VNA) and a parameter analyzer for DC biasing are used. The VNA captures the small-signal behavior of the device connected in a two-port configuration (e.g.: Port1: Gate-Source, Port2: Drain-Source) in the form of S-parameters in a wide frequency range starting from few MHz up to several GHz. Before starting the measurements the VNA needs to be calibrated up to the RF probe tips (= reference plane). There are several methodologies in use like SOLT, TRL, LRRM. Dummy structures (e.g. open and short) and sophisticated de-embedding methods are applied in order to get rid of parasitics introduced by the probe pads and the device connection.

4.7.3 Pulsed Measurements for HV MOS Modeling

This type of characterization has not yet been established itself as industry standard, though some companies offer such pulsed measurement systems (PMS) [5, 9, 13, 15, 30, 31, 45]. The primary purpose of PMS data is to determine the thermal resistance **RTH** and the thermal capacitance **CTH**.

Figure 4.20 shows a typical PMS setup which consists of a pulse generator to drive the gate voltage, a high-voltage generator to supply the drain voltage (could be pulsed or only buffered) and at least a two channel digital oscilloscope.

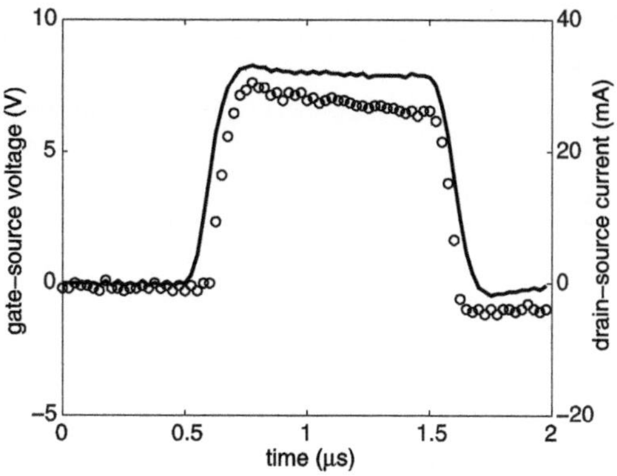

Fig. 4.21 Example pulse diagram of V_G (*solid*) and I_D (*symbols*) measured with an oscilloscope

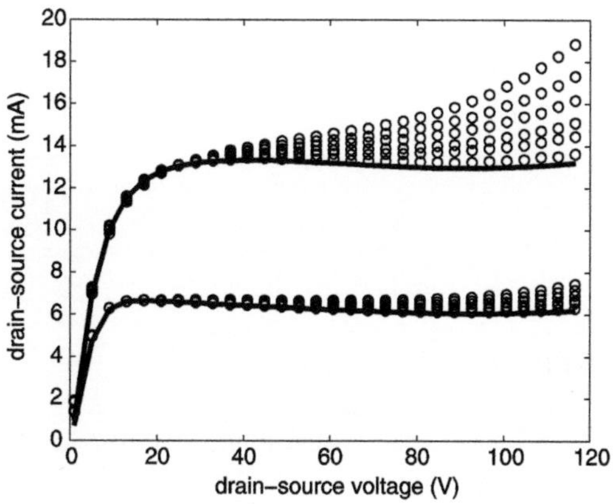

Fig. 4.22 MOSFET output data at $V_{GS} = 8.6$ V and 20 V recorded at pulse lengths of 0.5 μs to 100 μs (*symbols*). The *solid line* is the DC measurement

Figure 4.21 shows the voltage response of a HV MOSFET. The solid line is the pulsed gate voltage V_G, the symbols are the drain current I_D which can be calculated by the voltage drop on R_{DD}:

$$I_D = \frac{V_{DD} - V_D}{R_{DD}}. \tag{4.47}$$

Additional effects like trapping processes can influence the pulse shapes [41]. Figure 4.22 demonstrates the self-heating effect at two gate voltages as increasing current with decreasing pulse width. Table 4.2 lists figures of merit of an "ideal" PMS for modeling integrated HV transistors. Other types of devices e.g. bipolar transistors and resistors are also of interest. A thermal schematic [12, 22, 34, 46] and [10] which is typically used for thermal modeling is shown in Fig. 4.23.

Finally, alternative measurement and extraction methods have been developed like the ambient temperature measurement approach [40] and the AC output conductance approach [27].

Table 4.2 Figures of merit of an "ideal" PMS

$V_{GS} = \pm 20$ V (pulsed)
$V_{DS} = \pm 200$ V (pulsed); max. IDS > 100 mA @ 1 μA measurement resolution
pulse length variable: 10 ns [8]—1 s
bulk current monitoring

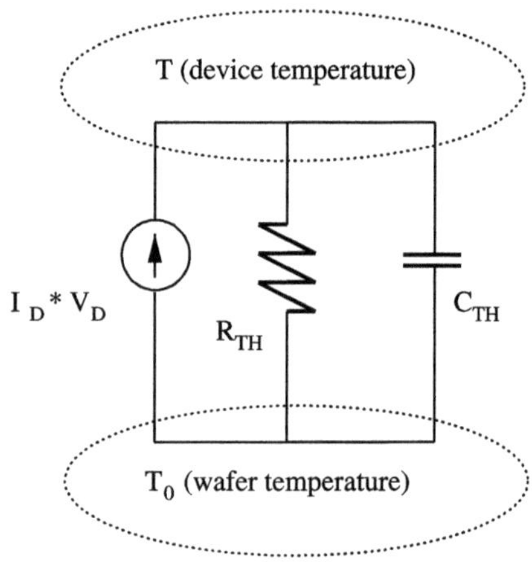

Fig. 4.23 A simple thermal model with a thermal resistance **RTH** and a thermal capacitance **CTH**

References

1. http://home.hiroshima-u.ac.jp/usdl/HiSIM_HV/index.html. Cited 2 December 2009
2. http://en-origin.nxp.com/models/hv_models. Cited 2 December 2009
3. Aarts, A.C.T., Kloosterman, W.J.: Compact modeling of high-voltage LDMOS devices including quasi-saturation. IEEE Trans. Elecron Devices **53**(4), 897–902 (2006)
4. Aarts, A.C.T., D'Halleweyn, N., van Langevelde, R.: A surface potential-based high-voltage compact LDMOS transistor model. IEEE Trans. Electron Devices **52**(5), 999–1007 (2005)
5. ACCENT Diva. http://findarticles.com/p/articles/mi_m0EIN/is_2001_July_9/ai_76390652. Cited 2 December 2009
6. Agilent 4285A. http://www.home.agilent.com/. Cited 2 December 2009
7. Agilent 415x or B1500A. http://www.home.agilent.com/. Cited 2 December 2009
8. Agilent, Ten nanosecond pulsed IV parametric test solution. Technical Overview (2006)
9. AGILENT B15xx. http://www.home.agilent.com/. Cited 2 December 2009
10. Anghel, C.: High voltage devices for standard MOS technologies—characterisation and modelling. Doctor Thesis, Lausanne, EPFL (2004)
11. Anghel, C., et al.: Investigations and physical modelling of saturation effects in lateral DMOS transistor architectures based on the concept of intrinsic drain voltage. In: ESSDERC, pp. 399–402, September 2001
12. Anghel, C., Gillon, R., et al.: Self-heating characterization and extraction method for thermal resistance and capacitance in high voltage MOSFETs. In: Automacs, EC Project
13. AURIGA AU4750. http://www.auriga-ms.com/charmain.shtml. Cited 2 December 2009
14. Baylis, C.P. II: Understanding pulsed IV measurement waveforms. In: IEEE EDMO 2003, p. 223 (2003)
15. Berkner, Modeling self heating using HICUM. In: HICUM Workshop TU Dresden (HP85124) (2003)

16. Canepari, A., Bertrand, G., Giry, A., Minondo, M., Blanchet, F., Jaouen, H., Reynard, B., Jourdan, N., Chante, J.-P.: LDMOS modeling for analog and RF circuit design. In: Proceedings of ESSDERC, Grenoble, France (2005)
17. Chauhan, Y.S., Krummenacher, F., Anghel, C., Gillon, R., Bakeroot, B., Declercq, M., Ionescu, A.M.: Analysis and modeling of lateral non-uniform doping in high-voltage MOSFETs. In: IEDM Tech. Dig., pp. 8.3.1–8.3.4, December 2006
18. Chen, J.F., Tian, K.-S., Chen, S.-Y., Wu, K.-M., Liu, C.M.: On-resistance degradation induced by hot-carrier injection in LDMOS transistors with STI in the drift region. IEEE Electron Device Lett. **29**(9), 1071–1073 (2008)
19. Contiero, C., Murari, B., Vigna, B.: Progress in power ICs and MEMS. In: Proceedings of 2004 International Symposium on Power Semiconductor Devices and ICs, Kitakyushu, pp. 3–12 (2004)
20. Efland, T., Malhi, S., Bailey, W., Kwon, O.K., Ng, W.T., Torreno, M., Keller, S.: An optimized RESURF LDMOS power device module compatible with advanced logic processes. In: IEDM Tech. Dig., pp. 237–240 (1992)
21. Enz, C., Krummenacher, F., Vittoz, E.: An analytical MOS transistor model valid in all regions of operation and dedicated to low-voltage and low-current applications. J. Analog Integr. Circuits Signal Process. 83–114 (1995)
22. Farhanah, A., et al.: Modeling 32 V asymmetric LDMOS using Aurora and Hspice Level 66. In: MOS AK (2007)
23. Frere, S.F., Moens, P., Desoete, B., Wojciechowski, D., Walton, A.J.: An improved LDMOS transistor model that accurately predicts capacitance for all bias conditions. In: IEEE International Conference on Microelectronics Test Structures, pp. 75–79, April 2005
24. Griffith, E.C., Kelly, S.C., Power, J.A., Bain, D., Whiston, S., Elebert, P., O'Neil, M.: Capacitance modelling of LDMOS transistors. In: IEEE Solid-State Device Research Conference, pp. 624–627, September 2000
25. Hussein, B., Declercq, M.: High Voltage Devices and Circuits in Standard CMOS Technologies. Kluwer Academic, Dordrecht (1999)
26. Jang, J., Arnborg, T., Yu, Z., Dutton, R.W.: Circuit model for power LDMOS including quasi-saturation. In: Proc. IEEE Int. Conf. on Simulation of Semiconductor Processes and Devices, SISPAD'99, pp. 15–18 (1999)
27. Jin, W., et al.: Self-heating characterization for SOI MOSFET based on AC output conductance. In: IEDM (1999)
28. Keithley 4200. http://www.keithley.com/. Cited 2 December 2009
29. Keithley 4200. http://www.keithley.com/. Cited 2 December 2009
30. Keithley PIV-A. http://www.keithley.com/. Cited 2 December 2009
31. Keithley PIV-Q. http://www.keithley.com/. Cited 2 December 2009
32. Knaipp, M., Roehrer, G., Minixhofer, R., Seebacher, E.: Investigations on the high current behavior of lateral diffused high-voltage transistors. IEEE Trans. Electron Devices **51**(10), 1711–1720 (2004)
33. Knaipp, M., Park, J.M., Vescolli, V., Roehrer, G., Minixhofer, R.: Investigations on an isolated lateral high-voltage n-channel LDMOS transistor with a typical breakdown of 150 V. In: Proceedings of the 36th European Solid-State Device Research Conference, ESSDRC 2006 (2006)
34. Labate et al.: Scalable electrical model for a SOI-RF-LDMOS including drain drift region resistance self-heating effects. In: MOS-AK (2006)
35. Liang, Y.C., Samudra, G.S.: Power Microelectronics, Devices and Process Technologies. World Scientific, Singapore (2009)
36. Ludikhize, A.: Kirk effect limitations in HV-ICs. In: Proc. ISPSD, pp. 249–252 (1994)
37. Miura-Mattausch, M., Sadachika, N., Miyake, M., Navarro, D., Ezaki, T., Mattausch, H.J., Ohguro, T., Iizuka, T., Taguchi, M., Miyamoto, S., Inagaki, R., Furui, Y., Fudanuki, N., Yoshida, T.: HiSIM2.4.0: Advanced MOSFET model for the 45 nm technology node and beyond. In: Proceedings of the NSTI-Nanotech 2007, pp. 479–484, Santa Clara (2007)
38. Oh, S.-Y., Ward, D.E., Dutton, R.W.: Transient analysis of MOS transistors. J. Solid-State Circuits **SSC-15**, 636–643 (1980)

39. Oritsuki, Y., Yokomiti, M., Sakuda, T., Sadachika, N., Miyake, M., Kajiwara, T., Kukuchihara, H., Yoshida, T., Feldmann, U., Mattausch, H.J., Miura Mattausch, M.: HiSIM-LDMOS/HV: A complete surface-potential based MOSFET model for high-voltage applications. In: Proceedings of the NSTI-Nanotech 2008, pp. 893–896, Boston (2008)
40. Paasschens, J.C.J., et al.: Dependence of thermal resistance on ambient and actual temperature. In: BCTM (2004)
41. Passant Baylis, C. II: Improved current-voltage methods for RF transistor characterization. Master Thesis
42. Perugupalli, P., Trivedi, M., Shenai, K., Leong, S.K.: Modeling and characterization of an 80 V silicon LDMOSFET for emerging RFIC applications. IEEE Trans. Electron Devices **45**(7), 1468–1478 (1998)
43. Posch, W.: Measurement and modelling of high-voltage MOS field effect transistors. Master Thesis, Institute of Solid-State Physics, Technical University Graz (2002)
44. Schrems, M., et al.: Scalable high voltage CMOS technology for smart power and sensor applications. E I Elektrotech. Informationstech. **125**(4), 109–117 (2008)
45. Sischka, F., Characterisation Handbook, p. 2, 8.10.01J
46. Skadron, K., et al.: A Short Tutorial on Thermal Modeling and Management (2008), coolchips08
47. Trivedi, M., Khandelwal, P., Shenai, K.: Performance modeling of RF power MOSFETS. IEEE Trans. Electron Devices **46**(8), 1794–1802 (1999)
48. Vestling, L.: Design and modeling of high-frequency LDMOS transistors. Ph.D. Thesis, Uppsala University, February (2002)
49. Yokomichi, M., Sadachika, N., Miyake, M., Kajiwara, T., Mattausch, H.J., Miura-Mattausch, M.: Laterally diffused metal oxide semiconductor model for device and circuit optimization. Jpn. J. Appl. Phys. **47**, 2560–2563 (2008)

Chapter 5
Physics of Noise Performance of Nanoscale Bulk MOS Transistors

R.P. Jindal

Abstract Detailed physical understanding of the noise mechanisms that exist in bulk MOS transistors is developed. These sources of noise consist of the intrinsic fluctuations which are inherent in the device structure and extrinsic fluctuations that are subject to optimization and elimination. While most noise sources are well understood, excess channel noise and $1/f$ noise continue to be areas of active research.

5.1 Introduction

Although the first patent for a field-effect device was filed by Julius Lilienfeld [1] back in 1925, its practical demonstration had to wait for 35 years due to the technological challenges associated with obtaining a clean semiconductor surface to observe this effect. To overcome this hurdle, a predecessor structure was first proposed by Warner [2] with substrate as the controlling electrode. Shortly thereafter, Kahng and Atalla published results [3] demonstrating what is known as a bulk MOSFET today. Once the device structure stabilized, noise research followed. Initial attempts at understanding the noise in a MOSFET were largely guided by the pioneering research of van der Ziel on JFETs [4, 5]. Based on this understanding of the intrinsic noise mechanisms the expected noise performance of MOSFET was considered very promising for integrated single-chip solutions for communication applications [6]. However, early MOS designs [7] performed poorly at the system level pointing to a serious gap in the understanding of the noise behavior of this nascent device. This motivated researchers at Bell Laboratories to examine the noise of the MOSFET in further detail in the early 80s. In this phase of intense research, several extrinsic noise mechanisms inherent in the bulk MOSFET structure were discovered and eliminated improving the noise performance almost by an

R.P. Jindal (✉)
W.H. Hall Board of Regents Eminent Scholar Endowed Chair, William Hansen Hall Department of Electrical and Computer Engineering, University of Louisiana at Lafayette, Lafayette, LA 70504, USA
e-mail: r.jindal@ieee.org

G. Gildenblat (ed.), *Compact Modeling*,
DOI 10.1007/978-90-481-8614-3_5,

order of magnitude [8]. Since then, over the last 25+ years, MOS has emerged as the technology of choice for high-frequency, high-performance, low-noise, and low-cost lightwave and wireless base-station and hand-held applications. This has been helped by the development of physics-based models describing the noise behavior of MOS devices. These models have been incorporated in compact MOS models used in circuit simulators to facilitate their usage by RF designers. Drawing from an earlier publication [9], in subsequent sections of this chapter, we will examine each of these noise mechanisms in detail developing a simple physical understanding of the manifestation of these noise sources in a MOSFET. This understanding is translated into model equations, where possible.

5.2 Preliminary Considerations

Before we plunge into the detailed noise mechanisms it is instructive to examine some basic aspects of the device. A bulk MOSFET is basically a four-terminal voltage-controlled resistor, the four terminals being the source, drain, gate and the substrate. The channel, connecting the source and the drain terminals, is the voltage controlled resistor. This voltage control can be experienced either from the top gate, as is typically done, or from the bottom gate or substrate. In this structure, the first intrinsic noise source is the resistive nature of the FET channel. In the context of resistors, based on thermodynamic considerations, Nyquist [10] has shown that irrespective of its physical structural details, a resistance must exhibit thermal noise. Hence, under thermodynamic equilibrium, all MOSFETs, irrespective of their channel length, gate oxide thickness or junction depths must exhibit only thermal noise. This does not preclude the presence of other fluctuations but they cannot be observed across the device terminals. Further, a substrate is needed to support the channel and a gate electrode is needed to produce and modulate the channel. The coupling of channel thermal noise to the gate and the substrate produces induced gate current noise and induced substrate current noise both of which are detrimental to the performance of the device.

However, noise mechanisms other than those outlined above also exist that further degrade the device performance. One can logically arrive at their manifestation by realizing that any device parameter that influences the channel charge and hence the channel current, if modulated, will result in channel noise. This would lead one to suspect that thermal noise generated by the distributed gate and substrate resistances, substrate current noise and gate current noise to contribute to MOSFET noise. These are termed as *extrinsic* noise mechanisms, since their existence is not inherent to the transistor action. We will next investigate the physics of each of these mechanisms referring to the original papers for the mathematical details.

Another aspect of MOSFET operation that has bearing on its noise behavior is the frequency of operation in relation to the transit frequency f_T. Typical MOSFET analyses are carried out in the quasi-static (QS) regime where the frequency of operation is small compared with f_T. When the operating frequency gets closer to f_T, the inertia of the current carriers in the MOS channel comes into play and the MOSFET enters the non-quasi-static (NQS) regime. This has a direct bearing on both its signal and noise performance. These effects will also be addressed in the following sections. Although the discussions that follow are valid for all type of MOSFETs, for ease of understanding, unless otherwise stated, we will refer to bulk n-channel MOSFETs.

Borrowing terminology from thermal noise in a resistor, we will describe the noise in terms of open-circuit noise voltage spectral density $S_V(f)$ with dimensions of V^2/Hz observed across the two terminals. Alternatively, we will also describe noise in terms of short-circuit noise current spectral density $S_I(f)$ with dimensions of A^2/Hz observed through an AC short placed across the device. Conceptually, this is nothing but the noise observed through a brick-wall filter of bandwidth of 1 Hz.

5.3 Intrinsic Fluctuations

In this section, we discuss the noise mechanisms that are inherent in the MOSFET device structure. These consist of the channel thermal noise, induced gate noise and induced substrate noise.

5.3.1 Channel Thermal Noise

As mentioned earlier, this noise arises due to the thermally generated random motion of carriers that constitute the FET channel. When the drain-to-source voltage V_{DS} is zero, the channel is a uniform resistor and hence the thermal noise current spectral density is simply given by the Nyquist relation.

$$S_{I_d}(f) = 4k_B T g_{d0} \tag{5.1}$$

where k_B is the Boltzmann constant, T is the absolute temperature of the sample and g_{d0} is the *equilibrium* drain conductance at zero drain-to-source voltage. The adjective "*equilibrium*" used here has a subtle consequence and will be revisited later in Sect. 5.3.4. At a finite V_{DS}, if the departure from equilibrium is not too drastic, as is the case for long-channel MOS devices, this expression is simply modified to take the channel non-uniformity into account. A detailed analysis of this effect in the context of the MOSFET built on lightly doped substrates was presented by Sah at a noise symposium held at the University of Minnesota and later at the Solid-State Device Research Conference in Boulder, Colorado [11], followed by an analysis by Jordan and Jordan [12]. The results can be expressed in the following form giving rise to a flat spectrum.

$$S_{I_d}(f) = 4k_B T \gamma_{lc} g_{d0}. \tag{5.2}$$

Here γ_{lc} is the long-channel bias-dependent noise parameter. We have introduced the subscript "lc" to denote the long-channel case where the departure from equilibrium is insignificant as far as the carrier thermal velocity distribution is concerned.

This is distinguished from γ for short-channel devices which will be a subject of extensive discussion in Sect. 5.5.2. According to this long-channel theory, the noise parameter approaches unity, as expected, when the drain-to-source voltage V_{DS} approaches 0. It decreases to a value of $2/3$ as the device enters the saturation regime and remains at that value when V_{DS} exceeds the saturation value. Let us carefully analyze the case for $V_{DS} = 0$ from a thermodynamic perspective. If we assume the ideal situation where there are no leakages or breakdown currents in the device, then, as V_{DS} approaches 0, all currents in the device stop flowing. The device is then in perfect equilibrium with no exchange of power between the device and the battery or the surroundings. Following Nyquist's derivation [10], if this FET channel is now connected to one end of a transmission line with characteristic impedance equal to the channel resistance, then in the steady-state, the noise power flowing out of the channel must equal that flowing into the channel. If this were not the case, then over time, the MOSFET would either cool or heat by itself. This is prohibited based on thermodynamic principles. Hence, at $V_{DS} = 0$, the FET channel must exhibit full thermal noise voltage fluctuations corresponding to the value of the channel resistance. As pointed earlier, this does not preclude the presence of other fluctuations. However, they cannot be felt by the external world. Examples of such fluctuations will be discussed in Sect. 5.3.4. Thus, under these conditions, the current noise fluctuations must correspond to noise generated by channel conductance of magnitude g_{d0} as predicted by the model.

The saturation value of $\gamma_{lc} = 2/3$ is valid only for long-channel MOS devices built on lightly doped substrates. The effect of the fixed bulk-charge on channel thermal noise was first pointed out by Sah et al. [13]. Klaassen and Prins [14] extended these analyses by allowing a finite realistic doping in the FET substrate. Then under the assumption of a constant surface-charge density they produced the following relation

$$S_{I_d}(f) = \frac{4k_BT}{I_{DS}L_{eff}^2} \int_0^{V_{DS}} g^2(V)dV \tag{5.3}$$

where I_{DS} is the drain-to-source current, L_{eff} is the metallurgical channel length, $g(V)$ is the local channel conductance of the channel and the rest of the symbols have been defined earlier. They pointed to values of γ substantially higher than the ideal can be observed due to substrate doping effects. As in the previous analyses, in this analysis also high-field effects such as mobility degradation and carrier heating were not included which is justifiable for long-channel devices. Recently Paasschens et al. [15] have critically examined the K-P analysis and identified some fundamental short-comings of the treatment and extended it to include velocity saturation effects. The reader is referred to the original paper for details.

As mentioned earlier, when the frequency of operation gets closer to the device f_T, the MOSFET enters the NQS regime. Recently, Deshpande and Jindal [16], following Shoji's [17] approach, have shown that in the NQS regime the channel thermal noise is no longer flat with respect to frequency. The modified expression is given by

$$S_{I_d}(f) = 4k_B T g_{d0}\left[\frac{2}{3} + \frac{76}{4725}\left(\frac{\omega}{\omega_T}\right)^2\right]. \tag{5.4}$$

Here ω_T is the angular transit frequency and ω is the operating angular frequency. It will be shown that departure from these theories also occurs as the channel lengths shrink. This will be a subject of detailed discussion in Sect. 5.5.2.

5.3.2 Induced Gate Noise

The thermal noise voltage fluctuations across an elementary section Δx of the FET channel lead to a readjustment of the potential distribution along the whole channel. The channel being one plate of the MOS capacitor, these voltage fluctuations generate a fluctuating charge across the gate capacitor of the FET. This rate of change of charge generated by these charge fluctuations is equivalent to a fluctuating gate current. The magnitude this gate current is directly proportional to the frequency of interest. Following the earlier work on induced gate noise by van der Ziel [5] in connection with JFETs, this effect was modeled by Shoji [17] for MOSFETs built on lightly doped substrates. Assuming MOSFET operation in the low channel electric field regime, Shoji evaluated the propagation of these fluctuations treating the MOS channel as an active distributed RC transmission line. A voltage fluctuation across a small section Δx at a point x in the channel of the device drives the two transmission lines of the channel one stretching from $x = 0$ to $x = x$ and the other stretching from $x = x$ to $x = L$, where L the electrical channel length of the device. Note that, above saturation, the electrical channel length becomes smaller than the metallurgical channel length. However, for long-channel length devices, this difference can be neglected. The gate current fluctuation is evaluated as the difference between the corresponding drain-side current fluctuation and the source-side current fluctuation. In the final analysis, Shoji retained only the leading term in a Bessel Function solution to this extremely complex calculation, and recovered the results for channel thermal noise [11, 12]. It turns out that this neglect of the higher order terms is equivalent to solving this problem under the quasi-static approximation. Under device saturation, again retaining the leading term, he developed an analytical expression for the gate current noise spectral density given below:

$$S_{I_g}(f) = \frac{64}{135}\omega^2 k_B T \frac{C^2}{g_{d0}}. \tag{5.5}$$

Here C is the total gate capacitance, ω is the frequency of operation and other symbols have been defined earlier. Analogous to the channel thermal noise, one can express [18] the induced gate noise by the expression

$$S_{I_g}(f) = 4k_B T \delta g_g. \tag{5.6}$$

Here $g_g = \omega^2 C_{gs}^2/5g_{d0}$ is the induced-gate noise conductance where $C_{gs} = (2/3)C$ is the gate-to-source capacitance under saturation. δ is a dimensionless parameter with value of 4/3 under saturation. The voltage fluctuations generated by this gate current across the gate impedance manifest themselves at the FET channel via the device transconductance. Based on an enhanced Klaassen and Prins approach, taking velocity saturation into account, van Langevelde et al. have shown [19] that the classical expression could underestimates the induced gate noise by up to 40%.

Since the device transconductance vanishes at $V_{DS} = 0$, the effect of these induced gate current fluctuations on the FET drain current is eliminated. Since the drain and gate current fluctuations arise from the same source, they must be correlated. Shoji showed that the cross-correlation spectrum of the gate and drain current fluctuation under saturation is given by

$$S_{I_g^* I_d}(f) = j\frac{4}{9}\omega k_B T C. \tag{5.7}$$

It should be noted that the correlation spectrum is purely imaginary. As shown by Shoji, for $V_{DS} = 0$ due to channel symmetry, the cross-correlation vanishes.

In the NQS regime the above expressions must be refined. Deshpande and Jindal [16] have shown that a more general expression taking NQS effects into account is given by

$$S_{I_g}(f) = \frac{64}{135} k_B T g_{d0} \left(\frac{\omega}{\omega_T}\right)^2 \left[1 - \frac{43}{2970}\left(\frac{\omega}{\omega_T}\right)^2\right]. \tag{5.8}$$

The expression for cross-correlation spectrum is given by

$$S_{I_g^* I_d}(f) = j\frac{4}{9} k_B T g_{d0}\left(\frac{\omega}{\omega_T}\right) + \frac{176}{945} k_B T g_{d0}\left(\frac{\omega}{\omega_T}\right)^2. \tag{5.9}$$

Here we see a fundamental departure from the classical expression given in (5.7). Instead of being purely imaginary, the cross-correlation spectrum now has a real part that increases as the square of the operating frequency. The traditional imaginary part increases only linearly with frequency. This has been suspected for a long time since the delay in the carrier transit through the channel gives rise to an in-phase component of induced gate noise. This effect has also been demonstrated based on Monte-Carlo/Hydrodynamic simulations of carrier transport by Jungemann et al. [20]. The real part of the cross-correlation has fundamental implications in terms of LNA design [21]. A more accurate expression for the cross-correlation coefficient taking the real part of the cross-correlation into account is given by [16]

$$c = \frac{\overline{i_g^* i_d}}{\sqrt{\overline{i_g i_g^*} \cdot \overline{i_d i_d^*}}} = j\frac{1}{4}\left(\frac{5}{2}\right)^{1/2} + \frac{11}{105}\left(\frac{5}{2}\right)^{1/2}\left(\frac{\omega}{\omega_T}\right). \tag{5.10}$$

Again, it should be emphasized that the real part is frequency dependent.

Vallur and Jindal [22] have derived analytical expressions to quantify the effect of electron temperature and velocity saturation on induced-gate noise in the NQS regime. Their results show an increase in the induced gate noise as a function of the constant electric field E in the channel normalized by the critical electric field E_C and are given below. This increase is due to the increase in the thermal voltage fluctuations in the channel. The increase in the thermal voltage fluctuations is due to (i) increase in the electron temperature which gives rise to a factor $(1 + E/E_C)$ (ii) an increase in the channel resistance due to a decrease in the carrier mobility introducing another factor $(1 + E/E_C)$. The second term in the square bracket is the NQS correction.

$$S_{I_g}(f)_{NQS} = \frac{64}{135} k_B T g_{d0}(1 + E/E_C)^2 \left(\frac{\omega}{\omega_T}\right)^2 \times \left[1 - \frac{43}{2970}(1 + E/E_C)^2 \left(\frac{\omega}{\omega_T}\right)^2\right] \Delta f. \quad (5.11)$$

Finally, it must be pointed out that the induced gate current fluctuation changes sign as a function of the position of the causing voltage fluctuation in the channel and therefore for a given position in the channel it must vanish. At $V_{DS} = 0$ this happens at $x = (1/2)L_{eff}$ [17, 23]. This is due to the symmetry in the channel about this half-way point. When the device is in saturation, this happens at precisely at $x = (5/9)L_{eff}$ [22]. This shift towards the drain is due to the fact that the resistance per unit length on the source side of the channel is less than the resistance per unit length on the drain side resulting in a larger voltage change near the drain side, thus causing a larger change in charge near the drain end.

5.3.3 Induced Substrate Noise

Analogous to the induction of gate current due to fluctuations in the channel potential, one expects induced substrate current noise in sympathy with thermal voltage fluctuations in the FET channel. The analysis runs parallel to that for induced gate noise and was first reported by Pu and Tsividis [24]. For a well designed device, since the substrate transconductance is much smaller than the gate transconductance, this noise is typically not important from low-noise circuit perspective. Also application of substrate bias [25] helps suppress this noise.

5.3.4 Equilibrium Noise

At $V_{DS} = 0$ the effect of induced gate and induced substrate noise current is decoupled from the drain terminal of the MOSFET due to vanishing gate and substrate transconductances. However, it has been shown by Jindal [23] and Paasschens et al. [15] that these fluctuations in the inversion layer charge give rise to channel conductance fluctuations with a flat spectrum. The noise spectrum associated with the fluctuations in the inversion layer charge is given by the following expression.

$$S_Q(f) = C^2 4 k_B T \left(\frac{1/g_{d0}}{12}\right) \quad (5.12)$$

where the symbols have been defined earlier The factor $1/12$ occurs due to the distributed nature of the channel resistance and will occur again in the context of distributed gate resistance noise to be discussed in Sect. 5.4.1. It is important to note that to preserve the validity of Nyquist's results [10], the noise current associated with these charge fluctuations must flow between the gate terminal and the source and drain terminals so as NOT to give rise to any fluctuating voltage across the FET channel. To sense these fluctuations, a sensing current must flow in the device. Such a situation is typical in case of transimpedance amplifiers [26] designed using a MOSFET as a tunable feedback resistive element.

However, excess channel noise measurements at zero drain bias have been reported as early as 1979 by Takagi and van der Ziel [27]. These results pertained to devices with gate oxide thickness of 1000 Å. The authors stated that this excess noise is due to the strong transverse electric field in the FET channel. They also found that this excess noise component decreases with decrease in sample temperature, which is contraindicative of hot-carrier effects. Recently Vadyala [28] has observed similar excess noise for devices with gate oxide thickness in the 18 Å range. A plausible explanation for this was provided by Jindal and van der Ziel [29] where, using a sub-layer channel conduction model based on an analytical solution to Boltzmann Transport Equation, they showed that the *equilibrium* channel conductance g_{d0} could be higher than the measured g_{d0} thus explaining this discrepancy.

5.3.5 Bulk Charge Effects

Considering a FET as a 4-terminal black-box it can be shown [30] that without making any assumption regarding the details of the device operation, the four small-signal device conductances are connected by the relation $g_s = g_d + g_m + g_{mb}$ Here g_d and g_s are the drain and source conductances respectively and g_m and g_{mb} are the gate and substrate transconductances respectively. Next consider the special case when $V_{DS} = 0$. Again without loss of generality, we can claim that $g_m = g_{mb} = 0$. Thus, $g_s = g_d \equiv g_{d0}$. Now if realize that for normal MOSFET device operation with the pinch-off occurring on the drain side of the channel, the value of g_s is fairly independent of other terminal voltages. Then, for *any* terminal voltages, in general, $g_{d0} = g_m + g_{mb} + g_d$. Finally, for well-designed MOSFET when the device enter saturation, $g_d \ll g_m$ or g_{mb} Thus under saturation we claim,

$$g_{d0} \approx g_m + g_{mb}. \tag{5.13}$$

We can therefore conclude [25] that a higher substrate doping will result in a larger substrate transconductance at the expense of the gate transconductance. Since the channel thermal noise is determined by g_{d0}, a fixed g_{d0} implies a fixed channel thermal noise. Thus, according to (5.13) a higher g_{mb} will imply a device with same output noise but lower g_m. This is undesirable for low-noise circuit design implying that a lightly-doped substrate is preferable. We will revisit the optimum substrate doping in Sect. 5.4.2 in connection with substrate resistance noise since this choice of dopant density has impact on substrate resistance noise. It should be noted that a higher substrate doping with mild substrate reverse bias [25] is a good solution by eliminating the substrate transconductance.

5.4 Extrinsic Fluctuations

Having discussed the intrinsic noise mechanisms, we will now focus on the extrinsic noise mechanisms. As the name implies, since they are not fundamental to the operation of the device, by suitable change of device structure and parameters, these fluctuations are amenable to reduction.

5.4.1 Gate Resistance Noise

As mentioned in Sect. 5.2, any fluctuations in the parameters that determine the channel resistance will manifest themselves as fluctuations at the device output as noise. One obvious suspect is the resistive MOSFET gate. To the reader today, this may appear as a trivial assertion. However the author has to confess that back in the 80s when this concept was enunciated for the first time, it took a lot of bold thinking followed by detailed mathematical analyses. Thornber first modeled this problem for a single-stripe gate and demonstrated [31] that this noise source becomes important when the device transconductance approaches the conductance of the resistive gate. However, for typical complex IC gate matrix layouts there are multiple gate stripes and resistive contacts. Further, the situation becomes even more complicated when one realizes that the gate resistance comes in two flavors. The gate runners that meander over thin-oxide produce transistor action where their distributed nature must be taken into account. Gate runners that run over thick-oxide serve only as resistive interconnects which can be treated as lumped elements. To take all these effects into account in a semi-automated manner Jindal [32] developed a generalized analytical formulation to quantify the contributions due to the gate resistors and the interconnect resistors in a complex gate matrix layout. The contributions were added with proper correlations taken into account. To better understand the above concepts, consider a relatively simple layout as shown in Fig. 5.1 marking the gate and interconnect resistances.

In general, noise voltage fluctuations across each interconnect and gate resistor produces fully correlated voltage fluctuations in every other resistor in the gate matrix. If the device has a constant channel length across the full device width, which is typically true, these effects can be modeled by a simple parameter $A_{v_{ij}}$. $A_{v_{ij}}$ is the average of the voltage fluctuation at the two ends of the ith distributed gate resistor due to a voltage fluctuation across the jth interconnect or gate resistor divided by the voltage fluctuation across this jth resistor. $A_{v_{ij}}$ can be calculated either by inspection or using standard circuit simulation tools. Under this simplification, the general

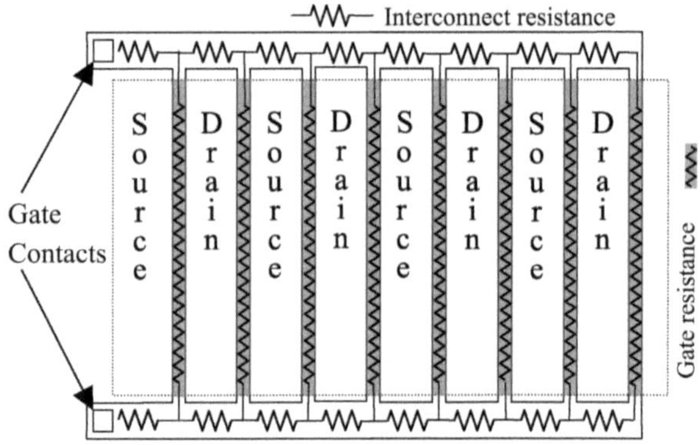

Fig. 5.1 A typical resistive gate matrix layout. (Reprinted from R.P. Jindal [9])

yet complex expression derived by Jindal [32] reduces to the following expression for equivalent noise resistance due to a resistive gate matrix.

$$R_n = \frac{1}{g_{m_T}^2}\left[\sum_{j=1}^{p}\left(\sum_{i=1}^{p} g_{m_i} A_{v_{ij}}\right)^2 R_j + \sum_{i=1}^{p} \frac{g_{m_i}^2}{12} R_i\right]. \tag{5.14}$$

Here, g_{m_i} is the transconductance associated with the ith resistor R_i and g_{m_T} is the total transconductance of the MOSFET. For simplicity and automation of computation, the summation over j varies from 1 to p includes all gate and interconnect resistors realizing that for an interconnect resistor, the associated transconductance g_{m_i} is zero while for a gate resistor g_{m_i} has a finite value. It can be easily shown that for a single gate stripe of resistance R connected to the signal at one end and open at the other, the effective noise resistance is $R/3$. When the signal is applied to both ends of the gate, the expression reduces to the celebrated result $R/12$. These results are frequently quoted in published literature and books as rules of thumb without reference to their origin,. As shown here, these results do emerge from a sound theoretical basis [32]. This analysis can be routinely applied to evaluate and refine actual gate layouts by strategically placing gate contacts. This technique has been used, but not published in open literature, to quantitatively evaluate the tradeoff between, often competing, cost constraints such as employing a certain resistivity gate material, contact area and lithographic rules.

At the time this analysis was first published, device channel lengths and widths were in the few micrometer range with the devices being primarily used for digital applications. This resulted in gate W/L ratios under 10 thus minimizing the gate resistance. As technology progressed, a demand for high speed performance has led to a continual shrinking of channel lengths which are now in the nanometer regime. At the same time the device widths grew to over 100 micrometers to control the channel thermal noise. Thus the gates became long and narrow increasing the W/L ratio to higher than 100. Using lower resistivity gate material such as salicides and multi-finger gates has helped contain the problem. However, it should be noted when a salicides is used the finite interfacial resistance between the salicides and the poly-silicon sets a lower limit on the gate sheet resistance. Thus, the effective gate resistance needs to be carefully modeled and controlled by using multi-finger

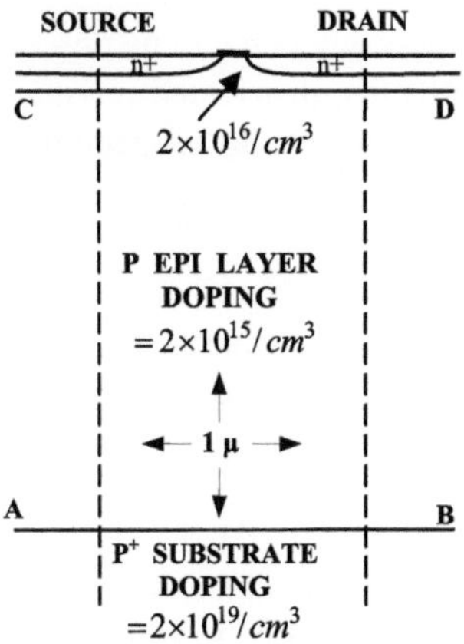

Fig. 5.2 Section of an n-channel FET. (Reprinted from R.P. Jindal [57])

gates and placing contracts throughout the layout strategically for a ultra-low-noise design. Since the publication of this work [32] in 1984, researchers worldwide have applied these concepts to RF CMOS front-end designs for lightwave receiver design and wireless cell phones and pager applications. Extensive research on this topic followed dealing with device applications [33, 34], experimental aspects [35–38], circuit simulation and compact models [39–44] and high-speed low-noise circuit designs [45–54]. At this stage of the technology development, it is one of the important considerations [55] in the design of any low-noise amplifier for lightwave and wireless applications.

5.4.2 Substrate Resistance Noise

Following the top gate noise analysis, the next logical step is to examine the noise arising from the resistive substrate or the bottom gate, first analyzed by Jindal [56]. For bulk MOSFETs, the bottom gate typically consists of an epitaxial layer grown on a low-resistivity silicon substrate with a contact to the back surface of the wafer. Alternatively, for the CMOS process, the transistor is located inside a tub which is connected to ground. In either case, the relatively lightly doped epitaxial layer contributes to a significant resistance between the back of the channel and the DC potential provided by the contact. Based on Nyquist's assertion [10] this resistance must generate thermal noise across its terminals. These thermal voltage fluctuations produce a fluctuating drain current through the substrate transconductance g_{mb}. Following the original analysis [56], as shown in Fig. 5.2, consider a vertical cross-section of a n-channel MOSFET fabricated on an epitaxial layer grown on a heavily doped p^+ substrate. The resistance generating the noise lies between the line AB representing the interface between the p^+ substrate and the p epitaxial layer and line CD representing the interface between the p epitaxial layer and the more heavily doped top channel-stop implant region. Above line CD the resistance is essentially zero. The two vertical lines are the lines of symmetry as dictated by the device layout.

The physical phenomenon can be described by a relatively simple expression. Let R_{sub} be the effective distributed resistance between the channel and the substrate. Then the channel current noise spectral density is given by

$$S_{I_d}(f) = 4k_B T R_{sub} g_{mb}^2 \quad (5.15)$$

where all the symbols have been defined earlier. Note that due to the distributed nature of the substrate resistance, its estimation may have to be done numerically. Due to its importance in the design of low-noise electronics, since its original publication [56], this noise source has been further researched at device level [36, 38, 58, 59], introduced in compact MOS models [41], and accounted for in circuit designs [46, 53, 60–64]. To avoid confusion, it needs to be pointed out that in some subsequent publications the phrase "substrate resistance noise" may not have been used by other authors. In fact in some text books [65] the name "epi-noise" has being used to refer to this very phenomenon.

As can be inferred from (5.15) a reduction in g_{mb} will have the desirable effect of reducing the contribution of this noise to the FET channel. This can be achieved by the application of a mild substrate reverse bias. This is a circuit technique of suppressing noise [25, 32] whereby a designer can improve the noise performance of an LNA.

5.4.3 Substrate Current Super-Shot Noise

In the previous section, we discussed thermally generated fluctuation of the substrate potential under equilibrium conditions which affect the device noise performance. Under non-equilibrium, fluctuations in the substrate potential can also occur by non-thermal means as well. At channel lengths in the micrometer range this is not an important issue. However, as the channel lengths shrink, the electric field in the channel increases and the carriers in transit from the source to the drain acquire enough energy from the electric field to cause band to band excitation of electrons also referred to as impact ionization. The generated electrons and holes are collected by the device terminals. Depending on the position of carrier generation, they can cause either a gate, substrate, source and/or drain current. For an n-channel MOSFET focusing on the substrate terminal, this results in a hole current collected by the substrate terminal. The fluctuations in this current produce a fluctuating potential across the substrate resistance which gets coupled to the FET channel. This excess noise mechanism was first reported by Kim [66] in the context of long-channel MOSFETs. His results showed that only one type of carrier (electrons for n-channel and holes for p-channel) was active in this ionization process. This gave rise to full shot noise. Rucker and van der Ziel [67] measured the noise associated with the gate current of a JFET and found it also exhibited full shot noise as expected for a weak avalanching situation. For MOSFETs with channel lengths in the sub-micrometer regime Jindal [68] also observed full shot noise at low drain-to-source voltages. However, as shown in Fig. 5.3, at higher drain-to-source voltages the noise steadily

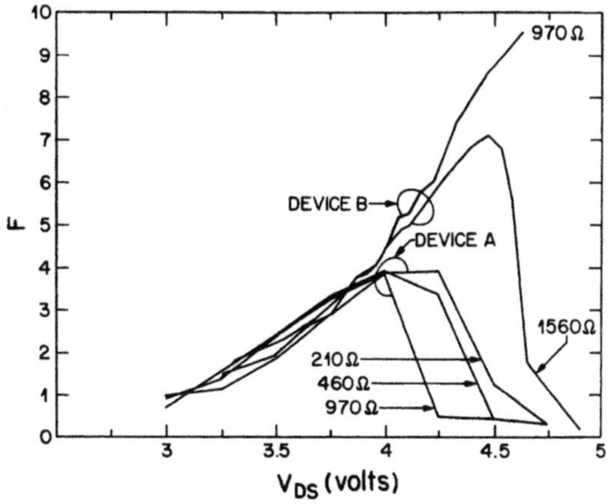

Fig. 5.3 Excess noise factor F versus drain-to-source voltage for devices A and B with external substrate resistance as a parameter. (Reprinted from R.P. Jindal [69])

increased above the shot noise level. Based on theoretical consideration, it can be easily shown that for impact ionization generated current, since each carrier is generated at random instants of time, one would expect full shot noise. The super-shot noise behavior is indicative of a more complex statistics of the carrier generation process and needs a deeper explanation. This excess noise was explained by Jindal [69] using a multi-step ionization process involving both holes and electrons. Each step consists of a single electron-hole pair production per traversal of the ionization region, which in this case is the high-field region of the FET channel, by the ionizing carrier. This ionizing carrier is an electron when drifting towards the drain diffusion and a hole when drifting away from the drain diffusion. Thus, the avalanche process has a gain larger than unity. For most of the supershot noise regime, the equivalent saturated diode noise current for substrate noise current is given by

$$I_{eq} = I_{sub}\sqrt{I_{sub}/I_{sub0}} \tag{5.16}$$

where I_{sub0} is the highest value of substrate current I_{sub} for which it exhibits full shot noise. For a detailed derivation of this effect and its behavior over an extended bias range, the reader is referred to the original publication [69]. The fluctuating substrate potential produced by the substrate current couples to the FET channel through the substrate transconductance by the following equation:

$$S_{I_d}(f) = 2qI_{eq}R_{sub}^2 g_{mb}^2 \tag{5.17}$$

where q is the electronic charge and the rest of the symbols have been defined earlier. In recent literature, a subset of this noise mechanism has been reported by Scholten et al. [44] under the name "Avalanche Noise". By application of mild substrate reverse bias, this noise mechanism can be essentially decoupled [25] from the channel and impact ionization is typically avoided by proper device design. This mechanism has received attention from device community [33, 70, 71] and extensively by the circuit design community far too numerous to list.

5.4.4 Gate Current Noise

Analogous to the substrate current, gate current is another source of potential noise in the MOSFETs. It can be generated in two ways. The first mechanism to generate the gate current is due to high energy carriers present near the drain end of the channel. Some of them are collected by the gate electrode as they travel over the barrier thus generating gate current. The second mechanism that generates gate current is Fowler-Nordheim tunneling through the gate oxide. This becomes more of an issue as the gate oxide thicknesses plunge into the 10's of Angstroms range. Both such mechanisms are expected to generate full shot noise.

5.5 Short-Channel Effects

Having discussed both the intrinsic and extrinsic noise mechanisms in a MOSFET in the long-channel regime, we will next focus our attention on the shrinking channel lengths and their impact on the noise behavior of these devices. While these short-channel effects directly or indirectly impact almost all noise mechanisms discussed in this chapter, they have a profound effect on the channel thermal noise. This is due to the fact that carrier transport in the FET channel acquires a different character in the nanometer regime.

5.5.1 Physical Origin

To improve the speed and performance, MOSFET channel lengths have continued to shrink over the last several decades. This closer spacing of the source and drain diffusions has introduced a host of phenomena discussed elsewhere [30]. They are broadly labeled as short-channel effects. In this section we will focus primarily on the following short-channel effects. (i) Channel-length modulation (ii) Carrier heating (iii) Velocity saturation (iv) Non-local quasi-ballistic effects.

Channel-Length Modulation With the MOSFET biased in the strong inversion, as we increase V_{DS}, ultimately the channel pinches off. As we further increase the drain voltage, the pinch-off point of the channel moves closer to the source diffusion. Thus the electrical channel length shrinks. This effect is referred to as channel-length modulation (CLM) [30].

Carrier Heating As one increases the electric field in the channel, the carriers gain higher energy from the electric field in-between the momentum randomizing collisions with the lattice. Hence their average energy increases. While this energy seldom follows a well-behaved distribution, it can be characterized by an effective temperature by approximating the distribution to be Maxwellian. This is justified when the electron-electron interaction is fairly strong. Thus the carrier temperature increases as a function of the electric field. There are several expressions for this effective carrier temperature. We will use the following expression [18] for electron temperature to facilitate further discussion.

$$T_e = T[1 + E(x)/E_C] \tag{5.18}$$

where $E(x)$ is the electric field at position x and E_C is the critical electric field.

Velocity Saturation As we continue to increase the electric field in the channel, the carriers get hotter. To reach a steady-state, they shed this excess energy acquired from the electric field in-between the collisions by the emission of high energy optical phonons in collision with the lattice. They thus reach a maximum drift velocity called the saturation drift velocity. At this stage any changes in the local electric field has very little effect on the carrier dynamics. Another way of looking at this phenomenon is that the carrier mobility decreases as the inverse of the electric field. Mimicking this behavior, we will use the following expression [72] for the electron mobility.

$$\mu_n(x) = \frac{\mu_{n0}}{1 + E(x)/E_C}. \tag{5.19}$$

Non-local Quasi-ballistic Effects As the channel length is shrunk further, ultimately a situation will be reached where the carriers will not spend enough time in the FET channel to get scattered. There traversal from the source to the drain is with only a few collisions with the lattice [73]. This is referred to as the quasi-ballistic regime of carrier transport.

5.5.2 *Effect On Channel Noise*

Researchers started to investigate hot-carrier effects soon after the intrinsic MOSFET channel thermal noise theory, as described in Sect. 5.3.1 was in place. However, these early studies on MOSFETs with channel lengths in the micrometer regime [74–76] yielded only a modest increase in noise at room temperature. This increase became more pronounced at 77°K. These findings were somewhat misleading in the context of short-channel MOSFETs and raised expectations [6] that MOS technology would be ideal for low-noise lightwave front-end applications. Initial attempts at single-chip repeater designs [7] proved that this enthusiasm was premature. This setback resulted in a renewed effort [8] at Bell Laboratories to explore, understand and possibly extend the limits of fine-line silicon MOS technology to meet the demands of ultra-sensitive lightwave and wireless communication systems. Due to this initiative, a multitude of noise mechanisms were discovered and eliminated [25] improving the FET performance by almost an order of magnitude [8]. Majority of these noise mechanisms have already been discussed in the preceding sections. In search of ultimate noise performance, new MOSFET structures with channel lengths

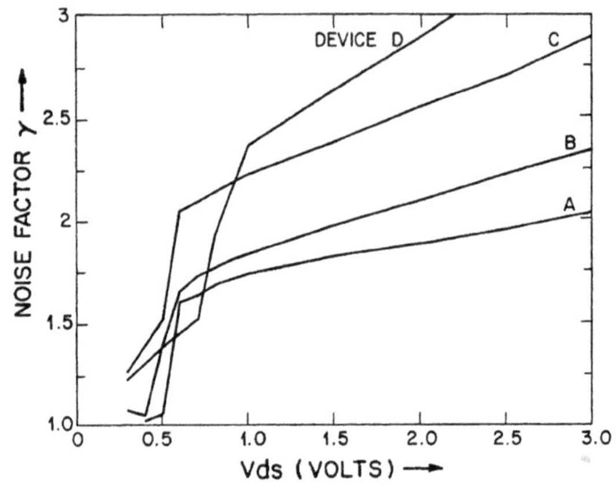

Fig. 5.4 Noise factor γ versus drain-to-source voltage at different gate-to-source voltages for devices A, B, C, and D. (Reprinted from Jindal, R.P. [25])

down to several hundred nanometers were investigated. From standard theory presented in Sect. 5.3.1, the expectation was for the channel thermal noise, the noise parameter γ would have value of unity and at $V_{DS} = 0$ and monotonically decrease to a value 2/3rd as the device approached saturation. However, Jindal [8] found that γ actually increased steadily with increasing V_{DS}. This was the first *room-temperature* experimental observation of the exacerbation of channel noise. These results were later published in the open literature [25, 77, 78]. The original figure is reproduced here as Fig. 5.4. For device A with $L_{eff} = 1250$ nm and device B with $L_{eff} = 750$ nm the initial expected dip in γ is there but barely perceptible. For device C with $L_{eff} = 500$ nm and device D with $L_{eff} = 250$ nm the dip is completely masked. Here L_{eff} denotes the metallurgical channel length as opposed to the electrical channel length taking CLM [30] into account. The initial rise instead of a decline in γ as a function of V_{DS} is believed to be due to carriers getting hotter [78]. As the device enters saturation this increase slows down pointing perhaps to a change in the noise mechanism. Since the publication of these results in 1985, this topic has been a focus of intense research activity in the global device community [36, 55, 79–113] over the last twenty-six years. Equally numerous papers have appeared from the circuit community as well. The resulting γ values [25, 36, 44, 79, 85, 91, 97, 107, 108] have been summarized in Fig. 5.5. Most experimental papers have largely confirmed the earlier findings [25, 77, 78] with one exception [79] in which case the observed values are significantly higher.

Researchers have attempted to develop a physical understanding of this excess noise over the last twenty-six years. However, convergence has yet to emerge. The research efforts can be classified into the "No excess noise group", "thermal noise" group and the "shot noise" group. We will next discuss each of them.

5.5.3 No Excess Noise School of Thought

One group of researchers [15, 19, 44, 109] claim to have successfully explained their measured values of the noise parameter γ for MOSFETs with gate-length down to

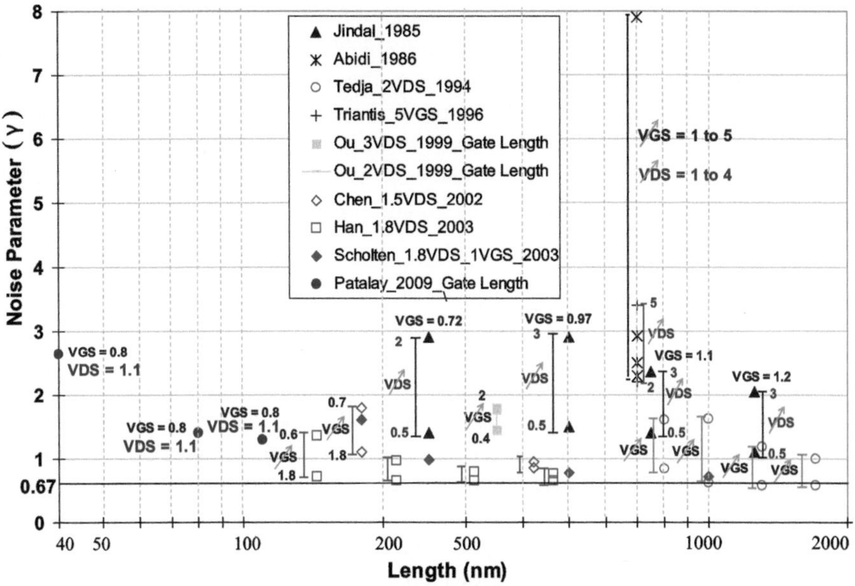

Fig. 5.5 Noise parameter γ versus channel length. (Reprinted from R.P. Jindal [111])

80 nm without having to invoke excess noise mechanism discussed in this section. However, they do observe an unexplained underestimation of noise by 20% at the shortest channel lengths.

5.5.4 Shot Noise School Of Thought

In the literature three distinct approaches have been suggested to explain excess noise observed in the FET channel to have a shot noise origin. In strong inversion, current flow in a MOSFET is primarily due to carrier drift which does not exhibit shot noise. However, it has been suggested by Obrecht et al. [92, 98] that a significant component of diffusion current exists in the FET channel near the source end of the device. They ascribe this excess channel noise to shot noise associated with carrier diffusion. An experimental verification of this theory is also claimed by Andersson and Svensson [99].

A second suggestion to explain excess channel noise arising from shot noise was proposed by Navid and Dutton [100]. They ascribe the existence of shot noise to non-equilibrium transport in the FET channel due to tens of collisions of the carriers as they traverse the distance from the source to the drain. Their theory is able to explain the earliest excess noise results reported by Jindal [25] in 1985 for 750 nm channel length devices. However, this theory was not tested against the 500 nm and 250 nm device published results [25] at the same time.

A third suggestion for the injection of shot noise in the FET channel was made by Sirohi [101] and Navid [102] where the suggested shot noise is generated due to the emission of carriers over the barrier as the they are injected from the source into the FET channel. Sirohi's semi-empirical theory is able to explain the noise behavior of MOS devices down to 250 nm [25]. Recently Devulapalli [103] has further developed this idea and presented a detailed analytical model explaining the excess noise down into the sub-100 nanometer regime.

5.5.5 Hot Carrier School of Thought

The argument in favor of hot carrier noise starts with the fact that as we decrease the device channel length, the lateral electric field in the channel increases with attendant increase in (i) the average carrier energy, (ii) a reduction in carrier mobility and (iii) a reduction in channel length due to the pinch off point moving closer to the source. The last effect is referred to as channel-length modulation (CLM) [30]. Let us go back to the experimental evidence. By making direct noise measurements on low-resistivity n-silicon layers grown epitaxially on high-resistivity silicon substrates Baechtold [75] has verified that indeed the noise does increase signifying an increase in the carrier temperature. A number of researchers [36, 85, 89] and [95] have developed theories to support this "hot-carrier" interpretation. However, as the carriers heat up, they also experience a reduction in their mobility. This reduction in mobility makes them less responsive to the fluctuating electric field produced by thermal noise voltage fluctuations in the FET channel. Hence, a reduction in carrier mobility implies a reduction in the fluctuating noise current. In the limit when the carrier velocity saturates, this noise must vanish. Thus, as pointed out by Chen and Deen [97] the pinched-off portion of the channel cannot contribute to the channel thermal noise rendering hot-carrier models questionable. Recently, using the transmission line analysis, Vallur and Jindal [22], have examined the effect of both carrier heating and mobility reduction on channel thermal noise. They have shown analytically that in the un-pinched off portion of the channel, for a specific dependence of the carrier temperature on electric field as described by (5.18) and mobility on electric field as described by (5.19) the two effects mask each other out. The channel noise is then given by

$$\overline{i_d i_d^*}_{NQS} = 4k_B T g_{d0}\left\{\frac{2}{3} + \frac{76}{4725}\left[\frac{\omega(1+E/E_C)}{\omega_T}\right]^2\right\}\Delta f. \tag{5.20}$$

This fortuitous cancellation of the two effects has perhaps led to significant confusion in the literature.

The effect of channel-length modulation can be explained as follows. CLM essentially produces an MOS device with effectively a larger g_{d0} implying a higher noise current. To make (5.20) consistent with channel-length modulation, g_{d0} should be interpreted as the drain-to-source conductance at zero V_{DS} for a device with channel length equal to electrical channel length as determined by CLM. Several models [44, 97, 106, 112] have been proposed taking the above effects into

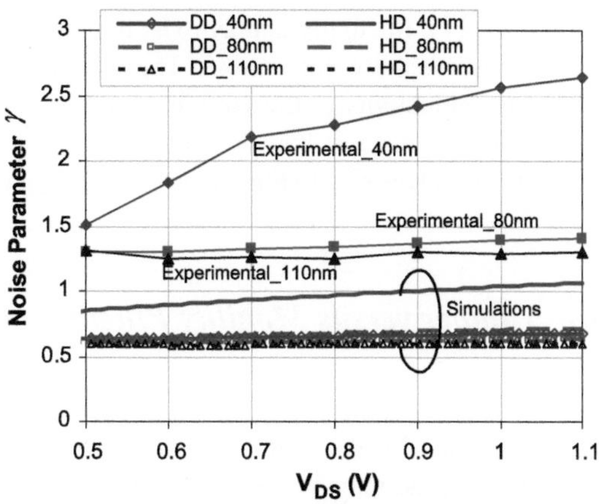

Fig. 5.6 Comparison of experimental and simulated drain noise parameter γ results for devices with gate lengths 40 nm, 80 nm, and 110 nm vs. V_{DS} at $V_{GS} = 0.8$ V (for experiments) and $V_{GS} = 0.7$ V (for simulations). (Reprinted from V.M. Mahajan et al. [113])

account albeit with different details each claiming a match between theory and experiment. We will not reproduce these equations here. The reader is referred to original publications for sometimes conflicting [110] details. Recent experimental results by Patalay et al. [108] have clearly demonstrated that at least for MOSFETs with channel lengths in the sub-100 nanometer regime the above theories cannot explain the measured noise.

Mahajan et al. have recently published [113] physics-based numerical investigation of channel noise using drift-diffusion (DD) and hydrodynamic (HD) transport models incorporating velocity saturation with and without carrier heating. Their results are reproduced in Fig. 5.6. The noise parameter γ is plotted as a function of the drain-to-source voltage with gate voltage as a parameter. The top three curves are the experimental data [108] for the 40 nm, 80 nm and 110 nm gate length devices. The simulation curves lie clustered at the bottom of the graph close to the ideal value. The only exception is the hydrodynamic simulation results for the 40 nm device shown by the solid line at the top of the simulation cluster. These results clearly show that in the presence of mobility degradation neither channel-length modulation nor carrier heating is sufficient to explain the measured noise. Thus, we must look for an alternate noise mechanism to explain this excess noise.

As the device enters the quasi-ballistic regime of carrier transport in the FET channel this excess channel noise will continue to morph itself and be an active area of research.

5.6 $1/f$ Noise

Of all the noise mechanisms know to us the one that has been perhaps most widely researched is $1/f$ noise. For the case of a MOSFET competing theories have survived over decades without a clear resolution. In the following sections we will explain this in greater detail. However, one must realize that $1/f$ noise is a non-equilibrium phenomenon. In other words, in any dissipative systems under equilibrium Nyquist theory [10] holds precluding the observation of $1/f$ noise across its terminals. $1/f$ fluctuations may exist but they are not revealed. One must pass a current to observe these fluctuations. These fluctuations therefore amount to fluctuations in the conductivity of the sample. However, since the conductivity is a product of the number of carriers and their mobility, the question whether it is number or mobility which gives rise to this disturbance is not easily answered.

5.6.1 Number versus Mobility Fluctuations Debate

The earliest explanation of $1/f$ noise goes back to 1950 given by van der Ziel [114], Du Pre [115] and McWhorter [116] due to trapping and de-trapping of carriers at the oxide-semiconductor interface of a MOSFET channel. Each trap is associated with a time constant which gives rise to a generation-recombination (g-r) Lorentzian spectrum. A suitably weighted superposition of Lorentzians is invoked to give rise to a $1/f$ spectrum. Therefore, $1/f$ noise was believed to be a surface-effect.

In 1969, Hooge published experimental data [117] showing that $1/f$ fluctuations were related to the number of carriers in the bulk of the sample thus claiming that it was a bulk effect. Since there are very few traps in the bulk there is no mechanism to produce fluctuations in the number of carriers in the sample. Hence it was suggested that the source of this noise is tied to mobility fluctuations. This started a multi-decade mobility versus number fluctuation debate among researchers as the source of $1/f$ noise in MOSFETs. It is interesting to note that while a well defined model existed for the number fluctuation theory, the mobility fluctuation idea did not have a physical model to base it on. The earliest models for bulk mobility fluctuations $1/f$ noise include the quantum $1/f$ noise by Handel [118] and phonon fluctuation model by Jindal and van der Ziel [119]. Since then there has been a healthy debate on how to model these mobility fluctuations as well.

5.6.2 Current Status

The first attempt to combine the number and mobility fluctuation models was made by Jayaraman and Sodini [121]. They suggested that due to Coulombic interaction the fluctuations in the trap occupancy at the oxide-semiconductor interface also produced correlated mobility fluctuations due to surface scattering. However, this still did not produce the desired results for p-channel devices. A possible resolution to this debate was presented by Ralls et al. [120]. By working with MOSFETs with ultra-small dimensions they were able to observer the random telegraph signal due to single electron trapping and de-trapping events with a characteristic time constant. For larger devices, collectively these fluctuations would give rise to a $1/f$ spectrum. Currently there is consensus that number fluctuations are the source of $1/f$ noise in n-channel MOS devices. For p-channel devices the situation is still not entirely clear. In the past, for high frequency applications $1/f$ noise was not an important consideration unless the circuit operation involved mixing. However, as channel lengths shrank, $1/f$ noise in MOS devices continued to increase [122] and needs to be taken into account.

From the compact modeling perspective, $1/f$ noise in MOSFETs has been simply incorporated in most commercial models and is given by the following equation [65]

$$\overline{i_n^2} = \frac{K}{f} \frac{g_m^2}{WLC_{ox}^2} \Delta f. \tag{5.21}$$

Here K is a technology dependent constant, W and L are the gate dimensions and C_{ox} is gate capacitance per unit area, f is the frequency and Δf is the bandwidth of observation.

5.7 Noise Capabilities of Compact MOS Models

Having developed a comprehensive understanding of the noise behavior of bulk MOSFET, it is instructive to examine the degree of penetration of this body of knowledge in the MOS compact models used in circuit simulators. At the present time there are four dominant compact MOS models that are being used in the world. These include the BSIM, EKV, HiSim and PSP. In Table 5.1 we list the status of implementation of various noise sources for the latest version of each of these models. This table is based on an informal exchange of information between the model developers and the author. No attempt has been made to benchmark these capabilities. The reader must note that it in some cases it is difficult to accurately capture the capabilities of a model by a simple "yes" or "no". Hence this table is to be viewed as a very high-level summary. Also a "no" entry in the table does not imply that one cannot simulate the effect due to that noise source at all. It is often possible to introduce noise by adding extra dissipative elements in the current path.

The large number of "No" entries in the table begs the question why? The first point to realize is that a model incorporating all these effects will be computationally less efficient than one which only has a subset of these effects built in. Often model developers focus on a specific customer set and tailor their model capabilities to the perceived current needs of the customer while at the same time keeping the model computationally efficient. Thus, one set of users may find these model capabilities entirely adequate while another set of users with different application might look for more.

While several models predict channel noise in the sub-threshold region of operation of the MOSFET a smooth transition from channel thermal noise to shot noise as the device bias is changed from above threshold to below threshold is an excellent indicator of the quality of the model.

Table 5.1 Summary of noise sources implemented in compact MOSFET models

NOISE SOURCE	BSIM 4.5	EKV 301.02	HiSIM 250	PSP
Channel thermal noise: Above threshold, QS including $V_{DS} = 0$	Yes	Yes	Yes	Yes
Channel thermal noise: Above threshold, NQS including $V_{DS} = 0$	No	No	No	No
Channel thermal noise: Sub-threshold, QS including $V_{DS} = 0$	No	Yes	Yes	Yes
Induced gate noise: Above threshold (QS charge estimation)	Yes	Yes	Yes	Yes
Induced gate noise: Above threshold (NQS charge estimation)	No	No	No	[a]
Induced gate noise at $V_{DS} = 0$	No	Yes	Yes	Yes
Correlation between induced gate noise and channel thermal noise (QS charge estimation)	Yes	Yes	Yes	Yes
Correlation between induced gate noise and channel thermal noise (NQS charge estimation)	No	No	No	No
Induced substrate noise: Above threshold (QS charge estimation)	No	Yes	No	No
Induced substrate noise: Above threshold (NQS charge estimation)	No	No	No	No
Correlation between induced substrate noise and channel thermal noise (QS charge estimation)	No	Yes	No	No
Correlation between induced substrate noise and channel thermal noise (NQS charge estimation)	No	No	No	No
Distributed gate resistance noise	No	Yes	No	No
Distributed substrate resistance noise	No	Yes	No	No
Substrate current super-shot noise	No	No	No	Yes
Flicker ($1/f$) noise	Yes	Yes	Yes	Yes
Drain/Source resistance noise	Yes	Yes	No	Yes
Drain/Source junction shot noise	No	No	Yes	Yes
Gate leakage current shot noise	Yes	Yes	Yes	Yes

[a]Frequency dependence needs validation

5.8 Conclusions

MOS technology continues to evolve in response to demand from a myriad of applications that impact our daily lives. The demand for a better understanding of noise has largely come from lightwave and wireless terminal and hand-held applications. Over the last 45 years, in addition to the intrinsic noise mechanisms several extrinsic noise sources have been found, understood and eliminated. As a result of this boost, MOS has emerged as the technology of choice with unprecedented levels of integration. As MOS technology marches into the nanometer regime challenges in terms of understanding the excess channel noise and $1/f$ noise continue to exist.

Acknowledgments This work is dedicated to Aldert van der Ziel whose fond memory continues to inspire and energize the activities of author's research group. The author would also like to acknowledge support received as William Hansen Hall Louisiana Board of Regents Eminent Scholar Endowed Chair and support from Semiconductor Research Corporation via contract 2007-VJ-1601. Interactions under NSF EPSCoR Award # EPS-0701491 are gratefully acknowledged.

References

1. Lilienfeld, J.E.: U.S. Patent 1,745,175. Filed 8 October 1926 (22 October 1925 in Canada). Issued 28 January 1930
2. Warner, R.M.: Microelectronics: its unusual origin and personality. IEEE Trans. Electron Devices **48**(11), 2457–2467 (2001)
3. Kahng, D., Atalla, M.M.: Silicon-silicon dioxide field induced surface devices. In: Proc. IRE Solid-State Devices Res. Conf., Pittsburgh, PA, Carnegie Inst. Technol. (1960)
4. van der Ziel, A.: Thermal noise in field effect transistors. Proc. IRE **50**(8), 1808–1812 (1962)
5. van der Ziel, A.: Gate noise in field effect transistors at moderately high frequencies. Proc. IEEE **51**(3), 461–467 (1963)
6. Ogawa, K.: Internal technical memorandum. Bell Laboratories, Holmdel, NJ (1980)
7. Fraser, D.L. Jr., Williams, G.F., Jindal, R.P., Kushner, R.A., Owen, B.: A single chip NMOS preamplifier for optical fiber receivers. In: IEEE Int. Solid-State Circuits Conf., vol. 26, pp. 80–81, February 1983
8. Jindal, R.P.: Internal technical memorandum. AT&T Bell Laboratories, Murray Hill, NJ (1984)
9. Jindal, R.P.: Compact noise models for MOSFETs. Invited paper in the special issue on Advanced Compact Models and 45-nm Modeling Challenges. IEEE Trans. Electron Devices **53**(9), 2051–2061 (2006)
10. Nyquist, H.: Thermal agitation of electric charge in conductors. Phys. Rev. **32**(1), 110–113 (1928)
11. Sah, C.T.: Theory and experiments on the $1/f$ surface noise of MOS insulated-gate field-effect transistors. IEEE Trans. Electron Devices **11**, 534 (1964)
12. Jordan, A.G., Jordan, N.A.: Theory of noise in metal oxide semiconductor devices. IEEE Trans. Electron Devices **12**(3), 148–156 (1965)
13. Sah, C.T., Wu, S.Y., Hielschler, F.H.: The effects of fixed bulk charge on the thermal noise in metal-oxide-semiconductor transistors. IEEE Trans. Electron Devices **13**(4), 410–414 (1966)
14. Klaassen, F.M., Prins, J.: Thermal noise in MOS transistors. Philips Res. Rep. **22**, 504–514 (1967)
15. Paasschens, J.C.J., Scholten, A.J., van Langevelde, R.: Generalizations of the Klaassen–Prins equation for calculating the noise of semiconductor devices. IEEE Trans. Electron Devices **52**(11), 2463–2472 (2005)
16. Deshpande, A., Jindal, R.P.: Modeling non-quasi-static effects in thermal noise and induced gate noise in MOS field effect transistors. Solid State Electron. **52**(5), 771–774 (2008)
17. Shoji, M.: Analysis of high frequency thermal noise of enhancement mode MOS field-effect transistors. IEEE Trans. Electron Devices **13**(6), 520–524 (1966)
18. van der Ziel, A.: Noise in Solid-State Devices and Circuits. Wiley, New York (1986)
19. van Langevelde, R., Paasschens, J.C.J., Scholten, A.J., Havens, R.J., Tiemeijer, L.F., Klaassen, D.B.M.: New compact model for induced gate current noise. In: IEDM Tech. Dig., pp. 36.2.1–36.2.4 (2003)

20. Jungemann, C., Neinhus, B., Nguyen, C.D., Meinerzhagen, B., Dutton, R.W., Scholten, A.J., Tiemeijer, L.F.: Hydrodynamic modeling of RF noise in CMOS devices. In: IEDM Tech. Dig., pp. 36.3.1–36.3.4 (2003)
21. Patel, A.: Design optimization of MOS amplifiers for high frequency ultra-low-noise wireless communication applications. Master's Thesis, Univ. of Louisiana at Lafayette, Lafayette (2007)
22. Vallur, S., Jindal, R.P.: Modeling short-channel effects in channel thermal noise and induced-gate noise in MOSFETs in the NQS regime. Solid State Electron. **53**(1), 36–41 (2009)
23. Jindal, R.P.: Effect of induced gate noise at zero drain bias in field-effect transistors. IEEE Trans. Electron Devices **52**(3), 432–434 (2005)
24. Pu, L.-J., Tsividis, Y.: Small-signal parameters and thermal noise of the four-terminal MOSFET in non-quasi-static operation. Solid-State Electron. **33**(5), 513–521 (1990)
25. Jindal, R.P.: High frequency noise in fine line NMOS field effect transistors. In: IEDM Tech. Dig., pp. 68–71 (1985) (invited paper)
26. Jindal, R.P.: Transimpedance preamplifier with 70 dB AGC range in fine line NMOS. IEEE J. Solid State Circuits **23**(2), 867–869 (1988)
27. Takagi, K., van der Ziel, A.: Drain noise in MOSFETs at zero drain bias as a function of temperature. Solid-State Electron **22**, 87–88 (1979)
28. Vadyala, R.: Experiments on the excess thermal noise in MOSFETs at zero drain bias. Master's Thesis, Univ. of Louisiana at Lafayette, Lafayette (2007)
29. Jindal, R.P., van der Ziel, A.: Effect of transverse electric field on Nyquist noise. Solid State Electron. **24**(10), 905–906 (1981)
30. Tsividis, Y.: Operation and Modeling of the MOS Transistor, 2nd edn. McGraw Hill, New York (1999)
31. Thornber, K.K.: Resistive-gate-induced thermal noise in IGFET's. IEEE J. Solid-State Circuits **16**(4), 414–415 (1981)
32. Jindal, R.P.: Noise associated with distributed resistance of MOSFET gate structures in integrated circuits. IEEE Trans. Electron Devices **31**(10), 1505–1509 (1984)
33. Shuang, X., Conn, D.R.: A low-noise gate structure for DMOS monolithic devices. IEEE Trans. Electron Devices **36**(7), 1393–1396 (1989)
34. Manku, T.: Microwave CMOS—Device physics and design. IEEE J. Solid-State Circuits **34**(3), 277–285 (1999)
35. Tedja, S., Williams, H.H., van der Spiegel, J., Newcomer, F.M., Vanberg, G.R.: Noise spectral-density measurements of a radiation hardened CMOS process in the weak and moderate inversion. IEEE Trans. Nucl. Sci. **39**(4), 804–808 (1992)
36. Tedja, S., van der Speigel, J., Williams, H.H.: Analytical and experimental studies of thermal noise in MOSFETs. IEEE Trans. Electron Devices **41**(11), 2069–2075 (1994)
37. Anelli, G., Faccio, F., Florian, S., Jarron, P.: Noise characterization of a 0.25 μm CMOS technology for the LHC experiments. Nucl. Instrum. Methods Phys. Res., Sect. A, Accel. Spectrom. Detect. Assoc. Equip. **457**, 361–368 (2001)
38. Re, V.I., Bietti, I., Castello, R., Manghisoni, M., Speziali, V., Svelto, F.: Experimental study and modeling of the white noise sources in submicron p- and n-MOSFETs. IEEE Trans. Nucl. Sci. **48**(4), 1577–1586 (2001)
39. Abou-Allam, E., Manku, T.: A small-signal MOSFET model for radio frequency IC applications. IEEE Trans. Comput.-Aided Des. Integr. Circuits Syst. **16**(5), 437–447 (1997)
40. Abou-Allam, E., Manku, T.: An improved transmission-line model for MOS transistors. IEEE Trans. Circuits Syst. II, Analog Digit. Signal Process. **46**(11), 1380–1387 (1999)
41. Tin, S.F., Osman, A.A., Mayaram, K., Hu, C.: A simple sub-circuit extension of the BSIM3v3 model for CMOS RF design. IEEE J. Solid-State Circuits **35**(4), 612–624 (2000)
42. Tsakas, E.F., Birbas, A.N.: Noise associated with interdigitated gate structures in RF submicron MOSFETs. IEEE Trans. Electron Devices **47**(9), 1745–1750 (2000)
43. Goo, J.-S., Choi, C.-H., Danneville, F., Morifuji, E., Momose, H.S., Yu, Z., Iwai, H., Lee, T.H., Dutton, R.W.: An accurate and efficient high frequency noise simulation technique for deep submicron MOSFETs. IEEE Trans. Electron Devices **47**(12), 2410–2419 (2000)

44. Scholten, A.J., Tiemeijer, L.F., van Langevelde, R., Havens, R.J., Zegers-van Duijnhoven, A.T.A., Venezia, V.C.: Noise modelling for RF CMOS circuit simulation. IEEE Trans. Electron Devices **50**(3), 618–632 (2003)
45. Steyaert, M., Chang, Z.: Low voltage BIMOS AM front-end amplifier. In: IEE Proc.-G Circuits Devices and Systems, vol. 137, pp. 57–60, February 1990
46. Chang, Z., Sansen, W.M.C.: Low-noise, low-distortion CMOS AM wide-band amplifiers matching a capacitive source. IEEE J. Solid-State Circuits **25**, 833–840 (1990)
47. Shaeffer, D.K., Lee, T.H.: A 1.5-V, 1.5-GHz CMOS low noise amplifier. IEEE J. Solid-State Circuits **32**(5), 745–759 (1997)
48. Zhou, J.J., Allstot, D.J.: Addition to Monolithic transformers and their application in a differential CMOS RF low-noise amplifier. IEEE J. Solid-State Circuits **34**, 1176 (1999)
49. Christoforou, Y., Rossetto, O.: GaAs preamplifier and LED driver for use in cryogenic and highly irradiated environments. Nucl. Instrum. Methods Phys. Res., Sect. A, Accel. Spectrom. Detect. Assoc. Equip. **425**(1), 347–356 (1999)
50. Terrovitis, M.T., Meyer, R.G.: Noise in current-commutating CMOS mixers. IEEE J. Solid-State Circuits **34**, 772–783 (1999)
51. Tsakas, E.F., Birbas, A.: Noise optimization for the design of a reliable high speed X-ray readout integrated circuit. Microelectron. Reliab. **40**(11), 1937–1942 (2000)
52. Tsakas, E.F., Birbas, A.N., Manthos, N., Kloukinas, K., Evangelou, I., Triantis, F.A., Van der Stelt, P.F., Speller, R.D.: Low noise high-speed X-ray readout IC for imaging applications. Nucl. Instrum. Methods Phys. Res., Sect. A, Accel. Spectrom. Detect. Assoc. Equip. **469**(1), 106–115 (2001)
53. Maschera, D., Simoni, A., Gottardi, M., Gonzo, L., Gregori, S., Liberali, V., Torelli, G.: An automatically compensated readout channel for rotary encoder system. IEEE Trans. Instrum. Meas. **50**(6), 1801–1807 (2001)
54. Goo, J.-S., Ahn, H.-T., Ladwig, D.J., Yu, Z., Lee, T.H., Dutton, R.W.: A noise optimization technique for integrated low-noise amplifiers. IEEE J. Solid-State Circuits **37**(8), 994–1002 (2002)
55. Scholten, A.J., Tiemeijer, L.F., van Langevelde, R., Havens, R.J., Zegers-van Duijnhoven, A.T.A., De Kort, R., Klaassen, D.B.M.: Compact modelling of noise for RF CMOS circuit design. IEE Proc.—Circuits Devices Syst. **151**(2), 167–174 (2004)
56. Jindal, R.P.: Distributed substrate resistance noise in fine line NMOS field effect transistors. IEEE Trans. Electron Devices **32**(11), 2450–2453 (1985)
57. Jindal, R.P.: A physical understanding of the noise performance of MOS transistors for wireless and lightwave applications in the giga-bit regime. In: 3rd SPIE Int. Symp. Fluctuation and Noise, Austin, TX, pp. 10–22, May 2005
58. Chang, C.Y., Su, J.G., Wong, S.C., Huang, T.Y., Sun, Y.C.: RF CMOS technology for MMIC. Microelectron. Reliab. **42**(4-5), 721–733 (2002)
59. Re, V., Svelto, F.: High accuracy measurement of low-frequency noise in front-end P-channel FETs. Nucl. Phys. B, Proc. Suppl. **44**, 607–612 (1995)
60. Kim, C.S., Park, J.-W., Yu, H.K., Cho, H.: Gate layout and bonding pad structure of a RF n-MOSFET for low noise performance. IEEE Electron Device Lett. **21**(12), 607–609 (2000)
61. Binkley, D.M., Rochelle, J.M., Paulus, M.J., Casey, M.E.: A low-noise, wide-band, integrated CMOS transimpedance preamplifier for photodiode applications. IEEE Trans. Nucl. Sci. **39**, 747–752 (1992)
62. Shaeffer, D.K., Shahani, A.R., Mohan, S.S., Samavati, H., Rategh, H.R., Hershenson, M.M., Xu, M., Yue, C.P., Eddleman, D.J., Lee, T.H.: A 115-mW, 0.5-μm CMOS GPS receiver with wide dynamic-range active filters. IEEE J. Solid-State Circuits **33**(12), 2219–2231 (1998)
63. Beuville, E., Borer, K., Chesi, E., Heijne, E.H.M.: Amplex, a low-noise, low-power analog CMOS signal processor for multi-element silicon particle detectors. Nucl. Instrum. Methods Phys. Res., Sect. A, Accel. Spectrom. Detect. Assoc. Equip. **288**(1), 157–167 (1990)
64. Gottardi, M., Gonzo, L., Gregori, S., Liberali, V., Simony, A., Torelli, G.: An integrated CMOS front-end for optical absolute rotary encoders. Analog Integr. Circuits Signal Process. **34**(2), 143–154 (2003)

65. Lee, T.H.: The Design of CMOS Radio-Frequency Integrated Circuits. Cambridge Univ. Press, Cambridge (1998)
66. Kim, C.S.: Avalanche multiplication and related noise in silicon MOSFETs. Ph.D. dissertation, Univ. of Florida (1971)
67. Rucker, L.M., van der Ziel, A.: Noise associated with JFET gate current resulting from avalanching in the channel. Solid-State Electron. **21**, 798–99 (1978)
68. Jindal, R.P.: Noise associated with substrate current in fine line NMOS field effect transistors. Presented at the 42nd Device Research Conference, pp. 18–20 (1984); see IEEE Trans. Electron Devices **31**(12), 1971 (1984)
69. Jindal, R.P.: Noise associated with substrate current in fine line NMOS field effect transistors. IEEE Trans. Electron Devices **32**(6), 1047–1052 (1985)
70. Jin, W., Chan, P.C.H., Fung, S.K.H., Ko, P.K.: Shot-noise-induced excess low-frequency noise in floating-body partially depleted SOI MOSFETs. IEEE Trans. Electron Devices **46**(6), 1180–1185 (1999)
71. Workman, G.O., Fossum, J.G.: Physical noise modeling of SOI MOSFETs with analysis of the Lorentzian component in the low-frequency noise spectrum. IEEE Trans. Electron Devices **47**(6), 1192–1201 (2000)
72. Caughey, D.M., Thomas, R.E.: Carrier mobilities in silicon empirically related to doping and field. Proc. IEEE **55**(12), 2192–2193 (1967)
73. Lundstrom, M.: Fundamentals of Carrier Transport, 2nd edn. Cambridge University Press, Cambridge (2000)
74. Klaassen, F.M.: On the influence of hot carrier effects on the thermal noise of field-effect transistors. IEEE Trans. Electron Devices **17**(10), 858–862 (1970)
75. Baechtold, W.: Noise temperature in silicon in the hot electron region. IEEE Trans. Electron Devices **18**(2), 1186–1187 (1971)
76. Takagi, K., Matsumoto, K.: Noise in Silicon and FET's at high electric fields. Solid-State Electron. **20**, 1–3 (1977)
77. Jindal, R.P.: Noise phenomena in submicron channel length silicon NMOS transistors. In: 8th Int. Conf. Noise Phys. Syst. 4th Int. Conf. $1/f$ Noise Conf., Rome, Italy, pp. 199–202 (1985)
78. Jindal, R.P.: Hot electron effects on channel thermal noise in fine line NMOS field effect transistors. IEEE Trans. Electron Devices **33**(9), 1395–1397 (1986)
79. Abidi, A.A.: High-frequency noise measurements on FET's with small dimensions. IEEE Trans. Electron Devices **33**(11), 1801–1805 (1986)
80. Sheu, B.J., Scharfetter, D.L., Ko, P.K., Jeng, M.C.: BSIM: Berkeley short-channel IGFET model for MOS transistors. IEEE J. Solid-State Circuits **22**(4), 558–566 (1987)
81. Cappy, A., Wolfgang, H.: High-frequency FET noise performance: A new approach. IEEE Trans. Electron Devices **36**(2), 403–409 (1989)
82. Danneville, F.: Microscopic noise modeling and macroscopic noise models: How good a connection? IEEE Trans. Electron Devices **41**(5), 779–786 (1994)
83. Wang, B., Hellums, J.R., Sodini, C.D.: MOSFET thermal noise modeling for analog integrated circuits. IEEE J. Solid-State Circuits **29**(7), 833–835 (1994)
84. Yasuhisa, O.: An improved analytical solution of energy balance equation for short-channel SOI MOSFET's and transverse-field-induced carrier heating. IEEE Trans. Electron Devices **42**(2), 301–306 (1995)
85. Triantis, D.P., Birbas, A.N., Kondis, D.: Thermal noise modeling for short-channel MOSFETs. IEEE Trans. Electron Devices **43**(11), 1950–1955 (1996)
86. Triantis, D.P., Birbas, A.N.: Optimal current for minimum thermal noise operation of submicrometer MOS transistors. IEEE Trans. Electron Devices **44**(11), 1990–1995 (1997)
87. Bonani, F., Ghione, G., Pinto, M.R., Smith, R.K.: An efficient approach to noise analysis through multidimensional physics-based models. IEEE Trans. Electron Devices **45**(1), 261–269 (1998)
88. Svelto, F.: Noise analysis of submicron PMOS in NWELL. Nucl. Phys. B **61**, 539–544 (1998)

89. Klein, P.: An analytical thermal noise model of deep submicron MOSFETs. IEEE Electron Device Lett. **20**(8), 399–401 (1999)
90. Scholten, A.J., Tromp, H.J., Tiemeijer, L.F., van Langevelde, R., Havens, R.J., de Vreede, P.W.H., Roes, R.F.M., Woerlee, P.H., Montree, A.H., Klaassen, D.B.M.: Accurate thermal noise model for deep-submicron CMOS. In: IEDM Tech. Dig., pp. 155–158 (1999)
91. Ou, J.J., Jin, X., Hu, C., Gray, P.R.: Submicron CMOS thermal noise modeling from an RF perspective. In: Symp. VLSI Tech. Dig., pp. 151–152 (1999)
92. Obrecht, M.S., Manku, T., Elmasry, M.I.: Simulation of temperature dependence of microwave noise in metal-oxide-semiconductor-field-effect transistors. Jpn. J. Appl. Phys. **39**, 1690–1693 (2000)
93. Jin, W., Chan, P.C.H., Lau, J.: A physical thermal noise model for SOI MOSFET. IEEE Trans. Electron Devices **47**(4), 768–773 (2000)
94. Goo, J.-S., Choi, C.-H., Abramo, A., Ahn, J.-G., Yu, Z., Lee, T.H., Dutton, R.W.: Physical origin of the excess thermal noise in short channel MOSFETs. IEEE Electron Device Lett. **22**(2), 101–103 (2001)
95. Knoblinger, G., Klein, P., Tiebout, M.: A new model for thermal channel noise of deep-submicron MOSFETs and its application in RF-CMOS design. IEEE J. Solid-State Circuits **36**(5), 831–837 (2001)
96. Rengel, R., Mateos, J., Pardo, D., Gonzalez, T., Martin, M.J.: Monte Carlo analysis of dynamic and noise performance of submicron MOSFETs at RF and microwave frequencies. Semicond. Sci. Technol. **16**(11), 939–946 (2001)
97. Chen, C.-H., Deen, M.J.: Channel noise modeling of deep submicron MOSFETs. IEEE Trans. Electron Devices **49**(8), 1484–1487 (2002)
98. Obrecht, M.S., Abou-Allam, E., Manku, T.: Diffusion current and its effect on noise in submicron MOSFETs. IEEE Trans. Electron Devices **49**(3), 524–526 (2002)
99. Andersson, S., Svensson, C.: Direct experimental verification of shot noise in short channel MOS transistors. Electron. Lett. **41**(15), 869–871 (2005)
100. Navid, R., Dutton, R.W.: The physical phenomena responsible for excess noise in short-channel MOS devices. In: Proc. Int. Conf. Simul. of Semicond. Processes and Devices, Kobe, Japan, pp. 75–78 (2002)
101. Sirohi, S.: Experiments and modeling of high-frequency noise in deep submicron MOSFETs. Master's Thesis, Univ. of Louisiana at Lafayette, Lafayette (2005)
102. Navid, R.: Amplitude and phase noise in modern CMOS circuits. Ph.D. Thesis Stanford Univ., Stanford (2005)
103. Devulapalli, V.: Modeling drain-current noise in short-channel MOS field-effect transistors. Master's Thesis, Univ. of Louisiana at Lafayette, Lafayette (2008)
104. Jungemann, C.: Hydrodynamic modeling of RF noise in CMOS devices. In: IEDM Tech. Dig., pp. 871–874 (2003)
105. Teng, H.F., Jang, S.L.: A non-local channel thermal noise model for nMOSFETs. Solid-State Electron. **47**(5), 815–819 (2003)
106. Han, K., Shin, H., Lee, K.: Analytical drain thermal noise current model valid for deep submicron MOSFETs. IEEE Trans. Electron Devices **51**(2), 261–269 (2004)
107. Han, K., Lee, K., Shin, H.: Thermal noise modeling for short-channel MOSFETs. In: Int. Conf. SISPAD, pp. 79–82 (2003)
108. Patalay, P.R., Jindal, R.P., Shichijo, H., Martin, S., Hou, F.-C., Trombley, D.: High-frequency noise measurements on MOSFETs with channel-lengths in sub-100 nm regime. In: Proc. 2nd IEEE Int. Workshop on Electron. Devices and Semicond. Tech., Mumbai, India (2009)
109. Scholten, A.J., Tiemeijer, L.F., Zegers-van Duijnhoven, A.T.A., Havens, R.J., de Kort, R., van Langevelde, R., Klaassen, D.B.M., Jeamsaksri, W., Velghe, R.M.D.S.: Modeling and characterization of noise in 90-nm RF CMOS technology. In: González, T., Mateos, J., Pardo, D. (eds.) Proc. 18th ICNF, pp. 735–740 (2005)
110. Roy, A.S., Enz, C.C.: Compact modeling of thermal noise in the MOS transistor. IEEE Trans. Electron Devices **52**(4), 611–614 (2005)

111. Jindal, R.P.: From millibits to terabits and beyond—over 60 years of innovation. Invited paper. In: Proc. 2nd IEEE Int. Workshop on Elect. Devices and Semicond. Tech., Mumbai, India (2009)
112. Asgaran, A.S., Deen, M.J., Chen, C.-H.: Analytical modeling of MOSFETs channel noise and noise parameters. IEEE Trans. Electron Devices **51**(12), 2109–2114 (2004)
113. Mahajan, V.M., Jindal, R.P., Shichijo, H., Martin, S., Hou, F.-C., Trombley, D.: Numerical investigation of excess RF channel noise in sub-100 nm MOSFETs. In: Proc. 2nd IEEE Int. Workshop on Elect. Devices and Semicond. Tech., Mumbai, India (2009)
114. van der Ziel, A.: On the noise spectra of semiconductor noise and of flicker effect. Physica **16**(4), 359–372 (1950)
115. Du Pre, F.K.: A suggestion regarding the spectral density of flicker noise. Phys. Rev. **78**(5), 615 (1950)
116. McWhorter, A.L.: $1/f$ noise and germanium surface properties. In: Semiconductor Surface Physics. Univ. of Pennsylvania Press, Philadelphia (1957)
117. Hooge, F.N.: $1/f$ noise is no surface effect. Phys. Lett. A **29**(3), 139–140 (1969)
118. Handel, P.H.: $1/f$ noise—an infrared phenomena. Phys. Rev. Lett. **34**(24), 1492–1495 (1975)
119. Jindal, R.P., van der Ziel, A.: Phonon fluctuation model for flicker noise in elemental semiconductors. J. Appl. Phys. **52**(4), 2884–2888 (1981)
120. Ralls, K.S., Skocpol, W.J., Jackel, L.D., Howard, R.E., Fetter, L.A., Epworth, R.W., Tennant, D.M.: Discrete resistance switching in submicrometer silicon inversion layers: individual interface traps and low-frequency $(1/f)$ noise. Phys. Rev. Lett. **52**(3), 228–231 (1984)
121. Jayaraman, R., Sodini, C.G.: A $1/f$ noise technique to extract the oxide trap density near the conduction band edge of silicon. IEEE Trans. Electron Devices **36**(9), 1773–1782 (1989)
122. Vandamme, L.K.J.: $1/f$ noise in MOSTs: Faster is noisier. In: Sikula, J., Levinshtein, M.E. (eds.) Advanced Experimental Methods for Noise Research in Nanoscale Electronic Devices, pp. 109–120. Kluwer, Dordrecht (2004)

Part II
Compact Models of Bipolar Junction Transistors

Chapter 6
Introduction to Bipolar Transistor Modeling

Colin C. McAndrew and Marcel Tutt

Abstract This chapter reviews the operation and modeling of bipolar junction transistors (BJTs) and heterojunction bipolar transistors (HBTs). The emphasis is on fundamental device physics and modeling; subsequent chapters give specific details of two advanced models, Mextram and HiCuM. We first give a synopsis of basic BJT device behavior and modeling, and then introduce the Gummel integral charge control relationship, which elegantly and physically encapsulates the core description of BJT operation. The development of approximations to this key relationship, which leads to the widely known and used SGP (SPICE Gummel-Poon) model, are detailed, as are modifications necessary for modeling III-V HBT devices.

6.1 Introduction

Bipolar transistors were the first type of transistor invented, by Walter Brattain and John Bardeen at AT&T Bell Laboratories in 1947 [1], and the basic physical analysis of bipolar transistor action (and an extremely large range of other semiconductor device behavior) was developed by William Shockley, also of AT&T Bell Laboratories [27]. For these pioneering scientific advances this trio was awarded the Nobel Prize in physics for 1956, and the explosive adoption and continuous (exponential) development of semiconductor technologies and microelectronics that this invention spawned continues to this day.

Although MOSFETs have supplanted bipolar junction transistors (BJTs) as the type of transistors most commonly used in modern microelectronic systems, BJTs are still widely used for many high frequency and/or high power applications, such as the RF portions of wireless communication systems. This chapter provides an

C.C. McAndrew (✉) · M. Tutt
Freescale Semiconductor, Tempe, AZ 85284, USA
e-mail: Colin.McAndrew@freescale.com

M. Tutt
e-mail: Marcel.Tutt@freescale.com

G. Gildenblat (ed.), *Compact Modeling*,
DOI 10.1007/978-90-481-8614-3_6,

overview of bipolar transistors and bipolar transistor modeling. Subsequent chapters provide details of two advanced BJT models, Mextram [20] and HiCuM [10], so the emphasis in this chapter is on fundamental operation and modeling, and not on specifics of these two models.

We start with a basic analysis of bipolar transistor action, and then move to development of the Gummel integral charge control relationship, which elegantly and succinctly captures the basic physics of bipolar transistor operation for homojunction devices, as incorporated in the Gummel-Poon (GP) BJT model. The assumptions and approximations that underlie the Gummel relationship are discussed, as are how it relates to the design and optimization of practical transistor structures, including how it naturally captures the Early effect and some aspects of high-level injection. The relationship between the so-called "ideality" factor for the collector-emitter transport current and the reverse Early voltage is discussed. A simplification of the GP model that leads to the so-called SPICE Gummel-Poon (SGP) model, the widely used BJT model from the pioneering SPICE circuit simulator, is presented, with details of how it differs from the original GP model. High-level injection effects in real transistors, most notably the Kirk effect and base push-out, differ from how they are embodied in the original Gummel-Poon model and approaches to modeling this are discussed.

Besides the primary collector-emitter current from transistor action, recombination and generation also lead to base current in BJTs, and models for this are derived.

Although historically most bipolar transistors were based on homojunction silicon technology, heterojunction bipolar transistors (HBTs) are now also common. III-V based HBTs, such as GaAs or InGaP HBTs, have been widely used for high frequency, high power applications, and the integration of germanium into standard silicon BiCMOS processes has led to SiGe HBTs being the transistor of choice for the RF portion of integrated circuits for many wireless applications. III-V HBTs have some significant differences in behavior compared to silicon-based BJTs, and these are reviewed and approaches to modeling them are presented. SiGe HBT models are generally based on models for silicon BJTs rather than on III-V HBT models.

6.2 Basic Bipolar Transistor Operation and Modeling

Figure 6.1 shows a cross-sectional view of a basic *npn* bipolar transistor. In normal operation the base-emitter junction is forward biased and the base-collector junction is reverse biased, with voltages across them of V_{BE} and V_{BC}, respectively. For now we will assume that the biases are such that the base region is not in high-level injection, that there is negligible generation and recombination in the base, that the depletion approximation holds, so there are sharply defined edges to the depletion regions around the base-emitter and base-collector *pn* junctions, and that the emitter, base, and collector regions have constant doping, of values N_E, N_B, and N_C, respectively. A somewhat idealized doping profile for the device of Fig. 6.1 is shown in Fig. 6.2, and it is clear that, except for the lightly doped epitaxial collector region (i.e. the "epi"), the doping is not uniform; but we will assume uniform doping for

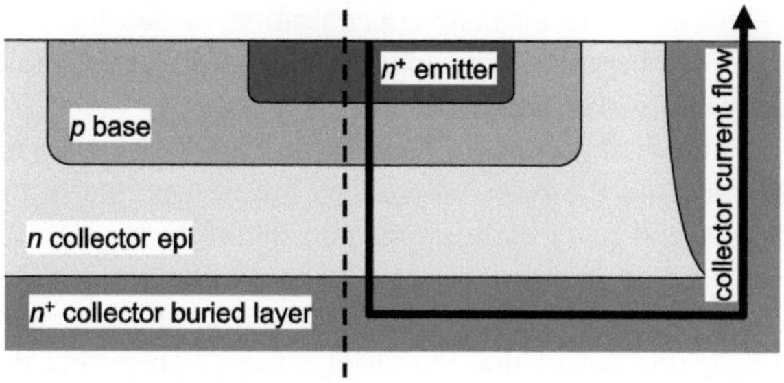

Fig. 6.1 Cross-section of a basic bipolar transistor

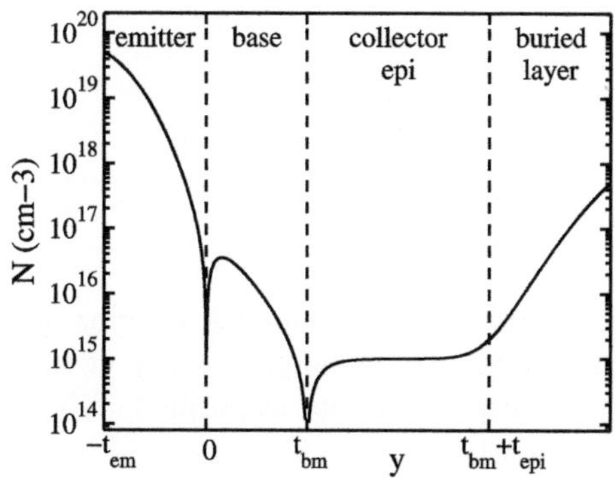

Fig. 6.2 Idealized doping profile of a vertical *npn* transistor (along dashed line of Fig. 6.1)

initial model development. The profile shown in Fig. 6.2 is along the vertical dashed line in Fig. 6.1. The basic operation of the transistor is 1-dimensional, in the vertical direction in Fig. 6.1 from the emitter down to the collector; we will denote this dimension as y and, as Fig. 6.2 shows, will select the origin of y as the position of the metallurgical junction between the emitter and base regions. The thickness of the base will be denoted t_{bm} (this is the distance between the base-emitter and base-collector metallurgical junctions, and is often called the "base width;" here we will use the terms "width" and "length" to denote lateral dimensions so we will adopt the term "thickness" to denote vertical dimensions), the distance between the silicon surface and the base-emitter metallurgical junction is t_{em}, and the thickness of the collector epi region is t_{epi}, measured from the base-collector metallurgical junction to the start of the buried n^+ collector layer.

The analysis below uses many standard results from basic *pn* junction theory; readers are referred to [24, 25, 29, 30] for details, but they are covered in most standard texts as well as those.

In the neutral base region, between the base-emitter and base-collector depletion regions, the electron concentration in equilibrium (i.e. with no biases applied to the transistor) is well approximated by

$$n_0 = \frac{n_i^2}{N_B} \tag{6.1}$$

where n_i is the intrinsic concentration, which has a value of about 10^{10} cm^{-3} at room temperature. For typical base doping levels, the equilibrium concentration of electrons is therefore low.

A forward bias on the base-emitter junction causes injection of electrons from the emitter into the base, where they are transported to the edge of the base-collector region and from there swept into the collector (at essentially their saturated drift velocity) by the large field in the base-collector depletion region. As just noted, the electron concentration in the base is low, and in general the base thickness is also small; this means that the electrons are transported through the base primarily by diffusion. Standard *pn* junction theory indicates that the position of the edge of the base-emitter junction depletion region in the base is given by

$$y = t_{j,E} = \frac{t_{0j,E}}{\sqrt{\phi_{bi,E}}}\sqrt{\phi_{bi,E} - V_{BE}} \tag{6.2}$$

where $\phi_{bi,E}$ is the built-in potential of the base-emitter junction and

$$t_{0j,E} = \sqrt{\frac{2\epsilon_s N_E}{qN_B(N_E + N_B)}\phi_{bi,E}} \tag{6.3}$$

is the thickness of the base-emitter junction depletion region in the base for zero applied bias. Here ϵ_s is the permittivity of silicon and q is the magnitude of the electronic charge. Similarly, the position of the edge of the base-collector depletion region in the base is

$$y = t_{bm} - t_{j,C} = t_{bm} - \frac{t_{0j,C}}{\sqrt{\phi_{bi,C}}}\sqrt{\phi_{bi,C} - V_{BC}} \tag{6.4}$$

where $\phi_{bi,C}$ is the built-in potential of the base-collector junction and

$$t_{0j,C} = \sqrt{\frac{2\epsilon_s N_C}{qN_B(N_C + N_B)}\phi_{bi,C}} \tag{6.5}$$

is the thickness of the base-collector junction depletion region in the base for zero applied bias.

The thickness of the base region between the base-emitter and base-collector depletion regions (the "neutral base" region) is therefore

$$t_b = t_{bm} - t_{j,E} - t_{j,C} = t_{bm} - \frac{t_{0j,E}}{\sqrt{\phi_{bi,E}}}\sqrt{\phi_{bi,E} - V_{BE}} - \frac{t_{0j,C}}{\sqrt{\phi_{bi,C}}}\sqrt{\phi_{bi,C} - V_{BC}}. \tag{6.6}$$

The electron concentration at the edge of the base-emitter depletion region in the base is

$$n_E = n_0 \exp\left(\frac{V_{BE}}{\phi_t}\right) = \frac{n_i^2}{N_B}\exp\left(\frac{V_{BE}}{\phi_t}\right) \tag{6.7}$$

where $\phi_t = kT/q$ is the thermal voltage, and similarly at the edge of the base-collector depletion region in the base the electron concentration is

$$n_C = n_0 \exp\left(\frac{V_{BC}}{\phi_t}\right) = \frac{n_i^2}{N_B}\exp\left(\frac{V_{BC}}{\phi_t}\right). \tag{6.8}$$

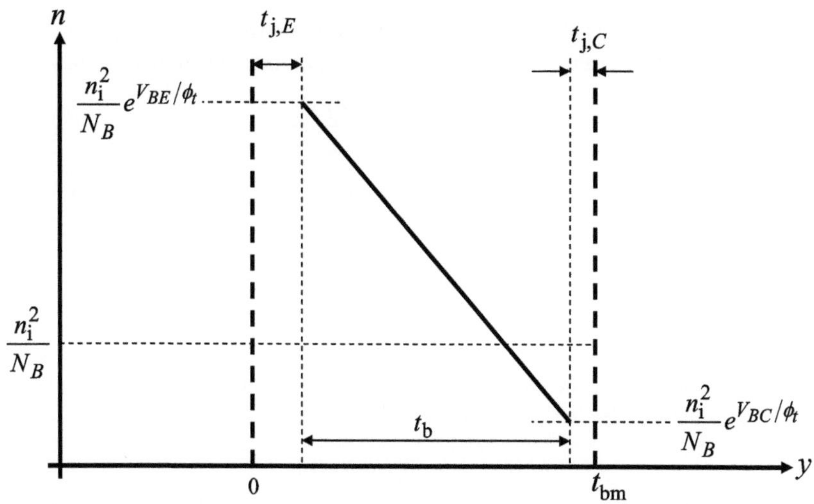

Fig. 6.3 Electron concentration in the base region under low-level injection conditions

This gives the situation shown in Fig. 6.3.

The electron concentration is linear in position because the diffusion current density, which is given by

$$J_{n,y} = qD_n \frac{\partial n}{\partial y} \tag{6.9}$$

is constant because we have assumed there is no generation or recombination in the base. Here D_n is the diffusion constant (taken to be $\phi_t \mu_n$ where μ_n is the mobility for electrons) and n is the mobile electron concentration. The constant current density, assuming a constant mobility, implies that $\partial n/\partial y$ must also be constant, as shown in Fig. 6.3. We note in passing that a similar situation occurs in an MOS transistor in weak inversion [31].

The collector-emitter current is therefore given by

$$I_{CE} = WLJ_{n,y} = qWLD_n \frac{\partial n}{\partial y} = WL\frac{qn_i^2 D_n}{N_B}\frac{\exp(V_{BE}/\phi_t) - \exp(V_{BC}/\phi_t)}{t_b} \tag{6.10}$$

where W and L are the width and length of the emitter region.

For dynamic operation of the transistor we need to consider the charges; these will also help reformulate I_{CE} of (6.10) in a more intuitive and convenient form. There are three components to the charge associated with the base Q_B; the base-emitter and base-collector depletion region charges (the "junction" charges) and the charge associated with the electrons in the neutral base, as in Fig. 6.3 (the so-called "transport" or "diffusion" charge). Because we have assumed constant doping, the junction charge of the base-emitter depletion region is

$$Q_{j,E} = -WLqN_Bt_{j,E} \tag{6.11}$$

(the base of an *npn* transistor is doped with acceptors, therefore the depletion charge in the base is negative) and differentiating this to get the capacitance we have

$$C_{j,E}(V_{BE}) = \frac{C_{0j,E}}{\sqrt{1 - V_{BE}/\phi_{bi,E}}} \tag{6.12}$$

where the zero-bias junction capacitance is, using a superscript prime to denote a quantity per unit area,

$$C_{0j,E} = WLC'_{0j,E} = WL\frac{qN_B t_{0j,E}}{2\phi_{bi,E}}. \tag{6.13}$$

Similarly, the base-collector junction charge (for the intrinsic portion of the base-collector junction under the emitter in Fig. 6.1; for total charge modeling the extrinsic portion must also be included) is

$$Q_{j,C} = -WLqN_B t_{j,C} \tag{6.14}$$

which gives a base-collector junction capacitance

$$C_{j,C}(V_{BC}) = \frac{C_{0j,C}}{\sqrt{1 - V_{BC}/\phi_{bi,C}}} \tag{6.15}$$

where

$$C_{0j,C} = WLC'_{0j,C} = WL\frac{qN_B t_{0j,C}}{2\phi_{bi,C}}. \tag{6.16}$$

Using these charge expressions the base thickness (6.6) is

$$t_b = t_{0b} + \frac{t_{0j,E}}{2\phi_{bi,E}}\left[q_{j,E}(V_{BE}) - q_{0j,E}\right] + \frac{t_{0j,C}}{2\phi_{bi,C}}\left[q_{j,C}(V_{BC}) - q_{0j,C}\right] \tag{6.17}$$

where t_{0b} is the base thickness in equilibrium,

$$q_{j,E} = \frac{Q_{j,E}}{C_{0j,E}}, \qquad q_{j,C} = \frac{Q_{j,C}}{C_{0j,C}} \tag{6.18}$$

are the base-emitter and base-collector junction charges normalized to the zero-bias junction capacitances (6.13) and (6.16), respectively, and $q_{0j,E}$ and $q_{0j,C}$ are the values of the normalized junction charges at $V_{BE} = 0$ and $V_{BC} = 0$, respectively.

Using the normalized charges I_{CE} can be rewritten as

$$I_{CE} = I_{t,E} - I_{t,C} \tag{6.19}$$

where $I_{t,E}$ is the forward transport current

$$I_{t,E} = I_S\frac{\exp(V_{BE}/\phi_t) - 1}{1 + \frac{q_{j,E} - q_{0j,E}}{V_{ER}} + \frac{q_{j,C} - q_{0j,C}}{V_{EF}}} \tag{6.20}$$

$I_{t,C}$ is the reverse transport current

$$I_{t,C} = I_S\frac{\exp(V_{BC}/\phi_t) - 1}{1 + \frac{q_{j,E} - q_{0j,E}}{V_{ER}} + \frac{q_{j,C} - q_{0j,C}}{V_{EF}}} \tag{6.21}$$

the forward and reverse Early voltages are

$$V_{EF} = \frac{2\phi_{bi,C}t_{0b}}{t_{0j,C}}, \qquad V_{ER} = \frac{2\phi_{bi,E}t_{0b}}{t_{0j,E}} \tag{6.22}$$

and the saturation current I_S is given by

$$I_S = WL\frac{qn_i^2 D_n}{t_{0b}N_B} = WL\frac{qn_i^2 D_n}{Q_0'}, \tag{6.23}$$

where Q_0' is the equilibrium dopant density in base (seen looking down vertically).

The above expressions show that the transport currents depend exponentially on the applied base-emitter and base-collector voltages, and that the modulation of the base thickness t_b by these voltages also affects the currents. Specifically, in normal operation, with positive V_{BE} and negative V_{BC}, $I_{t,C}$ is negligible so the collector-emitter current is overwhelmingly $I_{t,E}$. As the collector potential increases, V_{BC} will become increasingly negative, the base-collector depletion region will extend further into the base, $q_{j,C}$ will become increasingly negative, which will decrease the denominator of (6.20) and therefore increase I_{CE}. The is known as the Early effect [5]. Most important, from (6.23) we see that the dopant density per unit area in the base $Q_0' = t_{0b}N_B$ is a fundamental quantity that controls bipolar transistor action, and that modulation of the charge in the base is what controls the Early effect.

In equilibrium, the base region between the base-emitter and base-collector junction depletion edges is neutral; with biases applied the change (i.e. "excess") in the base transport charge Q_t is, from Fig. 6.3,

$$Q_t = WL\frac{qn_i^2}{N_B}\left\{\frac{1}{2}t_b\left[\exp\left(V_{BE}/\phi_t\right) + \exp\left(V_{BC}/\phi_t\right)\right] - t_{0b}\right\} \tag{6.24}$$

(although in Fig. 6.3 we show the electron concentration, from charge neutrality this is balanced by an equal change in the hole concentration, and it is holes that flow in from the base contact hence the sign of Q_t is positive; if it were taken to be negative the associated capacitance would have the wrong sign) and if we take $t_{0b} \approx t_b$ this can be reformulated as

$$Q_t = \tau_b\left(I_{t,E} + I_{t,C}\right) \tag{6.25}$$

where the so-called base transit time is

$$\tau_b = \frac{t_b t_{0b}}{2D_n} \approx \frac{t_{0b}^2}{2D_n} \tag{6.26}$$

and the factor of 2 arises from the linear $n(y)$ distribution in the base, which again follows because the transport currents are diffusion currents. (We use the subscript "t" and term "transport charge" to avoid ambiguity; if "d" were used it could be misinterpreted as denoting "depletion" rather than "diffusion.") Equation (6.25) highlights a key and standard relationship: transport charges are proportional to transport currents (and because the transport current is from diffusion these are also referred to as diffusion charges). Note that this is not an empirical relationship, it follows directly from the analysis. In practice, it is found that experimental data can be better fitted if the transit time is modified to be a function of the bias voltages or of the transport currents. This means that the transport charge is no longer directly proportional to the transport current; however, the concept of a transport current being modeled as a (bias dependent) transit time multiplied by a transport current is still useful and is widely used.

The model developed in this section shows how to formulate a basic model of BJT transistor operation. In practice, bipolar transistor behavior is more complicated. More advanced models are presented below and in the Chapters on the Mextram and HiCuM models; however, it is worth reviewing where the behavior of real devices can vary from the description provided above. We will not consider here aspects like 2- and 3-dimensional behavior (e.g. area and perimeter components of current), high-level injection in the base, modulation of the resistivity of the collector epi region (i.e. base push-out, or the Kirk effect [13]), or heterostructure (i.e. non-homogeneous material) effects. Rather, it is sometimes maintained that the collector-emitter current should (in the absence of the Early effect) vary *exactly* exponentially with V_{BE}/ϕ_t in forward operation, as predicted above, and we will list reasons why, even for the case of low-level injection, this is not the case.

- Doping profiles in real devices are not uniform, as we have assumed.
- The mobility, and hence D_n, is not constant, but varies with doping.
- The bandgap, and hence n_i, is not constant, but varies with doping.
- Self-heating causes temperature gradients that we have not taken into account.
- The depletion approximation is just that, an approximation.
- Details of transport effects such as drift and velocity saturation are ignored.

From these, it is not surprising that real devices may, even though the basic theory above is quite reasonable, deviate from exhibiting *perfectly* exponential behavior. We will address this further below.

Before progressing to more detailed models there are two features of the basic model that are useful to note. First, from (6.13), (6.16), and (6.22) we have a fundamental relationship between Early voltages, zero-bias junction capacitances per unit area, and the equilibrium dopant density per unit area in the base

$$C'_{0j,E} V_{ER} = C'_{0j,C} V_{EF} = q Q'_0 \tag{6.27}$$

which underscores the importance of Q'_0 in controlling bipolar transistor operation. Second, for small-signal modeling, in normal forward operation where $I_{t,C} \ll I_{t,E}$,

$$g_m \approx \frac{I_{CE}}{\phi_t} \tag{6.28}$$

where we have ignored the Early effect. The transconductance of a bipolar transistor can therefore be increased simply by increasing the collector-emitter current. The intrinsic transition (or "transit" or "unity gain transition") frequency, the frequency at which the magnitude of the capacitive (i.e. imaginary) component of the base current equals the magnitude of the transconductive (i.e. real) component of the collector current, is

$$\omega_T = \frac{g_m}{C_{j,E} + (A_C/A_E)\, C_{j,C} + C_t} \tag{6.29}$$

where we have explicitly scaled the base-collector junction capacitance by the ratio of total (not just intrinsic) base-collector area A_C to the emitter area $A_E = WL$, and the transport (more commonly "diffusion") capacitance is, assuming a bias independent base transit time,

$$C_t = \tau_b g_m. \tag{6.30}$$

From the previous expressions for the various quantities involved

$$\omega_T = \frac{1}{\tau_b}\frac{\frac{V_{EF}V_{ER}}{2\phi_t}(\frac{n_i}{N_B})^2\exp(\frac{V_{BE}}{\phi_t})}{\frac{V_{EF}}{\sqrt{1-\frac{V_{BE}}{\phi_{bi,E}}}} + \frac{A_C}{A_E}\frac{V_{ER}}{\sqrt{1-\frac{V_{BC}}{\phi_{bi,C}}}} + \frac{V_{EF}V_{ER}}{2\phi_t}(\frac{n_i}{N_B})^2\exp(\frac{V_{BE}}{\phi_t})} \tag{6.31}$$

which indicates that the transit frequency can also be increased, to a theoretical maximum value of $1/\tau_b$, by increasing V_{BE} and hence by increasing the collector-emitter current. Note that high-level injection effects are not included in the analysis so far, so the quantitative prediction of where the "flattening" of $\omega_T(I_{CE})$ occurs is not accurate in (6.31); but it predicts qualitatively correct behavior for low-level injection. In practice, ω_T peaks and then falls with increasing I_{CE} because the base transit time is not constant but increases with increasing current levels.

6.3 Base Current

Besides the bipolar transport current I_{CE} there are other currents that flow through the base-emitter and base-collector junctions. These currents arise from generation and recombination (which we have ignored in modeling the transport currents), and have several components. If the emitter is thick, then holes injected from the base into the emitter recombine in the emitter region. The current at the edge of the base-emitter depletion region in the emitter is a diffusion current, which transitions to a drift component far from the junction. The current is proportional to the net recombination rate, and for the Shockley-Read-Hall (SRH) process we have, near the junction on the emitter side, where $n \gg p$ and p is the mobile hole concentration, a net recombination rate

$$(R-G)_{\mathrm{SRH}} \approx \frac{p}{\tau_p} \tag{6.32}$$

where τ_p is the net recombination time constant for holes. The salient feature is that this rate, and hence the base-emitter component of base current, is proportional to p. Although not shown in Fig. 6.3, analogous to the electron concentration at the edge of the base side of the junction, the hole concentration at the edge of emitter side of the junction is $(n_i^2/N_E)\exp(V_{BE}/\phi_t)$, therefore the SRH recombination current is proportional to the exponential of the base-emitter voltage. There may also be Auger recombination, and near the junction on the emitter side we have

$$(R-G)_{\mathrm{AUG}} \approx C_n N_E^2 p \tag{6.33}$$

were C_n is a parameter; again this rate is proportional to p, and hence to $\exp(V_{BE}/\phi_t)$.

If the emitter is thin, then the hole transport through the emitter is from diffusion and there is negligible drift component. At the emitter contact the holes recombine, and the current density from this is conventionally modeled as being proportional to the excess hole density over that in equilibrium, with a proportionality constant S_p called the surface recombination velocity. If the hole density at the emitter contact is p_c, and this is much greater than the equilibrium concentration there, then equating the diffusion and surface recombination currents, assuming the base-emitter junction depletion region thickness in the emitter is much smaller than t_{em}, gives

$$qS_p p_c = q\phi_t \mu_p \frac{\frac{n_i^2}{N_E}\exp(\frac{V_{BE}}{\phi_t}) - p_c}{t_{em}} \tag{6.34}$$

where μ_p is the mobility for holes. This gives the hole concentration at the emitter contact

$$p_c = \frac{n_i^2 \exp(V_{BE}/\phi_t)}{N_E[1 + S_p t_{em}/(\phi_t \mu_p)]}. \tag{6.35}$$

The current is proportional to this concentration, so once again we have a component of base-emitter current that is proportional to $\exp(V_{BE}/\phi_t)$.

There is one final component that needs to be considered, from net recombination in the depletion region around the base-emitter junction. The Auger recombination process is negligible there (it is a 3-body process, the inverse of impact ionization, and in the depletion region the mobile carrier concentrations are low so such 3-body interactions are exceedingly rare), so the SRH mechanism dominates. Assuming the splitting of the electron and hole quasi-Fermi potentials is V_{BE} throughout the depletion region, the (position dependent) SRH net recombination rate is well approximated by

$$(R-G)_{\mathrm{SRH}} \approx \frac{n_i \exp(V_{BE}/\phi_t)}{\tau_p[\exp(\psi/\phi_t)+1] + \tau_n\{\exp[(V_{BE}-\psi)/\phi_t]+1\}} \tag{6.36}$$

where τ_n is the net recombination time constant for electrons and $\psi(y)$ is the electrostatic potential. The net recombination needs to be integrated over the depletion region to give the total current. However, because of the exponential dependence of (6.36) on ψ, it turns out that there is significant net recombination only over a very small portion of the depletion region. This occurs when the denominator of (6.36) is minimum, which happens when

$$\psi = \frac{1}{2}\left[V_{BE} - \phi_t \ln\left(\frac{\tau_p}{\tau_n}\right)\right] \tag{6.37}$$

and if $\tau_p \approx \tau_n$ this means that the peak net recombination rate, and therefore the associated component of current, is proportional to $\exp[V_{BE}/(2\phi_t)]$.

This analysis is somewhat subjective, but it indicates that a reasonable model for the base-emitter component of base current comprises two components, typically called ideal and non-ideal components,

$$I_{BE} = I_{BEI}\left\{\exp\left[V_{BE}/(N_{EI}\phi_t)\right] - 1\right\} + I_{BEN}\left\{\exp\left[V_{BE}/(N_{EN}\phi_t)\right] - 1\right\} \tag{6.38}$$

where $N_{EI} \approx 1$ is the so-called ideality factor of the ideal component of base-emitter current, $N_{EN} \approx 2$ is the ideality factor of the non-ideal component of base-emitter current, and I_{BEI} and I_{BEN} are the associated current factors. These parameters are determined from experimental data. Note that we have not indicated a geometric scaling relationship for these parameters. In general I_{BEI} scales approximately with the emitter area WL. However, because the non-ideal component is associated with net recombination in the base-emitter depletion region, and it could be expected that the net recombination rates there would be greatest at the surface where the silicon lattice is disrupted, I_{BEN} can be better modeled as scaling with the emitter perimeter. Some elegant studies based on flicker noise and stress have supported this [14].

An analogous model is used for the base-collector component of base current,

$$I_{BC} = I_{BCI}\left\{\exp\left[V_{BC}/(N_{CI}\phi_t)\right] - 1\right\} + I_{BCN}\left\{\exp\left[V_{BC}/(N_{CN}\phi_t)\right] - 1\right\} \tag{6.39}$$

where N_{CI} and N_{CN} are the ideality factors of the ideal and non-ideal components of base-collector current, and I_{BCI} and I_{BCN} are the associated current factors.

6.4 Gummel Integral Charge Control Relation

The most fundamental understanding of bipolar transistor action follows from the Gummel integral charge control relation (ICCR) [8]. It is instructive to understand in detail this relation; the derivation here is based on [4], and is provided for an *npn* transistor. Importantly, the ICCR extends the analysis of Sect. 6.2 to encompass high level injection.

The electron continuity equation is

$$q\frac{\partial n}{\partial t} - \nabla \cdot \mathbf{J}_n = q(G_n - R_n) \tag{6.40}$$

where $\mathbf{J}_n$ is the electron current density, and G_n and R_n are the electron generation and recombination rates, respectively. The electron current density consists of both drift and diffusion components

$$\mathbf{J}_n = -q\mu_n n \nabla\psi + qD_n \nabla n = -q\mu_n n \nabla\phi_n \tag{6.41}$$

where ϕ_n is the electron quasi-Fermi potential. In 1-dimensional steady-state operation, assuming there is negligible generation and recombination in the base, we have

$$J_{n,y} = -q\mu_n n \frac{\partial \phi_n}{\partial y} = \text{constant}. \tag{6.42}$$

Now,

$$\frac{\partial \phi_n}{\partial y} = -\phi_t \exp(\phi_n/\phi_t)\frac{\partial \exp(-\phi_n/\phi_t)}{\partial y} \tag{6.43}$$

and the electron concentration in the base is

$$n = \frac{n_i^2}{p} \exp\left(\frac{\phi_p - \phi_n}{\phi_t}\right) \tag{6.44}$$

where ϕ_p is the hole quasi-Fermi potential. So over any region where ϕ_p is constant we have

$$q\phi_t \frac{\partial \exp[(\phi_p - \phi_n)/\phi_t]}{\partial y} = \frac{p}{\mu_n n_i^2} J_{n,y}. \tag{6.45}$$

In the neutral base region, in the absence of generation and recombination, there is no current flow vertically due to holes (there is a lateral hole current flow, to provide the base current, which was addressed in Sect. 6.3), therefore ϕ_p is constant and does not depend on y. Also, the split between the hole and electron quasi-Fermi potentials at the base edge of the base-emitter junction is just V_{BE} and at the base edge of the base-collector junction is V_{BC}. So integrating (6.45) over the neutral base and scaling the resultant current density by the emitter area gives

$$I_{CE} = WLq\phi_t \frac{[\exp(V_{BE}/\phi_t) - \exp(V_{BC}/\phi_t)]}{\int_{t_{j,E}}^{t_{bm}-t_{j,C}} \frac{p(y)}{\mu_n(y)n_i(y)^2} dy} \tag{6.46}$$

where we have emphasized in the integral in the denominator that the hole concentration, electron mobility, and intrinsic concentration may be functions of position. Equation (6.46) is the Gummel integral charge control relation [8], and it succinctly embodies the basic principles of bipolar transistor operation. The basic exponential nature of the transport current is clear, as its dependence on the intrinsic concentration and on temperature (n_i also varies strongly with temperature). If the base doping is so high that it affects the mobility or n_i, the resultant effect on the transport current is also captured. The importance of the base doping is key, and under high-level injection conditions p will rise above the background doping level N_B, and therefore the transport current will decrease compared to extrapolation from low-level injection behavior. The scaling with emitter area is also apparent.

Note that for position independent p, μ_n, and n_i and operation in low-level injection (6.46) is exactly (6.10); however, the above result was not derived under the assumption of low-level injection and is also valid for the case of high-level injection in the base. In fact the above form is general and is applicable over any region where the hole quasi-Fermi potential is constant.

For the purpose of developing a compact model, we will now assume that p, μ_n, and n_i are constant, then (6.46) can be reformulated as

$$I_{CE} = \frac{I_{tf} - I_{tr}}{q_b} \tag{6.47}$$

where

$$I_{tf} = I_S\left[\exp(V_{BE}/\phi_t) - 1\right], \tag{6.48}$$

$$I_{tr} = I_S\left[\exp(V_{BC}/\phi_t) - 1\right] \tag{6.49}$$

(I_S is still given by (6.23)) and the normalized base charge is

$$q_b = \frac{1}{t_{0b}} \int_{t_{j,E}}^{t_{bm}-t_{j,C}} \frac{p}{N_B} dy. \tag{6.50}$$

Note that this is conventionally referred to as the base "charge" even though it is not a charge *per se*; the base region (between the depletion regions) is electrically neutral and the positive charge from the holes is balanced by the negative charge of the acceptor dopant atoms. Instead of using the theoretical formula (6.26) for the base transit time, if we introduce adjustable forward and reverse knee current parameters (the reason for this will become apparent), I_{KF} and I_{KR}, respectively, then the transport charge in the base, normalized to the integral of p over the neutral base region in equilibrium, can be modeled as [9]

$$\frac{Q_t}{WLqQ_0'} = \frac{q_2}{q_b} \tag{6.51}$$

where

$$q_2 = \frac{I_{tf}}{I_{KF}} + \frac{I_{tr}}{I_{KR}}. \tag{6.52}$$

The total normalized base charge then becomes the sum of the components from modulation of the neutral base thickness, i.e from the Early effect, and from the transport charge. The former has already been evaluated, it is t_b/t_{0b} which is the denominator of (6.20), so denoting this as q_1

$$q_1 = 1 + \frac{q_{j,E} - q_{0j,E}}{V_{ER}} + \frac{q_{j,C} - q_{0j,C}}{V_{EF}} \tag{6.53}$$

we then have for the normalized base charge

$$q_b = q_1 + \frac{q_2}{q_b} \tag{6.54}$$

which has the solution

$$q_b = \frac{q_1}{2} + \sqrt{\frac{q_1^2}{4} + q_2}. \tag{6.55}$$

This equation along with (6.47) through (6.49), (6.52), and (6.53) defines the Gummel-Poon (GP) model [9].

The Early effect enters the GP model through the q_1 term, and high-level injection effects enter via the q_2 component of the normalized base charge. Figures 6.4 and 6.5 show the qualitative characteristics of the GP model (the base current model of Sect. 6.3 is also included). At low bias, the non-ideal component of base current dominates, and $\beta = I_C/I_B$ drops compared to its peak value. In the ideal region, where both the collector-emitter transport current and the base current vary approximately as $\exp(V_{BE}/\phi_t)$, β is seen to be nearly constant; the slight decrease with increasing base bias is from the Early effect (associated with the base-emitter junction, so this is the reverse Early effect; the curves shown are for constant V_{BC}). At

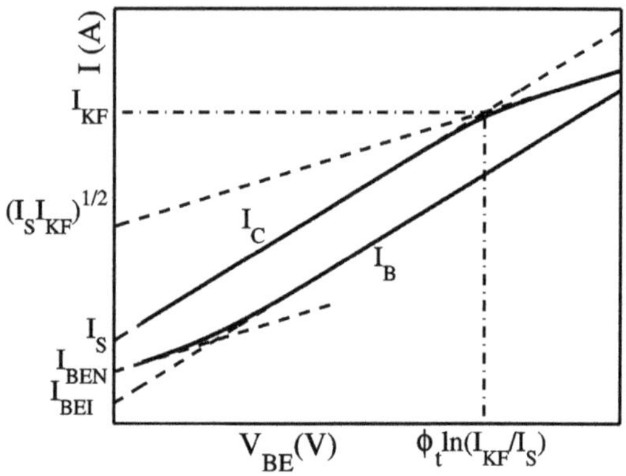

Fig. 6.4 Forward Gummel plot ($\log(I_C)$, $\log(I_B)$ vs. V_{BE}) from the GP model

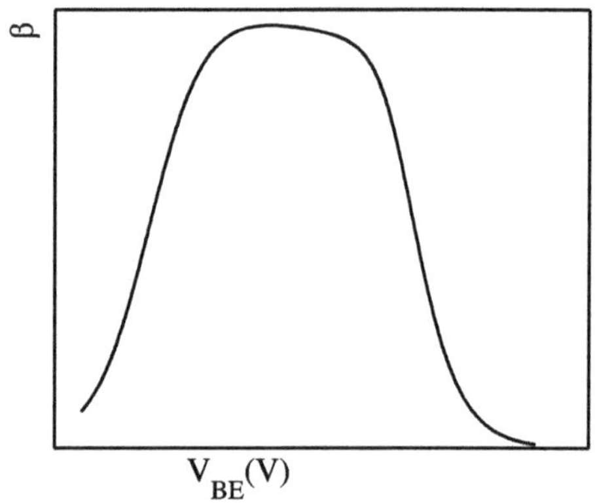

Fig. 6.5 $\beta = I_C/I_B$ of the GP model from the forward Gummel plot

very high forward bias, where high-level injection occurs, we have, ignoring the Early effect,

$$I_{CE} \approx \sqrt{I_S I_{KF}} \exp[V_{BE}/(2\phi_t)] \tag{6.56}$$

hence the slope of the $\log(I_C)$ vs. V_{BE} characteristic decreases by a factor of 2 going from ideal low-level to high-level injection behavior. Asymptotes for various components of the current are shown by dashed lines in Fig. 6.4, as are values for certain intercepts, in terms of the model parameters.

There is one important aspect of the Early effect, seen in Fig. 6.5 as the "droop" in β for increasing V_{BE} in the ideal region, that was explicitly pointed out in [9] but is often overlooked. From the discussion in Sect. 6.2 of imprecisions in the physical formulation of the "ideal" exponential model it is reasonable to expect that perhaps $I_S \exp[V_{BE}/(N_F\phi_t)]$ may be an appropriate model for the core transport current model (in forward operation, where the V_{BC} term is negligible), where N_F is an ideality factor parameter whose value should be very close to 1 (we will see below that N_F is used in common models, and is necessary for modeling III-V HBTs). From this model the ideality factor can be determined from data as

$$N_F = \frac{I_{CE}}{\phi_t g_m}. \tag{6.57}$$

Under low-level injection conditions, from the GP model we have

$$\begin{aligned} N_F &= \frac{1}{1 - \frac{\phi_t}{1+\frac{q_{j,E}(V_{BE})-q_{0j,E}}{V_{ER}}+\frac{q_{j,C}(V_{BC})-q_{0j,C}}{V_{EF}}} \frac{1}{V_{ER}} \frac{\partial q_{j,E}}{\partial V_{BE}}} \\ &\approx 1 + \phi_t \frac{C'_{j,E}}{q Q'_0}. \end{aligned} \tag{6.58}$$

If care is not taken, the (reverse) Early effect can therefore be mistakenly interpreted as a need to include a transport current ideality factor that differs from, and is slightly greater than, 1 (also, any difference between the measured and actual temperature at which data are taken can similarly be misinterpreted).

6.5 SPICE Gummel-Poon Model

The implementation of the Gummel-Poon model in the SPICE circuit simulator [23], which has been universally adopted and is referred to as the SGP model, differs in several respects from the theoretical model developed in the previous section. The most important of these are now detailed.

First, the normalized junction charge model is modified from the theoretical square-root voltage dependence, which holds only for uniformly doped junctions, to a parameterized form that gives a normalized capacitance of [19]

$$\frac{C_{j,E}}{C_{0j,E}} = \frac{1}{(1 - V_{BE}/\phi_{bi,E})^{m_{jE}}}, \qquad \frac{C_{j,C}}{C_{0j,C}} = \frac{1}{(1 - V_{BC}/\phi_{bi,C})^{m_{jC}}} \tag{6.59}$$

where m_{jE} and m_{jC} are parameters that are determined to best fit experimental data (when these parameters are 0.5 the model matches the square-root dependence given in the previous section). The charge models are just the integrals of these expressions, and are modified at high forward bias to avoid the singularity at the built-in potential [19].

Second, the solution for the normalized base charge (6.55) is modified to become

$$q_b = \frac{q_1}{2} + q_1 \sqrt{\frac{1}{4} + q_2} \tag{6.60}$$

as this enables the Early effect, which enters through the q_1 term, to remain active for the case of high-level injection, which is not the case for the form (6.55).

Third, rather than computing the transport charge based on the theoretical transit time (6.26) additional forward and reverse transit time parameters, τ_f and τ_r, respectively, are defined and the transport charge is calculated as

$$Q_t = \tau_f \frac{I_{tf}}{q_b} + \tau_r I_{tr} \tag{6.61}$$

(note that the second term does not include division by the factor q_b). An empirical bias dependence for τ_f is also included [19]. Comparison with (6.51) and (6.52) indicates that we should have (if the second term were divided by the factor q_b)

$$\tau_f \frac{I_{KF}}{WL} = \tau_r \frac{I_{KR}}{WL} = q Q_0' \tag{6.62}$$

but this is not enforced in the SGP model. Compare (6.62) to (6.27); the latter relationships are also not enforced, and the zero-bias junction capacitances and Early voltages are treated as separate model parameters. Although theoretically all of these parameters should be interrelated, to Q_0', and for the purpose of properly capturing parameter correlations for statistical modeling it is best to do this [3], remember that the models are approximations, so allowing the parameters to be separate can improve the accuracy of fitting experimental data. Specifically in this case, the Early voltages and knee current parameters affect the modeled collector-emitter current through the normalized base charge, and the zero-bias junction capacitances and transit times affect the modeled total base charge (which is *not* computed from the normalized base charge), hence this allows independent extraction of dc and small-signal (ac, or capacitance) parameters. This can be convenient in practice.

Fourth, the junction charge model, which gives the normalized junction capacitance (6.59), is not used when computing q_1 of (6.53). Rather, the change in the junction charge with respect to its zero-bias value is taken to just be the voltage across the junction, which is equivalent to assuming that the normalized junction capacitance is equal to 1 independent of bias, and the Early effect is assumed to be small so that the approximation $1 + x \approx 1/(1 - x)$ can be applied; this gives

$$q_1 = \frac{1}{1 - \frac{V_{BE}}{V_{ER}} - \frac{V_{BC}}{V_{EF}}}. \tag{6.63}$$

This approximation for q_1 *explicitly* makes the SGP model less accurate than the GP model. So why was it done? The reason is rather prosaic. With the approximation (6.63) in forward operation under low-level injection conditions the output conductance is

$$g_0 = \frac{\partial I_C}{\partial V_C} = \frac{I_{tf}}{V_{EF}} \tag{6.64}$$

and this predicts that g_0 is constant for a given V_{BE}. At the time that SPICE was developed the Ebers-Moll model, with a constant output conductance added empirically, was the standard for BJT modeling [7]; the more accurate GP formulation was not universally accepted as being more accurate. The degradation of the GP model to the SGP form, to be consistent with this, was done deliberately to conform to this perceived "standard" [21]. Figure 6.6 compares I_E/g_0 from measurements with the SGP model and results from the full GP model (reverse operation data are shown as this accentuates the problem). The "linear" nature of the SGP approximation is

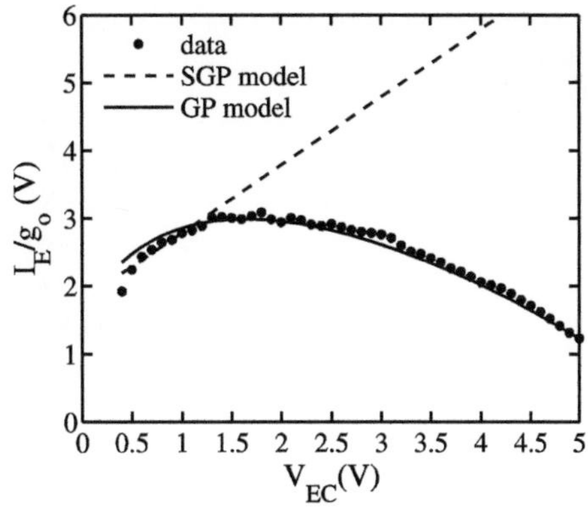

Fig. 6.6 I_E/g_0 showing imprecision in the SGP model

clear, as is the imprecision in fitting the data. The proper depletion charge formulation of the GP model is significantly more accurate; it is somewhat ironic that one of the "improvements" of the VBIC model [22] over the SGP model was to just revert to the original GP form for depletion charge modeling for I_{CE} and so drop the approximation (6.63).

Finally, as we have noted there are approximations involved in the model derivations above so it may be expected that in real devices there is some deviation from the ideal exponential behavior, and this can especially be true for HBTs. So-called forward and reverse non-ideality factors N_F and N_R are introduced to account for this, and the idealized forward and reverse currents are modified to become

$$I_{tf} = I_S \left\{\exp\left[V_{BE}/(N_F\phi_t)\right] - 1\right\}, \tag{6.65}$$

$$I_{tr} = I_S \left\{\exp\left[V_{BC}/(N_R\phi_t)\right] - 1\right\}. \tag{6.66}$$

Care needs to be exercised when using these parameters in practice. Ignoring base current, the Early effect, and high-level injection effects, in normal forward operation the power dissipated in a transistor is

$$P_{diss} = V_{CE} I_S \exp\left(\frac{V_{BE}}{N_F\phi_t}\right)\left\{1 - \exp\left[\frac{V_{BE}}{\phi_t}\left(\frac{1}{N_R} - \frac{1}{N_F}\right)\right]\exp\left(-\frac{V_{CE}}{N_R\phi_t}\right)\right\} \tag{6.67}$$

and for this to be positive we must have, noting that $V_{CE} > 0$ in forward operation,

$$\exp\left[\frac{V_{BE}}{\phi_t}\left(\frac{1}{N_R} - \frac{1}{N_F}\right)\right]\exp\left(-\frac{V_{CE}}{N_R\phi_t}\right) < 1. \tag{6.68}$$

Because V_{BE} is positive, and V_{CE} can be arbitrarily close to zero, passivity therefore requires that $N_F \leq N_R$. A similar analysis for reverse operation leads to the requirement that $N_R \leq N_F$, which implies that we should require that $N_F = N_R$. But this is not enforced.

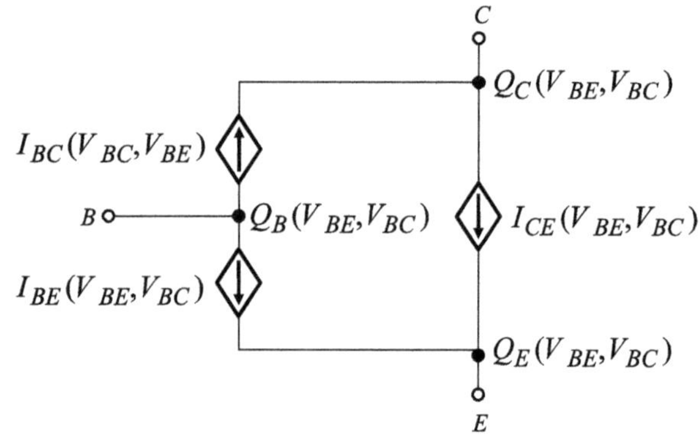

Fig. 6.7 General large-signal representation for a 3-terminal BJT

6.6 Small-Signal Model

Strictly, the small-signal representation of a large-signal model follows simply by linearizing the large-signal model; this is done automatically for models implemented in Verilog-A and so is not directly important for the purpose of compact modeling. The small-signal model is of interest to designers, for device and circuit analysis and understanding, so model developers need to define what small-signal information is to be reported by simulators. However, the conventional hybrid-π representation of the small-signal equivalent circuit for bipolar transistors is incomplete, in that some elements that can be important under certain operating conditions are omitted. Various equivalent network representations are possible, here we adapt the form presented in [31], which is rigorously correct for MOSFETs (actually, for any 4-terminal device), to a 3-terminal BJT; this requires 4 elements related to conductive currents and 4 elements related to charging ("capacitance") currents.

Figure 6.7 shows a general representation for a large-signal model for a 3-terminal bipolar transistor (the same topology is applicable to any 3-terminal device). There are three current sources, and the only extension of the models presented above is that we allow the base-emitter and base-collector components of base current to be functions of both V_{BE} and V_{BC}, and not just of the individual voltage difference across each junction; this can be the case when there is neutral base recombination for example, when altering V_{BC} modulates the base thickness through the Early effect, which then will (slightly) change I_{BE}. There are also three terminal charges in Fig. 6.7, and charge neutrality requires that

$$Q_E + Q_B + Q_C = 0. \tag{6.69}$$

As noted above, there are two components to the charges: the junction (depletion region) charge and the transport (or diffusion) charge. The junction charges can be associated with 2-terminal capacitances, because they depend only on the voltage difference across the junction. However, the transport charge cannot be represented like that: from (6.25) separate transport charges can be associated with the forward and reverse transport currents, but through the Early effect, and also through bias dependencies of the base transit time introduced in more complex models, the transport charge depends on all of the terminal voltages applied to a device. Additionally,

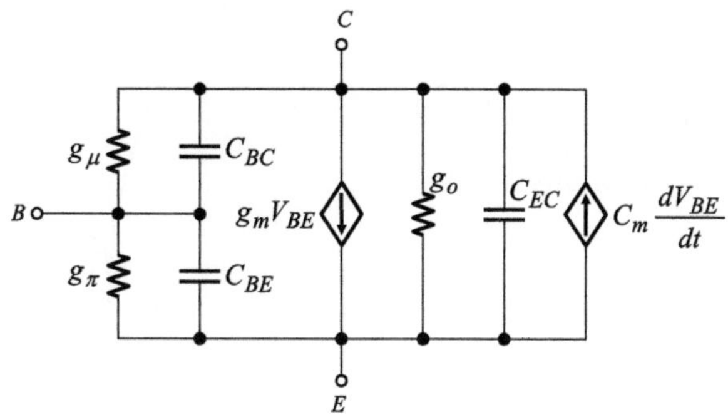

Fig. 6.8 Complete small-signal representation for a 3-terminal BJT corresponding to Fig. 6.7

besides the obvious choice of associating the forward and reverse transport current components of Q_t with the base-emitter and base-collector junctions more sophisticated partitioning schemes have been proposed [6]. This means that in general the terminal charges must be considered to be functions of all terminal voltages.

Linearization of the large-signal model of Fig. 6.7, following the practice of [31] for the charging components gives, after some rearrangement of the conductive components, the small-signal representation of Fig. 6.8. In this model the capacitances are given by $C_{KL} = -\partial Q_K / \partial V_L$, $K \neq L$, with all voltages other than V_L being held constant, and $C_{KK} = \partial Q_K / \partial V_K$, with all voltages other than V_K being held constant, therefore

$$C_{BE} = -\left(\frac{\partial Q_B}{\partial V_E}\right)_{V_B, V_C}, \tag{6.70}$$

$$C_{BC} = -\left(\frac{\partial Q_B}{\partial V_C}\right)_{V_B, V_E}, \tag{6.71}$$

$$C_{EC} = -\left(\frac{\partial Q_E}{\partial V_C}\right)_{V_B, V_E}, \tag{6.72}$$

$$C_m = C_{CB} - C_{BC} = \left(\frac{\partial Q_B}{\partial V_C}\right)_{V_B, V_E} - \left(\frac{\partial Q_C}{\partial V_B}\right)_{V_E, V_C}. \tag{6.73}$$

The conductance elements have values

$$g_\pi = \frac{\partial I_B}{\partial V_{BE}} = \frac{\partial I_{BE}}{\partial V_{BE}} + \frac{\partial I_{BC}}{\partial V_{BE}}, \tag{6.74}$$

$$g_\mu = \frac{\partial I_B}{\partial V_{BC}} = \frac{\partial I_{BC}}{\partial V_{BC}} + \frac{\partial I_{BE}}{\partial V_{BC}}, \tag{6.75}$$

$$g_m = \frac{\partial I_{CE}}{\partial V_{BE}} + \frac{\partial I_{CE}}{\partial V_{BC}} + \frac{\partial I_{BE}}{\partial V_{BC}} - \frac{\partial I_{BC}}{\partial V_{BE}}, \tag{6.76}$$

$$g_o = -\frac{\partial I_{CE}}{\partial V_{BC}} - \frac{\partial I_{BE}}{\partial V_{BC}}. \tag{6.77}$$

In most models I_{BE} is a function only of V_{BE} and I_{BC} is a function only of V_{BC}, so the "cross-derivatives" of the base current components are zero and can be omitted. However, the C_{EC} and C_m components are not necessarily zero. C_{EC} represents

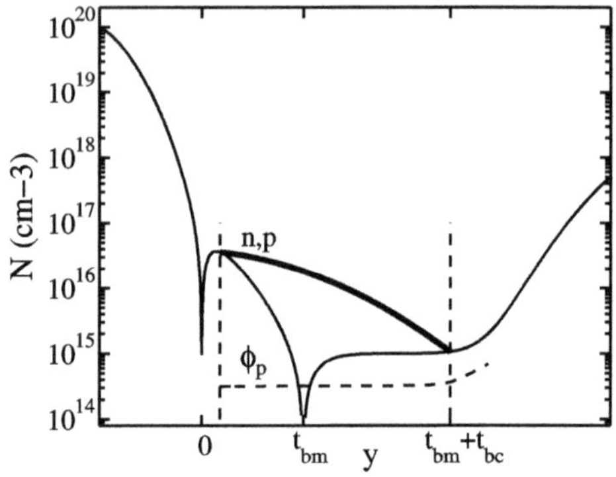

Fig. 6.9 Carrier concentrations and hole quasi-Fermi potential under base push-out conditions. The effective base region is between the *vertical dashed lines*

the effect of the collector voltage on the emitter charge; the junction component is not affected by the collector voltage, but the transport component (and generally the portion of transport charge associated with the forward transport current is partitioned completely to the emitter, none being partitioned to the collector) does depend on the collector voltage, both through the Early effect (which causes the forward transport current to change) and through the bias dependence of the transit time. Similarly, C_m represents the non-reciprocity of the base-collector capacitance; the junction component is reciprocal, and so does not contribute to C_m, but the collector voltage dependence of the transport charge causes C_m to be non-zero. Note that these effects are included in advanced bipolar compact models, but are sometimes not indicated in their small-signal representations. In most other studies of bipolar device operation C_{EC} and C_m are omitted from the small-signal equivalent circuit.

6.7 Kull-Nagel Model

The preceding analyses have omitted one very important aspect of bipolar transistor behavior: base push-out. When the base voltage is increased to such a level that the electron concentration in the base exceeds the doping level in the collector, the electrons from the base spill over ("push-out") into the collector. These electrons then modulate the conductivity of the collector region, and this effect is also referred to as quasi-saturation. An analysis of the effect was included in the original GP model [9] but this was not implemented in the SGP model. Other analyses were also proposed but were not adopted in any standard models [12, 32]. Two approaches have emerged as the preferred techniques for modeling the effect, and the development of [17], which is referred to as the Kull-Nagel model, has been used (with modifications) in both the Mextram and VBIC [22] models and will now be reviewed.

When the electrons push out from the base into the collector the hole concentration also, from neutrality, rises above the doping level in the base, and the hole quasi-Fermi potential remains constant through the portion of the collector into which the base region has encroached, see Fig. 6.9 (after [17]). The electron quasi-Fermi potential changes through this region, and if we take the thickness of the pushed out region to be t_{bc} then we denote voltage between the base edge of the base-collector junction depletion region to the end of the pushed out region, at $y = t_{bm} + t_{bc}$, as V_{BCC}, i.e. the voltage drop across the pushed out base region is $V_{BCC} - V_{BC}$. Neutrality in the pushed out region gives

$$np = (p + N_C)p = n_i^2 \exp\left(\frac{\phi_p - \phi_n}{\phi_t}\right) \tag{6.78}$$

and differentiating with respect to position y (remembering that ϕ_p is constant) and rearranging gives

$$n\frac{\partial \phi_n}{\partial y} = -\phi_t \left(2 + \frac{N_C}{p}\right)\frac{\partial p}{\partial y}. \tag{6.79}$$

Integrating the current density (6.42) across the pushed out base region, using (6.79) to change the variable of integration, gives the current flowing in the epitaxial collector (measured as positive in the same direction as I_{CE}) as

$$\begin{aligned} I_{epi} &= \frac{WLq\mu_n\phi_t}{t_{bc}} \int_{p_c}^{p_b} \left(2 + \frac{N_C}{p}\right) dp \\ &= \frac{WLq\mu_n\phi_t}{t_{bc}} \left[2(p_b - p_c) + N_C \ln\left(\frac{p_b}{p_c}\right)\right] \end{aligned} \tag{6.80}$$

where p_b and p_c are the hole concentrations at the ends of the pushed out region. Clearly, under conditions of base push-out the encroached-on part of the collector is in high-level injection, so we have for the electron concentration at the start and end of the pushed out region

$$n_b = 0.5N_C\left[1 + K_1(V_{BC})\right], \tag{6.81}$$

$$n_c = 0.5N_C\left[1 + K_1(V_{BCC})\right] \tag{6.82}$$

where

$$K_1(V) = \sqrt{1 + \frac{4n_i^2}{N_C^2}\exp\left(\frac{V}{\phi_t}\right)} \tag{6.83}$$

and charge neutrality implies that $p_b - p_c = n_b - n_c$ and we have

$$\frac{p_b}{p_c} = \frac{n_c}{n_b}\exp\left(\frac{V_{BC} - V_{BCC}}{\phi_t}\right). \tag{6.84}$$

Substituting these into (6.80) gives the final Kull-Nagel result

$$\begin{aligned} I_{epi} = \frac{1}{R_{CC}}\Bigg(V_{BC} - V_{BCC} + \phi_t \Bigg\{ & K_1(V_{BC}) - K_1(V_{BCC}) \\ & - \ln\left[\frac{1 + K_1(V_{BC})}{1 + K_1(V_{BCC})}\right]\Bigg\}\Bigg) \end{aligned} \tag{6.85}$$

where $R_{CC} = t_{bc}/(WLq\mu_n N_C)$ is the (low bias) resistance of the pushed out region in the collector.

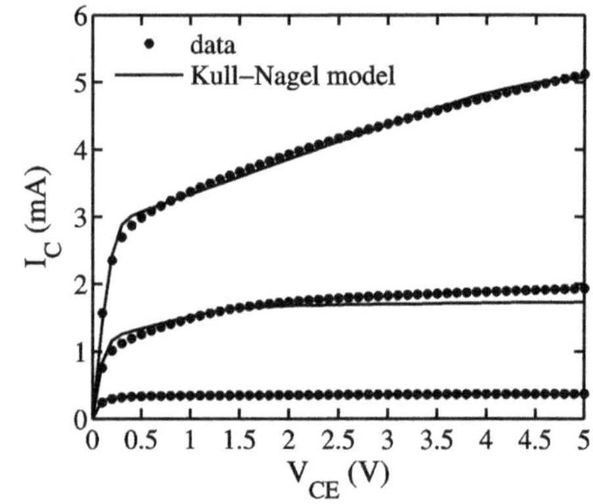

Fig. 6.10 Kull-Nagel model for base push-out (quasi-saturation)

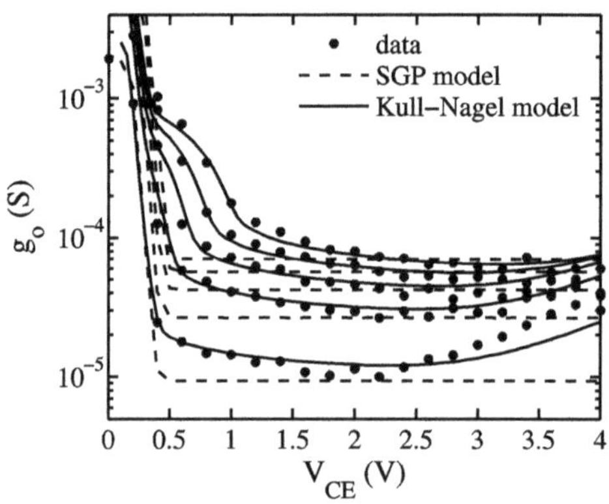

Fig. 6.11 Output conductance in the presence of base push-out

The Kull-Nagel model has proven effective for modeling high voltage transistors, see Fig. 6.10. Fitting conductances is often more difficult than fitting currents; Fig. 6.11 shows fitting of output conductance in the presence of base push-out (the "hump" toward the bottom left of each curve).

The original development of the Kull-Nagel model [17] included the effects of velocity saturation, and was applied across the whole thickness of the collector epi region, even though the hole quasi-Fermi potential is not constant across the portion of the collector where no base push-out occurs. For these reasons the model has been modified when used in Mextram and VBIC; but the basic form of the model, derived above, is maintained.

An alternative, adopted in HiCuM, is to note that when base push-out occurs the hole quasi-Fermi potential is constant through the "normal" base *and* through the portion of the collector where base push-out occurs. Therefore the Gummel ICCR is valid across this whole region. In fact, it is physically reasonable to *define* the base as the portion of an *npn* transistor where ϕ_p is constant. Use of this requires solving for the position in the collector to which the base pushes out, which is involved. However, in some respects this better represents the physics of the situation, rather than somewhat artificially splitting the effective base region into two. Both approaches have proven to represent experimental data reasonably.

6.8 III-V HBTs: Device Physics and Modeling Challenges

Silicon based bipolar transistors, historically made from silicon and over the past decade more usually SiGe HBTs for wireless transceiver applications, have been waning in their importance in the semiconductor industry, because of the ineluctable advances in CMOS process technology. However, for high power, high frequency applications bipolar transistors remain important, and generally these are heterostructures made from III-V compound semiconductors. Although there is some overlap, often the world's of "silicon" bipolar modeling and "RF" bipolar modeling are separate, with the companies and people, even within the same company, involved with the different areas participating in different conferences, reading different journals, using different terminologies, and following different approaches to modeling. III-V HBTs still share some common behaviors with Si based BJTs and HBTs, such as approximately exponential dependence of mobile carrier levels with applied potential, but also exhibit some different behaviors, especially deviations from supposedly "ideal" behavior.

The concept of building bipolar junction transistors using materials with different band gaps can be traced back to Shockley [28]. Kroemer [15, 16] has presented an analysis of these devices. Starting in the 1980s considerable work was dedicated to the development of HBTs based on III-V materials. Today, III-V HBT's are used in RF, microwave and millimeter wave applications including linear and non-linear applications such as low noise amplifiers, oscillators, receivers and power amplifiers. In addition to the use of III-V materials such as GaAs, InGaP, InP and AlGaAs the HBT's can have one or two heterojunctions. The majority of III-V HBT's use InGaP-GaAs with a single graded heterojunction at the base-emitter junction.

Typically, the devices are processed using epitaxial wafers grown using either metal-organic chemical vapor deposition (MOCVD) or molecular beam epitaxy (MBE). Devices are fabricated using etching techniques. The resulting device is referred to as a mesa structure. This approach yields a device where the contacts can be placed directly on the layer of interest. Figure 6.12 shows a cross-section of a single emitter device. An important feature of III-V HBT's is that all epitaxial layers are grown on a semi-insulating substrate, which, with the mesa structure, minimizes parasitics compared to planar silicon based technologies. The availability of through-substrate-vias allows connection of backside metalization to topside circuitry, which enables implementation of low loss transmission line structures which are essential for some RF applications and are not available in silicon based bipolar processes. In this section, the unique physics of these devices is reviewed and specific modeling issues presented by these devices are described.

The different band structure of III-V materials provides not only a wider bandgap, but also different carrier transport properties, compared to Si. Table 6.1 compares the energy gap values (E_g) of various semiconductors. $In_{0.51}Ga_{0.49}P$ and $Al_{0.3}Ga_{0.7}As$ are included because they are typically the wide bandgap material used in III-V HBT's. Figure 6.13 shows the velocity-field characteristics of n-type GaAs and Si. Several features are noteworthy. First, the low field mobility in GaAs is larger than that of Si (8500 compared to 1450 $cm^2/(V\,s)$). This translates to a lower access resistance in devices. Second, GaAs displays a negative differential mobility in part

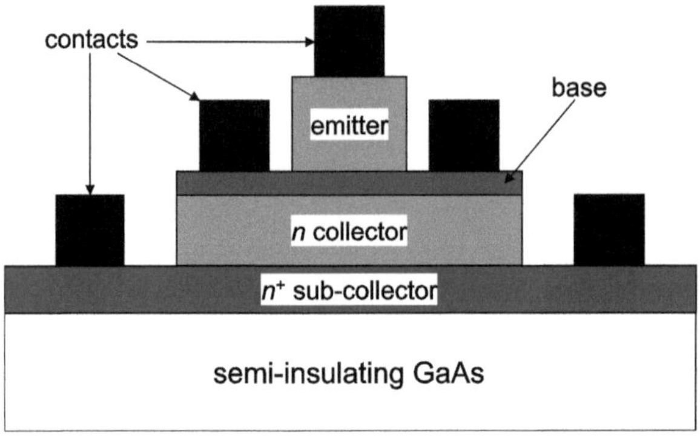

Fig. 6.12 Simplified cross-section of a single emitter mesa type HBT structure

Table 6.1 Energy bandgap for some common semiconductors

Material	E_g (eV)
Si	1.13
AlAs	2.16
GaP	2.24
GaAs	1.42
InP	1.26
InAs	0.35
$Al_{0.3}Ga_{0.7}As$	1.84
$In_{0.51}Ga_{0.49}P$	1.92

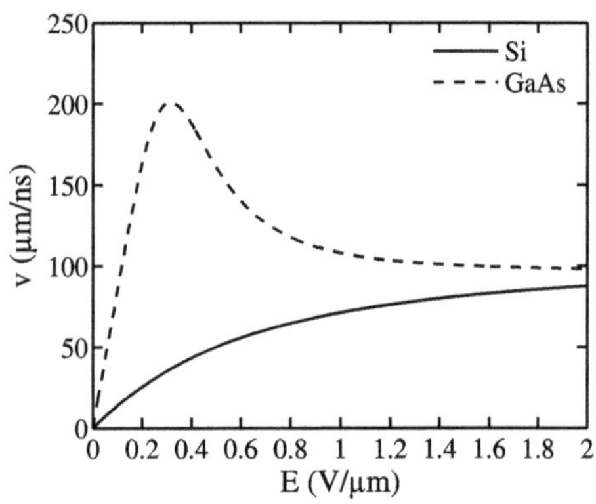

Fig. 6.13 Velocity-field profiles for n-type GaAs and Si

of this characteristic. The simplified energy band structure, shown in Fig. 6.14, explains this. At low energy, free electrons reside in the bottom of the Γ-valley. In the presence of an electric field, electrons can acquire sufficient energy to scatter into the L-valley, where they have reduced mobility (which is proportional to the second derivative of the $E - k$ relationship). Consequently, the average velocity will begin

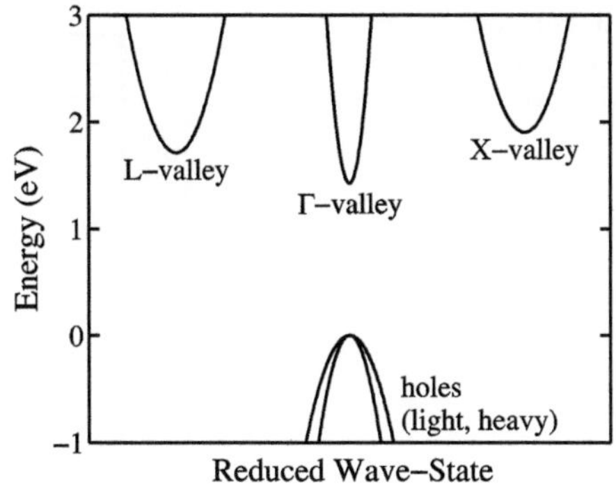

Fig. 6.14 Simplified energy band diagram for GaAs

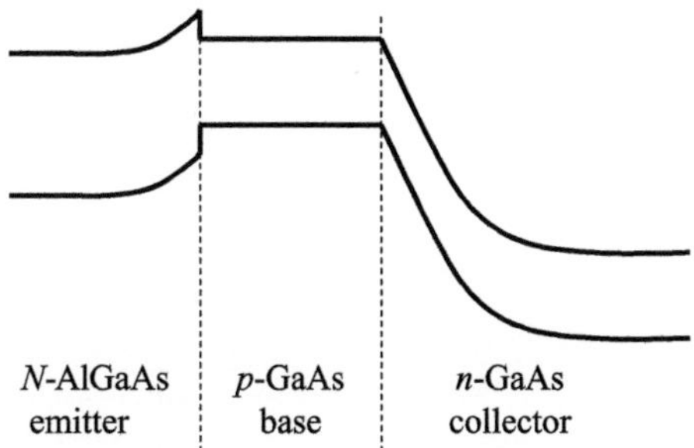

Fig. 6.15 Energy band diagram of a single heterojunction wide band-gap emitter HBT

to decrease. The separation between the top of the hole energy bands and the bottom of the Γ-valley is the energy gap. The negative differential mobility gives rise to ω_T "peaking" behavior, which will be discussed later. Third, at sufficiently high fields the saturation velocities of GaAs and Si are comparable at 100 μm/ns.

In most GaAs based HBT's, the base and collector are GaAs and the emitter is a wide bandgap material such as AlGaAs or InGaP. The use of the wide band gap emitter has several important effects. First, the discontinuity in the band gap provides a barrier to hole flow compared to electron flow. This enables the use of a very highly doped base layer and a relatively lowly doped emitter while maintaining good injection efficiency. The device is able to maintain very good current gain along with low base resistance and low collector capacitance. These latter two features are especially important for the high frequency operation of III-V HBT's. Reference [18] provides an extensive analysis of III-V HBT properties. Figure 6.15 depicts the energy band diagram of an abrupt *npn* single heterojunction bipolar transistor (SHBT). The upper case *N* denotes an *n*-doped wide bandgap semiconductor. The base and collector are *p*-type GaAs and *n*-type GaAs, respectively. In this case the device is forward biased.

Figure 6.16 shows the forward Gummel plot of a graded emitter InGaP/GaAs HBT. The base current displays three distinct regions. At low bias, the current is dominated by recombination current in the base-emitter depletion region. At moderate bias the base current has an ideality factor N_{EI} of 1.04, slightly different from the "theoretical" value of 1.0. At high bias, the current is limited by series resistance

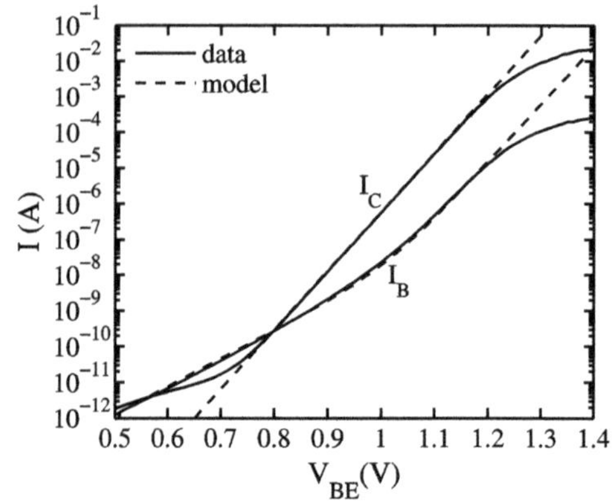

Fig. 6.16 Forward Gummel plot of a graded emitter InGaP/GaAs HBT

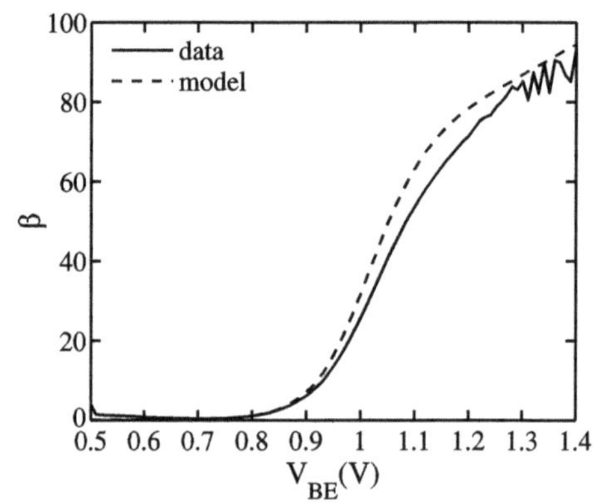

Fig. 6.17 $\beta = I_C/I_B$ of a graded emitter InGaP/GaAs HBT

(which was not included in the model results). In contrast, the collector current has an ideality factor N_F of 1.018; as discussed previously, although theoretically for silicon N_F should be close to 1, this is not observed for III-V HBTs, which is why it is useful in practice to have N_F available as a model parameter. The difference in ideality factors between base and collector currents indicates that at no point in the bias region is the device $\beta = I_C/I_B$ approximately constant as is seen in silicon devices. Figure 6.17 is a plot of β for this device displaying this fact. Accurate dc modeling of these devices requires that the base and collector currents be decoupled. Relating one current to the other through a constant current gain term, as was done in some older SPICE models, is inaccurate for III-V devices in general.

Figure 6.18 is a plot of the reverse Gummel characteristics of the same device. The base current is much greater than the emitter current yielding a reverse β much less than one. Moreover, the base current is entirely depletion region recombination current as reflected in its ideality factor N_{CN} of 1.95. The emitter current ideality factor is $N_R = 1.018$ (no series resistance was included in the model).

The presence of the differing semiconductor materials may cause one to wonder if reciprocity holds. That is, is the transport through the base the same when emitter

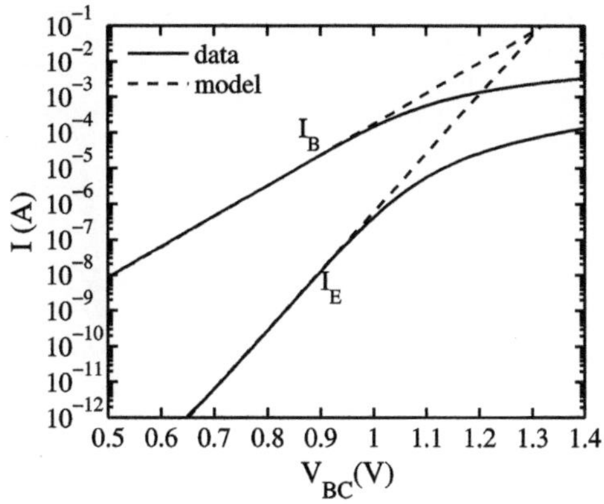

Fig. 6.18 Reverse Gummel plot of a graded emitter InGaP/GaAs HBT

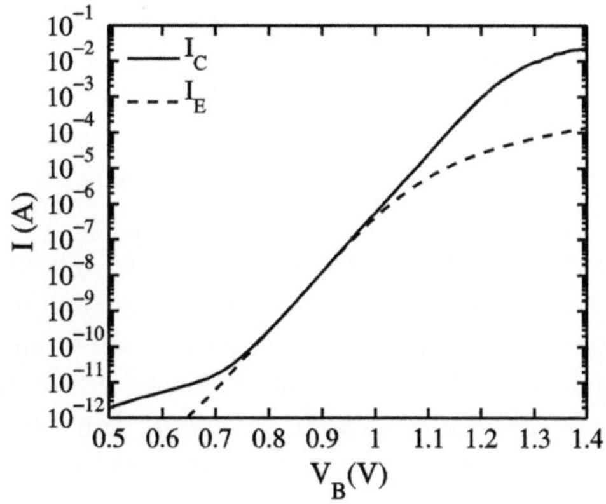

Fig. 6.19 Reciprocity of a graded emitter InGaP/GaAs HBT

and collector are reversed? This is a fundamental relationship in silicon homojunction transistors, expressed as [7, 24]

$$\alpha_R I_{CS} = \alpha_F I_{ES} = I_S \tag{6.86}$$

where α_F and α_R are the forward and reverse common base current gains and I_{ES} and I_{CS} are the so-called emitter and collector "diode" reverse saturation currents. Figure 6.19 compares the forward collector and reverse emitter currents for a graded emitter device. In the ideal region of operation these two current are indeed essentially equal. This does not hold for abrupt HBT's, because transport over the base-emitter junction differs from that in the case of the GaAs homojunction. In the latter case, there may be tunneling and thermionic components present in addition to the diffusion component. Only a diffusion component is present in the case of a homojunction. When the junction is graded, the behavior is more like that of homojunction device. The departure from reciprocity is observed in the forward and reverse Gummel characteristics. Modeling this behavior requires having separate forward and reverse saturation currents.

The difference in forward and reverse operation gives rise to another feature not observed in homojunction BJT's: the presence of an offset voltage in the output

Fig. 6.20 Output characteristics of an InGaP/GaAs HBT (I_B: 50 to 450 μA in 50 μA steps)

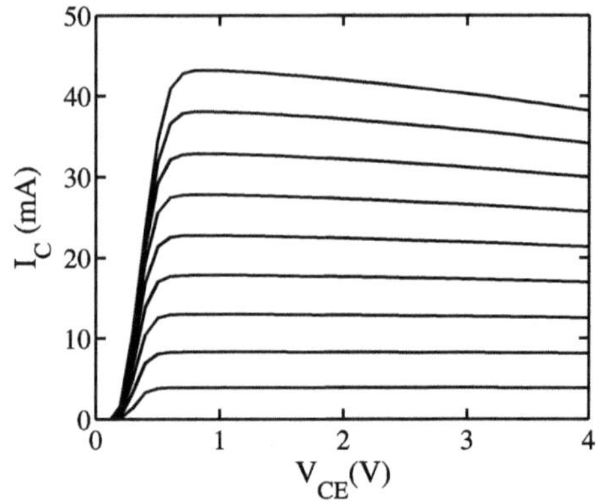

characteristics of GaAs HBTs. Figure 6.20 shows output characteristics of an InGaP/GaAs HBT where I_C is not zero at $V_{CE} = 0$ but rather for $V_{CE} \approx 0.12$ V. This offset voltage, V_{OS}, for graded junction HBT's is given by

$$V_{OS} = N_F \phi_t \ln\left(\frac{1}{\alpha_R}\right). \tag{6.87}$$

In the case of GaAs based HBT's α_R is on the order of 0.01, which is much lower than the approximately unity value seen in silicon.

Figure 6.20 also indicates two other important features of III-V devices. First, the high thermal resistance of the GaAs substrate gives rise to substantial device heating. This results in reduced current gain with increasing temperature, so self-heating is extremely important in these devices and cannot be neglected. Second, in regions where self-heating is not significant, the output characteristic is essentially flat indicating that the Early voltage V_{EF} is large. This is expected because when the base is highly doped compared to the collector, from the analysis of Sect. 6.2,

$$V_{EF} \approx \sqrt{\frac{2q\phi_{bi,C}}{\epsilon_s N_C}} t_{0b} N_B; \tag{6.88}$$

the very high base doping in these devices should lead to a large Early voltage, even though the base is relatively thin.

Figure 6.21 shows $f_T = \omega_T/(2\pi)$ as a function of collector current density. The unique feature displayed in this characteristic is the "peaking" behavior that is not seen in silicon homojunction devices. This behavior has been attributed to the transport properties of electrons in GaAs. Recall the velocity-field profile of Fig. 6.13. As the collector current is increased, the electrons compensate the depletion charge in the collector causing a reduction in electric field, which in turn causes the carrier velocity to increase. This is seen as a reduction in the transit time. A unique formulation for the transit time has been implemented to account for this [11]. Figure 6.22 shows the extracted base-collector capacitance C_{BC} as a function of collector current density. As I_C is increased, C_{BC} tends to decrease but then at some point it begins to rise again. This behavior has been explained in terms of charge compensa-

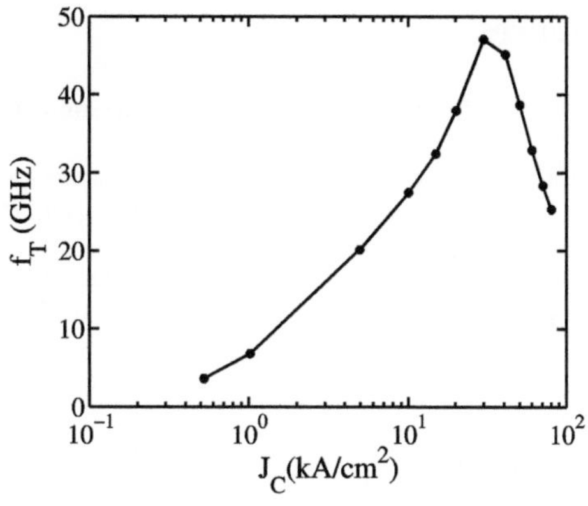

Fig. 6.21 f_T vs. collector current density

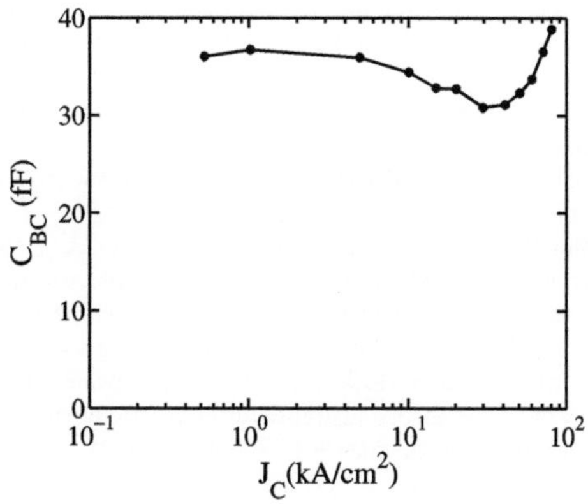

Fig. 6.22 C_{BC} vs. collector current density

tion in the collector [26], however, it has also been suggested that it is due to carrier velocity modulation [2].

6.9 Conclusions

In the chapter we have reviewed basic approaches to bipolar transistor modeling, including the physical basis of several common models. We have also shown where "non-ideal" characteristics are critical for modeling the non-ideal behavior observed in III-V HBTs, although silicon BJTs are often close to ideal. We have concentrated on analysis of the intrinsic transistor. Parasitics are also important, and these are treated in the Chapters on the Mextram and HiCuM models. The basis of these models is still the integral charge control relationship, although Mextram, adapting the Kull-Nagel approach, has separate handling of the base and base push-out regions, whereas in HiCuM these regions are treated in a unified manner.

References

1. Bardeen, J., Brattain, W.H.: The transistor, a semi-conductor triode. Phys. Rev. **84**(2), 230–231 (1948)
2. Betser, Y., Ritter, D.: Reduction of the base-collector capacitance in InP/GaInAs heterojunction bipolar transistors due to electron velocity modulation. IEEE Trans. Electron Devices **46**(4), 628–633 (1999)
3. Davis, W.F., Ida, R.T.: Statistical IC simulation based on independent wafer extracted process parameters and experimental designs. In: IEEE Bipolar Circuits and Technology Meeting (BCTM), pp. 262–265 (1989)
4. de Graaff, H.C., Klaassen, F.M.: Compact Transistor Modeling for Circuit Design. Springer, Berlin (1990)
5. Early, J.M.: Effects of space-charge layer widening in junction transistors. Proc. Inst. Radio Eng. (IRE) **40**(11), 1401–1406 (1952)
6. Fossum, J.G., Veeraraghavan, S.: Partitioned-charge-based modeling of bipolar transistors for non-quasi-static circuit simulation. IEEE Electron Device Lett. **7**(12), 652–654 (1986)
7. Getreu, I.: Modeling the Bipolar Transistor. http://www.lulu.com
8. Gummel, H.K.: A charge control relation for bipolar transistors. Bell Syst. Tech. J. **49**(1), 115–120 (1970)
9. Gummel, H.K., Poon, H.C.: An integral charge control model of bipolar transistors. Bell Syst. Tech. J. **49**(5), 164–174 (1970)
10. HiCuM group: HiCuM web site. http://www.iee.et.tu-dresden.de/iee/eb/hic_new/hic_start.html
11. Iwamoto, M., Root, D.E., Scott, J.B., Cognata, A., Asbeck, P.M., Hughes, B., D'Avanzo, D.C.: Large-signal HBT model with improved collector transit time formulation for GaAs and InP technologies. In: Digest IEEE Microwave Symposium (MTT-S), pp. 635–638 (2003)
12. Jeong, H., Fossum, J.G.: A charge-based large-signal bipolar transistor model for device and circuit simulation. IEEE Trans. Electron Devices **34**(1), 124–131 (1989)
13. Kirk, C.T.: A theory of transistor cutoff frequency f_T falloff at high current densities. IRE Trans. Electron Devices **9**(2), 809–814 (1962)
14. Koolen, M.C.A.M., Aerts, J.C.J.: The influence of non-ideal base current on $1/f$ noise behavior of bipolar transistors. In: IEEE Bipolar Circuits and Technology Meeting (BCTM), pp. 232–235 (1990)
15. Kroemer, H.: Theory of a wide-gap emitter for transistors. Proc. Inst. Radio Eng. (IRE) **45**(11), 1535–1537 (1957)
16. Kroemer, H.: Heterostructure bipolar transistors and integrated circuits. Proc. IEEE **70**(1), 13–25 (1982)
17. Kull, G.G., Nagel, L.W., Lee, S.W., Lloyd, P., Prendergast, E.J., Dirks, H.: A unified circuit model for bipolar transistors including quasi-saturation effects. IEEE Trans. Electron Devices **32**(6), 1103–1113 (1985)
18. Liu, W.: Handbook of III-V Heterojunction Bipolar Transistors. Wiley, New York (1998)
19. Massobrio, G., Antognetti, P.: Semiconductor Device Modeling with SPICE, 2nd edn. McGraw-Hill, New York (1983)
20. Mextram Group: Mextram website. http://mextram.ewi.tudelft.nl/
21. McAndrew, C.C., Nagel, L.W.: SPICE Early modeling. In: IEEE Bipolar/BiCMOS Circuits and Technology Meeting (BCTM), pp. 144–147 (1994)
22. McAndrew, C.C., Seitchik, J.A., Bowers, D.F., Dunn, M., Foisy, M., Getreu, I., McSwain, M., Moinian, S., Parker, J., Roulston, D.J., Schroter, M., van Wijnen, P., Wagner, L.W.: VBIC95, the vertical bipolar inter-company model. IEEE J. Solid-State Circuits **31**(10), 1476–1483 (1996)
23. Nagel, L.W.: SPICE2: A computer program to simulate semiconductor circuits. Memo ERL-M520 University of California, Berkeley (1975)
24. Roulston, D.J.: Bipolar Semiconductor Devices. McGraw-Hill, New York (1990)
25. Sah, C.T.: Fundamentals of Solid-State Electronics. World Scientific, Singapore (1991)

26. Samelis, A.: Modeling the bias dependence of the base-collector capacitance of power heterojunction bipolar transistors. IEEE Trans. Microw. Theory Tech. (MTT) **47**(5), 642–645 (1999)
27. Shockley, W.: The theory of p-n junctions in semiconductors and p-n junction transistors. Bell Syst. Tech. J. **28**, 435–489 (1949)
28. Shockley, W.: Circuit element utilizing semiconductive material. U.S. Patent 2,569,347 (1951)
29. Sze, S.M., Ng, K.K.: Physics of Semiconductor Devices, 3rd edn. Wiley, New York (2006)
30. Taur, Y., Ning, T.: Fundamentals of Modern VLSI Devices. Cambridge University Press, Cambridge (1998)
31. Tsividis, Y., McAndrew, C.: Operation and Modeling of the MOS Transistor, 3rd edn. Oxford University Press, London (2010)
32. Turgeon, L.J., Mathews, J.R.: A bipolar transistor model of quasi-saturation for use in computer aided design (CAD): In: IEDM Tech. Digest, pp. 394–397 (1980)

Chapter 7
Mextram

R. van der Toorn, J.C.J. Paasschens, W.J. Kloosterman, and H.C. de Graaff

Abstract We present the Mextram model, an industrial world standard compact model for bipolar transistors, showing the identity, philosophy and capabilities of the model. Mextram has been developed to capture all terminal characteristics of bipolar transistors that are relevant to industrial electronic circuit design of any Si or SiGe bipolar transistor, under all relevant practical circumstances. History, basic structure and features of the model are discussed, including simulation of heating effects, noise, geometrical scaling and statistical analysis. The relevance of the refined topology of its equivalent circuit, to simulation of advanced ac-characteristics of modern high-speed Si and SiGe transistors is extensively demonstrated.

7.1 Introduction

7.1.1 History

Mextram, currently a *TechAmerica/GEIA/Compact Model Council*[1] [25] world standard compact model for bipolar transistors, has a rich history. Four generations, or

[1]TechAmerica, formed by the merger of AeA (formerly the American Electronics Association), the Cyber Security Industry Alliance (CSIA), the Information Technology Association of America (ITAA) and the Government Electronics & Information Technology Association (GEIA), offers leading federal market research and standards development programs to the high-tech industry at large: www.geia.org/index.asp?bid=597.

R. van der Toorn (✉) · H.C. de Graaff
Faculty of Applied Mathematics, Electrical Engineering & Computer Science, Delft University of Technology, Mekelweg 4, 2628 CD Delft, The Netherlands
e-mail: R.vanderToorn@tudelft.nl

J.C.J. Paasschens · W.J. Kloosterman · H.C. de Graaff
NXP Semiconductors, Eindhoven, The Netherlands

G. Gildenblat (ed.), *Compact Modeling*,
DOI 10.1007/978-90-481-8614-3_7,

levels, of Mextram have been released since the release of the first industrial version in 1985 [11]. These have served computer-aided circuit design and simulation throughout multiple generations of semiconductor technology. Several generations of researchers have contributed to its development and utilization.

The first complete Mextram model definition was formulated by H.C. de Graaff at Philips Research laboratories in 1982. It was based on results of pioneering research in the field of bipolar transistor device physics carried out since the late 1960s [7, 14, 49]. The Ph.D. thesis [8] of De Graaff on the device physics of lightly doped collectors formed a stepping stone to the development of the model. The representation of physics of the electric field and charges in the intrinsic collector, often denoted as the *epilayer*, would form a key component of the Mextram model throughout the various generations of it. Indeed, one could say that Mextram's collector epilayer model forms part of the identity of model. A large part of the current chapter will be devoted to the role of and the ideas behind Mextram's epilayer model.

While it had been developed at Philips Research laboratories (Eindhoven, The Netherlands), and its first generations have been used and evolved within *Royal Philips Electronics* throughout the 1980s [12, 15, 32] and early 1990s [9, 35, 58], Mextram was brought in the public domain in 1994 [13, 33] which opened the way to close involvement of academia (Delft University of Technology, The Netherlands) in its development [16–18]. This can be viewed as a manifestation of a larger trend in semiconductor industry. With the ongoing down-scaling of electronic circuitry and devices, and the associated up-scaling of production and research facilities and costs came a trend towards commercialization of software development. Circuit simulation tools in general became core business of specialized software companies, that in turn served e.g. dedicated semiconductor manufacturers and fab-less circuits design houses. The general trend was directed away from company restricted, in-house software development and maintenance and towards an economy in which computer simulation and computer aided design (CAD) support was the domain of dedicated professional software vendor companies. In such context, large and general compact semiconductor models only had right of existence in the public domain, where they were shared among, and supported by, multiple companies in the semiconductor business.

One may consider as another manifestation of this trend the formation of inter-company organizations, through which semiconductor companies aim to deal with the necessity to actually coordinate developments in semiconductor business. A dominant exponent of this is the *International Technology Roadmap for Semiconductors (ITRS)*. A key organization with respect to compact model developments is the *Compact Model Council* (CMC) [25], in which a large group of leading companies in the semiconductor business, production companies, software vendors and associated serving parties, have organized themselves, with the aim to coordinate developments in computer-aided design tools for circuit design and compact semiconductor device modeling.

Considering bipolar applications, certainly in the late 20th century, trends have been in the direction of high frequency power amplifier applications, in the analog and mixed-signal domain, where distortion and noise are key quantities. Silicon bipolar technology has gone through revolutionary developments with the introduction of SiGe hetero-junction technology in the 1990s. To serve these trends in applications and technology, Mextram level 504 was developed at Philips Research by Paasschens and Kloosterman [34, 43, 44, 46]; Mextram 504 was released in 2000 and since then it has been subject of continued further research and development [39–41, 47, 48, 55].

Mextram was accepted as an industrial World Standard Compact Model for bipolar transistors by the CMC in 2004. By that time Mextram had been adopted by Delft University of Technology (The Netherlands), so that in the 21st century, Mextram was no longer only available in the public domain, but it has become fully part of it and it has found its home there.

7.1.2 Lumped-Element Modeling

Mextram is a physics-based compact model for bipolar transistors. The aim of Mextram is to capture all relevant terminal characteristics of any Si or SiGe bipolar transistor, under all relevant practical circumstances. In practice this means that Mextram supports accurate computer simulations of all relevant observable terminal quantities (voltages, currents, noise) as a function of applied terminal boundary conditions (e.g. biases, signals) and ambient temperature. It does so both for stationary and time-dependent boundary conditions.

As far as time-dependent signals are considered, one usually requires accurate simulation results only in the frequency range in which the transistor still acts as an active device. This allows for adoption of the *lumped-element* approach which brings a great reduction in computational complexity, as compared to, for example, solving partial differential equations describing all relevant physical fields inside the transistor.

In lumped-element models, a physical system is represented by an equivalent circuit. The equivalent circuit that is used in Mextram to represent the physical structure of a bipolar transistor is shown in Fig. 7.1.

On the one hand this equivalent circuit represents the essential physical topology of a bipolar device, and identifies the essential constituent parts of it. On the other hand it outlines the global mathematical structure of the model. The latter is basically formed by Kirchhoff's laws for voltages and currents and the conservation of charges in the network. Indeed, as a basis for time-independent (dc) simulations, Mextram is represented by a system of algebraic equations, whereas for elementary time-dependent simulations, it is represented by a system of ordinary differential equations. Once more, we recognize the very significant reduction in computational complexity, indeed by at least an order of magnitude, of lumped element models, compared to full physical modeling in space and time by partial differential equations.

Solving network equations still poses a considerable challenge. This challenge is fully common however with the problem of simulation of electronic circuits in

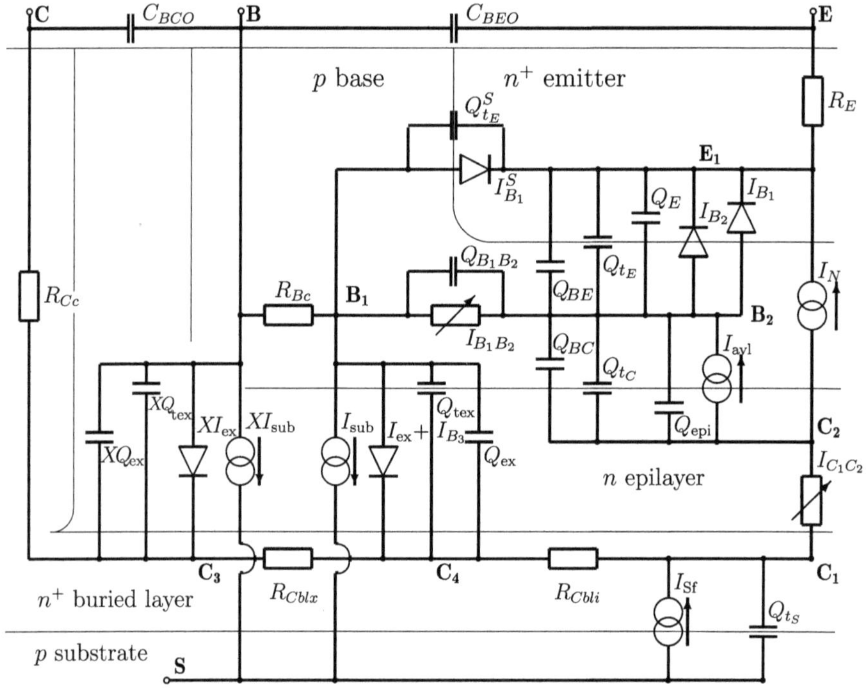

Fig. 7.1 Full equivalent circuit of the Mextram 504.7 model. The physical topology of a bipolar transistor is indicated by *thin lines*. The Mextram equivalent circuit is superimposed on this outline (Ref. [55] © [2007] IEEE)

general and lumped compact models such as Mextram can rely on the ample availability of so-called circuit simulators, i.e. software tools that have been especially developed to solve network equations. As a consequence, to understand Mextram, we have to understand its network and its constituents in the first place. Solving the mathematical equations that are thus implicitly formulated, can be left to circuit simulators.

7.1.3 Modeling Time-Dependence, Non-Linearity, Large Signals

Capturing the time-dependence of all observables is done by representing all relevant bias-dependent charges in the device; these charges are indicated by capacitor symbols in Fig. 7.1. The time-dependent processes are then basically the processes of charging and discharging these capacitors in the network.

The capacitors in the Mextram equivalent circuit are 'non-linear' network elements. This simply means that the physical charges that they represent depend on the biases in the network in a non-linear way, so that the associated capacitances would have a bias dependent value. In fact, most of the elements of the Mextram equivalent circuit are non-linear in this sense.

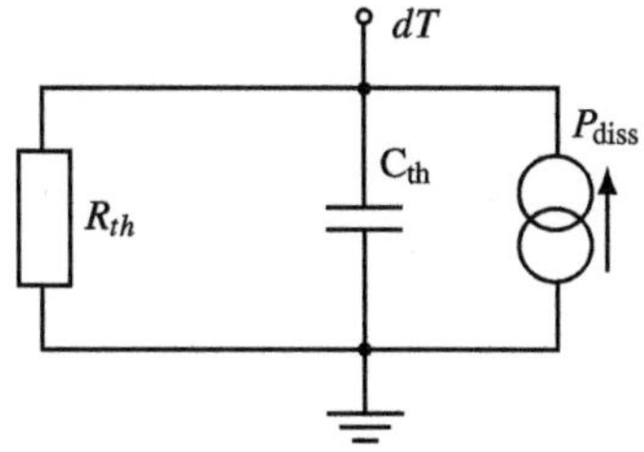

Fig. 7.2 Equivalent circuit for simulation of time-dependent self-heating of a device

By including the full non-linear bias dependence of its constituent elements, Mextram provides an appropriate basis to simulate the full, non-linear response of bipolar transistors to large signals. This forms the basis for simulation of distortion of electronic circuits involving bipolar transistors.

7.1.4 Temperature Dependence and Heating

To capture temperature dependence of the transistor characteristics, Mextram simply represents the temperature dependence of its network elements. As we shall see in more detail below, the network elements represent physical parts and processes within the transistor. Based on this, the temperature dependence of the network elements follows from physical principles.

In static situations, the temperature of the device itself can be calculated from the simple relation

$$T = T_{amb} + P_{diss} R_{th}. \tag{7.1}$$

This relation calculates the device temperature T from the ambient temperature T_{amb}, the total power P_{diss} dissipated in the device and the constant thermal resistance R_{th} of the device.

The fact that the device, through dissipation of power, increases its own temperature is known as *self-heating*.

In Mextram, relation (7.1) is encompassed by a small separate equivalent circuit, depicted in Fig. 7.2. While relation (7.1) is valid only for static situations, this circuit in fact represents the transistor as a first order linear time-dependent thermal system, with a thermal time constant $\tau_{th} = R_{th} C_{th}$, so that time-dependent self-heating can be addressed as well in circuit simulations. In Mextram, the thermal resistance itself depends on the ambient temperature [48].

In a circuit simulation setting, the increase in device temperature dT is accessible through the dedicated temperature terminal shown in Fig. 7.2. This provides the possibility to represent the thermal behaviour of a transistor by a more advanced network. This can be done by effectively disabling the internal self-heating network—by setting $R_{th} \to \infty$ and $C_{th} \to 0$—and connecting an alternative network to the thermal node.

By connecting temperature nodes of different instances of Mextram in a circuit, as depicted in Fig. 7.3, one can model the fact that transistors exchange heat and hence influence each other's temperature, a phenomenon known as *mutual heating*.

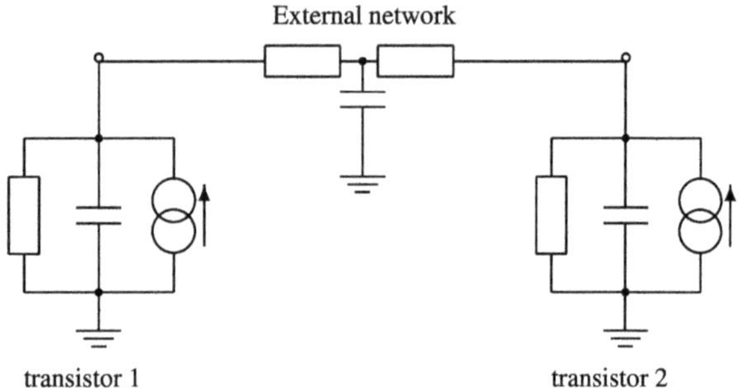

Fig. 7.3 Example of an equivalent circuit for simulation of time-dependent mutual heating of two devices

7.1.5 Noise Model

For modeling of noise, i.e. random fluctuations in currents and voltages, noise sources are associated with elements of the equivalent circuit [3, 5, 56, 57]. In Mextram, thermal noise sources are associated with resistors. In case of the variable base resistance, current crowding effects are taken into account [40]. Shot noise sources represent noise arising from currents that flow across junctions. The collector current shot noise model includes contributions from avalanche. The model for avalanche associated noise also covers the correlation between base and collector current noise due to avalanche [41]. So called $1/f$ noise sources in various parts of the transistors are also represented. Details of Mextram's noise model can be found in the model definition document [26].

7.1.6 Geometric Scaling and Statistical Modeling

In IC design it is important to have a geometrically scalable transistor model, i.e. a compact model that is able to describe the characteristics of a transistor as a function of its geometrical structure and size. The Mextram model has a sound physical base and describes the various physical parts of a transistor. As a result, the values of the model parameters of Mextram scale excellently with geometry [10, 45]. Actual complete and accurate geometrical scaling is only possible if the technological process and the geometrical layout of the transistors to be modeled is known in some detail. Therefore, in order to preserve general applicability, Mextram itself does not contain a geometrical scaling model. The geometrical scaling model is supposed to be added to it, as a shell (e.g., [59–61]).

On basis of Mextram, being a physics based model, supplemented with physically based geometry scaling of its parameters, it is possible to introduce statistical modeling [10, 45]. In this branch of modeling, besides the mean value, also the spread and correlation of compact model parameter values is determined. This may then serve as a basis for prediction of statistical variations in transistor characteristics. Statistical variations can be considered in one chip, on one wafer, in a batch (lot of wafers) or in a whole process during a period of time.

7.2 Model Structure and Components

7.2.1 Outline

7.2.1.1 Mextram Topology and Physical Topology of a Bipolar Transistor

We shall assume basic familiarity of the reader with the physical layout of bipolar transistors. Based on this, in the following we shall discuss the correspondence of network topology to the physical topology of a vertical bipolar transistor.

In the general topology of a vertical bipolar transistor one can always discern an intrinsic emitter layer, a base layer and a collector layer. Together, these form the intrinsic bipolar transistor, which actually implements the characteristic bipolar transistor action.

In Fig. 7.1, the physical topology of a bipolar transistor is indicated by thin lines. The Mextram equivalent circuit is superimposed on this outline. The intrinsic transistor is located between nodes E_1, B_1 and C_1. In this region, in the vertical direction, from top to bottom, one recognizes the physical emitter, base and collector (epi-) layers. In planar technology, all of the transistor may find itself on top of a semiconducting substrate layer, with which the buried collector layer forms a pn-junction.

The buried collector layer itself, in planar bipolar technologies serves to contact the intrinsic transistor from below. The buried collector in turn is contacted from above by means of a vertical collector plug, which is indicated at the left of the figure and which is represented in Mextram by the resistor R_{Cc}.

7.2.1.2 Intrinsic Transistor and Resistances

Node E_1 of Mextram's circuit (Fig. 7.1) corresponds to the emitter of the intrinsic transistor. It is connected to the external emitter contact E by the emitter resistor R_E. The base node B is connected to the internal node B_2 via two resistances. These are the constant resistor R_{Bc} and the variable resistor R_{Bv}; in Fig. 7.1 R_{Bv} is represented by the current source I_{B1B2}. This non-linear current source I_{B1B2}, describes DC current crowding under the emitter. The effective resistance of this current source at zero bias conditions, denoted by R_{Bv}, is a parameter of the Mextram model. Between the two base resistors an extra internal node is present: B_1.

Starting from the external collector node C, we meet a constant resistor R_{Cc} representing the collector plug and two resistors R_{Cblx} and R_{Cbli} that represent the resistance of the collector buried layer. These resistors form a network with the various extrinsic collector-base junction capacitances XQ_{tex}, XQ_{ex}, Q_{ex}, Q_{tex}. The filter network thus formed enables complete, simultaneous modeling of ac-characteristics—e.g. high frequency current-, voltage- and power gain—for advanced BiCMOS technologies. We shall come back to this in Sect. 7.2.2. Here, we just mention that the resistances R_{Cblx} and R_{Cbli} can have zero value in Mextram, in which case the internal nodes C_3 and C_4 disappear and merge into node C_1. The simplified circuit that thus appears brings reduced computational load and while it will be still accurate for dc modeling.

The epilayer has its own 'resistance'. At high current densities many effects take place in the epilayer, to be discussed in more detail in Sect. 7.3. In Mextram the epilayer is modeled by a controlled current source $I_{C_1C_2}$.

7.2.1.3 Junction Currents, Main Current

The main transistor current is represented by the current source I_N in Fig. 7.1. In Mextram the description of this current is based on the Gummel's charge control relation [21]. This means that the deviations from an ideal transistor current are given in terms of the charges in the intrinsic transistor and quasi Fermi levels at the relevant nodes of the intrinsic transistor. The main current depends (even in the ideal case) on the voltages of the internal nodes E_1, B_2 and C_2.

In addition to the main current, a bipolar transistor carries base currents. In the forward mode the base current shows an ideal and a non-ideal regime. In Mextram, these regimes are represented by current contributions I_{B_1} and I_{B_2} respectively, each represented by a diode in Fig. 7.1.

In reverse mode Mextram also has an ideal and a non-ideal base current. Because commonly the extrinsic part of the collector-base interface is geometrically much larger that the intrinsic part under the emitter, the reverse collector base currents are mainly determined by the extrinsic base-collector pn-junction. On these grounds, in Mextram 504 they are neglected in the intrinsic transistor.

The last current source in the intrinsic transistor is the avalanche current I_{avl}. This current describes the generation of electrons and holes in the collector epi-layer due to impact ionization, and is therefore proportional to the current $I_{C_1C_2}$. The avalanche model of Mextram 504 accurately describes avalanche in the so-called weak avalanche regime. It captures current dependence (Kirk effect) of avalanche, as well as the consequences of the finite length of the collector epilayer [34].

7.2.1.4 Charges in the Intrinsic Transistor

Charges stored in the transistor are represented in Fig. 7.1 by capacitances. The charges Q_{t_E} and Q_{t_C} represent depletion charges resulting from the base-emitter and base-collector pn-junctions. The extrinsic regions will have similar depletion capacitances. The two diffusion charges Q_{BE} and Q_{BC} are related to the built-up of charge in the base due to the main current: electrons traversing the base and hence adding to the total charge. The charge Q_{BE} is related to forward operation and Q_{BC} to reverse operation. In hard saturation both are present. The charge Q_E is related to the built-up of holes in the emitter. Its bias dependence is similar to that of Q_{BE}. The charge Q_{epi} describes the built-up of charge in the collector epilayer.

7.2.1.5 Extrinsic Transistor, Parasitic Currents and Capacitances

The extrinsic regions of a transistor are simply defined as the regions outside the intrinsic region. In planar technologies on a semiconducting substrate, in the extrinsic region, we recognize a parasitic PNP-transistor and other, more simple parasitics.

We mention the current $I^S_{B_1}$ in the junction formed by the side wall of the emitter. Since the pn-junction between base and emitter is not only present in the intrinsic region below the emitter, a part of the ideal base current will flow through the sidewall. This part is given by $I^S_{B_1}$. Similarly, the sidewall has a depletion capacitance with charge $Q^S_{t_E}$.

The extrinsic base-collector region has the same elements as the intrinsic transistor. We already mentioned the base currents. For the base-collector region these are the ideal base current I_{ex} and the non-ideal base current I_{B_3}. Directly connected to these currents is the diffusion charge Q_{ex}. The depletion capacitance between the base and the collector is split up in three parts. We have already seen the charge Q_{t_C} of the intrinsic transistor. The charge Q_{tex} is the junction charge between the base and the epilayer. Mextram models also the charge XQ_{tex} between the outer part of the base and the collector plug.

The collector-substrate junction has, as any pn-junction, a depletion capacitance given by the charge Q_{t_S}. Furthermore, the base, collector and substrate together form a parasitic PNP transistor. This transistor has a main current of itself, given by I_{sub}. This current runs from the base to the substrate. In Mextram 504, the reverse mode of this parasitic transistor is not really modeled, since it is assumed that the potential of the substrate is the lowest in the whole circuit. However, to give a warning signal when this is no longer true a substrate failure current I_{Sf} is included.

Finally, the overlap capacitances C_{BEO} and C_{BCO} that model the constant capacitances between base and emitter or base and collector, due to for instance overlapping metal layers are shown in Fig. 7.1.

7.2.2 Relevance of Model Structure to Modeling Results

7.2.2.1 Introduction

Since their introduction in the 1990s, hetero-junction bipolar transistors (HBT's) have become the active devices of choice for e.g. modern low-noise power amplifier modules for wireless RF applications. To meet the tight specifications posed by such applications on noise, linearity and power efficiency for example, sophisticated circuit design schemes are applied [54]. In order to reduce development costs and time-to-market, modern industrial electronic design efforts rely on circuit simulations. Therefore the need for an industrially supported, advanced compact bipolar transistor model capable of describing relevant characteristics of modern HBT's in the relevant regimes of operation is required. In this section, we shall discuss the relevance of the distributed capacitance and resistance effects in this respect, thus

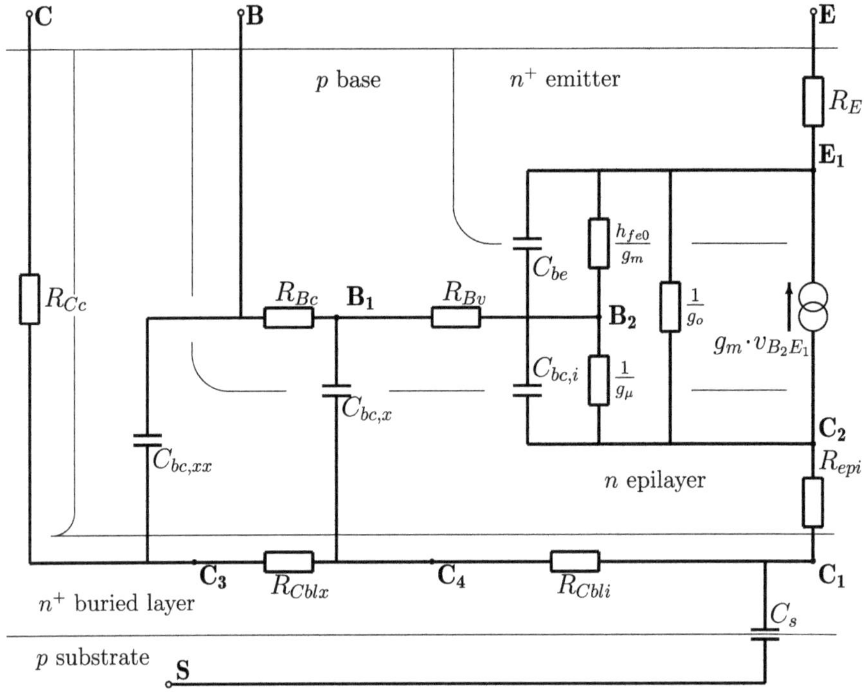

Fig. 7.4 Small-signal equivalent circuit of the Mextram 504.7 model with distributed collector resistance, as used as a starting point for small-signal analytical calculations (Ref. [55] © [2007] IEEE)

addressing the actual relevance of the refined structure of the Mextram equivalent circuit.

We shall focus on consequences for the modeling of common-emitter admittance-parameters and related functions, e.g. cut-off frequency f_T, unilateral gain G_u and available bandwidth f_a [28, 29]. For modern advanced RF-circuit design, adequate modeling of the real parts of admittance- ("y"-) parameters of the transistor, viewed as a two-port, deserves attention, even though at low frequencies y_{11}, y_{12} and y_{22} are relatively small quantities: y_{11}, y_{12} and y_{22} show *second order* ($O(\omega^2)$) dependence on frequency ω. The real parts of y- parameters play a role in *unilateralization* [54], stability [6, 50] and in the unilateral power gain G_u [37] of amplifier stages.

7.2.2.2 Analysis of Spatial Distribution Effects

Mextram supports extensive modeling of RC-distribution effects. Base resistance (R_{Bv} ("$I_{B_1B_2}$"), R_{Bc}) is distributed across collector capacitance. Two resistors R_{Cblx} and R_{Cbli} represent the resistance in the buried layer underneath the extrinsic base-collector junction and underneath the intrinsic transistor respectively; see Figs. 7.1 and 7.4.

In the present section we focus on the consequences of the distribution of resistances for small-signal characteristics. To address this issue we bear on analytical expressions for the small-signal admittance parameters, or y-parameters, of the circuit shown in Fig. 7.4; this circuit represents those features of the full equivalent circuit of Fig. 7.1 that are essential to the problem at hand.

A symbolic representation of y-parameters as algebraic functions of circuit elements reveals the influence of circuit elements on the small-signal characteristics. Below, we shall present the admittance parameters, or y-parameters, of the circuit in common-emitter configuration, with substrate node S connected to the emitter node E. When v_{be} and v_{ce} denote the complex input and output small-signal voltages, and i_b and i_c the corresponding complex input and output small-signal currents, the y- parameters are defined by

$$\begin{pmatrix} i_b \\ i_c \end{pmatrix} = \begin{pmatrix} y_{11} & y_{12} \\ y_{21} & y_{22} \end{pmatrix} \begin{pmatrix} v_{be} \\ v_{ce} \end{pmatrix}. \tag{7.2}$$

Calculation of the y-parameters as functions of the elements of the equivalent circuit technically comes down to algebraic manipulation of 2-dimensional matrices, the elements of which are complex rational functions. In the course of such manipulations formal Taylor expansions in the angular frequency (ω) can be applied in order to reduce the algebraic complexity of the intermediate and final expressions obtained. This means that the results presented here are to be interpreted as Taylor expansions in ω, about $\omega = 0$, and hence are approximate results in a formal low frequency limit. Furthermore, again to reduce the algebraic complexity, we apply the limits $h_{fe0} \to \infty$ and $g_o \to 0$. In this way we arrive at the following results.

For the transit time

$$\tau_T = \frac{1}{2\pi f_T}, \tag{7.3}$$

where f_T is the cut-off frequency (of the small-signal current gain $h_{fe} = |y_{21}/y_{11}|$), we find [55]

$$\begin{aligned} \tau_T = {} & \frac{C_{be} + C_{bc}}{g_m} + C_{bc,xx}(R_{Cc} + R_E) \\ & + C_{bc,x}(R_{Cc} + R_{Cblx} + R_E) \\ & + C_{bc,i}(R_{Cc} + R_{Cblx} + R_{Cbli} + R_{epi} + R_E); \end{aligned} \tag{7.4}$$

we have used the abbreviation $C_{bc} = C_{bc,i} + C_{bc,x} + C_{bc,xx}$. Relation (7.4) shows that, for a given fixed distribution of collector-base capacitance ($C_{bc,xx}$, $C_{bc,x}$, $C_{bc,i}$), the nearer the collector resistance contributions are to the collector contact C, the more they contribute to the transit time [28]. For collector-base capacitance, the opposite holds. In short: for high maximum f_T, first priority would be to have low $C_{bc,i}$ and low R_{Cc}.

In full Mextram simulations, the effect of distribution of collector resistance on the maxima of $f_T(I_C, V_{CE})$ is clearly recognizable in case of, for example, a modern QUBiC4X [19] SiGe HBT of emitter dimensions $0.4 \times 1.0\ \mu\text{m}^2$, as is shown in Fig. 7.5a. Note that at the corresponding bias conditions, g_m is high and hence the contribution of the first terms of expression (7.4) is relatively small.

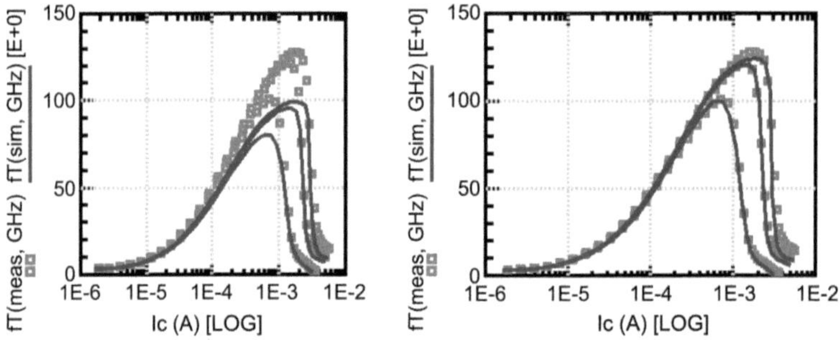

Fig. 7.5 Measured (symbols) and simulated (curves) cut-off frequency f_T as a function of collector current I_c, for several fixed values of the collector-emitter voltage: $V_{CE} = 0.5$, 1.0, 1.5 V, at an ambient temperature of 25°C. Cut-off frequency is calculated as $f_T = f_{meas} \cdot |Y_{21}/Y_{11}|$, from y-parameters measured at $f_{meas} = 5.337$ GHz. Fig. 7.5d shows results obtained when all of the collector resistance $R_{Cc} + R_{Cblx} + R_{Cbli}$ is assigned to R_{Cc}, while keeping all other model parameters unchanged; Sub-fig. a: Mextram simulations, with intentionally manipulated collector resistance distribution: all of the collector buried layer resistance included in the collector contact resistance R_{Cc}. Sub-fig. b: Full Mextram simulations (Ref. [55] © [2007] IEEE)

While relation (7.4) shows that distribution of collector resistance does affect device characteristics, the following well-known relation is not sensitive to it, at leading order in ω,

$$\mathrm{Im}(y_{12}) = -\omega C_{bc} + \mathrm{O}(\omega^2). \tag{7.5}$$

This relation can therefore be used to extract parameters for bias dependent modeling of the total collector-base capacitance. One should realize in this context, that in the representation of Fig. 7.4, the intrinsic capacitances $C_{bc,i}$ and C_{be} not only represent bias dependent junction depletion capacitances, but also diffusive charge storage in the intrinsic transistor. This latter remark also applies to the imaginary part of y_{11}, for which we find

$$\begin{aligned}\mathrm{Im}(y_{11}) = \omega(C_{bc} + \tilde{C}_{be} + \tilde{g}_m(C_{bc}R_{Cc} + C_{bc,x}R_{Cblx} \\ + C_{bc,i}(R_{Cbli} + R_{Cblx} + R_{epi}))),\end{aligned} \tag{7.6}$$

where

$$\tilde{g}_m = \frac{g_m}{1 + g_m R_E}, \tag{7.7}$$

and

$$\tilde{C}_{be} = \frac{C_{be}}{1 + g_m R_E}. \tag{7.8}$$

At low current densities, whenever g_m is sufficiently low to make the contribution from the corresponding terms negligible, one may use relation (7.6) to measure base-emitter depletion capacitance as a function of bias conditions. At higher current densities however, one not only observes the dependence on collector-resistance distribution, but also effects of diffusive base and emitter charges, corresponding to parameters like τ_B and τ_E in full Mextram simulations.

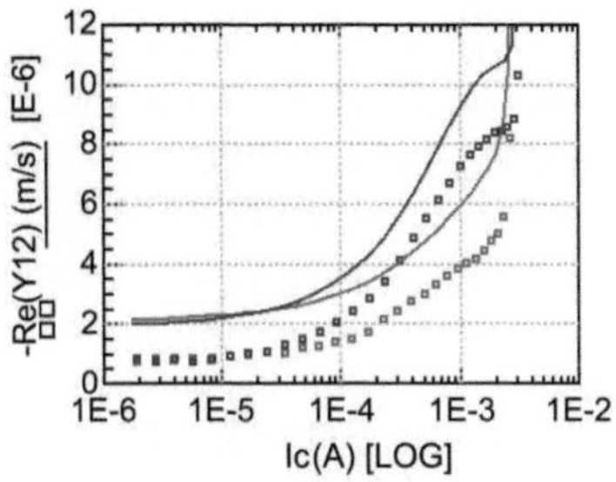

(a) Suppressed collector resistance distribution effects

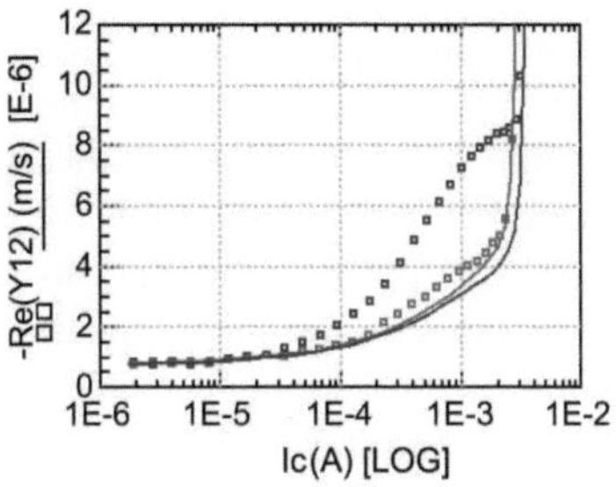

(b) Suppressed avalanche modeling.

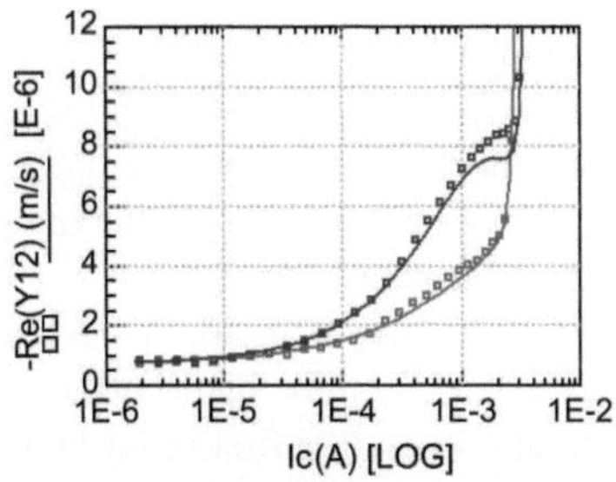

(c) Suppressed mixed Kirk- and avalanche effects. ("EXAVL = 0")

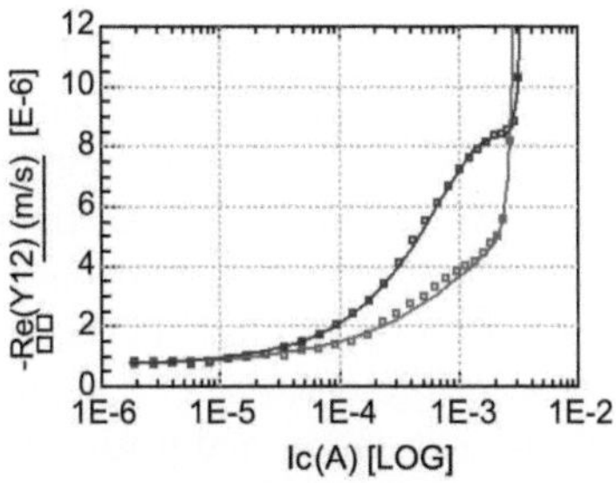

(d) Full Mextram Simulations.

Fig. 7.6 Measured (*markers*) and Mextram-simulated (*curves*) values of minus the real part of the 12-admittance parameter, $-\mathrm{Re}(Y_{12})$, as a function of collector current (I_C), for two values of collector emitter voltage V_{CE}: 1.5 V (lower measurements) and 2.0 V (upper measurements). The measurements were taken at $f_{meas} = 5.3$ GHz on an HBT, featuring $f_{T,peak} = 130$ GHz, from an industrial SiGe BiCMOS process. Besides showing the modeling capabilities of Mextram with respect to $\mathrm{Re}(Y_{12})$ (d), this sequence of plots serve to demonstrate the significance of resistance distribution (a), avalanche (b) and combined Kirk/avalanche (c) effects

In analogy to relation (7.5) one finds [38]

$$\mathrm{Re}(y_{12}) = -g_{\mu} + O(\omega^2). \tag{7.9}$$

Firstly, this expression shows that the real part of the y_{12} admittance parameter provides a suitable observable to study avalanche effects in the context of RF/ac conditions. Secondly, it shows that in addition to that, higher order ($O(\omega^2)$), RC-distribution effects are relevant to the quantity $\mathrm{Re}(y_{12})$. This is illustrated in Figs. 7.6a to d, which show measured (markers) and Mextram-simulated (curves) values of minus the real part of the 12-admittance parameter, $-\mathrm{Re}(Y_{12})$, as a function of collector current (I_C), for two values of collector emitter voltage V_{CE}: 1.5 V (lower measurements) and 2.0 V (upper measurements). In the various subplots, simulation results are shown for various intentionally manipulated parameter settings of the Mextram model. In this way, the significance of various aspects of Mextram is demonstrated.

Figure 7.6d demonstrates that Mextram is indeed quite capable of, in this case, simulating the details of the real part of the admittance parameter y_{12}. Figure 7.6b demonstrates the significance of avalanche conductance—the leading term in expression (7.9)—in this respect: for the simulations shown in this figure, the Mextram model parameters have been modified in such a way so as to suppress avalanche effects altogether. Apparently, to the observed values for $-\mathrm{Re}(Y_{12})$ at $V_{CE} = 2.0$ V avalanche was highly significant, as *without* taking avalanche into account, the model does not reproduce these measured values. At the frequency $f_{meas} = 5.3$ GHz at which these measurements were taken however, far below the characteristic cut-off frequency $f_{T,peak} = 130$ GHz of the device, the second order RC-distribution effects represented by the $\mathrm{O}(\omega^2)$ term in expression (7.9), turns out to be as significant as avalanche effects. Indeed, the Mextram simulation results presented in Fig. 7.6a were produced with a modified Mextram parameters set that changes just the distribution of the total buried layer collector resistance over the components R_{Cbli}, R_{Cblx} and R_{Cc}. The increase in $-\mathrm{Re}(Y_{12})$ for $V_{CE} = 1$ V (lower curve) with increasing current is dominantly due to diffusions charges in the base and emitter, Q_{BE}, Q_{BC} and Q_E in Fig. 7.1. Distinctive modeling of these three charges is relevant to the details of various components of the admittance parameters.

The real part of y_{11} turns out to be a complicated function of circuit elements, too extensive to be presented here. It significantly depends on the distribution of base- and collector resistances and it vanishes up to and including first order in a Taylor expansion in ω.

For the real part of y_{21} we find

$$\mathrm{Re}(y_{21}) = \tilde{g}_m + O(\omega^2), \tag{7.10}$$

where $O(\omega^2)$ denotes terms that are proportional to ω^2. At low frequencies, for example at the frequency of 5.3 GHz at which the measurements of Fig. 7.5 have been taken, $\tilde{g}_m$ fully dominates $\mathrm{Re}(y_{21})$ and emitter resistance R_E can be found or verified by confronting simulated to measured values of $\mathrm{Re}(y_{21})(I_C, V_{CE})$. At high frequencies however—in this case above 10 GHz, as shown in Fig. 7.7a—the higher order terms in frequency in $\mathrm{Re}(y_{21})$ are significant. The *available bandwidth* f_a for example, is defined as the -3 dB point, as a function of frequency, of the voltage gain G_v [29],

$$G_v = \left| \frac{-y_{21} Z_L}{1 + y_{22} Z_L} \right|, \tag{7.11}$$

where Z_L is chosen such that the low frequency value of G_v has a specified reference value, e.g. 10. For the QUBiC4X SiGe HBT [19], f_a (Fig. 7.7b) turned out to be determined by the high frequency drop-off of $\mathrm{Re}(y_{21})$ as a function of frequency. Indeed, assuming $C_s, C_{bc} \ll C_{be}$ one may derive

$$\mathrm{Re}(y_{21}) = \frac{g_m(1 + g_m R_E)}{(1 + g_m R_E)^2 + C_{be}^2 (R_{Bc} + R_{Bv} + R_E)^2 \omega^2}. \tag{7.12}$$

Equation (7.12) shows that, given R_E (see e.g. below (7.10)), one may obtain or verify base resistance from the frequency dependence of $\mathrm{Re}(y_{21})$, comparing simulated and measured values of $\mathrm{Re}(y_{21})$ as a function of frequency; consistency with the assumptions $C_s, C_{bc} \ll C_{be}$ may then be checked from the model simulations.

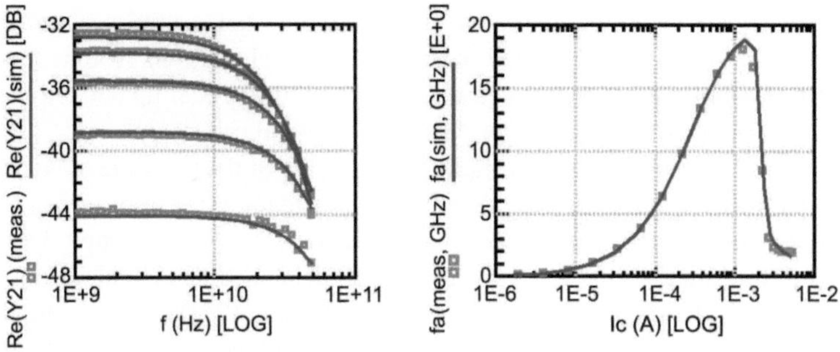

(a) $Re(y_{12})$ as a function of frequency, for, from upper to lower curves, a series of base-emitter bias conditions: $V_{BE} = 960, 930, 900, 870, 840\,\text{mV}$

(b) Available bandwidth

Fig. 7.7 Results (**b**) for measured (*symbols*) and Mextram simulated (*curve*) available bandwidth f_a as a function of collector current I_C, for a fixed value of the collector-base voltage ($V_{CB} = 0$ V), at an ambient temperature of 25°C. The bandwidth is determined as the -3 dB point in a frequency sweep measurement of the voltage gain of the transistor in common emitter configuration with a single load resistance. The low-frequency voltage gain was kept fixed at a value of 10. (**a**) shows measured and simulated values of $\text{Re}(y_{21})$, which largely determines f_a, as a function of frequency for $V_{CE} = 1.5$ V and, from upper to lower curves, $V_{BE} = 960, 930, 900, 870, 840$ mV, at an ambient temperature of 25°C (Ref. [55] © [2007] IEEE)

The imaginary part of y_{21} is a complicated function of collector- and base-capacitance and resistance distribution. However, for low currents, or low g_m one finds simply

$$\lim_{g_m \to 0} \text{Im}(y_{21}) = -C_{bc}\omega. \tag{7.13}$$

The real part of y_{22} is of $O(\omega^2)$ and depends on the modeling of substrate capacitance and resistance. The imaginary part of y_{22} may serve verification of the modeling of substrate capacitance C_s:

$$\text{Im}(y_{22}) = \left(C_{bc} + C_s + \tilde{g}_m(C_{bc,i} R_B + C_{bc,x} R_{Bc})\right)\omega + O(\omega^2). \tag{7.14}$$

7.2.2.3 More Advanced RF Characteristics

In this section we shall confront full-model (Fig. 7.1) small-signal simulations with measured small-signal ac-characteristics of a modern QUBiC4X [19] BNX-type SiGe HBT of emitter dimensions $0.4 \times 1.0\ \mu\text{m}^2$. In such simulations, the network elements of our small-signal equivalent circuit (Fig. 7.4) are to be considered as bias-dependent, but our algebraic results of Sect. 7.2.2.2 still indicate dependence of ac-characteristics on circuit elements.

The model parameter values used during Mextram simulations, as presented in the present section, were obtained by a combination of standard parameter extraction techniques[2] [45] supplemented by interpretation of measured y-parameter characteristics along the lines sketched in Sect. 7.2.2.2, calculations from technological data of the QUBiC4X process and analysis of geometrical scaling behavior of device characteristics and extracted parameters for individual transistors.

The cut-off frequency characteristic $f_T(I_C, V_{CE})$ of the example device was presented in Fig. 7.5 and discussed in Sect. 7.2.2.2. The total transit time τ_T (7.3) is the sum of contributions from many parts of the model. As was indicated by the result (7.4) from the small-signal analysis, these include R-C times from various parts of the Mextram network. A full analysis reveals (see, e.g. [27]) that in addition transit times from the neutral emitter, base and collector epilayer will contribute to the total transit time. In full Mextram simulations, these latter contributions are implicitly represented by partial derivatives of various charges with respect to currents. Avoiding details here, we summarize this by stating that in Mextram, RF device characteristics such as $f_T(I_C, V_{CE})$ are synthesized as the result of contributions from various physical processes. One demonstration of this was presented in Fig. 7.5a, which demonstrated for example the relevance of collector resistance distribution to the $f_T(I_C, V_{CE})$ device characteristic.

Figure 7.8d shows measured (symbols) and simulated (curves) unilateral gain G_u [37],

$$G_u = \frac{|\gamma_{21} - \gamma_{12}|^2}{4(\mathrm{Re}(\gamma_{11}) \cdot \mathrm{Re}(\gamma_{22}) - \mathrm{Re}(\gamma_{12}) \cdot \mathrm{Re}(\gamma_{21}))}, \tag{7.15}$$

where the gamma's are immittance parameters. G_u is a figure of merit for RF power gain [37]. Expression (7.15) gives the same numerical result for the value of G_u, independent of whether admittance (y_{ij}) or impedance (z_{ij}) parameters are substituted for the immittance parameters γ_{ij}. In this sense, the quantity G_u is a property of the device that transcends the arbitrariness that is intrinsic to the choice of representing a device by either admittance or impedance parameters. In this well-defined sense, G_u is a *device invariant* [22, 52].

Figures 7.8a to c illustrate how various significant physical contributions together synthesize the total unilateral power gain. In Fig. 7.8b, for example, we show results for $G_u(I_C, V_{CE} = 1.0, 1.5\ \mathrm{V})$ that are obtained if distribution of collector resistance would *not* be taken into account: for the simulations in this figure, all of the resistance $R_{Cc} + R_{Cblx} + R_{Cbli}$ was assigned to the plug resistance R_{Cc} [55].

Comparison of Figs. 7.8b and d, in turn, demonstrates the impact of weak avalanche effects in the collector base junction, as well as Mextram's ability to model this phenomenon. Indeed, as was also mentioned in Sect. 7.2.2.2, Mextram incorporates an advanced model for weak avalanche [34]; among other things, this avalanche effect takes into account the modulation of the electric field in the collector epilayer, by charge carriers taking part in the collector current (Kirk effect). Due to this effect, and due to the finite length of the collector epilayer, at high current densities the peak value of the electric field will occur at the buried layer end of the

[2]For the most recent model descriptions, source code, and documentation, see the web-site http://mextram.ewi.tudelft.nl.

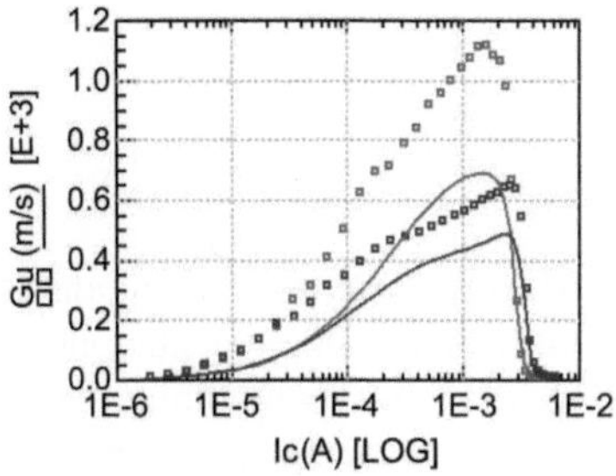

(a) Suppressed collector resistance distribution effects

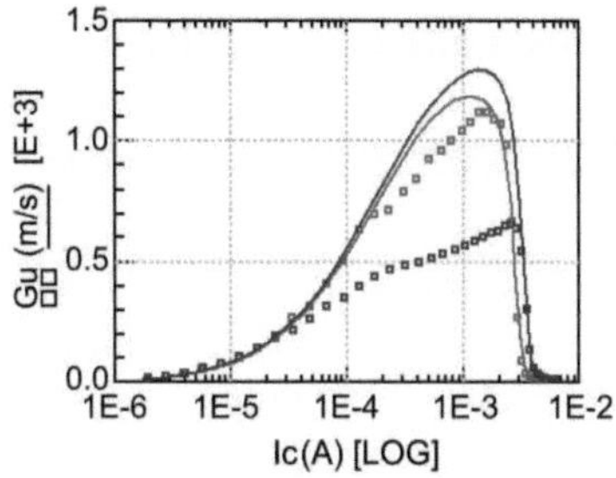

(b) Suppressed avalanche modeling.

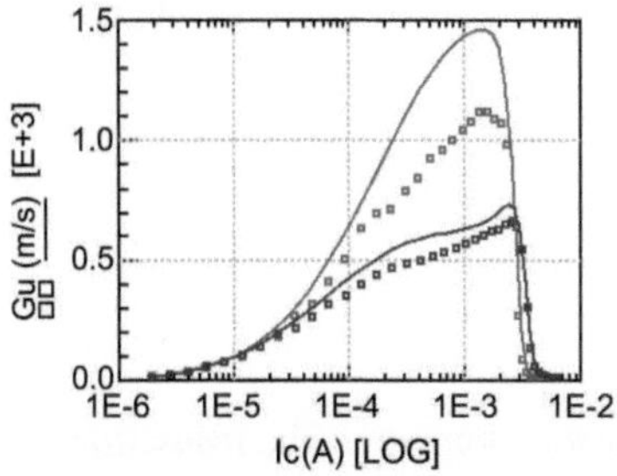

(c) Evenly distributed base resistance. ($R_{Bc} = R_{Bv}$)

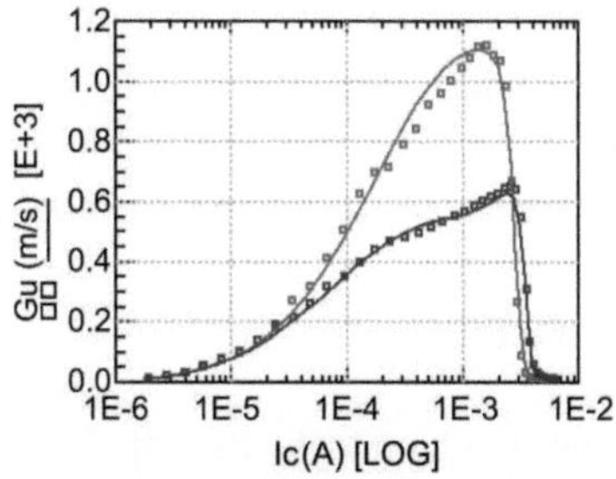

(d) Full Mextram Simulations.

Fig. 7.8 Measured (*markers*) and Mextram-simulated (*curves*) values of the unilateral power gain, G_u, as a function of collector current (I_C), for two values of collector emitter voltage V_{CE}: 1.5 V (upper measurements) and 2.0 V (lower measurements). The measurements were taken at $f_{meas} = 5.3$ GHz on an HBT, featuring $f_{T,peak} = 130$ GHz, from an industrial SiGe BiCMOS process. **d** demonstrates the modeling capabilities of Mextram with respect to G_u. The full sequence of plots demonstrates the significance, to modeling of G_u, of parasitic collector resistance distribution (**a**), avalanche (**b**) and base resistance distribution (**c**) effects

collector epilayer. At such a state, the electric peak value, and hence the avalanche effects, will *increase* with increasing collector current. This effect is demonstrated under RF conditions in Fig. 7.6c.

7.2.2.4 Conclusion

In Sect. 7.2 we have discussed the refined topology of the Mextram model and we have analyzed the significance of this topology to the capabilities of the model. We showed that the refined network topology captures higher order frequency effects, that are physically due to spatial distribution of charges, resistances and currents inside bipolar transistors and we showed that these are significant to RF small-signal representations of devices such as admittance parameters as well as to various well-known RF figures of merit such as cut-off frequency, available bandwidth and unilateral power gain.

7.3 Mextram Philosophy

7.3.1 Introduction

In this section we outline the philosophy of Mextram. We avoid going deeply into the technical details that are typical to compact model implementations. Instead we shall present an outline of the structure of Mextram, highlighting the physical content of the model and discuss its main components and key parameters in terms of bipolar transistor physics.

The conceptual understanding thus developed in practice turns out to be useful to those who apply the model, e.g. during parameter extraction or circuit simulation.

7.3.2 Main Transistor Current Model

7.3.2.1 Outline

The central unique feature of bipolar transistors, known as the *transistor action*, is embodied by the so called *main current* that is formed by charge carriers flowing through the base layer, from emitter to collector. Modeling this current as a function of electrical boundary conditions and temperature belongs to the core tasks of compact models for bipolar transistors.

The main current in the base of a bipolar transistor is well-described by a relation known as Gummel's *Charge Control Relation* (CCR) [21], in terms of (1) boundary conditions in terms of the quasi Fermi potentials at two suitably selected points in the transistor and (2) an integral property of the transistor known as the *Gummel number* (G_B).

The Gummel number can be viewed as a weighted integral of the majority carrier concentration in the base. From a natural viewpoint known as the *charge control principle*, which prescribes the relation between charges and currents in a bipolar transistor, the Gummel number depends on the main current itself. The CCR therefore tends to form an *implicit* equation for the main transistor current, rather than an explicit expression. When the CCR is applied in a compact model to calculate the main transistor current, this implicitness of the CCR presents an issue to be resolved (I).

During the derivation of the CCR [21, 27], in case of an npn transistor, it is assumed that the hole quasi Fermi potential φ_p is spatially constant. This is easily justified (II) in the base and its vicinity, on basis of the general transport equation for holes

$$\mathbf{J}_p = -qp\mu_p \nabla\varphi_p, \tag{7.16}$$

and the facts that the hole current $|\mathbf{J}_p|$ will be relatively small, while the hole concentration p is relatively large. In relation (7.16), q is the elementary charge and μ_p the hole mobility.

Modern bipolar transistors, especially *hetero-junction bipolar transistors* (HBT's), such as realized in high performance silicon-germanium technologies (SiGe), have a profoundly non-homogeneous spatial structure, as the material composition of emitter, base and collector can differ significantly. In terms of compact modeling, this suggest to represent emitter, base and collector regions by separate model components (III), so that differences in material properties, e.g. intrinsic carrier concentration and mobility, can be naturally represented by compact model parameters associated with different regions. Moreover, as we will discuss in Sects. 7.3.2.3 to 7.3.2.8, the collector epilayer hosts a remarkably wide range of physical states and phenomena.[3] This also suggests a separate part of a compact transistor model to be dedicated to the epilayer.

The three considerations (I–III) outlined above are reflected by the structure of Mextram's model for the main transistor current in the intrinsic transistor, which is formed by the current sources I_N and $I_{C_1C_2}$ between nodes E_1, C_2 and C_1 of Mextram's circuit (see Fig. 7.1).

The current source I_N between nodes E_1 and C_2 represents the main current in the base region, modeled by the CCR. Hence, the CCR is applied to model the main transistor current in a region where application of the CCR is indeed firmly justified (II).

Especially under time-independent conditions, when charging currents are absent, and outside the avalanche regime, so that $I_{avl} = 0$, it will hold that $I_N = I_{C_1C_2}$. That is, this equality will hold for any physically sound state, or "valid solution", of the transistor model. The fact that the main transistor current in the collector *epilayer*, $I_{C_1C_2}$, is nevertheless represented in Mextram as an *independent* variable resolves the issue (I) of the implicitness of the CCR. Indeed, especially in the *forward* ($I_{C_1C_2} > 0$) transistor mode of operation, in Mextram the current $I_{C_1C_2}$ is effectively just parameterized in terms of the nodal voltage at node C_2 and thus put under direct control of the circuit simulator. As this nodal voltage is not used otherwise in the model in forward mode ($I_{C_1C_2} > 0$), a circuit simulator solver will, through V_{C_2} effectively treat $I_{C_1C_2}$ as an independent variable which (a) has to obey the Kirchhoff current laws for the circuit and on which (b) other model components may depend. Since a value of the current $I_{C_1C_2}$ is thus well-defined at any stage during the computational process of circuit simulation, including model evaluation, it can be used to evaluate e.g. the Gummel number, so that the CCR can be evaluated as an explicit expression for I_N. Thus we see that the mere existence of $I_{C_1C_2}$ as an independent quantity within Mextram indeed resolves the issue of the implicitness of the CCR (I).

[3]This is also apparent from the extensive literature that is devoted to the physics of the epilayer: see e.g. references [1, 2, 4, 7, 9, 13, 14, 20, 23, 30, 36, 46, 51, 53].

7.3.2.2 Main Current I_N: Charge Control Relation

In Mextram, the main transistor current in the base, I_N, is basically modeled by the Charge Control Relation (CCR)

$$I_N = \frac{q n_{i0}^2 A_{em}}{G_B} \left(\exp\left(\frac{V_{B_2E_1}}{V_T}\right) - \exp\left(\frac{V^*_{B_2C_2}}{V_T}\right)\right). \tag{7.17}$$

In this relation, A_{em} is the effective emitter area, n_{i0} is a suitable constant reference value for the intrinsic carrier concentration $n_i(x)$ and

$$V_T = \frac{k_B T}{q} \tag{7.18}$$

is the thermal voltage, which is here expressed in terms of Boltzmann's constant k_B, absolute temperature T and elementary charge q.

For an npn transistor, the Gummel number G_B is defined as

$$G_B = \int_{E_1}^{C_2} \frac{p(x) n_{i0}^2}{\mu_n(x) V_T n_i(x)^2} dx, \tag{7.19}$$

where in the context Mextram the integration over the spatial coordinate x is performed from the position of the node E_1 at the emitter-base interface until the position of the node C_2 at the base-collector interface.

Expression (7.19) shows that Mextram's Gummel number G_B can be viewed as a weighted integral of the hole concentration in the base region. This hole concentration[4] in itself is given firstly by the base doping and secondly by additional holes in the neutral base that compensate the charge of the electrons that take part in the main current (Webster- or knee current-effect). In Mextram, the Webster effect is indeed directly represented through the Gummel number and by a model parameter I_K, called the *knee current*.

The integral (7.19) is furthermore determined by the width of the neutral part of the base, i.e. the part of the base between the emitter-base (EB) and base-collector (BC) depletion regions. The width of these depletion regions, and hence the width of the neutral base and therefore the Gummel number, is modulated by the applied EB- and BC- bias voltages. The associated effects on the transistor characteristics are know as the *reverse* (EB) and *forward* (BC) *Early effects*, modeled indeed in Mextram through the Gummel number and represented by model parameters V_{ER} and V_{EF} respectively.

In modern SiGe HBT technologies, the base of the transistor consists of a silicon-germanium alloy of possibly variable composition. Addition of germanium to silicon lowers the bandgap and strongly increases the intrinsic carrier concentration n_i and its representative value n_{i0}. This tends to increase the main current level (7.17) of the transistor. From the point of view of technology, the germanium content thus provides an independent degree of freedom for device optimization.

[4] Again we focus on npn transistors; for pnp transistors the Gummel number would be formulated in terms of electron concentrations.

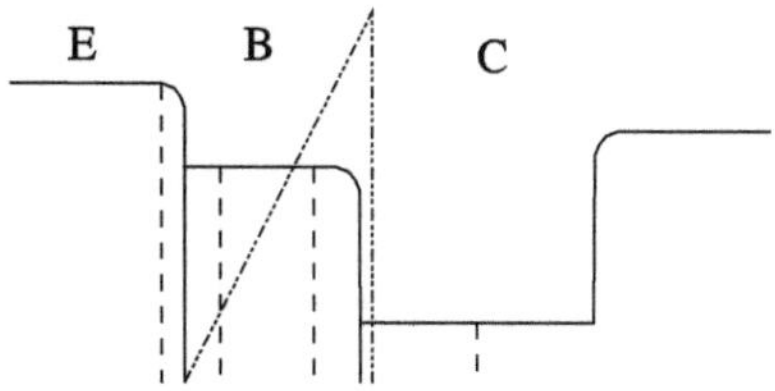

Fig. 7.9 Schematic doping profile of a SiGe bipolar transistor which has a graded Ge content. The depletion layers are schematically shown with *dashed lines*. The triangular Ge-profile is shown with a *dash-dotted line*. (Ref. [44] © [2001] IEEE)

In Mextram, the model parameter V_{g_B} represents the bandgap (voltage) in the base of the transistor. The amount of germanium in the base of a transistor will therefore affect the value of this parameter. In fact, it will affect the value of various model parameters. In many practical cases, when Mextram is applied to model a SiGe hetero-junction bipolar transistor, the effect of germanium on the device characteristics can just be captured through the influence of the germanium on the model parameter values. Effectively this means that in many cases, as far as application of Mextram to model their characteristics is concerned, transistors made in SiGe technology are not different in any significant sense compared to transistors made in Si technologies: the difference will only be reflected in the model parameter values.

A case in which this in not true, and which is therefore supported by Mextram through dedicated modeling features, is illustrated schematically in Fig. 7.9. In this case, a gradient in the germanium content induces a significant gradient in the intrinsic carrier concentration $n_i(x)$. In such cases, the Gummel number (7.19) may be sensitive to spatial gradients in the germanium fraction in the base. This is of significance to the forward and reverse Early effects in SiGe HBT's [31, 44]. The description of this effect in Mextram is based on the assumption that the intrinsic carrier concentration can be represented as

$$n_i^2 \propto \exp\left(\frac{x}{W_{B0}}\frac{\Delta E_g}{kT}\right). \tag{7.20}$$

Basically it is assumed [24] that the bandgap decreases linearly with spatial position: $E_g = E_{g0} - \Delta E_g x / W_{B0}$. The model parameter ΔE_g is the difference in bandgap between the neutral edges of the base, at zero bias. In practice, the value of ΔE_g can be estimated using process knowledge.

In the CCR (7.17) the voltages $V_{B_2E_1}$ and $V^*_{B_2C_2}$ represent the local differences between hole- and electron-quasi-Fermi potentials. In case of $V^*_{B_2C_2}$ this fact is emphasized by the star superscript which distinguishes the model variable $V^*_{B_2C_2}$ from the nodal voltage difference $V_{B_2C_2}$. This latter quantity, or rather the nodal voltage V_{C_2}, we discussed in Sect. 7.3.2.1, in forward mode of operation takes the form of an arbitrary parameterization of the model variable $I_{C_1C_2}$, which in turn represents the main current in the epilayer. The Fermi level difference $V^*_{B_2C_2}$ at the collector base interface indeed forms an interface variable between the main current (I_N) model in the base region and the model for the main current ($I_{C_1C_2}$) in the collector epilayer.

In Sect. 7.3.2.7 below we shall discuss how $V^*_{B_2C_2}$ is calculated from the model of the epilayer.

7.3.2.3 The Epilayer Current

In bipolar transistor terminology, *reverse mode of operation* denotes a state in which the base-collector junction is forward biased ($V_{BC} > 0$), and the base-emitter junction is reverse biased ($V_{BE} < 0$). In such a state, electrons in the base will flow from collector to emitter and the electric current will be directed from emitter to collector. In Mextram, a negative sign is assigned to the epilayer current under such conditions: $I_{C_1C_2} < 0$. Forward biasing of the internal base-collector junction [36, 49] of the intrinsic transistor not only occurs in the reverse mode of operation, it is also characteristic for modes of operation denoted as *saturation* and *quasi-saturation*, where $I_{C_1C_2} > 0$.

Whenever the base-collector junction is forward biased, the collector epilayer is invaded by majority carriers—holes in an npn transistor—from the base. In the Mextram model, this state of the collector epilayer is described by a model known as the Kull model [36]. In terms of functions,[5] K_j,

$$K_j = \sqrt{1 + 4\exp\left(\frac{V_{B_2C_j} - V_{dC}}{V_T}\right)}, \tag{7.21}$$

that have a direct interpretation in terms of associated hole concentrations p_j

$$K_j = 2p_j + 1, \tag{7.22}$$

and a voltage

$$E_C = V_T\left(K_2 - K_1 - \ln\left(\frac{K_2+1}{K_1+1}\right)\right) \tag{7.23}$$

the current $I_{C_1C_2}$ in the epilayer is then calculated as

$$I_{C_1C_2} = \frac{V_{C_1C_2} + E_C}{R_{Cv}}. \tag{7.24}$$

In expression (7.21) the quantity V_{dC} is the built-in diffusion voltage of the base-collector junction.

Expression (7.24) for the epilayer current is applied under all conditions. When the base-collector junction is forward biased ($I_{C_1C_2} < 0$) the Kull model is assumed to be valid throughout the collector and relations (7.21) to (7.24) can be combined with the CCR (7.17), to form an adequate model for the main current in the intrinsic transistor. The set of equations thus formed is closed by the identity

$$V_{B_2C_2} = V^*_{B_2C_2} \tag{7.25}$$

[5] In the notation used here, the functions K_j are associated with nodes: K_1 is associated with node C_1 of the equivalent circuit, Fig. 7.1. In much of the Mextram literature and documentation, the value of the function K_1 is notated as K_W (and K_2 is denoted as K_0).

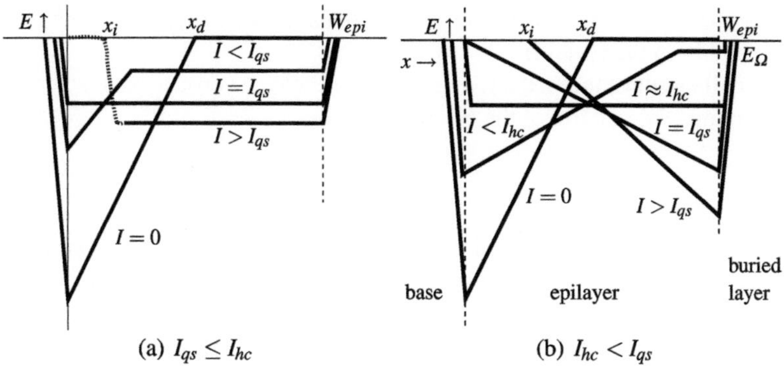

Fig. 7.10 Electric field in an n-type collector epilayer as function of current $I_{C_1C_2} = I$. In **a** the width of the depletion layer decreases because ohmic voltage drop is the dominant effect. In **b** the width of the depletion layer increases because velocity saturation is dominant (Kirk effect). At $I = I_{qs}$ quasi-saturation starts (see text). In this epilayer model, the dopant concentration is assumed to be homogeneous. The epilayer ends at position W_{epi}, where the highly doped buried layer starts

which holds in this mode of operation; this identity indeed just follows from the assumption that the Kull model is valid throughout the collector.

In forward mode of transistor operation ($I_{C_1C_2} > 0$) relation (7.24) is still applied, but relation (7.25) is not. As we discussed in Sect. 7.3.2.1, in this mode of operation the nodal voltage V_{C_2} is not addressed anywhere else in the model, so that relation (7.24) reduces to a mere parameterization of $I_{C_1C_2}$ that serves to make the main current in the epilayer accessible to the circuit simulator as a model variable.

7.3.2.4 State of the Epilayer (I)

Although they are represented in the model formulation in a smoothly unified manner, in forward mode of transistor operation ($I_{C_1C_2} > 0$) Mextram distinguishes several qualitative states of the collector epilayer. Figure 7.10 presents an overview.

In thermal equilibrium ($I_{C_1C_2} = 0$), a depletion layer will exist at the base-collector junction. If this depletion layer does not extend throughout the epilayer, the epilayer will furthermore show a neutral part, between the depletion layer and the buried layer.

At low, non-zero current levels, this thermal equilibrium state is modified in two ways. Firstly, the space charge density in the depletion layer is modulated by the charge of the carriers that take part in the current (*Kirk effect*). In Fig. 7.10, this would result in a clockwise rotation of the slope of the electric field. Secondly, ohmic carrier transport through the neutral part of the epilayer requires a non-zero homogeneous electric field E_Ω,

$$E_\Omega = \frac{I_{C_1C_2}}{q N_{epi} \mu_{epi} A_{em}}; \tag{7.26}$$

which is proportional to the current density (*Ohm's law*); μ_{epi} is the low field majority mobility in the epilayer.

Under all circumstances, the total electrostatic voltage drop across the epilayer is equal to minus the integral of the electric field, i.e.

$$V_{C_1B_2} + V_{dC} = -\int_{epi} E(x)dx. \tag{7.27}$$

Depending on the relative significance of the Kirk effects and the Ohmic effect, with increasing current levels the thickness of the depletion region x_d may either decrease or increase.

When the Ohmic effect is dominant—see Fig. 7.10a—the neutral part will increase in width and the depletion region will shrink until it finally vanishes altogether; this will happen at a critical current level $I_{C_1C_2} = I_{qs}$. For even higher current levels E_Ω will still increase. Relation (7.27) then requires that the length of the Ohmic region will decrease again, the electric field at the base-collector junction will collapse and holes from the base will invade the collector epilayer. The transistor is then entering a state known as *quasi-saturation*. From this scenario, and relation (7.27), it is also clear that the current value I_{qs} at which quasi-saturation starts depends on the applied collector base voltage, $V_{C_1B_2}$, as well as on the built in voltage, V_{dC}.

When the Kirk effect is dominant—see Fig. 7.10b—the depletion layer width x_d will increase until the depletion region extends throughout the epilayer, a state known as *total depletion*. At a somewhat higher current level known as the *hot carrier current* I_{hc},

$$I_{hc} = qN_{epi}v_{sat}A_{em}, \tag{7.28}$$

the charge density in the base-collector depletion layer formed by the carriers that take part in the main current becomes equal to the dopant level in the epilayer (N_{epi}), hence the total space charge density vanishes and the electric field $E(x)$ is spatially homogeneous. At a certain current level *somewhat higher than* I_{hc}, a state is reached in which the electric field has been tilted that far that it vanishes altogether at the base-collector interface. For even higher current levels, holes from the base will invade the epilayer and, again, a state of quasi-saturation is entered. Again it is clear that the current level I_{qs} at which quasi-saturation starts depends on the sum of the applied collector base voltage, $V_{C_1B_2}$, and the built in voltage, V_{dC}.

7.3.2.5 Critical Current for Onset of Quasi-Saturation I_{qs}

In the scenario depicted in Fig. 7.10a, quasi-saturation starts, at the current level I_{qs}, when the voltage drop over the Ohmic part in the epilayer, $W_{epi}E_\Omega$, is equal to the total voltage drop over the epilayer. In the Mextram model, the resistance of the epilayer in this state is used as a model parameter, R_{Cv}, so that

$$W_{epi}E_\Omega|_{I_{C_1C_2}=I_{qs}} = I_{qs}R_{Cv}. \tag{7.29}$$

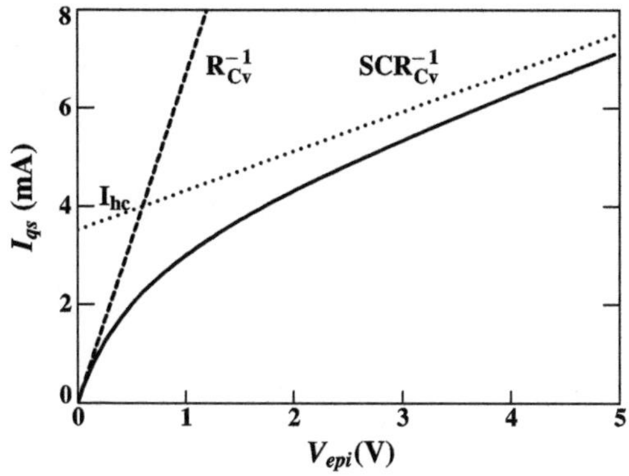

Fig. 7.11 The current at onset of injection as function of the total voltage drop over the epilayer $V_{epi} = V_{dC} - V_{B_2C_1}$ for the default parameter set [42]. We have also shown the two limiting cases, that intersect exactly at I_{hc}, which here equals 4 mA. This plot can be interpreted as a part of the state space of the transistor: the part of the quadrant above the $I_{qs}(V_{epi})$ curve represents states of (quasi-) saturation, the part below the curve represents states of normal forward operation

It follows that

$$I_{qs}|_{\text{Ohmic}} = \frac{V_{C_1B_2} + V_{dC}}{R_{Cv}}. \tag{7.30}$$

In case of quasi-saturation induced by the Kirk effect, i.e. in the scenario depicted in Fig. 7.10b, is can be shown that the critical current for onset of quasi-saturation can be formulated analogously, in terms of the effective resistance, SCR_{Cv}, of the space charge region

$$I_{qs}|_{\text{Kirk}} = \frac{V_{C_1B_2} + V_{dC}}{SCR_{Cv}}. \tag{7.31}$$

Figure 7.11 shows how the critical current for onset of quasi-saturation I_{qs} depends on the total voltage drop over the epilayer $V_{epi} = V_{dC} - V_{B_2C_1}$ according to the two expressions (7.30) and (7.31). The key quantity to interpret this graph is the hot carrier current I_{hc} (7.28), which in Mextram is available as a model parameter. If a state of quasi-saturation is entered at current levels lower than I_{hc}, according to the scenarios depicted in Fig. 7.10 this can only be a case of quasi-saturation induced by the Ohmic voltage drop across the neutral part of the epilayer, because in the scenario of quasi-saturation induced by Kirk effect, I_{qs} must certainly be larger than I_{hc}. Conversely, if a state of quasi-saturation is entered at current levels higher than I_{hc}, this can only be a case of quasi-saturation induced by the Kirk effect, for at current levels above I_{hc}, the Ohmic region will have vanished. Based on these conclusions, in Mextram the critical current for onset of quasi-saturation is represented by a smooth unification of the two expressions (7.30) and (7.31) [13]. This unification is represented by the curve in Fig. 7.11.

Figure 7.11 can be interpreted as a part of the state space of the transistor: the part of the quadrant above the $I_{qs}(V_{epi})$ curve represents states of (quasi-) saturation, the part below the curve represents states of normal forward operation.

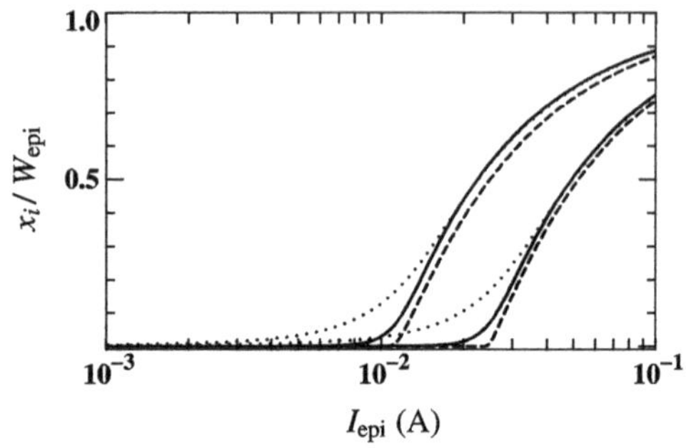

Fig. 7.12 The normalized thickness of the injection region x_i / W_{epi} as function of the current in the epilayer I_{epi} for $V_{C_1B_2} = 1, 3$ V. *Dashed line*: abrupt transition at onset of quasi-saturation. *Solid line* ($a_{xi} = 0.1$) and dotted line ($a_{xi} = 0.3$): results from the Mextram level 504 model [46] (Ref. [43] © [2000] IEEE)

7.3.2.6 Width of the Injection Region x_i

As is indicated in Figs. 7.10a and b, for current levels $I_{C_1C_2} > I_{qs}$ a quasi-neutral *injection region* will exist at the base side of the epilayer. In an npn-type transistor, this layer contains, firstly, holes that invade the epilayer from the base and, secondly, electrons to account for the epilayer current as well as to maintain quasi-neutrality.

In the regime of quasi-saturation, the width x_i of this layer is modeled based on relation (7.27) and the assumption that the voltage drop over the injection region is small, so that the voltage drop V_{epi} over the epilayer is to a good approximation equal to the voltage drop over either the Ohmic part (Fig. 7.10a) or the depleted part of the epilayer (Fig. 7.10b).

Outside the quasi-saturation regime ($I_{C_1C_2} < I_{qs}$), x_i should vanish. As was noted by Paasschens et al. [43, 46], the behavior of x_i in the transition from forward operation to quasi-saturation deserves special care. Model implementations which incorporate a rather abrupt increase of x_i as a function of current at the onset of quasi-saturation tend to lack accuracy in modeling of quantities involving derivatives, such as output conductance and cut-off frequency. For that reason, Mextram level 504 adopted a model for x_i that is based on the assumptions outlined above and that furthermore implements a gradual transition of x_i across $I_{C_1C_2} = I_{qs}$. The model has a parameter a_{xi} that controls the degree of abruptness of the transition, as is demonstrated in Fig. 7.12.

7.3.2.7 Internal Collector-Base Junction Bias

Relation (7.24) can be written as

$$V_{C_1C_2} = I_{C_1C_2} R_{Cv} - E_C. \tag{7.32}$$

On the other hand, in the regime of quasi-saturation induced by an Ohmic voltage drop across the Ohmic part of the epilayer—Fig. 7.10a—according to the reasoning in Sect. 7.3.2.6 the voltage drop across the epilayer is approximately due to the voltage drop across the Ohmic part ($x_i < x < W_{epi}$) of the epilayer, so that

$$V_{C_1C_2} \approx I_{C_1C_2} R_{Cv} \left(1 - \frac{x_i}{W_{epi}}\right); \tag{7.33}$$

this expression is based on the interpretation of R_{Cv} as the resistance of the collector epilayer if it is in a fully Ohmic state (7.29). In the interpretation of Kull's formalism represented by expressions (7.32) and (7.33) it holds that

$$E_C \approx I_{C_1C_2} R_{Cv} \frac{x_i}{W_{epi}}. \tag{7.34}$$

For a given value of x_i/W_{epi}, the combination of this expression with expressions (7.21), (7.22) and (7.23) implies an equation for the hole concentrations at the intrinsic collector-base interface (p_{2^*}) and at the end of the collector epilayer, p_1

$$V_T \left(2p_{2^*} - 2p_1 - \ln\left(\frac{p_{2^*}+1}{p_1+1}\right)\right) \approx I_{C_1C_2} R_{Cv} \frac{x_i}{W_{epi}}. \tag{7.35}$$

The hole concentration p_1 is well defined by relations (7.21) and (7.22) at any stage of model evaluation in the context of a circuit simulation, because the nodal voltage $V_{B_2C_1}$ is a model variable that is defined by the circuit simulator. In Mextram 504, relation (7.35) is interpreted and approximately solved as an equation for the hole concentration at the base-collector interface. Using relations (7.22) and (7.21), from the hole concentration p_{2^*} the difference of the quasi-Fermi potentials at the base-collector interface, $V_{B_2C_{2^*}} = V^*_{B_2C_2}$ is calculated.[6] This difference in quasi Fermi potentials at the collector base junction forms the interface between the Kull model for the physics in the epilayer and the charge control relation, which describes the main transistor current in the base region.

7.3.2.8 Charges; Depletion Capacitance, Injected Charge

Both to model time dependent behavior of transistors and in view of the charge control relation (7.17), it is essential to have an adequate representation of the charges stored inside transistors. Most of the modeled pn-junction regions in the Mextram model are represented by voltage dependent depletion charges and diffusion charges that are in turn based on regional transit time models. We shall not discuss these models in detail here but refer to the Mextram literature and documentation for further details. Some parts of the charge model deserve to be highlighted here though.

---

[6] In the notation of expression (7.21), the voltage $V_{B_2C_{2^*}}$ appears in the expression for K_{2^*}. In much of the Mextram literature and documentation, this voltage $V_{B_2C_{2^*}}$ is notated as $V^*_{B_2C_2}$.

In Mextram, the current dependence of the charge stored in the depletion region of the collector epilayer is taken into account. The physical origin of this dependence is again the Kirk effect (Fig. 7.10b): the thickness x_d of the depletion region is modulated by the main current and as a result the depletion capacitance

$$C_{CB} = \frac{\epsilon A}{x_d(I_{C_1C_2})},$$

depends on the current. In Mextram, in consistency with the definition of the model nodes, the holes associated with the base-collector depletion charge are taken into account in the Gummel number (7.19) for the main current I_N. Implicitness (in terms of the main current itself) of the expressions for the main current I_N thus formed by relations (7.17) and (7.19) is prevented by the fact that the current dependence of the depletion charge is expressed in terms of $I_{C_1C_2}$, not I_N.

In Mextram, the holes stored in the injection layer in saturation or quasi-saturation do not contribute to the Gummel number (7.19), because the upper limit of the integral, the node C_2, is chosen at the base side of this injection region. In Mextram, instead of through the Gummel number, the physical effect of quasi-saturation on the main current is taken into account by the calculation of the difference $V^*_{B_2C_2}$ of the quasi Fermi potentials at the base-collector interface.

To take into account the effect of quasi-saturation on the time-dependent behavior of transistors, e.g. on the cut-off frequency characteristics, the charge Q_{epi} stored in the injection region is calculated according to [1]

$$Q_{epi} \approx \tau_{epi} \left(\frac{x_i}{W_{epi}} \right)^2 I_{C_1C_2}. \tag{7.36}$$

Interestingly, this expression follows from again a charge control relation like (7.17), but now applied to the epilayer instead of to the base region.

7.4 Conclusion

In this chapter we have presented an introduction to the Mextram model, an industrial world standard compact model for bipolar transistors. Mextram has been developed to capture all terminal characteristics of bipolar transistors that are relevant to industrial electronic circuit design of any Si or SiGe bipolar transistor, under all relevant practical circumstances. It supports accurate computer simulations of all relevant observable terminal quantities (voltages, currents, noise) as a function of applied terminal boundary conditions (biases, signals) and ambient temperature. It does so both for stationary and time-dependent boundary conditions.

In Sect. 7.1 we discussed the history and basic structure and features the model. Simulation of heating effects, noise, geometrical scaling and statistical analysis was briefly discussed. In Sect. 7.2 we presented a detailed discussion of the relevance of the structure of the model, especially the refined topology of its equivalent circuit, to simulation of advanced ac-characteristics of modern high-speed Si and SiGe transistors.

Section 7.3 presented the main underlying ideas of Mextram. The discussion focused on how the central problem of modeling the transistor action in widely different regimes of operation is addressed in Mextram. It was explained how the central approach is reflected by the overall structure of the model.

In conclusion we can say that we have presented several different views on the Mextram model, aiming to show the identity, philosophy and capabilities of the model. We have not aimed for a complete and fully detailed overview of the model. Given the ongoing developments of the model, any detailed description of the model could be complete only temporarily anyhow. As we have seen however, the overall structure, approach and identity of the Mextram model has been recognizable over decades now and can be expected to remain so for the foreseeable future.

Acknowledgments The authors are indebted to the IEEE Intellectual Property Rights Office for granted permission to re-use figures of IEEE copyrighted publications and to Daniel P. Vidal (Delft University) for reading the penultimate version of this manuscript.

References

1. Beale, J.R.A., Slatter, J.A.G.: Equivalent circuit of a transistor with a lightly doped collector operating in saturation. Solid-State Electron. **11**, 241–252 (1968)
2. Berkner, J.: Kompaktmodelle für Bipolartransistoren. Praxis der Modellierung, Messung und Parameterbestimmung—SGP, VBIC, HICUM und MEXTRAM (Compact Models for Bipolar Transistors. Practice of Modelling, Measurement and Parameter Extraction—SGP, VBIC, HICUM und MEXTRAM). Expert, Renningen (2002) (In German)
3. Bonani, F., Ghione, G.: Noise in Semiconductor Devices. Springer, Berlin (2001)
4. Bowler, D.L., Lindholm, F.A.: High current regimes in transistor collector regions. IEEE Trans. Electron. Devices **ED-20**, 257–263 (1973)
5. Buckingham, M.J.: Noise in Electronic Devices and Systems. Ellis Horwood, Chichester (1983)
6. Chu, G.Y.: Unilateralization of junction transistor amplifiers at high frequencies. Proc. IRE **43**(8), 1001–1006 (1955)
7. de Graaff, H.C.: Collector models for bipolar transistors. Solid-State Electron. **16**, 587–600 (1973)
8. de Graaff, H.C.: Electrical behaviour of lightly doped collectors in bipolar transistors. PhD thesis, Eindhoven University of Technology (1975)
9. de Graaff, H.C., Klaassen, F.M.: Compact Transistor Modelling for Circuit Design. Springer, Wien (1990)
10. de Graaff, H.C., Klaassen, F.M.: Compact Transistor Modelling for Circuit Design. Springer, Berlin (1990)
11. de Graaff, H.C., Kloosterman, W.J.: Mextram a new bipolar transistor model. Report 6038, Philips Natl. Lab. (1985)
12. de Graaff, H.C., Kloosterman, W.J.: New formulation of the current and charge relations in bipolar transistor modeling for CACD purposes. IEEE Trans. Electron. Devices **ED-32**, 2415 (1985)
13. de Graaff, H.C., Kloosterman, W.J.: Modeling of the collector epilayer of a bipolar transistor in the Mextram model. IEEE Trans. Electron. Devices **ED-42**, 274–282 (1995)
14. de Graaff, H.C., van der Wal, R.J.: Measurement of the onset of quasi-saturation in bipolar transistors. Solid-State Electron. **17**, 1187–1192 (1974)

15. de Graaff, H.C., Kloosterman, W.J., Geelen, J.A.M., Koolen, M.C.A.M.: Experience with the new compact Mextram model for bipolar transistors. In: Proc. of the Bipolar Circuits and Technology Meeting, pp. 246–249 (1989)
16. de Vreede, L.C.N., de Graaff, H.C., Mouthaan, K., de Kok, M., Tauritz, J.L., Baets, R.G.F.: Advanced modeling of distortion effects in bipolar transistors using the Mextram model. In: Proc. of the Bipolar Circuits and Technology Meeting, pp. 48–51 (1994)
17. de Vreede, L.C.N., de Graaff, H.C., Mouthaan, K., de Kok, M., Tauritz, J.L., Baets, R.G.F.: Advanced modeling of distortion effects in bipolar transistors using the Mextram model. IEEE J. Solid-State Circuits **31**, 114–121 (1996)
18. de Vreede, L.C.N., de Graaff, H.C., Tauritz, J.L., Baets, R.G.F.: Extension of the collector charge description for compact bipolar epilayer models. IEEE Trans. Electron. Devices **ED-42**, 277–285 (1998)
19. Deixler, P., Rodriguez, A., De Boer, W., Sun, H., Colclaser, R., Bower, D., Bell, N., Yao, A., Brock, R., Bouttement, Y., Hurkx, G., Tiemeijer, L., Paasschens, J., Huizing, H., Hartskeerl, D., Agrarwal, P., Magnee, P., Aksen, E., Slotboom, J.: QUBiC4X: An $f_t/f_{max} = 130/140$ GHz SiGe:C-BiCMOS manufacturing technology with elite passives for emerging microwave applications. In: Proc. BCTM, pp. 233–236. IEEE, New York (2004)
20. Getreu, I.E.: Modeling the Bipolar Transistor. Elsevier, Amsterdam (1978)
21. Gummel, H.K.: A charge control relation for bipolar transistors. Bell Sys. Tech. J. 115–120 (1970)
22. Gupta, M.: Power gain in feedback amplifiers, a classic revisited. IEEE Trans. Microw. Theory Tech. **40**(5), 864–879 (1992)
23. Hassan, M.M.S.: Modelling of lightly doped collector of a bipolar transistor operating in quasi-saturation region. Int. J. Electron. **86**, 1–14 (1999)
24. Hong, G.B., Fossum, J.G., Ugajin, M.: A physical SiGe-base HBT model for circuit simulation and design. In: IEDM Tech. Digest, pp. 557–560 (1992)
25. http://en.wikipedia.org/wiki/Compact_Model_Council
26. http://mextram.ewi.tudelft.nl
27. Hurkx, G.A.M.: Bipolar and Bipolar-MOS Integration. Elsevier, Amsterdam (1994). Chap. 3
28. Hurkx, G.A.M.: The relevance of f_T and f_{max} for the speed of a bipolar CE amplifier stage. IEEE Trans. Electron. Devices **44**, 775–781 (1997)
29. Hurkx, G.A.M., Agarwal, P., Dekker, R., van der Heijden, E., Veenstra, H.: RF figures-of-merit for process optimization. IEEE Trans. Electron. Devices **51**(12), 2121–2128 (2004)
30. Jeong, H., Fossum, J.G.: A charge-based large-signal bipolar transistor model for device and circuit simulation. IEEE Trans. Electron. Devices **ED-36**, 124–131 (1989)
31. Joseph, A.J., Cressler, J.D., Richey, D.M., Jaeger, R.C., Harame, D.L.: Neutral base recombination and its influence on the temperature dependence of early voltage and current gain-early voltage product in UHV/CVD SiGe heterojunction bipolar transistors. IEEE Trans. Electron. Devices **44**, 404–413 (1997)
32. Kloosterman, W.J., de Graaff, H.C.: Avalanche multiplication in a compact bipolar transistor model for circuit simulation. IEEE Trans. Electron. Devices **ED-36**, 1376–1380 (1989)
33. Kloosterman, W.J., Geelen, J.A.M., Klaassen, D.B.M.: Efficient parameter extraction for the Mextram model. In: Proc. of the Bipolar Circuits and Technology Meeting, pp. 70–73 (1995)
34. Kloosterman, W.J., Paasschens, J.C.J., Havens, R.J.: A comprehensive bipolar avalanche multiplication compact model for circuit simulation. In: Proc. of the Bipolar Circuits and Technology Meeting, pp. 172–175 (2000)
35. Koolen, M.C.A.M., Aerts, J.C.J.: The influence of non-ideal base current on $1/f$ noise behaviour of bipolar transistors. In: Proc. of the Bipolar Circuits and Technology Meeting, pp. 232–235 (1990)
36. Kull, G.M., Nagel, L.W., Lee, S., Lloyd, P., Prendergast, E.J., Dirks, H.: A unified circuit model for bipolar transistors including quasi-saturation effects. IEEE Trans. Electron. Devices **ED-32**(6), 1103–1113 (1985)

37. Mason, S.J.: Power gain in feedback amplifier. Trans. IRE **1**(2), 20–25 (1954)
38. Milovanović, V., Van der Toorn, R.: Rf small signal avalanche characterization and repercussions on bipolar transistor circuit design. In: The IEEE Region 8 EUROCON. The International Conference on Computer as a Tool, IEEE, Saint Petersburg, Russia (2009)
39. Milovanović, V., van der Toorn, R., Humphries, P., Vidal, D., Vafanejad, A.: Compact model of Zener tunneling current in bipolar transistors featuring a smooth transition to zero forward bias current. In: Proc. of the Bipolar Circuits and Technology Meeting (2009)
40. Paasschens, J.C.J.: Compact modeling of the noise of a bipolar transistor under DC and AC current crowding conditions. IEEE Trans. Electron. Devices **51**, 1483–1495 (2004)
41. Paasschens, J.C.J., de Kort, R.: Modelling the excess noise due to avalanche multiplication in (heterojunction) bipolar transistors. In: Proc. of the Bipolar Circuits and Technology Meeting, pp. 108–111 (2004)
42. Paasschens, J.C.J., Kloosterman, W.J.: The Mextram bipolar transistor model, level 504. Unclassified Report NL-UR 2000/811, Philips Natl. Lab. (2000). See footnote 2
43. Paasschens, J.C.J., Kloosterman, W.J., Havens, R.J., de Graaff, H.C.: Improved modeling of output conductance and cut-off frequency of bipolar transistors. In: Proc. of the Bipolar Circuits and Technology Meeting, pp. 62–65 (2000)
44. Paasschens, J.C.J., Kloosterman, W.J., Havens, R.J.: Modelling two SiGe HBT specific features for circuit simulation. In: Proc. of the Bipolar Circuits and Technology Meeting, pp. 38–41 (2001)
45. Paasschens, J.C.J., Kloosterman, W.J., Havens, R.J.: Parameter extraction for the bipolar transistor model Mextram, level 504. Unclassified Report NL-UR 2001/801, Philips Natl. Lab. (2001). See footnote 2
46. Paasschens, J.C.J., Kloosterman, W.J., Havens, R.J., de Graaff, H.C.: Improved compact modeling of output conductance and cutoff frequency of bipolar transistors. IEEE J. Solid-State Circuits **36**, 1390–1398 (2001)
47. Paasschens, J.C.J., Havens, R.J., Tiemeijer, L.F.: Modelling the correlation in the high-frequency noise of (heterojunction) bipolar transistors using charge-partitioning. In: Proc. of the Bipolar Circuits and Technology Meeting, pp. 221–224 (2003)
48. Paasschens, J.C.J., Harmsma, S., van der Toorn, R.: Dependence of thermal resistance on ambient and actual temperature. In: Proc. of the Bipolar Circuits and Technology Meeting, pp. 96–99 (2004)
49. Pals, J.A., de Graaff, H.C.: On the behaviour of the base-collector junction of a transistor at high collector current densities. Philips Res. Rep. **24**, 53–69 (1969)
50. Reisch, M.: High-Frequency Bipolar Transistors. Advanced Microelectronics, vol. 11. Springer, Berlin (2003). Chap. B.2
51. Rey, G., Dupuy, F., Bailbe, J.P.: A unified approach to the base widening mechanism in bipolar transistors. Solid-State Electron. **18**, 863–866 (1975)
52. Rollett, J.: Stability and power-gain invariants of linear twoports. In: IRE Trans. on Circuit Theory, pp. 29–32 (1962)
53. Schröter, M., Lee, T.Y.: Physics-based minority charge and transit time modeling for bipolar transistors. IEEE Trans. Electron. Devices **ED-46**, 288–300 (1999)
54. Van der Heijden, M.P., De Vreede, L.C., Burghartz, J.N.: On the design of unilateral dual-loop feedback low-noise amplifiers with simultaneous noise, impedance, and iip3 match. IEEE J. Solid-State Circuits **39**(10), 1727–1736 (2004)
55. van der Toorn, R., Dohmen, J.J., Hubert, O.: Distribution of the collector resistance of planar bipolar transistors: Impact on small signal characteristics and compact modeling. In: Proc. of the Bipolar Circuits and Technology Meeting, pp. 184–187 (2007)
56. van der Ziel, A.: Noise. Sources, Characterization, Measurement. Prentice-Hall, Englewood Cliffs (1970)
57. van der Ziel, A.: Noise in Solid-State Devices and Circuits. Wiley-Interscience, New York (1986)
58. Versleijen, M.P.J.G.: Distributed high frequency effects in bipolar transistors. In: Proc. of the Bipolar Circuits and Technology Meeting, pp. 85–88 (1991)

59. Wu, H.: A scalable mextram model for advanced bipolar circuit design. Ph.D. thesis, Delft University of Technology (2007)
60. Wu, H., Mijalkovic, S., Burghartz, J.: Parameters extraction of a scalable Mextram model for high-speed SiGe HBTs. In: Proc. BCTM, pp. 140–143. IEEE, New York (2004)
61. Wu, H., Mijalkovic, S., Burghartz, J.: A referenced geometry based configuration scalable mextram model for bipolar transistors. In: Proc. IEEE Int. Behavioral Modeling and Simulation Workshop, pp. 50–55. IEEE, New York (2006)

Chapter 8
The HiCuM Bipolar Transistor Model

Michael Schröter and Bertrand Ardouin

Abstract This chapter provides an overview on the advanced compact bipolar transistor model HiCuM in terms of the modelling approach with respect to circuit design. The relevant physical effects occurring in modern heterojunction bipolar transistors (HBTs) are briefly described with the main focus on SiGe HBTs. Geometry scaling, a statistical design methodology, and parameter extraction methods in an industrial environment are discussed. These methods are applied step by step to a selected advanced SiGe HBT process technology with a transit frequency beyond 200 GHz. The corresponding results for a consistent set of important device characteristics exhibit excellent agreement over bias, temperature and geometry, and demonstrate the suitability of the model for such high-frequency bipolar process technologies.

8.1 Introduction

The push for higher operating frequencies, such as 60 GHz for wireless communications, 160 GHz for imaging applications, 120 Gb/s for fiber-optic data transmission, or 77 GHz for automotive radar, creates a continuous demand for high-speed circuits and devices. At these *circuit* frequencies, bipolar transistors have distinctive advantages due to their inherently higher speed and better suitability for analog circuits

M. Schröter (✉)
Chair for Electron Devices and Integrated Circuits (CEDIC), University of Technology Dresden, Dresden, Germany
e-mail: mschroter@ieee.org

M. Schröter
ECE Dept, University of California at San Diego, La Jolla, USA

B. Ardouin
XMOD Technologies Inc., Bordeaux, France
e-mail: ardouin@xmodtech.com

G. Gildenblat (ed.), *Compact Modeling*,
DOI 10.1007/978-90-481-8614-3_8,

(e.g. [1, 2]). This is especially true for heterojunction bipolar transistors (HBTs). While III/V HBTs provide the biggest potential for speed *and* power handling capability (due to their relatively large breakdown voltage) SiGe HBTs have become a serious contender for high-speed applications [3]. This is a result of the integration into CMOS technology that on one hand has led to a tremendous boost in footprint reduction and on the other hand enables power efficient system-on-chip solutions.

Overall, there is a large variety of HBT device types and designs resulting from the different properties of the materials used (e.g. AlGaAs/GaAs, InGaP/GaAs, InGaAs/InP, GaAsSb/InP, SiGe, SiGeC). While in III/V processes single heterojunction bipolar transistors (S-HBTs) were built initially, the more recent trend appears to be heading towards double HBTs (D-HBTs), in which emitter *and* collector region consist of wider bandgap material. For SiGe HBTs this has always been the case due to lattice stability constraints. Nevertheless, the first-order and also many second-order effects are the same regardless of the particular HBT type, especially when it comes to production technologies [2]. As a consequence, it is attractive from a circuit design and EDA point of view to have a compact transistor model that covers all relevant HBT technologies. Most important requirements for such a model are: (i) accuracy over a wide bias, temperature, frequency, and geometry range; (ii) computational efficiency and reliability; (iii) as simple as possible parameter determination based on standard equipment; (iv) modularity and easy extension to new features. These requirements have been addressed by HiCuM from the beginning of its development in the early eighties [4–11]. The version at that time filled a growing gap for accurate simulation of large-signal high-speed fiber-optic circuits. From then on, "highspeed" and "high-frequency" (HF) kept being the main theme of further model development and extensions up to the version presently existing in circuit simulators. Besides its availability in circuit simulators [12] a parameter extraction infrastructure has been built to support user-friendly deployment of the model (e.g. [13]).

The limited space for this chapter does not permit a detailed presentation of all model equations or even their physics-based derivation. For this information the reader is referred to the literature already mentioned above and also given throughout the text below. Instead, this chapter provides a high-level overview on the approaches used for the model formulations and the reasons for their selection. For this, the important physical effects in SiGe and III/V HBTs and how they are taken into account in HiCuM will be concisely described. Finally, a rough overview on the parameter extraction approach and application examples will be given.

8.2 Model Fundamentals

Figure 8.1 exhibits the equivalent circuit (EC) of HiCuM/Level2 version 2.23, which—admittedly—looks fairly complicated. However, each EC element represents a particular device region and physical effect. This will be explained in the subsequent sections. As mentioned above, HiCuM development started out with the goal of improving the dynamic description of BJTs by putting the main emphasis on

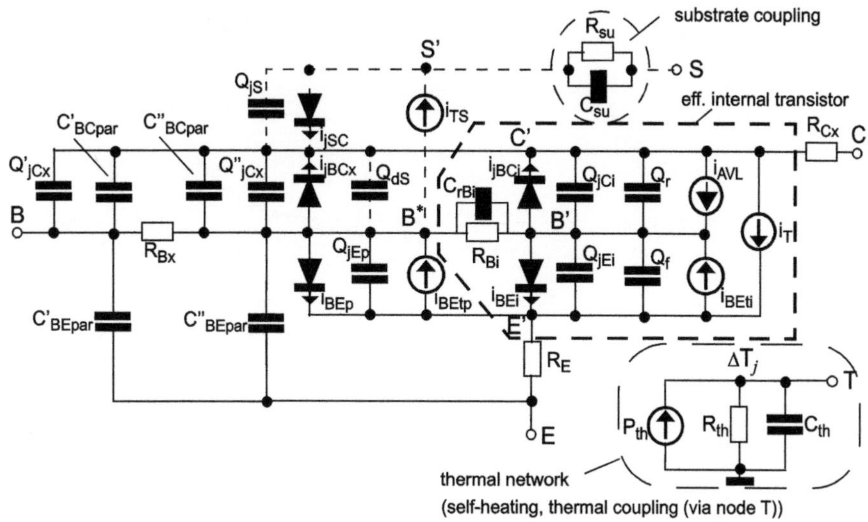

Fig. 8.1 Complete equivalent circuit of HiCuM/Level2 with adjunct network for self-heating. Substrate related elements are indicated by *dashed lines*

accurately modelling charge storage. This approach has turned out to be extremely successful and, with speed advantage as the main reason for the survival of BJT and HBT technology, is expected to do so in future. Thus, the model overview will start with charge modelling. The internal portion of these charges will also be needed for an elegant description of the transfer current. Afterwards, DC currents, series resistances and other important effects will be discussed, also with respect to their relevance in different types of HBTs.

Within this text, an npn transistor is assumed, but the model can be (and already has been) employed also for pnp transistors. (This requires the addition of n-well related elements.) The EC in Fig. 8.1 can be constructed from considering first a one-dimensional (intrinsic) transistor, then extending it to an (effective) internal transistor defined by the region under the emitter, and finally adding the elements for the parasitic external regions. The effective internal transistor, indicated by the dashed box in Fig. 8.1, already includes emitter perimeter injection into the base that is found in Si BJTs and SiGe HBTs. This is accomplished by defining a (measurable) effective emitter width w_E and area A_E that are usually somewhat larger than their window counterparts (cf. Sect. 8.3.1.2, [14, 15]). This approach makes the model scalable using just a single-transistor EC (i.e. a single transfer current source i_T with the associated minority charge elements Q_f and Q_r) instead of requiring a two-transistor EC. The discussion below will focus on HiCuM/Level2 (Fig. 8.1). From its parameters the simplified HiCuM/Level0 [16, 17] and the distributed model HiCuM/Level4 [2, 18] can be generated, forming a hierarchy of compact models that allows adapting accuracy and simulation time to different design tasks.

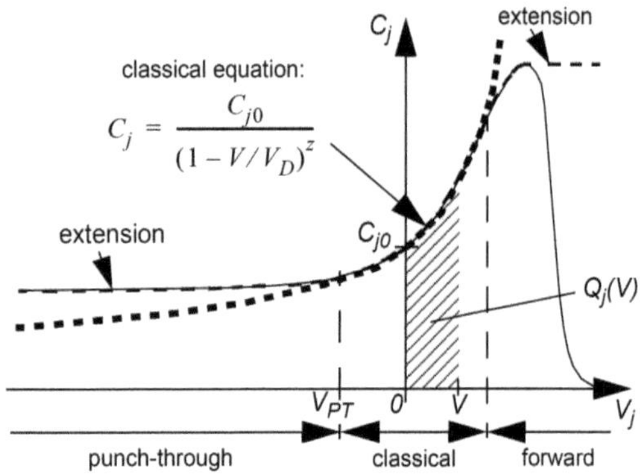

Fig. 8.2 Typical voltage dependence of a depletion capacitance C_j (*solid line*) and its generalized classical description (*dotted line*). The behavior modeled in HiCuM is indicated by the *dashed line*. C_{j0} is the zero-bias capacitance, V_D the built-in voltage, and z an exponent factor. The hatched area represents the depletion charge Q_j at a voltage V

8.2.1 Charges

Fundamentally, the stored charge under non-equilibrium conditions can be divided into a depletion component Q_j, which represents the space-charge mostly around the junctions, and the mobile component Q_m, which consists mostly of minority carriers. This partitioning can be easily performed in device simulation, providing a clear reference for developing accurate compact charge expressions [19–21]. In HiCuM the small-signal quantities, such as depletion capacitances and transit times, are described accurately and numerically smooth as a function of bias (i.e. voltage and current). The charges are then obtained by analytical integration. This way, the relevant state variable "charge" is guaranteed to be continuously differentiable, and the derivative is guaranteed to be accurate. This is important for simulator convergence and accurate small-signal analysis. However, during model development finding integrable expressions for the derivatives can sometimes be difficult.

8.2.1.1 Depletion Charges

At reverse bias or low injection across a pn junction the minority charge is negligible, and only the depletion charge needs to be taken into account. The typical voltage dependence of a depletion capacitance is illustrated in Fig. 8.2 along with the classical expression. The latter is extended in HiCuM for all junctions towards forward operation in order to ensure numerical stability and to provide an accurate description of the transit frequency f_T at medium current densities (below and at the peak). For the BC capacitance components C_{jCi} and C_{jCx} the punch-through of the respective space-charge region (SCR) to the buried layer is taken into account by an additional smooth extension of the already modified analytical expression. The corresponding punch-through voltage V_{PT} depends on the width w_{Ci} and doping N_{Ci} of the internal collector. In all cases, the depletion charge Q_j is calculated analytically from integrating the corresponding depletion capacitance as shown in Fig. 8.2.

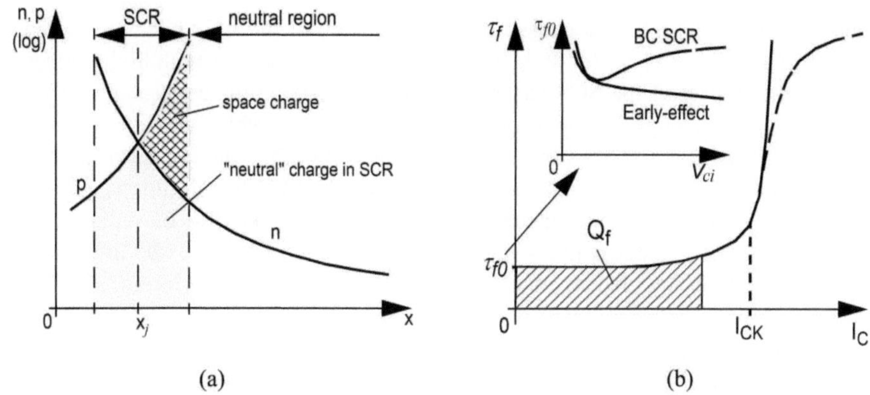

Fig. 8.3 Minority carrier storage: **a** Schematic carrier density in a SCR and adjacent neutral region; x_j is the junction depth, **b** Typical current dependence of the forward transit time for BJTs and SHBTs (*dashed line*) and D-HBTs (*solid line*). The hatched area represents the associated minority charge. The *inset* shows the possible cases of voltage dependence for τ_{f0}

8.2.1.2 Mobile Charges

At forward bias carriers are injected across the junctions. The corresponding excess density (i.e. the difference between the non-equilibrium and the zero-bias carrier density) leads to *minority* carriers and an associated mobile charge storage. In neutral regions majority carriers from the transistor terminals fully compensate the injected minorities, while in SCRs only partial compensation exists as shown in Fig. 8.3a. Although the carrier density itself is higher in SCRs compared to neutral regions, the associated "neutral" charge is usually negligible due to the small width of the SCR. It can nevertheless be analytically modeled if necessary [2]. With increasing forward bias the neutral charge in the SCR even decreases, while beyond some injection level the minority charge in the neutral regions, which increases exponentially according to Boltzmann statistics, becomes significant. Figure 8.3b illustrates the typical behavior of the associated forward transit (or storage) time τ_f in BJTs and HBTs. τ_f is the sum of various components, each related to a transistor region:

$$\tau_f = \tau_{Ef} + \tau_{BE} + \tau_B + \tau_{BC} + \tau_{pC}. \tag{8.1}$$

Here, τ_{Ef} is the emitter contribution, τ_{BE} is BE space charge contribution, τ_B is the base contribution, τ_{BC} is the storage time related to the BC space charge region, and τ_{pC} is the portion of the neutral collector contribution occurring at high current densities. From a compact modelling point of view, τ_f is a function of the collector current and of the base-collector (or collector-emitter) voltage.

Neglecting the component related to the neutral charge in the BE SCR, τ_f is current independent at low injection, but is still a function of the effective collector voltage $V_{ci} = V_{B'C'} - V_{DCi} \approx V_{C'E'} - V_{CEs}$, with V_{CEs} as internal CE saturation voltage and V_{DCi} as built-in voltage of the internal BC junction. The voltage dependence (cf. Fig. 8.3) is caused by the expansion of the BC SCR with increasing reverse voltage $V_{C'B'}$ (or $V_{C'E'}$). For high-speed transistors with high collector doping the impact of the SCR extension into the base dominates through the base transit time τ_{Bf} (roughly proportional to base width w_B^2), which corresponds to the Early-effect. In contrast, for power transistors with low N_{Ci} the impact of the SCR extension into the collector dominates through the BC SCR transit time τ_{BC} ($\sim$ BC SCR width). Notice that in HiCuM the time constant related to the internal collector resistance is automatically included in the measurement of (the low-current component of) τ_f and, hence, being taken into account in its description.

At medium injection τ_f starts to increase due to the increase in neutral base width and mobile charge within the BC SCR. In D-HBTs also the conduction band barrier starts to form [2, 22], leading to a stronger increase in charge storage at the end of the neutral base as compared to BJTs and S-HBTs. With increasing current density, the collector space charge is more and more compensated by the mobile charge in the BC SCR, leading to a reduction of the electric field there (Kirk-effect [23]). Once the field has collapsed to a critical value E_{lim}, which defines the boundary between velocity saturation and the ohmic region in the carrier velocity vs. field relation, an additional diffusion current is necessary to carry the collector current increase [2, 19, 24]. The resulting minority carrier gradient in the collector leads to additional charge storage at high injection that increases rapidly with current. The critical current [15, 24],

$$I_{CK} = \frac{V_{ci}/r_{Ci0}}{[1+(V_{ci}/V_{lim})^{\delta}]^{1/\delta}}\left[1+\frac{u+\sqrt{u^2+a_{ickpt}}}{2}\right] \tag{8.2}$$

with $u = (V_{ci} - V_{lim})/V_{PT}$ indicates the onset of this high-current effect as function of V_{ci} and is used in HiCuM as part of a physics-based analytical formulation for the transit time behavior described above [19]. The current, voltage and temperature dependent model contains the relevant structural parameters of the transistor through the model parameters $r_{Ci0} = w_{Ci}/(q\mu_{nCi0}N_{Ci}A_E f_{cs})$, $V_{PT} = (qN_{Ci}w_{Ci}^2)/(2\varepsilon)$, $V_{lim} = E_{lim}/w_{Ci}$. In addition, the impact of collector current spreading at narrow emitter widths is taken in to account via the factor f_{cs} and a bias dependent extension of the charge formulation [15].

8.2.2 *Transfer Current*

Solving the combined continuity and transport equation in the neutral base region results, under certain simplifications, in a closed-form analytical expression for the spatially dependent minority carrier distribution. From this classical theory of drift transistors [25–27] then the transfer current I_T is calculated. Due to the simplifications necessary for obtaining a closed-form solution the variation of the SCR widths (i.e. the Early effect) has to be added empirically, and the expression is only valid for low-injection, while for high injection (in the base) the differential equation is somewhat different and yields a different $V_{B'E'}$ dependence [28]. In order to link the low- and high-injection solution in the bias range relevant for circuit design, an empirical smoothing function can be used (e.g. [29]). However, the theory is still valid only for the neutral base, so that the impact of high-current effects in the collector needs to be added empirically.

Recognizing that the impact of neutral base recombination on transfer current is negligible in modern BJTs and HBTs, a direct integration of the transport equation yields a closed-form solution for the entire bias range [4, 30, 31],

$$I_T = I_{Tf} - I_{Tr} = I_S \frac{\exp(V_{B'E'}/V_T) - \exp(V_{B'C'}/V_T)}{1 + \Delta Q_{p,T}/Q_{p0,T}} \tag{8.3}$$

with the saturation current

$$I_S = (A_E^2 q^2 V_T \overline{\mu_{nB} n_{iB}^2})/Q_{p0,T} \sim A_E. \tag{8.4}$$

If for $Q_{p0,T}$ the zero-bias hole charge Q_{p0} and for $\Delta Q_{p,T}$ the actual *base* charge ΔQ_p are inserted, and if the series resistances are neglected, the Integral Charge-Control Relation (ICCR) first derived by Gummel [32, 33] is obtained. The charge term in the denominator now determines the detailed bias dependence of I_T. Obviously, the task of linking low- and high-injection region has now been shifted to the charge modelling. However, for BJTs already a simple current independent transit time τ_f leads to a reasonable approximation of $I_T(V_{B'E'})$ at medium current densities. Thus, and because an accurate description for the charge is required in any way, (8.3) is a much more attractive approach than classical theory. For instance, assuming low-injection (in a BJT) gives $\Delta Q_{p,T} = Q_{jEi} + Q_{jCi}$. Hence, the "forward" Early effect is automatically modeled as function of bias by Q_{jCi}/Q_{p0}.

A current independent τ_f is not sufficient, however, for practical applications in which transistors are required to be operated at high current densities. Also, access to the potential differences across the BE and BC junction is not possible so that the integration interval has to extended to the emitter and collector contact ($x = 0$ and $x = x_{C'}$, respectively) of the intrinsic transistor. In this case, integration of the transport equation without any simplifications leads to the Generalized ICCR (GICCR) [21, 30, 31]. Neglecting recombination then gives for the transfer current related excess charge, subdivided into the respective depletion and minority charge,

$$\Delta Q_{p,T} = Q_{j,T} + Q_{m,T} = q \int_{x_{je}}^{x_{jc}} h_g \Delta(p-n)dx + q \int_0^{x_{C'}} h_g \Delta m dx, \tag{8.5}$$

in which the excess carrier densities are weighted with the function

$$h_g = \overline{\mu_{nB} n_{iB}^2}/(\mu_n n_i^2)$$

that has been conveniently normalized to an average value in the base region. Since the intrinsic carrier density n_i depends exponentially on bandgap its influence dominates h_g compared to the mobility μ_n. The spatially different weighting becomes important in HBTs due to the bandgap variation there. Defining an average weight factor for the regional charge components allows to keep using the actual charge terms and to maintain the elegant solution (8.3) for the transfer current characteristic. For the latest most advanced HBTs though some weight factors exhibit a significant bias dependence [2]. Note, that due to the integration between the internal contacts of the transistor the influence of the strongly bias dependent internal collector resistance on the transfer current (and its derivatives) is automatically included in the GICCR presented above.

Compared to other approaches for modelling I_T, the GICCR has the distinctive advantage that it is based on an exact solution of the transport equation. In other words, any simplification or assumption towards a compact formulation can be evaluated quantitatively for an arbitrary set of transistor structures by device simulation before a new model equation is derived. This makes model development much more reliable and less empirical. It is even possible to include recombination and to extend the GICCR concept to the 2D/3D case leading to the GICCR master equation [34]. As a consequence, HiCuM has been based on this approach from the beginning which so far has turned out to be quite successful.

8.2.3 Base Current Components

The base current portion controlled by $V_{B'E'}$ represents the back injection of holes into the emitter (I_{jBEi}), the recombination within the BE space-charge region (I_{jREi}), and the excess recombination in the neutral base (I_{Bhrec}) due to the BC barrier effect (e.g. [22, 35]) at high collector current densities:

$$I_{BEi} = \underbrace{I_{BEiS}\left[\exp\left(\frac{V_{B'E'}}{m_{BEi}V_T}\right) - 1\right]}_{I_{jBEi}} + \underbrace{I_{REiS}\left[\exp\left(\frac{V_{B'E'}}{m_{REi}V_T}\right) - 1\right]}_{I_{jREi}} + \underbrace{\frac{\Delta Q_{fB}}{\tau_{Bhrec}}}_{I_{Bhrec}}. \tag{8.6}$$

The saturation currents I_{BEiS} and I_{REiS}, the ideality coefficients m_{BEi} and m_{REi}, and the lifetime τ_{Bhrec} are model parameters; ΔQ_{fB} is the excess minority charge in the neutral base. For the emitter perimeter related component I_{BEp} in Fig. 8.1 a similar expression is used with the corresponding model parameters and $V_{B*E'}$, but with the last component being dropped since there is no barrier effect outside of the *effective* emitter. For the BC and CS junction related currents (I_{jBCi}, I_{jBCx}, I_{jSC} in Fig. 8.1) only the injection component (first term in (8.6)) is taken into account with the respective parameters and controlling voltages since the recombination there is negligible for practical applications.

With increasing HBT speed the collector breakdown voltage is decreasing. Therefore, the corresponding avalanche current I_{AVL} needs to be taken into account during circuit design using a sufficiently accurate model description. The presently existing expression in HiCuM is based on a first-order approximation of the local electric field distribution around its peak [36, 37] and represents the weak avalanche effect. The current dependence at high injection is neglected since it leads to 3D pinch-in which cannot be modeled properly in any way with a lumped model but requires a discretized multi-transistor model [38]. For advanced SiGe HBTs (and III/V HBTs) though the finite distance carriers have to travel in order to acquire sufficient energy for successful impact ionization [39] needs to be taken into account in order to obtain physics-based parameters [2].

The DC current gain B as well as the AC current gain β in a bipolar transistor result from the physically *independent* mechanisms governing base and collector current. This is even more the case once the contributions of the external HBT regions are included. Hence, an accurate physics-based description of the current gain can only be obtained if the base current is modeled independently of the transfer (collector) current. In other words, the requirement from circuit designers for providing a simple current gain parameter value and, at the same time, an accurate bias dependent description for the current gain cannot be met.

Finally, tunneling in the reverse biased BE SCR is taken into account by simple physics-based expressions for I_{BEti} and I_{BEtp} in Fig. 8.1 [37]. Whether the internal (index i) or perimeter (index p) component dominates depends on the device design and doping profile. Trap-assisted tunnelling has not been included yet.

8.2.4 Series Resistances

The distributed internal base current flow causes a voltage drop across the lateral base region due to the finite ohmic resistance there. This voltage drop is described by the internal base resistance R_{Bi}. The exact definition and value of R_{Bi} depend on the transistor operating mode (such as DC, AC, or transient). For DC operation the solution of the non-linear differential equation for the distributed current yields the compact element expression

$$R_{Bi} = r_{SBi}(V_{B'E'}, V_{B'C'})\frac{w_E}{l_E} g_i(w_E, l_E)\psi_{dc}(I_{Bi}, r_{SBi}, w_E, l_E). \qquad (8.7)$$

Here r_{SBi} is the bias dependent internal base sheet resistance [40], g_i is a geometry function [41–45], l_E is the emitter length, and ψ_{dc} is the bias and geometry dependent DC emitter current crowding function (e.g. [41, 42, 46]). In contrast to dynamic current crowding, DC current crowding can be neglected in most HBTs so that $\psi_{dc} = 1$.

Both low- and high-frequency small-signal (AC) operation lead to a linear differential equation, allowing to define a lumped impedance that depends also on frequency [46, 47]. Although truncation after the first frequency term enables a simple network-based solution (R_{Bi} and C_{rBi} in Fig. 8.1) [48], this is *not* suitable for *large-signal non-linear* transient operation. For the latter the shape of the input signal would have to be known in order to be able to derive any closed-form solution.

The *internal* collector resistance is strongly bias dependent and, hence, needs to be described within the *internal* transistor model. In HiCuM its influence is taken into account for DC behavior by the GICCR formulation and for dynamic behavior by the transit time τ_{f0}.

The internal transistor is connected to the device terminals through ohmic regions that can be represented by lumped series resistance elements. Common to all structures is the emitter resistance R_E, the external base resistance R_{Bx}, and the external collector resistance R_{Cx}. Each of these in turn depends on components that are given by the doping, structure, and dimensions of a specific region, including contacts. These resistances are usually defined for DC operation. The influence of distributed effects in the external base at high frequencies are taken into account to first order by the Π equivalent circuit consisting of R_{Bx}, Q_{jCx}, C_{BCpar} (representing shallow trench isolation and metalization), and using the already existing nodes B and B^*. The ratio of the capacitances (cf. Fig. 8.1) can be adjusted according to the frequency behavior of the complete external base region based on a chain matrix representation of each subregion [49]. Similarly, the parasitic BE capacitance C_{BEpar}, representing spacer isolation and metalization, is split across R_{Bx}. This element partitioning improves the modelling of mm-wave transistors.

8.2.5 NQS Effects

For fast changes of the controlling voltages the carriers react delayed the farther they are away from the controlling junction. For instance, for a fast change of $V_{B'E'}$ the carriers in the BC SCR that constitute the transfer current show a larger delay than those in the base region that constitute the base minority charge. These so-called non-quasi-static (NQS) effects are taken into account in HiCuM for both I_T and Q_f in a form that is consistent between time and frequency domain [2, 50]. The respective expressions are based on a second-order truncation of the solution of the combined continuity and transport equation and allow to accurately describe over a wide bias range not only the phase shift but also the slight drop in magnitude of transfer current and minority charge as a function of frequency (e.g. [2, 9, 51]). The characteristic time constants τ_1 and τ_2 can be shown to be a certain fraction of the (base) transit time, depending on the drift field in the neutral base [2, 9, 51, 52]. Note that HiCuM does not use an inconsistent non-physical ideal delay expression in form of $\exp(-j\omega\tau)$, which is often found elsewhere in the literature (e.g. [53]) and also certain circuit simulators (incl. SPICE). Moreover, accounting for the NQS effect in both I_T *and* Q_f provides an accurate and correct phase shift for the current gain.

8.2.6 Substrate Effects

SiGe HBTs use for isolation purposes the collector substrate junction and, possibly, a deep trench. As a consequence, HiCuM includes the substrate depletion capacitance C_{jS}. In addition, at high-frequencies the bulk substrate node needs to be coupled to its contact on the chip surface by a more or less complicated impedance network [54]. High-frequency substrate coupling is a complicated 3D problem for which only very limited compact solutions are presently existing. In HiCuM, a simple parallel circuit is included consisting of resistance R_{su} and capacitance C_{su} representing the substrate resistivity and the relatively high permittivity of silicon, respectively. If a more sophisticated solution is needed, the above built-in elements can be turned off and replaced by any customized network. Finally, the base, collector and substrate region in a SiGe HBT can act as a parasitic transistor with its own transfer current I_{TS} and diffusion capacitance C_{dS} (cf. dashed connections in Fig. 8.1).

Note that Si-based SOI technologies, due to the thin bulk oxide, only eliminate the parasitic substrate transistor but not C_{jS} and the coupling network. In contrast, III/V HBTs are fabricated on semi-insulating substrates so that all substrate related elements can be dropped in a compact model.

8.2.7 Temperature Effects

Transistor operation generally causes power to be dissipated in regions where voltage drop and current density are in phase with each other. As a result, heat is generated in these dissipative regions, leading to an increase in the lattice temperature within the device and a shift in device characteristics that, in the worst case, can cause destruction. This self-heating effect is taken into account by the adjunct network in Fig. 8.1 consisting of the thermal resistance R_{th} and the thermal capacitance C_{th}. The power source P_{th} is calculated as a sum of the power dissipated in the I_T current source, the diodes and resistors. The (default) network represents the most important thermal time constant. For larger structures, such as power transistors, a more sophisticated thermal network or even a distributed model may be required in order to properly describe the different temperature in each of the cells or emitter fingers. This thermal coupling, which is also present between separate structures on a chip, can be simulated with HiCuM by linking the thermal adjunct networks of each finger or device at the externally available node ΔT_j in Fig. 8.1 (e.g. [55]).

In a physics-based compact model like HiCuM a large part of the temperature description is automatically built-in via those model parameters that depend on, e.g., intrinsic carrier density, mobility or saturation velocity. The corresponding temperature coefficients correspond to regionally averaged values and are defined as model parameters. Note that the device temperature is given by the sum of the wafer temperature and the possible increase ΔT_j from the thermal network in Fig. 8.1.

8.2.8 Noise

Low-frequency noise is represented by a Hooge-type flicker noise model in the back injection current component I_{jBE} (e.g. [56, 57]). High frequency noise is described by adding shot noise current sources to diode and transfer currents as well as a thermal noise source to resistors. Furthermore, the correlation between transfer current and neutral base charging current noise, which can become important at high frequencies, is included to first-order [58]. Finally, avalanche current noise is modeled according to [59].

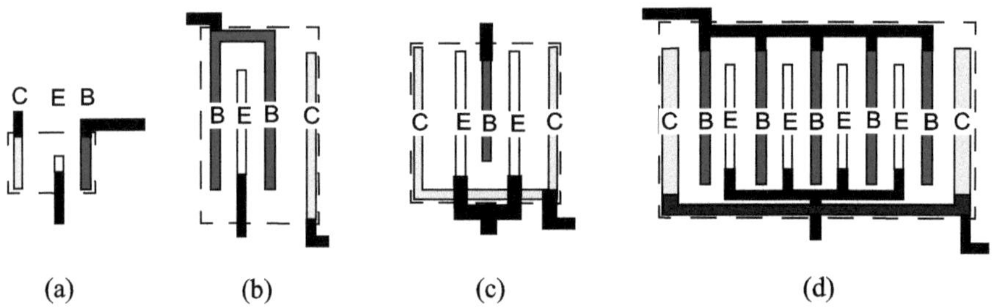

Fig. 8.4 Subset of BJT and HBT transistor configurations typically found in process design kits (PDKs): **a** minimum size type with single base contact; **b** double base standard type (e.g. for LNA and mixer design); **c** double-emitter power cell with 2 collectors; **d** multi-finger power transistor of double-base-per-emitter type

8.2.9 Geometry Dependence

Circuit design and optimization (e.g. [60]) is based on proper device sizing and, thus requires a geometry scalable compact transistor model. As additional advantages, such a model enables matching and statistical modelling as well as process debugging through parameter extraction. From a design point of view, it is desirable to be able to select the transistor *configuration*, defined by the emitter window width w_{E0} and length l_{E0} together with the number and spatial arrangement of contact slots (i.e. stripes) for base, emitter and collector. Figure 8.4 displays a small subset of the many possible configurations used in practical applications the variations of which by far exceed those possible with MOS transistors.

Except for the thermal and substrate coupling network, a compact analytical description as function of device configuration exists for each element of the HiCuM equivalent circuit in Fig. 8.1. While this is fairly simple for some elements it can become quite *complicated* for others. For instance, the elements of the internal transistor scale simply with the effective emitter area A_E (which has to be calculated though from measuring the perimeter injection; cf. Sect. 8.3.1.2), whereas a compact geometry dependent model for R_{Bi} and R_{Bx} is significantly more complicated, especially in Si-based transistors [42–45]. Even more complicated is the calculation of the thermal and substrate coupling elements since it requires the solution of a distributed 3D problem. Although compact expressions exist for special cases (e.g. [61]) the most flexible approach appears to be a series or Green's function based solution of the underlying differential equations (e.g. [62, 63]). This not only allows to handle arbitrary geometries but also to generate distributed thermal and electrical models for, e.g., power amplifier applications (e.g. [64]). Both tasks are difficult to achieve with, e.g., a simulator-specific or a Verilog-A preprocessor.

Due to the large variety of bipolar processes and structures as well as due to the sometimes complicated geometry calculations mentioned before all existing geometry equations for HiCuM (and also the SGPM) that were developed and gathered over the past 20 years have been implemented in a program named TRADICA [49]. This GUI driven tool not only enables the generation of model libraries with thousands of different configurations within seconds (e.g. [65]) but also offers device

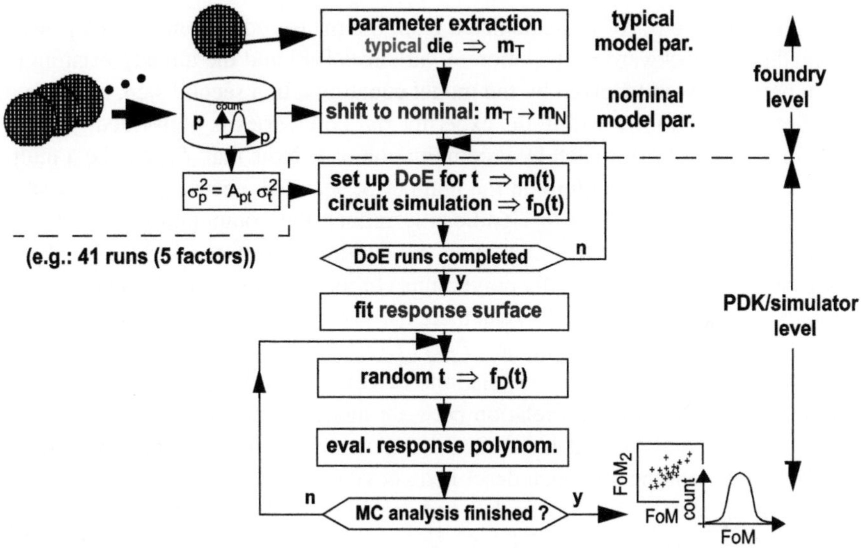

Fig. 8.5 Flow-chart for generic and computationally efficient physics-based statistical simulation of devices *and* circuits

sizing algorithms as well as predictive and statistical modelling capability (see next section). Compared to a circuit simulator, it also provides to modelling, design and process engineers a quick overview on DC and AC device characteristics as function of not only transistor configuration but also technology parameters such as some vertical dimensions and doping concentrations. TRADICA was linked to a design kit in cooperation with Atmel and Cadence [66].

8.2.10 Statistical and Predictive Modelling

The continuous trend towards smaller devices increases the importance of taking process tolerances into account. A physics-based compact model such as HiCuM contains all necessary ingredients to address predictive and statistical modelling. The latter is considered a small-signal perturbation of the former, which deals with the large-signal and possibly non-linear transition between process generations. Many, especially the key, model parameters of HiCuM depend on technology parameters such as dimensions and (average) doping concentrations, establishing a relation between the model parameter vector $\boldsymbol{m}$, the technology parameter vector $\boldsymbol{t}$, and the process control monitor (PCM) vector $\boldsymbol{p}$. The latter contains the set of data acquired in the fab through in-line measurements on production wafers. The procedure for statistical modelling with HiCuM and TRADICA (or a PDK) is shown in Fig. 8.5 and has been described in more detail in [67].

Based on a (possibly early) extracted *typical* model parameter set $\boldsymbol{m}_T$ first a shift to *nominal* parameters is performed using the predictive feature of TRADICA.

Also, the variance σ_t^2 of $\boldsymbol{t}$ is calculated from the measured variance σ_p^2 of $\boldsymbol{p}$ using the method of backward propagation of variances [68] and the already existing relation between $\boldsymbol{p}$ and $\boldsymbol{t}$ given by the model equations. In a second step, a design of experiment (DoE) for $\boldsymbol{t}$ within the experimental boundaries is performed, yielding a figure of merit $\boldsymbol{f}_D$ for each DoE combination of $\boldsymbol{t}$. Note that $\boldsymbol{f}_D$ can be a multi-dimensional vector depending on the number of selected electrical characteristics. The purpose of the DoE is to significantly reduce the computational effort (versus a full Monte-Carlo (MC) simulation) by building a response surface for $\boldsymbol{f}_D(\boldsymbol{t})$ based on a multi-variate (usually polynomial) fit. In a third step, a MC analysis with random $\boldsymbol{t}$ values is performed to calculate the statistical distribution of the desired figures of merit from the continuous $\boldsymbol{f}_D(\boldsymbol{t})$ DoE-results.

Employing the physics-based relations of HiCuM model parameters $\boldsymbol{m}$ with $\boldsymbol{t}$ and $\boldsymbol{p}$ ensures the proper correlation between model parameters in contrast to the built-in MC methods of circuit simulators, which directly use $\boldsymbol{m}$ as random variable [69–71]. Note that the approach described above is applicable directly to complete circuits of all classes, i.e. analog HF and digital. It both obviates the need of ill-defined Worst Case (WC) model parameter sets and allows in fact to define WC parameters based on a particular application. In addition it can be employed for any device type, enabling to even *include* the *physical correlations* between the various device types fabricated on the same chip.

8.3 Parameter Extraction

Compact models provide an approximate representation of the electrical characteristics of electronic devices under given bias, frequency and temperature conditions based on a set of parametrized analytical equations. The determination of the model parameters from electrical device measurements is called *parameter extraction*. Only the combination of an adequate model with a set of properly determined model parameters will provide the desired accuracy and confidence in the simulation results to allow efficient computer aided design (CAD) of integrated circuits, especially in the domains of analog and high-frequency applications. For circuit designers, the term *model* (as another indication of the importance of parameter extraction), commonly refers to the combination of compact model *and* its associated parameter set as if the two can not be distinguished from each other.

Very often, especially in the III-V HBT area, pure optimization strategies are pursued for parameter determination that ignore physical background and result in parameter sets for a single device geometry. The strategies pursued here can be divided into "slopes and intercept" or "direct" methods on one hand and "brute force" nonlinear fitting on the other hand. In the first case, model equations are suitably reformulated in terms of expressions with measured terminal variables. Then, plotting one expression vs. the other allows to extract the desired unknowns (model parameters) from the slope and (y-axis) intercept of a linear fit of the resulting curve. The second method just boils down to a nonlinear multivariate optimization problem.

Both approaches are practically feasible only for sufficiently simple models (i.e. EC and formulation). Especially for the second one the results depend strongly on the measured characteristics selected for optimization. In addition, for instance direct measurement of series resistances on a single transistor usually leads to results that depend on the transistor model assumed and, hence, neither yield necessarily the actual resistance nor can easily be compared among each other. This is true particularly for the base resistance.

Since the effort for determining a unique set of physically meaningful parameters for a single transistor structure can be quite elaborate, especially for advanced compact models, the above mentioned "classical" approaches provide models for only a *very limited number* of different devices (usually $3\ldots10$) for circuit design. This is obviously inadequate for circuit optimization and, in particular, for foundries or process departments that have to support a large user base with a wide range of circuit applications and, hence, a wide range of desired transistor configurations (cf. Fig. 8.4). Furthermore, geometry scaling and statistical modelling (i.e. including process variations in circuit design) becomes very difficult or often impossible.

The opposite and fundamentally different approach is a physics- and geometry-based parameter extraction methodology. It yields a significantly better overall process description due to the inherent smoothing of measurement errors, of local statistical variations and of extraction and/or optimization noise. At the same time, it allows to deliver geometry scalable models and has several advantages: (a) more geometry configurations can be made available to designers enabling more design freedom and, in particular, circuit optimization through transistor sizing, (b) model parameter extraction can be performed on a small subset of transistors, thus reducing this effort, and (c) test chip area can be saved. Finally, geometry scalable models ensure that important model parameters (like base resistance) behave properly with variations of geometry and technology parameters. Therefore, such a physics-based scalable approach has been developed for HiCuM over many years. It allows to determine almost all regional components separately from each other from well-defined test structures (e.g. [67, 72–76]). This approach will be described in this chapter since it exemplifies the state-of-the-art that has been achieved in the silicon-based bipolar transistor community and that has led to successful model deployment and circuit design.

Figure 8.6 provides an overview on the extraction flow that was used for obtaining the results shown in the remainder of this and in the next chapter. In this chapter the different steps are briefly presented along with selected graphical representations of the data in order to show the suitability of each method. As much as possible, we will present *analytical* parameter extraction methods, whenever they exist, and use *local* optimization only (i.e., a minimum number of model parameters having a localized influence on a well-defined subset of the measured data). However, presenting an *exhaustive* description of each parameter extraction method is beyond the scope of this chapter and fills an entire book on its own. Looking at Fig. 8.6 it is interesting to note the final fine tuning of the model parameters. This latter step becomes increasingly important for very advanced technologies like the

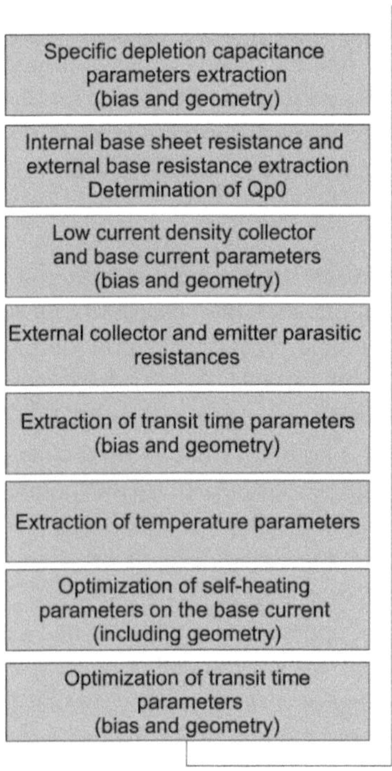

Fig. 8.6 Overview on the complete extraction flow

one studied here because many different physical effects interact at medium and high current densities. In order to solve this problem, it appears that once a good estimate of the first order parameters is obtained, a global fitting of a small subset of relevant parameters with respect to different transistor geometries is relatively efficient, and in most cases leads to a unique (physics-based) solution. This procedure is, at the present state-of-the-art in parameter extraction, the best compromise.

8.3.1 Parameter Extraction Methods

The presented parameter extraction approach and methods are illustrated by applying them to the most recent experimental data obtained for an Infineon SiGe HBT process that is based on its 200 GHz production technology B7HF200. This process is presently under development towards higher maximum oscillation frequency and smaller gate delays. The development is performed within the DOTFIVE research project [77] supported by the European commission through the seventh framework program.

8.3.1.1 Junction Capacitance Parameters

In HiCuM, the Early effect (or base width modulation) is described via the voltage dependence of the intrinsic (weighted) depletion charges stored in the internal BE and BC junction regions (cf. (8.3)). This approach, which provides an inherently consistent modelling of the DC and AC behavior, also adds conditions for capacitance modelling and extraction: (i) Internal and external capacitances must be partitioned properly, since only the internal charges are relevant for the Early effect modelling. (ii) The junction capacitance parameters have to be extracted *before* completing the DC parameter extraction (contrary to the Spice Gummel-Poon model).

Although various attempts have been made in the literature [78, 79] to split junction capacitances based on S parameter measurements of a single transistor, it turns out that the best method to do so is to rely on geometry information. As a side effect of capacitance partitioning between intrinsic and extrinsic part, a more physics-based description of the capacitances can be obtained. Actually, the intrinsic and extrinsic BE and BC regions have different doping profiles, thus requiring in any way two distinct sets of capacitance parameters (including built-in potential and grading coefficient). Therefore, the voltage dependence of the total capacitance is more accurately described with respect to geometry when using proper partitioning.

Depletion capacitances can be measured with two different methods. The first one is based on a precision capacitance meter (CV method) and the second one is based on "cold S parameters measurements" with a vector network analyzer (VNA). In our parameter extraction flow, the "cold S parameters" method will be preferred since determining the model parameters from measurements of actual transistor structures makes modelling more consistent. This method will be described in detail below. The CV method though is still useful for measuring large area BE, BC and SC diodes and therefore to access directly different junction regions and their associated parameter sets. It can also be used as a backup solution when the "cold S parameters" method yields to inaccurate results. However, in advanced SiGe HBT technologies the doping profile often varies with emitter width so that the capacitance per area in large area structures differs from that of actual transistors. This provides another reason for using the "cold S parameters" method.

In "cold S parameters" methods the transistor is biased such that DC current flow is negligible. This means that forward bias across each junction is limited to typically less than 0.5 V in silicon. The S parameters measured under these conditions are then converted into Y parameters from which, after proper deembedding, the depletion capacitances can be derived according to the following formulas:

$$C_{BE} = \Im \frac{(Y_{11} + Y_{12})}{\omega}, \tag{8.8}$$

$$C_{BC} = -\Im \frac{(Y_{12} + Y_{21})}{2\omega}, \tag{8.9}$$

$$C_{CS} = \Im \frac{(Y_{22} + Y_{12})}{\omega}. \tag{8.10}$$

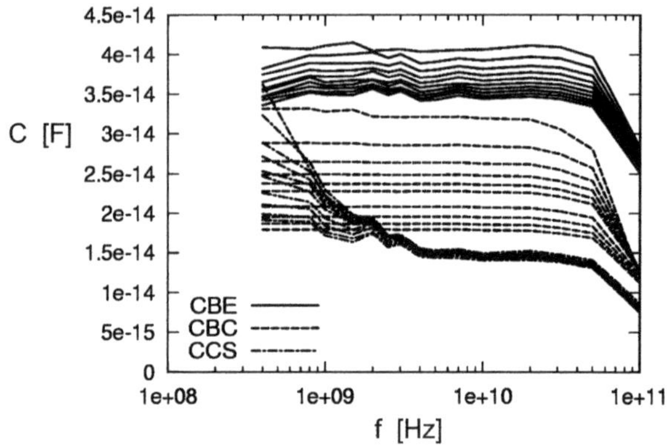

Fig. 8.7 Capacitance of Base-Emitter (*solid lines*), Base-Collector (*dashed lines*) and Collector-Substrate (*dash-dotted lines*) junction versus frequency and bias determined from "cold S parameters" measurements

The equations above are valid only at low frequencies. Above a certain frequency limit, the distributed nature of the transistor structure has to be accounted for, and a more sophisticated equation has to be used [78]. This effect can be seen in Fig. 8.7, where the extracted capacitance decreases at higher frequencies. Also, at very low frequency, the extracted capacitance terms can be quite noisy because the reactance term $1/(\omega C)$ is very large. Therefore, one has to define a suitable frequency range at which the junction capacitances can be extracted accurately. In order to further increase the accuracy, the extracted capacitance can be averaged in the selected frequency range.

Next we investigate the capacitance versus geometry relations. The preferred practice is to select transistors with emitter width variations and sufficiently large emitter lengths in order to eliminate the influence of 3D effects. Moreover, this type of configurations will lead to a larger spread in the perimeter over area ratio (P/A), which improves the extraction accuracy. In order to retrieve the depletion capacitances it is necessary to subtract the parasitic isolation capacitances (including those related to the metal backend) from the measured values. The bias independent BE and BC isolation capacitances C_{BEPAR} and C_{BCPAR} can be estimated from simple geometry (process and layout) considerations, since the main purpose here is to (a) have a reasonable estimate of the capacitance splitting range and (b) to obtain the bias dependent depletion portion of the capacitances. Note that the total capacitance stays unchanged with this procedure since C_{BEPAR} and C_{BCPAR} are taken into account later in the model. Therefore we have:

$$C_{jE} = C_{BE} - C_{BEPAR}, \tag{8.11}$$

$$C_{jC} = C_{BC} - C_{BCPAR}. \tag{8.12}$$

From a scaling point of view a depletion capacitance can be partitioned into an internal and an external part as follows:

$$C_j = C_{jA}A + C_{jP}P. \tag{8.13}$$

Here, C_j is the total junction capacitance, C_{jA} and C_{jP}, respectively, are the area and perimeter specific junction capacitances, A is the area of the junction and P

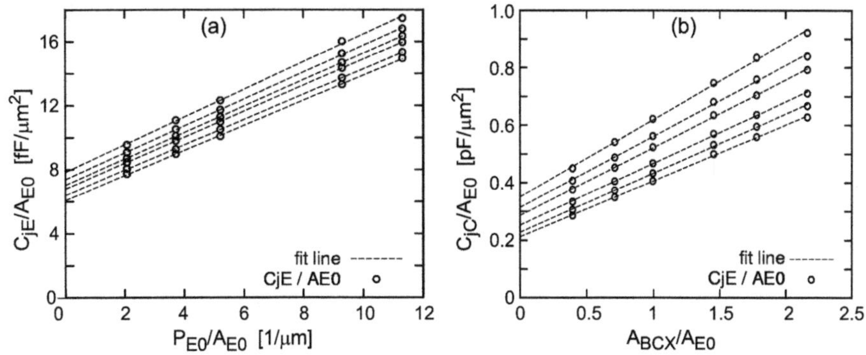

Fig. 8.8 **a** C_{jE}/A_{E0} versus P_{E0}/A_{E0} at $V_{BE} = (-1, -0.6, -0.2, 0, 0.2, 0.4)$ V. Emitter (contact) window dimensions are $(0.18, 0.22, 0.40, 0.57, 1.03) \times 10$ μm². $C_{jE} = C_{BE} - C_{BEPAR}$. (**b**) C_{jC}/A_{E0} versus A_{BCX}/A_{E0} for $V_{BE} = (-1.4, -1, -0.6, -0.20, 0.2)$ V. Emitter window dimensions are $(0.18, 0.22, 0.27, 0.40, 0.57, 1.03) \times 10$ μm². $C_{jC} = C_{BC} - C_{BCPAR}$

its perimeter. A and P have to be determined according to actual junction dimensions (not layout) including mask bias and spacers. By dividing all terms of above equation by the area, one obtains:

$$\frac{C_j}{A} = C_{jA} + C_{jP}\frac{P}{A}. \tag{8.14}$$

Therefore, by plotting the l.h.s. as a function of the P/A ratio for several geometries, C_{jA} and C_{jP} can be determined from the y axis intercept and the slope, respectively, of the obtained linear regression. Figure 8.8a shows an example for the BE depletion capacitance. In this figure, A_{E0} and P_{E0} are the actual emitter window area and perimeter, respectively, obtained by subtracting spacer width and other mask bias from the drawn dimensions.

Figure 8.8b shows a similar example for the BC depletion capacitance. Note that the junction perimeter has been replaced by the area $A_{BCX} = A_{BC} - A_{E0}$, where A_{BC} is the area of the total BC junction. This distinction is made to account for the selectively implanted collector (SIC), which leads to two distinct BC junction regions: the internal region below the emitter, and the external region with much lower doping outside the emitter. A possible perimeter part of the external capacitances can be taken into account, although this is typically not necessary due to the STI being the junction boundary.

Figure 8.9 shows the resulting extraction of the BE area and perimeter specific depletion capacitances (for consistency with Fig. 8.8, a reduced number of bias points is shown). Similar results are obtained for the base-collector specific capacitances (not shown here). Once the specific capacitances are obtained, the related specific model parameters can be determined by a simple non-linear fit of the well-known depletion capacitance equation shown as inset in Fig. 8.2.

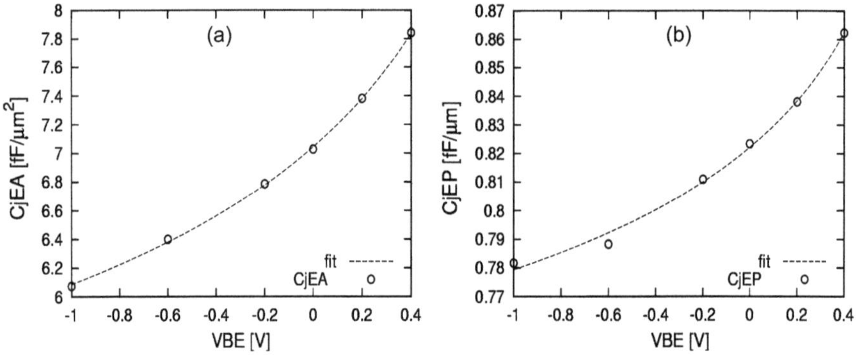

Fig. 8.9 Extracted C_{jEA} *(left)* C_{jEP} *(right)* versus V_{BE} (Reduced number of data points for clarity)

8.3.1.2 Collector and Base Current Parameters at Low Injection

At low current densities (e.g., below the onset of high-current effects and without voltage drop across series resistances, the collector current can be split into an internal and perimeter portion related to the emitter *window* dimensions:

$$I_C = I_{CA} A_{E0} + I_{CP} P_{E0}, \tag{8.15}$$

which can be conveniently reformulated as:

$$I_C = I_{CA} A_{E0} \left(1 + \frac{I_{CP} P_{E0}}{I_{CA} A_{E0}} \right) \tag{8.16}$$

or

$$I_C = I_{CA} A_{E0} \left(1 + \gamma_C \frac{P_{E0}}{A_{E0}} \right) = I_{CA} A_E. \tag{8.17}$$

Here γ_C is the perimeter specific to area specific collector current ratio, and A_E is the effective electrical emitter area. Therefore, according to the equation above, when plotting I_C/A_{E0} versus P_{E0}/A_{E0} one can extract the area specific collector current parameter I_{CA} and γ_C from the y intercept and the slope of the curves. According to Fig. 8.10b the scaling equations used describe fairly well the collector current behavior, although the discrepancies are magnified compared to the $I_C(w_E)$ plot in Fig. 8.10a. It has to be noted though, that the emitter length scaling may need more elaborate scaling equations due to, e.g., proximity effects.

A similar procedure is applied to the base current in order to determine the related specific parameters. Note that this has to be done for the ideal and non-ideal (recombination) base current components. Fig. 8.11a and b show the results obtained for the studied process regarding the ideal base current. Similar results can be obtained for the base recombination current, although those are typically noisier. Actually, the base recombination current is highly subject to (i) process statistical variations, (ii) measurement noise or inaccuracy, (iii) erratic behavior due to the stress applied to the transistors during measurements.

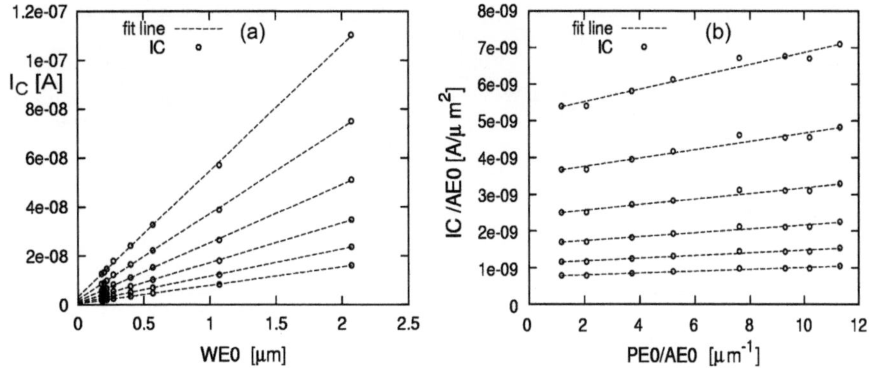

Fig. 8.10 **a** I_C vs. w_{E0} for $V_{BE} = 0.45$ to 0.5 V in steps of 10 mV. **b** I_C/A_{E0} vs. A_{E0}/P_{E0} for the same bias. The emitter window dimensions are $(0.18, 0.22, 0.27, 0.4, 0.57, 1.03, 2.03) \times 10\ \mu m^2$

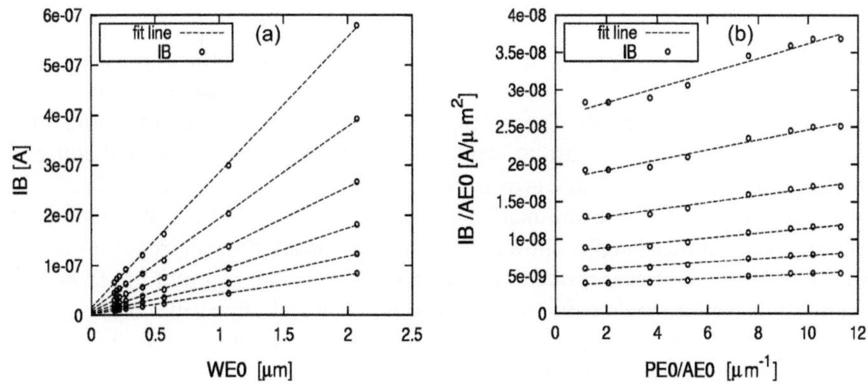

Fig. 8.11 **a** I_B vs. w_{E0} for $V_{BE} = 0.65$ to 0.7 V in steps of 10 mV (ideal region), **b** I_B/A_{E0} vs. A_{E0}/P_{E0} for the same bias. The emitter window dimensions are $(0.18, 0.22, 0.27, 0.4, 0.57, 1.03, 2.03) \times 10\ \mu m^2$

8.3.1.3 Base Resistance

The determination of parameters for the external and internal base resistance is one of the most challenging topics of bipolar model parameter extraction. The literature provides an extensive number of methods for base resistance extraction from the measurements of a single transistor, based on DC measurements [80, 81] and S parameters measurements [78, 82–84]. Most of these have limited accuracy or do not allow to distinguish the internal from the external part of the resistance. Following the physics/based approach of the HiCuM model we will present a method based on specific test structures (tetrodes or ring emitter structures) in order to determine the internal base sheet resistance and the length specific external base resistance. The actual model parameter values are then determined by scaling equations. The tetrode structure consists of a transistor with two base terminals B_1 and B_2 sepa-

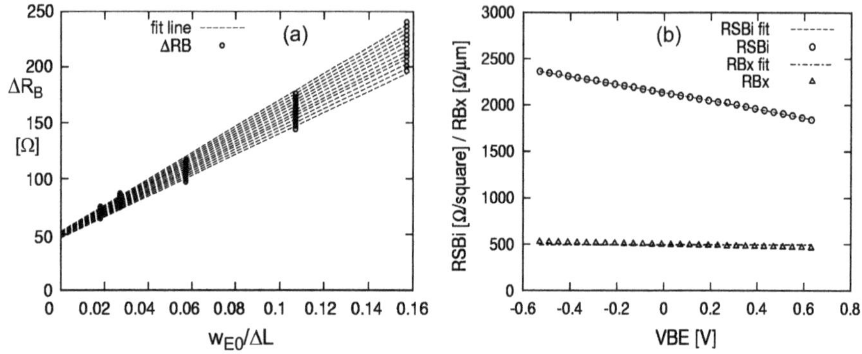

Fig. 8.12 **a** Measured (corrected) resistance between the 2 tetrode base terminals versus $w_{E0}/\Delta L$ for different BE voltages. $\Delta L = l_{E02} - l_{E01} = 10\ \mu\text{m}$, and the emitter window width is (0.18, 0.27, 0.57, 1.03, 1.57) μm. **b** Extracted internal base sheet resistance r_{SBi} and external base resistance per unit length (R_{Bx}) versus $V_{BE} = V_{BC}$ ranging from -0.5 to 0.6 V in steps of 100 mV

rated by a ring emitter that forces the current to flow through the internal base below the emitter [72, 73]. The base resistance can be measured directly between the two base terminals by applying a voltage ΔV with an offset V_1 ranging from -0.6 to $+0.6$ V (i.e. negligible current across the junctions), with emitter and collector being grounded. $\Delta V = 20$ mV is usually a good compromise between accuracy and symmetry across the base terminals. Subtracting the currents from structures with two different emitter lengths allows removing the impact of the current flow across the corners and emitter foreside. Measuring structures with several (3 to 6) emitter widths then leads to the results shown in Fig. 8.12a, which shows the corrected base resistance versus emitter width normalized to the length difference ΔL. As it can be observed the investigated process is fairly scalable.

Next, we can extract the internal base sheet resistance r_{SBi} and the length specific external base resistance R_{Bx} as a function of average bias from the slope and the intercept of the fitting curves. The result is shown in Fig. 8.12b. One can observe that the external base resistance is fairly constant with respect to bias as expected, while the internal base sheet resistance has a noticeable voltage dependence due to the base width modulation following the equation (valid for $V_{CE} = 0$):

$$r_{SBi} = r_{SBi0} \frac{Q_{p0}}{Q_{p0} + Q_{jEi} + Q_{jCi}}. \tag{8.18}$$

Here Q_{jEi} and Q_{jCi} are the BE and BC depletion charges respectively which are known from the depletion capacitance model parameters. QP0 is a HiCuM model parameter which can be quite accurately determined from $r_{SBi}(V_{BE}, V_{BC})$ and the optimization of (8.18). The result is also displayed in Fig. 8.12b.

8.3.1.4 External Collector Resistance

Since in HiCuM the modelling of the epitaxial (internal) collector region is handled by the GICCR [8, 34] there is no resistor element in the equivalent circuit related

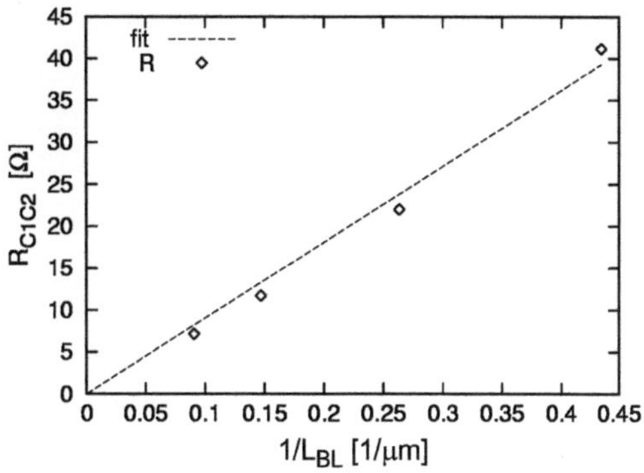

Fig. 8.13 Measured resistance R_{C1C2} between the two collectors $(C1, C2)$ of a special test structure versus reciprocal buried layer length $L_{BL} = 2.3, 3.8, 6.8, 11\ \mu m$

to the internal collector. Only the external collector series resistance R_{Cx} consisting of the contribution from buried layer, sinker and contact has to be determined. Although many different methods exist for the determination of the collector resistance in the literature, all being based on measurements of a single transistor [85], none of these give satisfactory results. Therefore, we will present here a specific test structure approach [86], which allows determining unambiguously the various components of the external collector resistance components. The actual model parameter value R_{Cx} will be computed according to the extracted component values and a set of scaling rules. The used scaling rules for R_{Cx} can become fairly complicated depending on the investigated collector contact configurations (see examples in Fig. 8.4, contact(s) at foreside, etc.) and presenting them goes beyond the scope of this chapter.

The external collector resistance determination test structure consists of a transistor structure with two separate collector contacts connected to external metal pads in Kelvin (Force/Sense) configuration [86]. Different buried layer widths are required and two different lengths can be used to correct for current-spreading effects. For the process investigated here though, only a fixed width but different lengths were available on the existing test chip. Therefore, only the sum of all components could be extracted resulting in an overall length dependent resistance.

Figure 8.13 shows the measured resistance between the two collectors of the test structure as a function of the inverse of the buried layer length. Note that the buried layer extension ($L_{BL} > l_E$) has been taken into account leading to a fitting curve intercepting the y axis at 0. From the slope of the curve the length specific collector resistance can be extracted. The discrimination between buried layer and sinker plus contact contribution had to be realized later by investigation of transistor output characteristics in saturation, which gives suboptimal accuracy though.

8.3.1.5 Emitter Resistance

The emitter resistance parameter extraction is one of the most challenging steps. Many different methods have been proposed in the literature [80, 82, 87, 88]. Most of these lead to questionable results. Especially the methods based on open-collector

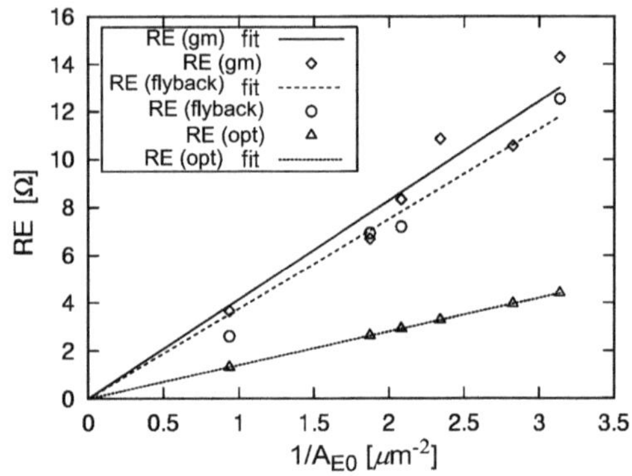

Fig. 8.14 Extracted emitter resistance R_E versus $1/A_{E0}$ for transistors with emitter window dimensions of $(0.18, 0.21) \times 1.9\ \mu m^2$ and $(0.16, 0.2, 0.4) \times 2.8\ \mu m^2$. The parasitic probe resistances have been subtracted

measurements extract the emitter resistance R_E in hard saturation, which does not correspond to the normal transistor operation mode and the typical vertical current flow through the emitter regions. Other methods are facing either limitations due to approximations of the transistor equations or inaccuracy due to limited measurement precision. Figure 8.14 presents the extracted R_E for different transistor geometries as a function of the inverse of the emitter window area for the "g_m" method and for the "flyback" method. As it can be seen the results are subject to extraction noise. This can be attributed to both the limitations of the methods and the use of a structure with several transistors in parallel, which reduces the emitter resistance measurement sensitivity. Finally, Fig. 8.14 also presents the recalculated emitter resistance after final optimization of the model parameters (opt). As it can be seen, the final value used for R_E is much smaller than the extracted value, which was mandatory to obtain a good global agreement. At the time of writing this text the cause of this difference is not well understood. We hypothesize that for very advanced SiGe HBT processes, like the one investigated here, high current and barrier effects as well as self-heating have an increased impact on the transistor characteristics, thus making an accurate emitter resistance parameter extraction more difficult.

8.3.1.6 Transit Time Parameters

The transit frequency f_T is defined as the frequency at which the extrapolated small-signal current gain becomes unity. It is an important performance metric that, from a modelling point of view, has the advantage of being unambiguously defined and easy to measure (via S parameters), but also allows conveniently to determine the transit time via

$$\tau_f = \frac{1}{2\pi f_T} - \frac{C_{BE} + C_{BC}}{g_m} - (R_{Cx} + R_E)C_{jC}. \tag{8.19}$$

In HiCuM the current and voltage dependence of τ_f is split into a current independent low-current component, which only depends on voltage $V_{B'C'}$, and a current and voltage dependent high current component [2, 19],

$$\tau_f = \tau_{f0}(V_{B'C'}) + \Delta\tau_f(V_{B'C'}, I_T). \tag{8.20}$$

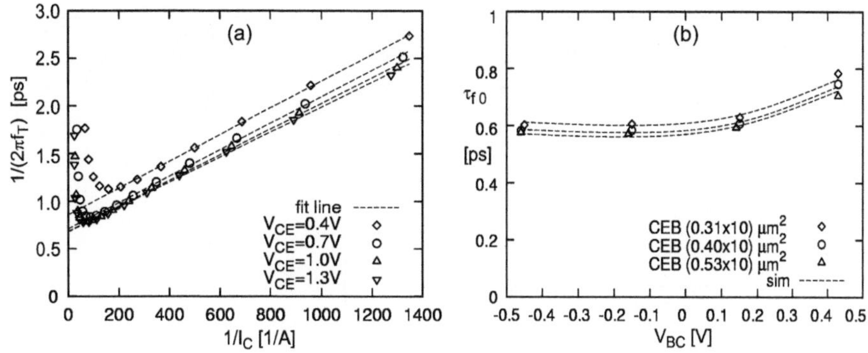

Fig. 8.15 **a** $1/(2\pi f_T)$ vs. $1/I_C$ for a drawn emitter size of $0.18(0.31) \times 10\ \mu m^2$. **b** Low-current transit time $\tau_{f0}(V_{BC})$ for drawn emitter sizes of $(0.31, 0.4, 0.53) \times 10\ \mu m^2$

Low-Current Transit Time The first step of the transit time parameter extraction method consists in determining the transit frequency from S-parameter measurements at various collector current densities and various voltages V_{CE} or V_{BC}. The transit frequency f_T is determined using the current gain bandwidth product by, e.g., choosing a spot frequency in the -20 dB/decade roll-off region of the current gain. The second step consists in determining the low-current transit time τ_{f0} and the transit time increase $\Delta\tau_f$ at high current densities from the reciprocal transit frequency (cf. Fig. 8.15a). According to (8.19), plotting $1/(2\pi f_T)$ versus $1/g_m$ or $1/I_C$ at a constant voltage V_{CE} or V_{BC} yields a curve, which can be approximated by a straight line at low current densities with a slope that is proportional to $(C_{BE} + C_{BC})$. Extrapolating this curve towards the y axis (e.g., towards high current densities) gives as intercept the low-current transit time τ_{f0} plus the last term on the r.h.s. of (8.19). With the terms R_E, R_{Cx} and C_{BC} already known, τ_{f0} can then be determined. More advanced methods also exist [89, 90] with the possible drawback of being more complicated and less robust. Figure 8.15a also shows the results for the fitting lines used.

By applying this method for different V_{CE} (or V_{BC}), one can obtain $\tau_{f0}(V_{BC})$ as it is presented in Fig. 8.15b for several transistor geometries. From these data one can determine the HiCuM model parameters T0, TVBL and DT0H from nonlinear optimization of the $\tau_{f0}(V_{BC})$ equation or from using a direct method as described in [91]. Also the geometry dependence of the transit time can be determined by fitting the following equation to the extracted transit time:

$$\tau_0 = \tau_{f0i} \frac{1 + (\tau_{f0p}/\tau_{f0i})\gamma_C P_{E0}/A_{E0}}{1 + \gamma_C T_{E0}/A_{E0}}. \tag{8.21}$$

Here, τ_{f0i} and τ_{f0p}, respectively, are the area and perimeter specific low-current transit time; γ_C, A_{E0} and P_{E0} have already been defined earlier for the collector (transfer) current. As can be seen from Fig. 8.15b, the low-current transit time can be described quite well as a function of voltage and geometry.

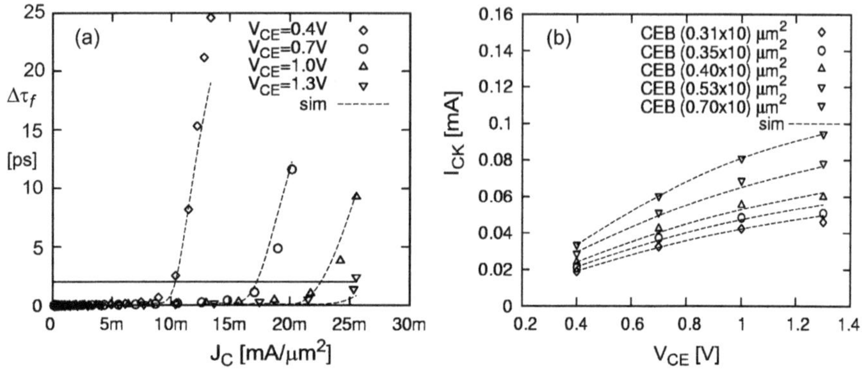

Fig. 8.16 **a** Transit time increase $\Delta\tau_f$ at high current densities for the example of a transistor with emitter window size of $0.18 \times 10\ \mu m^2$. **b** Critical current $I_{CK}(V_{CE})$ for drawn emitter sizes of $(0.31, 0.35, 0.40, 0.53, 0.70) \times 10\ \mu m^2$

High-Current Transit Time The next step of the transit time parameter extraction consists in the determination of the transit time increase $\Delta\tau_f$ at high current densities. The most simple description available in HiCuM reads

$$\Delta\tau_f = THCSw^2\left[1 + \frac{2I_{CK}}{I_{Tf}\sqrt{(i^2 + AHC)}}\right] \tag{8.22}$$

with the normalized collector injection width

$$w = \frac{i + \sqrt{(i^2 + AHC)}}{1 + \sqrt{(i^2 + AHC)}} \quad \text{and} \quad i = 1 - \frac{I_{CK}}{I_{Tf}}. \tag{8.23}$$

From Fig. 8.15, $\Delta\tau_f$ is calculated from the difference between the measured data and the straight lines extrapolated from low injection. The results are shown in Fig. 8.16a as a function of collector current (or collector current density) for several V_{CE} voltages. From this we can determine the critical current I_{CK} (see (8.2)), which indicates the onset of high-current effects in the collector and base. Several methods exist [91] for that purpose, but we will present here only a simple method that has the advantage of being very robust. This method consists of defining a constant threshold value below the steep increase (typically $K^*\tau_0$, where K is between 2 and 5) and taking the intercept with the extracted $\Delta\tau_f$ curves (after proper interpolation). From this the $I_{CK}(V_{CE})$ curve can be obtained (see Fig. 8.16b). It has to be noted that the bias range for S parameter measurements has to be carefully chosen in order to apply this methodology since the following conditions have to be satisfied: (a) obtain data up to high current densities (for $\Delta\tau_f$ determination) and (b) limit the power dissipation in the transistor to avoid stress and/or device destruction.

From the critical current $I_{CK}(V_{CE})$ curves one can extract the HiCuM model parameters RCI0, VLIM and VPT by nonlinear optimization of (8.2) or by using a direct method as described in [19]. Also the geometry dependence of the parameter RCI0 can be determined using the following scaling equations:

$$RCI0 = r_{Ci0,A}/(A_E f_{cs}) \tag{8.24}$$

$$f_{cs} = \begin{cases} \frac{(\zeta_b - \zeta_l)}{\ln[(1+\zeta_b)/(1+\zeta_l)]}, & l_{E0} > w_{E0}, \\ 1 + \zeta_b, & l_{E0} > w_{E0}, \end{cases} \tag{8.25}$$

$$\zeta_b = 2\frac{w_{Ci}}{w_E}\tan\delta_C, \qquad \zeta_l = 2\frac{w_{Ci}}{l_E}\tan\delta_C. \tag{8.26}$$

Here, $r_{Ci0,A}$ is the area-specific low-field internal collector resistance, w_{Ci} is the collector width under the emitter, and δ_C is the current spreading angle in the collector. As it can be seen from Fig. 8.16b, both bias and geometry dependence of the critical current are very well modelled by the analytical scaling equations.

Finally, the two HiCuM model parameters THCS and AHC in (8.22) can be determined from the $\Delta\tau_f$ curves of Fig. 8.16 by nonlinear optimization of (8.22) or by a direct method as described in [19]. In order to obtain a better robustness for the latter step, one can plot $\Delta\tau_f$ as a function of I_C/I_{CK}, in which case the V_{CE} dependence disappears and all curves will overlay on top of each other. This is also a good criterion to evaluate the accuracy of the I_{CK} determination. In Fig. 8.16, a good accuracy can be observed for the $\Delta\tau_f$ description.

8.3.1.7 Temperature Related Parameters

Accurate transistor modelling as a function of temperature is mandatory for accurately predicting integrated circuit behavior in a wide variety of environments, such as consumer, automotive or space applications. Moreover, the increasing impact of self-heating in modern SiGe HBTs increases the importance of accurate temperature modelling. In HiCuM/L2 v2.23 the temperature equations for the collector current parameters read:

$$c_{10}(T) = c_{10}(T_0)\left(\frac{T}{T_0}\right)^{ZETACT} \exp\left[\frac{VGB}{V_T}\left(\frac{T}{T_0} - 1\right)\right], \tag{8.27}$$

$$Q_{p0}(T) = Q_{p0}(T_0)\left[2 - \left(\frac{V_{DEi}(T)}{V_{DEi}(T_0)}\right)^{z_{Ei}}\right]. \tag{8.28}$$

Furthermore, the temperature dependence of the emitter backinjection component of the base current is described by the saturation current

$$I_{BEiS}(T) = I_{BEiS}(T_0)\left(\frac{T}{T_0}\right)^{ZETEBET} \exp\left[\frac{VGE}{V_T}\left(\frac{T}{T_0} - 1\right)\right]. \tag{8.29}$$

In above equations, ZETACT, VGB, ZETABET and VGE are model parameters to be determined. Note that the temperature dependence of Q_{p0} is fixed by the previously determined capacitance parameters, which emphasizes again the importance of accurate depletion capacitance determination and partitioning. In order to extract the model parameters, collector and base current measurements with respect to temperature are employed. Gummel plots at $V_{BC} = 0$ V are well suited for this purpose.

The next step consists in nonlinear optimization by applying (8.27) and (8.28) to the collector current measurements and (8.29) to the base current measurements.

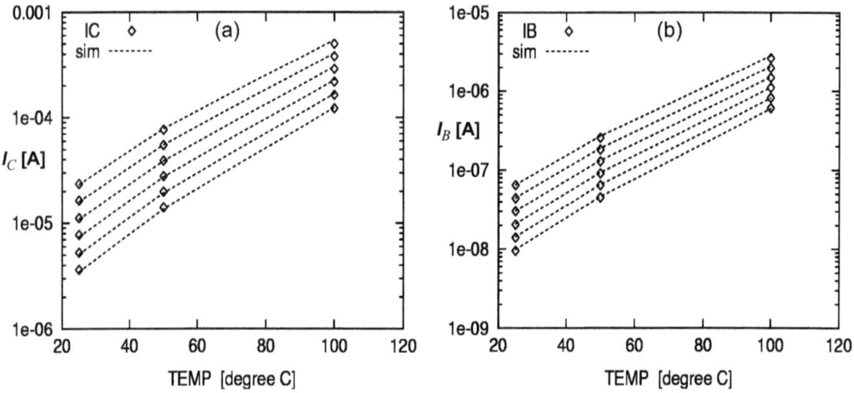

Fig. 8.17 Temperature dependence of **a** collector current and **b** base current for $V_{BE} = 0.65$ to 0.7 V in steps of 10 mV and at $V_{BC} = 0$ V

Alternatively, VGB and VGE can be estimated directly from the measured data by assuming ZETACT = ZETABET = 3 and neglecting the QPO temperature dependence:

$$VG = V_T \ln\left[\frac{P(T)/P(T_0)}{\exp[(T/T_0) - 1]}\left(\frac{T_0}{T}\right)^3\right]. \tag{8.30}$$

Here $P(T)$ is the parameter whose temperature dependence is to be extracted at a temperature T. Figure 8.17 exhibits the resulting comparisons between measurement and simulation for the collector and base current.

Similar methods can be applied to the other BE and BC junction related current components. Then by applying the methods described for room temperature on measured data taken at other temperatures the temperature coefficients can be determined for the remaining parameters (Parasitic resistances, transit time, etc.). These temperature related parameters can play a significant role even at room temperature for advanced technologies for which the self-heating effect becomes increasingly important.

8.4 Application Examples

Production HiCuM model parameters have been generated for a variety of processes and are also available from a number of foundries (e.g. [69–71, 92, 93]).[1] A more detailed overview, especially on circuit results obtained with HiCuM-based transistor libraries in design kits, is given in [2]. Below, the model playbacks are presented for the selected SiGe HBT process for which the parameter extraction was performed

[1]Production HICUM parameters have been available from BiCMOS *foundries* such as Atmel, IBM, ffIP, Jazz, ST, Toshiba, TSMC.

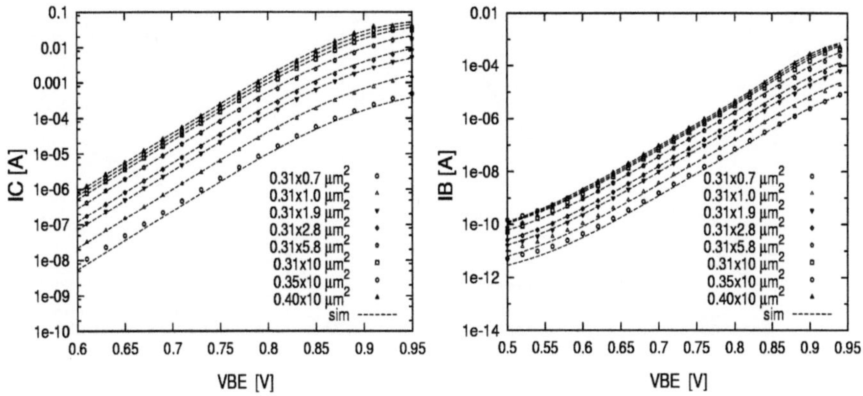

Fig. 8.18 Collector current (*left*) and base current (*right*) vs. V_{BE} at $V_{BC} = 0$ for different drawn emitter lengths and widths

Fig. 8.19 DC current gain vs. collector current density at $V_{BC} = 0$ for different emitter lengths

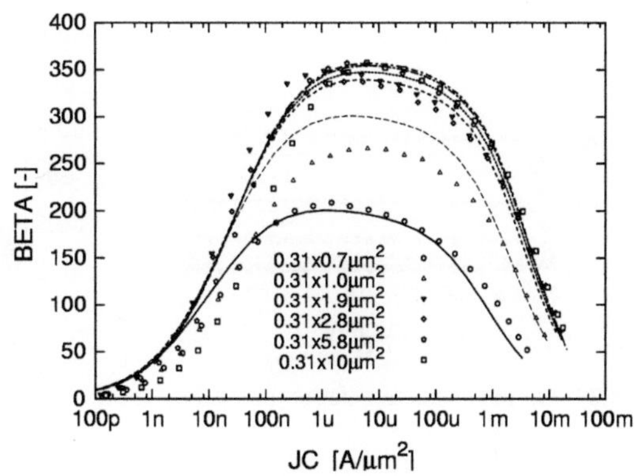

in the previous chapter. Note that parameters were not extracted for all structures presented; i.e. their parameters are generated based on the specific electrical data obtained during the extraction and on the design rules of the process.

Figure 8.18 exhibits the collector and base current versus BE voltage for different emitter dimensions. As can be seen the agreement over both bias and geometry is very good, except perhaps for the smallest transistor with a drawn emitter size of $0.31 \times 0.7\ \mu m^2$ (i.e. window size of $0.18 \times 0.57\ \mu m^2$), which suffers from variations in emitter mask and vertical doping concentrations and is typical at the beginning of (a new) process development. Nevertheless, the process already scales quite well in both length and width direction.

Figure 8.19 shows the corresponding DC current gain as a function of collector current. Good agreement can be observed for different emitter lengths, except for a few deviations due to the nonideal base current component, which is intrinsically more subject to process variations.

Figure 8.20 presents the output characteristics for forced V_{BE} for different transistor geometries. Note that the avalanche parameters have not been extracted for

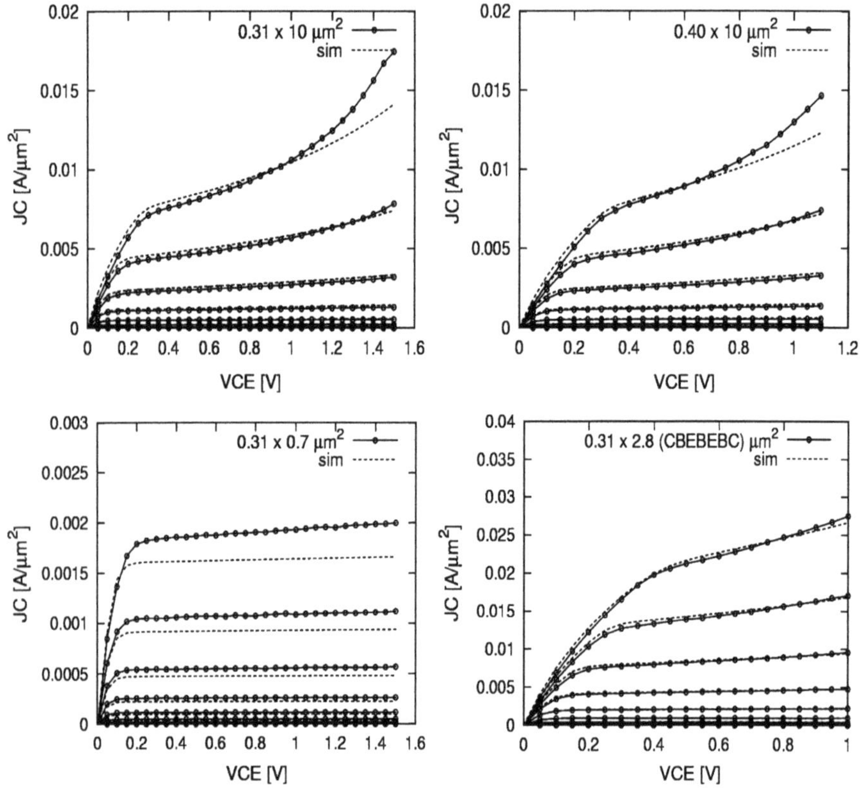

Fig. 8.20 Output characteristics at forced V_{BE} for different emitter sizes

this process. Again good agreement is obtained even in the presence of very high self-heating, except for the smallest transistor (for the reasons mentioned above).

The output characteristics at forced I_B in Fig. 8.21 show excellent agreement for different transistor geometries. This also holds, according to Fig. 8.22, for both the BE and BC depletion capacitance as well as, according to Fig. 8.23, the transit frequency over a wide range of emitter sizes and bias conditions (cf. Fig. 8.24).

The results shown above represent a standard set of characteristics. Model comparisons and verifications for special characteristics, such as distortion and noise can be found in the literature (e.g., [10, 11, 59, 64, 69, 74, 93]). Publicly available results obtained for circuits have been compiled in [2].

8.5 Conclusions

The first part of this book chapter has provided "HiCuM in a nutshell"—a concise overview on the approach and the physical effects taken into account as well as on relevant literature containing theory, derivations and results. Most important fea-

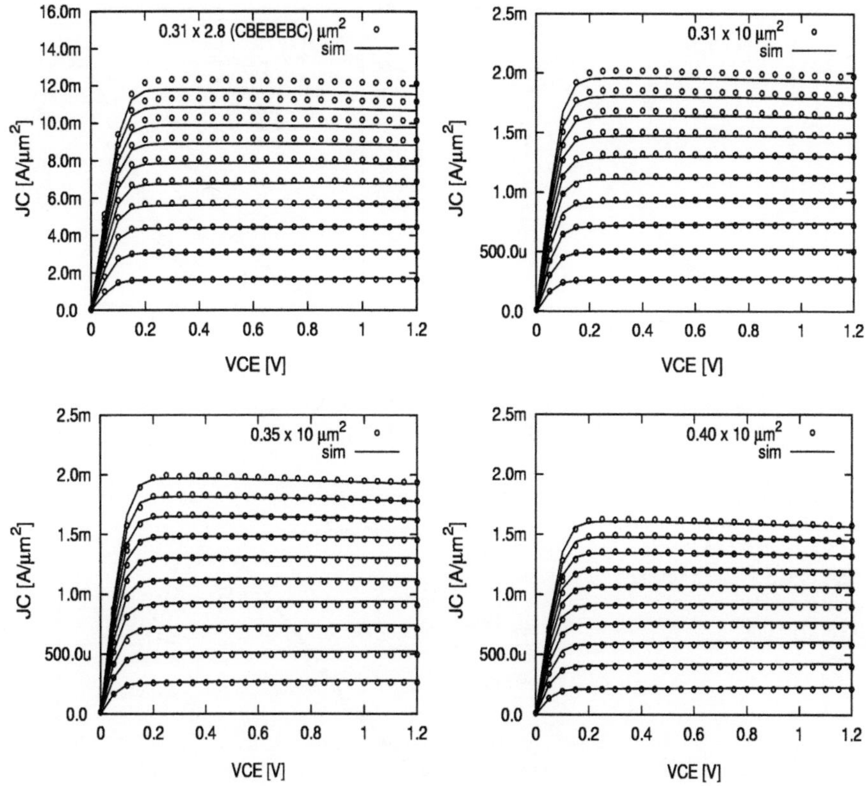

Fig. 8.21 Output characteristics at forced I_B for different emitter sizes

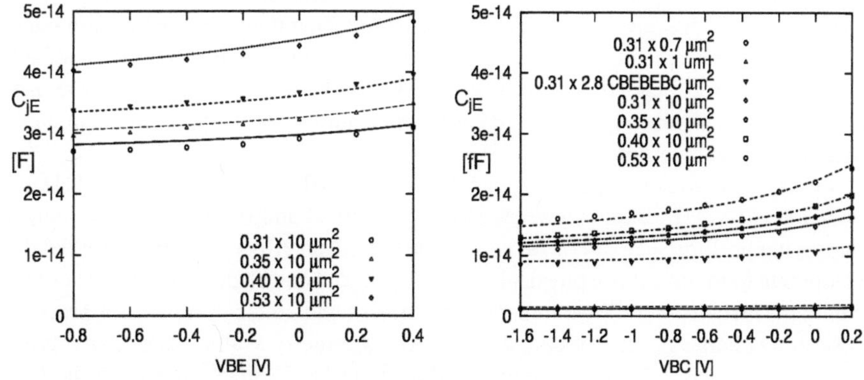

Fig. 8.22 Voltage and geometry dependence of **a** BE depletion capacitance and **b** BC depletion capacitance

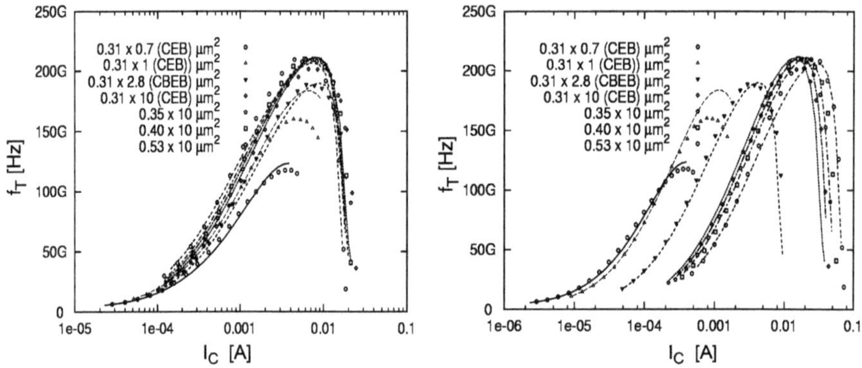

Fig. 8.23 Transit frequency at $V_{CE} = 1$ V vs. collector current *density* (*left*) and vs. collector current (*right*) for different (drawn) emitter dimensions

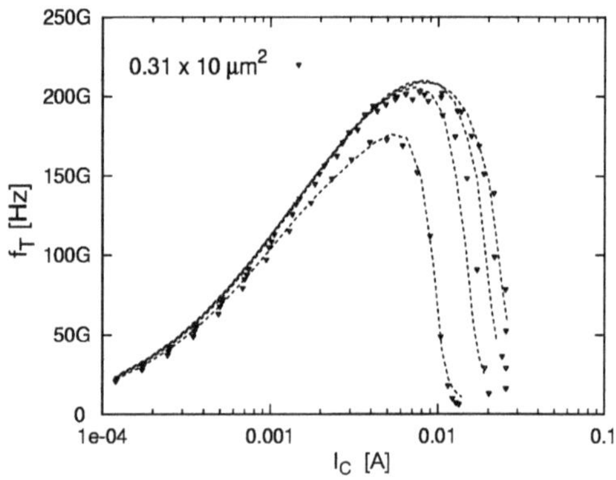

Fig. 8.24 Transit frequency vs. collector current density for $V_{CE} = 0.4$ to 1.3 V in steps of 0.3 V and a selected emitter window size of $0.18 \times 10\ \mu m^2$

tures of the model include an accurate and consistent physics-based description of charges and currents over a wide bias, temperature, and geometry range, an accurate description of non-quasi-static effects, noise correlation and electrothermal effects. The second part has given an overview on the model parameter extraction approach, which was also implemented into a commercial package. Finally, for a selected advanced 200 GHz SiGe HBT process the model has been compared to experimental characteristics, exhibiting results that are representative for the accuracy obtainable in a production environment.

Future development especially for SiGe and InP HBTs is targeting (low-)THz applications [77, 94]. In this view, the already observed and further expected rapidly improving performance of HBTs requires *inparallel* an effort on compact model development to include those physical effects that have been neglected so far. These include, for instance, non-local transport in the base and collector region for properly describing charge storage and breakdown. Also geometry-scaling related effects are becoming increasingly more complicated to describe. Besides improving the process technology and compact models, significant work is also required for providing the associated infrastructure such as parameter determination methods, suitable test structures (also for geometry scaling), and accurate methods for de-embedding data from HF measurements up to 110 GHz (and even 220 GHz). Without this parallel development effort, a process technology can not be fully exploited and will not generate the maximum return on the ever increasing investment. For instance, the soaring mask cost, which by far exceed US $1 Million for a 90 nm-(Bi)CMOS process node, and the resulting pressure to reduce the overall design cost have raised over the past decade the importance of compact modelling. Saving just one design cycle and its associated mask cost quickly pays off model support and development cost!

According to present progress and first extrapolations towards next process generations in existing research projects [77, 94], HBTs show great potential for addressing the needs of future ultra-high-speed electronics applications. These developments appear to proceed significantly faster than predicted by the ITRS [95]. HBT technology is an excellent candidate for achieving the goal of process technology development to eventually combine in a modular fashion analog HF and digital signal processing along with, e.g., MEMS, on a single chip in order to maximize system and chip functionality and performance at minimum power dissipation. In this view of addressing a very diverse application market, pure CMOS system-on-chip solutions appear to be just a stepping stone towards this goal and to be economically suitable only for a few very specialized very high-volume products.

Acknowledgments We would like to thank Magali De Matos, Thomas Zimmer (IMS Lab, University Bordeaux I, France), Christian Raya (XMOD), and Paulius Sakalas (CEDIC) for their help with measurements, Julia Krause, Hung Iran, and Anindya Mukherjee (all with CEDIC, TUD) for carefully proofreading this book chapter as well as Klaus Aufinger and Thomas Meister (both with Infineon Technologies) for providing wafers and related process information. Furthermore, we are indebted to the European Commission for financial support of the integrated research project DOT-FIVE through their seventh framework program. We also like to express our sincere gratitude to the many individuals that have supported our bipolar transistor modelling and, in particular, HiCuM development effort over the past two decades. Finally, the first author (MS) likes to acknowledge financial support from the German Research Foundation (DFG).

References

1. Long, J.: SiGe radio frequency ICs for low-power portable communication. Proc. IEEE **93**, 1598–1623 (2005)
2. Schröter, M., Chakravorty, A.: Compact Hierarchical Bipolar Transistor Modelling with HICUM. World Scientific, Singapore (to appear)
3. Cressler, J. (ed.): Silicon Heterostructure Handbook. CRC Press, Boca Raton (2005)
4. Rein, H.-M., Stübing, H., Schröter, M.: Verification of the integral charge-control relation for high-speed bipolar transistors at high current densities. IEEE Trans. Electron Devices **32**, 1070–1076 (1985)
5. Stuebing, H., Rein, H.-M.: A compact physical large-signal model for high-speed bipolar transistors at high current densities—Part I: One-dimensional model. IEEE Trans. Electron Devices **ED-34**, 1741–1751 (1987)
6. Rein, H.-M., Schröter, M.: A compact physical large-signal model for high-speed bipolar transistors at high current densities—Part II: Two-dimensional model and experimental results. IEEE Trans. Electron Devices **34**, 1752–1761 (1987)

7. Koldehoff, A., Schröter, M., Rein, H.-M.: A compact bipolar transistor model for very high-frequency applications with special regard to narrow stripes and high current densities. Solid-State Electron. **36**, 1035–1048 (1993)
8. Schröter, M., Friedrich, M., Rein, H.-M.: A generalized integral charge-control relation and its application to compact models for silicon based HBT's. IEEE Trans. Electron Devices **40**, 2036–2046 (1993)
9. Schröter, M., Rein, H.-M.: Investigation of very fast and high-current transients in digital bipolar circuits by using a new compact model and a device simulator. IEEE J. Solid-State Circuits **30**, 551–562 (1995)
10. Schröter, M., Pehlke, D.R., Lee, T.-Y.: Compact modelling of high-frequency distortion in Si integrated bipolar transistors. IEEE Trans. Electron Devices **47**, 1529–1539 (2000)
11. Schröter, M.: High-frequency circuit design oriented compact bipolar transistor modelling with HICUM (invited paper), IEICE Trans. Electron. **E88-C**(6), 1098–1113 (2005). Special Issue on Analog Circuit and Device Technologies. See also: M. Schröter, "Advanced compact bipolar transistor models—HICUM", Chapter 8.4 (pp. 807–823) in "Silicon heterostructure Handbook", ed. by J. Cressler, CRC Press NY, 2005. For the latest HICUM version and related references see: www.iee.tu-dresden.de/iee/eb
12. Compact modelling Council (CMC): http://www.eigroup.org/cmc
13. XMOD Technologies: www.xmodtech.com
14. Rein, H.-M.: A simple method for separation of the internal and external (peripheral) currents of bipolar transistors. Solid-State Electron. **27**, 625–632 (1984)
15. Schröter, M., Walkey, D.J.: Physical modelling of lateral scaling in bipolar transistors. IEEE J. Solid-State Circuits **31**, 1484–1491 (1996). **33**, 171 (1998)
16. Schröter, M., Lehmann, S., Fregonese, S., Zimmer, T.: A computationally efficient physics-based compact bipolar transistor model for circuit design—Part I: Model formulation. IEEE Trans. Electron Devices **53**, 279–286 (2006)
17. Fregonese, S., Lehmann, S., Zimmer, T., Schröter, M., Celi, D., Ardouin, B., Beckrich, H., Brenner, P., Kraus, W.: A computationally efficient physics-based compact bipolar transistor model for circuit design—Part II: Parameter extraction and experimental results. IEEE Trans. Electron Devices **53**, 287–295 (2006)
18. Schröter, M.: Compact bipolar transistor modelling—Issues and possible solutions. In: Proc. WCM, International NanoTech Meeting, San Francisco (CA), pp. 282–285 (2003)
19. Schröter, M., Lee, T.-Y.: Physics-based minority charge and transit time modelling for bipolar transistors. IEEE Trans. Electron Devices **46**(2), 288–300 (1999)
20. Schröter, M., Tran, H.: Charge-storage related parameter calculations for Si and SiGe bipolar transistors from device simulation. In: Proc. WCM, International NanoTech Meeting, Boston (MA), pp. 735–740 (2006)
21. Tran, H.: Physics-based analytical modelling of SiGe heterojunction bipolar transistors for high-speed integrated circuits. Ph.D. thesis, Chair for Electron Dev. and Integr. Circuits, TU Dresden (2006)
22. Tiwari, S.: A new effect at high currents in heterostructure bipolar transistors. IEEE Electron Devices Lett. **9**, 142–144 (1988)
23. Kirk, C.T.: A theory of transistor cutoff frequency (fT) falloff at high current densities. IEEE Trans. Electron Devices **ED-9**, 164–174 (1962)
24. Schröter, M., Rein, H.-M.: Transit time of high-speed bipolar transistors in dependence on operating point, technological parameters, and temperature. In: Proc. Bipolar Circuits and Technology Meeting, Minneapolis, pp. 250–253 (1989)
25. Kroemer, H.: Zur Theorie des Diffusions- und Drifttransistors. Arch. Elektr. Uebertrag. (AEUe) **8**, 223–228 (1954)
26. Kroemer, H.: Zur Theorie des Diffusions- und Drifttransistors. Arch. Elektr. Uebertrag. (AEUe) **8**, 363–369 (1954)
27. Kroemer, H.: Zur Theorie des Diffusions- und Drifttransistors. Arch. Elektr. Uebertrag. (AEUe) **8**, 499–504 (1954)
28. Schröter, M.: A survey of present compact models for high-speed bipolar transistors. FREQUENZ **47**, 178–190 (1993)

29. de Graaff, H.C., Kloosterman, W.J.: New formulation of the current and charge relations in bipolar transistor modelling for CACD purposes. IEEE Trans. Electron. Devices **ED-32**, 2415–2419 (1985)
30. Schröter, M., Friedrich, M., Rein, H.-M.: A generalized integral charge-control relation and its application to compact models for silicon based HBT's. IEEE Trans. Electron Devices **40**, 2036–2046 (1993)
31. Schröter, M.: Integral charge-control relations. In: Cressler, J. (ed.) Silicon Heterostructure Handbook, pp. 1181–1208. CRC Press, Boca Raton (2005). Chap. A3
32. Gummel, H.K.: A charge-control relation for bipolar transistors. BSTJ **49**, 115–120 (1970)
33. Gummel, H.K., Poon, H.C.: An integral charge-control model for bipolar transistors. BSTJ **49**, 827–852 (1970)
34. Schröter, M., Tran, H.: Two-/three-dimensional GICCR for Si/SiGe bipolar transistors. In: Proc. WCM, International NanoTech Meeting, Anaheim (CA), pp. 99–104 (2005)
35. Schröter, M.: HICUM status overview. In: HICUM Workshop 2004, Bordeaux, France (2004)
36. DeGraaff, H.C., Kloosterman, W.J.: The MEXTRAM bipolar transistor model. Philips Report Nr. 006/94 (level 503.2) (1995)
37. Schröter, M., Yan, Z., Lee, T.-Y., Shi, W.: A compact tunneling current and collector breakdown model. In: Proc. IEEE Bipolar Circuits and Technology Meeting, Minneapolis, pp. 203–206 (1998)
38. Rickelt, M., Rein, H.-M.: A novel transistor model for simulating avalanche-breakdown effects in Si bipolar circuits. IEEE J. Solid-State Circuits **37**, 1184–1197 (2002)
39. Slotboom, J., Streutker, G., van Dort, M., Woerlee, H., Pruijmboom, A., Gravesteijn, D.: Nonlocal impact ionization in silicon devices. In: Proc. IEDM, pp. 127–130 (1991)
40. Rein, H.-M., Schröter, M.: Experimental determination of the internal base sheet resistance of bipolar transistors under forward-bias conditions. Solid-State Electron. **34**, 301–308 (1991)
41. Rein, H.-M., Schröter, M.: Base spreading resistance of square emitter transistors and its dependence on current crowding. IEEE Trans. Electron Devices **36**, 770–773 (1989)
42. Schröter, M.: Simulation and modelling of the low-frequency base resistance of bipolar transistors in dependence on current and geometry. IEEE Trans. Electron Devices **38**, 538–544 (1991)
43. Schröter, M.: Modelling of the low-frequency base resistance of single base contact bipolar transistors. IEEE Trans. Electron Devices **39**, 1966–1968 (1992)
44. Schröter, M., Krause, J., Lehmann, S., Celi, D.: Compact layout and bias dependent base resistance modelling for advanced SiGe HBTs. IEEE Trans. Electron Devices **55**(7), 1693–1701 (2008)
45. Lehmann, S., Schröter, M.: Improved layout dependent modelling of the base resistance in advanced HBTs. In: Proc. WCM, International NanoTech Meeting, Boston, pp. 795–800 (2008)
46. Ghosh, H.N.: A distributed model of junction transistor and its application in the prediction of the emitter-base diode characteristic, base impedance, and pulse response of the device. IEEE Trans. Electron Devices **ED-12**, 513–531 (1965)
47. Pritchard, R.L.: Two-dimensional current flow in junction transistors at high frequencies. Proc. IRE **46**, 1152–1160 (1958)
48. Versleijen, M.: Distributed high frequency effects in bipolar transistors. In: Proc. Bipolar/BiCMOS Circuits and Technology Meeting (BCTM), Minneapolis, pp. 85–88 (1991)
49. Schröter, M., Rein, H.-M., Rabe, W., Reimann, R., Wassener, H.-J., Koldehoff, A.: Physics- and process-based bipolar transistor modelling for integrated circuit design. IEEE J. Solid-State Circuits **34**, 1136–1149 (1999). See also: TRADICA User's Guide and www.xmodtech.com
50. Weil, P.B., McNamee, L.P.: Simulation of excess phase in bipolar transistors. IEEE Trans. Circuits Syst. **CAS-25**, 114–116 (1978)
51. Koldehoff, A., Schröter, M., Rein, H.-M.: A compact bipolar transistor model for very high-frequency applications with special regard to narrow stripes and high current densities. Solid-State Electron. **36**, 1035–1048 (1993)
52. TeWinkel, J.: Extended charge-control model for bipolar transistors. IEEE Trans. Electron Devices **ED-20**, 389–394 (1973)

53. Rudolph, M.: Limitations of current compact transit-time models for III-V-based HBTs. In: Proc. IMS, pp. 487–490 (2008)
54. Pfost, M., Rein, H.-M., Holzwarth, T.: Modelling substrate effects in the design of high-speed Si-bipolar ICs. IEEE J. Solid-State Circuits **31**, 1493–1497 (1996)
55. Walkey, D.J., et al.: Equivalent circuit modelling of static substrate thermal coupling using VCVS representation. IEEE J. Solid-State Circuits **37**, 1198–1206 (2002)
56. Chen, X.Y., Deen, M.J., Yan, Z.X., Schröter, M.: Effects of emitter dimensions on low-frequency noise in double-polysilicon BJTs. Electron. Lett. **34**(2), 219–220 (1998)
57. Deen, J., Rumyantsev, S., Schröter, M.: On the origin of $1/f$ noise in polysilicon emitter bipolar transistors. J. Appl. Phys. **85**(2), 1192–1195 (1999)
58. Sakalas, P., Chakravorty, A., Herricht, J., Schröter, M.: Compact modelling of high frequency correlated noise in HBTs. In: Proc. Bipolar/BiCMOS Circuits and Technology Meeting (BCTM), Maastricht (Belgium), pp. 279–282 (2006)
59. Sakalas, P., Ramonas, M., Schröter, M., Jungemann, C., Shimukovitch, A., Kraus, W.: Impact ionization noise in SiGe HBTs: comparison of device and compact modelling with experimental results. IEEE Trans. Electron Devices **56**(2), 328–336 (2009)
60. Rein, H.-M., Möller, M.: Design considerations for very-high-speed Si-bipolar ICs operating up to 50 Gb/s. IEEE J. Solid-State Circuits **31**, 1076–1090 (1996)
61. Vanhoucke, T., Hurkx, G.A.M.: A new analytical model for the thermal resistance of deep-trench bipolar transistors. IEEE Trans. Electron Devices **53**, 1379–1388 (2006)
62. Özisik, M.: Heat Conduction. Wiley, New York (1993). ISBN 0-471-53256-8
63. Walkey, D.J., Smy, T.J., Reimer, C., Schröter, M., Tran, H., Marchesan, D., Jackson, M., Kleckner, T.: Modelling thermal resistance in trench-isolated bipolar technologies including trench heat flow. Solid-State Electron. **46**(1), 7–17 (2002)
64. Sakalas, P., Schröter, M., Kraus, W., Kornau, L.: Modelling of SiGe power HBT intermodulation distortion using HICUM. In: Proc. ESSDERC, Lisboa (Portugal), pp. 311–314 (2003)
65. Schröter, M., Berntgen, J., Kraus, W., Wassener, H.-J., Strobel, H.-J.: Scalable HICUM model for SiGe. In: ECOMPASS Workshop, Bonn (Germany), 17–18 April 2002
66. Final Report on DETAILS: HF design technology for precise analog IP-based front-end solutions in highly-integrated data transfer systems (in German). A Joint Industry Project Funded by the German Ministry for Education and Research, pp. 103–104 and pp. 122–123 (2007)
67. Schröter, M., Wittkopf, H., Kraus, W.: Statistical modelling of bipolar transistors. In: Proc. Bipolar/BiCMOS Circuits and Technology Meeting (BCTM), Santa Barbara (CA), pp. 54–61 (2005)
68. McAndrew, C., et al.: Efficient statistical BJT modelling, why is beta more than IC/IB. In: Proc. BCTM, pp. 28–31 (1997)
69. Mallardi, R., Newton, K., Schröter, M.: Development and design kit integration off a scalable and statistical high current model for advanced SiGe HBTs. In: Proc. WCM, International NanoTech Meeting, Boston (MA), pp. 729–734 (2006)
70. Schneider, W., Schröter, M., Kraus, W., Wittkopf, H.: Statistical simulation of high-frequency bipolar circuits. In: Proc. DATE07, Nice, pp. 1397–1402 (2007)
71. Yoshitomi, S., Kawakyu, K., Teraguchi, T., Kimijina, H., Yonemura, K.: Modification of compact bipolar transistor model for DFM (Design for Manufacturing) applications. In: Proc. MIXDES, pp. 125–130 (2006)
72. Rein, H.-M., Schröter, M.: Experimental determination of the internal base sheet resistance of bipolar transistors under forward-bias conditions. Solid-State Electron. **34**, 301–308 (1991)
73. Schröter, M., Lehmann, S.: The rectangular bipolar transistor tetrode structure and its application. In: Proc. ICMTS, Tokyo, Japan, pp. 206–209 (2007)
74. Lee, T.-Y., Schröter, M., Racanelli, M.: A scalable model generation methodology for bipolar transistors for RF IC design. In: Proc. Bipolar/BiCMOS Circuits and Technology Meeting (BCTM), pp. 171–174 (2001)
75. Zimmermann, Y., Schröter, M., Zampardi, P., Klimashov, A., Tkachenko, G.: III-V HBT modelling, scaling and parameter extraction using TRADICA and HICUM. In: IEEE Workshop on Power Amplifiers, San Diego, CA (2003)

76. Hu, J., Zampardi, P.J., Cismaru, C., Kwok, K., Yang, Y.: Physics-based scalable modelling of GaAs HBTs. In: Proc. Bipolar/BiCMOS Circuits and Technology Meeting (BCTM), pp. 176–179 (2007)
77. EU project targets 0.5-THz SiGe bipolar transistor. EE Times Europe Print Edition Covering 17 March–6 April 2008. See also DOTFIVE website: http://www.dotfive.eu/
78. Ardouin, B., Zimmer, T., Fouillat, P.: Direct method for base-emitter and base-collector capacitance splitting using high frequency measurements. In: Proc. Bipolar/BiCMOS Circuits and Technology Meeting (BCTM), pp. 114–117 (2001)
79. Berger, D., et al.: Extraction of the BC capacitance splitting along the base resistance using HF measurements. In: Proc. Bipolar/BiCMOS Circuits and Technology Meeting (BCTM), pp. 180–183 (2000)
80. Ning, T.H., Tang, D.D.: Method for determining the emitter and base series resistances of bipolar transistors. IEEE Trans. Electron Devices **ED-31**(4), 409–412
81. Zimmer, T., Meresse, A., Cazenave, Ph., Dom, J.P.: Simple determination of BJT extrinsic base resistance. Electron. Lett. **27**(21) (1991)
82. Kloosterman, W.J., Paasschens, J.C.J., Klaassen, D.B.M.: Improved extraction of base and emitter resistance from small signal high frequency measurements. In: Proc. Bipolar/BiCMOS Circuits and Technology Meeting (BCTM), pp. 93–96 (1999)
83. Nakadai, T., Hashimoto, K.: Measuring the base resistance of bipolar transistors. In: Proc. Bipolar/BiCMOS Circuits and Technology Meeting (BCTM), pp. 200–203 (1991)
84. Huszka, Z., Seebacher, E., Pflanzl, W.: An extended two-port method for the determination of the base and emitter resistance. In: Proc. Bipolar/BiCMOS Circuits and Technology Meeting (BCTM), pp. 188–191 (2005)
85. Berkner, J.: Principles, strategies and methods for SGP, VBIC, HICUM and MEXTRAM models and model parameter extraction. In: BCTM Short Course, Minneapolis, September 2001
86. Schröter, M.: Methods for extracting parameters of geometry scalable compact bipolar transistor models. Internal Report (2000)
87. Morizuka, K., Hidaka, O., Mochizuki, H.: Precise extraction of emitter resistance from an improved floating collector measurement. Trans. Electron Devices **42**(2) (1995)
88. Tran, H., Schröter, M., Walkey, D.J., Marchesan, D., Smy, T.J.: Simultaneous extraction of thermal and emitter series resistances in bipolar transistors. In: Proc. Bipolar/BiCMOS Circuits and Technology Meeting (BCTM), pp. 170–173 (1997)
89. Malorny, M., Schröter, M., Celi, D., Berger, D.: An improved method for determining the transit time of Si/SiGe bipolar transistors. In: Proc. BCTM, pp. 229–232 (2003)
90. Raya, C.: Moderation et optimisation de transistors bipolaires a heterojonction Si/SiGeC ultra rapides pour applications millimetriques. Ph.D. Thesis (in French), University Bordeaux I, July 2008
91. Ardouin, B., Zimmer, T., Berger, D., Celi, D., Mnif, H., Burdeau, T., Fouillat, P.: Transit time parameter extraction for the HICUM bipolar compact model. In: Bipolar Circuits and Technology Meeting (BCTM), Minneapolis, pp. 106–109, 1–2 October 2001
92. Pfeiffer, U., Garcia, A.-V.: Millimeter-wave design considerations for power amplifiers in a SiGe process technology. IEEE Trans. Microw. Theory Tech. **54**, 57–64 (2006)
93. HICUM Workshop, Digest of Presentations, 2001–2008 at www.iee.tu-dresden.de/iee/eb/
94. Technology for Frequency Agile Digitally Synthesized Transmitters (TFAST), US DoD Project; See: http://www.darpa.mil/mto/programs/tfast/index.html
95. International Roadmap for Semiconductors (ITRS), 2008 edn.

Part III
Compact Models of Passive Devices

Chapter 9
Integrated Resistor Modeling

Colin C. McAndrew

Abstract This chapter details models for resistors. Although resistors may seem to be simple devices, in practice the effects of velocity saturation, depletion pinching, and self-heating cause the dc $I(V)$ characteristics to be nonlinear, and for this nonlinearity to vary with geometry and temperature. Even slight nonlinearities in resistors can contribute significantly to distortion and harmonics in highly linear analog and RF circuits, so accurate modeling of the nonlinearities is needed. Parasitics and self-heating also cause the ac behavior of resistors to vary with frequency. Basics of resistor modeling are reviewed, and a physically based 3-terminal resistor model is derived, and shown to be applicable to both poly resistors and diffused resistors (which are really JFETs). The model includes geometry and temperature dependence, and also has statistical variability, including mismatch, built in. Details of some useful parameter extraction procedures are provided.

9.1 Introduction

One of the first principles of physics taught to aspiring electrical and electronics engineers is Ohm's law for resistors: $V = IR$ where R is the resistance of the resistor, I is the current forced through the resistor, and V is the potential difference that is developed across the resistor as a consequence of the current flow through the resistor. Unfortunately, resistors in the real world, especially in the real world of integrated circuit (IC) manufacturing processes, exhibit significantly more complex behavior than that embodied by Ohm's "law." The relationship between V and I is nonlinear; in other words the resistance R depends on the applied bias. R can even depend on biases in addition to those directly applied across the resistor itself. R also depends on the geometric layout of the resistor and on temperature, and these

C.C. McAndrew (✉)
Freescale Semiconductor, Tempe, AZ 85284, USA
e-mail: Colin.McAndrew@freescale.com

G. Gildenblat (ed.), *Compact Modeling*,
DOI 10.1007/978-90-481-8614-3_9, © Springer Science+Business Media B.V. 2010

dependencies can be complex. Further, both for circuit simulation, which uses modified nodal analysis (MNA–based on solving Kirchoff's current law at each node in a circuit with the nodal voltages, not branch currents, as the independent variables [18]), and to align better with common biasing schemes, in which the applied biases are voltages and not currents, it is often more convenient to analyze resistor behavior through the conductance form $I = GV$, where $G = 1/R$ is the conductance, rather than through the resistance form $V = IR$.

This chapter presents details of accurate modeling of resistors in semiconductor technologies.

First, basic aspects of resistors and resistor models are presented. This includes the concept of sheet resistance, analysis of the geometry dependence of resistance and its temperature variation, including end and contact effects, and phenomenological modeling of the bias dependence of resistance, including a review of the `r2_cmc` model (CMC indicates the Compact Model Council).

Second, physical models of semiconductor resistors are presented. There are two main classes of resistors available in semiconductor processes: diffused resistors and polysilicon (or "poly") resistors. These are both 3-terminal, rather than 2-terminal, resistors, where the potential of the semiconductor region in or above which the body of the resistor sits modulates the resistance of the body of the resistor. Although diffused and poly resistors are structurally different and operate by seemingly different physical principles—the former is really a junction field-effect transistor (JFET) and the latter is in effect an "upside down" insulated-gate field-effect transistor (IGFET)—the main physical sources of nonlinearity are the same: depletion pinching (which reduces the effective cross-sectional area through which the current in the resistor flows, thereby increasing the effective resistance), velocity saturation, and self-heating. We show how a compact model for the depletion pinching effect can be derived based on the principle of symmetric linearization, and how velocity saturation and self-heating can be added to the basic depletion pinching model; this is the basis of the `r3_cmc` model.

Third, other important aspects of resistor modeling are detailed. This includes modeling of parasitic currents (from junction leakage for diffused resistors) and capacitances (from junction depletion capacitance for diffused resistors and from dielectric capacitance for poly resistors), noise, and both global and local (i.e. mismatch) statistical variations.

Fourth, techniques for parameter extraction are developed. Although there are simple algorithms that can be used to determine many resistor model parameters, separating the effects of self-heating and velocity saturation can be difficult because they exhibit qualitatively similar behavior over bias and geometry; we show that the frequency dependence of the small-signal resistance can be used to separate the influence of velocity saturation (which does not depend on frequency) from that of self-heating (which does depend on frequency).

Finally, techniques to implement resistor models are discussed, with emphasis on handling very small valued resistors for which the $V = IR$ formulation is numerically more robust than the $I = GV$ formulation.

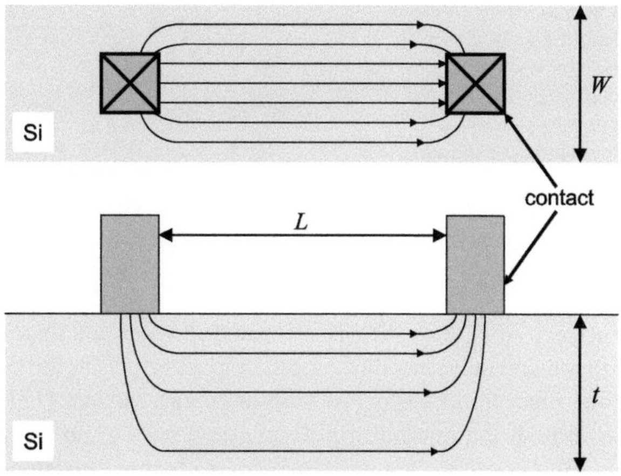

Fig. 9.1 Top and side views of a basic semiconductor resistor

9.2 Semiconductor Resistors

Resistors in semiconductor manufacturing technologies are in essence "thin film" resistors, because the current flow is confined to a thin conducting layer. Figure 9.1 shows an idealized representation of current flow through a slab of semiconducting material of width W and thickness t, where voltage is applied to the slab via the two contacts, which are spaced a distance L apart. The thin lines with arrows represent flow paths of current through the semiconductor. If we ignore, for now, the spreading of the current flow at the ends near the contacts, and assume that the doping level N in the slab is uniform and that the current flow lines are parallel and uniform throughout the height and width of the slab, then the resistance of the resistor is

$$R = \rho \frac{L}{tW} = \frac{1}{q\mu_0 N} \cdot \frac{L}{tW} \tag{9.1}$$

where ρ is the resistivity of the semiconductor, q is the magnitude of the electronic charge, μ_0 is the low-field mobility of the conducting carriers in the semiconductor (taken to be constant) and we have assumed that all dopants are activated so that the mobile charge carrier density is equal to N. This is more conveniently written as

$$R = \rho_s \frac{L}{W} \tag{9.2}$$

where $\rho_s = 1/(q\mu_0 Nt)$ is called the "sheet resistance" of the slab. The unit of ρ_s is strictly Ohm, but as it is the resistance of one "square" of the material, when looked at from above, it is commonly quoted in $\Omega/\square$. Increasing the number of squares W/L of the resistor increases its resistance. Equation (9.2) is the basic relationship for modeling (idealized) semiconductor resistors.

In practice, semiconductor layers in IC manufacturing processes are often formed by ion implantation, which leads to vertical non-uniformity in the doping profile; N decreases from the surface with depth into the semiconductor. The current thus preferentially flows through paths of higher doping (i.e. lower resistivity), which is depicted in Fig. 9.1 by the current flow lines being squeezed closer together near the

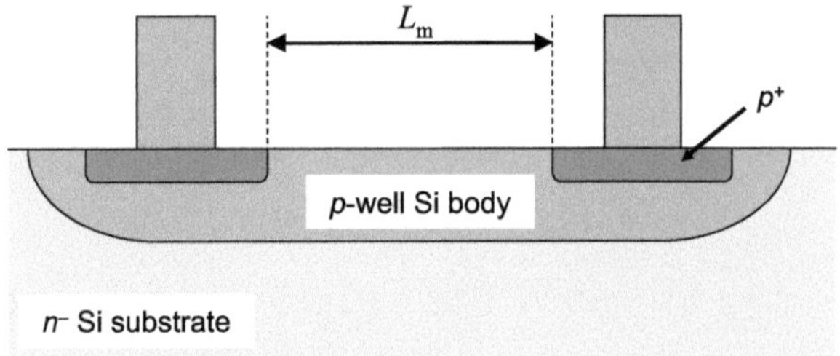

Fig. 9.2 Mask (design) length L_m for a p-well resistor with enhanced contact doping (lateral out-diffusion of the p^+ implant is not shown)

surface. Also, polysilicon is formed from an amalgamation of separate crystals of silicon and there are surface and contact effects at the boundaries between the grains that affect the transport of mobile charge carriers [13]. However, at a macroscopic level both the nonuniform doping and poly grain boundary effects can be ignored and the concept of a uniform film with sheet resistance ρ_s is still applicable, and universally used, in practice.

The contacts in Fig. 9.1 add resistance over and above that of the body of the resistor, and the current spreading (both vertical and lateral) near the contacts also causes the effective total resistance to increase. As we will see below, some of these contributions may be modeled as separate parasitic resistances in series with the resistance of the body of the resistor; but they may also be modeled by considering the effective electrical length of the resistor, which we will denote by L, as being different from the mask (or design) length L_m. What constitutes L_m depends on the type of resistor being modeled, for example Fig. 9.2 shows L_m for a diffused resistor with a lowly doped body, such as a well resistor, that has an added implant to ensure there is a low resistance between the contact and the semiconductor. Similarly, poly resistors can have unsalicided body regions, to give a high sheet resistance (and therefore as small a resistor area as possible for a given target resistance), but the ends of the resistor are salicided to ensure there is a good contact to the poly. In these situations L_m is defined as the distance between the mask edges that define the body of the resistor, and not the distance between the contacts. The effective electrical width W of a resistor can, from several physical mechanisms, be different from the mask width W_m.

9.2.1 Effective Resistor Geometry and Total Resistance

The effective resistor length can be different from L_m by a fixed amount, e.g. from biasing (i.e. sizing) of masks to compensate for various manufacturing effects or from out-diffusion (not shown) of the contact enhancement region implants in Fig. 9.2 past the edges of the mask that define the implant. From the top view in Fig. 9.1 the current spreading in the width direction occurs at the edges of the device, not along the center of the device in the width direction, so assuming the contact size scales with width (which is true of large geometry technologies; for fine geometry technologies the contacts are usually of a fixed size and multiple contacts are placed for wider devices) it is apparent that to first order the relative effect

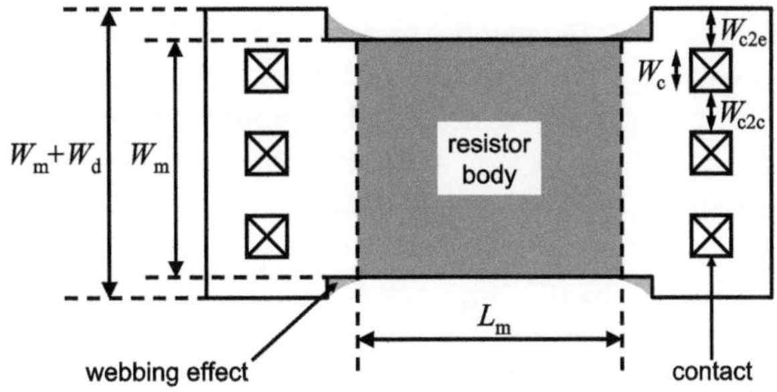

Fig. 9.3 Top view of resistor layout including dogbone end regions

of the current spreading on total resistance, which is modeled as part of the length offset, should decrease reciprocally as the width increases. The effective resistor length is therefore modeled as

$$L = L_m + \Delta_L + \frac{\Delta_{LW}}{W_m}. \tag{9.3}$$

Here Δ_L and Δ_{LW} are model parameters, and the width dependence is modeled using W_m rather than W, to avoid circular definitions of the effective geometries.

Effective width modeling is more involved [15]. From mask biasing there can be a fixed component of offset (between mask and effective electrical width). For LOCOS processes there is also a reciprocal width dependence to the width offset [15]. In addition, well resistors receive deep, fairly low dose implants, and these can outdiffuse to such an extent that for very narrow resistors the doping in the middle of a resistor is less than it is in the middle of a wide resistor. Analysis shows that this can be physically modeled through an exponential dependence of effective width on W_m and is referred to as the finite dopant source effect [15]. Finally, resistors are not always laid out in nice rectangular patterns but, especially for narrow resistors, can have end regions that are wider than the resistor body; see Fig. 9.3. This is often referred to as a "dogbone" shape, and we denote as W_d the amount the end region is wider than the body of the resistor. The corners of the dogbone are generally not perfectly formed on-wafer, but become rounded and "webbed." This shows up as an effective width that increases as the resistor length decreases. Putting all of these effects together gives [15]

$$W = \frac{W_m + \Delta_W + (\Delta_{WW}/W_m) + \Delta_{WFD}[1 - \exp(-W_m/W_{FD})]}{1 - (W_{WE}W_d)/(L_m W_m)} \tag{9.4}$$

where Δ_W and Δ_{WW} are model parameters for the fixed and width dependent width offsets, respectively, Δ_{WFD} and W_{FD} are parameters of the finite dopant effect model, and W_{WE} is the parameter that controls the webbing effect model.

The total resistance comprises the body resistance R plus resistance from the contacts and the end regions. If the resistance per contact is R_C and the resistance of a unit width of the end region is R_{EW} the total resistance is

$$R_{tot} = 2\frac{R_C}{N_C} + 2\frac{R_{EW}}{W_m} + R \tag{9.5}$$

where N_C is the number of contacts at each end and, for simplicity, the end region resistance is taken to scale with mask rather than effective width. Note that for fixed size contacts N_C scales roughly proportionally to width, so we can consider the quantity $R_{EWC} = R_{EW} + R_C(W_m/N_C)$ to be a combined effective end-plus-contact resistance for a unit width, therefore

$$R_{tot} \approx 2\frac{R_{EWC}}{W_m} + R = 2\frac{R_{EWC}}{W_m} + \rho_s\frac{L}{W} \approx \rho_s\frac{L + 2R_{EWC}/\rho_s}{W}. \qquad (9.6)$$

The end resistance can therefore be implicitly modeled by including it as part of the Δ_L component in (9.3).

9.2.2 Resistor Temperature Dependence

To first order the length and width offsets do not change with temperature T, so the majority of the temperature dependence of resistors comes from the temperature dependence of ρ_s, which in turn comes from the temperature dependence of μ_0. A simple, single parameter T^{X_T} mobility temperature dependence has been proposed for silicon [10] and used for resistor modeling [4]. However, this form is monotonic with respect to T, and poly resistors can exhibit nonmonotonic variation in resistance over temperature, so the quadratic form

$$R(T) = R(T_{ref})\left(1 + T_{C1}\delta T + T_{C2}\delta T^2\right) \qquad (9.7)$$

is preferred, where $\delta T = T - T_{ref}$, T_{ref} is the reference temperature, and T_{C1} and T_{C2} are first and second order temperature coefficients of resistance. This form has the added benefit of being well understood by IC designer engineers.

The contact and end region resistances can have different temperature coefficients than those of the body resistance. If we add B and E to subscripts, to denote body and end-plus-contact respectively (the contact and end regions may themselves have a different dependence on temperature but they can be modeled using composite first and second order temperature coefficients), then

$$\begin{aligned} R_{tot}(T) = {} & 2\frac{R_{EWC}}{W_m}\left(1 + T_{C1E}\delta T + T_{C2E}\delta T^2\right) \\ & + \rho_s\frac{L}{W}\left(1 + T_{C1B}\delta T + T_{C2B}\delta T^2\right) \end{aligned} \qquad (9.8)$$

gives the total resistance at temperature T. Rearranging this expression, and assuming $2R_{EWC} \ll \rho_s L$, gives

$$\begin{aligned} \frac{R_{tot}(T)}{R_{tot}(T_{ref})} = {} & 1 + \left[T_{C1B} + \frac{2R_{EWC}(T_{C1E} - T_{C1B})}{\rho_s L}\right]\delta T \\ & + \left[T_{C2B} + \frac{2R_{EWC}(T_{C2E} - T_{C2B})}{\rho_s L}\right]\delta T^2 \end{aligned} \qquad (9.9)$$

so if the end and contact resistances are included via Δ_L, and not modeled separately (with separate body and end-plus-contact resistance temperature coefficients as in (9.8)), then any difference in temperature coefficients between the body and the end-plus-contact regions can be modeled through a geometry dependence of the temperature coefficients; specifically, from (9.9), by a reciprocal dependence on length.

There is one situation where the contact resistance cannot be subsumed into Δ_L: Contacts to p-type silicon can have a resistance that decreases with temperature and can be affected by self-heating [1]. As we will see, the degree of self-heating varies with frequency, yet this is not included in our "effective temperature coefficient" analysis. Worse, the contact resistance can decrease so much with temperature that it becomes negligible, but lumping together the body and end-plus-contact resistance temperature coefficients does now allow explicit limiting of the contact resistance to a minimum of zero, and can implicitly have the contact resistance become negative, which is physically incorrect. If modeling of the frequency dependence is important, and the contact resistance can drop to very low values over temperature, then the contact resistances (and their temperature coefficients) should be modeled explicitly.

9.3 2-Terminal Resistor Models

So far we have considered the resistance to be constant, i.e. independent of bias. Real resistors exhibit a bias dependence of their resistance. The simplest way to model this is using the form

$$R(V) = R_0\left(1 + V_{C1}V + V_{C2}V^2\right) \tag{9.10}$$

where R_0 is the value of the resistance with zero applied bias and V_{C1} and V_{C2} are empirical first and second order voltage coefficients. There is a wide variation in how resistor nonlinearity is modeled in different simulators. Some extend the quadratic form (9.10) to arbitrary order polynomials, others have polynomial forms for conductance rather than resistance, and other empirical fitting relations can be implemented using controlled sources or Verilog-A. Besides lack of standardization, forms like (9.10) have two practical problems. First, as specified it allows the resistance to become zero or negative, and so clamping needs to be implemented in the model code to prevent this. Second, of the three main sources of nonlinearity in integrated resistors two, velocity saturation and self-heating, depend on the electric field $E = V/L$ rather that the voltage directly, and so have a geometry dependence that is not included in (9.10).

To standardize the form of bias dependence for 2-terminal resistors, and avoid the effort of having to characterize different models for different simulators, the CMC has provided the `r2_cmc` model. This model improves on (9.10) in that it avoids numerical problems and formulates the nonlinearity as a function of E. The basic form of the nonlinearity modeling was presented in [2], and these notes are included as part of the `r2_cmc` documentation available at [7].

The bias dependence of r2_cmc is given by

$$R(E) = R_0 \left(1 - P_3 - P_2 + P_3\sqrt[3]{1 + |Q_3 E|^3} + P_2\sqrt{1 + Q_2^2 E^2}\right) \quad (9.11)$$

where P_2, P_3, Q_2, and Q_3 are parameters. For $P_2 = 0$, if $|Q_3 E|$ is somewhat greater than 1 the resistance (9.11) becomes

$$R_{linear}(E) \approx R_0 \left[1 + P_3 \left(|Q_3 E| - 1\right)\right] \quad (9.12)$$

so the model approximates a linear (first order) dependence of resistance on field. For $P_3 = 0$, if $|Q_2 E|$ is somewhat less than 1 the resistance (9.11) becomes

$$R_{quadratic}(E) \approx R_0 \left(1 + 0.5 P_2 Q_2^2 E^2\right) \quad (9.13)$$

so the model approximates a quadratic (second order) dependence of resistance on field. As long as $0 \leq P_3 < 1$ and $0 \leq P_2 < 1 - P_3$ the resistance R can never become negative, so the form (9.11) approximates the empirical first and second order voltage coefficient model (9.10) but has no numerical problems and is more accurate, being based on electric field rather than voltage. Note the form (9.11) has a singularity at $V = 0$ because of the use of the absolute value function. This makes the r2_cmc model singular. However, this is "buried" in the model, and the derivative of the current through the resistor with respect to V up to third order exists; the fourth order derivative does not exist at $V = 0$; its left and right limits are not equal there.

Despite being useful, and improving on simple polynomial models by being numerically well behaved and having the nonlinearity be a function of electric field rather than voltage, the r2_cmc model is neither physical nor does it enable modeling of the depletion pinching effect. Specifically, because the amount of depletion pinching depends on the bias of the well or substrate that a resistor sits in (for diffused resistors) or above (for poly resistors), it cannot be comprehended in a 2-terminal model; modeling of the depletion pinching effect requires a 3-terminal formulation, which we now describe.

9.4 Physical 3-Terminal Resistor Model

The nonlinearities in semiconductor resistors can depend on more than just the voltage across the resistor; the potential of the well or substrate in which a diffused resistor is formed, or the semiconductor region above which a poly resistor sits, modulates the thickness of the conducting region. This can be modeled using empirical linear voltage coefficients of resistance [9], including for the well or substrate bias, or using the SPICE JFET model [9, 14, 15]. However, both of these approaches can have significant inaccuracies in modeling experimental data. We will develop physical models for the depletion pinching bias dependence for both diffused and polysilicon resistors in this section, and then we will add models for velocity saturation and self-heating, which can also significantly affect resistor nonlinearity.

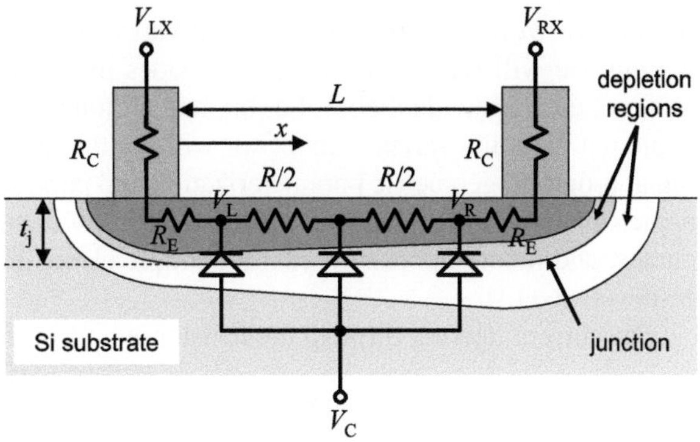

Fig. 9.4 Diffused resistor cross-section, with equivalent network overlay

9.4.1 Diffused Resistor (JFET) Depletion Effect Model

Figure 9.4 shows the cross-section of a diffused resistor. The bias V (measured with respect to the substrate, which is at the potential V_C) varies with position x along the body of the resistor, and therefore so does the extent of the depletion region that forms around the *p-n* junction between the body of the resistor and the well or substrate in which it is formed. This depletion region "pinches" the conducting thickness of the resistor body, and from (9.1) therefore modulates its resistance; this is precisely the physical mechanism of operation of a JFET, in which the gate is the substrate and the source and drain are the left and right terminals. Although not shown, the depletion region extends from the bottom of the resistor up the sidewalls to the surface, and therefore also pinches the resistor from each side, in the width direction.

The conducting thickness t_c of the resistor body is, assuming a uniformly doped *p*-type substrate and *n*-type resistor body,

$$t_c(V) = t_{c0} - \frac{t_{d0}}{\sqrt{\phi_{bi}}}\left[\sqrt{\phi_{bi} + V(x)} - \sqrt{\phi_{bi}}\right] \tag{9.14}$$

where ϕ_{bi} is the built-in potential of the junction, t_{d0} is the width of the depletion region in the resistor body with zero applies bias, and $t_{c0} = t_j - t_{d0}$ is the zero-bias conducting depth of the resistor body where t_j is the metallurgical junction depth (see Fig. 9.4). The conducting width of the resistor is

$$W_c(V) = W - 2\frac{t_{d0}}{\sqrt{\phi_{bi}}}\left[\sqrt{\phi_{bi} + V(x)} - \sqrt{\phi_{bi}}\right] \tag{9.15}$$

where we have taken the effective electrical width W to define the conducting width with zero applied bias. The factor of 2 enters because the depletion region pinches the width direction from both sides, and we have taken the built-in potential and zero-bias depletion region width to be the same for the sidewalls as for the bottom of the resistor body, i.e. we have assumed that the doping levels for the body and substrate around the edges of the resistor are the same as for the bottom. For ion implanted resistors this is not strictly true; the doping density decreases with depth into the semiconductor and so is not even constant down the sidewalls of the resistor. However, the built-in potential varies only weakly (logarithmically) with doping, we will see that it is the total (sides plus bottom) pinching that is important, not each individual contribution, and although the ratio of sidewall to bottom depletion pinching varies with W we will see that this can be accounted for by using appropriate geometric parameterization and parameter extraction. Therefore the approximation is acceptable in practice, does not significantly affect modeling accuracy, and most important it enables simple closed form modeling expressions to be developed [4].

The current flowing through the resistor at a position x is

$$I(x) = q\mu W_c(x) N t_c(x) \frac{dV}{dx} \tag{9.16}$$

where μ is the mobility (which may vary with bias and thus differ from μ_0), and N is the (assumed uniform) doping of the resistor body. Introducing (9.14) and (9.15) gives

$$\begin{aligned} I(x) = q\mu W N t_{c0} & \left[1 - \frac{t_{d0}}{t_{c0}\sqrt{\phi_{bi}}}\left(\sqrt{\phi_{bi}+V} - \sqrt{\phi_{bi}}\right)\right] \\ & \times \left[1 - \frac{2t_{d0}}{W\sqrt{\phi_{bi}}}\left(\sqrt{\phi_{bi}+V} - \sqrt{\phi_{bi}}\right)\right]\frac{dV}{dx}. \end{aligned} \tag{9.17}$$

This equation can be integrated along the body of the resistor to give an exact but somewhat complex expression for the current, which can be significantly simplified by approximation and by assuming that the build-in potentials of the bottom and sidewalls are the same [4]. We present here a different and simpler derivation, based on the symmetric linearization procedure of [6], that leads to an identical result. Denoting $V_{LC} = V_L - V_C$, $V_{RC} = V_R - V_C$, and $V_r = V_R - V_L$, then $\sqrt{\phi_{bi} + V}$ can be approximated by linearizing it about the bias $V_m = (V_{LC} + V_{RC})/2$,

$$\sqrt{\phi_{bi} + V(x)} \approx \sqrt{\phi_{bi} + V_m} + \frac{V(x) - V_m}{2\sqrt{\phi_{bi} + V_m}}. \tag{9.18}$$

Although it may not seem that a linear expansion of $\sqrt{\phi_{bi} + V}$ in V may be accurate for small ϕ_{bi} it turns out that this first order expansion is as accurate as using a third order expansion to simplify the directly integrated solution. Substituting (9.18) into (9.17) and integrating the latter along the resistor body gives, after some manipulation,

$$\begin{aligned} I = \frac{q\mu W N t_{c0}}{L} & \left[1 - \frac{t_{d0}}{t_{c0}\sqrt{\phi_{bi}}}\left(\sqrt{\phi_{bi}+V_m} - \sqrt{\phi_{bi}}\right)\right] \\ & \times \left[1 - \frac{2t_{d0}}{W\sqrt{\phi_{bi}}}\left(\sqrt{\phi_{bi}+V_m} - \sqrt{\phi_{bi}}\right)\right] V_r. \end{aligned} \tag{9.19}$$

If the depletion pinching effect is not too large, which is obviously true for a device that is meant to approximate a linear resistor, the terms in square brackets can be multiplied out and the product of the thickness and width depletion pinching terms can be ignored, giving

$$I = \frac{1}{R_0}\left[1 - \frac{t_{d0}}{\sqrt{\phi_{bi}}}\left(\frac{1}{t_{c0}} + \frac{2}{W}\right)\left(\sqrt{\phi_{bi}+V_m} - \sqrt{\phi_{bi}}\right)\right] V_r \tag{9.20}$$

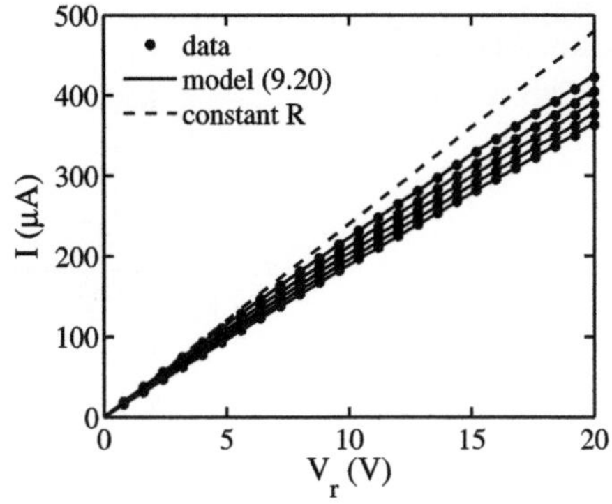

Fig. 9.5 $I(V)$ characteristics for a wide/long diffused resistor. Curves from top to bottom are for V_C from 0 to 20 V

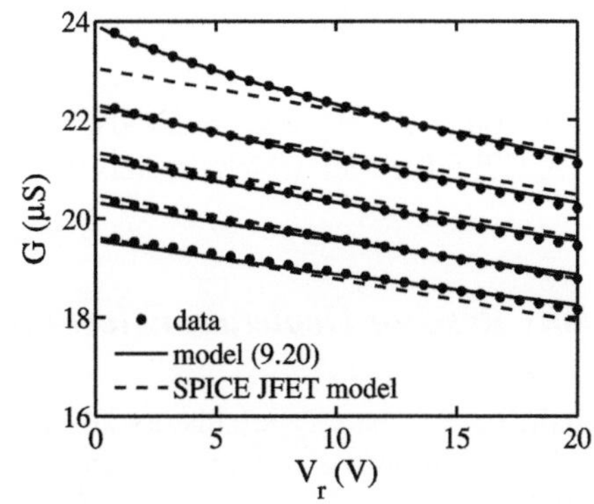

Fig. 9.6 $G(V)$ characteristics for a wide/long diffused resistor. Curves from top to bottom are for V_C from 0 to 20 V

where we have introduced (9.2), with t defined by t_{c0} and the sheet resistance considered to define the zero-bias resistance.

Equation (9.20) presents an intuitive view of the depletion pinching effect for diffused resistors: the amount of depletion pinching depends on the ratio of the (zero-bias) depletion region to conducting region depths, and it varies inversely proportionally to width.

Figure 9.5 shows the model (9.20) fitted to measured $I(V)$ data from a p-type body diffused resistor; the resistor is long so that the nonlinearity in $I(V_r)$ is predominantly from depletion pinching and not from velocity saturation or self-heating. Also shown is a fit from a linear resistor model, with its conductance matched to the measured data at zero bias. The accuracy of the model (9.20) is apparent; however this is not the best plot to use to evaluate how well a model matches experimental data: for highly linear resistors any model visually appears to fit $I(V)$ data well. To evaluate resistor modeling accuracy it is better to plot the conductance $G = I/V_r$ vs. V_r than it is to plot I vs. V_r directly (taking care to exclude data for low bias where G is "noisy" because of measurement inaccuracy and, of course, for zero bias where G cannot be calculated). Figure 9.6 shows the model (9.20) fitted to $G(V)$ data; also shown for comparison is the SPICE JFET model [14]. The "depletion" nature of the bias dependence is clear in the leftmost part of the upper curve (this is where V_m is lowest; from (9.20), for higher V_C the $G(V)$ curve should become

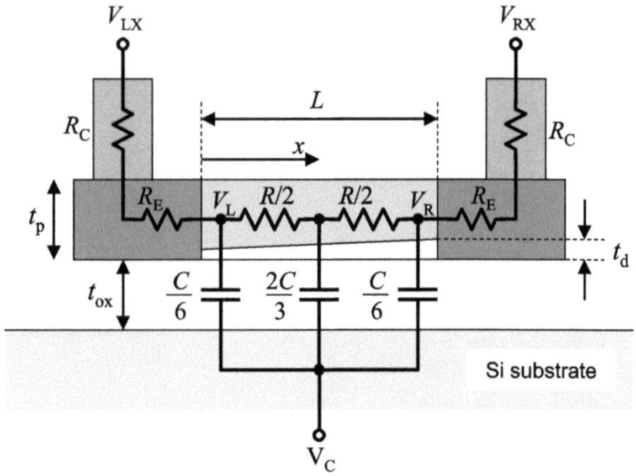

Fig. 9.7 Poly resistor cross-section, with equivalent network overlay

more linear around $V_r = 0$, as is observed). This behavior is not modeled accurately by the SPICE JFET model, which predicts that the $G(V)$ curves should be linear.

9.4.2 Poly Resistor Depletion Effect Model

Although diffused resistors were historically widely used for IC design, poly resistors are often commonly used today. Figure 9.7 shows the basic structure of a poly resistor. Because the resistor body is not formed in a well or substrate region of the silicon, there is no junction depletion effect. This makes poly resistors generally more linear than diffused resistors. In addition, because they are formed on top of a thick oxide layer the parasitic capacitance is lower than for diffused resistors, and multiple implants can be used not just to tailor the sheet resistance of the poly film but also to adjust its temperature coefficient; poly resistors can therefore have much lower variation over temperature than diffused resistors.

There are two potential drawbacks to poly resistors. First, because of the polycrystalline nature of the material, the transport of carriers is highly dependent on what happens at the boundaries between the grains of crystalline silicon [13]. There is typically a depletion region that forms at the boundaries between the grains, and this can lead to nonlinearities in the dc $I(V)$ behavior of poly resistors, especially at very low current densities. However, although in small resistors variability due to the randomness of the number and size of grains can be observed, in practical devices there are a sufficient number of grains so that the randomness averages out, and the poly film can be modeled using the concept of sheet resistance detailed in Sect. 9.2. Second, because the dielectric under the poly film has a much larger thermal resistivity than silicon, poly resistors are affected by self-heating more than diffused resistors; often self-heating is the dominant cause of nonlinearity in poly resistors.

The structure in Fig. 9.7 looks like an upside-down MOS system, but rather than the voltage on the "gate" (i.e. the silicon substrate, or control, terminal) inducing an inversion charge in the poly, it changes the thickness t_d of the depletion region at the bottom of the polysilicon film. The modulation of t_d by the potential of the silicon substrate changes the cross-sectional area through which current flows in the body of the poly resistor, thereby altering its resistance. Although the physical t_d modulation mechanism is based on MOS device physics rather than p-n junction physics, as for diffused resistors, we shall see that the ensuing modeling equations are similar.

If the potential at the interface between the bottom of the poly and the dielectric it is formed on is $\psi_s(x)$, referenced to the potential $V(x)$ in the poly (which is the "body" potential of the upside-down MOS structure) then [20]

$$t_d(x) = \sqrt{\frac{2\epsilon_s}{qN}\psi_s(x)} \tag{9.21}$$

where ϵ_s is the permittivity of silicon and N is the poly doping level, taken to be p-type. For the device structures and biases encountered in practice the bottom of the poly is neither inverted nor accumulated, therefore [20]

$$\psi_s = \frac{\gamma^2}{4}\left[\sqrt{1+\frac{V_C - V(x) - V_{FB}}{\gamma^2/4}} - 1\right]^2 \tag{9.22}$$

where V_{FB} is the flatband voltage of the silicon-poly system (in essence the work-function difference between the two materials), $\gamma = \sqrt{2q\epsilon_s N}/C'_{ox}$, and $C'_{ox} = \epsilon_{ox}/t_{ox}$ is the dielectric capacitance per unit area, where ϵ_{ox} is the permittivity of the dielectric between the silicon substrate and the poly and t_{ox} is its thickness.

Using (9.22) in (9.21) gives the thickness of the conducting part of the poly as

$$t_c(V) = t_p - \frac{\epsilon_s}{C'_{ox}}\left[\sqrt{1+\frac{V_C - V(x) - V_{FB}}{\gamma^2/4}} - 1\right]. \tag{9.23}$$

Assuming that the doping density in the poly is uniform, the current flowing in the poly at position x is

$$I(x) = q\mu W N t(x)\frac{dV}{dx} \tag{9.24}$$

where this differs from (9.16) in that the width does not depend on bias. Again applying the principal of symmetric linearization [6] with respect to the mid-point voltage $V_m = (V_{CL} + V_{CR})/2$, where $V_{CL} = V_C - V_L$ and $V_{CR} = V_C - V_R$, and integrating along the body of the resistor from $x = 0$ to $x = L$ gives

$$\begin{aligned} I = {} & \frac{q\mu W N t_p}{L}\left(1+\frac{\epsilon_s}{t_p C'_{ox}}\right) \\ & \times\left[1 - \frac{2\epsilon_s}{\gamma C'_{ox}(t_p + \epsilon_s/C'_{ox})}\sqrt{0.25\gamma^2 - V_{FB} + V_m}\right] V_r. \end{aligned} \tag{9.25}$$

Although developed using a different approach for a different physical system, (9.25) has the same structure as the basic depletion pinching formulation (9.20), which was derived based on reference [4] for diffused resistors and JFETs. This result can also be shown to be valid for the case where a field plate above a diffused resistor modulates a MOS depletion region in the silicon under the field plate, and pinches the diffused resistor from above [21].

Note that from (9.25) the voltage nonlinearity from depletion pinching is (ignoring V_{FB}, as it is negligible in practice) proportional to

$$\frac{\epsilon_{ox}}{qNt_{ox}(t_p + t_{ox}\epsilon_s/\epsilon_{ox})} \tag{9.26}$$

which indicates that to minimize the nonlinearity of a poly resistor the dielectric thickness, poly thickness, and poly doping should all be as large as possible, which physically makes sense: increasing the dielectric thickness reduces the electric field across the dielectric, therefore decreasing the modulation of t_d by V_C; increasing the poly thickness reduces the relative change in the conducting depth of the poly for a given amount of depletion region variation; and increasing the poly doping reduces the variation in t_d needed to accommodate a specific charge change, thereby reducing the relative modulation of t_c.

9.4.3 Unified Depletion Effect Model

The depletion pinching formulations (9.20) and (9.25) are qualitatively the same and with rearranging can be unified into the single model

$$I = g_f\left(1 - d_f\sqrt{d_p + V_{LC} + V_{RC}}\right)V_r \tag{9.27}$$

where

$$g_f = \frac{1}{R_0\left(1 - d_f\sqrt{d_p}\right)} \tag{9.28}$$

is a conductance factor, d_p is an effective depletion potential (note that this is twice the depletion potentials of (9.20) and (9.25) because we have factored out the one half in the definition of V_m to simplify the expressions and eliminate unnecessary multiplications by 2 during model evaluation). The quantity d_f is an effective depletion factor which has a geometry dependence

$$d_f = d_{f\infty} + \frac{d_{fW}}{W} + \frac{d_{fL}}{L} + \frac{d_{fWL}}{WL} \tag{9.29}$$

where $d_{f\infty}$, d_{fW}, d_{fL}, and d_{fWL} are parameters. The first of these is the depletion factor for a wide and long resistor, the reciprocal width dependence term follows theoretically from (9.20), and the remaining terms are added empirically to help with fitting experimental data over geometry. The model (9.27) along with (9.29) is the core of the `r3_cmc` model, and from the derivation is applicable to both poly resistors and diffused resistors (or JFETs). Note that the polarity of the bias voltages with respect to the substrate need to be reversed depending on whether the resistor body is n-type or p-type material.

9.4.4 Velocity Saturation

Velocity saturation is an important physical source of nonlinearity, and although we may be accustomed to thinking of it as only being important for modeling short-channel MOSFETs, for accurate modeling of distortion in highly linear resistors its effect is noticeable in surprisingly long devices, up to around 50 μm. Conventionally, velocity saturation is analyzed based on the velocity-field $v(E) = \mu(E)E$ dependence. However, for the purposes of analyzing data, highlighting certain aspects of velocity saturation, and modeling, it is more instructive to consider velocity saturation as an effective mobility reduction factor

$$\mu_{red} = \frac{\mu_0}{\mu(E)}. \tag{9.30}$$

The most common form of velocity saturation model, expressed in terms of μ_{red}, is [5, 19]

$$\mu_{red} = \left[1 + \left(\frac{\mu_0}{v_{sat}} \cdot \frac{|V|}{L} \right)^{\beta} \right]^{1/\beta} \tag{9.31}$$

where v_{sat} is the saturated carrier speed at high fields, V is the voltage across the resistor body, and β is an adjustable parameter. It is commonly held that for holes $\beta = 1$ is suitable and for electrons $\beta = 2$ is better [5]. The form with $\beta = 1$ [19] is widely used because it leads to simple closed form solutions. However, neither analytic simplicity nor qualitatively correct behavior are sufficient for our purpose: accurate modeling of distortion. For this we need a quantitatively correct model.

To derive a quantitatively correct mobility reduction model, we need to have good experimental data, preferably from actual diffused resistors rather than isolated bars of semiconductor material (models always have inaccuracies, which can often be "corrected" by appropriate determination of parameters of the model; if the goal is to model a device for the purpose of IC design, rather than to do basic physical characterization, then it is always preferable to base a model and its parameters on data from a real device). As we have seen, part of the nonlinearity in real resistors comes from depletion pinching, so we cannot just look at $I(V)$ data from one device and separate the influences of depletion pinching and velocity saturation. However, from (9.20) the effect of depletion pinching does not depend on the length of a resistor, whereas from (9.31) the effect of velocity saturation depends strongly on length, so by comparing the degree of nonlinearity of short and long resistors we can infer the effect of velocity saturation alone. (Self-heating also causes nonlinear $I(V)$ behavior, however for velocity saturation analysis we will use data from high sheet resistance devices, which minimizes the influence of self-heating. In addition, velocity saturation is more important for diffused than for poly resistors, so we will use data from diffused resistors).

Figure 9.8 shows measured dc $I(V)$ data from three high sheet resistance diffused resistors, of the same (wide) width but of different lengths, normalized to the conductance G_0 at low field for each device. The increasing nonlinearity of the data

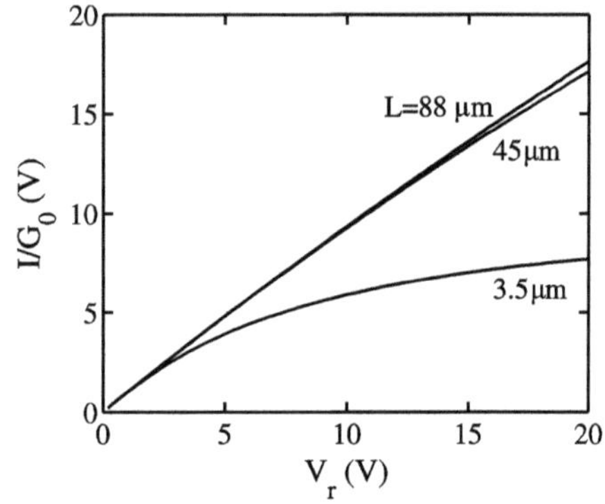

Fig. 9.8 Normalized $I(V)$ characteristics from measured data

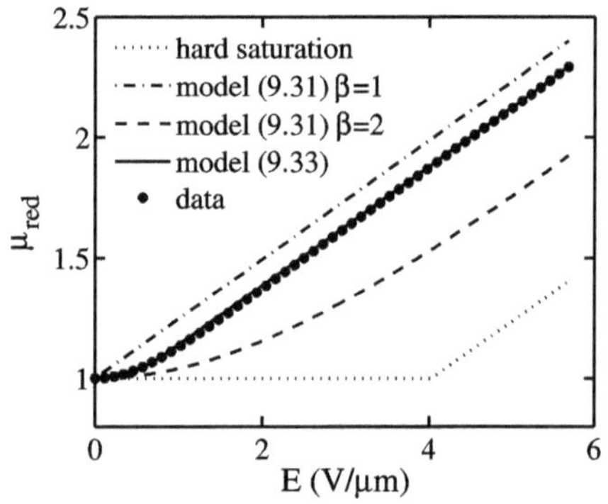

Fig. 9.9 Measured and modeled μ_{red} from velocity saturation

as the length becomes smaller is predominantly from velocity saturation. It is apparent that the effect of velocity saturation is noticeable for the $L = 45$ μm long device compared to the $L = 88$ μm long device and has become the dominant cause of nonlinearity for the $L = 3.5$ μm long device. By forming the ratio of the (normalized) currents from different length devices the depletion region modulation component of nonlinearity is canceled, thereby giving μ_{red}, see Fig. 9.9. Also shown in Fig. 9.9 are fitting results from the model (9.31) for both $\beta = 1$ and $\beta = 2$. It is clear that neither model represents the measured data well (adjusting the parameters of the models does not help). Further, because physically the effect of velocity saturation must be symmetric, $\mu_{red}(-E) = \mu_{red}(E)$, it is apparent from Fig. 9.9 that for $\beta = 1$ the model (9.31) is singular, which is known to cause problems for accurate distortion modeling [3, 12].

Inspection of the measured μ_{red} data in Fig. 9.9 shows that it has the following characteristics: it has value 1 for, and is symmetric about, $E = 0$; it has slope $d\mu_{red}/dE$ zero for $E = 0$; it asymptotically approaches $1 + (E - E_{co})/E_{cr}$ for large fields, where E_{co} and E_{cr} are corner and critical field parameters, respectively, that describe the offset and slope of the asymptote. For low fields the $\beta = 2$ model gives

$$\mu_{red} \approx 1 + \frac{\mu_0^2}{2v_{sat}^2}\left(\frac{V}{L}\right)^2 \tag{9.32}$$

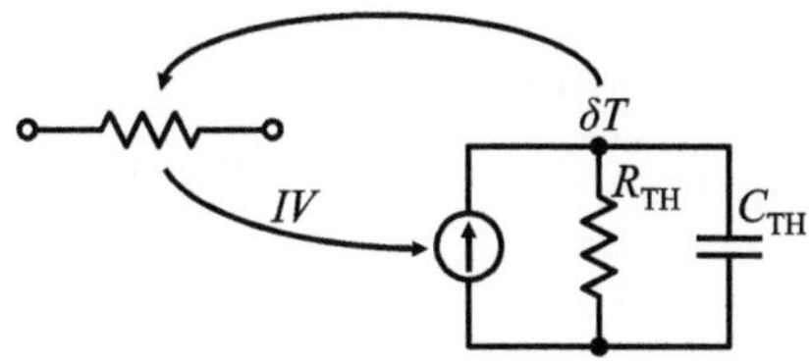

Fig. 9.10 Thermal network for self-heating modeling

and this symmetric, quadratic behavior around $V = 0$ is apparent in the data in Fig. 9.9. An empirical model that has these characteristics, and retains the approximate quadratic behavior for low fields, is

$$\mu_{red} = 1 + \sqrt{\left(\frac{E - E_{ce}}{2E_{cr}}\right)^2 + \frac{d_\mu E_{ce}}{E_{cr}}} + \sqrt{\left(\frac{E + E_{ce}}{2E_{cr}}\right)^2 + \frac{d_\mu E_{ce}}{E_{cr}}} - \sqrt{\left(\frac{E_{ce}}{E_{cr}}\right)^2 + \frac{4d_\mu E_{ce}}{E_{cr}}} \quad (9.33)$$

and Fig. 9.9 shows that this model fits the measured data well. Here d_μ is a fitting parameter for the "hardness" of the transition around E_{co} and

$$E_{ce} = \sqrt{E_{co}^2 + 4d_\mu^2 E_{cr}^2} - 2d_\mu E_{cr}. \quad (9.34)$$

9.4.5 Self-Heating

Self-heating can be important for both diffused and poly resistors, but is especially significant in the latter. This is modeled in the standard way after [22]: the power dissipation is calculated in each non-energy storage element of the equivalent network and the sum of all of these is fed into a thermal network comprising a thermal resistance and thermal capacitance. The state variable for the thermal network is the local temperature rise of the resistor above the ambient temperature. This temperature rise is fed back into the temperature dependence equations for the $I(V)$ model. The simulator then self-consistently solves for the current through the resistor in the presence of self-heating. Figure 9.10 shows the equivalent network for self-heating modeling.

There is one often under appreciated aspect of self-heating modeling: it has a strong dependence on frequency. At dc and low frequencies the temperature variation in a device tracks changes in the bias applied to a device. At high frequencies the temperature change cannot "keep up" with rapid changes in bias, so the effect of self-heating disappears (it still affects the quiescent operating point of course). Thermal time constants for integrated devices are on the order of 0.1 to 10.0 μs, so signals with frequencies in the 0.1 to 10.0 MHz range can have their ac behavior changed by the frequency response of self-heating. Figure 9.11 shows measured and modeled ac resistance of a poly resistor as a function of frequency and bias; the significance of the self-heating effect and its frequency dependence is clear.

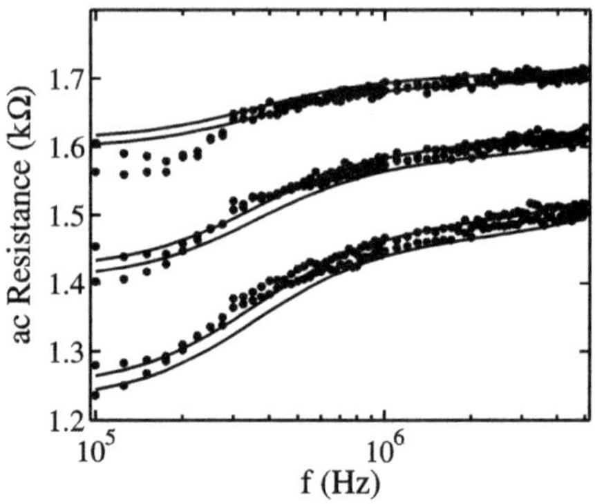

Fig. 9.11 Frequency dependence of self-heating. Symbols are data, lines are model. V_r is ±6, ±9, and ±12 V from top to bottom. The separation in each pair of curves is from the difference in depletion pinching

9.4.6 Complete 3-Terminal Resistor and JFET Model

Depletion pinching, velocity saturation, and self-heating are all important in practice, and so need to be combined to give a complete 3-terminal resistor model. Self-heating requires no special attention and is added to any basic model as discussed in Sect. 9.4.5. Although pedantically velocity saturation can affect the position dependence of V along a resistor body, a self-consistent handling of the local effects of depletion pinching and velocity saturation is not possible, and is not really needed. The mobility reduction factor (9.33) is directly used to scale down the current from the unified depletion effect model (9.27).

There is one final aspect that needs to be handled in a complete model: pinch-off. Although most resistors are intended to be as nearly linear as possible, so pinch-off should not be encountered, it is possible to see pinch-off in lowly doped well resistors. Moreover, the model we have derived is applicable to JFETs, which can pinch-off. And any model should behave as accurately and physically as possible for all biases, even if outside the normal region of operation. Therefore it is important to model pinch-off properly, and the biggest issue is to avoid the problem of negative output conductance, which was a problem with some early MOSFET models.

Strictly the conducting body thickness becomes zero when the control to resistor body voltage is

$$V_P = \phi_{bi}\left[\left(\frac{t_j}{t_{d0}}\right)^2 - 1\right] = \frac{1}{2d_f^2} - \frac{d_p}{2} \tag{9.35}$$

and this is the value used in [4]. However, at this bias the output conductance is negative. Without considering velocity saturation it is straightforward to calculate a saturation voltage for V_r such that the output conductance at that point is zero [16]. However, this is not useful as velocity saturation is important in practice. Unfortunately there is no closed form solution for the zero output conductance point for a general velocity saturation model. Fortuitously, it can be shown that if an approximate μ_{red} model is used that has a value of $d\ln(\mu_{red})/dE$ less than that of the actual μ_{red} model then it will predict a saturation voltage V_{sat} that is guaranteed to

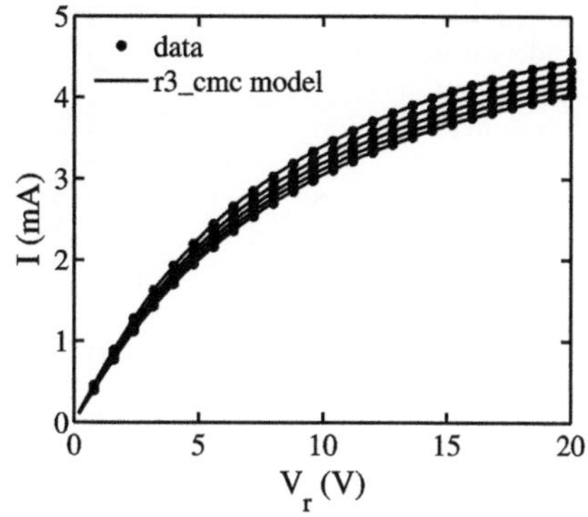

Fig. 9.12 $I(V)$ data and r3_cmc model fits for a short diffused resistor. Curves from top to bottom are for V_C from 0 to 20 V

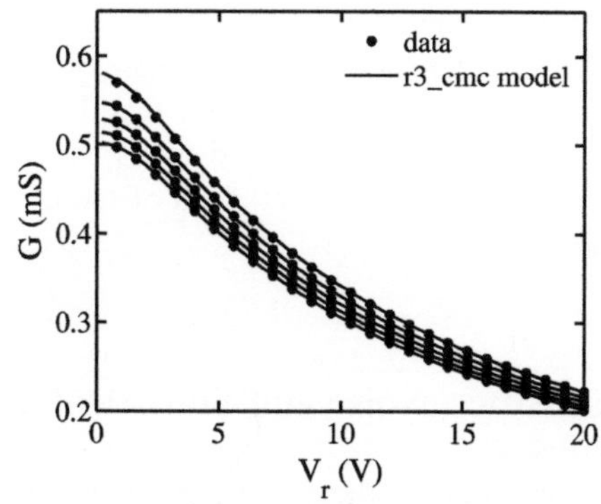

Fig. 9.13 $G(V)$ data and r3_cmc model fits for a short diffused resistor. Curves from top to bottom are for V_C from 0 to 20 V

be at a point where the output conductance is not negative. The asymptote of the μ_{red} model (9.33) precisely provides such an approximate model, therefore V_{sat} is computed as the value of V_r that makes the derivative of

$$\frac{1 - d_f\sqrt{d_p + 2V_{LC} + V_r}}{1 + [(V_r/L) - E_{co}]/E_{cr}} V_r \tag{9.36}$$

with respect to V_r zero. This gives a fourth order equation that has an analytic solution (although it is a complex solution; most of the code for the r3_cmc model is devoted to the calculation of V_{sat}). V_r is smoothly limited to V_{sat} using the symmetric limiting form

$$\frac{2V_r V_{sat}}{\sqrt{(V_r - V_{sat})^2 + 4A_{TS}^2} + \sqrt{(V_r + V_{sat})^2 + 4A_{TS}^2}} \tag{9.37}$$

where A_{TS} is a model parameter.

Figures 9.12 and 9.13 show $I(V)$ and $G(V)$ data and fits from the complete r3_cmc model. This device, a diffused resistor, is significantly affected by both depletion pinching and velocity saturation. The accuracy of the model is apparent. The velocity saturation model (9.31) with $\beta = 1$ is not able to accurately represent the "soft" onset of velocity saturation for low biases seen in Fig. 9.13.

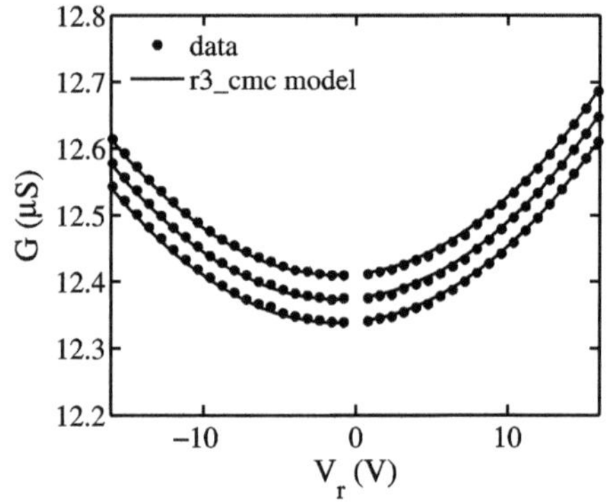

Fig. 9.14 $G(V)$ data and r3_cmc model fits for a long p-type poly resistor. Curves from top to bottom are for V_C from -16 to 0 V

Figure 9.14 shows $G(V)$ data and r3_cmc model fits for a polysilicon resistor (the $I(V)$ characteristics of this device are essentially indistinguishable from linear). The main source of nonlinearity for this device is from self-heating, however the conductance is clearly modulated by V_C, and there is a slight asymmetry in the characteristics that is apparent; both of these phenomena are caused by depletion pinching.

9.5 Parasitics, Noise and Statistical Modeling

For poly resistors the dielectric capacitance to the substrate is the main parasitic effect, this is modeled in r3_cmc using area and perimeter component bias independent capacitances. For diffused resistors a simple p-n junction model is used, including area and perimeter components for ideal diode currents and the standard SPICE depletion capacitance model [14].

Thermal noise is modeled using standard $4kTG$ components for the body and end resistances, where k is Boltzmann's constant. For diffused resistors there is shot noise $2qI$ for each parasitic diode current. The standard SPICE resistor flicker noise model does not have physically correct scaling with bias and geometry, so r3_cmc uses the geometry scaling derived in [17]

$$i_{flicker}^2 = K_{FN} \cdot \left(\frac{I}{W}\right)^{A_{FN}} \cdot \frac{W}{L} \cdot \frac{1}{f^{B_{FN}}} \tag{9.38}$$

where K_{FN}, A_{FN}, and B_{FN} are parameters.

The r3_cmc model has both global and local (i.e. mismatch) statistical variation modeling included as part of the basic model formulation; these do not need to be added on top of the basic model in model parameter files. Physically, statistical variation affects the sheet resistance and the effective electrical length and width. Because the sheet resistance variation depends on variation in the doping level, this averages over the area of the device hence its mismatch variance should be inversely proportional to area. The effective width averages over the length of the resistor, hence its mismatch variance should vary inversely with length, and similarly the mismatch variance of the effective length averages over width. The statistical variations are specified by standard deviations, which are model parameters, and the number of standard deviations of variation to apply, which are model parameters for global variation and instance parameters for local variation:

$$W = W_{nom} + N_{SW}\sigma_W + \frac{N_{mmSW}\sigma_{mmW}}{\sqrt{mL}}, \tag{9.39}$$

$$L = L_{nom} + N_{SL}\sigma_L + \frac{N_{mmSL}\sigma_{mmL}}{\sqrt{mW}}, \tag{9.40}$$

$$\rho_s = \rho_{s,nom}\exp\left[0.01\left(N_{SR}\sigma_R + \frac{N_{mmSR}\sigma_{mmR}}{\sqrt{mWL}}\right)\right] \tag{9.41}$$

where m is the multiplicity parameter (the number of devices connected in parallel), the added subscript "nom" signifies the nominal value of the parameter, the N are number of standard deviations of variation for each parameter, with added subscripts "mm" for the mismatch component, and the σ parameters are the standard deviations for each parameter. The above forms have the correct physical scaling for mismatch over geometry. The exponential mapping for the sheet resistance variation is included to prevent the resistance unphysically becoming negative for devices that have a large statistical variation in resistance; for such devices the observed distribution in resistance is generally log-normal, and the exponential mapping transitions from giving a normal distribution for small variations to becoming log-normal for large variations. The factor of $1/100$ is included for convenience, to allow the sheet resistance variation to be specified in terms of percentage.

9.6 Parameter Extraction

As noted previously, it is often better to analyze the "large-signal" conductance $G = I/V_r$ rather than $I(V)$ data, and this is also useful for parameter extraction. Specifically, assuming measurements are done with $V_L = 0$ for convenience, then $G(V_r, V_C)$ at $V_r = 0$ is useful for the purpose of parameter extraction because it is affected by neither velocity saturation nor self-heating. For diffused resistors, $G(0, V_C)$ can be relatively easily determined from extrapolation, see Fig. 9.6. For poly resistors the situation is a little more tricky, since poly resistors are highly linear and so G can be quite "noisy." However, for poly resistors the depletion pinching effect is manifest as an essentially linear variation of G with bias, and as Fig. 9.14 shows self-heating, which is the main cause of nonlinearity for poly resistors, causes a characteristic quadratic variation of G with bias. Therefore fitting $G(V)$ data around $V_r = 0$ with a quadratic polynomial in V_r not only aligns with the expected physical behavior, it averages out the effects of measurement noise and, as we will see, the coefficients of the polynomial can be directly used to give initial estimates of some parameters (in practice final parameter values are often determined using numerical optimization procedures based on the complete resistor model; here we discuss techniques to derive reasonable values for parameters based

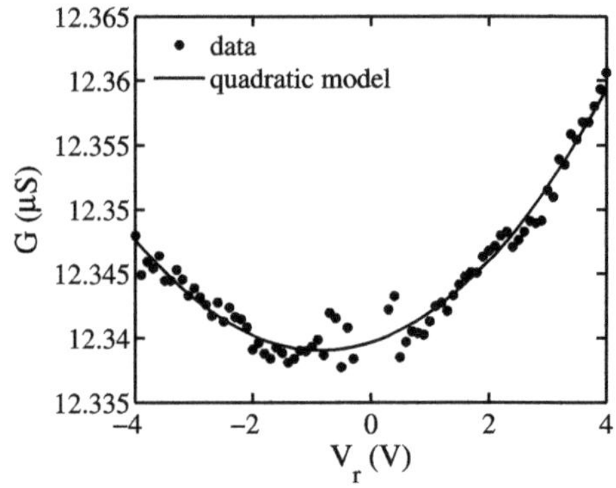

Fig. 9.15 Quadratic polynomial fit to $G(V)$ data

on simplified models and analysis). Figure 9.15 shows a quadratic polynomial fit to $G(V)$ data from a poly resistor. The measurement "noise" is apparent, so using any single data point to calculate parameter values is problematic, but the regression used to fit the data gives reliable values. Note the predominant parabolic shape is from self-heating, and the asymmetry is from depletion pinching.

Let the regression fit be

$$G(V_r, V_C) = b_0 + b_1 V_r + b_2 V_r^2 \tag{9.42}$$

where the coefficients are functions of V_C. From this the zero-bias resistance R_0 can be calculated as $1/b_0$, and from R_0 over temperature and geometry the first and second order temperature coefficients are easily calculated.

The parameters of the effective geometry offset models can be determined by looking at R_0 over length for a fixed (wide) width and over width for a fixed (long) length. The finite dopant effect model is only needed for well resistors, so ignoring this and considering only non-dogbone shaped devices, the effective geometry models of Sect. 9.2.1 can be manipulated to give

$$\begin{bmatrix} 1 & -\frac{R_{01}}{L_{m1}} & -\frac{R_{01}}{W_{m1}L_{m1}} & \frac{1}{L_{m1}} & \frac{1}{W_{m1}L_{m1}} \\ 1 & -\frac{R_{02}}{L_{m2}} & -\frac{R_{02}}{W_{m2}L_{m2}} & \frac{1}{L_{m2}} & \frac{1}{W_{m2}L_{m2}} \\ 1 & -\frac{R_{03}}{L_{m3}} & -\frac{R_{03}}{W_{m3}L_{m3}} & \frac{1}{L_{m3}} & \frac{1}{W_{m3}L_{m3}} \\ 1 & -\frac{R_{04}}{L_{m4}} & -\frac{R_{04}}{W_{m4}L_{m4}} & \frac{1}{L_{m4}} & \frac{1}{W_{m4}L_{m4}} \\ 1 & -\frac{R_{05}}{L_{m5}} & -\frac{R_{05}}{W_{m5}L_{m5}} & \frac{1}{L_{m5}} & \frac{1}{W_{m5}L_{m5}} \end{bmatrix} \begin{bmatrix} \rho_s \\ \Delta_W \\ \Delta_{WW} \\ \rho_s \Delta_L \\ \rho_s \Delta_{LW} \end{bmatrix} = \begin{bmatrix} R_{01}\frac{W_{m1}}{L_{m1}} \\ R_{02}\frac{W_{m2}}{L_{m2}} \\ R_{03}\frac{W_{m3}}{L_{m3}} \\ R_{04}\frac{W_{m4}}{L_{m4}} \\ R_{05}\frac{W_{m5}}{L_{m5}} \end{bmatrix} \tag{9.43}$$

which can be used to directly solve for the parameters from five or more arbitrary geometry resistors (if more than five geometries are used (9.43) is solved in a least squares sense using the Moore-Penrose pseudo-inverse [8]).

For diffused resistors a reasonable initial estimate for d_p is 1.5 (twice a typical p-n junction built-in voltage). For poly resistors it can be difficult to determine d_p from data as the depletion pinching effect gives an essentially linear variation of G with bias and any large value of d_p is consistent with this. However, if d_p is too large then this also causes d_f to be large, and the depletion pinching factor in

(9.27) can become negative. From (9.25) a reasonable estimate for d_p is $q\epsilon_s N/C_{ox}^{\prime 2}$ however because of incomplete dopant activation it can be better to evaluate this as $\epsilon_s/(\mu_0 t_p \rho_s C_{ox}^{\prime 2})$.

From (9.27)

$$\frac{1}{G(V_r, V_C)}\left(\frac{\partial G}{\partial V_r}\right)_{V_r=0, V_C=0} = -\frac{d_f}{2\sqrt{d_p}} = \frac{b_1}{b_0} \tag{9.44}$$

so d_f can be calculated from the coefficients of the fitted quadratic polynomial once d_p is known. Alternatively, if G at $V_r = 0$ is known at two values of V_C then

$$d_f = \frac{G(0, V_{C2}) - G(0, V_{C1})}{G(0, V_{C2})\sqrt{d_p + 2V_{C1}} - G(0, V_{C1})\sqrt{d_p + 2V_{C2}}}. \tag{9.45}$$

Equation (9.27) can also be rearranged to form

$$\frac{G^2}{d_f^2 g_f^2} - \frac{2G}{d_f^2 g_f} + \frac{1}{d_f^2} = d_p + V_{LC} + V_{RC} \tag{9.46}$$

so from measured G for three different biases this can be directly solved for g_f, d_f, and d_p.

For devices that are affected by self-heating, analysis of the self-heating effect, assuming that it is small, gives

$$\frac{G(V_r, 0)}{G(0, 0)} \approx \frac{1}{1 + T_{C1} R_{TH} V_r I} \approx 1 - \frac{T_{C1} R_{TH} W L}{\rho_s}\left(\frac{V_r}{L}\right)^2 \tag{9.47}$$

therefore b_2/b_0 from the quadratic model allows the thermal resistance to be calculated, given T_{C1} and ρ_s from previous extraction steps. Note that thermal resistance scales approximately inversely with area, so the quadratic variation with electric field in (9.47) is qualitatively the same as the low field variation of velocity saturation seen in (9.32) and Fig. 9.9. It can therefore be difficult to separate the effects of self-heating and velocity saturation in measured data. However, velocity saturation does not depend in frequency but self-heating does, see Fig. 9.11, so the frequency response of the small-signal resistance is useful to separate the two effects.

9.7 Details of Model Implementation

Code for the `r2_cmc` and `r3_cmc` models described in this chapter are available at the Compact Model Council (CMC) web site [7]. The models are defined in Verilog-A, which is the *de facto* standard language for compact modeling. Verilog-A makes it particularly easy to implement self-heating, because the most time-consuming and error-prone part of adding self-heating to a model, the generation of all derivatives, is handled automatically. Although Figs. 9.4 and 9.7 depict a 2-section resistor body model, with $1/6 : 2/3 : 1/6$ partitioning of the parasitic capacitance (this gives optimal modeling of the frequency response for a 2-section model [4]), the actual number of sections needed to ensure a specified accuracy up to a given frequency for small-signal modeling depends on the length of the resistor. The models are therefore implemented using a single element for the body of the resistor and the contact and end resistances, which are depicted as separate elements in Figs. 9.4 and 9.7, are combined. For accurate modeling up to a frequency f

$$N_S = \sqrt{\frac{2\pi f \rho_s C_A}{3}} f \tag{9.48}$$

sections are needed, where C_A is the parasitic capacitance per unit area. The internal end-plus-contact resistances are then set to zero when the N_S sections are connected in series.

Model implementation is fairly straightforward, however there is one aspect that requires special attention: how to handle small or zero values for the end-plus-contact resistance. This can be directly handled in Verilog-A as follows (using a simple linear resistor as an example, with line numbers indicated for reference):

```
01 `include "disciplines.h"
02 `define  rMin 0.001
03 module res_va(p,m);
04 inout       p, m;
05 electrical p, m;
06 parameter  real R = 1.0 from[0.0:inf);
07 analog begin : analogBlock
08   if (R<`rMin) begin
09     V(p,m)   <+ I(p,m)*R;
10   end else begin
11     I(p,m)   <+ V(p,m)/R;
12   end
13 end
14 endmodule
```

In this form if the resistance is less than `rMin` then the model switches from the $I = GV$ formulation of line `11` to the $V = IR$ formulation of line `09`. This can be somewhat tricky to implement when the Verilog-A code is mapped to C code, and the voltage contribution form line `09` is not "natural" for the MNA formulation, which is based on currents that are controlled by node voltages. Further, although it appears that the above model only depends on two system variables (i.e. the voltages on nodes `p` and `m`), implicitly a third variable is added to the MNA system of unknowns: the current through the element, which is necessary to handle the voltage contribution form.

A better approach for handling zero and small valued resistors was reported in [11]. This implementation recognizes that an additional system variable is needed, the current flowing through the element, and it adds an internal node `I_pm` that is used to explicitly include the current:

```
`include "disciplines.h"
`define  rMin 0.001
module res_va(p,m);
```

```
04 inout      p, m;
05 electrical p, m, I_pm;
06 parameter  real R=1.0 from[0.0:inf);
07 analog begin : analogBlock
08   I(I_pm) <+ V(p,m)-1.0e-6*R*V(I_pm);
09   if (R<`rMin) begin
10    I(p,m)  <+ 1.0e-6*V(I_pm);
11   end else begin
12     I(p,m)  <+ V(p,m)/R;
13   end
14 end
15 endmodule
```

Note that this form includes only current contributions, lines `08`, `10`, and `12`, it has no voltage contributions.

If the resistance is greater than or equal to `rMin` then line `12` is activated, which is the normal $I = GV$ formulation. For this case line `08` defines a current contribution to the node `I_pm`, but this is the only current contribution assigned to that node, and the sum of all currents flowing into a node must be zero, so it must be zero. The right hand side of the contribution operator in line `08` will therefore be forced to be zero at the converged solution of the simulator, which gives

$$\mathtt{V(I\_pm)} = \mathtt{1.0e6} * \mathtt{V(p,m)/R} \tag{9.49}$$

and this makes the voltage on the node `I_pm` equal to the current through the resistor, scaled up by 10^6. This scaling is necessary because this model formulation is storing a current value as the voltage on the node `I_pm`. The convergence tolerance for Kirchoff's current law is 10^{-12} by default in Verilog-A and in most simulators, but the convergence tolerance for voltages is only 10^{-6}, so if the current "stored" on the node `I_pm` were not scaled by 10^6 it would cause accuracy problems.

If the resistance is less than `rMin` then lines `08` and `10` are active. At the converged simulator solution, `I(p,m)` is the current flowing through the resistor. Line `10` therefore forces the voltage on the `I_pm` node to be the resistor current, scaled up by 10^6. Summation of the currents flowing into the node `I_pm` being zero forces

$$\mathtt{V(p,m)} = \mathtt{1.0e-6} * \mathtt{R} * \mathtt{V(I\_pm)} = \mathtt{R} * \mathtt{I(p,m)} \tag{9.50}$$

which is just the $V = IR$ formulation. For a resistance of zero this just forces `V(p,m)` to be zero, as expected.

So without explicitly including a voltage contribution this implementation elegantly and numerically stably handles resistors of any value, large and small and even zero. Some slightly more concise, but equivalent, formulations of current contribution only formulation for zero or small valued resistances are provided in [11].

9.8 Conclusions

In this chapter we have reviewed models for integrated resistors. This has included basic modeling of linear resistors, including effective geometry and temperature variation, empirical 2-terminal models of nonlinear resistors, and a physically based 3-terminal model that includes the effects of self-heating, velocity saturation, and depletion pinching for both diffused and polysilicon resistors. Models for parasitics and for statistical variation were also detailed. Some useful "tricks" and approaches for parameter extraction were described, although a good method to determine the depletion potential parameter for poly resistors is still wanting. It was shown that it can be difficult to separate the effects of velocity saturation and self-heating, and although it is at present not standard practice the best way to do this appears to be from the frequency dependence of the small-signal resistance; the measurements needed to enable this are somewhat time-consuming compared to those for dc and capacitance characterization, but if self-heating is important these measurements are necessary anyway to characterize the thermal capacitance. Finally, we have provided some useful information to aid accurate and numerically stable model implementation.

References

1. Banerjee, K., Amerasekera, A., Dixit, G., Hu, C.: Temperature and current effects on small-geometry-contact resistance. In: IEDM Tech. Digest, pp. 115–118 (1997)
2. Banoo, K., Gummel, H.K., Singhal, K.: Modelling resistor voltage coefficients. Agere Systems Notes (2004)
3. Bendix, P., Rakers, P., Wagh, P., Lemaitre, L., Grabinski, W., McAndrew, C.C., Gu, X., Gildenblat, G.: RF distortion analysis with compact MOSFET models. In: Proc. IEEE Custom Integrated Circuits Conf., pp. 9–12 (2004)
4. Booth, R.V.H., McAndrew, C.C.: A 3-terminal model for diffused and ion-implanted resistors. IEEE Trans. Electron Devices **44**(5), 809–814 (1997)
5. Caughey, D.M., Thomas, R.E.: Carrier mobilities in silicon empirically related to doping and fields. Proc. IEEE **55**(12), 2192–2193 (1967)
6. Chen, T.-L., Gildenblat, G.: Symmetric bulk charge linearisation in charge-sheet MOSFET model. Electron. Lett. **37**(12), 791–793 (2001)
7. Compact Model Council web site: http://www.geia.org/index.asp?bid=597
8. Forsythe, G.E., Malcolm, M.A., Moler, C.B.: Computer Methods for Mathematical Computations. Prentice-Hall, New York (1977)
9. Ito, A.: Modeling of voltage-dependent diffused resistors. IEEE Trans. Electron Devices **44**(12), 2300–2302 (1997)
10. Jacoboni, C., Canali, C., Ottaviani, G., Alberigi Quaranta, A.: A review of some charge transport properties of silicon. Solid-State Electron. **20**(2), 77–89 (1977)
11. Lemaitre, L., McAndrew, C.C.: Voltage-controlled-current-source-only Verilog-A resistor model for $R \geq 0$. In: Proc. IEEE Behavioral Modeling and Simulation Workshop, pp. 93–95 (2008)
12. Li, X., Jha, A., Gildenblat, G., van Langevelde, R., Smit, G.D.J., Scholten, A.J., Klaassen, D.B.M., McAndrew, C.C., Watts, J., Olsen, C.M., Coram, G.J., Chaudhry, S., Victory, J.: Benchmark tests for MOSFET compact models with application to the PSP model. IEEE Trans. Electron Devices **56**(2), 243–251 (2009)
13. Lu, N.C.C., Gerzberg, L., Lu, C.Y., Meindl, J.D.: Modeling and optimization of monolithic polycrystalline silicon resistors. IEEE Trans. Electron Devices **ED-28**(7), 818–830 (1981)
14. Massobrio, G., Antognetti, P.: Semiconductor Device Modeling with SPICE, 2nd edn. McGraw-Hill, New York (1983)

15. McAndrew, C.C., Sekine, S., Cassagnes, A., Wu, Z.: Physically based effective width modeling of MOSFETs and diffused resistors. In: IEEE International Conf. on Microelectronic Test Structures, pp. 169–174 (2000)
16. McAndrew, C.C.: R3, an accurate JFET and 3-terminal diffused resistor model. In: Tech. Proc. Workshop on Compact Modeling, pp. 86–89 (2004)
17. McAndrew, C.C., Coram, G., Blaum, A., Pilloud, O.: Correlated noise modeling and simulation. In: Tech. Proc. Workshop on Compact Modeling, pp. 40–45 (2005)
18. Nagel, L.W.: SPICE2: A computer program to simulate semiconductor circuits. Memo ERL-M520 University of California, Berkeley (1975)
19. Trofimenkoff, F.N.: Field-dependent mobility analysis of the field-effect transistor. Proc. IEEE **53**(11), 1765–1766 (1965)
20. Tsividis, Y., McAndrew, C.: Operation and Modeling of the MOS Transistor, 3rd edn. Oxford University Press, London (2010)
21. Victory, J., McAndrew, C.C., Hall, J., Zunino, M.: A 4-terminal compact model for high voltage diffused resistors with field plates. In: IEEE Bipolar/BiCMOS Circuits and Technology Meeting (BCTM), pp. 139–142 (1997)
22. Vogelsong, R., Brzezinski, C.: Simulation of thermal effects in electrical systems. In: IEEE Applied Power Electronics Conference and Exposition (APEC), pp. 353–356 (1989)

Chapter 10
The JUNCAP2 Model for Junction Diodes

A.J. Scholten, G.D.J. Smit, R. van Langevelde, and D.B.M. Klaassen

Abstract The physics-based junction model for CMOS, called JUNCAP2, is described. It contains single-piece formulations for the Shockley-Read-Hall generation/recombination current and the trap-assisted tunneling current which are valid both in forward and reverse mode of operation. Moreover, the trap-assisted tunneling model extends the previous model (Hurkx et al. in IEEE Trans. Electron Devices 39(9), 2090–2098, 1992) to the high electric fields encountered in today's CMOS technologies. Furthermore, the model contains expressions for junction capacitance, ideal current, band-to-band tunneling current, avalanche breakdown, and junction shot noise. The parameter extraction, experimental verification and a calculationally efficient evaluation procedure (JUNCAP2 Express) are also discussed in this chapter. The JUNCAP2 model is incorporated in the PSP model for bulk MOSFET's (see Chap. 1 of this book) and in the PSP-SOI model for SOI MOSFET's (see Chap. 2 of this book).

10.1 Introduction

Low-power CMOS applications are more and more affected by off-state leakage currents as CMOS technology is scaling down into the sub-100-nm era [2, 3]. Most

A.J. Scholten (✉) · G.D.J. Smit · D.B.M. Klaassen
NXP-TSMC Research Center, High Tech Campus 4, 5656 AE Eindhoven, The Netherlands
e-mail: andries.scholten@nxp.com

G.D.J. Smit
e-mail: gert-jan.smit@nxp.com

D.B.M. Klaassen
e-mail: d.b.m.klaassen@nxp.com

R. van Langevelde
Philips Research Europe, High Tech Campus 37, 5656 AE Eindhoven, The Netherlands
e-mail: ronald.van.langevelde@philips.com

G. Gildenblat (ed.), *Compact Modeling*,
DOI 10.1007/978-90-481-8614-3_10, © Springer Science+Business Media B.V. 2010

well-known leakage mechanisms in this context are gate leakage, increasing with decreasing insulator thickness, and source-to-drain leakage, increasing with decreasing channel length. To keep the source-drain leakage under control, pocket implants are commonly used in today's CMOS technologies. These highly-doped pocket implants are located closely to the highly doped source/drain junctions, causing high electric fields across these junctions. These high fields lead to increased junction leakage due to the effects of trap-assisted tunneling and band-to-band tunneling [2].

The junction modeling of older compact MOSFET models (e.g. NXP's JUNCAP model [4] or the BSIM3 junction model [5]) is mostly restricted to junction capacitance, ideal junction current, and a very simple Shockley-Read-Hall generation/recombination model. Only the BSIM4.4 model [5] contains equations for trap-assisted tunneling. These equations, however, have limited physical background and do not always accurately reproduce experimental data.

In this chapter, the physics-based model, JUNCAP2, is described. The JUNCAP2 model is capable of modeling the junction leakage currents in advanced CMOS technologies in detail. The model has been introduced in Ref. [6]. The physical background of the model equations and the parameter extraction procedure have been published in Ref. [7]. The calculationally efficient evaluation procedure (JUNCAP2 Express) was presented in Ref. [8]. The JUNCAP2 model is incorporated in the PSP model for bulk MOSFET's (see [9, 10] and Chap. 1 of this book) and in the PSP-SOI model for SOI MOSFET's (see [11] and Chap. 2 of this book).

10.2 Model Derivation

In this section, the derivation of the model equations for the JUNCAP2 junction model will be presented. Basic electrostatics and junction capacitance are recapitulated in Sect. 10.2.1, and the ideal junction current is briefly recalled in Sect. 10.2.2. Next, the derivation of more accurate compact formulations for the Shockley-Read-Hall and trap-assisted tunneling currents will be presented in Sects. 10.2.3 and 10.2.4, respectively. The model is completed with band-to-band tunneling (Sect. 10.2.5), avalanche breakdown (Sect. 10.2.6), junction noise (Sect. 10.2.7), and geometrical scaling (Sect. 10.2.8).

10.2.1 Capacitance

The cross-section of a junction is given in Fig. 10.1. Donor and acceptor concentrations are denoted with N_D and N_A, respectively. A symmetrical junction is considered with net doping concentration $N_{D,\mathit{eff}} = N_D - N_A$ given by:

$$N_{D,\mathit{eff}} = N_0 \cdot \left(\frac{x}{x_0}\right)^m \qquad \text{for } x \geq 0, \tag{10.1}$$

$$N_{D,eff} = -N_0 \cdot \left(\frac{-x}{x_0}\right)^m \quad \text{for } x < 0 \tag{10.2}$$

where the x is the direction perpendicular to the junction (see Fig. 10.1). The net doping concentration is parameterized by the parameters N_0, x_0 and m. The latter quantity is directly related to the so-called grading coefficient $P \equiv 1/(m+2)$. Different values of m, or alternatively P, correspond to different net doping profiles. Well-known examples are $P = 1/3$ (linear grading), $P = 1/2$ (step junction), and $P > 1/2$ (hyper-abrupt junction).

The Poisson equation for the junction is solved under the total-depletion approximation, and using the boundary conditions that (i) the electrostatic field F at the depletion-region edges is zero and that (ii) the electrostatic potential difference $\Delta\psi$ is equal to $\Delta\psi = V_{AK} - V_{bi}$, see Appendix 1. Here, V_{AK} is the applied bias between anode A and cathode K, and V_{bi} is the built-in voltage of the junction. This leads to the following expressions for the electrostatic potential ψ and field F:

$$\psi = \frac{V_{bi} - V_{AK}}{2 \cdot (1-P)} \cdot \left[-P \cdot \left(\frac{2 \cdot x}{W_{dep}}\right)^{1/P} + \frac{2 \cdot x}{W_{dep}} \right], \tag{10.3}$$

$$F = F_{max} \cdot \left[1 - \left(\frac{2 \cdot x}{W_{dep}}\right)^{(1-P)/P} \right] \tag{10.4}$$

where F_{max} is the maximum electrostatic field and W_{dep} is the depletion-region width given by:

$$F_{max} = \frac{V_{bi} - V_{AK}}{W_{dep} \cdot (1-P)}, \tag{10.5}$$

$$W_{dep} = W_{dep,0} \cdot \left(1 - \frac{V_{AK}}{V_{bi}}\right)^P. \tag{10.6}$$

Here, $W_{dep,0}$ is the zero-bias depletion region width which may be written in terms of N_0, x_0, m, and V_{bi}. From (10.6) the junction depletion capacitance per unit area,

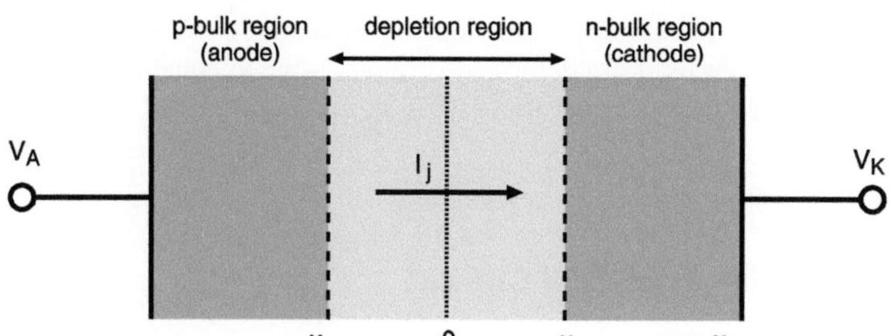

Fig. 10.1 Schematic cross section of p-n junction

C'_j, is straightforwardly derived, using $C'_j = \varepsilon_{Si} / W_{dep}$. This leads to the well-known equation:

$$C'_j = \frac{C_{jo}}{(1 - \frac{V_{AK}}{V_{bi}})^P} \tag{10.7}$$

where C_{jo}, the zero-bias junction depletion capacitance per unit area, is given by:

$$C_{jo} = C_{JOR} \cdot \left(\frac{V_{BIR}}{V_{bi}} \right)^P . \tag{10.8}$$

Here, two adjustable model parameters C_{JOR} and V_{BIR} have been introduced. C_{JOR} is the zero-bias capacitance per unit area at the reference temperature, and V_{BIR} is the built-in voltage at the reference temperature. The temperature dependence of the junction capacitances is governed by the temperature dependence of V_{bi}, which is discussed in Appendix 1.

10.2.2 Ideal Current

The ideal current density through a p-n junction, I'_D, caused by drift and diffusion of charge carriers through the depletion region, is modeled by the Shockley equation:

$$I'_D = I_{DSAT} \cdot \left[\exp\left(\frac{V_{AK}}{\phi_{TD}} \right) - 1 \right] \tag{10.9}$$

where ϕ_{TD} is the thermal voltage. The saturated ideal current density I_{DSAT} depends on temperature as

$$I_{DSAT} = I_{DSATR} \cdot F_{TD}^2 \tag{10.10}$$

where the adjustable model parameter I_{DSATR} is the saturation current density at the reference temperature, and the quantity F_{TD} (which is proportional to the intrinsic carrier density n_i) is introduced in Appendix 1. In the JUNCAP2 model, the temperature dependence of I_{DSATR} can be adjusted to fit the data by tuning the band gap voltage ϕ_G, see Appendix 1.

It should be noted that (10.9) has *not* been extended with a so-called "ideality factor", which is often seen in compact models for junctions. This "ideality factor" has no physical meaning and is simply used to fit deviations from the ideal behavior predicted by (10.9) in the forward mode of operation. In the JUNCAP2 model, instead, non-ideal forward currents are described by physics-based equations for Shockley-Read-Hall recombination current (Sect. 10.2.3), and trap-assisted tunneling current (Sect. 10.2.4).

10.2.3 Shockley-Read-Hall Current

Here, an improved expression for the Shockley-Read-Hall (SRH) current in a junction is presented. It builds forth on the theory behind Diode Level 500 [1, 13]. This

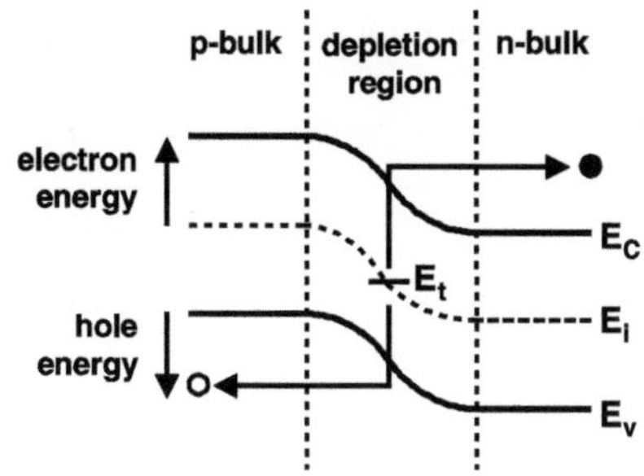

Fig. 10.2 Schematic representation of the Shockley-Read-Hall generation process

improved model, however, is valid for any grading coefficient P, whereas the theory in [1] uses the $P = 0.5$ approximation. In addition, the JUNCAP2 model features a single-piece description of the current which is valid both in the forward and reverse mode of operation. This is in contrast with Ref. [1], which uses separate expressions in the forward and reverse, and an additional parameter to tune the forward current.

The SRH generation (see Fig. 10.2) and recombination of charge carriers at depletion layer traps gives rise to deviations from the ideal behavior discussed in Sect. 10.2.2. In the forward mode of operation, the net SRH generation-recombination rate U_{SRH} is positive, leading to net recombination and non-ideal additional current at low forward bias. In the reverse mode of operation, U_{SRH} is negative, corresponding to a net generation rate and causing additional leakage.

Assuming Boltzmann statistics, mid-gap traps, and equal capture cross section σ for electrons and holes, and adopting the commonly used approximation that the quasi-Fermi levels of electrons (ϕ_n) and holes (ϕ_p) remain constant throughout the depletion layer ($\phi_p = V_{AK}/2$ and $\phi_n = -V_{AK}/2$), U_{SRH} is given by [14]:

$$U_{SRH} = \sigma \cdot v_{th} \cdot N_t \cdot n_i \cdot \left[\exp\left(\frac{V_{AK}}{\phi_{TD}}\right) - 1\right] \cdot f(x) \qquad (10.11)$$

where

$$f(x) = \left[\exp\left(\frac{\psi(x) - \phi_n}{\phi_T}\right) + \exp\left(\frac{\phi_p - \psi(x)}{\phi_T}\right) + 2\right]^{-1}. \qquad (10.12)$$

Here, n_i is the intrinsic carrier density, v_{th} is the thermal velocity and N_t the trap density.

The SRH current density I'_{SRH} is given by the integral of $U_{SRH}(x)$ over the depletion region, so that

$$I'_{SRH} = \sigma \cdot v_{th} \cdot N_t \cdot n_i \cdot \left[\exp\left(\frac{V_{AK}}{\phi_{TD}}\right) - 1\right] \cdot \int_{-x_p}^{x_n} f(x) \cdot \mathrm{d}x. \qquad (10.13)$$

To proceed further, the integral of $f(x)$ has to be evaluated. The function $f(x)$ is plotted in Fig. 10.3 for several bias conditions. It is seen that $f(x)$ is a symmetrical function changing from box-shaped in reverse bias to triangularly shaped in the

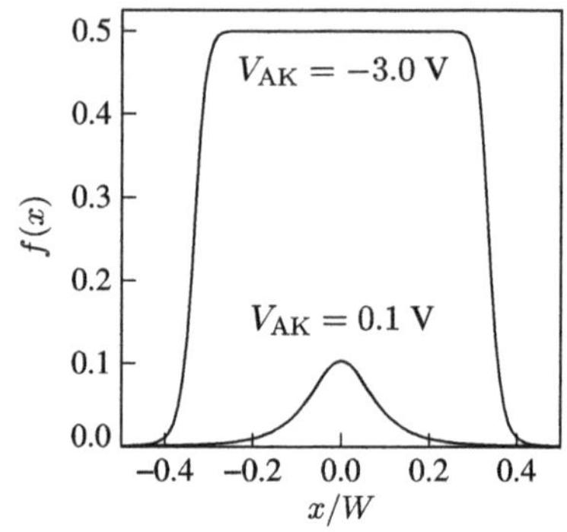

Fig. 10.3 The function $f(x)$ plotted for different bias conditions

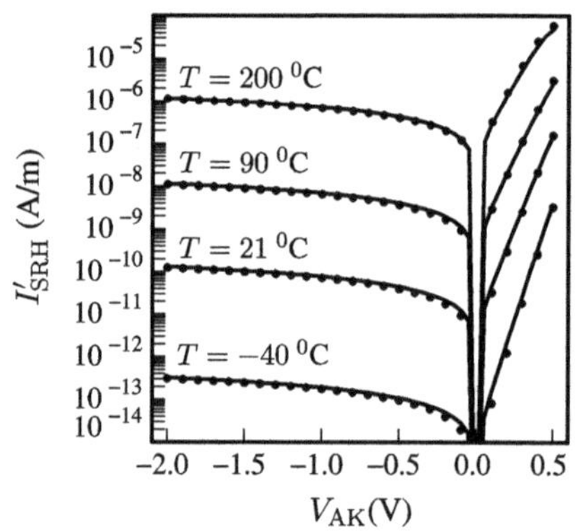

Fig. 10.4 Calculation of the SRH current using numerical integration of $f(x)$ (markers), compared with the approximation used in the compact model (*solid lines*). The calculation is done at various temperatures as indicated in the graph

forward mode of operation. In either case, the integral can be suitably approximated by

$$\int_{-x_p}^{x_n} f(x) \cdot \mathrm{d}x \approx W_{SRH} \cdot \max[f(x)] \tag{10.14}$$

where W_{SRH} is the full-width-at-half-maximum of the function $f(x)$, and $\max[f(x)]$ is the maximum value of $f(x)$, which is reached at $x = 0$:

$$\max[f(x)] = \frac{1}{2 \cdot [\exp(\frac{V_{AK}}{2 \cdot \phi_{TD}}) + 1]}. \tag{10.15}$$

Details on the calculation of w_{SRH} are given in Appendix 2. Having evaluated w_{SRH}, the SRH current can be rewritten as (see (10.13)):

$$I'_{SRH} = C_{SRH} \cdot F_{TD} \cdot \left[\exp\left(\frac{V_{AK}}{2 \cdot \phi_{TD}}\right) - 1\right] \cdot w_{SRH} \cdot W_{dep} \tag{10.16}$$

where C_{SRH} has been introduced, which is an adjustable model parameter, theoretically equal to $\sigma \cdot v_{th} \cdot N_t/[2 \cdot n_i(T_{KR})]$. The quantity W_{dep} is derived from the capacitance using $W_{dep} = \varepsilon_{Si}/C'_j$.

In Fig. 10.4 the SRH current based on the approximation (10.14) is compared with the current obtained when $f(x)$ is integrated numerically. Both results agree

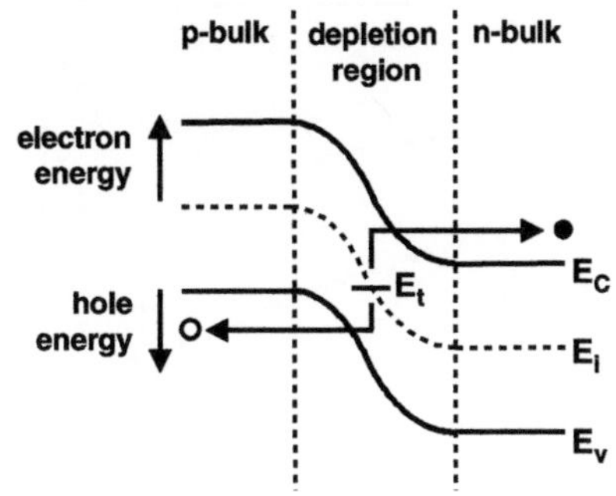

Fig. 10.5 Schematic representation of the trap-assisted tunneling process

very well both in the reverse and in the low forward bias regime, thereby justifying the approximations in the analytical calculation.

10.2.4 Trap-Assisted Tunneling Current

The JUNCAP2 formulation of the trap-assisted tunneling (TAT) current builds forth on the theory behind the Diode Level 500 model [1]. In contrast to Ref. [1], the improved model (i) does not use the $P = 1/2$ approximation, (ii) features a single-piece description of the current instead of separate forward and reverse descriptions, and, most importantly, (iii) contains both low-field as well as high-field behavior of trap-assisted tunneling.

In the presence of a high electric field, the SRH recombination rate is enhanced w.r.t. (10.11) due to trap-assisted tunneling. This process is depicted schematically in Fig. 10.5. Treating, as before, electrons and holes on equal footing, the total recombination rate U_{traps} is given by [12]:

$$\begin{aligned} U_{traps} &= (1+\Gamma)\cdot U_{SRH} \\ &= U_{SRH} + U_{TAT} \end{aligned} \tag{10.17}$$

where Γ is the so-called field-enhancement factor, and where $U_{TAT} = \Gamma \cdot U_{SRH}$ is the TAT recombination rate. The expression for $\Gamma(x)$ reads [12]:

$$\Gamma(x) = a_{TAT} \cdot \int_0^1 \exp\left(a_{TAT}\cdot u - K \cdot u^{3/2}\right)\cdot \mathrm{d}u \tag{10.18}$$

where $a_{TAT} = \Delta E/(k_B \cdot T)$, and $K = F_{nor}/|F|$, where

$$F_{nor} = \frac{4}{3}\cdot\frac{2\cdot m_{eff}\cdot \Delta E^3}{q\cdot\hbar}. \tag{10.19}$$

Here, m_{eff} is the effective mass for trap-assisted tunneling, which is an adjustable model parameter expected to be close to $0.25m_0$ [12]. For the bias condition of interest, i.e. reverse and low-forward bias, ΔE is, in case of electrons, given by the energy difference between conduction band and traps. Assuming, as before, mid-gap traps, one has $\Delta E = q\cdot\phi_{GD}/2$, where ϕ_{GD} is the band gap voltage at the device temperature. For details on the definition of the quantity ΔE, please refer to [12].

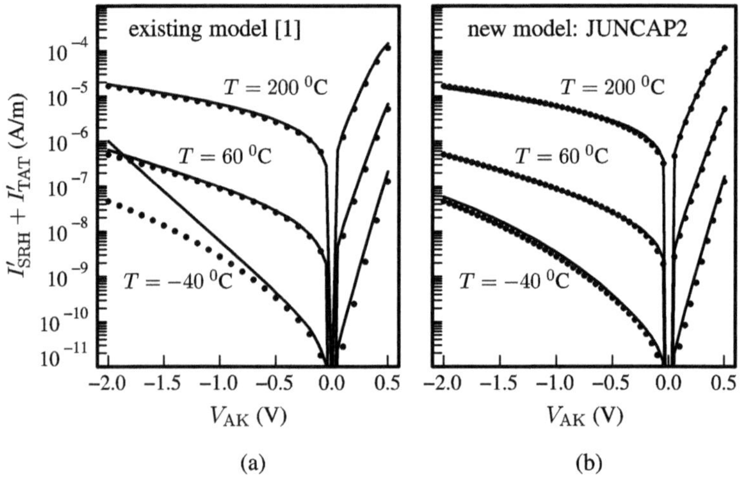

Fig. 10.6 Comparison of numerical calculation of TAT current density, indicated by markers, with (**a**) old model [1] and (**b**) improved model. *Solid lines* represent model curves. The calculation is done at various temperatures as indicated in the graph. The same model parameters are used as in Fig. 10.11

The TAT current density is found by integration of U_{TAT} over the depletion region:

$$I'_{TAT} = \int_{-x_p}^{x_n} \Gamma(x) \cdot U_{SRH}(x) \cdot \mathrm{d}x. \tag{10.20}$$

This is a cumbersome task, since already the integral defining Γ [see (10.18)] cannot be performed analytically. In Diode Level 500 [1], a low-field approximation for Γ [12] has been adopted, and the subsequent integration of U_{TAT} over the depletion region has been done for the specific case of a symmetrical step junction ($P = 0.5$). The low-field approximation of Γ is valid for electric fields smaller than $F_{lim} = 1.5 F_{nor}/a_{TAT}$, and severely overestimates Γ at fields in excess of F_{lim}, leading to problematic model behavior at high reverse bias, as shown in Fig. 10.6(a). As F_{lim} is smallest at low temperatures, the modeling problems are first encountered at low temperatures.

Here, an improved model is described that is valid for both low and high fields. Just as in the preceding section, the approximation that the value of the integral is equal to the maximum of the integrand times the "width" of the integrand. The function $\Gamma(x) \cdot U_{SRH}(x)$ has a maximum at $x = 0$. Depending on the bias condition, its width is determined by either $\Gamma(x)$ or $U_{SRH}(x)$. This leads to the approximation:

$$I'_{TAT} = \Gamma_{max} \cdot U_{SRH,max} \cdot W_{TAT} \tag{10.21}$$

where W_{TAT} is given by

$$W_{TAT} = \frac{W_{SRH} \cdot W_{\Gamma}}{W_{SRH} + W_{\Gamma}} \tag{10.22}$$

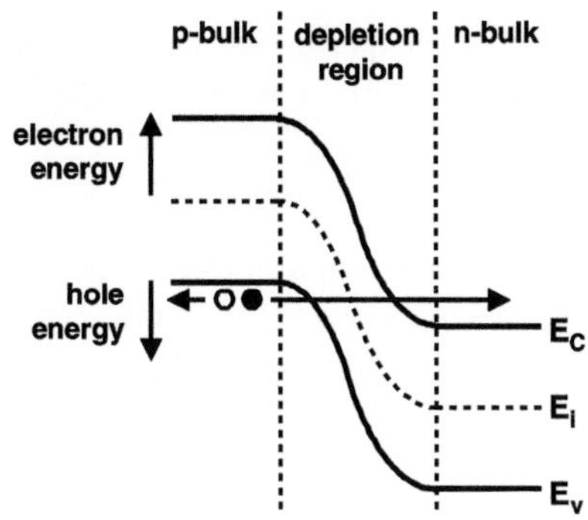

Fig. 10.7 Schematic representation of the band-to-band tunneling process

with W_Γ given by

$$W_\Gamma = \frac{\int_{-x_p}^{x_n} \Gamma(x) \cdot \mathrm{d}x}{\Gamma_{max}}. \tag{10.23}$$

Details of the calculation of W_Γ and Γ_{max} are given in Appendices 3 and 4, respectively. The resulting equation for the trap-assisted tunneling current becomes:

$$I'_{TAT} = C_{SRH} \cdot F_{TD} \cdot \left[\exp\left(\frac{V_{AK}}{2 \cdot \phi_{TD}} \right) - 1 \right] \cdot \Gamma_{max} \cdot W_{TAT}. \tag{10.24}$$

Finally, to gain model flexibility and in line with the Diode level 500 model, the parameter C_{SRH} in this equation is replaced by a new, separately adjustable model parameter C_{TAT}. Theoretically, the equality $C_{TAT} = C_{SRH}$ should hold.

In Fig. 10.6(b) the improved model is compared with the results obtained when (10.18) and (10.20) are integrated numerically. It can be observed that the improved model stays very close to the result from numerical integration, proving the validity of the approximations involved. Further note that the problematic model behavior at high reverse bias, as depicted in Fig. 10.6(a), has been remedied.

10.2.5 Band-to-Band Tunneling Current

For band-to-band tunneling (see Fig. 10.7) the following expression is adopted from Ref. [15]:

$$I'_{BBT} = C_{BBT} \cdot V_{AK} \cdot {F_{max}}^2 \cdot \exp\left(-\frac{F_0}{F_{max}} \right) \tag{10.25}$$

where C_{BBT} and F_0 are parameters. The band-to-band tunneling current I'_{BBT} depends on temperature because F_{max} depends on temperature through V_{bi}, and because F_0 depends on temperature. The first effect prevails at low biases and would give a decrease in the BBT current with increasing temperature; the second effect prevails at higher negative bias and gives an increase in BBT current with temperature. This can lead to crossing model curves for different temperatures, an effect which is not seen in experiments. Instead, experiments show an increase of the BBT current with temperature for all biases. Therefore (10.25) is slightly modified into

$$I'_{BBT} = C_{BBT} \cdot V_{AK} \cdot F_{max,r}{}^2 \cdot \exp\left(-\frac{F_{BBT}}{F_{max,r}}\right) \tag{10.26}$$

where $F_{max,r}$ is the maximum electric field at the reference temperature. All temperature dependence is empirically absorbed by F_{BBT} using:

$$F_{BBT} = F_{BBTR} \cdot \left[1 + S_{T,FBBT} \cdot (T_{KD} - T_{KR})\right]. \tag{10.27}$$

In the BBT current model, F_{BBTR}, $S_{T,FBBT}$, and C_{BBT} are adjustable model parameters.

10.2.6 Avalanche Breakdown Current

The starting point of this section is the Diode Level 500 [1] description of avalanche current. In Diode Level 500 the total diode current density I' is given by:

$$I' = \frac{(I'_{BBT} + I'_{TAT}) \cdot e^{-\mu_{av}} + (I'_{SRH} + I'_D) \cdot \frac{1+e^{-2\mu_{av}}}{2}}{1 - 2 \cdot \mu_{av} \cdot [1 + \exp(-2 \cdot \mu_{av})]}. \tag{10.28}$$

Here

$$\mu_{av} = 0.3295 \cdot \left(\frac{F_{max}}{F_{max,br}}\right)^2 \cdot \exp\left(b_n \cdot \frac{F_{max} - F_{max,br}}{F_{max} \cdot F_{max,br}}\right) \tag{10.29}$$

where b_n is a temperature-dependent parameter, see [1]. Breakdown occurs when F_{max} reaches the value $F_{max,br}$, because at that point the value of μ_{av} becomes 0.3295, and the denominator of (10.28) becomes zero.

Because the quantity μ_{av} varies over a fairly limited range [0, 0.3295], (10.28) can be simplified considerably by using Taylor expansions:

$$I' = \left(I'_D + I'_{SRH} + I'_{TAT} + I'_{BBT}\right) \cdot \frac{1}{1 - 3 \cdot \mu_{av}}. \tag{10.30}$$

To maintain the correct definition of $F_{max,br}$, the factor 0.3295 (10.29) is slightly modified to the value of 1/3. Furthermore μ_{av} is approximated by:

$$\mu_{av} = \frac{1}{3} \cdot \left(\frac{F_{max}}{F_{max,br}}\right)^{2+b_n/F_{max,br}}. \tag{10.31}$$

For the large reverse voltages under consideration and for $P = 0.5$, the electric field F scales with $\sqrt{-V_{AK}}$, leading to

$$\mu_{av} = \frac{1}{3} \cdot \left|\frac{-V_{AK}}{V_{BR}}\right|^{P_{BR}} \tag{10.32}$$

where V_{BR} is the breakdown voltage and $P_{BR} = 1 + 0.5 \cdot b_n \cdot F_{max,br}$. Combining this μ_{av} expression with (10.30) one arrives at:

$$I' = \left(I'_D + I'_{SRH} + I'_{TAT} + I'_{BBT}\right) \cdot \frac{1}{1 - |\frac{-V_{AK}}{V_{BR}}|^{P_{BR}}}. \tag{10.33}$$

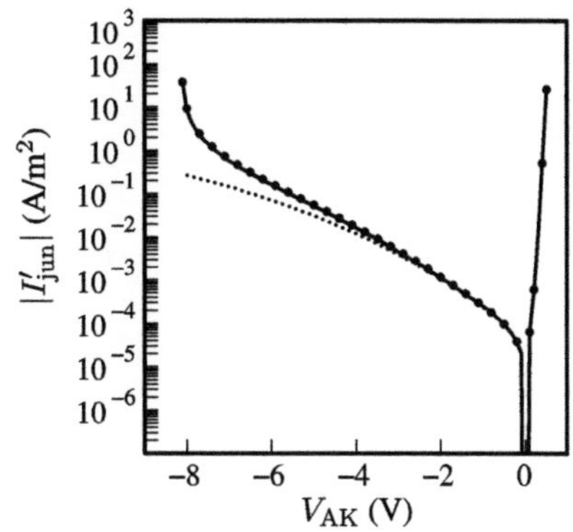

Fig. 10.8 Comparison of (10.28) and (10.29) (*markers*) and the simplified implementation (10.33) (*solid line*) of avalanche. *Dotted line*: no avalanche

The above formula for breakdown is a well-known (but so far empirical) formula for avalanche and breakdown in junctions and can be found in textbooks, e.g., Ref. [16]. To the best of the authors' knowledge this derivation published in Ref. [7], is the first to establish a relation between the empirical formula and the theoretical expressions [1] and can be regarded as a derivation of the former.

Now the starting point, (10.28) and (10.29) will be compared with the simplified formula (10.33). In order to do this V_{BR} is expressed in terms of $F_{max,br}$ using the electrostatics in Sect. 10.2.1. It is shown in Fig. 10.8 that the simplified approach stays very close to the more sophisticated starting formulas, corroborating the presented derivation.

10.2.7 Noise

The JUNCAP2 junction model is equipped with shot noise. This is important for the modeling of MOSFETs fabricated in partially depleted silicon-on-insulator (PDSOI) technology. Here, junction shot noise, together with the shot-like noise of the MOSFET avalanche current, is the origin of the kink-related excess Lorentzian noise observed in the drain current of PDSOI transistors [17].

10.2.8 Geometrical Scaling

Just as in existing junction models for MOSFETS, e.g., JUNCAP [4] or the BSIM3/4 junction models [5], the JUNCAP2 junction model distinguishes a bottom component, an STI-edge component, and a gate-edge component of the junction capacitances:

$$C_j = A_B \cdot C'_{j,bot} + L_S \cdot C'_{j,sti} + L_G \cdot C'_{j,gat} \tag{10.34}$$

where $C'_{j,bot}$ is the bottom component (in F/m^2), $C'_{j,sti}$ is the STI-edge component (in F/m) and $C'_{j,gat}$ is the gate-edge component (in F/m) of the junction capacitance.

Fig. 10.9 Schematic top view of a MOSFET. For the drain junction, the meaning of A_B, L_S, and L_G is indicated

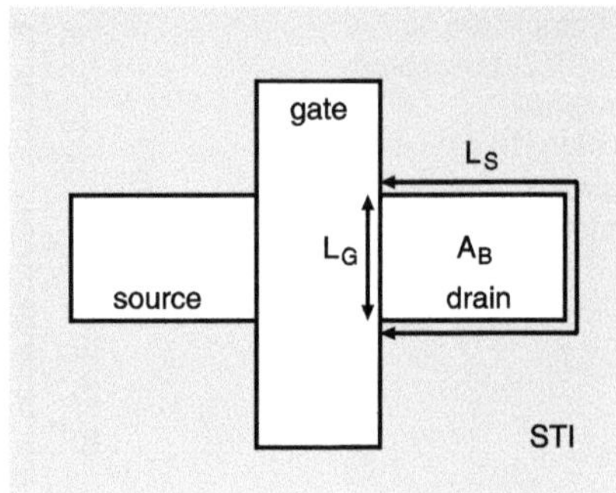

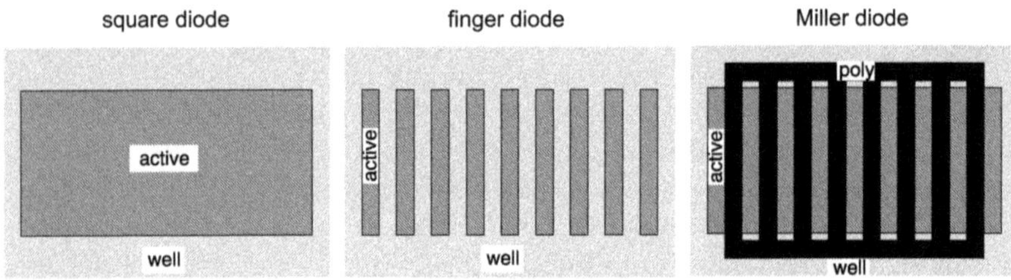

Fig. 10.10 Schematic representation of three test structures needed for parameter extraction

Furthermore A_B, L_S, and L_G are the junction area, STI-edge, and gate-edge, as illustrated in Fig. 10.9. Similarly, for the junction currents one can write

$$I_j = A_B \cdot I'_{j,bot} + L_S \cdot I'_{j,sti} + L_G \cdot I'_{j,gat} \tag{10.35}$$

where $I'_{j,bot}$ is the bottom component (in A/m^2), $I'_{j,sti}$ is the STI-edge component (in A/m, usually very small) and $I'_{j,gat}$ is the gate-edge component (in A/m) of the junction currents.

10.3 Parameter Extraction

10.3.1 Test Structures

For extraction of JUNCAP2 parameters, one uses the same three test structures that are commonly used for junction characterization (see Fig. 10.10). The first structure is a simple, square diode, which has a large bottom component, a relatively small STI-edge component, and no gate-edge component. The second structure is a finger diode, which has a much larger STI-edge component, and no gate-edge component. The third structure is a Miller diode, which is a multi-fingered MOSFET with source and drains tied together. It has a relatively small STI-edge component, and a significant gate-edge component. The test structures should be sufficiently large so that currents and capacitances are easily measurable.

For extraction of JUNCAP2 parameters, one needs both CV and IV measurements over a large range of temperatures. All measurements are carried out on the three test structures and are used to extract the three components. For the capacitances this is done using

$$C'_{j,bot} = \frac{L_{S,finger} \cdot C_{j,square} - L_{S,square} \cdot C_{j,finger}}{L_{S,finger} \cdot A_{B,square} - L_{S,square} \cdot A_{B,finger}}, \tag{10.36}$$

$$C'_{j,sti} = \frac{A_{B,square} \cdot C_{j,finger} - A_{B,finger} \cdot C_{j,square}}{L_{S,finger} \cdot A_{B,square} - L_{S,square} \cdot A_{B,finger}}, \tag{10.37}$$

$$C'_{j,gat} = \frac{C_{j,Miller} - A_{B,Miller} \cdot C'_{j,bot} - L_{S,Miller} \cdot C'_{j,sti}}{L_{G,Miller}} \tag{10.38}$$

for each bias point and temperature. Similar formulas apply to the junction currents, resulting in the current components $I'_{j,bot}$, $I'_{j,sti}$, and $I'_{j,gat}$.

10.3.2 Extraction of CV Parameters

From the CV curves, the parameters C_{JOR}, V_{BIR}, and P are extracted for each capacitance component. First, the parameters are initialized to reasonable initial values: C_{JOR} is set to the measured zero-bias capacitance closest to the chosen reference temperature, V_{BIR} is set to 1 V, and P is set to 0.5. Now the three parameters are optimized to fit the CV curves at all temperatures simultaneously.

10.3.3 Extraction of IV Parameters

The remaining parameters are determined from the IV curves. First initial values are derived for the parameters that determine the ideal current, i.e., I_{DSATR} and ϕ_G. This is done as follows: those measurement points from the forward current are selected for which the measured ideality factor

$$n = \phi_{TD} \cdot \frac{\partial I'_{j,gat}}{\partial V_{AK}} \tag{10.39}$$

is reasonably close to 1 (e.g. $0.9 < n < 1.1$). Here the gate-edge component has been taken, but the same applies to the other current components. The band gap should be close to 1.16 eV, so ϕ_G is initialized to 1.16. Now I_{DSATR} is set to 1 and the modeled current is calculated. The average ratio between measured and modeled currents for the selected measurement points is now the starting value for I_{DSATR}. Please note once more that the ideality factor employed here is only a quantity directly derived from measurements. It is not a model parameter as in many other junction models.

To initialize the parameters for SRH and TAT the activation energy is used

$$E_{act} = \frac{\partial \ln(I'_{j,bot})}{\partial \phi_{TD}^{-1}}. \tag{10.40}$$

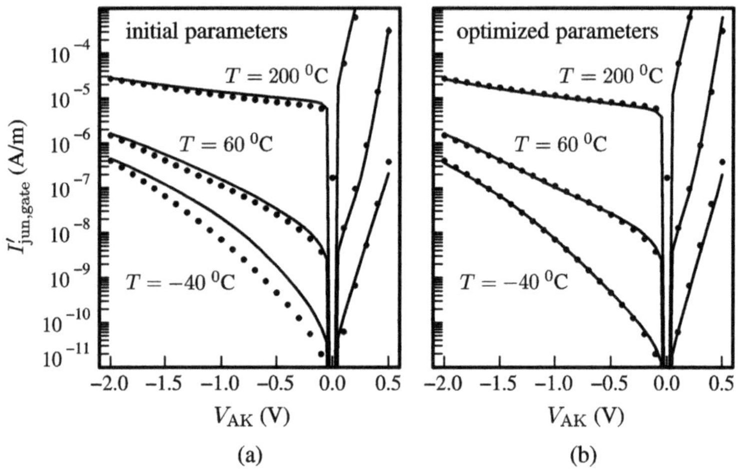

Fig. 10.11 Comparison of model curves (*solid lines*) with the experimental data (*markers*) after parameter initialization (**a**) and after parameter optimization (**b**). The experimental data are the gate-edge contribution of a n^+-pwell junction in 0.12 μm technology. Only three temperatures are shown for clarity, but measurements and parameter extraction have been done for a much larger set of temperatures

If the activation energy is around half the band gap ($0.3\ \mathrm{eV} < E_{act} < 0.7\ \mathrm{eV}$) the current is dominated by SRH and TAT. The parameter m_{eff} is initialized to $0.25 \cdot m_0$, and at first C_{SRH} and C_{TAT} are set to 1. The average ratio between measured and modeled currents for the selected measurement points is now the starting value for $C_{SRH} = C_{TAT}$.

A similar procedure is followed for the BBT parameters. Here the selection criterion is $E_{act} < 0.2$ eV. The parameters F_{BBTR} and S_{FBBT} are initialized to 1×10^9 V/m and -1×10^{-3}, respectively. C_{BBT} is first set to 1, and the average ratio between measured and modeled currents for the selected measurement points is now the starting value for C_{BBT}.

In Fig. 10.11(a) the model result after parameter initialization is compared with measured data. Here the gate-edge contribution of a n^+-pwell junction in 0.12 μm technology is taken as an example, because the ideal current, SRH, TAT, and BBT are all of importance. The model curves after parameter initialization are already quite close to the measured data. The next step is a least-squares optimization of the model to the data, using the scheme in Table 10.1, resulting in an excellent fit as depicted in Fig. 10.11(b). Note that the improved SRH and TAT model gives a good description of excess leakage current in the reverse mode of operation and of the non-ideality in the forward mode of operation simultaneously. A more detailed comparison between model and experimental data will be presented in Sect. 10.4.

Table 10.1 Summary of adjustable model parameters and measurements used to determine them

Measurement	Physical effect	Parameters
CV	capacitance	C_{JOR}, V_{BIR}, P
forward IV	ideal current	I_{DSATR}, ϕ_G
forward and reverse IV	SRH, TAT	C_{SRH}, C_{TAT}, m_{eff}
reverse IV	BBT	C_{BBT}, F_{BBTR}, S_{FBBT}

10.4 Model Verification

10.4.1 Capacitances

Junction capacitances and currents have been measured over a large temperature range on NMOS and PMOS junctions in a 0.12 μm CMOS technology. Parameter extraction has been done using the test structures and procedures outlined in Sect. 10.3. In Fig. 10.12, the CV curves of the bottom, STI-edge and gate-edge components of the NMOS junction are shown as a function of bias voltage for various temperatures. The model gives an accurate description of the capacitances for all temperatures.

10.4.2 Currents

In Fig. 10.13 the bottom and gate-edge contributions of the NMOS junction current are shown. As expected, the STI edge contribution to the current was found to be negligible. The bottom component is dominated by the ideal current, but also shows SRH and TAT contributions. The gate-edge component, in addition, clearly shows a BBT contribution in reverse. TAT contributions are seen in both forward and reverse. This is illustrated by switching off TAT in the model, see Fig. 10.14, in which case both the simulated forward and reverse currents are far below the experimental data. The model parameters not only give an excellent description of the three devices used for parameter extraction (not shown here), but also of a fourth independent device with a different geometry, see Fig. 10.15. Next, in Fig. 10.16, the IV characteristics for the PMOS junction are shown. Finally, in Fig. 10.17, the gate-edge contribution of an NMOS junction in 65-nm technology is shown. Compared to the 0.12 μm technology (cf. Fig. 10.13), the importance of BBT has increased. Note that BBT has a much stronger bias dependence and a much weaker temperature dependence than TAT. Consequently, attempts to capture both effects in one model equation are bound to fail (see e.g. [5]).

10.5 JUNCAP2 Express

To calculate the current in a MOSFET at a given bias condition, the junction currents in all three components (bottom, gate-edge, and isolation edge) of both the

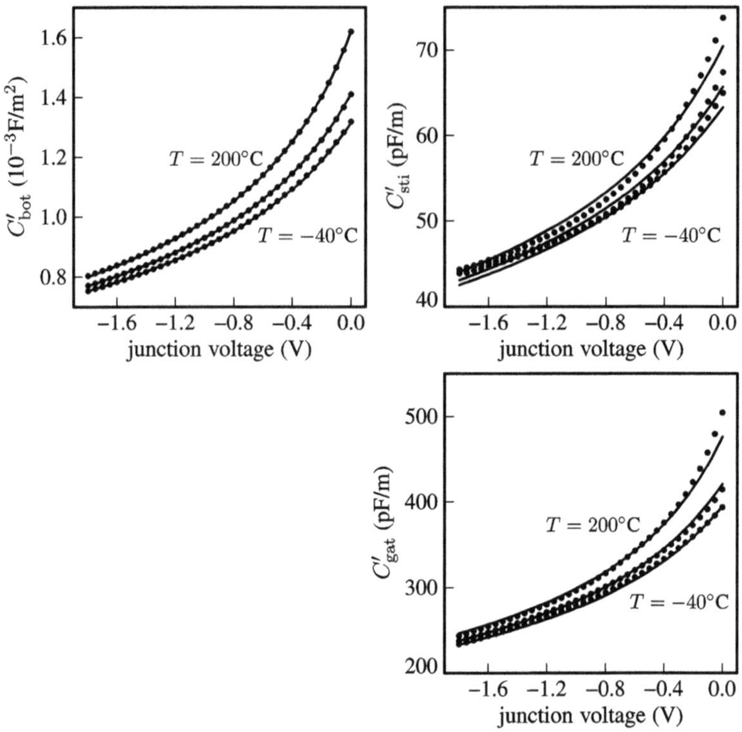

Fig. 10.12 *Bottom* (*upper-left figure*), STI-edge (*upper-right figure*), and gate-edge (*lower figure*) components of the NMOS junction capacitance as a function of junction voltage in a 0.12 μm CMOS technology. Temperatures are −40°C (*bottom curve*), 60°C, and 200°C (*top curve*). *Markers* are measurements, *lines* are the JUNCAP2 model

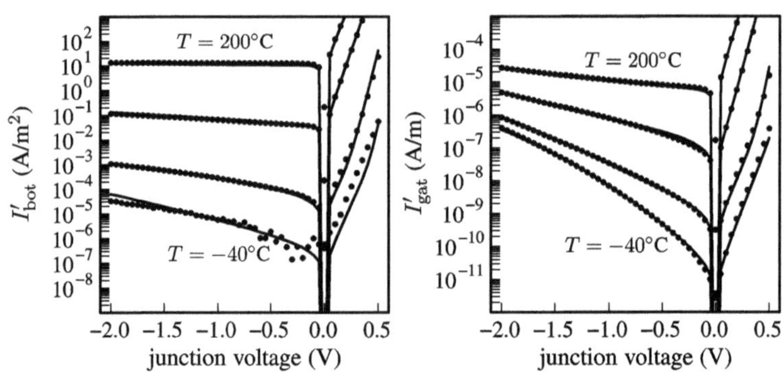

Fig. 10.13 *Bottom* (*left figure*) and gate-edge (*right figure*) components of the NMOS junction current as a function of the voltage across the junction in a 0.12 μm CMOS technology. Temperatures are −40°C (*bottom curve*), 21°C, 125°C, and 200°C (*top curve*). *Markers* are measurements, *lines* are the JUNCAP2 model

Fig. 10.14 Same measurements as in right figure of Fig. 10.13, but now the trap-assisted tunneling (TAT) has been turned off in the model

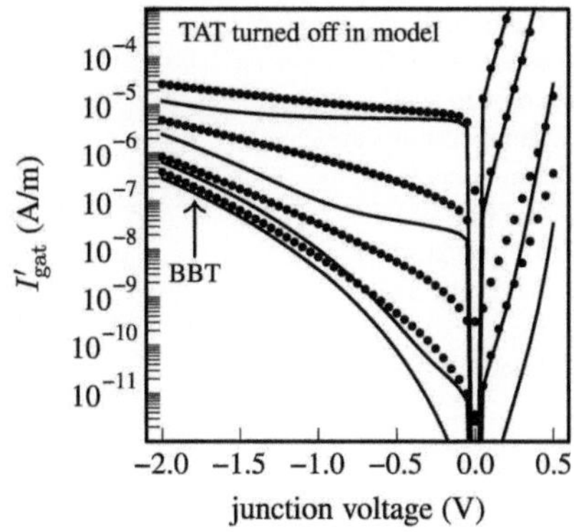

Fig. 10.15 NMOS junction current in a junction diode with different geometry than the three diodes used for parameter extraction (see Sect. 10.4.1)

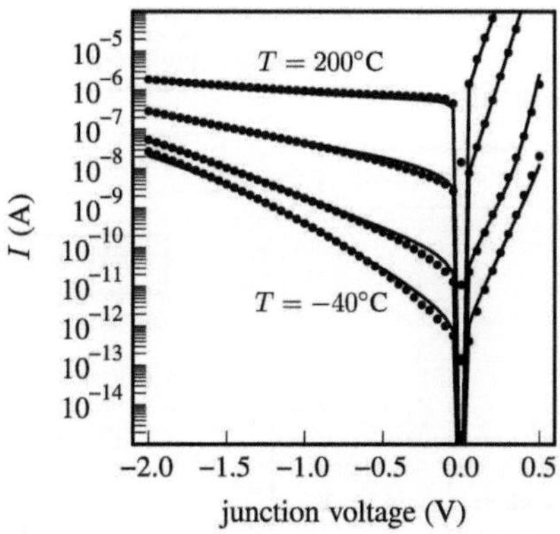

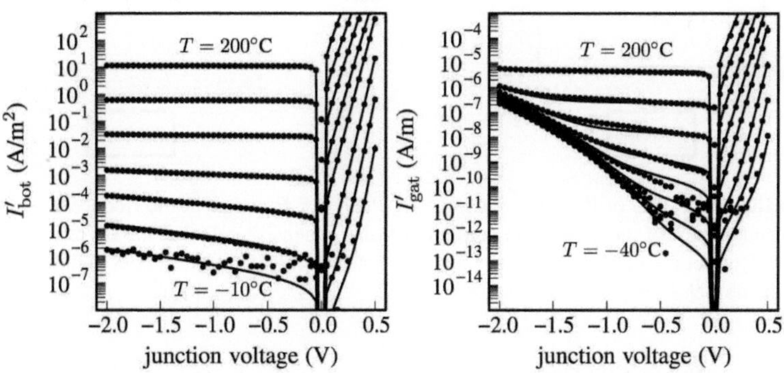

Fig. 10.16 *Bottom* (*left figure*) and gate-edge (*right figure*) components of the PMOS junction current as a function of the voltage across the junction in a 0.12 µm CMOS technology. Temperatures are −40°C (*bottom curve*), −10°C, 21°C, 60°C, 90°C, 125°C, 160°C, and 200°C (*top curve*). *Markers* are measurements, *lines* are the JUNCAP2 model

source and drain junctions have to be calculated. So, while the expression for the channel current is evaluated once, the expression for the junction current including the various mechanisms, is evaluated *six* times. To alleviate this calculational

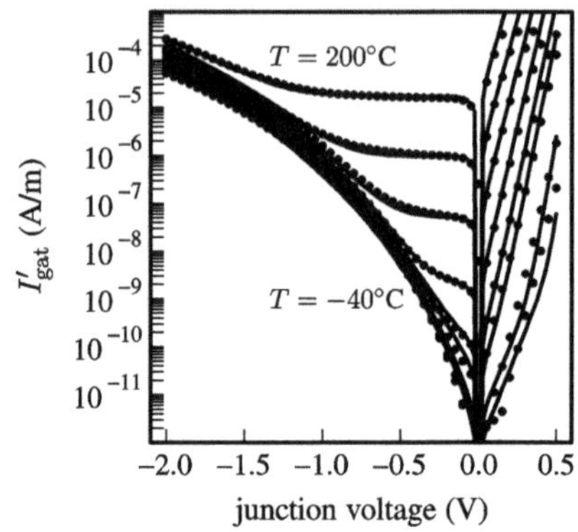

Fig. 10.17 Gate-edge component of the NMOS junction in a 65-nm CMOS technology. Temperatures are −40°C (*bottom curve*), −10°C, 21°C, 60°C, 90°C, 125°C, 160°C, and 200°C (*top curve*). *Markers* are measurements, *lines* are the JUNCAP2 model

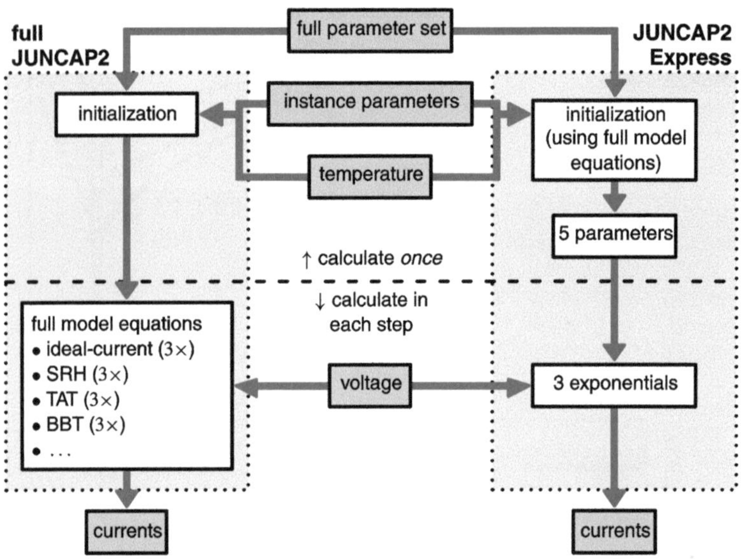

Fig. 10.18 Illustration of the basic idea behind JUNCAP Express: in the frequently called voltage-dependent part of the model, the computationally expensive calculations are replaced by simple ones. This set of simple equations is mapped onto the full JUNCAP model in the initialization part of the model

burden, the "JUNCAP2 Express" option has been developed; see Fig. 10.18. In the bias-independent part of the model (only called in the initialization phase), the full expressions for all three components are used to calculate the coefficients of a very simple expression. Now, in the bias-dependent part, which is called repeatedly by the circuit simulator in all the bias- or time-steps of the circuit simulation, only this very simple expression needs to be evaluated. Details of the JUNCAP Express calculations are given in Appendix 6. In Fig. 10.19, the JUNCAP2 Express simulations are shown to be in close agreement with the original full JUNCAP2 simulations, validating the JUNCAP2 Express approach.

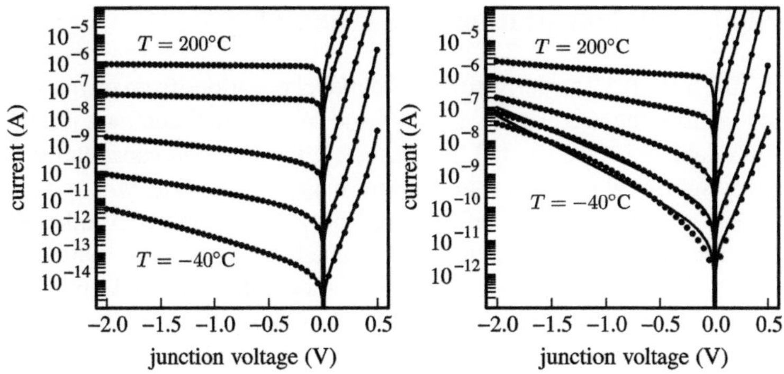

Fig. 10.19 Junction current I_j as a function of the voltage V_j across n^+/p-well junctions in a 0.12 μm CMOS technology for temperatures of −40, 21, 90, 160, and 200°C. Symbols represent simulations with the full JUNCAP2 model, while *lines* represent the corresponding JUNCAP Express results. *Left picture*: large square junction with a bottom area of 70000 μm^2 and an STI edge of 1100 μm. *Right picture*: multi-fingered junction with a bottom area of 23100 μm^2 and gate edges of 77000 μm

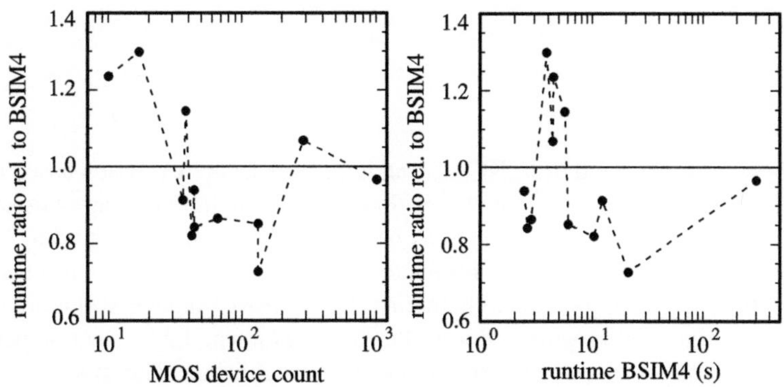

Fig. 10.20 Runtimes of PSP relative to BSIM4 for 12 representative digital standard cells as a function of MOS device count (*left*) and BSIM4 runtime (*right*) using the circuit simulator Eldo from Mentor Graphics [18]. *Dashed lines* serve to guide the eye

Using the JUNCAP2 Express option, the junction calculation overhead in PSP is reduced by a factor of 5–15 (dependent on simulator and analysis type). As a consequence, the junction calculation now only consumes ∼ 10% of the total PSP runtime, while its accuracy is hardly sacrificed at all. In Fig. 10.20, the PSP runtime is benchmarked against BSIM4 for transient analyses on a set of representative circuits with varying transistor count. The circuit simulator used was Eldo from Mentor Graphics [18]. In PSP, the JUNCAP Express option was switched on. As can be seen, the PSP/BSIM4 runtime ratio is on average close to 1. Moreover, as expected, there is *no* systematic increase/decrease in this number with transistor count.

10.6 Model Implementation and Availability

In Sects. 10.2 and 10.5, the derivation of the model equations from physics has been presented. Before these equations can be used in an actual compact model, some numerical adaptations are needed. For instance, the junction capacitance formula (10.7) gives numerical problems at a forward bias of $V_{AK} = V_{bi}$, where the approximations that are used to arrive at (10.7) are not valid anymore. Another example is the junction breakdown formula, (10.33) which in the form presented here not only gives the breakdown in the reverse mode of operation (wanted), but also in the forward mode of operation (unwanted). The detailed treatment of these and other numerical issues is beyond the scope of this paper; it suffices here to mention that they are taken care of in the actual implementations of JUNCAP2 that can be found in Refs. [4, 9].

Source code (C-code as well as in Verilog-A code) and documentation of JUNCAP2 can be downloaded from the internet [4, 9]. JUNCAP2 is available as stand-alone model, and is also part of the PSP MOSFET model [4, 9, 10]. At present, it is available in commercial circuit simulators ADS (Agilent) and Spectre (Cadence) via a dynamically linked library called "SiMKit" [4], and in various other commercial circuit simulators.

10.7 Conclusion

The physical background of JUNCAP2, an improved accurate physics-based model for MOSFETs, has been presented. For Shockley-Read-Hall and trap-assisted tunneling, the previous description [1] has been extended to arbitrary grading coefficient, and a single-piece description for forward and reverse currents has been derived. Moreover, for trap-assisted tunneling, the improved models capture both the "low" and "high" field regimes. For avalanche breakdown, the connection between theory and a so-far empirical formula has been established. The model is shown to give a very accurate description of junction leakage currents, and a simple, effective parameter extraction strategy has been presented. JUNCAP2 is available as stand-alone model and is also part of the Compact Model Council (CMC, see [20]) standard bulk MOSFET model PSP and of the PSP-SOI model for SOI MOSFET's.

Acknowledgments This work was partly sponsored by the European project MEDEA+ T206. The authors acknowledge Laurent Lemaitre (Freescale) and Marjan Driessen and co-workers (NXP/Corporate I&T/DTF) The results of Fig 10.20 were first presented by A. Juge (ST Microelectronics) at the CMC meeting on October 4, 2007, Boston, MA, and have been reproduced with permission of the author.

Appendix 1: Built-in Voltage

The electrostatic potential difference $\Delta\psi$ between both edges of the depletion region, i.e. $-x_p$ and x_n (see Fig. 10.1), is derived from carrier statistics. In the nondegenerate case and assuming total ionization, the Fermi potentials on the n-side and p-sides of the junction are:

$$\phi_n(x_n) = \psi(x_n) - \phi_{TD} \cdot \ln\left[\frac{N_{D,\mathit{eff}}(x_n)}{n_i}\right], \tag{10.41}$$

$$\phi_p(-x_p) = \psi(-x_p) + \phi_{TD} \cdot \ln\left[\frac{N_{A,\mathit{eff}}(-x_p)}{n_i}\right] \tag{10.42}$$

where $N_{A,\mathit{eff}} \equiv -N_{D,\mathit{eff}}$ is the net acceptor concentration.

Assuming zero voltage drop in the bulk regions, one has $V_{AK} = \phi_p(-x_p) - \phi_n(x_n)$, so that (10.41) and (10.42) yield:

$$\Delta\psi \equiv \psi(-x_p) - \psi(x_n) = V_{AK} - V_{bi} \tag{10.43}$$

where V_{bi} is the built-in voltage, given by

$$V_{bi} = \frac{k_B \cdot T}{q} \cdot \ln\left[\frac{N_{D,\mathit{eff}}(x_n) \cdot N_{A,\mathit{eff}}(-x_p)}{n_i^2}\right]. \tag{10.44}$$

In the special case of a step junction this formula reduces to the well-known textbook formula: $V_{bi} = (k_B \cdot T/q) \cdot \ln(N_D \cdot N_A/n_i^2)$. In that special case the built-in voltage is independent of the depletion region width and thus independent of applied bias. In most literature, it is tacitly assumed that this is also the case for more general junctions with arbitrary grading coefficient. However, from inspection of the more general expression (10.44) it can be seen that this is strictly speaking not true. Nevertheless, motivated by the weak (logarithmic) dependence of V_{bi} on depletion region width, the approximation can be made that the built-in voltage is a constant for junctions with arbitrary grading coefficient.

The temperature dependence of V_{bi} is governed by the exponential dependence of n_i on temperature. The expressions for n_i are well known [14] and lead to the model equations:

$$V_{bi} = V_{BIR} \cdot \frac{T_{KD}}{T_{KR}} - 2 \cdot \phi_{TD} \cdot \ln F_{TD}, \tag{10.45}$$

$$F_{TD} = \left(\frac{T_{KD}}{T_{KR}}\right)^{1.5} \cdot \exp\left(\frac{\phi_{GR}}{2 \cdot \phi_{TR}} - \frac{\phi_{GD}}{2 \cdot \phi_{TD}}\right). \tag{10.46}$$

Here, T_{KD} and T_{KR} are the device temperature and reference temperature in Kelvin, respectively, ϕ_{TD} and ϕ_{TR} are the corresponding thermal voltages, F_{TD} is $n_i(T_{KD})/n_i(T_{KR})$, and the adjustable model parameter V_{BIR} is the built-in voltage at the reference temperature. Finally, ϕ_{GD} and ϕ_{GR} are the band gap voltage (i.e. the band gap divided by the elementary charge) at the device temperature and reference temperature, respectively. They, in turn, are given by:

$$\phi_G(T) = \phi_G - \frac{7.02 \times 10^{-4} \cdot T^2}{1108.0 + T}. \tag{10.47}$$

Here the adjustable model parameter ϕ_G has been introduced, which is the zero-Kelvin band gap voltage. It will be used to tune the temperature dependence of the ideal current.

Appendix 2: Evaluation of W_{SRH}

In order to determine W_{SRH}, the equation $f(x) = \max[f(x)]/2$ has to be solved. Using the definition of $f(x)$, (10.12), one finds that $f(x) = \max[f(x)]/2$ is fulfilled when ψ takes the values $\pm\psi^*$, where ψ^* given by:

$$\psi^* = \phi_{TD} \cdot \text{arcosh}\left[2 + \exp\left(\frac{-V_{AK}}{2 \cdot \phi_{TD}}\right)\right]. \tag{10.48}$$

To arrive at W_{SRH}, the corresponding x-values are needed. The x-values are each others opposites due to the assumed symmetry of the junction. Therefore (10.3) is solved for $\psi = \psi^*$ and $x = w_{SRH} \cdot W_{dep}/2$, to obtain w_{SRH}, which is the relative (dimensionless) width of $f(x)$ divided by the total width W_{dep} of the depletion region. An analytical solution for w_{SRH} is only obtained in case $P = 1/2$ (step junction):

$$w_{SRH,step} = 1 - \sqrt{1 - \frac{2 \cdot \psi^*}{V_{bi} - V_{AK}}}. \tag{10.49}$$

A first-order perturbational approach is used to approximate the deviation of w_{SRH}, called Δw_{SRH}, for values of P deviating from $1/2$:

$$\Delta w_{SRH} = \frac{-2 \cdot w_{SRH,step} \cdot [1 + w_{SRH,step} \cdot \ln(w_{SRH,step})]}{1 - w_{SRH,step}} \cdot \left(P - \frac{1}{2}\right). \tag{10.50}$$

Appendix 3: Evaluation of W_Γ

Using the expression (10.18) for Γ, and changing the order of integration, one finds for $\int \Gamma(x) \cdot \mathrm{d}x$:

$$\int_{-x_p}^{x_n} \Gamma(x) \cdot \mathrm{d}x = a_{TAT} \cdot \int_0^1 \exp(a_{TAT} \cdot u) \cdot \xi \cdot \mathrm{d}u, \tag{10.51}$$

$$\xi = 2 \cdot \int_0^{W_{dep}/2} \exp\left(-\frac{F_{nor} \cdot u^{3/2}}{|F(x)|}\right) \cdot \mathrm{d}x. \tag{10.52}$$

The integral ξ is now approximated by its maximum (obtained for $F = F_{max}$) multiplied by the x value for which the integrand has dropped to $1/e$ times its maximum. Using (10.4) one arrives at:

$$\xi = W_{dep} \cdot \left(b_{TAT} \cdot u^{3/2} + 1\right)^{P/(P-1)} \cdot \exp\left(-b_{TAT} \cdot u^{3/2}\right) \tag{10.53}$$

where the abbreviation $b_{TAT} = F_{nor}/F_{max}$ has been used.

Returning to the integral $\int \Gamma(x) \cdot \mathrm{d}x$, substitution of (10.53) into (10.51) leads to:

$$\int_{-x_p}^{x_n} \Gamma(x) \cdot \mathrm{d}x = a_{TAT} \cdot W_{dep} \cdot \int_0^1 \left(b_{TAT} \cdot u^{3/2} + 1\right)^{P/(P-1)} \cdot \exp\left(a_{TAT} \cdot u - b_{TAT} \cdot u^{3/2}\right) \cdot du. \tag{10.54}$$

Because the exponent in this integral changes much more rapidly than the other part of the integrand, the following approximation can be made:

$$\int_{-x_p}^{x_n} \Gamma(x) \cdot \mathrm{d}x = \Gamma_{max} \cdot a_{TAT} \cdot W_{dep} \cdot \left(b_{TAT} \cdot {u_{max}}^{3/2} + 1\right)^{P/(P-1)} \tag{10.55}$$

where Γ_{max} is the field enhancement factor evaluated for $F = F_{max}$. If the upper boundary of the integral in (10.54) would have been infinity, u_{max} would have been equal to the value for u where the exponent in the integrand of (10.54) reaches its maximum

$$u_{max} = \left(\frac{2 \cdot a_{TAT}}{3 \cdot b_{TAT}}\right)^2. \tag{10.56}$$

However, because the integral over u runs from 0 to 1, u_{max} has to be limited to 1. Consequently, u_{max} is replaced by

$$\sqrt{\frac{{u_{max}}^2}{{u_{max}}^2 + 1}}. \tag{10.57}$$

From (10.55) and (10.23) one can now straightforwardly derive the expression for W_Γ:

$$W_\Gamma = W_{dep} \cdot \left(b_{TAT} \cdot {u_{max}}^{3/2} + 1\right)^{P/(P-1)}. \tag{10.58}$$

Appendix 4: Evaluation of Γ_{max}

In Ref. [12], approximations for Γ in both "low" and "high" field regimes are given. Gluing them together using a smoothing function is a cumbersome task because of the strong (exponential) dependence of Γ on F. Therefore, a different approach will be used here, leading to a single-piece expression for Γ_{max}, valid for both low and high fields. The starting point is the equation for Γ, i.e. (10.18), for $F = F_{max}$:

$$\Gamma_{max} = a_{TAT} \cdot \int_0^1 \exp\left(a_{TAT} \cdot u - b_{TAT} \cdot u^{3/2}\right) \cdot du. \tag{10.59}$$

As in the derivation of the low-field Γ in Ref. [12], the function $h(u) = a_{TAT} \cdot u - b_{TAT} \cdot u^{3/2}$ is expanded around $u = u_{max}$. Note that the derivative $h'(u)$ does not necessarily have to be zero, since u_{max} represents the maximum of $h(u)$ on the interval [0, 1], and is therefore limited to 1 by (10.57). One can write:

$$h(u) = -\left[k_{TAT} \cdot (u - l_{TAT})\right]^2 + m_{TAT} \tag{10.60}$$

with k_{TAT}, l_{TAT}, and m_{TAT} given by:

$$k_{TAT} = \sqrt{\frac{3}{8} \cdot b_{TAT} \cdot u_{max}{}^{-1/2}}, \tag{10.61}$$

$$l_{TAT} = \frac{4 \cdot a_{TAT}}{3 \cdot b_{TAT}} \cdot u_{max}{}^{1/2} - u_{max}, \tag{10.62}$$

$$m_{TAT} = \frac{2 \cdot a_{TAT}{}^2}{3 \cdot b_{TAT}} \cdot u_{max}{}^{1/2} - a_{TAT} \cdot u_{max} + \frac{b_{TAT}}{2} \cdot u_{max}{}^{3/2}. \tag{10.63}$$

Using the 2nd-order Taylor expansion of $h(u)$, and replacing the lower limit of integration by $-\infty$, (10.59) becomes:

$$\Gamma_{max} = \frac{a_{TAT} \cdot \exp(m_{TAT}) \cdot \mathrm{erfc}[k_{TAT} \cdot (l_{TAT} - 1)] \cdot \sqrt{\pi}}{2 \cdot k_{TAT}}. \tag{10.64}$$

In the treatment of trap-assisted tunneling, no distinction has been made so far between "low-field" and "high-field" regimes, as in Ref. [12]. Actually, the expression (10.64) for Γ_{max} is valid for both regimes. In the model presented above, the transition from "low-field" to "high-field" regime is embedded in (10.56) and (10.57). The condition $u'_{max} = 1$ corresponds exactly to the point of transition between low and high-field approximations in Ref. [12]. Thus, it is (10.57) that actually takes care of the smooth transition between both regimes. The reason that this works so well is because u_{max} varies only between 0 and 1. If $u_{max} \ll 1$, the above expression (10.64) for Γ_{max} reduces to the low-field expression in Ref. [12]. In case $u_{max} \approx 1$, the expression (10.64) reduces to the high-field expression in Ref. [12].

In (10.64) the erfc-function has been used. A suitable approximation of this function will be derived in Appendix 5.

Appendix 5: Approximation of the erfc-Function

Analytical approximations for erfc(y) exist [19], but they have to be handled with care. They are usually only valid for positive arguments y, while for negative arguments, $\mathrm{erfc}(y) = 2 - \mathrm{erfc}(-y)$ has to be applied. In this construction, continuity of odd-order derivatives of the approximated erfc(y) function is warranted, but continuity of the function itself and its even-order derivatives is not automatically fulfilled. This is relevant for the JUNCAP2 model, since both positive and negative arguments of the erfc function can occur. Therefore now an erfc(y) approximation will be constructed which is not extremely accurate, but which has the desired property of C_3 continuity in $y = 0$. The starting point is the expression found in [19]:

$$\mathrm{erfc}(y) \approx \left(a_{\mathrm{erfc}} \cdot t + b_{\mathrm{erfc}} \cdot t^2 + c_{\mathrm{erfc}} \cdot t^3\right) \cdot \exp(-y^2) \tag{10.65}$$

with

$$t = \frac{1}{1 + p_{\mathrm{erfc}} \cdot y}. \tag{10.66}$$

In the above, a_{erfc}, b_{erfc}, c_{erfc}, and p_{erfc} are parameters which are given in [19], and which will be adjusted to meet the requirements. As mentioned above, for $y < 0$ $\mathrm{erfc}(y) = 2 - \mathrm{erfc}(-y)$ will be used. The requirement of continuity in $y = 0$ leads to the condition:

$$c_{\mathrm{erfc}} = 1 - a_{\mathrm{erfc}} - b_{\mathrm{erfc}} \tag{10.67}$$

which is fulfilled by the parameters given in [19]. Continuity of the second derivative leads to the condition:

$$b_{\mathrm{erfc}} = \frac{6 - 5 \cdot a_{\mathrm{erfc}} - p_{\mathrm{erfc}}{}^{-2}}{3} \tag{10.68}$$

which is not fulfilled by the parameters given in [19]. The last condition follows from a requirement for the limiting behavior of Γ, which can be derived directly from its definition, (10.18):

$$\lim_{b_{TAT} \to 0} \Gamma = \exp(a_{TAT}) - 1 \approx \exp(a_{TAT}) \tag{10.69}$$

where the last approximation is justified because typically $a_{TAT} > 10$. One can show that the latter condition is fulfilled by (10.64) when in the erfc approximation one uses:

$$p_{\mathrm{erfc}} = \sqrt{\pi} \cdot a_{\mathrm{erfc}}. \tag{10.70}$$

Now one has three conditions, (10.67), (10.68), and (10.70) to determine the four parameters of the erfc approximation. This leaves one parameter that can be fitted to achieve maximal accuracy. This was done by a least-square curve fit of the approximate erfc function to the exact result, where the relative error was minimized on the interval $0 \le y \le 5$ by adjusting the parameter a_{erfc}. This leads to the result: $a_{\mathrm{erfc}} = 0.29214664$; $b_{\mathrm{erfc}} = 0.26992880$; $c_{\mathrm{erfc}} = 0.43792456$; $p_{\mathrm{erfc}} = 0.51781644$.

Appendix 6: JUNCAP2 Express

In JUNCAP Express, the full JUNCAP2 equations are mapped onto a single, much simpler equation $I_j = I_{for1} + I_{for2} + I_{rev}$, where I_{for1}, I_{for2}, and I_{rev} model the ideal current, the non-ideal forward current, and the remaining non-ideal reverse current, respectively, and are given by the equations

$$I_{for1} = g(V_{AK}, \mathbf{ISATFOR1}, 1), \tag{10.71}$$

$$I_{for2} = g(V_{AK}, \mathbf{ISATFOR2}, \mathbf{MFOR2}), \tag{10.72}$$

$$I_{rev} = -g(-V_{AK}, \mathbf{ISATREV}, \mathbf{MREV}), \tag{10.73}$$

where the function g is given by

$$g(V, I_0, m) = I_0 \cdot \left[\exp(V \cdot m/\phi_{TD}) - 1.0\right]. \tag{10.74}$$

In the JUNCAP Express pre-processing step, the coefficients **ISATFOR1**, **ISATFOR2**, **MFOR2**, **ISATREV**, and **MREV** are determined in such a way that the original JUNCAP2 IV-curves are reproduced as accurately as possible.

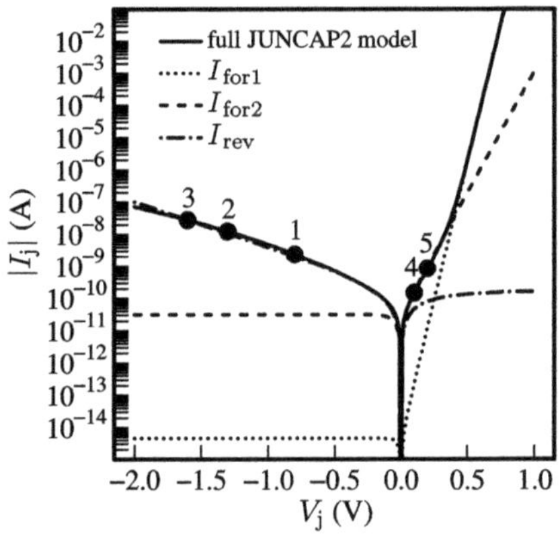

Fig. 10.21 *Solid line*: example of an IV characteristic calculated with the full JUNCAP2 model; points 1, 2, 3, 4, and 5 on this line are used in the JUNCAP2 Express calculations. *Dotted line*: JUNCAP2 Express ideal current contribution. *Dashed line*: JUNCAP2 Express non-ideal forward current contribution, calculated using points 4 and 5. *Dash-dotted line*: JUNCAP2 Express reverse current contribution, calculated using points 1, 2, and 3

Ideal Current The coefficient **ISATFOR1** is easily found: for the three junction components (bottom, STI-edge, and gate edge), the voltage dependence of the ideal current is the same, so that the total ideal current of JUNCAP Express, (10.71), can be mapped directly onto that of the full JUNCAP2 model. This immediately leads to

$$\mathbf{ISATFOR1} = AB \cdot I_{DSAT,bot} + LS \cdot I_{DSAT,sti} + LG \cdot I_{DSAT,gat}, \quad (10.75)$$

where $I_{DSAT,bot}$, $I_{DSAT,sti}$, and $I_{DSAT,gat}$ are the saturation current densities of the three junction components after temperature scaling.

Non-ideal Forward Current Determination of the coefficients **ISATFOR2** and **MFOR2** is less straightforward. Here, first the full-JUNCAP2 currents I_4 and I_5 are determined. They are calculated at low forward voltages $V_4 = 0.1$ V and $V_5 = 0.2$ V, respectively (points '4' and '5' in Fig. 10.21). Next, the ideal contributions to these currents are subtracted to arrive at the non-ideal contributions $I_{4,cor}$ and $I_{5,cor}$:

$$I_{i,cor} = I_i - g(V_i, \mathbf{ISATFOR1}, 1), \quad (10.76)$$

where $i = 4, 5$. Now everything is ready to determine **ISATFOR2** and **MFOR2** by demanding that the I_{for2} curve crosses the points (V_4, I_4) and (V_5, I_5). This leads to a set of equations that cannot be solved analytically. However, using the fact that **MFOR2** will always be close to $1/2$, so that both $\mathbf{MFOR2} \cdot V_4$ and $\mathbf{MFOR2} \cdot V_5$ are well in excess of ϕ_{TD}, it is easy to find that in good approximation:

$$\mathbf{MFOR2} = \frac{\phi_{TD} \cdot \ln(I_{4,cor}/I_{5,cor})}{V_4 - V_5}, \quad (10.77)$$

and

$$\mathbf{ISATFOR2} = \frac{I_{4,cor}}{[\exp(V_4 \cdot \mathbf{MFOR2}/\phi_{TD}) - 1]}. \quad (10.78)$$

Non-ideal Reverse Current The determination of the coefficients **ISATREV** and **MREV** is similar to that of **ISATFOR2** and **MFOR2**, but is somewhat more involved. The reason is that the value of **MREV** can be much smaller than **MFOR2** so that the approximation used in the determination of **MFOR2** is no longer sufficient. First, three voltages are chosen, $V_1 = -0.4 \cdot \mathbf{VJUNREF}$, $V_2 = -0.65 \cdot \mathbf{VJUNREF}$, and $V_3 = -0.8 \cdot \mathbf{VJUNREF}$ in the reverse mode of operation, where **VJUNREF** is an adjustable model parameter that determines the range of validity of the JUNCAP Express approach. The corresponding full JUNCAP2 currents are I_1, I_2, and I_3, respectively (points '1', '2', and '3' in Fig. 10.21). The ideal and non-ideal forward contributions to these currents are subtracted:

$$I_{i,cor} = I_i - g(V_i, \mathbf{ISATFOR1}, 1) - g(V_i, \mathbf{ISATFOR2}, \mathbf{MFOR2}), \quad (10.79)$$

where $i = 1, 2, 3$. Points "1" and "2" are used for the determination of **MREV**. Similarly to (10.77), the value of **MREV** is (to a first-order approximation) given by

$$m_0 = \frac{\phi_{TD} \cdot \ln(\alpha_{rev})}{V_2 - V_1}, \quad (10.80)$$

where $\alpha_{rev} = I_{1,cor}/I_{2,cor}$. Using a Newton-Raphson step one obtains a first-order result:

$$\mathbf{MREV} = m_0 + \phi_{TD} \cdot \frac{(\alpha_{rev} - 1) \cdot (\alpha_{rev}^{\frac{V_2}{V_2 - V_1}} - 1)}{\alpha_{rev} \cdot V_1 - V_2 + (V_2 - V_1) \cdot \alpha_{rev}^{\frac{V_1}{V_1 - V_2}}}. \quad (10.81)$$

Having fixed **MREV**, finally the reverse-current prefactor **IDSATREV** can be determined by demanding JUNCAP Express to cross the point '3', which leads to:

$$\mathbf{ISATREV} = \frac{-I_{3,cor}}{[\exp(V_3 \cdot \mathbf{MREV}/\phi_{TD}) - 1]}. \quad (10.82)$$

References

1. Hurkx, G.A.M., de Graaff, H.C., Kloosterman, W.J., Knuvers, M.P.G.: A new analytical diode model including tunneling and avalanche breakdown. IEEE Trans. Electron Devices **39**(9), 2090–2098 (1992); see also Ref. [4]
2. Solomon, P.M., Frank, D.J., Jopling, J., D'Emic, C., Dokumaci, O., Ronsheim, P., Haensch, W.E.: Tunnel current measurements on p/n junction diodes and implications for future device design. In: IEDM Tech. Dig., pp. 233–236 (2003)
3. Montree, A.H., van Brandenburg, A.C.M.C., Klaassen, D.B.M., Peset Llopis, R., Ponomarev, Y.V., Roes, R.F.M., Scholten, A.J., van Veen, R.S.: Limitations to adaptive back bias approach for standby power reduction in deep sub-micron CMOS. In: Proc. Eur. Solid-State Device Research Conf., pp. 580–583 (1999)
4. www.nxp.com/models
5. www-device.eecs.berkeley.edu
6. Scholten, A.J., Smit, G.D.J., Durand, M., van Langevelde, R., Dachs, C.J.J., Klaassen, D.B.M.: A new compact model for junctions in advanced CMOS technologies. In: IEDM Tech. Dig., pp. 209–212 (2005)

7. Scholten, A.J., Smit, G.D.J., Durand, M., van Langevelde, R., Klaassen, D.B.M.: The physical background of JUNCAP2. IEEE Trans. Electron Devices **53**(9), 2098–2107 (2006)
8. Scholten, A.J., Smit, G.D.J., De Vries, B.A., Tiemeijer, L.F., Croon, J.A., Klaassen, D.B.M., van Langevelde, R., Li, X., Wu, W., Gildenblat, G.: The new CMC standard compact MOS model PSP: advantages for RF applications. IEEE J. Solid-State Circuits **44**(5), 1415–1424 (2009)
9. pspmodel.asu.edu
10. Gildenblat, G., Li, X., Wu, W., Wang, H., Jha, A., van Langevelde, R., Smit, G.D.J., Scholten, A.J., Klaassen, D.B.M.: PSP: An advanced surface-potential-based MOSFET model for circuit simulation. IEEE Trans. Electron Devices **53**(9), 1979–1993 (2006)
11. Wu, W., Li, X., Gildenblat, G., Workman, G.O., Veeraraghavan, S., McAndrew, C.C., van Langevelde, R., Smit, G.D.J., Scholten, A.J., Klaassen, D.B.M., Watts, J.: PSP-SOI: An advanced surface potential based compact model of partially depleted SOI MOSFETs for circuit simulations. Solid-State Electron. **53**, 18–29 (2009)
12. Hurkx, G.A.M., Klaassen, D.B.M., Knuvers, M.P.G.: A new recombination model for device simulation including tunneling. IEEE Trans. Electron Devices **39**(2), 331–338 (1992)
13. Calzolari, P.U., Graffi, S.: A theoretical investigation on the generation current in silicon *p-n* junctions under reverse bias. Solid-State Electron. **15**, 1003–1011 (1972)
14. Sze, S.M.: Physics of Semiconductor Devices. Wiley, New York (1981)
15. Hurkx, G.A.M.: On the modelling of tunnelling currents in reverse-biased p-n junctions. Solid-State Electron. **32**(8), 665–668 (1989)
16. Lindmayer, J., Wrigley, C.Y.: Fundamentals of Semiconductor Devices. van Nostrand, Princeton (1965)
17. Jin, W., Chan, C.H., Fung, S.K.H., Ko, P.K.: Shot-noise-induced excess low-frequency noise in floating-body partially depleted SOI MOSFET's. IEEE Trans. Electron Devices **46**(6), 1180–1185 (1999)
18. Eldo: Mentor Graphics. http://www.mentor.com
19. Abramovitz, M., Stegun, I.A.: Handbook of Mathematical Functions, p. 299. Dover, New York (1970)
20. www.geia.org/CMC—Council

Chapter 11
Surface-Potential-Based MOS Varactor Model

**Zeqin Zhu, Gennady Gildenblat, James Victory,
and Colin C. McAndrew**

Abstract A surface-potential-based scalable model for MOS varactors was developed jointly by Arizona State University, Sentinel IC Technologies, Jazz Semiconductor, and Freescale Semiconductor to facilitate RF CMOS design. We give details of the model, which is based on PSP, and show how it fits key device characteristics, including capacitance, gate current, and quality factor as functions of voltage, frequency, and geometry, for several technologies. Recent advances in the parameter extraction procedure are also reviewed. The model is implemented in Verilog-A and provides a robust and accurate description of MOS varactors, including their RF performance. A VCO design application is presented to illustrate the capabilities of the new model.

11.1 Introduction

Geometry scaling and low power consumption enable the continued advance of analog and RF CMOS technologies, in which MOS varactors are now standard components. These devices are embedded in integrated voltage-controlled oscillators

Z. Zhu (✉) · G. Gildenblat
Department of Electrical Engineering, Ira A. Fulton School, Arizona State University, 650 E. Tyler Mall, Tempe, AZ 85281, USA
e-mail: zeqin.zhu@asu.edu

G. Gildenblat
e-mail: gildenblat@asu.edu

J. Victory
Sentinel IC Technologies, 401 Glenneyre Street, #D, Laguna Beach, CA 92651, USA
e-mail: james@sentinel-ic.com

C.C. McAndrew
Freescale Semiconductor, 2100 E. Elliot Road, Tempe, AZ 85284, USA
e-mail: Colin.McAndrew@freescale.com

G. Gildenblat (ed.), *Compact Modeling*,
DOI 10.1007/978-90-481-8614-3_11,

(VCOs) [1, 5, 12, 24, 34, 53, 60] as part of phase-locked loops and frequency synthesizers [48], low noise and parametric amplifiers [13, 15, 38, 46], and filters [9]. Novel designs of both bulk [28, 30, 33, 62] and SOI [10, 16, 18, 63] MOS varactors have been reported to provide better performance in terms of die area, tuning range, quality factor Q, and phase noise based on various test structures and de-embedding techniques [43]. The continued evolution of MOS varactors, and their expanding use in integrated circuit (IC) design, have led to increasing interest in, and expectations of, compact models of MOS varactors. Prior compact MOS varactor models were either threshold-voltage-based models [11, 29] or behavioral models [8, 36, 40, 45, 47, 50] which were validated over a relatively narrow range of geometry and frequency [2, 11, 29, 50]. The new model, called MOSVAR, was developed at the request of leading semiconductor companies and is the first completely physical model of the MOS varactor based on the surface-potential approach [59]. It includes all relevant physics and provides an accurate description both of the behavior of the ideal, intrinsic varactor and of the parasitic, extrinsic elements that are unavoidable in a real MOS varactor structure. The Verilog-A model and user's manual are available at the PSP model webpage [37].

The recent change from a threshold-voltage-based to a surface-potential-based approach to compact modeling of both bulk and SOI MOS devices not only has provided innovative solutions for several long-standing compact modeling problems but also has facilitated physical correlation and natural consistency between MOS device models integrated into a process design kit (PDK). With a high physical content inherited from the PSP model [19, 37], MOSVAR enables a scalable and robust description of key MOS varactor characteristics.

11.2 Device Technology

The MOSVAR model is designed to be unified with the industry standard PSP MOSFET model as much as possible. Nevertheless, it includes a few unique features such as a dynamic inversion model, which captures the time and frequency dependence of the inversion layer inertia [20, 57], a bias-dependent parasitic resistance model, which is critical for accurate modeling of Q, and a gate current model, which plays an important role in modeling of both Q and device reliability [41]. Figure 11.1 shows a three-dimensional view of a typical n^+-poly/n-well MOS varactor with its equivalent network embedded in the physical layout [58]. Apart from the ideal charge model, MOSVAR includes many secondary effects such as parasitic capacitances, finite polysilicon doping, quantum mechanical (QM) effects, gate current, and the parasitic resistances associated with the gate and well regions.

MOS varactors can be built with varying combinations of gate and well doping polarity. The resulting variation in capacitance-voltage behavior opens up different circuit design architectures, particularly for VCOs. The most common structure is n^+-poly gate over n-well, which is included in most PDKs for modern day RF CMOS and BiCMOS technologies. With the availability of deep n-well in some RF CMOS technologies a p^+-poly gate over isolated p-well sitting in deep n-well

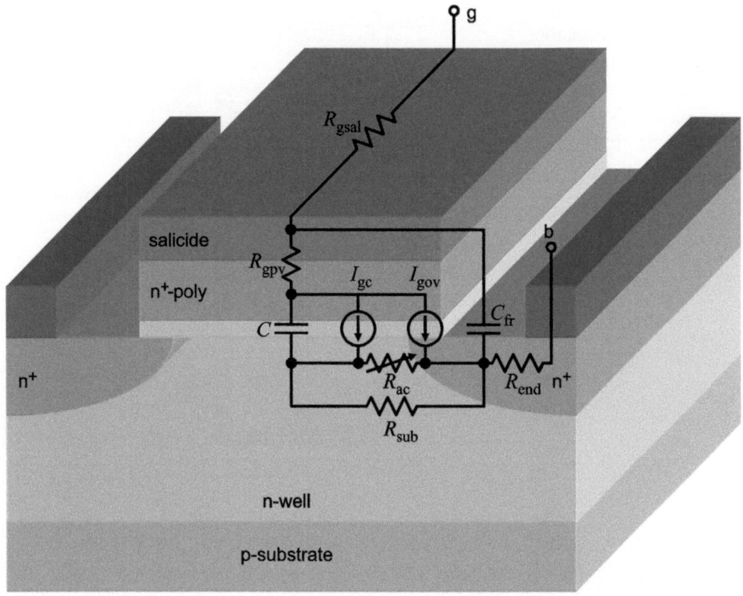

Fig. 11.1 Vertical cross section of an n$^+$-poly/n-well MOS varactor

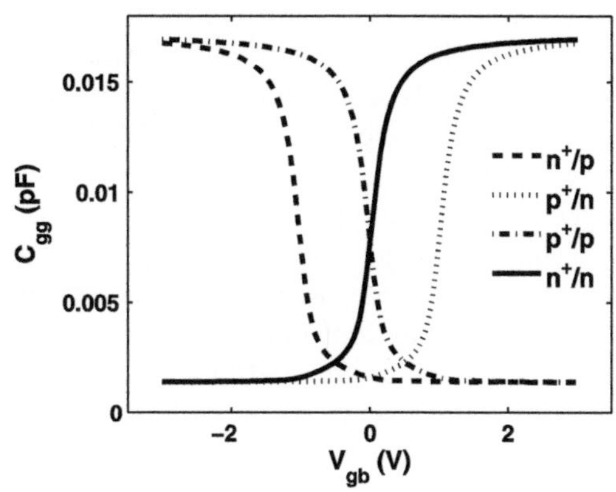

Fig. 11.2 High-frequency C-V curves for the four possible polarity combinations of gate and well layers in MOS varactors

structure can be used to form a varactor with the substrate isolation necessary for RF applications. MOSVAR includes independent treatment of the polarities of the varactor gate and well, which enables simulation of all possible MOS varactor polarity combinations, see Fig. 11.2.

The MOSVAR model includes intrinsic charge and parasitic elements for a unit cell, where only length L and width W of the gate are scaled.[1] In practice MOS varactors consist of many interconnected unit cells built in a common well. As such,

[1]We denote the effective and drawn dimensions as L, W and L_g, W_g, respectively.

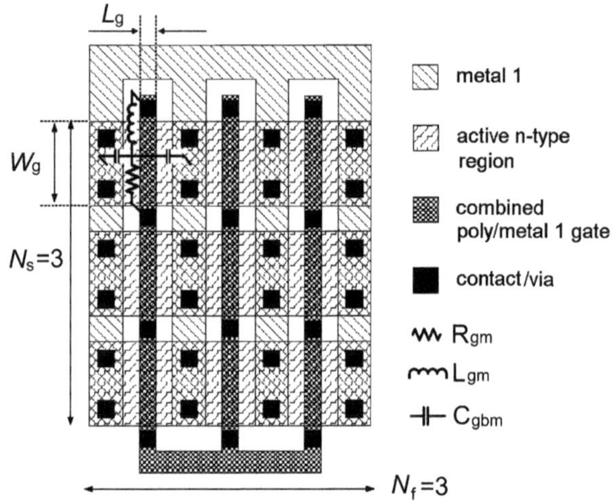

Fig. 11.3 An example of an n$^+$-poly/n-well MOS varactor layout. Parasitic resistance, capacitance, and inductance associated with the metal routing of the cells are shown for the top left unit cell

to model a MOS varactor accurately the well-to-substrate junction parasitic diode and the resistance, capacitance, and inductance associated with the metal routing of the cells need to be added. These parasitic elements can significantly influence the performance of a varactor, including the tuning range, Q, and high frequency behavior. A standard approach is to build a subcircuit or macro model which contains the MOSVAR model at its core, which scales with the number of segments in series (N_s) and in parallel (N_f), along with parasitic elements for the well and metal parasitics, that accurately scale with the specific varactor layout used, by defining these parasitics as appropriate functions of L, W, N_s, and N_f for example.

Figure 11.3 shows a typical device layout where nine individual gate segments are arrayed to form the complete varactor element. The active (well) region is broken between series segments to enable metal 1 contact to the poly gate. This reduces the deleterious effect of the poly resistance by eliminating long and narrow poly gate segments and by providing parallel resistance paths. In the layout presented, parallel metal 1 routes are used to contact the n-well and the gate. Metal 2 (not shown in Fig. 11.3) can be placed orthogonally to the metal 1 to reduce the metal resistance, at the expense of adding metal 1 to metal 2 capacitance, which degrades the tuning range. Parasitics R_{gm}, L_{gm}, and C_{gbm} associated with metal routing of the individual cell are not included in the MOSVAR model. A common practice is to include them in the encapsulating subcircuits.

11.3 Intrinsic Device Model

Since MOS varactors are usually biased from accumulation through weak inversion neither of the common regional analyses of the MOS system, for operation in depletion or in strong inversion, are sufficiently accurate to reproduce experimental data. Hence the surface potential based formulation, which is applicable to all regions of operation, is natural for the development of compact MOS varactor models. The classical analysis that underlies the MOS C-V curve can be found in [32] and forms the basis of the MOSVAR model. An even more physical formulation involves simultaneous solution of the coupled Poisson and Schrödinger equations. However, this approach is too expensive computationally to be incorporated into a compact model. Hence we adopt the classical approach and include QM effects through properly parameterized corrections.

There are two charges in a MOS varactor, the charge on the gate Q_g and the charge in the silicon $Q_s = -Q_g$. The silicon charge comprises two components, the inversion layer charge Q_i plus the well charge Q_b, which includes both accumulation charge and the charge from uncovered dopant atoms in the substrate, i.e. depletion charge. For the purpose of modeling it is convenient to normalize the charges with respect to WLC_{ox}, where C_{ox} is the oxide capacitance per unit area, so, for example, $q_g = Q_g/(WLC_{ox})$. Because there is no direct source of inversion charge (i.e. minority carriers) in a MOS varactor this charge increases and decreases through the processes of generation and recombination. This produces a difference between the actual value of the terminal charge, and their equilibrium values denoted as q_{g0} and $q_{s0} = -q_{g0}$.

Standard semiconductor theory gives the normalized, quasi-equilibrium silicon charge as [17]

$$q_{s0} = -\operatorname{sgn}(\psi_{s0}) \cdot \gamma_t \cdot \left\{ \exp(-\beta\psi_{s0}) + \beta\psi_{s0} - 1 + \exp(-2\beta\phi_F)\left[\exp(\beta\psi_{s0}) - \beta\psi_{s0} - 1\right]\right\}^{1/2} \tag{11.1}$$

where ψ_{s0} is the surface potential (assumed constant along the silicon surface), $\beta = q/k_B T$ and the normalized body factor $\gamma_t = \sqrt{2qp_b\varepsilon_s/\beta}/C_{ox}$ where p_b is the well concentration of the majority carriers (holes), ε_s denotes the permittivity of silicon, and ϕ_F is the so-called "Fermi potential" [35, 52]. Charge balance between the gate and silicon charges gives

$$V_{gb} - V_{fb} - \psi_{s0} = -q_{s0} \tag{11.2}$$

where V_{gb} and V_{fb} are the intrinsic gate-to-well voltage and the flatband voltage, respectively; this is an implicit equation that is solved for ψ_{s0} using a highly accurate analytical approximation [19, 21].

The normalized quasi-equilibrium capacitance is given by $C_{qe} = dq_{g0}/dV_{gb}$ or, equivalently,

$$C_{qe} = \frac{C_{sqe}}{C_{sqe} + 1} \tag{11.3}$$

where $C_{sqe} = -dq_{s0}/d\psi_{s0}$. As is common in modern compact models, in order to assure charge conservation the charge q_{s0} is implemented in the model code while the capacitance C_0 is calculated by the simulator based on the derivative, which is automatically computed from the Verilog-A definition of the model.

11.4 Inversion Layer Inertia

The quasi-equilibrium approximation is adequate for varactor operation near flat-band. However, if the applied dc bias and/or ac signal cause a device to enter the inversion region of operation, the inertia in the formation and collapse of the inversion layer must be included in a compact varactor model to capture properly the frequency dependence of the behavior of the device.

11.4.1 Relaxation Time Approximation

A simple approach to model the inversion layer inertia is based on the relaxation time approximation (RTA) in which the inversion charge q_i satisfies the approximation [57]

$$\frac{dq_i}{dt} = -\frac{q_i - q_{i0}}{\tau} \tag{11.4}$$

where τ is the relaxation time (model parameter **TAU**[2]) and q_{i0} denotes the quasi-equilibrium value of q_i. These charges are related to the quasi-equilibrium surface potential ψ_{s0} through the normalized bulk charge:

$$q_{b0} = -\operatorname{sgn}(\psi_{s0}) \cdot \gamma_t \cdot \left[\exp(-\beta\psi_{s0}) + \beta\psi_{s0} - 1\right]^{1/2} \tag{11.5}$$

and

$$q_{i0} = q_{s0} - q_{b0}. \tag{11.6}$$

Once ψ_{s0} is determined from (11.2), q_{i0} is evaluated from (11.5) and (11.6). Hence q_{i0} can be regarded as a known function of the instantaneous terminal voltages. Consequently, the RTA (11.4) can be solved with respect to q_i, which is done directly by the circuit simulator. The next step is to determine the actual (rather than quasi-equilibrium) surface potential ψ_s. For this purpose consider the well charge

$$q_b = -\operatorname{sgn}(\psi_s) \cdot \gamma_t \cdot \left[\exp(-\beta\psi_s) + \beta\psi_s - 1\right]^{1/2}. \tag{11.7}$$

Combining (11.7) with Gauss's law (E_{ox} denotes the oxide field)

$$\varepsilon_{ox} E_{ox} = -C_{ox}(q_i + q_b) \tag{11.8}$$

or, equivalently

$$V_{gb} - V_{fb} - \psi_s = -q_i - q_b \tag{11.9}$$

yields ψ_s as an implicit function of $V_{gbi} = V_{gb} + q_i$

$$V_{gbi} - V_{fb} - \psi_s = \operatorname{sgn}(\psi_s) \cdot \gamma_t \cdot \left[\exp(-\beta\psi_s) + \beta\psi_s - 1\right]^{1/2}. \tag{11.10}$$

[2]MOSVAR model parameters are denoted by capital bold font.

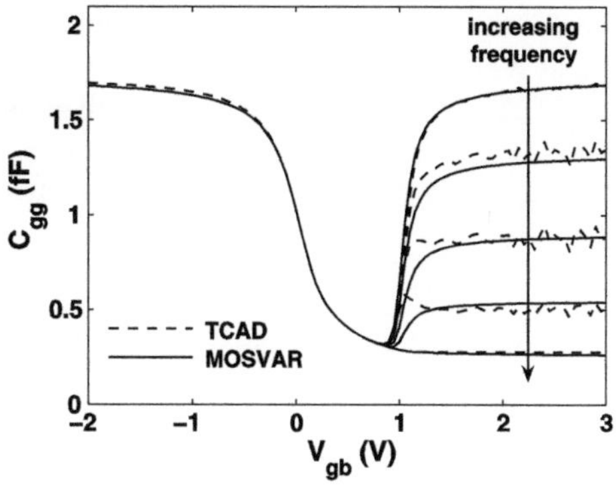

Fig. 11.4 Comparison of simulation results from MOSVAR and TCAD. The parameter **TAU** = 0.27 s and the frequency range is 0.01–100 Hz (after [57])

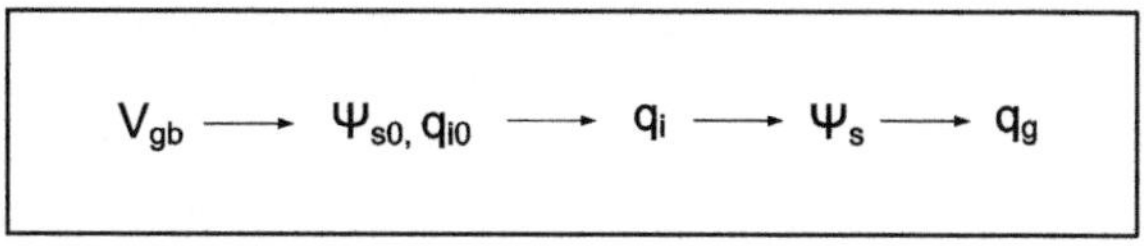

Fig. 11.5 RTA solution sequence

Note that the inertia of the ψ_s response to V_{gb} enters (11.10) through the q_i term in V_{gbi}. This equation has already been encountered in a different physical context in the compact modeling of MOS transistors. Indeed, apart from the change of V_{gb} into a V_{gbi}, this is a standard form of the surface potential in which the minority carrier contribution has been omitted. The surface potential equation is solved to obtain the surface potential in the source and drain overlap regions where the semiconductor surface is never inverted and the minority carrier concentration is negligible.

This enables the use of another accurate analytical approximation for $\psi_s(V_{gbi})$ [58]. With ψ_s determined the gate charge is evaluated as

$$q_g = V_{gb} - V_{fb} - \psi_s. \tag{11.11}$$

With q_g determined one can evaluate the non-equilibrium capacitance C_{gg}, or $C = C_{gg}/(WLC_{ox})$. The results of Fig. 11.4 illustrate both the inertia in the formulation of the inversion layer and a good agreement with TCAD simulations. The remaining difference is most likely associated with the assumption of the bias-independent relaxation time.

The solution procedure discussed in this section is summarized in Fig. 11.5

11.4.2 Analytical Solution for the Small-Signal Case

For the important special case of small-signal excitation, the equations of the RTA allow analytical solution, which provides additional insight into the underlying physics [20].

Let the voltage V_{gb} applied to the gate of a MOS varactor be given by

$$V_{gb} = V_Q + \mathrm{Re}\left(\Delta V_{gb} e^{j\omega t}\right) \tag{11.12}$$

where V_Q is the dc bias and ΔV_{gb} is the complex amplitude of the small harmonic component with angular frequency ω. It is convenient to set $q_i = q_{iQ} + \Delta q_i$ and $q_{i0} = q_{iQ} + \Delta q_{i0}$ where q_{iQ} is the quiescent point value of q_i corresponding to a dc bias of $V_{gb} = V_Q$; Δq_i and Δq_{i0} are the complex amplitudes of the harmonic components of q_i and q_{i0}, respectively . It follows from (11.4) that (in phasor notation)

$$\Delta q_i = \frac{\Delta q_{i0}}{1 + j\omega\tau} \tag{11.13}$$

while from the quasi-equilibrium nature of q_{i0}

$$\Delta q_{i0} = \frac{\partial q_i}{\partial V_{gb}} \Delta V_{gb} \tag{11.14}$$

where, here and below, all derivatives are evaluated at the quiescent point.

Apart from the relaxation time approximation (11.4), the essential physics of the method developed in [57, 58] is that inertia of the inversion layer formation is transferred to the response of the surface potential ψ_s through (11.10).

The well charge corresponding to the surface potential determined from (11.10) is given by (11.7) [57, 58]. From (11.7) and (11.10), one concludes that $q_b = q_b(\psi_s)$ while $\psi_s = \psi_s(V_{gb}, q_i)$. Hence

$$\Delta q_b = -C_{hf} \Delta V_{gb} + B \Delta q_i \tag{11.15}$$

where

$$C_{hf} = -\frac{dq_b}{d\psi_s} \frac{\partial \psi_s}{\partial V_{gb}} \tag{11.16}$$

and

$$B = \frac{dq_b}{d\psi_s} \frac{\partial \psi_s}{\partial q_i}. \tag{11.17}$$

Combining (11.14) and (11.15) one finds that the gate charge phasor

$$\Delta q_g = -\Delta q_i - \Delta q_b \tag{11.18}$$

is given by $\Delta q_g = \tilde{C} \Delta V_{gb}$ where

$$\tilde{C} = C_{hf} - \frac{1 + B}{1 + j\omega\tau} \frac{\partial q_i}{\partial V_{gb}}. \tag{11.19}$$

The normalized capacitance $C = \mathrm{Re}\,\tilde{C}$ is

$$C = C_{hf} - \frac{1 + B}{1 + (\omega\tau)^2} \frac{\partial q_i}{\partial V_{gb}}. \tag{11.20}$$

In particular, the quasi-equilibrium capacitance $C_{qe} = \lim_{\omega \to 0} C$ is

$$C_{qe} = C_{hf} - (1 + B) \frac{\partial q_i}{\partial V_{gb}} \tag{11.21}$$

so that from (11.20) one finds

$$C = C_{hf} + \frac{C_{qe} - C_{hf}}{1 + (\omega\tau)^2} \tag{11.22}$$

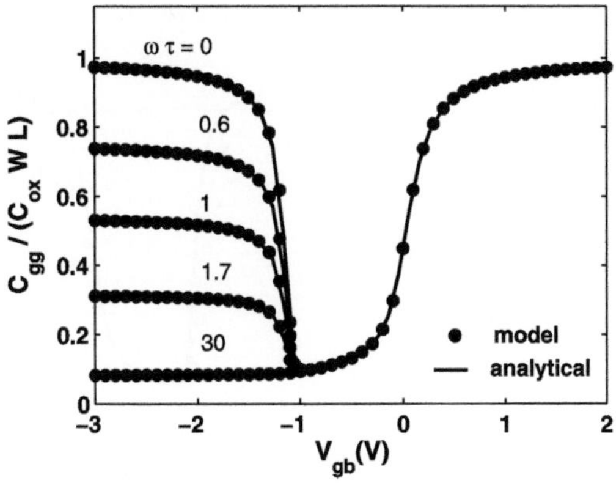

Fig. 11.6 Comparison of the analytical solution (11.22) with circuit simulator results for an n-well varactor

where, since q_i and q_b are both normalized to WLC_{ox}, so are C, C_{hf} and C_{qe}.

Hence $C_{hf} = C|_{\omega\tau\gg 1}$ can be interpreted as the "high-frequency capacitance." The expression for C_{hf} follows from (11.7) and (11.10) as

$$C_{hf} = \frac{1 - \exp(-\beta\psi_{s0})}{1 - \exp(-\beta\psi_{s0}) - [2q_{b0}/(\beta\gamma_t^2)]}. \tag{11.23}$$

The expression for C_{qe} is given by (11.3), or, more explicitly by [32, 42]

$$C_{sqe} = \frac{\beta\cdot\gamma_t\cdot \mathrm{sgn}(\psi_s)\{1 - \exp(-\beta\psi_{s0}) + \exp(-2\beta\phi_F)[\exp(\beta\psi_{s0}) - 1]\}}{2\{\exp(-\beta\psi_{s0}) + \beta\psi_{s0} - 1 + \exp(-2\beta\phi_F)[\exp(\beta\psi_{s0}) - \beta\psi_{s0} - 1]\}^{1/2}}. \tag{11.24}$$

Figure 11.6 compares the analytic solution (11.22) with the results of SPICE simulation, where the substrate doping is $2\cdot 10^{23}$ m^{-3} and the oxide thickness is 2 nm.

The curve for $\omega\tau = 30$ is very nearly $C_{hf}(V_{gb})$. As expected, a perfect agreement is achieved. Physically, the frequency dependence of MOS capacitance manifests itself after the development of the inversion layer since the response of the majority carriers is practically instantaneous. This is captured explicitly by (11.22) since in accumulation and depletion operation $C_{qe} \simeq C_{hf} \simeq C$ do not depend on $\omega\tau$. We note in passing that while the expressions for C_{hf} and C_{qe} assume uniform doping, no such assumption is made in deriving (11.22). Similarly, while comparison of (11.22) with the results of numerical simulation is made for the case of constant τ, the small-signal expression (11.22) remains valid for bias dependent τ. On the other hand, the inertia of the inversion layer formation is only one of the factors responsible for the frequency dependence of the varactor capacitance [58].

11.5 The Effects of Finite Polysilicon Doping and Quantum Mechanical Corrections

A more complete description of charge in the polysilicon region is one area where MOSVAR differs significantly from the PSP model. In MOSVAR, the polysilicon

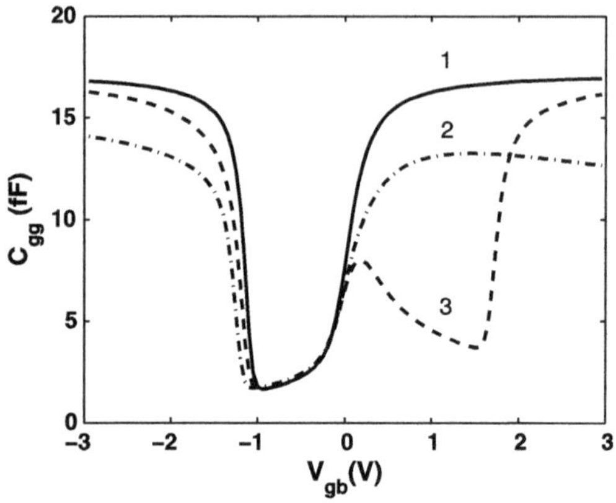

Fig. 11.7 Further illustration of the capabilities of MOSVAR for an n^+-poly over n-well structure; Curve "*1*": the ideal MOS capacitance; Curve "*2*": with poly depletion and QM corrections; Curve "*3*": the somewhat unrealistic case of an inversion layer forming in the polysilicon. **TOXO** = 2 nm, **NSUBO** = $3 \cdot 10^{23}$ m^{-3}, **NPO** = $1 \cdot 10^{27}$, $2 \cdot 10^{26}$ and $2 \cdot 10^{24}$ m^{-3} for curves *1*, *2* and *3* respectively. These are low frequency simulations, to allow the inversion charge to respond and highlight poly doping and QM effects in inversion

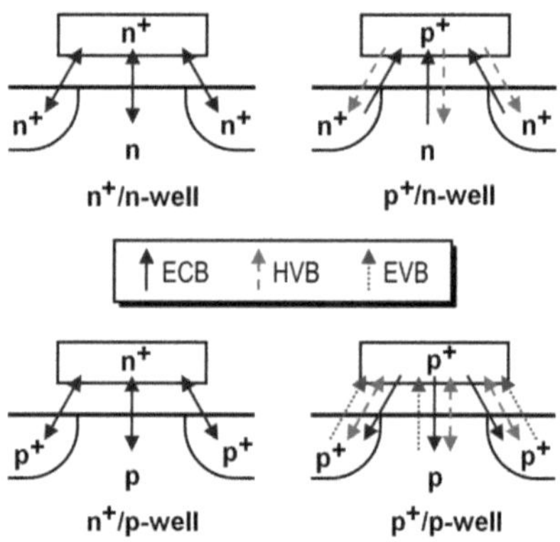

Fig. 11.8 Four possible MOS varactor configurations and the dominant gate current components

gate is allowed to enter all regions of operation (accumulation, depletion, and inversion, cf. Fig. 11.7), whereas in PSP only polysilicon depletion is taken into account. Another significant difference compared to PSP is that instead of the two polarity combinations of the gate and active regions encountered in bulk MOSFETs, MOSVAR allows four such combinations, see Fig. 11.8. The polarity of the gate doping is controlled by the model parameter **TYPEP** and that of the active region by the parameter **TYPE**. For example, the combination of **TYPEP** = 1 with **TYPE** = −1 describes p-type gate with an n-type active region.

In the most general case, MOSVAR solves two coupled surface potential equations in the active region and in the polysilicon gate. For example, in the quasi-

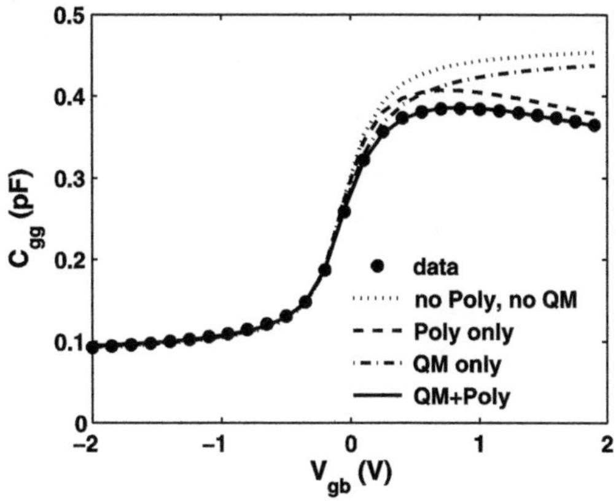

Fig. 11.9 QM correction and polysilicon effects in silicon data (1 GHz, de-embedded) for an n^+-gate/n-well device in a 130 nm process from Jazz Semiconductor. **TOXO** $= 2.21$ nm, **NSUBO** $= 4.6 \cdot 10^{23}\ m^{-3}$, **NPO** $= 1 \cdot 10^{26}\ m^{-3}$, and **QMC** $= 0.4$

equilibrium case the surface potentials ψ_{s0} and ψ_{p0} (polysilicon) are determined from

$$V_{gb} - V_{fb} - \psi_{s0} - \psi_{p0} = -q_{s0}\left(\psi_{s0}\right) \tag{11.25}$$

and

$$q_{s0} + q_{p0} = 0 \tag{11.26}$$

where q_{s0} is given by (11.1) and q_{p0} by

$$\begin{aligned} q_{p0} = &- \operatorname{sgn}(\psi_{p0}) \cdot \gamma_{pt} \cdot \big\{\exp(-\beta\psi_{p0}) + \beta\psi_{p0} - 1 \\ &+ \exp(-2\beta\phi_p)\big[\exp(\beta\psi_{p0}) - \beta\psi_{p0} - 1\big]\big\}^{1/2}. \end{aligned} \tag{11.27}$$

Here $\gamma_{pt} = \sqrt{2qN_p\varepsilon_s/\beta}/C_{ox}$ where N_p and ϕ_p are the doping concentration and "Fermi potential" of the polysilicon gate, respectively. These equations are written for a varactor with an n^+-poly gate over p-well.

Figure 11.7 shows typical results for relatively lightly doped polysilicon gates, to illustrate the model capabilities. Comparison with experimental data (cf. Fig. 11.9) highlights the essential role of the polysilicon depletion effect; in particular, for accumulation in the silicon region polysilicon depletion significantly reduces the device capacitance. Since in real applications a varactor's quiescent operating point is often selected near the boundary between depletion and accumulation ($V_{gb} = V_{fb}$), the effect of polysilicon depletion must be taken into account to accurately model the sensitivity dC/dV of the device.

One simplification made in MOSVAR concerning the polysilicon finite doping effect is that even when the frequency dependent inversion layer formation is operational in the varactor model for the active region, the polysilicon surface potential is still computed using the quasi-static approximation.

The reduction of the varactor capacitance associated with the formation of a polysilicon space charge region is automatically included in MOSVAR if the polysilicon doping level **NPO** $< 10^{27}\ m^{-3}$. For higher doping levels the effect is considered negligible and ψ_{p0} is set to zero.

The QM correction in MOSVAR is conceptually the same as that in MM11 [56] and PSP. Since first principles modeling of quantum effects, based on simultaneous solution of the coupled Poisson and Schrödinger equations, is at present impractical for the purpose of circuit simulation, all major quantum effects (inversion charge centroid shift [55, 56], surface quantization [55, 56], and surface degeneracy [4, 22]) are introduced through a bias dependent oxide thickness and one adjustable parameter—**QMC**. In this way, the oxide capacitance including quantum correction becomes

$$C_{ox,qm} = \begin{cases} C_{ox}, & q_q = 0 \\ \frac{C_{ox}}{1+q_q(q_{eff}^2+q_{lim2})^{-1/6}}, & q_q > 0 \end{cases} \tag{11.28}$$

where

$$q_{eff} = |q_b| + 0.5|q_i|, \tag{11.29}$$

$$q_q = 0.4 \cdot Q_M \cdot C_{ox}^{2/3} \cdot \mathbf{QMC}. \tag{11.30}$$

Q_M = 5.951993 and 7.448711 $\mathrm{F}^{-2/3}\,\mathrm{m}^{4/3}\,\mathrm{V}^{1/3}$ for p-well and n-well varactors, respectively [31]. The small constant $q_{lim2} = 100\beta^{-2}$ is included to avoid a singularity at $V_{gb} = V_{fb}$ when $q_{eff} = 0$.

The above functional dependence on q_{eff} follows from the expression [14, 51]

$$\Delta E = \frac{3}{2} \cdot \left(\frac{3 \cdot q \cdot \hbar}{2 \cdot \sqrt{m^*} \cdot \varepsilon_s} \right)^{2/3} \cdot \frac{|q_b| + (55/96)|q_i|}{[|q_b| + (11/32)|q_i|]^{1/3}} \tag{11.31}$$

for the shift of the first allowed energy level relative to the conduction band edge. References [56] and [3] explain the approximations made in going from (11.31) to (11.28). Typical results showing the relative magnitude of the quantum and polysilicon depletion effects for a realistic case are shown in Fig. 11.9 for an n^+-gate/n-well structure. Unlike the polysilicon depletion effect, QM correction does not introduce any non-monotonicity in the $C_{gg}(V_{gb})$ dependence. Note also that the model simulations depicted in Fig. 11.9 include the inertia of the inversion layer formation described in Sect. 11.4.

11.6 Gate Tunneling Current

Engineering models of the tunneling current in MOSFETs have been widely discussed, in particular in [6, 23, 56] and [61]. The tunneling current model in MOSVAR differs from these in several significant aspects. First, there are additional combinations of possible polarities of gate and silicon dopants, leading to different possible tunneling current components, such as electron conduction band (ECB), electron valence band (EVB) and hole valence band (HVB), see Fig. 11.8. A detailed physical analysis is presented in [6]. It is clear that HVB tunneling, which is not significant in bulk MOSFETs, becomes essential for varactors. Second, the surface potential in varactors is position-independent, which simplifies the tunneling current model. Both channel and overlap tunneling current densities are computed using the Tsu-Esaki formula [54]

$$J_g = \frac{q \cdot m_{ox}^* \cdot k_B \cdot T}{2 \cdot \pi^2 \cdot \hbar^3} \int D(E_x) \cdot F_s(E_x) \cdot dE_x \tag{11.32}$$

where $D(E_x)$ is the transmission coefficient, $F_s(E_x)$ is the supply function, and E_x denotes the kinetic energy associated with motion in the direction normal to the potential barrier.

For the purpose of compact modeling, we incorporate the mono-energetic approximation of the integral in (11.32), as developed in [23, 61] and used in PSP [19]. In the mono-energetic approximation $E_x = qV_{ox}$ (V_{ox} denotes the oxide voltage) for $V_{ox} > 0$ and $E_x = 0$ for $V_{ox} < 0$, with a smoothing function used to provide non-singular behavior at $V_{ox} = 0$. The resulting expressions for the different tunneling current components differ primarily in the supply function. For example, in the case of the p$^+$-poly/p-well structure shown in Fig. 11.8, the HVB current density is

$$J_{g,HVB} \approx J_{g0,HVB} \cdot D_{HVB} \cdot F_{s,HVB} \tag{11.33}$$

where $J_{g0,HVB} = 10^{12} \cdot \mathbf{IGCHVLW}$ is a model variable, theoretically equal to $qm_{ox}^*(k_BT)^2/(2\pi^2\hbar^3)$, the parameter **IGCHVLW** is the HVB current pre-factor for 1 μm^2 area, and the transmission coefficient is estimated using the WKB method as

$$D_{HVB} = \exp\left[-B_{HVB}\left(\frac{3}{2} - \mathbf{GC2HVO}z_g - \mathbf{GC3HVO}z_g^2\right)\right]. \tag{11.34}$$

Here,

$$B_{HVB} = \frac{4t_{ox}}{3\hbar}\sqrt{2 \cdot q \cdot m_{ox}^* \cdot \chi_B^{HVB}} \tag{11.35}$$

the normalized oxide voltage

$$z_g = \frac{V_{ox}}{\chi_B^{HVB}} \tag{11.36}$$

and $\chi_B^{HVB} = \mathbf{CHIBPO}$ (a model parameter) is the band offset between the valence bands of Si and SiO_2, as shown in Fig. 11.11 ($\chi_B = \mathbf{CHIBO}$ which represents the band offset between the conduction bands of Si and SiO_2 is also shown but is only used in the ECB model). The model parameters **GC2HVO** and **GC3HVO** compensate for the use of the WKB approximation and for uncertainty in the electron effective mass in the oxide, m_{ox}^*. The surface potential based HVB supply function is

$$F_{s,HVB} = \ln\left[\frac{1 + \Delta_{HVB}\exp(\beta V_{gb})}{1 + \Delta_{HVB}}\right] \tag{11.37}$$

where

$$\Delta_{HVB} = \exp\left[-\beta(\psi_s + E_g/q - \alpha_b + E_x/q)\right]. \tag{11.38}$$

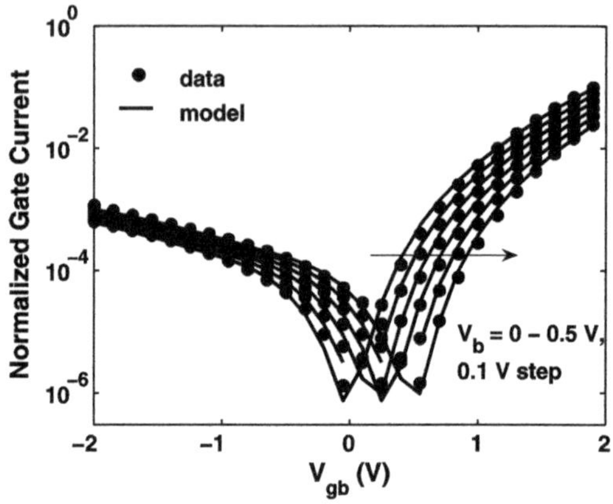

Fig. 11.10 Comparison of measured and simulated I_g for an n$^+$-poly/n-well MOS varactor from Jazz Semiconductor's 130 nm process

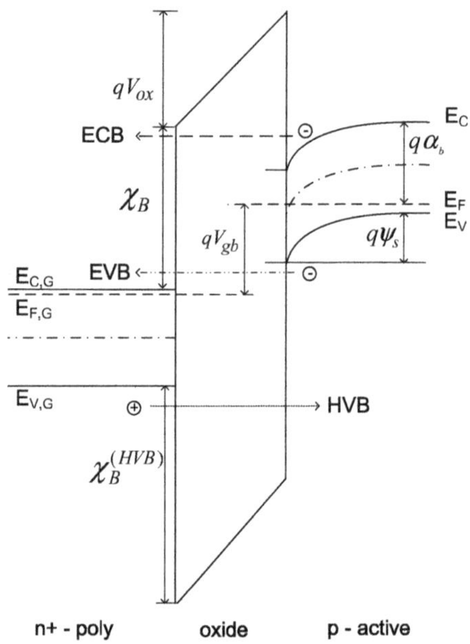

Fig. 11.11 Band diagram

Here E_g is the silicon bandgap, and $\alpha_b = (E_C - E_F)/q$ in the well. The remaining components of the tunneling current are evaluated similarly. Typical results for an n$^+$-poly/n-well structure are shown in Figs. 11.10 , 11.12 and 11.13. Nine devices with different lengths and widths were fit simultaneously using the same parameter set, with no scaling parameters (other than the instance parameters $\mathbf{L} = L_g$ and $\mathbf{W} = W_g$) for the gate current. The contributions from both the active and overlap regions were included in the model (contributions for each relevant mechanism for the type of device were included as detailed in Fig. 11.8).

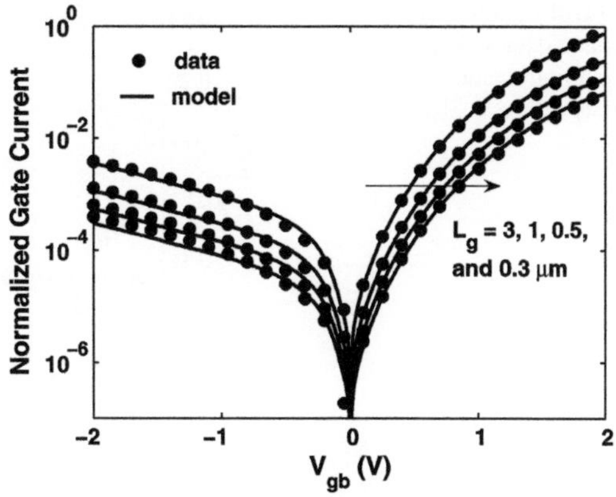

Fig. 11.12 Comparison of measured and simulated I_g for an n^+-poly/n-well MOS varactor from Jazz Semiconductor's 130 nm process

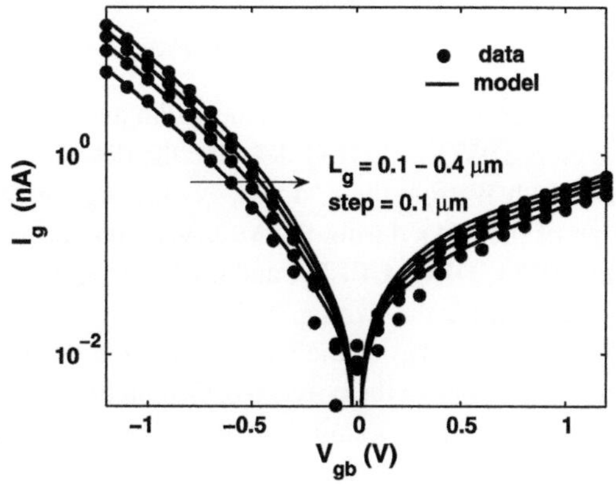

Fig. 11.13 Comparison of measured and simulated I_g for a p^+-poly/p-well MOS varactor from a Fujitsu 65 nm process

11.7 Parasitic Elements

11.7.1 Parasitic Capacitance C_{fr}

The total capacitance in the MOSVAR model is given by[3]

$$C_{gg}(V_{gb}) = \left[C_0(V_{gb}) \cdot L \cdot W + C_{fr}\right] \cdot m \tag{11.39}$$

where C_0 is the bias-dependent capacitance of the intrinsic device per unit area, length

$$L = L_g + \mathbf{DLQ} \tag{11.40}$$

width

$$W = W_g + \mathbf{DWQ} \tag{11.41}$$

[3]As all modern compact models, MOSVAR is charge based. However, it is convenient to discuss parasitic elements and parameter extraction in terms of capacitances.

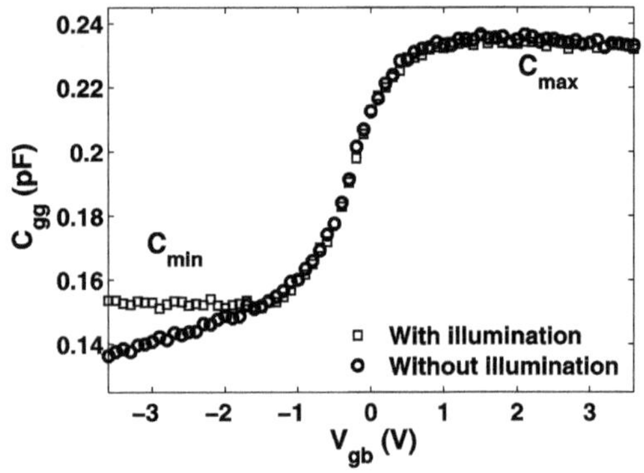

Fig. 11.14 Comparison of measured data with and without illumination for n^+-poly over n-well varactor

and m denotes the multiplicity factor. The parasitic capacitance

$$C_{fr} = 2 \cdot \mathbf{CFRW} \cdot W + 2 \cdot \mathbf{CFRL} \cdot L \tag{11.42}$$

includes components from both the width and length dimensions of the device. The parameters **DWQ** and **DLQ** describe the deviations of the effective channel length L and width W from their "drawn" values L_g and W_g.

One of the tasks during the MOSVAR model parameter extraction process is to obtain **DWQ**, **DLQ**, **CFRW**, and **CFRL** from test data. In the MOSVAR model, a separate term for the overlap region is not included, so **CFRW** encompasses both the poly gate overlap of the well contact region (n^+ in the case of n-well) and the fringe capacitance of the poly edge. Originally, the parameters **DWQ** and **DLQ** were extracted assuming for simplicity that **CFRL** = 0 [58]. The **CFRL** term models capacitance associated with the poly gate extension onto STI and fringe capacitances of the poly edge along the length. As in [58, 64] extraction is based on C_{min} and C_{max} values for the high-frequency capacitance, C_{hf}, obtained from S-parameters with proper de-embedding. To assure identical conditions for measurements and simulations, special attention is paid to the frequency behavior of C in the inversion region, which directly influences the C_{min} value. Figure 11.14 shows typical $C(V_{gb})$ curves with and without illumination. Without illumination, the MOS varactor at least partially enters the deep depletion regime producing erratic values of C_{min} and complicating comparison with simulated results. Under illumination, the time required for the inversion layer formation is reduced and classical high-frequency MOS $C(V_{gb})$ curve is experimentally observed. Thus the sample should be illuminated to produce consistent measurements for different devices.

The extraction technique relies on regression fitting of C_{max} and C_{min} values over geometry. From (11.39) and (11.41), $C_{min}(W_g) = C_{max}(W_g)$ for $W_g = -\mathbf{DWQ}$. Hence, the intercept point of $C_{min}(W_g)$ and $C_{max}(W_g)$ lines for fixed L_g (shown in Fig. 11.15) is $(XINT_{wg}, YINT_{cwg})$ where

$$XINT_{wg} = -\mathbf{DWQ}, \tag{11.43}$$

$$YINT_{cwg} = 2 \cdot \mathbf{CFRL} \cdot (L_g + \mathbf{DLQ}) \cdot m. \tag{11.44}$$

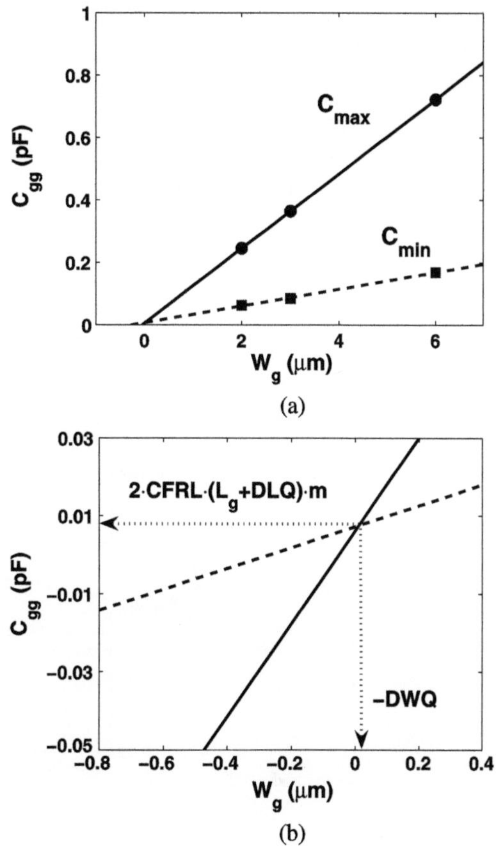

Fig. 11.15 **a** Measured $C_{min}(W_g)$ and $C_{max}(W_g)$ of the Jazz 180 nm RFCMOS technology for $L_g = 0.5$ μm, $m = 20$ and $f = 500$ MHz, and **b** Zoom-in of (**a**) near the intercept point

Similar extrapolation of the $C_{min}(L_g)$ and $C_{max}(L_g)$ for fixed W_g plots shown in Fig. 11.16 yields an intercept point $(XINT_{lg}, YINT_{clg})$, with

$$XINT_{lg} = -\mathbf{DLQ}, \tag{11.45}$$

$$YINT_{clg} = 2 \cdot \mathbf{CFRW} \cdot (W_g + \mathbf{DWQ}) \cdot m. \tag{11.46}$$

The parameters **CFRL** and **CFRW** are computed from (11.44) and (11.46), respectively, based on **DLQ** and **DWQ** from (11.43) and (11.45).

11.7.2 Gate Tunnel Current in the Overlap Region

With **DLQ** and **DWQ** determined, it becomes possible to find the relative magnitude of the gate tunneling currents in the channel, I_{gc}, and overlap region, I_{gov}, contributing to the total gate current

$$I_g = I_{gc} + I_{gov} = i_{gc} \cdot L \cdot W \cdot m + i_{gov} \cdot W \cdot m. \tag{11.47}$$

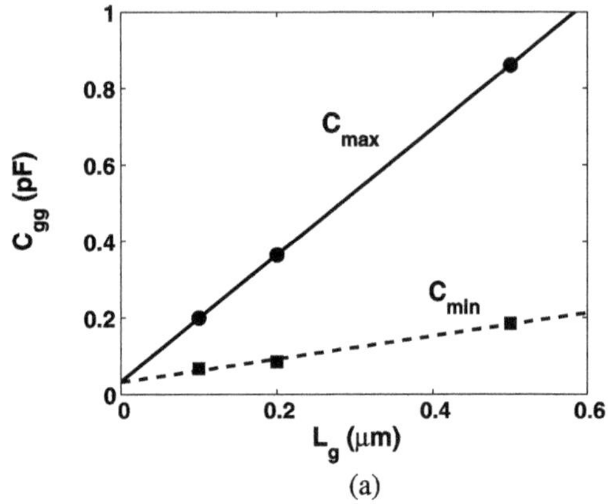

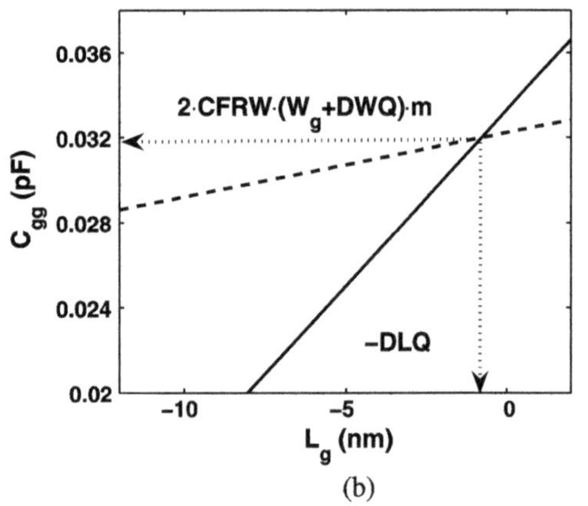

Fig. 11.16 **a** Measured $C_{min}(L_g)$ and $C_{max}(L_g)$ of the Jazz 180 nm RFCMOS technology for $W_g = 3$ μm, $m = 20$ and $f = 500$ MHz, and **b** Zoom-in of (**a**) near the intercept point

This involves extracting the parameters **IGCHVLW** and **IGOVHVW** in expressions [59]

$$i_{gc}(V_{gb}) = \mathbf{IGCHVLW} \cdot D(V_{gb}) \cdot F(V_{gb}) \tag{11.48}$$

and

$$i_{gov}(V_{gb}) = 2 \cdot \mathbf{IGOVHVW} \cdot \mathbf{LOV} \cdot D_{ov}(V_{gb}) \cdot F_{ov}(V_{gb}) \tag{11.49}$$

where $D(V_{gb})$ and $D_{ov}(V_{gb})$ are the tunneling transmission coefficients in the channel and overlap regions, respectively, while $F(V_{gb})$ and $F_{ov}(V_{gb})$ are the supply functions [54] in these regions. The overlap length **LOV** = 10 nm used in this example was obtained from technology information.

To decouple the gate current parameters for the channel region from those for the overlap region, $I_g(V_{gb})$ data from devices of maximum and minimum drawn channel length, L_{gmax} and L_{gmin}, are used. The difference in the tunneling currents is [cf. Fig. 11.17(a)]

$$\Delta I_g = \mathbf{IGCHVLW} \cdot D(V_{gb}) \cdot F(V_{gb}) \cdot (L_{gmax} - L_{gmin}) \cdot W \cdot m \tag{11.50}$$

and does not depend on the gate current in the overlap region [cf. Fig. 11.17(b)].

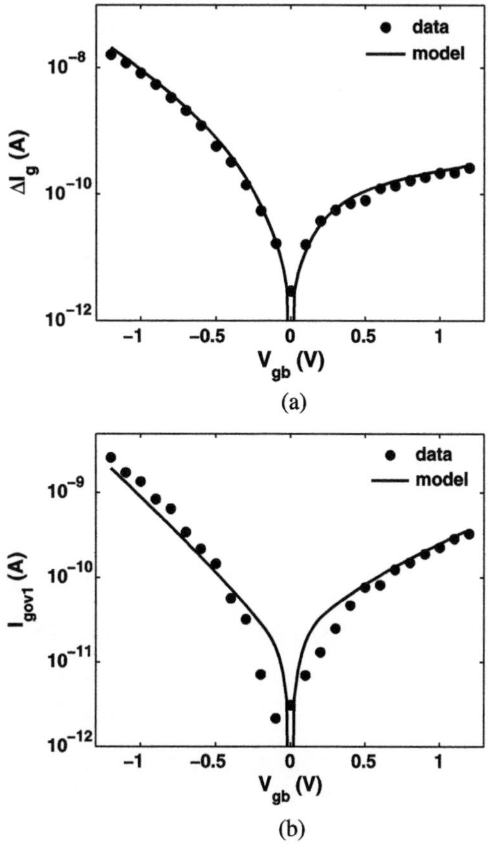

Fig. 11.17 **a** Measured and simulated $\Delta I_g(V_{gb})$ in (11.50) for $L_{gmax} = 0.5$ µm, $L_{gmin} = 0.1$ µm, $W_g = 3$ µm and **b** Measured and simulated $I_{gov1}(V_{gb})$ in (11.49) for $L_{gmin} = 0.1$ µm, $W_g = 3$ µm. Both (**a**) and (**b**) are of Fujitsu 65 nm RFCMOS technology

11.7.3 Parasitic Resistances

The well under the gate oxide is described as a parallel combination of a bias independent well resistance and a bias dependent accumulation resistance. The assumption of bias independence of the well resistance in depletion follows from the high doping of the surface regions in contemporary MOSFET technologies; variation of the thin depletion region under the oxide has a negligible effect on the resistance of the well. Measurements validate this assumption; they show no appreciable well resistance variation with gate bias. The bias-independent well resistance (cf. Fig. 11.1) is

$$R_{sub} = \frac{\mathbf{RSHS} \cdot L_g}{12 \cdot W_g \cdot m} \tag{11.51}$$

where **RSHS** is a model parameter (the sheet resistance of the well under the gate), and the coefficient $1/12$ accounts for the two-sided nature of the contact to the active region [58].

As the gate bias increases, an accumulation charge forms in the active region of an n^+-poly/n-well device and an additional component of the well conductance

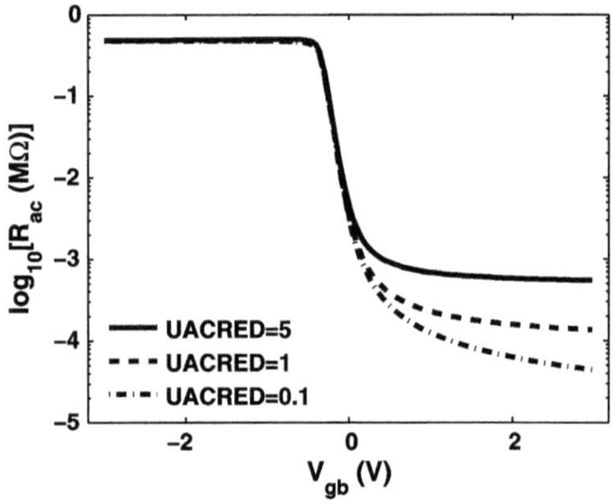

Fig. 11.18 Bias dependence of R_{ac} for an n^+-poly/n-well MOS varactor; **UAC** $= 0.05\ m^2/V/s$, $W_g = L_g = 1\ \mu m$

needs to be included. The resistance of this additional conductance path, R_{ac} (cf. Fig. 11.1), is inversely proportional to the accumulation change Q_{ac}

$$R_{ac} = \frac{L_g}{12 \cdot W_g \cdot \mu_{acv} \cdot Q_{ac} \cdot m} \tag{11.52}$$

where the mobility variation with gate bias is modeled as [58]

$$\mu_{acv} = \frac{\mathbf{UAC}}{1 + \mathbf{UACRED} V'_{gb}}. \tag{11.53}$$

Here **UAC** is the accumulation mobility value for low values of the vertical field, the parameter **UACRED** describes the mobility degradation associated with the V_{gb} increase and the effective gate bias $V_{gb} - \mathbf{VFBO}$ is softly clamped to zero when the device is operating outside of the accumulation region:

$$V'_{gb} = 0.5\left[\mathbf{VFBO} - V_{gb} + \sqrt{(\mathbf{VFBO} - V_{gb})^2 + \varepsilon^2}\right] \tag{11.54}$$

where $\varepsilon = 0.2$.

The accumulation charge in (11.52) is approximated by

$$Q_{ac} = \gamma_t \cdot C_{ox} \cdot \exp(-\beta\psi_{s0}/2). \tag{11.55}$$

Implementation of this equation in the MOSVAR code includes further modifications to avoid overflow problem for large ψ_s. The bias dependence of R_{ac} is illustrated in Fig. 11.18.

A significant portion of the MOS varactor resistance is from the well end-resistance, formed by the source/drain contact regions, the salicided contact diffusion and the contact resistance. This resistance does not change with gate length and is the factor that limits the maximum Q for minimum L_g. The end resistance is given by

$$R_{end} = \frac{\mathbf{REND}}{2 \cdot W_g \cdot m} \tag{11.56}$$

where **REND** is the end resistance per unit width and the division by 2 is from the symmetry of the gate segment (cf. Fig. 11.1).

The salicided poly gate resistance is given by the well-known expression

$$R_{gsal} = \frac{\mathbf{RSHG} W_g}{3 \cdot L_g \cdot \mathbf{NGCON}^2 \cdot m} \tag{11.57}$$

where the parameters **RSHG** and **NGCON** are the sheet resistance and number of gate contacts to the polysilicon, respectively.

The silicide to poly contact resistance as described by [27] and implemented into a MOSFET gate resistance model in [44], is given by

$$R_{gpv} = \frac{\mathbf{RPV}}{W_g \cdot L_g \cdot m} \tag{11.58}$$

where **RPV** is a model parameter. The combination of the two gate resistance components R_{gp} and R_{gpv} enables accurate and scalable modeling of total gate resistance.

11.8 Silicon Data Validation of RF Model

Robust measurement and extraction procedures for predominantly process and geometry based parameters across a wide bias and frequency space for 0.18 μm, 0.13 μm, and 65 nm RFCMOS technology were presented in [58, 59, 64]. At relatively low frequencies (500 MHz), the total gate capacitance is given by $\mathrm{Im}(y_{11})/\omega$. $Q = \mathrm{Im}(y_{11})/\mathrm{Re}(y_{11})$ values are extracted in the GHz range where $\mathrm{Re}(y_{11})$ is within the dynamic range of the network analyzer.

Figures 11.19 and 11.20 show $C(V_{gb})$, $Q(V_{gb})$, and $Q(f)$ for n^+-poly/n-well devices with varying L_g and W_g. The model accurately predicts the reduced tuning range C_{max}/C_{min} with decreasing L_g due to the influence of the parasitic capacitance C_{fr}. Further, the significant decrease in Q with increasing L_g due to increasing well resistance is accurately modeled. The model also accurately predicts the decrease in Q with increasing W_g due to increased R_{gsal} (not shown). Qualitatively similar behavior accurately modeled by MOSVAR is observed for oxide thickness of 1.9–7 nm [58, 59].

11.9 Circuit Applications Examples

Modern day RF standards, such as WLAN, WiMAX, UWB, and high speed optical communications systems use high frequency VCOs for frequency synthesis and system timing [7, 25, 49]. The oscillation frequencies of these VCOs are generally higher than that of the local oscillator due to the use of integer and fractional frequency dividers to improve synthesizer performance and capacity, and to lower phase noise. In a typical VCO, a tank circuit, consisting of an integrated inductor and a MOS varactor, controls the oscillation frequency.

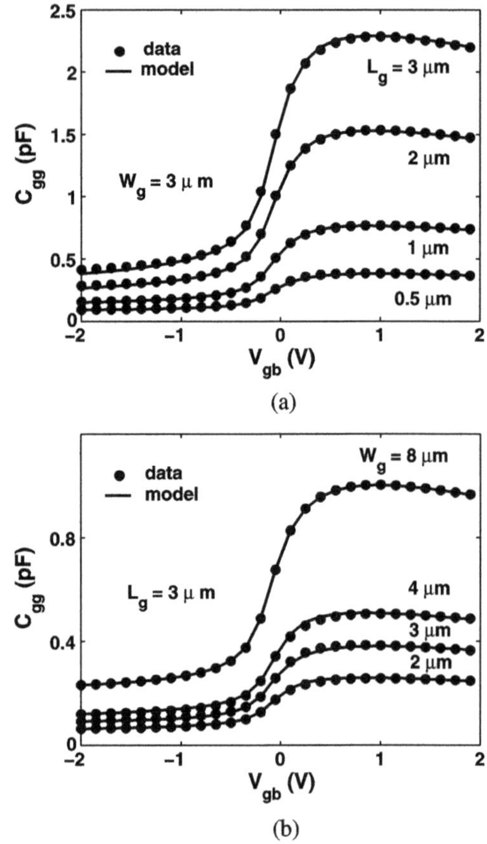

Fig. 11.19 **a** Measured and simulated $C_g(V_{gb})$ for L_g scaling and **b** Measured and simulated $C_g(V_{gb})$ for W_g scaling. Both (**a**) and (**b**) are from a Jazz 130 nm technology; $f = 1$ GHz

Figure 11.21 shows an RF CMOS VCO, based on the negative resistance principle, using complementary cross coupled NFET and PFET pairs. This topology of VCO offers a larger available transconductance, decreased switching time, and tighter output symmetry than a standard NFET-only topology [39]. This gives improved phase noise and simplifies design. The VCO oscillation frequency is set by the differential inductor (L_{tank}) and MOS varactors (C_1 and C_2) tank. Resistive tank losses are compensated by the negative resistance of the cross coupled pairs, enabling sustained oscillation.

The tank quality factor (Q_{tank}), which directly affects the phase noise of the VCO, is

$$Q_{tank} = \frac{Q_L Q_C}{Q_L + Q_C} = \frac{\omega L_{tank}}{R_L + \omega^2 \cdot L_{tank} \cdot C_{tank} \cdot R_C} \tag{11.59}$$

which is based on the approximations for the tank components [39]:

$$Q_L = \frac{\omega L_{tank}}{R_L}, \tag{11.60}$$

$$Q_C = \frac{1}{\omega \cdot C_{tank} \cdot R_C}. \tag{11.61}$$

R_C and R_L model the resistive losses for varactor and inductor, respectively, in the series tank path. Equation (11.59) shows that as frequency increases, Q_{tank} approaches Q_C. Figure 11.22 shows Q for three inductors designed to peak at 2, 5, and 10 GHz, respectively. The inductor Q deviates from the ideal Q_L at high frequencies, peaking and decreasing due to skin effect and capacitive losses. The inductance is decreased by lowering the conductor length, which reduces capacitive losses. This, in turn, extends the range of ideal Q_L behavior with frequency, giving a higher peak Q for higher frequency applications. Varactors scaled through parallel multiplicity of fixed geometry unit cells, while keeping $C_{tank} R_C$ constant, show a steady decrease in Q with frequency as shown in Fig. 11.22. At 2 GHz, Q_L is roughly an order of magnitude lower than Q_C and hence dominates Q_{tank}.

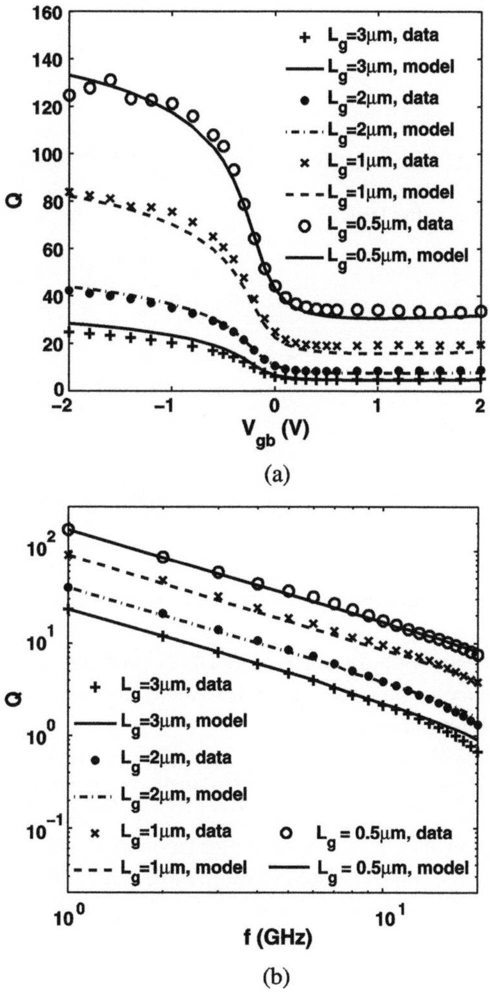

Fig. 11.20 **a** Measured and simulated $Q(V_{gb})$ and **b** Measured and simulated $Q(F)$ for L_g scaling with $W_g = 3$ μm. Both are from a Jazz 0.13 μm technology. After [59]

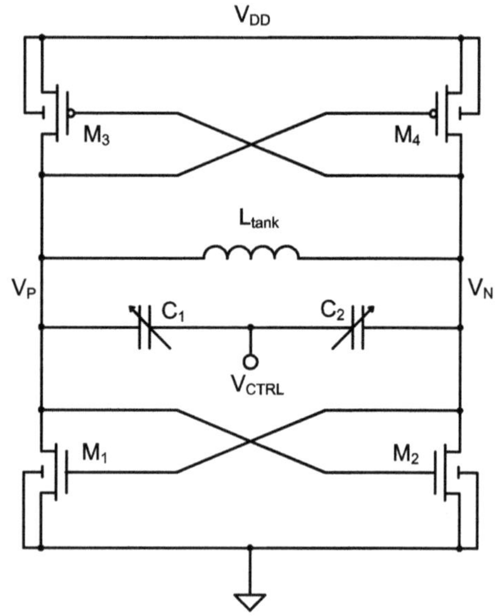

Fig. 11.21 A VCO based on cross coupled NFET and PFET pair

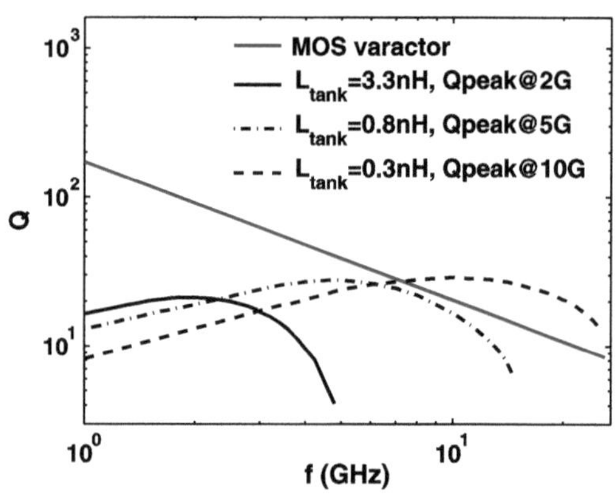

Fig. 11.22 Inductor and MOS varactor Q over frequency. After [59]

At 5 GHz and above, Q_C is of the same order or lower than Q_L, clearly affecting Q_{tank}.

Leeson's phase noise model [26] provides additional insight into the impact of the varactor on phase noise. It relates the phase noise transfer function $H(j\omega)$ to the oscillator parameters by

$$|H(j\omega)|^2 = \frac{1}{4Q_{tank}^2}\left(\frac{\omega_o}{\Delta\omega}\right)^2 \tag{11.62}$$

where ω_o is the oscillator center frequency and $\Delta\omega$ is the offset frequency from ω_o where the phase noise measurement is taken. At frequencies where Q_C dominates, (11.59) and (11.62) show that $|H(j\omega)|^2 \propto R_C^2$. Figure 11.23 shows phase noise simulations of the VCO with the MOSVAR model. The phase noise spectral density increases with L_g, as expected due to the increased R_C. Furthermore, the geometry

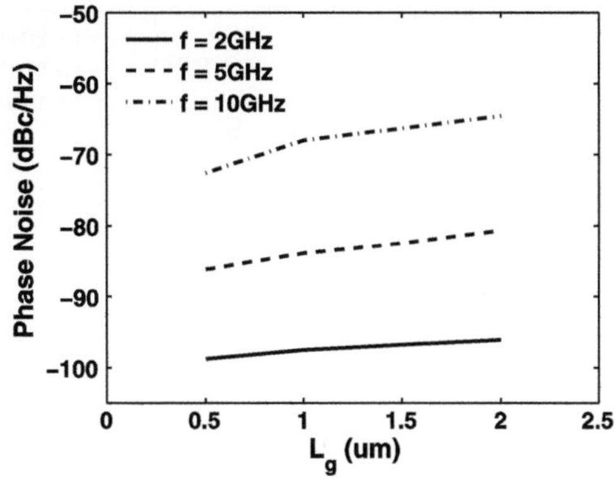

Fig. 11.23 Simulated phase noise at $\Delta f = 100$ kHz. After [59]

sensitivity of the phase noise increases at higher frequencies due to the increased influence of Q_C compared to Q_L.

In addition to phase noise, VCO tuning dynamics are commonly benchmarked through the VCO gain, KVCO, defined as

$$K_{VCO} = \left|\frac{d\omega}{dV_{CTRL}}\right|. \tag{11.63}$$

For the case of the LC circuit, the center frequency is given by

$$\omega_o = 1/\sqrt{L_{tank}C_{tank}}. \tag{11.64}$$

The voltage controlled varactor capacitance is given by (11.39). The derivative of C with respect to voltage is

$$\frac{dC(V_{gb})}{dV_{CTRL}} = L \cdot W \cdot \left|\frac{dC_0(V_{gb})}{dV_{gb}}\right|. \tag{11.65}$$

Combining (11.63), (11.64) and (11.65) yields

$$K_{VCO} = \frac{\omega \cdot L \cdot W}{2[L \cdot W \cdot C_0(V_{gb}) + W\mathbf{CFRW} + L\mathbf{CFRL}]} \cdot \left|\frac{dC_0(V_{gb})}{dV_{gb}}\right| \tag{11.66}$$

which for small L approaches

$$K_{VCO} = \frac{\omega}{2} \cdot \frac{L}{\mathbf{CFRW}} \cdot \left|\frac{dC_0(V_{gb})}{dV_{gb}}\right| \tag{11.67}$$

and for large L saturates to

$$K_{VCO} = \frac{\omega}{2C_0(V_{gb}) + \mathbf{CFRL/W}} \cdot \left|\frac{dC_0(V_{gb})}{dV_{gb}}\right|. \tag{11.68}$$

Equations (11.67) and (11.68) provide direct insight into the behavior of K_{VCO} over varactor length. Simulations of the K_{VCO} vs. L_g, shown in Fig. 11.24, verify the physical accuracy of the MOSVAR model, facilitating evaluation of critical VCO design tradeoffs. This aspect of MOS varactor modeling was originally reported in [59].

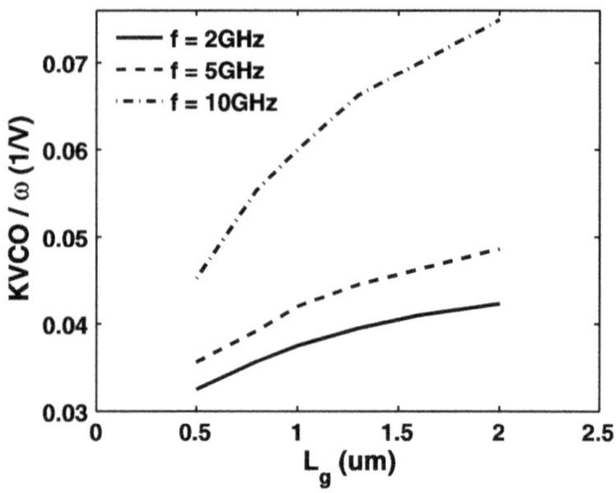

Fig. 11.24 Simulated KVCO over frequency and L. After [59]

11.10 Conclusions

A surface-potential-based approach has been used to develop a physical varactor model MOSVAR, that includes both the intrinsic device and the parasitic elements. Secondary effects such as gate tunneling current are modeled by further developing the techniques originally established for the PSP bulk MOSFET model. MOSVAR inherits from PSP its extremely accurate analytical approximations for the surface potential, both with and without the effects of minority carriers (inversion charge). The new varactor model is designed to maximize compatibility with the PSP model; it inherits many parameters from PSP, so once these have been characterized for a MOSFET the same values can be used for an initial MOSVAR model for a varactor made from the same basic MOS system (i.e. same gate, gate dielectric, and well). In addition to the features available in PSP, MOSVAR includes various effects specific to MOS varactors. These include different polarity combinations of the semiconductor layers, an additional tunneling current mode, and inertia in the formation of the inversion layer.

Extensive, scalable models for parasitic resistance and capacitance are included in MOSVAR in order to accurately model Q. A new parameter extraction procedure is described and verified against experimental data from several technology nodes. Circuit applications of MOSVAR are illustrated using a typical VCO architecture.

Acknowledgments The authors would like to thank S. Chaudhry, L. Dong and J. Zheng (*Jazz Semiconductor*), I. Amano and R. Hiroyuki (*Fujitsu Microelectronics Limited*) for providing the measured data, Z. Yan (*Jazz Semiconductor*) and J. Cordovez (*Sentinel-IC Technologies*) for model verification and testing. We are grateful to A. Scholten (*NXP-TSMC Research Center*) and G. Coram (*Analog Devices*) for their help in the model code development and to W. Wu (*Arizona State University*) for reading the manuscript and valuable comments. Special thanks are extended to the Compact Model Council (CMC) varactor subcommittee for verification of the MOSVAR model. This work was supported in part by CMC and Semiconductor Research Corporation (SRC).

References

1. Ainspan, H., Plouchart, J.O.: A comparison of MOS varactors in fully-integrated CMOS LC VCO's at 5 and 7 GHz. In: Proc. European Solid-State Circuits Conf., pp. 447–450 (2000)
2. Andreani, P., Mattisson, S.: On the use of MOS varactors in RF VCO's. IEEE J. Solid-State Circuits **35**, 905–910 (2000)
3. Bhattacharyya, A.B.: Compact MOSFET Models for VLSI Design. Wiley, Singapore (2009)
4. Brown, D.M., Gray, P.V.: Si–S_iO_2 fast interface state measurements. J. Electrochem. Soc. **115**, 760–766 (1968)
5. Bunch, R.L., Raman, S.: Large-signal analysis of MOS varactors in CMOS-GLC VCOs. IEEE J. Solid-State Circuits **38**, 1325–1332 (2003)
6. Cai, J., Sah, C.T.: Gate tunneling currents in ultrathin oxide metal-oxide-silicon transistors. J. Appl. Phys. **89**, 2272–2285 (2001)
7. Cao, C., Ding, Y.: Kenneth, K.O.: A 50-GHz phase-locked loop in 130-nm CMOS. In: Proc. IEEE Cust. Integr. Circuits. Conf., pp. 21–24 (2006)
8. Chatterjee, S., Musah, T., Tsividis, Y., Kinget, P.: Weak inversion MOS varactors for 0.5 V analog integrated filters. In: Symposium on VLSI Circuits, Dig. of Technical Pap., pp. 272–275 (2005)
9. Chatterjee, S., Tsividis, Y., Kinget, P.: 0.5-V analog circuit techniques and their application in OTA and filter design. IEEE J. Solid-State Circuits **40**, 2373–2387 (2005)
10. Chen, K.M., Huang, G.W., Wang, S.C., Yeh, W.K., Fang, Y.K., Yang, F.L.: Characterization and modeling of SOI varactors at various temperatures. IEEE Trans. Electron Devices **51**, 427–433 (2004)
11. Chen, Z., Wong, W., Cheng, J., He, D.: A new MOS varactor BSIM4 model with temperature effect. In: The 9th Int. Conf. on Solid-State and Integr.-Circuit Technology, pp. 527–529 (2008)
12. Dai, L., Harjani, R.: Design of Higher-Performance CMOS Voltage Controlled Oscillators, 1st edn. Springer, Assinippi Park (2002)
13. Dunwell, D., Frank, B.: Accumulation-mode MOS varactors for RF CMOS low-noise amplifiers. In: Topical Meet. on Silicon Monolithic Integr. Circuits in RF Syst., pp. 145–148 (2007)
14. Fang, F.F., Howard, W.E.: Negative field effect mobility on (100) Si surface. Phys. Rev. Lett. **16**, 797–799 (1966)
15. Feng, H., Wu, Q., Guan, X., Zhan, R., Wang, A.: A 2.45 GHz wide tuning range VCO using MOS varactor in 0.35/spl mu/m SiGe BiCMOS technology. In: IEEE Int. Symposium on Microw., Antenna, Propag. and EMC Technologies for Wirel. Commun., pp. 10–13 (2005)
16. Fong, N.H.W., Plouchart, J.O., Zamdmer, N., Liu, D., Wagner, L.F., Plett, C., Tarr, N.G.: A 1-V 3.8–5.7-GHz wide-band VCO with differentially tuned accumulation MOS varactors for common-mode noise rejection in CMOS SOI technology. IEEE Trans. Microw. Theory Tech. **51**, 1952–1959 (2003)
17. Garrett, C.G.B., Brattain, W.H.: Physical theory of semiconductor surfaces. Phys. Rev. **99**, 376–387 (1955)
18. Geynet, L., De Foucauld, E., Vincent, P., Jacquemod, G.: Fully-integrated multi-standard VCOs with switched LC tank and power controlled by body voltage in 130 nm CMOS/SOI. In: IEEE Radio Freq. Integr. Circuits (RFIC) Symposium, pp. 11–13 (2006)
19. Gildenblat, G., Li, X., Wu, W., Wang, H., Jha, A., van Langevelde, R., Smit, G.D.J., Scholten, A.J., Klaassen, D.B.M.: PSP: An advanced surface-potential-based MOSFET model for circuit simulation. IEEE Trans. Electron Devices **53**, 1979–1993 (2006)
20. Gildenblat, G., Zhu, Z., Wu, W.: Analytical expression for the bias and frequency-dependent capacitance of MOS varactors. IEEE Trans. Electron Devices **54**, 3107–3108 (2007)
21. Gildenblat, G., Zhu, Z., McAndrew, C.C.: Surface potential equation for bulk MOSFET. Solid-State Electron. **53**, 11–13 (2009)
22. Gray, P.V., Brown, D.M.: Freeze-out characteristics of the MOS varactor. Appl. Phys. Lett. **13**, 247–248 (1968)

23. Gu, X., Chen, T.L., Gildenblat, G., Workman, G., Veeraraghavan, S., Shapira, S., Stiles, K.: A surface potential-based compact model of n-MOSFET gate-tunneling current. IEEE Trans. Electron Devices **51**, 127–135 (2004)
24. Gutiérrez, I., Meléndez, J., Hernández, E.: Design and Characterization of Integrated Varactors for RF Applications. Wiley, Chichester (2006)
25. Han, Y., Larson, L.E., Lie, D.Y.C.: A low-voltage 12 GHz VCO in 0.13 μm CMOS for OFDM applications. In: Silicon Monolithic Integr. Circuit in RF Syst., pp. 379–382 (2006)
26. Leeson, D.B.: A simple model of feedback oscillator noise spectrum. Proc. IEEE **54**, 329–330 (1966)
27. Litwin, A.: Overlooked interfacial silicide-polysilicon gate resistance in MOS transistors. IEEE Trans. Electron Devices **48**, 2179–2181 (2001)
28. Maget, J., Tiebout, M., Kraus, R.: Influence of novel MOS varactors on the performance of a fully integrated UMTS VCO in standard 0.25 μm CMOS technology. IEEE J. Solid-State Circuits **37**, 953–958 (2002)
29. Molnar, K., Rappitsch, G., Huszka, Z., Seebacher, E.: MOS varactor modeling with a subcircuit utilizing the BSIM3v3 model. IEEE Trans. Electron Devices **49**, 1206–1211 (2002)
30. Morandini, Y., Larchanche, J.F., Gaquiere, C.: High frequency characterization of compact n^+-poly/n-well varactor using waffle-layout. In: Silicon Monolithic Integr. Circuits in RF Syst., pp. 167–170 (2008)
31. MOSVAR Manual: PSP-Based MOSVAR v1.1.0 (2008). http://pspmodel.asu.edu/downloads/MOSVAR1p1p0.pdf
32. Nicollian, E.H., Brews, J.R.: MOS (Metal Oxide Semiconductor) Physics and Technology. Wiley, New York (1982)
33. Oehm, J., Pham-Stabner, D.: Linear controlled temperature independent varactor circuitry. In: Proc. European Solid-State Circuits Conf., pp. 143–146 (2002)
34. Oh, Y., Kim, S., Lee, S., Rieh, J.S.: The island-gate varactor—a high-Q MOS varactor for millimeter-wave applications. IEEE Microw. Wirel. Compon. Lett. **19**, 215–217 (2009)
35. Pao, H.C., Sah, C.T.: Effects of diffusion current on characteristics of metal-oxide (insulator)-semiconductor transistors. Solid-State Electron. **9**, 927–937 (1966)
36. Porret, A., Melly, T., Enz, C.C., Vittoz, E.A.: Design of high-Q varactors for low-power wireless applications using a standard CMOS process. IEEE J. Solid-State Circuits **35**, 337–345 (2000)
37. PSP model website. http://pspmodel.asu.edu
38. Ranganathan, S., Tsividis, Y.: Discrete-time parametric amplification based on a three-terminal MOS varactor: analysis and experimental results. IEEE J. Solid-State Circuits **38**, 2087–2093 (2003)
39. Razavi, B.: RF Microelectronics. Prentice Hall, Upper Saddle River (1998)
40. Rustagi, S.C., Leung, C.C.C.: Accumulation mode MOS varactor SPICE model for RFIC applications. Electron. Lett. **36**, 1735–1736 (2000)
41. Sadat, A., Yang, H., Xiao, E., Yuan, J.S.: Breakdown effects on MOS varactors and VCOs. In: IEEE 57th Freq. Control Symposium, pp. 556–559 (2003)
42. Sah, C.: Theory of the metal oxide semiconductor capacitor. Tech. rep., Solid State Electron Lab Tech. Rep., Univ. of Illinois (1964)
43. Sameni, P., Siu, C., Iniewski, K., Hamour, M., Mirabbasi, S., Djahanshahi, H., Chana, J.: Modeling of MOS varactors and characterizing the tuning curve of a 5–6 GHz LC VCO. IEEE Int. Symposium on Circuits and Syst., pp. 5071–5074 (2005)
44. Scholten, A.J., Tiemeijer, L.F., van Langevelde, R., Havens, R.J., Duijnhoven, A.T.A.Z., Venezia, V.C.: PSP: Noise modeling for RF CMOS circuit simulation. IEEE Trans. Electron Devices **50**, 618–632 (2001)
45. Senapati, B., Ehwald, K., Winkler, W., Furnhammer, F.: High performance MOS varactor SPICE model. Electron. Lett. **38**, 1416–1417 (2002)
46. Shin, S.H., Yoo, H.J.: A multistandard RF front-end using varactor controlled tunable interstage matching network. In: IEEE Radio and Wirel. Symposium, pp. 181–184 (2007)

47. Shin, H., Kang, I.M., Jeon, J.W., Gil, J.: Active and passive RF device compact modeling in CMOS technologies. In: Int. Conf. on Simul. of Semicond. Process. and Devices, pp. 17–22 (2006)
48. Shin, S., Lee, K., Kang, S.M.: Low-power 2.4 GHz CMOS frequency synthesizer with differentially controlled MOS varactors. In: IEEE Int. Symposium on Circuits and Syst., pp. 553–556 (2006)
49. Simon, M., Laaser, P., Filimon, V., Geltinger, H., Friedrich, D., Raman, Y., Weigel, R.: An 802.11a/b/g RF transceiver in an SoC. In: Proc. IEEE Int. Solid-State Circuits Conf., pp. 562–622 (2007)
50. Song, S., Shin, H.: An RF model of the accumulation-mode MOS varactor valid in both accumulation and depletion regions. IEEE Trans. Electron Devices **50**, 1997–1999 (2003)
51. Stern, F., Howard, W.E.: Properties of semiconductor surface inversion layers in the electric quantum limit. Phys. Rev. **163**, 816–835 (1967)
52. Sze, S.M.: Physics of Semiconductor Devices, 3rd edn. Wiley, New York (2006)
53. Tiebout, M.: Low Power VCO Design in CMOS. Springer, Berlin (2006)
54. Tsu, R., Esaki, L.: Tunneling in a finite superlattice. Appl. Phys. Lett. **22**, 562–564 (1973)
55. van Dort, M.J., Woerlee, P.H., Walker, A.J., Juffermans, C.A.H., Lifka, H.: Influence of high substrate doping levels on the threshold voltage and the mobility of deep-submicron MOSFETs. IEEE Trans. Electron Devices **39**, 932–938 (1992)
56. van Langevelde, R., Scholten, A.J., Klaassen, D.B.M.: Physical background of MOS model 11. http://www.nxp.com/acrobat_download/other/models/nl_tn2003_00239.pdf
57. Victory, J., McAndrew, C.C., Gullapalli, K.: A time-dependent, surface-potential based compact model for MOS capacitors. IEEE Electron Device Lett. **22**, 245–247 (2001)
58. Victory, J., Yan, Z.X., Gildenblat, G., McAndrew, C.C., Zheng, J.: A physically based, scalable MOS varactor model and extraction methodology for RF applications. IEEE Trans. Electron Devices **52**, 1343–1353 (2005)
59. Victory, J., Zhu, Z., Zhou, Q., Wu, W., Gildenblat, G., Yan, Z., Cordovez, J., McAndrew, C.C., Anderson, F., Paassches, J.C.J., van Langevelde, R., Kolev, P., Cherne, R., Yao, C.: PSP-based scalable MOS varactor model. In: IEEE Custom. Integr. Circuits Conf., pp. 495–502 (2007)
60. Wong, W., Hui, P.S., Chen, Z., Shen, K., Lau, J., Chan, P., Ko, P.: A wide tuning range gated varactor: Wide tuning-range MOS varactors. IEEE J. Solid-State Circuits **5**, 773–779 (2000)
61. Wu, W., Li, X., Gildenblat, G., Workman, G., Veeraraghavan, S., McAndrew, C.C., van Langevelde, R., Smit, G.D.J., Scholten, A.J., Klaassen, D.B.M.: A compact model for valence-band electron tunneling current in partially depleted SOI MOSFETs. IEEE Trans. Electron Devices **54**, 316–322 (2007)
62. Xu, H., Kenneth, K.O.: High-Q thick-gate-oxide MOS varactors with subdesign-rule channel lengths for millimeter-wave applications. IEEE Electron Device Lett. **29**, 363–365 (2008)
63. Yan, T., Zhang, G., Shi, H., Huang, R., Zhang, X.: Wide tuning-range MOS varactors based on SOI. In: The 7th Int. Conf. on Solid-State and Integr. Circuits Technology, vol. 1, pp. 206–208 (2004)
64. Zhu, Z., Victory, J., Chaudhry, S., Dong, L., Yan, Z., Zheng, J., Wu, W., Li, X., Zhou, Q., Kolev, P., McAndrew, C.C., Gildenblat, G.: Improved parameter extraction procedure for PSP-based MOS varactor model. In: Int. Conf. on Microelectron. Test Struct., pp. 148–153 (2009)

Chapter 12
Modeling of On-chip RF Passive Components

Zhiping Yu

Abstract The challenge for accurate modeling of on-chip RF passive components, $L/C/R$, lies on the proper consideration of undesired parasitics and capturing of distributed nature of the structure at high frequency. This is especially true for inductive components, namely inductors and transformers including baluns, because of the open capacitive and (long-range) magnetic couplings between the top widing metal layers and the lossy silicon substrate.

This chapter is mostly focused on the equivalent circuit modeling (i.e., compact modeling) of spiral inductors and transformers, while the modeling of on-chip resistors and capacitors are briefly mentioned. Among the modeling of spirals inductive components, the emphasis is on the calculation and extraction of three key parameters: inductance L (and for transformers also the magnetic coupling coefficient k between the primary and secondary windings), quality factor Q, and self-resonant frequency SRF (due to inevitable parasitic capacitance across the port of the inductor). Both one-π and two-π circuit topologies for inductor modeling are reviewed with tilt toward the former because of its renaissance recently. The discussion on transformer modeling is largely based on the seminal work by Long (IEEE J. Solid-State Circuits 35(9):1368, 2000). Up-to-date circuit applications of spiral transformers are provided as a motivation of studying this important subject.

12.1 Introduction

All RF circuits use inductance (L) and capacitance (C) for the resonant tanks in selecting frequencies. Besides, inductance can be used in the source de-generation circuit for providing resistive input impedance in common-source amplifiers, as a choking device for high frequency, etc. On the other hand, capacitance is frequently

Z. Yu (✉)
Institute of Microelectronics, Tsinghua University, Beijing 100084, China
e-mail: yuzhip@tsinghua.edu.cn

G. Gildenblat (ed.), *Compact Modeling*,
DOI 10.1007/978-90-481-8614-3_12,

used as a DC (direct current) path blocking, to reduce the fluctuation of power supply, and etc.

The realization of inductance in IC technology commonly uses the same principle as in discrete circuit elements, i.e., use metal wires as the turns in the inductors. But different from solenoid structure, turns in IC cannot be stacked on top of each other in a straight fashion. Instead, they are layouted as planar spirals, in which the magnetic coupling between turns is not as strong as stacked wires. Plus, the on-chip spiral inductors are built over the lossy silicon substrate. All these structure change and environmental influence make the modeling of spiral inductors quite different from the textbook classical inductors.

The modeling of capacitors on-chip is relatively simply compared to the modeling of inductors. But there are also some special issues which need attention. For example, the parasitic capacitance between the plate(s) of a capacitor with substrate ground.

In this chapter, we'll discuss the modeling of on-chip passive components, mainly the inductors.

12.2 Circuit Requirement and Applications for On-chip RF Passive Components

The requirements for CMOS on-chip resistor/inductor/capacitor ($R/L/C$) are mainly the passiveness, i.e., they won't generate energy, rather they consume. This passiveness is very important for the stability of the circuits.

Then, there are a few common requirements for RF passive components:

1. Minimum loss of DC energy, which are due to the series resistance in inductors and parallel leakage resistance to capacitors.
2. Minimum undesired (yet often unavoidable) parasitic components such as the shunt capacitance between the two terminals of an inductor and wire inductance for the capacitor leads.
3. Form factor, or the foot-print of the component. Especially, spiral inductors often occupy large chip area and it is desirable to have small die area for the same inductance value.

12.3 *R* and *C* Realization in RF CMOS

12.3.1 IC Resistors

At high frequencies, even a simple resistor exhibits complex behavior. For one, there will be inductance associated with the leads of the resistor. Then, between two contacts of the resistor, there is a parasitic capacitance across the resistor. A simple equivalent circuit for resistors is then as shown in Fig. 12.1. It can be seen that at

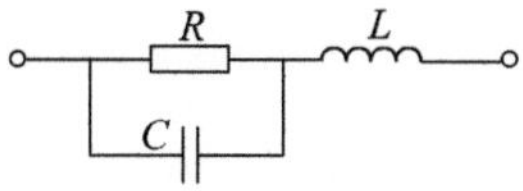

Fig. 12.1 Equivalent circuit for a resistor in high frequency

the low frequency, the parasitic series inductance, L, plays more role than C does, while at high frequency the shunt capacitance, C, decreases the overall impedance.

In CMOS process, metal and polysilicon lines can all be used as planar resistors with polysilicon a better choice because of its large resistivity. Moreover, using the so-called silicidation, the resistance of polysilicon line can be reduced too.

The resistivity of a conductive line is usually characterized by using the sheet resistance, which is defined by taking a rectangular sheet of conductor with depth of d (or called thickness), length l, and width w. The resistance seen from the two side planes (of $d \times w$) is then

$$R = \rho \frac{l}{wd} \tag{12.1}$$

where ρ is the resistivity of the conductor. For $w = l$, or a square sheet, the resistance is independent upon the value of l (or equivalently, w), and can be termed as the sheet resistance

$$R_{\square} = \rho \frac{1}{d}. \tag{12.2}$$

That is, the sheet resistance is inversely proportional to the thickness of the conductor sheet. Note that the units for the sheet resistance is the same as for an ordinary resistance, i.e., in ohms (Ω) (while for resistivity, ρ, in units of, say, Ω cm). And, using $R_{\square}$, the resistance of a conductor region of length l and width w becomes

$$R = \frac{l}{w} R_{\square} \tag{12.3}$$

The sheet resistance for silicided polysilicon line is commonly $5 \sim 10$ ohms. Besides the value of resistance, there are two resistor parameters, which are important in IC design:

1. The precision or variation of the resistance, $\Delta R/R$, usually in percentage
2. The temperature coefficient, TC, which is defined as

$$\mathrm{TC} = \frac{R(T) - R(T_0)}{R(T_0)} \frac{1}{T - T_0} = \frac{1}{R} \frac{\partial R}{\partial T} \tag{12.4}$$

   where T_0 is a fixed reference temperature, usually taken as the room temperature, $T_0 = 300$ K. The units of TC is commonly taken as ppm/°C, i.e., parts per million per centigrade.

For polysilicon resistors, depending on the doping in the poly, the TC may be around 1000 ppm/°C.

In advanced bipolar process, the self-aligned polysilicon emitter is often used for IC resistors.

In CMOS process, the source/drain (S/D) diffusion regions can be used for resistors. Its resistivity and TC are similar to those of silicided polysilicon (within a range of 2). If the S/D regions are heavily doped, the TC can be even lower. This type of resistors are named diffusion resistors. Due to the existence of the pn junction between S/D regions and the substrate, there exist sizable junction capacitance and large voltage coefficient, defined as $\Delta R/\Delta V$ where V is the applied voltage across the resistor and caused by the dependence of the width of the junction depletion region on the bias. The junction capacitance will limit the frequency range for the diffusion resistors, while the voltage dependence will cause nonlinearity, thus limiting the dynamic range.

The wells, within which the FETs are formed in CMOS process, can also be used for resistors. The doping level in wells is considerably lower than that in S/D regions, so the sheet resistance is high, typically at $R_{\square} = 1\text{–}10\ \text{k}\Omega$. For the same reason as with S/D diffusion regions, the junction capacitance is large (even larger because of the big area for the bottom of the well). Also, the TC is high (because of the light doping) with typical value of 3000–5000 ppm/°C.

An MOS transistor can also be made a resistor by applying a proper gate voltage to turn the transistor on and to let the transistor work in the linear region, where the drain bias is small and the drain current is proportional to the drain voltage. But the precision of this resistor (due to channel resistance) is not that good because the mobility for the inversion carriers is a strong function of temperature. Also, the nonlinearity is high.

For small resistance value, one can use metal interconnect such as aluminum (Al) for resistors. For most interconnect lines, it can be considered that $R_{\square} \sim 50\ \text{m}\Omega$, hence to get a resistor of resistance around 10 Ω is feasible.

For more discussion about the IC resistors available in CMOS process, refer to [16]. Also, in [25], there is detailed discussion on the modeling of on-chip polysilicon resistors.

12.3.2 Capacitors in IC Process

The best-quality capacitors in IC processes are still the plate capacitors, which are easily implemented using the back-end of line (BEOL) process, i.e., metal interconnect layers and their interlayer dielectrics (ILD). These capacitors are called MIM for metal-insulator-metal. The drawback of MIM capacitors is that they usually occupy large area.

The ILD thickness is relatively large (in the order of 0.5–1 μm), hence the unit area capacitance value is small (with typical value of 5×10^{-5} pF/μm^2). Another issue one must pay attention to during the use of MIM capacitors is that the bottom plate of the capacitor has parasitic capacitance with the substrate. This parasitic capacitance can be as large as 10–30% of the nominal capacitance of the MIM capacitor.

The standard capacitance formula,

$$C \approx \epsilon \frac{A}{H} = \epsilon \frac{W \cdot L}{H} \tag{12.5}$$

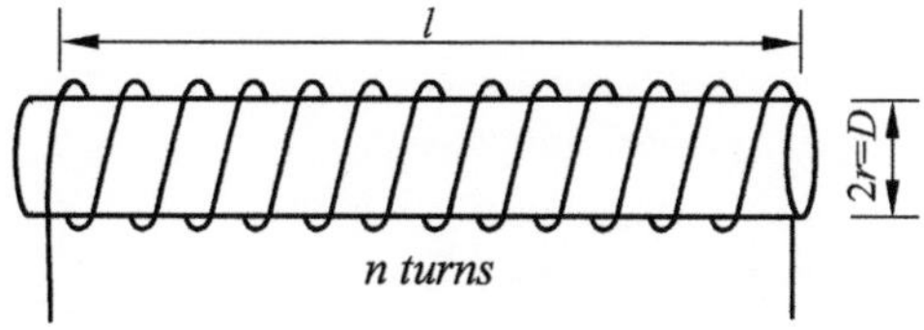

Fig. 12.2 Single-layer solenoid inductor

where A is the plate area with length L and width W, and H is the separation between the two plates, underestimates the capacitance because it does not take into account the fringing effects. The formula is accurate as long as the plate dimensions are much larger than the plate separation H. The rough correction in considering the fringing capacitance is to extend the plate dimensions using H, as follows

$$C = \epsilon \frac{(W+2H)\cdot(L+2H)}{H} \approx \epsilon \left[\frac{WL}{H} + 2W + 2L\right]. \tag{12.6}$$

The above approximation is obtained by neglecting the term of $4H$, which can be considered small compared to term $2(W+L)$. The TC of MIM capacitors is usually low (in the range of 30–50 ppm/°C).

The capacitance per unit area is a concern for IC design in order to reduce the die size. One way to increase the capacitance density (or area efficiency) is to exploit lateral flux between adjacent metal lines within a given interconnect layer. A simple pattern which illustrates the two terminals of the capacitor are distinguished by two sets of interleaving fingers. The two plates of the capacitor are constructed out of the same metal layer using the line sidewall, an alternative to exploit the lateral electric field flux.

Another way of obtaining capacitors in planar CMOS process is to use, as an MOS capacitor, simply the gate capacitance of an ordinary MOSFET transistor. Capacitance per unit surface area depends on the gate dielectric thickness, but is typically in the range of 1–5 fF/μm^2, or roughly 20-100 times larger than capacitors made of interconnect layers. Generally, a MOS capacitor will have a small, positive temperature coefficient (on the order of 30 ppm/°C).

As is in the case for IC resistors, readers are referred to [16] for a more detailed discussion on IC capacitors.

12.4 Inductors and Transformers

12.4.1 Non-planar Inductors: Solenoid

The most effective way to make inductors is to use a solenoid, on which the metal wires are wound (see Fig. 12.2). The structural parameters for this single-layer, i.e., all turns of the winding(s) having the same diameter, solenoid inductor are: number of turns n; the diameter (radius) of the solenoid D (r); and the length of the coil[1]

[1] In this chapter, we use coil and winding alternatively.

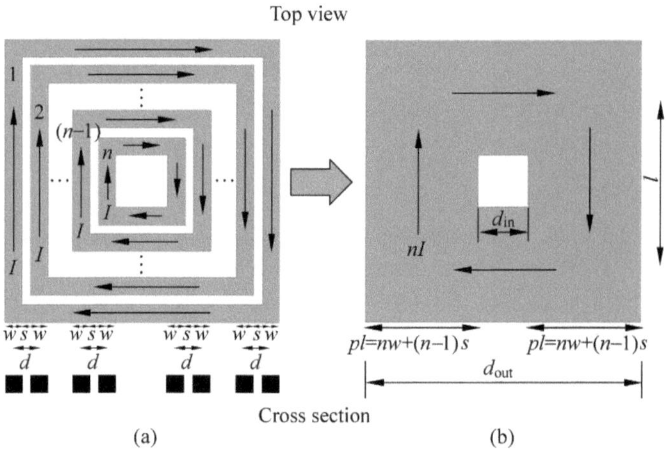

Fig. 12.3 Current sheet model for spiral inductors. (**a**) With metal line width and spacing; (**b**) with spacing $s = 0$, it becomes a current sheet [19]

along the solenoid l. It is assumed that the turns are tightly packed (no spacing between turns in the axial direction of the solenoid), and the inductance according to Wheeler in the late 1920s [24] is

$$L \approx \frac{\mu_0 n^2 \pi r^2}{l + 0.9r} \tag{12.7}$$

where SI units are used, i.e., L in henry (H), the permeability of free space $\mu_0 = 4\pi \times 10^{-7}$ H/m, and all length units are in meters (m). Note that if the length of the solenoid l is comparable to (or smaller than, which is the case for planar spiral inductors) the radius, r, of the cross-section, the inductance L is not proportional to the area of the solenoid cross-section ($\propto r^2$), but to r. We will see later that this rule applies to the spiral inductors as well. As long as $l > (2/3)r$, the above formula has a good accuracy of better than 1%. When the length is small compared to the radius (say $l = 0.4r$), this formula (12.7) underestimates the true inductance.

12.4.2 Spiral Inductors from Current Sheet

Before discussing the modeling of on-chip planar spiral inductors (also called monolithic inductors), we examine another extreme case, i.e., the inductor consists of a current sheet, or windings without spacing between adjacent turns. Assuming that we have a spiral inductor, as shown in Fig. 12.3(a), with n turns, and the winding parameters of width w and spacing s (the pitch for the winding is thus $d = w + s$). A current sheet model can be constructed as in Fig. 12.3(b). To characterize the overall geometry of this square shaped spiral, we can view the conductor as a single wire with width equal to

$$W = nw + (n - 1)s$$

(cf. Fig. 12.3(b)) and the length for each side of the square for the wire is l. It can be seen that

$$l = (d_{out} + d_{in})/2, \qquad d_{out} = d_{in} + 2W$$

where d_{in}, d_{out} are inner and outer diameters, respectively, of the hollowed square for the current sheet. It is convenient to define a "filling factor"

$$\rho = \frac{W}{l} = \frac{nw + (n-1)s}{l}. \tag{12.8}$$

It can easily be seen that $0 \leq \rho \leq 1$ and

$$\rho = \frac{d_{out} - d_{in}}{d_{out} + d_{in}}. \tag{12.9}$$

It is indeed a fitted name for ρ for when $d_{in} = 0$, i.e., for a full square, $\rho = 1$, while for a thin wire (no width or $d_{in} = d_{out}$) $\rho = 0$.

In the following discussion of this hollowed square current sheet, we use d_{in} and d_{out} as original geometric parameters.

It can be derived that the inductance for this inductor made of "fat" conductor, i.e., the current sheet, is [19]

$$L_{sq\_sheet} = \frac{2\mu_0 l}{\pi}\left[\ln\left(\frac{2.067}{\rho}\right) + 0.18\rho + 0.125\rho^2\right]. \tag{12.10}$$

Now, consider partitioning the metal conductor of width $W = (d_{out} - d_{in})/2$ into closely packed n strips (with no inter-strip spacing), the inductance can be written [19]

$$L_{n\_strips} \approx \frac{\mu_0 n^2 l}{2\pi}\left[\ln\left(\frac{2}{\rho}\right) + 0.5 + \frac{\rho}{3} - \frac{\rho^2}{24}\right]. \tag{12.11}$$

Note that L now also depends on a new parameter, the number of turns n, and $L \propto n^2$. It will be seen that this squared dependence of the inductance on the number of turns holds for all spiral inductors.

Further consider the non-zero spacing, s, and non-zero thickness, t, of the conductor winding, a more complete formula with parameters l, ρ which is based on d_{in} and d_{out}, and s, t (p. 50 in [19])

$$\begin{aligned} L_{sq\_spacing} = {} & \frac{\mu_0 n^2 l}{2\pi}\left[\ln\left(\frac{2}{\rho}\right) + 0.5 + \frac{\rho}{3} - \frac{\rho^2}{24}\right] \\ & + \frac{\mu_0 n^2 l}{2\pi}\left[\frac{(n-1)s^2}{2(\rho l)^2} + \frac{(n-1)s}{3nl} - \frac{1}{n}\ln\left(\frac{w+t}{w}\right)\right]. \end{aligned} \tag{12.12}$$

Note that the dependence of L on strip width w also appears.

The above inductors made of spiral conductors are in the category of planar spiral inductors, or simply spiral inductors, or even spirals. They can have shapes different from a square. Despite of the perception of round-shape spirals may have bigger L and better quality-factor Q, which is to be defined later, with the same die area, the fact is that both Q and L are not that sensitive (i.e., the 2nd-order function) to the shape. Octagonal or circular spirals are moderately better than squares (typically on the order of 10%).

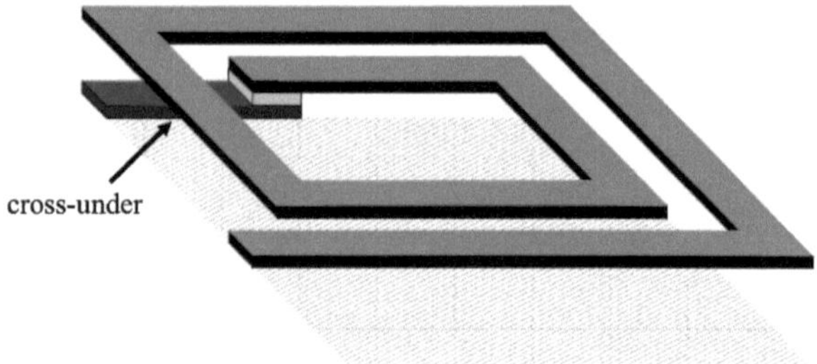

Fig. 12.4 3-D view of square spiral with 1.75 turns

The most common realizations for the spiral inductors is to use the top-most metal layer for the main part of the inductor (often also with two or more levels strapped together to reduce sheet resistance of the winding). The connection to the center end of the spiral is implemented with a cross-under at some lower level of metal (see Fig. 12.4). For CMOS process, there are always parasitics such as capacitance between the winding of the spiral and the conductive substrate. Maximizing the distance by using the top metal layers for winding can minimize this parasitic capacitance.

12.4.3 CMOS Spiral Inductors

Now we consider realistic spiral inductors made using planar CMOS process.

For general shapes such as shown in Fig. 12.5, the characterizing parameters are n, inner and outer diameters (or called dimensions), d_{in}, d_{out}, and metal winding parameters w, s, t. Now using d_{avg} instead of l for the average dimension

$$d_{avg} = (d_{out} + d_{in})/2.$$

The simplest formula for estimating the inductance of a square spiral, is

$$L_{sq} = (0.6 \sim 1) \times 10^{-3} n^2 d_{avg} \tag{12.13}$$

where multiplying factor 1 is given by Voorman [23], and the units for L is nH and μm for d_{avg}. The primary features of this formula is that the inductance of a spiral is proportional to

1. squared number of turns, n^2
2. dimension represented by the average dimension, d_{avg}, but not the area of the spiral

These features are the same as the formula for solenoid inductors, (12.7), when the length is much smaller than the radius of the cross-section. In fact, a planar spiral can be viewed as the special case of $l = 0$ while single-layer stacking becomes a spread winding on a plane. As pointed out by T. Lee [16] that (12.13) typically yields numbers on the high side.

For shapes other than the square spirals, multiply the value given by the square spiral formulas by the square root of the area ratio between the particular shape and

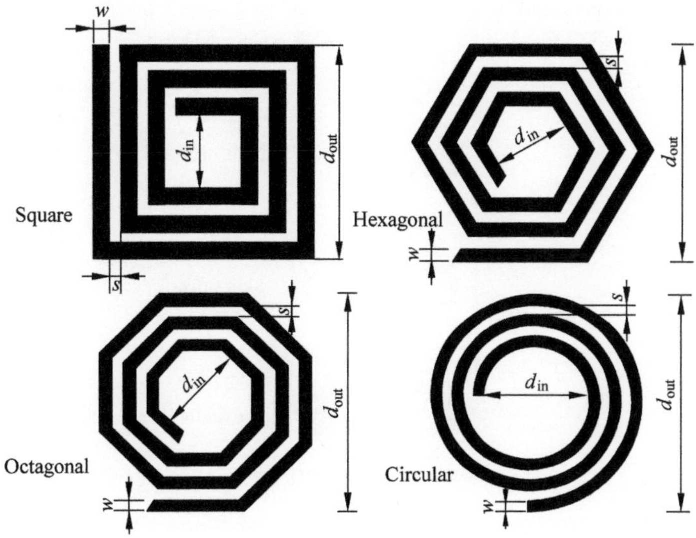

Fig. 12.5 Planar spiral inductors with different shapes

Table 12.1 Coefficients for current-sheet inductance formula with four different shapes [20]

Shape	c_1	c_2	c_3	c_4
Square	1.27	2.07	0.18	0.13
Hexagon	1.09	2.23	0.00	0.17
Octagon	1.07	2.29	0.00	0.19
Circle	1.00	2.46	0.00	0.20

the square (of the side d_{out}) to obtain the crude estimate of the correct value. For example, for circular spirals, multiply the square-spiral value by $\sqrt{\pi/4} \approx 0.89$, and for octagonal spirals by 0.91.

There is also a Wheeler formula for a hollow square spiral inductor, which reads

$$L \approx \frac{9.375\mu_0 n^2 d_{avg}^2}{11d_{out} - 7d_{avg}} \tag{12.14}$$

where L is in units H and d_{avg}, d_{out} in meters.

In a generalized form to (12.10), the inductance of planar spirals of all regular shapes can be cast in the following form [20],

$$L \approx \frac{\mu_0 n^2 d_{avg} c_1}{2}\left[\ln\left(\frac{c_2}{\rho}\right) + c_3\rho + c_4\rho^2\right]. \tag{12.15}$$

The various coefficients c_n, $n = 1, 2, 3, 4$ are a function of geometry and they are given in Table 12.1 for four representative shapes: square, hexagon, octagon, and circle.

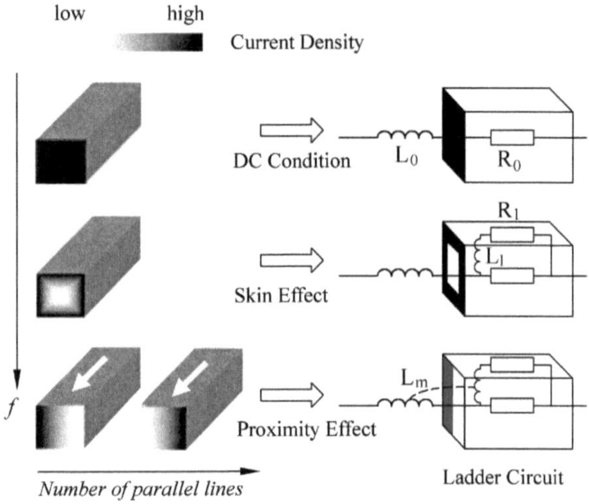

Fig. 12.6 Skin and proximity effects in the conductors of the spiral winding [2]

Aside from the large area potentially consumed, another serious problem with CMOS spiral inductors is their relatively large loss. There are two types of loss, all caused by the resistive Joule heat in the conductor of windings and conductive substrate.

For windings, the DC resistive loss is exacerbated by the skin and proximity effects within the conductor of windings. These two effects are all caused by the interplay between the electric and magnetic fields and result in the current through the cross-section of the conductor distributed unevenly: concentrated in the surface layer of the conductor, thus leading to larger AC resistance than that at DC (cf. Fig. 12.6). In addition to the winding resistive loss, the capacitive and magnetic coupling between the metal layer(s) for windings and the substrate causes the conductive currents in the form of both eddy current, which is circular, and straight path to the substrate contact. All these resistive losses (within winding conductor and in the substrate) reduce the Q of the spiral inductors.

Another issue is the parasitic capacitance between the spiral inductor and the substrate. For silicon CMOS, the distance between the winding metal layer and the conductive substrate is typically in the order of 2–5 μm (cf. Fig. 12.33), creating a parallel plate capacitor that resonates with the inductor. The resonant frequency of the LC combination represents the upper frequency limit of the inductor (called SRF for self-resonant frequency).

One additional parasitic capacitance is the shunt capacitance across the inductor that arise mainly from the overlap of the cross-under with the rest of the spiral (see Fig. 12.4), and also due to the turn-to-turn capacitances (between neighboring metal lines). However, the lateral capacitance from turn to turn usually has a negligible overall effect because it is the series connection of these capacitances that ultimately appears across the terminals of the inductor.

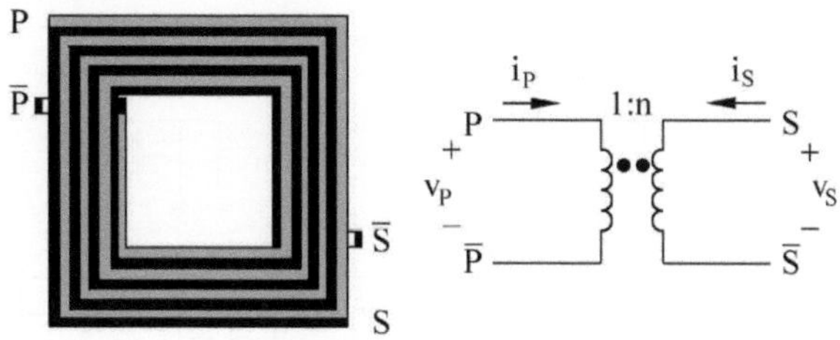

Fig. 12.7 Monolithic transformer of (4-turns each for primary/secondary windings) and its circuit symbol [18]

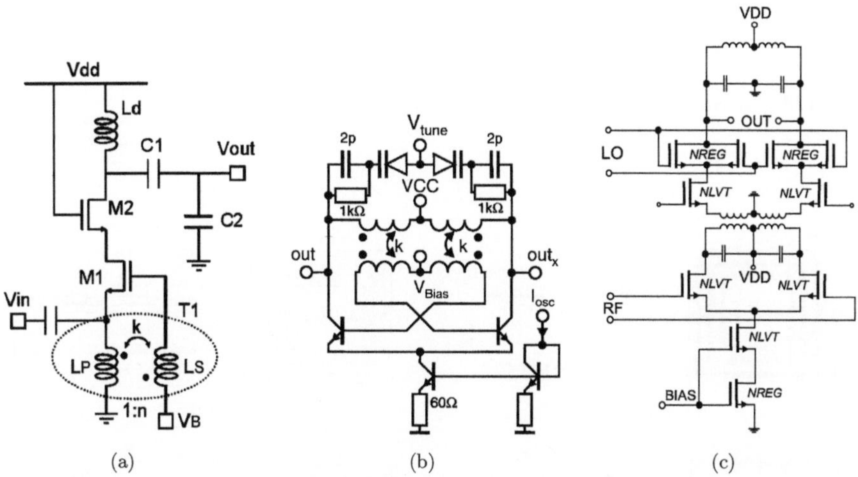

Fig. 12.8 (**a**) Transformer-coupled g_m-boosted common-gate LNA topology [17], (**b**) transformer-coupled VCO [26], (**c**) transformer-coupled 17 GHz low-voltage mixer [21]

12.4.4 Planar Transformers

CMOS transformers are increasingly used for impedance transformation and interstage coupling. They have advantages especially at high operation frequency to provide high Q (compared to capacitive coupling) and to save chip area (compared to transmission lines) [5].

The operation of a passive transformer is based on the mutual inductance between two or more windings. It is designed to couple alternating current from one winding to the other without a significant loss of power. A typical planar spiral transformer (in this case, with interwound windings) and its circuit symbol are shown in Fig. 12.7.

Transformers are used in amplifiers (low-noise amplifiers, or LNAs) (Fig. 12.8(a)), oscillators (Fig. 12.8(b)), mixers (Fig. 12.8(c)), and power amplifiers (Fig. 12.9).

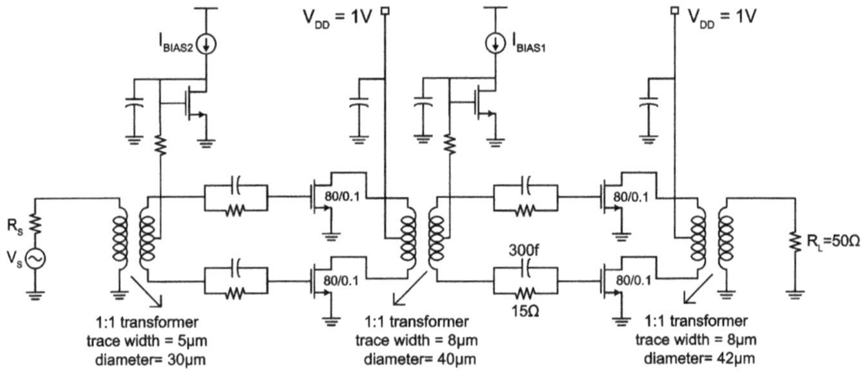

Fig. 12.9 Transformer-coupled two-stage 60 GHz PA [5]

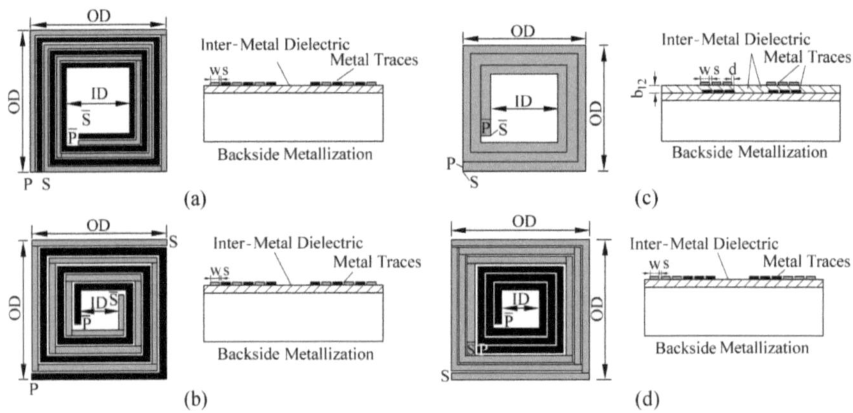

Fig. 12.10 Different configuration of planar transformers: (**a**) parallel conductor (Shibata) winding, (**b**) interwound (Frlan) winding, (**c**) overlay (i.e., stacking) (Finlay) winding, (**d**) concentric spiral winding [18]

12.4.5 Monolithic Spiral Transformers: Structures

Different configurations of planar transformers are shown in Fig. 12.10 and they include: parallel, interwound, overlay (or stacking), and concentric windings.

A physical layout for a stacking transformer used for 60 GHz operation is shown in Fig. 12.11 with top layer a differential winding and bottom layer a single circle. Since the required inductance is in the order of several tens pH, only one turn with varying radius is enough.

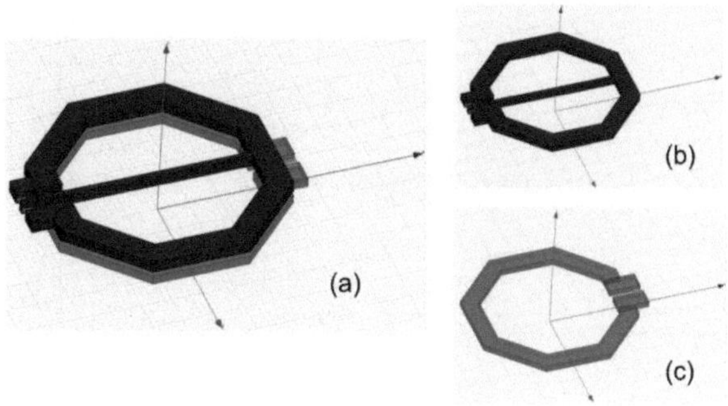

Fig. 12.11 3D perspective of stacked transformer (balun) used in 60 GHz PA. (**a**) Entire transformer, (**b**) top layer with center tab as differential winding, (**c**) bottom layer for a single turn

12.5 Modeling of Spiral Inductors and Transformers

For Spice circuit simulation, it is preferred to have equivalent circuit models for spiral inductors and transformers. Furthermore, it is desired to have frequency-independent element values in the model. Two less ideal approaches are (1) with terminal characteristics of the component represented by frequency-dependent S-parameters, and (2) using frequency-dependent values for elements in the equivalent circuits to model effects like skin and proximity. Besides, one of the most important requirements for equivalent circuits of any passive components is the preservation of the passitivity, i.e., no signal energy can emerge from the passive components at all time.

12.5.1 Characterization of Spiral Inductors

Before discussing the modeling of on-chip spiral inductors, we first look at the characterization of the spiral inductors. From measurement point of view, the spiral inductor is a two-port system with the substrate contact as the common ground. For high-frequency applications, it is preferred to measure S-parameters of the system. But for the analysis, it is more often to convert S-parameters to either Y- or Z-parameters.

In Fig. 12.12(a), the measurement setup (two-port system) for a spiral inductor is shown, together with a basic, bare-bone equivalent circuit model. Based on the Y-parameters obtained, the interpretation of the series resistance R_s and inductance L_s in the equivalent circuit for the spiral inductor is

$$R_s = \mathrm{Re}\left(-\frac{1}{Y_{21}}\right), \tag{12.16}$$

$$L_s = \frac{1}{\omega}\,\mathrm{Im}\left(-\frac{1}{Y_{21}}\right). \tag{12.17}$$

And the quality factor

$$Q = \frac{\omega L_s}{R_s}. \tag{12.18}$$

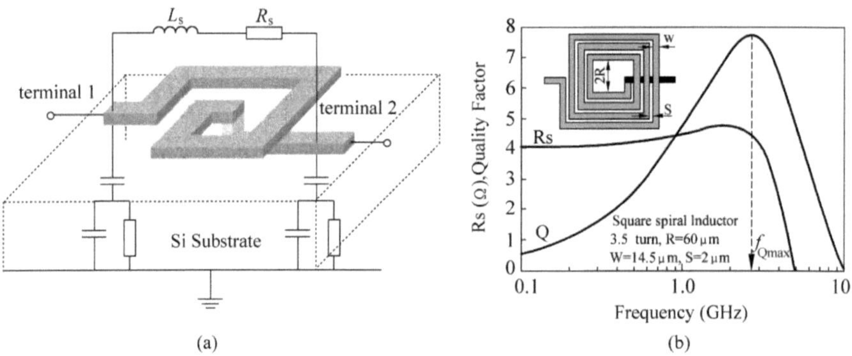

Fig. 12.12 (**a**) The basic circuit model for on-chip spiral inductor and its measurement setup (cf. [2]). (**b**) The interpretation of measured Y-parameters for frequency dependent series resistance R_s and Q-factor for a 3.5 turns spiral inductor [7]

It can be seen from Fig. 12.12(b) that R_s initially increases due to skin/proximity effects and then drops sharply above the self-resonant frequency f_{SR} (or SRF), which roughly corresponds to where Q peaks.

12.5.2 1-π Model for Spiral Inductors

One step behind the basic model shown in Fig. 12.12(a) is the so-called 1-π model, which is among the simplest models for a monolithic spiral inductor on some type of substrate (insulating/semi-insulating/semiconductor). Any reasonably accurate model for on-chip spiral inductors must include the circuit elements which can represent:

1. Inductance and its series resistance to reflect loss of the winding conductor
2. Capacitance appearing across the terminals of the component to "trigger" the intrinsic self-resonant frequency, beyond which the inductive component shows capacitive nature
3. To certain extent, the capacitive coupling and accompanying resistive loss between the inductor winding and the substrate has to be modeled.

To summarize the above main and parasitic mechanisms for on-chip spiral inductors, we can come up with an equivalent circuit as shown in Fig. 12.13(b). Because of the π-shape of the topology, it is so named as 1-π model.

Some key features of this model are

1. The model is symmetric, i.e., $C_{ox1} = C_{ox2}$, $R_{Si1} = R_{Si2}$, and $C_{Si1} = C_{Si2}$.
2. The main part of the equivalent circuit at low frequency is just an inductance L_s in series with a resistance R_s, representing the DC loss of the spiral winding. In order to model the frequency-dependent skin/proximity effects for winding conductor, a ladder circuit consisting of a resistor, R_{sk}, in series with an inductor,

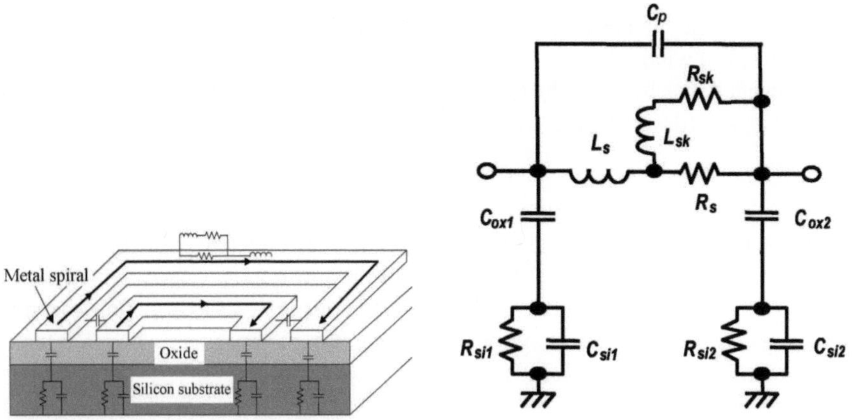

(a) 3D perspective of on-chip spiral inductor [14]

(b) Equivalent circuit of 1-π model of spiral

Fig. 12.13 1-π model of on-chip spiral inductors

L_{sk} (the subscript *sk* for skin), is added in shunt with R_s. At low frequency (e.g., at DC case),

$$R_{dc} = R_s \parallel R_{sk} = R_s R_{sk}/(R_s + R_{sk})$$

which is smaller than R_s. As frequency becomes large, the effective resistance of this ladder approaches R_s, a qualitatively correct representation of the resistance increase due to the increase of the operation frequency. Note that this ladder circuit can be further expanded to higher orders by replacing R_{sk} with another L-R ladder, and so on.

3. The shunt capacitance across the terminals of the inductor, C_p, is mainly due to the overlapping plate capacitor between the under-cross and the main winding layer(s) (see Fig. 12.4). But the capacitance between adjacent turns of the winding and the capacitive coupling between the winding and substrate through the oxide layer also contribute to this shunt capacitance. It should be pointed out this shunt capacitance is critical in having the SRF and cannot be neglected at all time, even for discrete inductors (working at high frequency).
4. The substrate loss due to the eddy current, which is caused by the magnetic flux perpendicular to the plane of the spiral inductor, is partly modeled by the ladder circuit in the main branch, which consists of L_s and R_s.
5. The capacitive coupling between the winding and substrate contact is taken care of by two circuit section of C_{ox} in series with $(R_{si} \parallel C_{si})$ from the terminal to the substrate contact on each terminal of the inductor.

The shunt capacitance C_p, considering only the overlapping part between the cross-under and the main windings (cf. Fig. 12.4), is

$$C_p = n \cdot w^2 \cdot \frac{\epsilon_{ox}}{t_{ox,u}} \tag{12.19}$$

where $t_{ox,u}$ is the thickness of the oxide between the metal layer of the cross-under and main spiral winding.

The capacitance between the spiral and substrate, C_{ox1} and C_{ox2} in Fig. 12.13(b), is calculated as (designated as C_{ox})

$$C_{ox} = w \cdot l \cdot \frac{\epsilon_{ox}}{t_{ox}} \tag{12.20}$$

where l is the total length of the winding and t_{ox} is the effective oxide thickness for the dielectric between the spiral and the top of the substrate.

Resistances R_{Si1}, R_{Si2} (all designated as R_{Si}) in Fig. 12.13(b) models the substrate's resistive loss, which is simply due to the current capacitively coupled to the substrate through C_{ox}. The value is given

$$R_{Si} \approx \frac{2}{w \cdot l \cdot G_{sub}} \tag{12.21}$$

where G_{sub} is a fitting parameter that has the dimensions of conductance per area. It is constant for a given substrate resistivity and the distance between the spiral and substrate. A typical value of G_{sub} is about 10^{-7} S/μm^2 [16].

The eddy current in the substrate due to the magnetic coupling between the spiral and the substrate forms the image of the current in the spiral (but circulating in the opposite direction), thus effectively reduces the inductance of the spiral as a whole. One rough way to partially model this inductive effect of the image current is through C_{Si1}/C_{Si2} (hereinafter designated as C_{Si}) in Fig. 12.13(b), which mainly reflects the capacitance of the substrate and has the value of

$$C_{Si} \approx \frac{w \cdot l \cdot C_{sub}}{2} \tag{12.22}$$

where similar to G_{sub}, C_{sub} is a fitting parameter with typical range of 10^{-3}–10^{-2} fF/μm^2.

To get the values of L_{sk}, R_{sk} in the above model, we follow the approach proposed in [2]:

1. Based on the geometry of the spiral, first calculate the DC quantities of the inductor, namely R_{dc} and L_{dc} (e.g., using (12.15)), and a critical frequency defined as

$$\omega_{crit} = \frac{3.1}{\mu_0} \frac{w+s}{w^2} R_{\square} \tag{12.23}$$

where $R_{\square}$ is the sheet resistance for the metal line of the winding.[2]

[2] As shown in [14], with ω_{crit} and DC resistance R_{dc}, the frequency dependent resistance of the inductor can be expressed as

$$R(f) = R_{dc} \cdot \left[1 + 0.1 \left(\frac{\omega}{\omega_{crit}} \right)^2 \right]. \tag{12.24}$$

2. To find the ratio of $\alpha = R_{sk}/R_s$ from solving the following equation:

$$\frac{0.315R_{dc}^2}{(240/n^{1.23})\omega_{crit}^2 L_{dc}^2} \cdot \alpha^2 + \frac{l^2 \epsilon_{ox}}{6C_{ox}L_{dc}c^2} \cdot \alpha - 1 = 0 \quad (12.25)$$

where l is the length of the winding for the spiral, n is the number of turns in the winding, C_{ox} is the oxide capacitance per unit surface area, and c is the speed of light.
3. In the DC case, one can solve for L_s, R_{sk}, L_{sk} from the following three equations

$$\frac{R_s R_{sk}}{R_s + R_{sk}} = \frac{R_{sk}}{1+\alpha} = R_{dc}, \quad (12.26)$$

$$L_{sk} = \frac{L_s}{0.315}\alpha, \quad (12.27)$$

$$L_s + \left(\frac{1}{1+\alpha}\right)^2 \cdot L_{sk} = L_{dc}. \quad (12.28)$$

And finally to calculate R_s from α and R_{sk}.

12.5.3 2-π Model for Spiral Inductors

The winding of a spiral inductor is essentially a distributed system with respect to the substrate contact. 1-π model as shown in Fig. 12.13(b) partitions the capacitive coupling between the winding and substrate contact into two pieces and assigns them to the terminals of the 2-port equivalent circuit. An improved approach to modeling this distributed nature is to partition the series inductive branch into two equal parts and add a capacitive leg between the newly-created intermediate node and the ground, as shown in Fig. 12.14(a). For symmetric configuration,

$$L_{s1} = L_{s2}, \qquad R_{s1} = R_{s2}, \qquad R_1 = R_2, \qquad L_1 = L_2, \qquad C_{p1} = C_{p2},$$

$$C_{sub3}/2 = C_{sub1} = C_{sub2}, \qquad 2R_{sub3} = R_{sub1} = R_{sub2}.$$

Note the change of the symbols in this case: L_1 is for L_{sk1} and C_{sub1} for C_{Si1} in Fig. 12.13(b), etc.

A critical observation for 2-π model is that as shown in Fig. 12.14(a), there is no mutual coupling between L_{s1} and L_{s2}. It means that the 2-π model is not from the physical partitioning of the winding. Instead, the segments are simply derived by dividing the overall inductance into two. That is, $L_{s1} + L_{s2} = L_s$, without mutual coupling between them. The capacitances C_{p1}, C_{p2} are to model the turn-to-turn capacitance, and capacitance C_o is to model the overlapping capacitor between the winding and the under-cross.

The above model has two variations. One is shown in Fig. 12.14(b) [2], which models explicitly the proximity effects in a multi-turn spiral by introducing the mutual coupling between the main inductor (L_0 in Fig. 12.14(b)) in one segment with

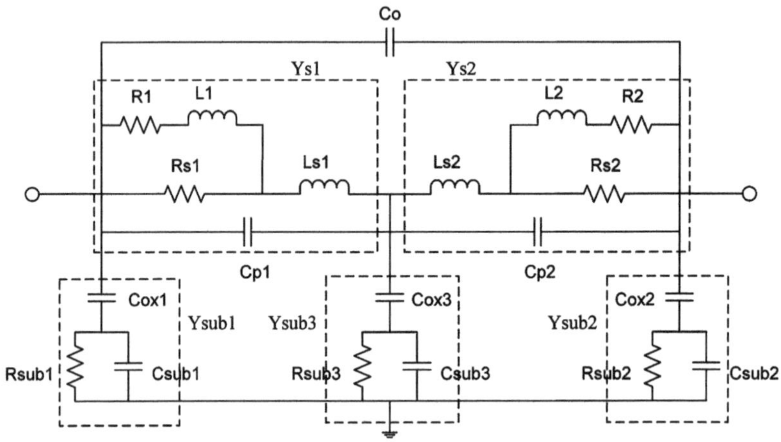

(a) General 2-π model for spiral inductors [14]

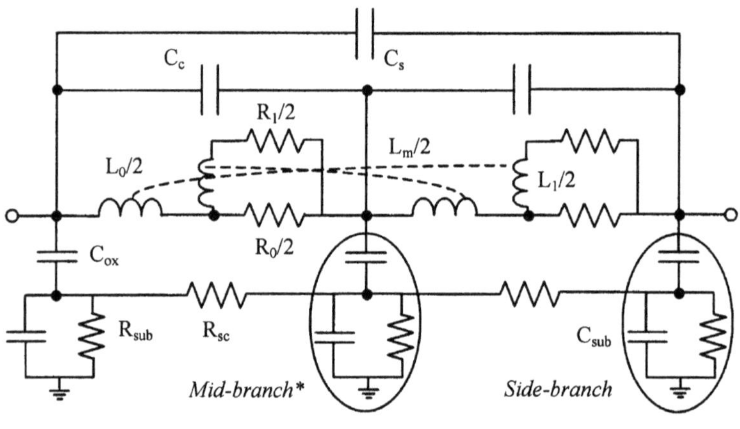

(b) 2-π model with proximity effects explicitly considered for spiral inductors [2]

Fig. 12.14 2-π models for spiral inductors

the inductor in the ladder on the other segment (L_1 in Fig. 12.14(b)). The underlying physics is explained in [14]. A brief explanation is provided in [2] as follows. In addition to the skin effect, the magnetic field generated by neighboring lines further changes the current distribution and results in a higher current density at the edges of the metal lines. This is described as the proximity effect and has a greater impact than the skin effect on the increase of resistance and degradation of Q in present-day spiral inductor designs. Since the inductance is induced by the magnetic field in the adjacent space of the line, we model the magnetic interaction between the external field and internal current by adding the mutual inductance L_m in Fig. 12.14(b).

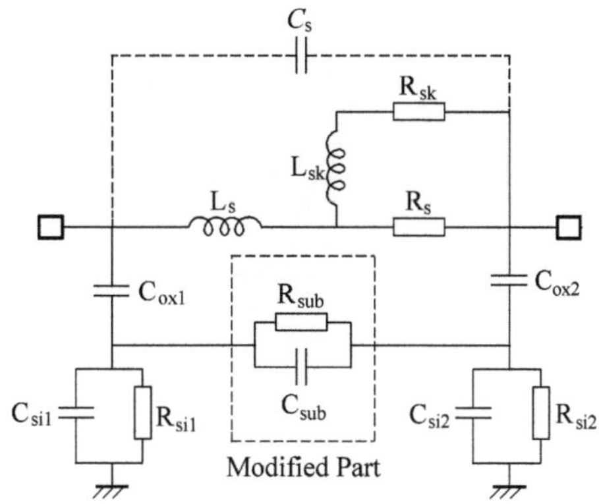

Fig. 12.15 Single-π model considering the lateral substrate RC coupling [7]

Another 2-π model is motivated from the experimental results that Y_{11} and Y_{22} for a spiral inductor is not the same, and it is attributed to the asymmetry of the structure. Thus arises a so-called asymmetric 2-π model as shown in Fig. 12.14(a), where corresponding elements on two segments are not necessarily the same. The model, including parameter extraction is discussed in [11] and is shown in Fig. 12.14(a).

12.5.4 Improved 1-π Models for Spiral Inductors

There is a resurgence recently of 1-π models for spiral inductors because of two reasons: simplicity and good enough accuracy as compared to 2-π models. Besides, the easiness of parameter extraction is also a plus.

We in the following examine several circuit topologies for this 1-π approach, including the T-models for spiral inductors.

12.5.4.1 Single-π Model with Lateral Substrate RC Coupling

This model, featuring a shunt RC branch placed within the substrate and separated from the inductor terminals by the oxide capacitors (Fig. 12.15), is proposed by H. Shin and his co-workers in Seoul National University (SNU) in Korea, in 2003 [7] (and [13] in 2005). The purpose of this lateral substrate RC branch is to model the spiral lateral coupling with and through the substrate. The adjacent metal strips in winding have signal coupling in the silicon substrate via the oxide capacitance. To represent the lateral substrate coupling, R_{sub} and C_{sub} are introduced in this 1-π model. The shunt capacitance, C_s, across two terminals of the inductor, which appears in almost all models for spiral inductors, can be neglected because of the existence of R_{sub}/C_{sub} branch.

As the number of turns of the spiral inductor increases, the substrate coupling between the metal strips becomes larger, resulting in larger C_{sub}. The model validation is shown in Fig. 12.16, which compares the fitting from the model with the

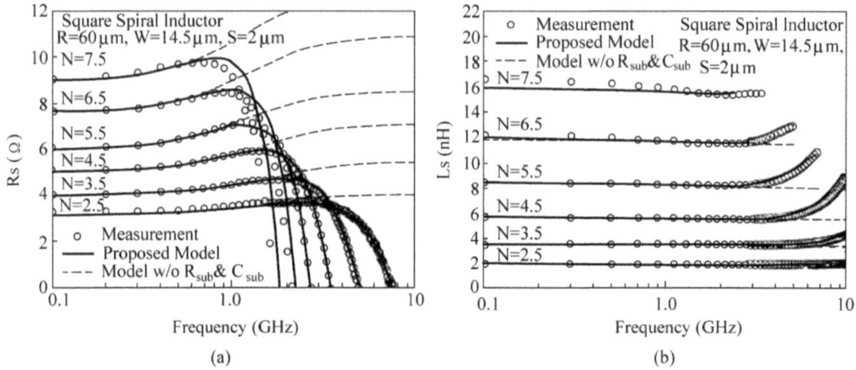

Fig. 12.16 Comparison between the measurement and model. (**a**) Series resistance, and (**b**) series inductance as a function of frequency and the number of turns [7]

measured data for model with and without R_{sub}/C_{sub} branch. It can be seen that the addition of the lateral substrate RC branch improves the model accuracy at high frequencies significantly.

The procedure of parameter extraction for this model is outlined in [13].

Further, an article following Shin's work is published in 2008 [4], systematically describing the procedure of parameter extraction for this type of model. It is summarized below.

The extraction starts with extracting the series inductance and resistance at low frequencies. Then, the oxide capacitance is evaluated in an intermediate frequency range. Afterward, the substrate effects including the substrate resistance and capacitance, as well as coupling, are extracted at higher frequencies. All the lumped circuit element values are analytically determined by the network analysis from the measured network parameters (S- or Y-parameters).

12.5.4.2 Single-π Model with Substrate Inductance

This model is also called the substrate-eddy-π model and was proposed by M. Fujishima and his co-workers in University of Tokyo [6, 15], as shown in Fig. 12.17(b). It introduces a substrate inductor branch in magnetic coupling with the top inductor due to the metal winding. This model considers the losses generated in both vertical and horizontal directions, hence capable of explaining the reduction in equivalent resistance between inductor terminals with increasing frequency. The eddy current is considered by the mutual coupling between the inductances of the winding and the substrate. The circuit elements enclosed with the dotted line have improved the conventional 1-π model. C_{ox1} and C_{ox2} are the capacitances between the wiring and the substrate, and C_{Si1} and C_{Si2} are the substrate capacitances, and R_{Si1} and R_{Si2} are the substrate resistances in the vertical direction. L_s and L_{sk} are the wiring inductances, and R_s and R_{sk} are the wiring resistances describing the skin and proximity

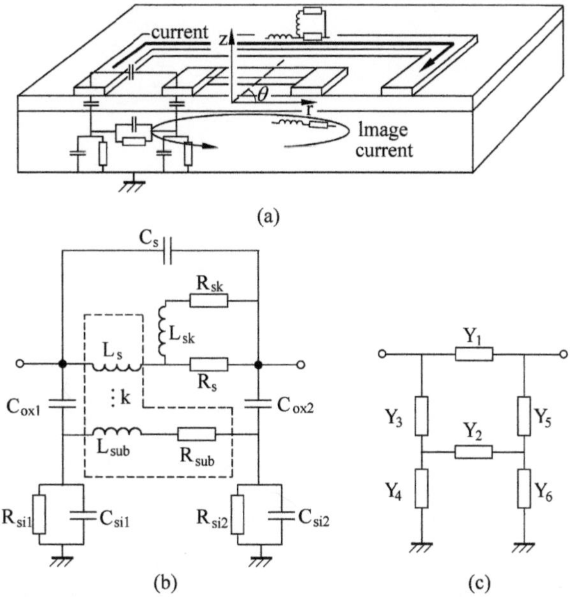

Fig. 12.17 1-π model of on-chip spiral inductors with substrate eddy current. (**a**) 3D perspective for eddy current in the substrate; (**b**) Equivalent circuit of 1-π model with substrate-eddy current; (**c**) lumped circuit block representation [15]

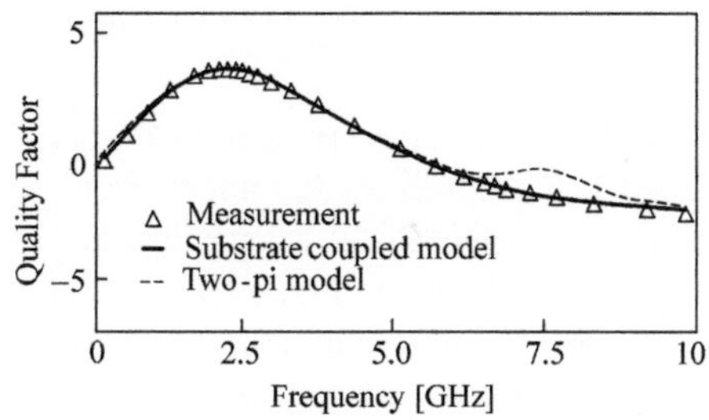

Fig. 12.18 Q-factor predicted by the measured, substrate-coupled model and the 2-π model of an inductor in [15]

effects, and C_s is the capacitance between terminals. Moreover, R_{sub} and L_{sub} denote the substrate resistance and inductance in the circular direction in the substrate. And k is the magnetic coupling coefficient between L_s and L_{sub}.

The circuit parameters are extracted based on the block diagram shown in Fig. 12.17(c) and the procedure can be found in [15].

One advantage of this substrate inductively coupled model is shown in Fig. 12.18. It can be seen that the Q-factor of the substrate-coupled model corresponds closely to the measurement results. The sample data shown is taken from the 0.35-μm-process inductor. The 2π model, on the other hand, deviates from the measured result at the singular point near 7.2 GHz.

12.5.4.3 T-model Considering Substrate Coupling/Loss

This model has a T-shape in circuit topology and consists of two RLC branches to account for the spiral winding, substrate loss, and their mutual interaction, as shown

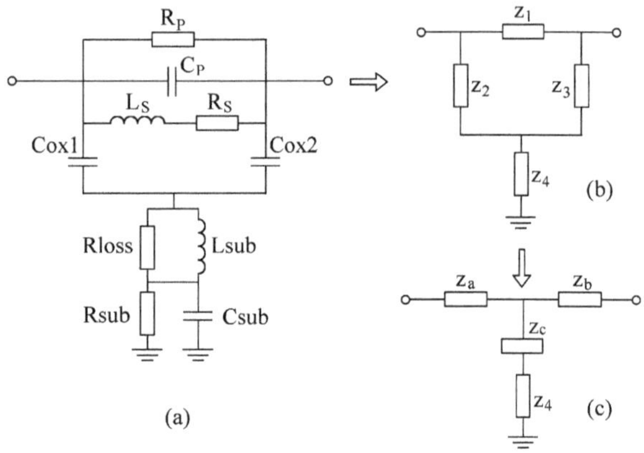

Fig. 12.19 T-section model for transmission-line nature of spiral inductor [8]

in Fig. 12.19(a). It was proposed by J.-C. Guo of NCTU, Taiwan, in [8] based on the work of Horng (also in Taiwan) in [10].

There are two RLC networks, each consisting of four elements, which are linked through C_{ox} to account for the coupling between the spiral winding and the lossy substrate. In Fig. 12.19(a), R_P is a new parameter created to model the series resistance in the winding originated from the lossy substrate return path. L_{sub} and R_{loss} are two more new elements to describe the substrate loss induced by the eddy current.

The model parameter extraction is done through the block diagram in Fig. 12.19(b). Z_1 represents the upper RLC network for the spiral inductor, and Z_4 is the one representing the lossy substrate. $Z_2 = Z_3 = 1/(j\omega C_{ox})$.

The upper Δ-network in Fig. 12.19(b) is further transformed to a Y-network in Fig. 12.19(c) to facilitate the parameter extraction in a more close fashion to the Y-parameters from the 2-port measurement.

For the detailed procedure of parameter extraction, refer to [8]. In general, measured (indirectly) Z- or Y-parameters are not enough to uniquely determine the model parameter values. The approximation under very low or very high frequency can be made to remove some unknown elements and to extract the others at the first pass. And then in the second pass, values for remaining parameters can be obtained. Finally, a global optimization step is needed to get all parameter values to best fit the measured S-parameters, $L(\omega)$, $\mathrm{Re}(Z_{in})$, and $Q(\omega)$.

12.5.4.4 Modified T-Model with Large Bandwidth out of Transmission-Line Nature

A modified T-section equivalent circuit (Fig. 12.20) [9, 10] is proposed to model on-chip spiral inductors with extremely large bandwidth. This model combines a five-element modified T-section to include up to half-wavelength long transmission-line effects, together with one parallel and one series feedback resonators to account for higher-order resonances present in the coil-coupling and ground-return paths. The

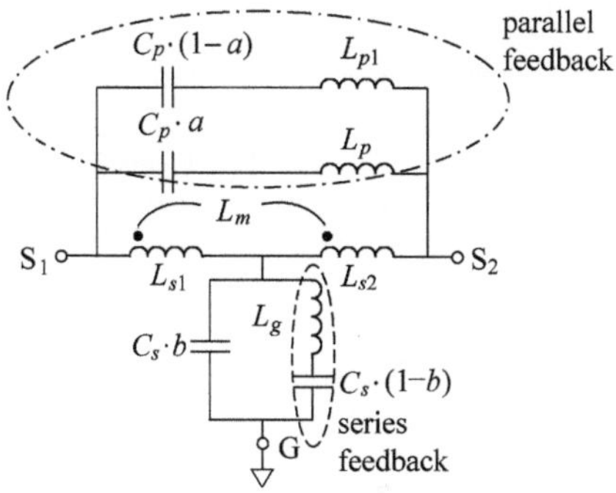

Fig. 12.20 Extended T-model to include parallel/series resonators for considering the coil and substrate couplings [9]

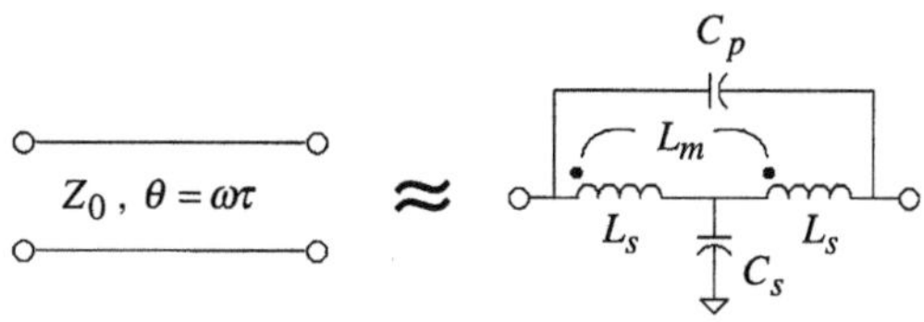

Fig. 12.21 Five-element equivalent circuit for modeling a lossless T-line [9]

modeled S-parameters show excellent agreement with measured results over the entire measurement frequency range of 20 GHz and 50 GHz for a series of planar spiral inductors implemented on silicon and InGaAs substrate, respectively. The effective modeling bandwidth is at least several times larger than in a conventional π-section model [10].

The model shown in Fig. 12.20 [9] for spiral inductors is developed based on an LTCC (low-temperature cofired-ceramic) substrate, but can be applied to CMOS on-chip spiral inductors as well. The feature of this model is to have a remarkably large bandwidth. It combines a five-element modified T-model (Fig. 12.21, where including the mutual inductance L_m, there are five circuit elements) to include the effects of a transmission line with an electrical length up to π (i.e., half-wavelength), and several resonators, $L_g/C_s(1-b)$ for series feedback and $L_p/C_p \cdot a$, $L_{p1}/C_p \cdot (1-a)$ for parallel feedback in Fig. 12.20 to account for the self and ground resonant phenomena.

Lump the shunt capacitance $C_s \cdot b$ with the series feedback resonator as impedance Z_g, Y_p for admittance of the parallel feedback resonator, one can write down $Y_{11}, Y_{22}, Y_{12} = Y_{21}$ in terms of L_{s1}, L_{s2}, L_m, and Z_g, Y_p. The elements in the equivalent circuit as shown in Fig. 12.20 can then be extracted from the measured S-parameters, following procedures outlined in [9].

To consider the frequency-dependent losses, resistances need to be added to and between terminals for either T- or π-shape models for spiral inductors (Fig. 12.22).

The model parameters can directly be extracted from the measured S-parameters.

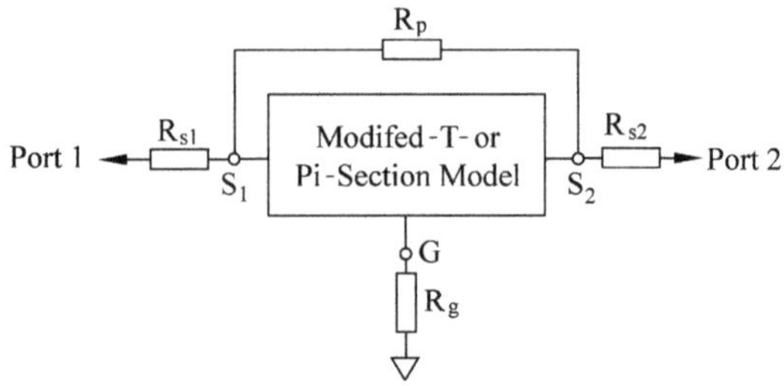

Fig. 12.22 Resistances added to the modified-T or π-section models for inclusion of the frequency-dependent losses [9]

Table 12.2 Model evaluation for spiral inductors with CMOS process

Technology	t (μm)	w (μm)	s (μm)	t_{ox} (μm)	d_{in} (μm)	n's	L's (nH)	ρ_{sub} (Ω cm)
0.13 μm (8M, Cu) [8]	3	15	2		60	2.5–5.5	1.96–8.66	
0.18 μm (1P6M, Al) [4]	2 (M6)	14.5	2	7	60	4.5		10
[10]		10	2		60	5.5	8.64	
90 nm (1P9M) [12]	3 (M9)	12	3		95	3	8.64	

12.5.4.5 The Process Validation of Selected Models for Spiral Inductors

We now summarize some of the model application and validation with selected CMOS process. Table 12.2 cites the models for (1) T-model considering the substrate loss [8]; (2) 1-π model with substrate RC lateral coupling [4]; (3) T-model with transmission-line characterization [10]; (4) 90 nm CMOS transformer [12] to be discussed later in this chapter.

12.5.5 Models for Transformers and Baluns

Spiral transformers can be categorized in three types: unbalanced (or called single-sided, e.g., Fig. 12.7) , balanced (e.g., Fig. 12.8(b)), and balun (short for balanced-unbalanced, e.g., Fig. 12.32).

We first look at the modeling of unbalanced transformers.

There are three independent physical parameters for the transformer and they are:

- The inductance for primary (L_p) and secondary (L_s) windings, respectively.
- The mutual inductance between the primary and secondary windings, M.

Then taking i_p, i_s in Fig. 12.7 as input variables, from definition of L_p, L_s, M, we have the following relationship.

$$v_p = L_p \frac{di_p}{dt} + M \frac{di_s}{dt}, \tag{12.29}$$

$$v_s = L_s \frac{di_s}{dt} + M \frac{di_p}{dt} \tag{12.30}$$

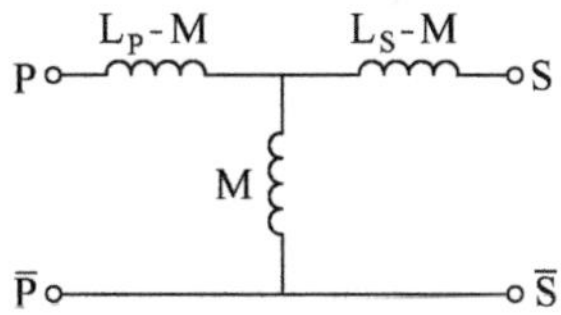

Fig. 12.23 T-shaped model for unbalanced transformers

which can be re-cast as

$$v_p = (L_p - M)\frac{di_p}{dt} + M\frac{d(i_p + i_s)}{dt}, \qquad (12.31)$$

$$v_s = (L_s - M)\frac{di_s}{dt} + M\frac{d(i_p + i_s)}{dt}. \qquad (12.32)$$

The mutual inductance M has the following relationship with L_p, L_s,

$$M = k_m\sqrt{L_p L_s} \qquad (12.33)$$

where k_m is called magnetic coupling coefficient and has the value between 0 and 1. Note that among M and k_m, only one is independent and either one can be used as the model parameter for transformers.

To draw an equivalent circuit for the above equations, one readily obtains the model for transformer with a common ground as shown in Fig. 12.23.

For $k_m = 1$, we can actually derive a constant ratio of the voltage appearing on the secondary port, v_s, vs. that on the primary port, v_p. Multiplying (12.31) with $\sqrt{L_s}$ and (12.32) with $\sqrt{L_p}$, and let $M = \sqrt{L_p L_s}$, one finds (cf. Fig. 12.7),

$$\frac{v_s}{v_p} = \sqrt{\frac{L_s}{L_p}} \equiv n \qquad (12.34)$$

n is conventionally called the turns ratio of the transformer for the following reason. For a transformer with windings on a solenoid, since the cross-section of the windings is the same for both windings, and because of $L \propto N^2$, where N is the number of turns for a single winding (see Sect. 12.4.2), we naturally have

$$\sqrt{\frac{L_s}{L_p}} = \frac{N_s}{N_p} = n. \qquad (12.35)$$

Note that the last equality holds only for the case of solenoid. Thus we can conclude that in the ideal case, the electrical quantity of $\sqrt{L_s/L_p}$ represents the physical quantity of turns ratio between the secondary and primary windings.

Now we discuss how to construct an equivalent circuit using controlled sources (both voltage and current sources) based on the representation shown in Fig. 12.24, which can be used as the basis for a more general case where $k_m < 1$. We need two additional controlled current sources to accomplish this task as shown in Fig. 12.25.

Another way to consider cases of $k_m < 1$ (or $M < \sqrt{L_p L_s}$) is to introduce the concept of leakage inductance, which will not contribute to the coupling between the primary and secondary windings, but acts as an ordinary self-inductance. Designate the leakage inductances as L_{kp}, L_{ks} for two windings, respectively, the modeling requirement is (refer to Fig. 12.26),

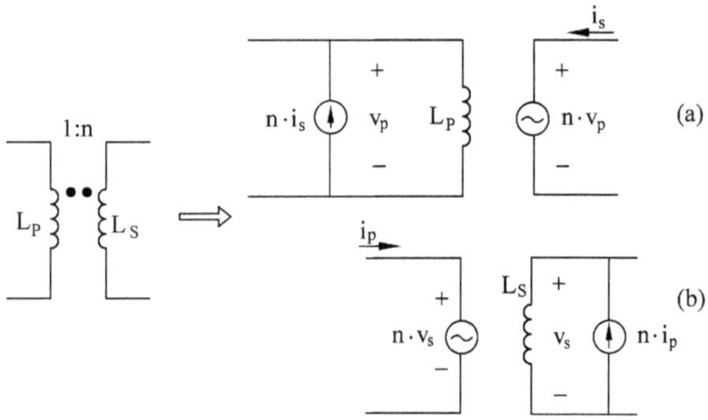

Fig. 12.24 Equivalent circuit for transformers driven from either the primary (**a**) or secondary (**b**) side. Note that in drawing this way, it has implicitly been assumed that $M = \sqrt{L_p L_s}$ and hence $n = \sqrt{L_s/L_p}$

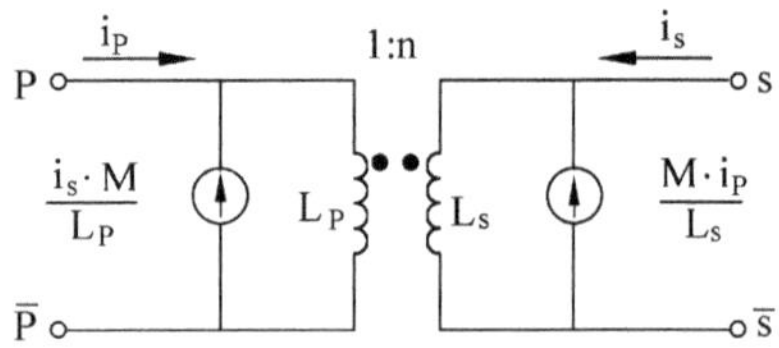

Fig. 12.25 Equivalent circuit for transformers with $k_m < 1$

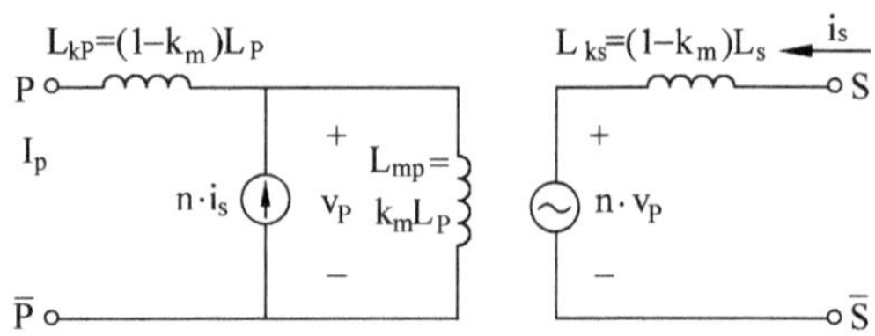

Fig. 12.26 General model for unbalanced transformer. The series resistance of the windings is not included [18]

1.

$$L_p = L_{kp} + L_{mp} = (1 - k_m)L_p + k_m L_p, \tag{12.36}$$

$$L_s = L_{ks} + L_{ms} = (1 - k_m)L_s + k_m L_s. \tag{12.37}$$

2. The coupling coefficient for the magnetic inductances is a unity, i.e.,

$$M_m = \sqrt{L_{mp} L_{ms}} = M = k_m \sqrt{L_p L_s} \tag{12.38}$$

where M_m is the mutual inductance between L_{mp} and L_{ms}.

3. The turns ratio for the coupled L_{mp} and L_{ms} remains unchanged, i.e., $n = \sqrt{L_{ms}/L_{mp}} = \sqrt{L_s/L_p}$.

The equivalent circuit now looks as in Fig. 12.26. Note that L_{ms} is not shown in the figure. Instead, the controlled sources as shown in Fig. 12.24(a) is used for the ideal transformer made of L_{mp} and L_{ms}.

The equivalent circuit in using the leakage inductance is not unique, however. This is because from (12.36–12.38) there are four unknowns, $L_{kp}, L_{mp}, L_{ks}, L_{ms}$ from three knowns, L_p, L_s, and M or k_m. Then one has the freedom in choosing one from either L_{kp} or L_{ks} under the constraint that the leakage inductance cannot be bigger than the self-inductance for the corresponding winding. For example, one

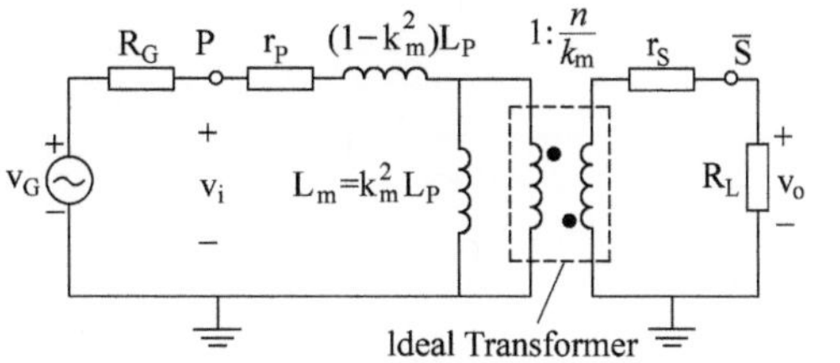

Fig. 12.27 Equivalent circuit for transformer with all leakage inductance put on the primary side and considering the winding resistance [18]

can put the leakage inductance all with one winding of the transformer. Taking the primary winding, because now one must maintain

$$M_m^2 = (L_p - L_{kp})L_s = k_m^2 L_p L_s \tag{12.39}$$

then

$$L_{kp} = (1 - k_m^2)L_p$$

and

$$n_m = \sqrt{\frac{L_s}{k_m^2 L_p}} = \frac{1}{k_m} n$$

where $n = \sqrt{L_s/L_p}$.

To realize the above relationship and to include the series resistance with the windings, we obtain the equivalent circuit as shown in Fig. 12.27. One more comment before we move on: for an ideal transformer as is included in Fig. 12.27, it is assumed that the self-inductance for the primary and secondary windings is infinitely large, and the label over the ideal transformer is just the voltage ratio for the secondary and primary ports.

In the equivalent circuit shown in Fig. 12.26, the voltage-controlled-voltage-source (VCVS) $(n \cdot v_p)$ and the current-controlled-current-source (CCCS) $(n \cdot i_s)$ have parameter n as the proportionality coefficient. Note that, the controlling voltage for the VCVS is the voltage drop across the magnetic inductance, L_m, and is reduced from the input voltage across the terminals P, $\bar{P}$ by the voltage drop on the leakage inductance L_{kp}.

For a more realistic situation, one should add series resistance to both the primary and secondary sides of the equivalent circuit of the transformer as shown in Fig. 12.28. Also, this model works in both directions, i.e., apply the signal source on either port of the transformer.[3]

[3] One question might arise when $k_m = 1$ and then L_s won't appear in the equivalent circuit. The answer is that in this case ($M = \sqrt{L_p L_s}$), the turn ratio n has implicit information on L_s, knowing L_p. That is, $L_s = n^2 L_p$.

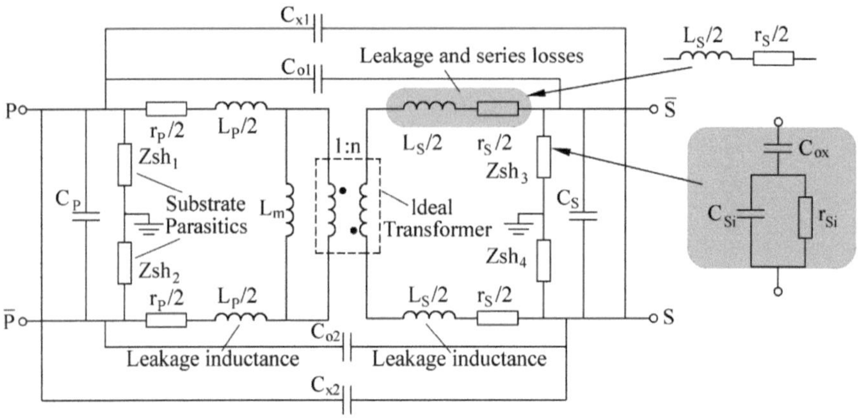

Fig. 12.28 Complete transformer model using leakage inductance and primary to secondary driving [18]. Note that in this figure L_p and L_s are actually leakage inductances, i.e., L_p for L_{kp} and L_s for L_{ks}

12.5.5.1 Complete Transformer Model with Resistive Loss and Capacitive Parasitics

Figure 12.28 shows a complete model for an unbalanced transformer with all resistive loss and capacitive parasitics (with the substrate) considered. The features in the model are

1. Using the leakage inductance to model the non-unity k_m (i.e., $k_m < 1$) in magnetic coupling of the original transformer.
2. Using symmetric topology for each winding, i.e., splitting the leakage inductance and resistance into two equal parts in series with the winding terminals.
3. Using ideal transformer to represent coupling from the primary side to the secondary.
4. The capacitive coupling among the terminals of the transformer and coupling between the top windings and the substrate are included.

12.5.5.2 High Frequency Behavior of Monolithic Transformers

It is clear from Fig. 12.28 that the shunt inductor in the primary, L_m, affects the low end of the frequency response by bypassing the signal which can be transferred to the secondary side, while the series element, L_{kp}, blocks transmission of the signal from primary to secondary as the operating frequency increases. The terms low, mid, and high frequency used in some discussion of transformer frequency response are in the relative sense implied by the effects of these two elements.

Some qualitative observations on the behavior of monolithic transformers can be made from the high-frequency equivalent circuit. Assume in Fig. 12.28, $C_{oi} = C_{xi} = C_o, i = 1, 2$, then reflected to the secondary, one has the equivalent circuit as shown in Fig. 12.29(b).

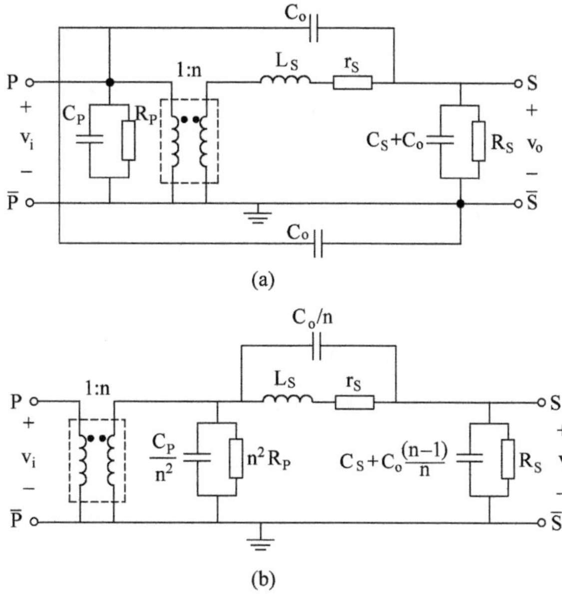

Fig. 12.29 High-frequency transformer model: (**a**) before reflected to the secondary; (**b**) after reflected to the secondary winding [18]

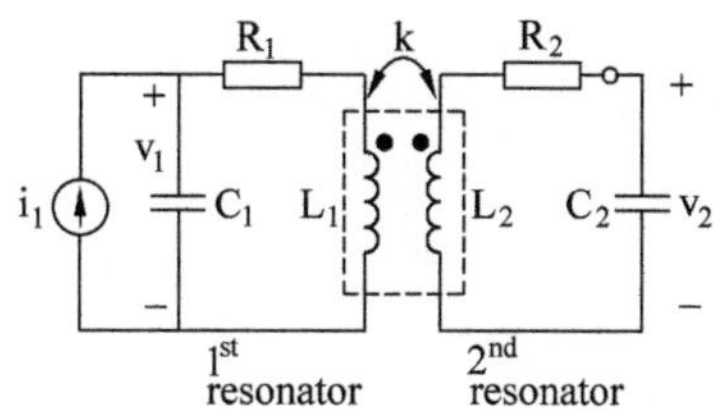

Fig. 12.30 Transformer Q-factor defined using two coupled LC resonators [1]

Finally, we discuss two metrics for transformers: quality factor Q and self-resonant frequency.

1. To define Q for a transformer, we consider the transformer as two coupled LC resonators, as shown in Fig. 12.30. One can easily define Q-factor for each resonator as

$$Q_p = \frac{\omega_{0p} L_p}{R_p}, \qquad Q_s = \frac{\omega_{0s} L_s}{R_s}$$

where ω_{0p} and ω_{0s} are resonant frequencies for primary and secondary resonators, respectively. To get a rough idea how a transformer-coupled resonators can have a boosted Q-factor, we assume two identical LC resonators, i.e., $L_p = L_s = L_0$, etc. Then, there is only one unique resonant frequency of $\omega_0 = 1/\sqrt{L_0 C_0}$. And the Q-factor for each resonator is defined using ω_0 as $Q_0 = \sqrt{L_0/C_0}/R_0$.

We consider now the trans-impedance between v_2 and i_1 in Fig. 12.30, which can be evaluated as

$$Z_{21}(s) = \frac{v_2}{i_1} = \frac{skL_0}{[(1+k)s^2/\omega_0^2 + sR_0C_0 + 1][(1-k)s^2/\omega_0^2 + sR_0C_0 + 1]} \tag{12.40}$$

where k is the magnetic coupling coefficient of the transformer. The resonant (or called oscillation) frequency for this coupled resonator can be defined and found from the poles of Z_{21}, and the results are

$$\omega_S = \frac{\omega_0}{\sqrt{1+k}}, \tag{12.41}$$

$$\omega_I = \frac{\omega_0}{\sqrt{1-k}} \tag{12.42}$$

for $k < 1$. ω_S, ω_I represents, respectively, so-called superior and inferior modes [1]. It can be seen from (12.41–12.42) that as the value of k increases, the superior mode is shifted to lower frequency, while the inferior mode moves to higher frequency. Further, it is shown in [1] that the amplitude corresponding to ω_I decreases drastically when k is increased (the exact reason why it is called inferior).

The quality factor of a transformer-coupled resonator is defined by exploring the phase dependence on frequency for the input impedance of a resonator as follows [1]

$$Q = \frac{\omega}{2}\left|\frac{\partial\phi}{\partial\omega}\right|. \tag{12.43}$$

It is found that

$$Q_S = Q_0\sqrt{1+k}, \tag{12.44}$$

$$Q_I = Q_0\sqrt{1-k}. \tag{12.45}$$

The conclusion is thus for the superior mode, the quality factor is enhanced by a factor of $\sqrt{1+k}$ while the resonant frequency becomes lower than the original resonant frequency for separate resonators.

2. Self-Resonant Frequency (SRF)

   The SRF can be obtained by treating the input port of the transformer (with the output port often open circuited) as an inductor with its effective inductance drawn w.r.t. frequency. Once the effective inductance passes the zero value (becoming negative from positive), the corresponding frequency is termed as the SRF. One example is shown in Fig. 12.31 where the dependence of SRF on the polarity of magnetic coupling coefficient, k, is shown (positive k results in a bigger SRF) [3].

12.5.6 Parameter Extraction for Transformer Model

Taking a balun as an example, the parameters to be extracted include the inductance for the primary and secondary windings, and the magnetic coupling coefficients, k's, etc.

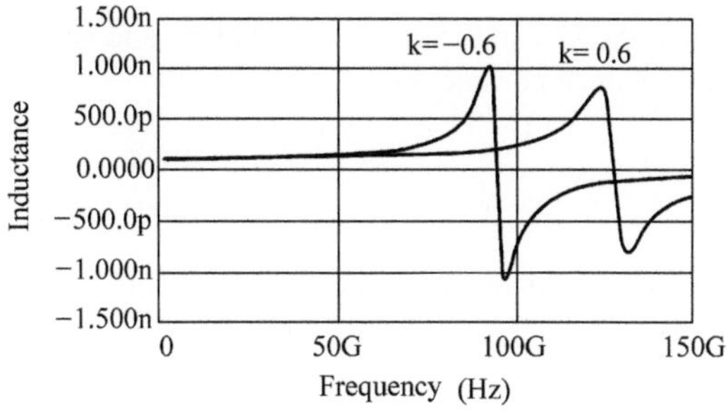

Fig. 12.31 Self-Resonant Frequency (SRF) for transformer [3]

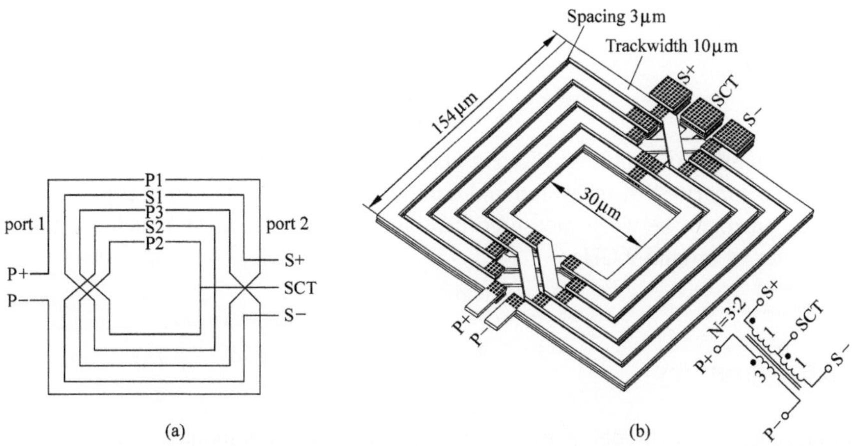

Fig. 12.32 Structure for 3 : 2 balun used for model parameter extraction [22]

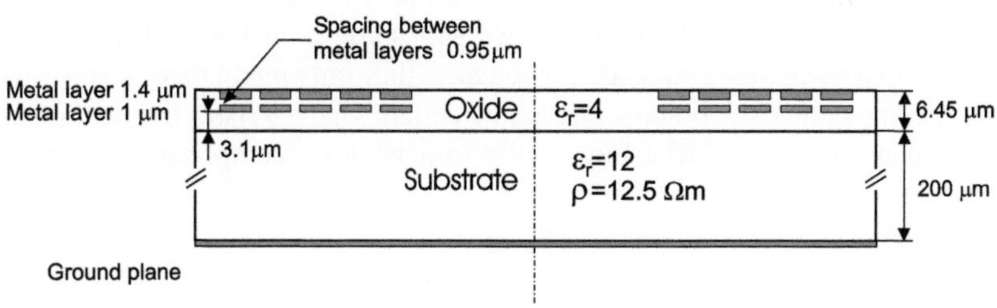

Fig. 12.33 Substrate structure for 3 : 2 balun used for model parameter extraction [22]

The structure in Fig. 12.32 is from Thüringer of TU Wien [22]. The silicon process of the balun is shown in Fig. 12.33 [22].

The measurement setup is to leave the center tap of the secondary winding for the balun in Fig. 12.32 open, and the primary terminals P^+, P^- are taken as port 1, while the secondary terminals S^+, S^- as port 2. The terminals P^-, S^- are grounded. This two-port system is measured for S-parameters, S_{11}, etc., and then converted to Z- and Y-parameters.

Then the self inductances for primary and secondary windings are derived from the impedance parameters

$$L_P = \mathrm{Im}(Z_{11})/\omega, \tag{12.46}$$

$$L_S = \mathrm{Im}(Z_{22})/\omega. \tag{12.47}$$

The mutual inductance can be extracted from the impedance and admittance parameters as

$$M = \sqrt{\left(Y_{11}^{-1} - Z_{11}\right) Z_{22}/\omega^2}. \tag{12.48}$$

The magnetic coupling coefficient between the primary and secondary windings is

$$k = \sqrt{\frac{(Y_{11}^{-1} - Z_{11})Z_{22}}{\mathrm{Im}(Z_{11})\,\mathrm{Im}(Z_{22})}}. \tag{12.49}$$

The equivalent circuit and its extracted parameters are shown in Fig. 12.34(a) and the comparison of measured and computed S_{11}, S_{22} is shown in Fig. 12.34(b). Good agreement between measured data and model calculation is achieved up to frequency of around 9 GHz.

12.6 Summary

In this chapter, the modeling of planary on-chip passive components—resistors, capacitors, and inductors/transformers—is discussed. The structures of these components, realized using CMOS process, are also introduced.

All monolithic electronic components, especially those built on the lossy silicon substrate, suffer from undesired parasitics. For spiral inductors and transformers, due to the existence of the parasitic capacitance, they only retain their intrinsic properties, i.e., as inductive components, below certain characteristic frequency, called self-resonant frequency (SRF). Beyond that frequency, the component behaves more like a resonant circuit.

The challenge for an accurate modeling of spiral inductors not only involves the extraction of various capacitive parasitics, but also need to elegantly and concisely account for the interplay between the magnetic fields generated by the originating conductive current and their induced current, the so-called eddy current. The superimposition of these two types of currents: originally driving current and eddy-current, results in the tendency of AC current concentrating near the surface layer within the conductor. The surface-approaching phenomena of AC currents are called skin and proximity effects, which are frequency-dependent. It should be noticed that the skin effect has square-root dependence on frequency (the larger the operation frequency, the smaller the skin depth), while the proximity effect has squared frequency dependency. Since proximity effects only show up with multi-turn spirals, so its proper modeling is important for large spiral inductance (in the range of nH order) since multiple turns are required in CMOS spirals to achieve these values.

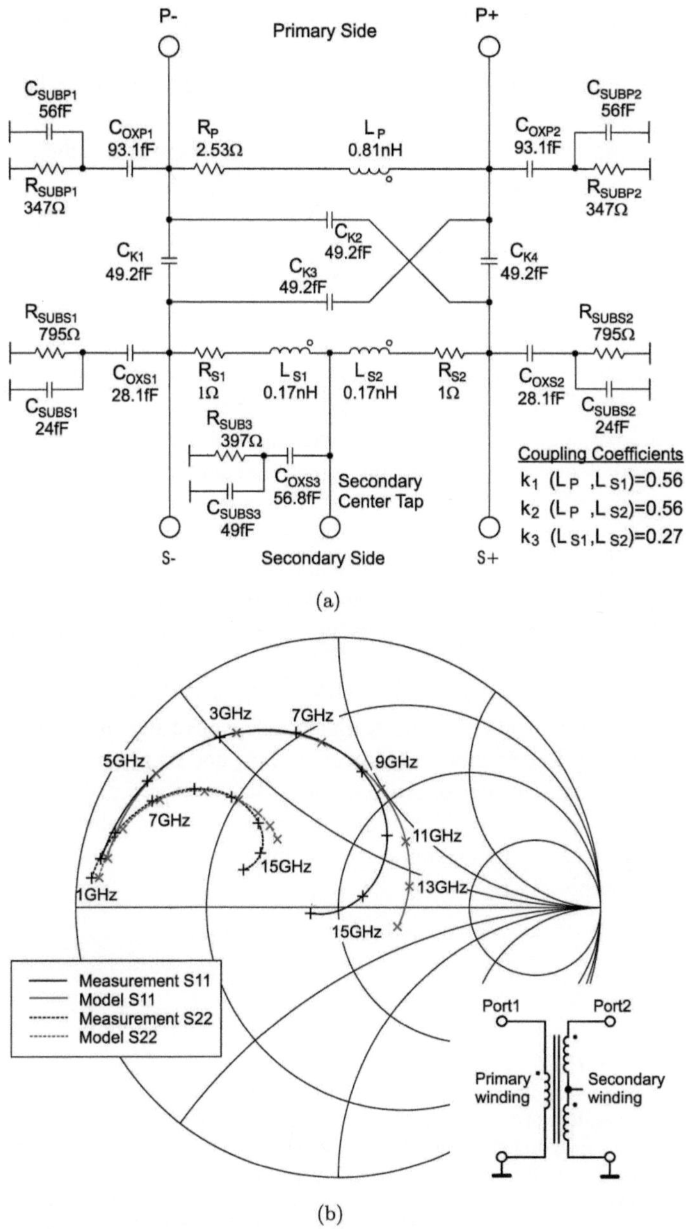

Fig. 12.34 Equivalent circuit and extracted parameters (**a**), and comparison of computed and measured S_{11}, S_{22} (**b**) for 3 : 2 balun [22]

The modeling of transformers, including baluns, is always intimidating. This chapter devotes quite a space to this subject. The key point for the transformer modeling is to realize that the physical turns ratio, n, is not a fundamental parameter. The electrical parameters of the self-inductance for the primary/secondary windings and their mutual inductance are. Strictly speaking, the turns ratio is an electrical concept, which is defined through the ratio (square-root) of the inductances for the secondary and primary windings.

Understanding this non-obvious issue about n helps appreciate the introduction of the leakage inductance when the magnetic coupling coefficient, k, between primary/secondary windings is less than 1. Also, how to use controlled voltage/current sources in substituting the ideal transformer is critical to bridge the circuit with non-Spice elements to a fully Spice-compatible circuitry.

The last issue with transformer modeling is the concept of quality factor Q, which is defined based on the phase-frequency response of one-port system (12.43) and can avoid the mathematical singularity problem in applying the conventional formula.

The model-parameter extraction is an extremely important step in any modeling and/or equivalent-circuit configuring process. From the measured S-parameters to model parameters usually involves two stages: (1) processing of the raw data through de-embedding to obtain the information on the intrinsic part of the device-under-test (DUT), (2) to partition the entire measurement frequency range into several sub-ranges: low-mid-high frequencies, and within each frequency sub-range to determine the most sensitive parameters in this range. Finally, using global optimization to smooth the parameter values and to achieve maximum model accuracy.

Another issue which is not touched in this chapter is the structure synthesis of on-chip passive components from the design targets. For readers interested, the book by Prof. Thomas H. Lee of Stanford University [16] and its sister book, *Planar microwave engineering: a practical guide to theory, measurement, and circuits*, Cambridge University Press, 2004, are good references, including discussion of some other important passive components, such as transmission lines and antennas, which is not covered in this chapter.

References

1. Baek, D., Song, T., Ko, S., Yoon, E., Hong, S.: Analysis on resonator coupling and its application to CMOS quadrature VCO at 8 GHz. IEEE RFIC '03, p. 85 (2003)
2. Cao, Y., Groves, R.A., Huang, X., Zamdmer, N.D., Plouchart, J.O., Wachnik, R.A., King, T.-J., Hu, C.: Frequency-independent equivalent-circuit model for on-chip spiral inductors. IEEE J. Solid-State Circuits **38**(3), 419 (2003)
3. Cheema, H.M., Sakian, P., Janssen, E., Mahmoudi, R., van Roermund, A.: Monolithic transformers for high frequency bulk CMOS circuits. In: IEEE Topical Meeting on SiRF (Silicon Monolithic IC in RF Systems) '09, p. 1 (2009)
4. Chen, H.-H., Zhang, H.-W., Chung, S.-J., Kuo, J.-T., Wu, T.-C.: Accurate systematic model-parameter extraction for on-chip spiral inductors. IEEE Trans. Electron Devices **55**(11), 3267 (2008)
5. Chowdhury, D., Reynaert, P., Niknejad, A.M.: Design considerations for 60 GHz transformer—coupled CMOS power amplifiers. J. Solid-State Circuits **44**(10), 2733 (2009)
6. Fujishima, M., Kino, J.: Accurate subcircuit model of an on-chip inductor with a new substrate network. In: IEEE VLSI Circuits, p. 376, June 2004

7. Gil, J., Shin, H.: A simple wide-band on-chip inductor model for silicon-based RF ICs. IEEE Trans. Microw. Theory Tech. **51**(9), 2023 (2003)
8. Guo, J.-C., Tan, T.-Y.: A broadband and scalable model for on-chip inductors incorporating substrate and conductor loss effects. IEEE Trans. Electron Devices **53**(3), 413 (2006)
9. Horng, T.S., Wu, J.M., Yang, L.Q., Fang, S.T.: A novel modified equivalent circuit for modeling LTCC embedded inductors with a large bandwidth. IEEE Trans. Microw. Theory Tech. **51**(12), 2327 (2003)
10. Horng, T.S., Jau, J.K., Huang, C.H., Han, F.Y.: Synthesis of a super broadband model for on-chip spiral inductors. In: IEEE RFIC Symp. Dig., p. 453 (2004)
11. Huang, F., Lu, J., Jiang, N., Zhang, X., Wang, Y.: Frequency-independent asymmetric double-π equivalent circuit for on-chip spiral inductors: physics-based modeling and parameter extraction. J. Solid-State Circuits **41**(10), 2272 (2006)
12. Hus, H.-M., Tsai, M.-C., Huang, K.-H.: An op-chip transformer in silicon based technology. J. Micromech. Microeng. **17**, 1504–1510 (2007)
13. Kang, M., Gil, J., Shin, H.: A simple parameter extraction method of spiral on-chip inductors. IEEE Trans. Electron Devices **52**(9), 2005 (1976)
14. Kuhn, W.B., Ibrahim, N.M.: Analysis of current crowding effects in multiturn spiral inductors. IEEE Trans. Microw. Theory Tech. **49**, 3138 (2001)
15. Lai, I.C.H., Fujishima, M.: A new on-chip substrate-coupled inductor model implemented with scalable expressions. J. Solid-State Circuits **41**(11), 2491 (2006)
16. Lee, T.H.: The Design of CMOS Radio-Frequency Integrated Circuits, 2nd edn. Cambridge University Press, Cambridge (2004)
17. Li, X., Shekhar, S., Allstot, D.J.: G_m-boosted common-gate LNA and differential collpitts VCO/QVCO in 0.18-μm CMOS. J. Solid-State Circuits **40**(12), 2609 (2005)
18. Long, R.: Monolithic transformers for silicon RF IC design. IEEE J. Solid-State Circuits **35**(9), 1368 (2000)
19. Mohan, S.S.: The design, modeling and optimization of on-chip inductor and transformer circuits. Ph.D. thesis, Stanford University (1999)
20. Mohan, S.S., del Mar Hershenson, M., Boyd, S.P., Lee, T.H.: Simple accurate expressions for planar spiral inductances. J. Solid-State Circuits **34**(10), 1419 (1999)
21. Simbürger, W., et al.: Silicon-based RF ICs up to 100 GHz: research trends and applications. In: ICSICT, Beijing, China (2004)
22. Thüringer, R.: Characterization of integrated lumped inductors and transformers. MS Thesis, TU Wien (2002)
23. Voorman, J.O.: Continuous-Time Analog Integrated Filters. IEEE Press, New York (1993)
24. Wheeler, H.A.: Simple inductance formulas for radio coils. Proc. IRE **16**(10), 1398 (1928)
25. Yu, Z., McAndrew, C.C.: RF CMOS is more than CMOS: modeling of RF passive components. In: CICC (Custom Integrated Circuits Conference), p. 407, San Jose, September 2009
26. Zannoth, M., Kolb, B., Fenk, J., Weigel, R.: A fully integrated VCO at 2 GHz. J. Solid-State Circuits **33**(12), 1987 (1998)

Part IV
Modeling of Multiple Gate MOSFETs

Chapter 13
Multi-Gate MOSFET Compact Model BSIM-MG

Darsen Lu, Chung-Hsun Lin, Ali Niknejad, and Chenming Hu

Abstract As the scaling of conventional planar CMOS is reaching its limits, multiple-gate CMOS structures will likely take up the baton. To facilitate circuit simulation in such advanced technologies, we have developed BSIM-MG: a versatile compact model for multi-gate MOSFETs. In this chapter separate formulations for common multi-gate and independent multi-gate MOSFETs are presented. The core I-V and C-V models are derived and agree well with TCAD simulations without using fitting parameters, reflecting the predictivity and scalability of the model. Physical effects such as volume inversion, short channel effects and quantum mechanical effects are included in the model. We verify BSIM-MG against triple-gate SOI FinFET experimental data. The model fits data very well across a wide range of biases, gate lengths and temperatures. It is also computationally efficient and suitable for simulating large circuits. Finally, several multi-gate circuit simulation examples are presented to demonstrate the use of the model.

13.1 Introduction

Over the past several decades, the size reduction of CMOS circuits has fueled the growth of the microelectronics industry. The manufacturing cost of integrated circuits has decreased exponentially, making countless new applications available to the general public.

D. Lu (✉) · A. Niknejad · C. Hu
EECS, University of California, Berkeley, CA, USA
e-mail: darsen@eecs.berkeley.edu

A. Niknejad
e-mail: niknejad@eecs.berkeley.edu

C.-H. Lin
IBM Thomas J. Watson Research Center, Yorktown Heights, NY, USA
e-mail: linc@us.ibm.com

G. Gildenblat (ed.), *Compact Modeling*,
DOI 10.1007/978-90-481-8614-3_13,

In the past forty years, CMOS technology advances relied on the improving manufacturing capabilities. The minimum manufacturable line width has decreased year by year. Surprisingly, the basic structure of the MOSFET has not changed much.

Simple scaling has become more and more challenging due to fundamental device-physics reasons [1]. At the device gate length (L) less than 100 nm, further reduction in L has yielded limited improvements in performance due to velocity saturation [2] and source velocity limit [3]. For this reason, the strained silicon technology has been put into production [4]. The scaling of gate dielectric also poses a challenge. As the physical thickness of the SiO_2 gate dielectric (T_{ox}) is scaled beyond 1.2 nm, quantum mechanical tunneling current from the gate into the channel becomes significant [5]. Further reduction in T_{ox} will result in large static leakage current and large power consumption even when the device is turned off. Therefore at around the 45 nm node, high-K gate dielectric is used to scale down the equivalent oxide thickness (*EOT*) without increasing the gate tunneling current. Metal gate electrodes are also used to eliminate the unwanted poly-silicon gate depletion effect [6]. Even with these advances, there is little room left for EOT scaling. This means the gate control of the channel can not be made much stronger, therefore channel length can not be made much shorter lest the drain exerts a proportionately large control leading to excessive short-channel effects and high off-state transistor leakage. A new approach is needed to allow future reduction of channel length. The multi-gate structure is a promising approach [7].

Another roadblock for transistor size reduction is the random fluctuation of the number of dopant atoms in the MOSFET channel, called random dopant fluctuation (*RDF*) effect, which increases the variation in device threshold voltage (V_{th}) [1, 8]. Since the standard deviation of V_{th} due to RDF is inversely proportional to $\sqrt{WL}$, circuits with small device dimensions such as the SRAM cell is especially susceptible to RDF.

RDF is a fundamental source of variation, which can not be eliminated as long as the channel is highly doped. Perhaps the only way to fight RDF is to reduce the channel doping. This is not possible for conventional bulk MOSFETs since heavy doping is required to control the short channel effects and subthreshold leakage. However, ultra thin-body (UTB) devices such as silicon-on-insulator (UTBSOI) MOSFETs [9] and especially UTB Multi-Gate Devices like the FinFET [10] have much better short-channel behavior. In fact reducing the body thickness presents a new scaling path to allow future device size reduction in addition or in lieu of the past dependence on EOT reduction and channel doping escalation.

In other words, the main advantage of the multi-gate devices is the improved short channel effects. Since the channel (body) is controlled electrostatically by the gate from multiple sides, the channel is better-controlled by the gate than in the conventional transistor structure. Unwanted leakage components are reduced and a small transistor can be used to continue the cost reduction through miniaturization. Improved gate control also provide lower output conductance, i.e., smaller $\frac{dI_{ds}}{dV_{ds}}$ in the current saturation region. This provides greater voltage gain, which is beneficial to analog circuits as well as to the noise tolerance of digital circuits.

A second advantage of the multi-gate devices is the improved on-state drive current (I_{on}) and therefore faster circuit speed. I_{on} improvement arises from several reasons [11]. Reduction of channel doping reduces impurity Coulombic scattering. Reduced channel doping reduces the electric field normal to the SiO_2 interface and therefore reduces the surface roughness scattering. Finally, a promising multi-gate structure, FinFET, provides a larger channel width with a small footprint area. This raises I_{on}, which is handy for driving a large capacitive load such as long interconnect. The advantages of multi-gate devices is well known and experimentally demonstrated by many researchers [12–15].

Given these advantages, multi-gate devices will likely be used in future CMOS technologies. However, a production-worthy multi-gate compact model (SPICE model) which allows efficient circuit design is yet to be developed.

The BSIM (Berkeley Short-channel IGFET Model) series compact models have served the industry for 20 years [16, 17]. BSIM3 and BSIM4 industry standard models have been widely used for the simulation of planar bulk MOSFETs. As technology advances, new compact models are developed to support new device architectures and incorporate new device physics. BSIMSOI [18, 19] was developed to model partially-depleted, fully-depleted and dynamically-depleted SOI devices. Recently, BSIM-MG [20, 21] was introduced for multi-gate circuit simulation.

In this chapter we will discuss multi-gate MOSFET compact model focusing on BSIM-MG.

Besides BSIM-MG, other multi-gate compact models have also been developed. Interested readers may refer to [22–27].

13.1.1 Various Flavors of Multi-gate MOSFET

There are many different flavors of multi-gate MOSFETs. Several examples are shown in Fig. 13.1.

A well-known example is the FinFET [10]. The FinFET consists of a thin silicon body (the fin) and a gate wrapping around its top and two sides. The ITRS [28] considers it the candidate to replace planar MOSFETs for the aforementioned benefits of multi-gate transistor and because a FinFET is relatively easy to fabricate. FinFETs can be made on either bulk or SOI substrates, creating the bulk FinFET (Fig. 13.1(a)) or the SOI FinFET (Fig. 13.1(b)). In some FinFET processes the oxide hard mask on top of the fin is not removed, creating the double-gate FinFET (Fig. 13.1(c)). In double-gate FinFETs the top surface of the fin does not conduct current, whereas in triple-gate FinFETs (Figs. 13.1(a), (b)) the side surfaces and the top surface all conduct current.

Another example of multi-gate MOSFET is the all-around gate device (Fig. 13.1(d)). It consists of a pillar-like body surrounded by the gate dielectric and the gate. The nanowire MOSFET [29] is one example of all-around gate devices. Depending on the fabrication process, the channel may be either vertically or horizontally oriented.

Optionally, a FinFET can have two separated gates that are independently biased. This can be achieved by removing the top portion of the gate of a regular FinFET

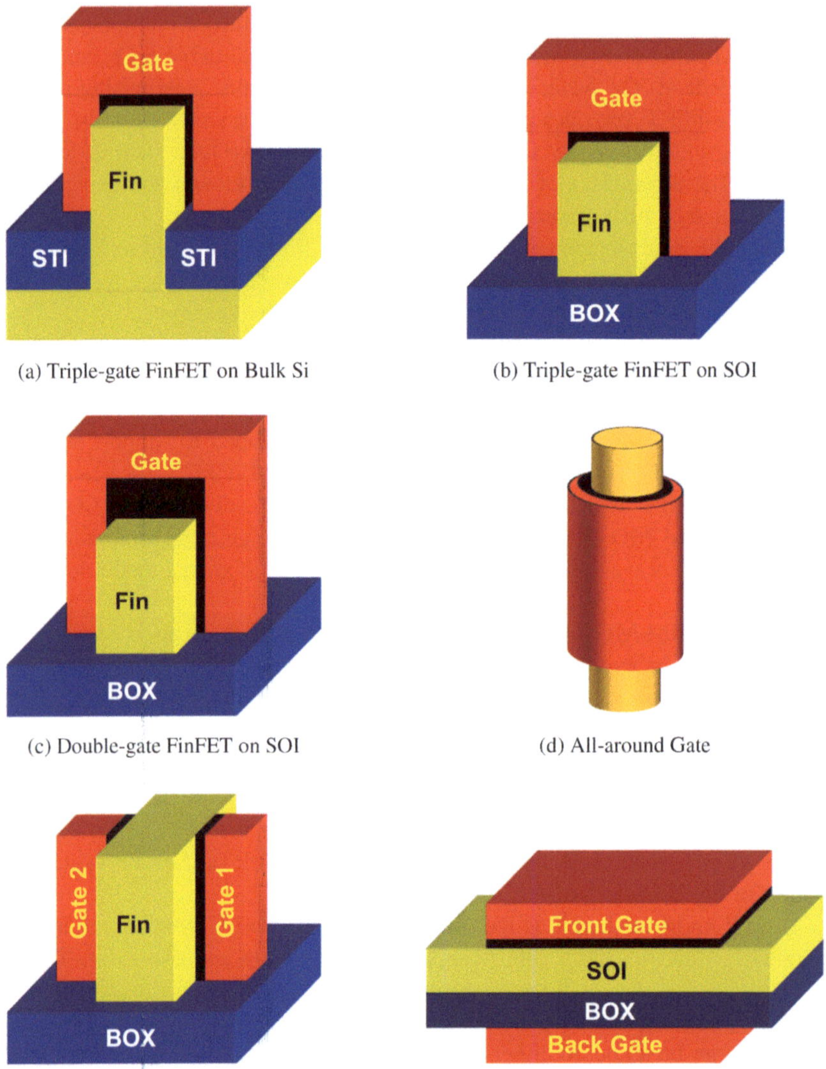

(a) Triple-gate FinFET on Bulk Si
(b) Triple-gate FinFET on SOI
(c) Double-gate FinFET on SOI
(d) All-around Gate
(e) Independent Double-gate FinFET on SOI
(f) Planar Double-gate SOI

Fig. 13.1 Illustration of various flavors of multi-gate FETs (SOI: silicon-on-insulator layer; BOX: buried oxide)

using chemical mechanical polishing, forming the independent double-gate FinFET (Fig. 13.1(e)) [30].

Independent double-gate MOSFETs may also be made as a planar device [31]. The planar double-gate SOI (Fig. 13.1(f)) is essentially a planar SOI MOSFET with a thin buried oxide (labeled as BOX). A heavily-doped region in silicon under the buried oxide acts as the back-gate. Unlike the front-gate, the back-gate is primarily used for tuning the device V_{th}. The buried oxide is usually not thin enough to induce an inversion layer at the back surface. V_{th} tuning can be used to compensate for variability in IC manufacturing from chip to chip or even circuit to circuit within the same chip. Doing so improves the IC speed and power consumption. It can also be used to dynamically raise or lower V_{th} circuit by circuit within a chip in response to the need for less leakage or more speed. This is a very effective means of managing power consumption.

13.1.2 BSIM-IMG and BSIM-CMG

It is likely that more than one flavor of multi-gate MOSFETs will be used in production. Therefore the compact model should ideally cover as many of these flavors as possible. We have classified multi-gate MOSFETs into two main categories: independent multi-gate (IMG) and common multi-gate (CMG) MOSFETs.

IMG refers to independent double-gate MOSFETs with two separate gates. The front- and back-gate stacks are allowed to have different gate workfunctions, biases, dielectric thicknesses and materials. The independent-gate FinFET (Fig. 13.1(e)) and the planar double-gate SOI (Fig. 13.1(f)) belong to this category.

CMG refers to a special case where the gates are "on and the same." The gate stacks of CMG MOSFETs have identical gate workfunction, bias and dielectric thickness and material. Regular FinFETs and all-around gate MOSFETs (Figs. 13.1(a)–(d)) fall into to this category.

Two separate compact models BSIM-IMG and BSIM-CMG are developed for IMG and CMG devices, respectively. BSIM-IMG has 4 or 5 terminals: front-gate, back-gate, drain, source and an optional substrate terminal. BSIM-CMG has one less terminal compared to BSIM-IMG because there is only one gate terminal.

13.2 Core Model for the Independent Double-gate MOSFET

In this section we present the core model of BSIM-IMG [21]: the calculation of surface potential and the derivation of the basic long channel I-V and C-V models.

13.2.1 Basic Modeling Framework

Figure 13.2 illustrates the basic modeling framework for the independent double-gate MOSFET. This view corresponds to a horizontal cross section of the independent-gate FinFET (Fig. 13.1(e)) or a vertical cross section along the length direction of the planar double-gate SOI MOSFET (Fig. 13.1(f)). The silicon body (with thickness T_{si}) is sandwiched between the front- and back-gates and gate dielectrics. The front and back sides are asymmetric: the two gate stacks may have different gate work functions (Φ_{fg}, Φ_{bg}), materials (metal or heavily doped semiconductor), dielectric thicknesses (T_{ox1}, T_{ox2}), and dielectric constants (ε_{ox1}, ε_{ox2}).

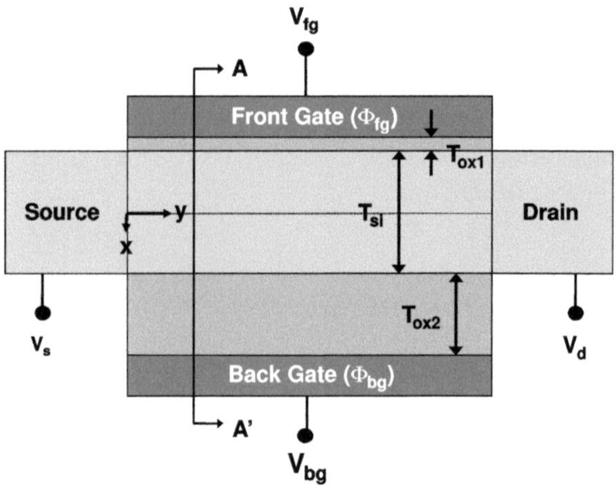

Fig. 13.2 Basic framework for modeling independent double-gate MOSFETs

We assume the silicon body is lightly-doped and fully-depleted. Independent double-gate MOSFETs will likely have a lightly-doped body to minimize RDF. Since IMG design provides a raised V_{th} with the back gate bias, it is particularly useful for transistors with lightly doped body (low V_{th}).

In BSIM-IMG and BSIM-CMG, the electric potential definition is referenced to the electron quasi Fermi level at the N+ source:[1]

$$\psi(x) = -\frac{E_c(x) - E_f(source)}{q} \tag{13.1}$$

$E_f(source)$ is the electron quasi Fermi level at the N+ source.

In a fully-depleted double-gate structure, the substrate is no longer visible or important in the 2D cross section (Fig. 13.2).[2] Therefore we use the N+ source instead of the substrate as a potential reference point. Figure 13.3 shows the energy band diagram at flat-band condition for the vertical cutline along $A - A'$ in Fig. 13.2.

13.2.2 Surface Potential Calculation

BSIM-IMG is a surface-potential based model. Namely, the I-V and C-V are expressed in terms of electric potentials at the silicon/oxide interfaces. Surface potentials must be calculated at both the source end and the drain end. We start our analysis based on the one-dimensional Poisson's equation:

$$\frac{d^2\psi(x)}{dx^2} = \frac{qN_c}{\varepsilon_{si}} \cdot \exp\left[\frac{q(\psi(x) - V_{ch})}{kT}\right] \tag{13.2}$$

[1]In this chapter we present equations for N-type devices.

[2]A substrate still exists underneath the buried oxide or the shallow trench isolation (Figs. 13.1(a)–(c)), but has little influence on the electrostatics in the body.

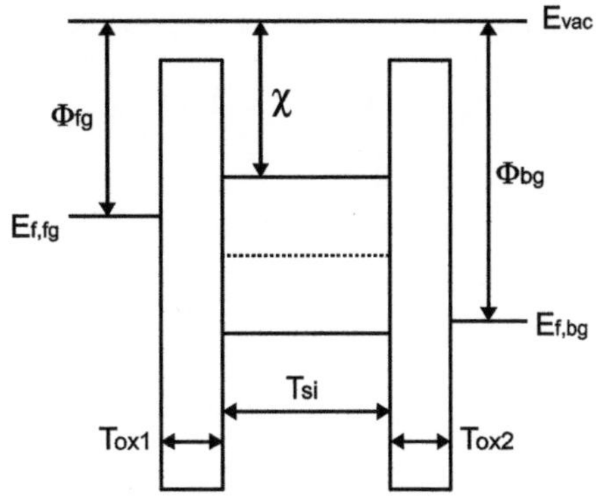

Fig. 13.3 Energy band diagram at flat-band condition corresponding to the $A - A'$ cutline in Fig. 13.2. χ is the electron affinity

where N_c is the density of states of the conduction band; V_{ch} is the channel potential, or the quasi Fermi level relative to that at the source. $V_{ch} = 0$ at the source end and $V_{ch} = V_{ds}$ at the drain end.

Two boundary conditions must be satisfied to ensure electric field continuity at the body and gate dielectric interfaces:

$$V_{fg} - \Delta\Phi_1 - \psi_{s1} = -\frac{\varepsilon_{si}}{C_{ox1}} \cdot \mathcal{E}_1, \tag{13.3}$$

$$V_{bg} - \Delta\Phi_2 - \psi_{s2} = -\frac{\varepsilon_{si}}{C_{ox2}} \cdot \mathcal{E}_2 \tag{13.4}$$

where $\Delta\Phi_1$ is the work function of the front gate (Φ_{fg}) relative to that of the N+ source (χ). Similarly, $\Delta\Phi_2$ is the work function of the back gate (Φ_{bg}) relative to χ. $\psi_{s1} = \psi(x = -T_{si}/2)$; $\psi_{s2} = \psi(x = +T_{si}/2)$; $\mathcal{E}_1 = \frac{d\psi}{dx}|_{x=-T_{si}/2}$; $\mathcal{E}_2 = \frac{d\psi}{dx}|_{x=+T_{si}/2}$.

The analytical solution to this problem is known [32, 33]. However, it requires solving a set of two coupled non-linear equations:

$$V_{fg} - \Delta\Phi_1 - V_{ch} = r_0 + \frac{2kT}{q} \ln\left[\frac{2\beta}{\sinh(\alpha - \beta)}\right] + r_1\beta\coth(\alpha - \beta), \tag{13.5}$$

$$V_{bg} - \Delta\Phi_2 - V_{ch} = r_0 + \frac{2kT}{q} \ln\left[\frac{2\beta}{\sinh(\alpha + \beta)}\right] - r_2\beta\coth(\alpha + \beta) \tag{13.6}$$

r_0, r_1 and r_2 are constants; α and β are unknowns. Unfortunately this set of coupled equations is very difficult to solve. Iterative techniques are either computationally expensive or have convergence problems [34]. A non-iterative technique is therefore desirable for a compact model.

We have found that by assuming that the inversion carrier density at the back surface is much smaller than that at the front surface,[3] a closed-form analytical

[3]Because of this assumption, BSIM-IMG does not model the case when then back surface enters into strong inversion. This is currently being investigated.

solution can be obtained. The assumption is:

$$N_c \exp\left[\frac{q(\psi_{s1} - V_{ch})}{kT}\right] \gg N_c \exp\left[\frac{q(\psi_{s2} - V_{ch})}{kT}\right]. \tag{13.7}$$

Integrating the Poisson's equation (13.2), we obtain

$$\mathcal{E}_1^2 - \mathcal{E}_2^2 = \frac{2N_c kT}{\varepsilon_{si}}\left\{\exp\left[\frac{q(\psi_{s1} - V_{ch})}{kT}\right] - \exp\left[\frac{q(\psi_{s2} - V_{ch})}{kT}\right]\right\}. \tag{13.8}$$

The last term in (13.8) can be dropped as long as (13.7) holds. We also approximate $\mathcal{E}_2$ as follows:

$$\mathcal{E}_2^{(0)} = \frac{(V_{fg} - \Delta\Phi_1) - (V_{bg} - \Delta\Phi_2)}{\frac{\varepsilon_{si}}{\varepsilon_{ox}}(T_{ox1} + T_{ox2}) + T_{si}}. \tag{13.9}$$

Quantities with superscript (0) are approximate and will be refined later through a perturbation step. Substituting (13.3) and (13.9) in (13.8) and dropping the last term in (13.8), we obtain

$$\left[\frac{C_{ox1}(V_{fg} - \Delta\Phi_1 - \psi_{s1}^{(0)})}{\varepsilon_{si}}\right]^2 - \left[\frac{(V_{fg} - \Delta\Phi_1) - (V_{bg} - \Delta\Phi_2)}{\frac{\varepsilon_{si}}{\varepsilon_{ox}}(T_{ox1} + T_{ox2}) + T_{si}}\right]^2 = \frac{2N_c kT}{\varepsilon_{si}}\left\{\exp\left[\frac{q(\psi_{s1}^{(0)} - V_{ch})}{kT}\right]\right\}. \tag{13.10}$$

The above equation is solved to obtain the front surface potential $\psi_{s1}^{(0)}$. Although (13.10) is a transcendental equation, a closed-form expression may be obtained through an analytical approximation method without iteration. Mathematical details of the approximation can be found in [34].

To further improve the accuracy of the solution, a perturbation step is performed. $\psi_{s1}^{(0)}$ is used to compute the back-side electric field $\mathcal{E}_2^{(1)}$.

$$\mathcal{E}_2^{(1)} = \frac{\psi_{s1}^{(0)} - (V_{bg} - \Delta\Phi_2)}{\frac{\varepsilon_{si}}{\varepsilon_{ox}}T_{ox2} + T_{si}}. \tag{13.11}$$

Equation (13.11) and the boundary condition (13.3) are substituted into (13.8) to compute a refined front surface potential $\psi_{s1}^{(1)}$. Again, the last term of (13.8) is dropped:

$$\left[\frac{C_{ox1}(V_{fg} - \Delta\Phi_1 - \psi_{s1}^{(1)})}{\varepsilon_{si}}\right]^2 - \left[\frac{\psi_{s1}^{(0)} - (V_{bg} - \Delta\Phi_2)}{\frac{\varepsilon_{si}}{\varepsilon_{ox}}T_{ox2} + T_{si}}\right]^2 = \frac{2N_c kT}{\varepsilon_{si}}\left\{\exp\left[\frac{q(\psi_{s1}^{(1)} - V_{ch})}{kT}\right]\right\}. \tag{13.12}$$

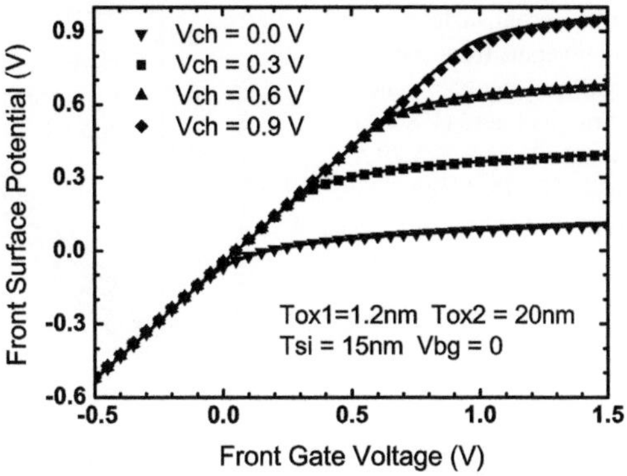

Fig. 13.4 Front surface potential versus front-gate bias with varying channel voltage ($\Phi_{g1} = 4.17$ V, $\Phi_{g2} = 5.29$ V; *Symbols*: TCAD; *Lines*: Model)

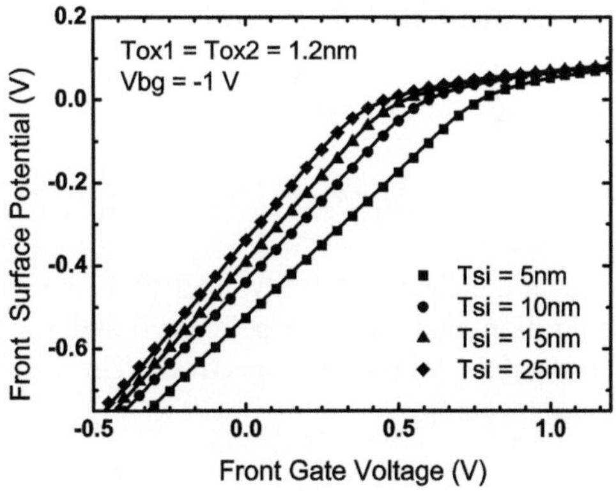

Fig. 13.5 Front surface potential versus front-gate bias with varying body thickness ($V_{ch} = 0$, $\Phi_{g1} = 4.4$ V, $\Phi_{g2} = 4.4$ V; *Symbols*: TCAD; *Lines*: Model)

Solving (13.12) we obtain $\psi_{s1}^{(1)}$.

Other quantities that need to be calculated are the inversion carrier density (Q_i) and the back-side surface potential (ψ_{s2}). Q_i is calculated by integrating from the front surface to the back surface:

$$Q_i = \int_{-T_{si}/2}^{T_{si}/2} qN_c \exp\left\{\frac{q[\psi(x) - V_{ch}]}{kT}\right\} dx. \tag{13.13}$$

Q_i can be expressed in closed form as a function of ψ_{s1}. The full expression for Q_i is quite complex and consists of 3 different solution forms depending on the boundary conditions. Therefore we do not present it here. Interested readers may refer to [34]. ψ_{s2} is calculated as a function of ψ_{s1} and Q_i from Gauss' Law.

Figures 13.4–13.6 compare the calculated surface potential with TCAD. TCAD simulations are performed for long-channel devices. To show the intrinsic properties of the model we do not use any fitting parameters in the verification of the core model. In Fig. 13.4, ψ_{s1} is plotted versus V_{fg} for different values of V_{ch}, showing good agreement. Each V_{ch} corresponds to a different position along the channel. Figures 13.5 and 13.6 demonstrate the scalability of the model with T_{si} and T_{ox2}.

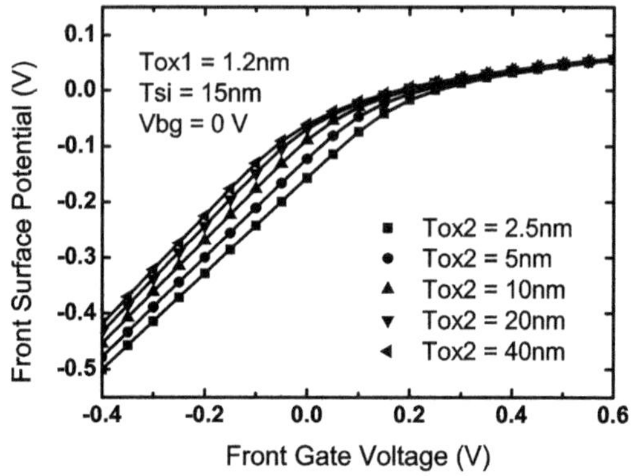

Fig. 13.6 Front surface potential versus front-gate bias with varying back-gate dielectric thickness ($V_{ch} = 0$, $\Phi_{g1} = 4.17$ V, $\Phi_{g2} = 5.29$ V; *Symbols*: TCAD; *Lines*: Model)

13.2.3 Drain Current Model

The drain current is derived based on drift and diffusion [35]:

$$I_{ds} = \mu W \left[Q_i(y) \frac{d\psi_{s1}(y)}{dy} - \frac{kT}{q} \frac{dQ_i(y)}{dy} \right] \tag{13.14}$$

where μ is the carrier mobility[4] and W is the channel width of the MOSFET. y is the position along the channel. $y = 0$ and $y = L$ correspond to the source and drain ends of the MOSFET, respectively. The inversion carrier density $Q_i(y)$ can be expressed (according to Gauss' Law) as

$$Q_i(y) = C_{ox1} \left[V_{fg} - \Delta\Phi_1 - \psi_{s1}(y) \right] - \varepsilon_{si} \mathcal{E}_2(y). \tag{13.15}$$

The back-side electric field $\mathcal{E}_2(y)$ is not known at all positions of y. To make (13.14) integrable, we linearize $\mathcal{E}_2(y)$:

$$\mathcal{E}_2(y) = \frac{\mathcal{E}_2(0) + \mathcal{E}_2(L)}{2} + \frac{\mathcal{E}_2(L) - \mathcal{E}_2(0)}{\psi_{s1}(L) - \psi_{s1}(0)} \left[\psi_{s1}(y) - \frac{\psi_{s1}(0) + \psi_{s1}(L)}{2} \right]. \tag{13.16}$$

Equations (13.15) and (13.16) are substituted into (13.14) and integrated from $y = 0$ (source end) to $y = L$ (drain end) to obtain the drain current expression. The result is:

$$I_{ds} = \mu \frac{W}{L} \left[\frac{Q_{is} + Q_{id}}{2} (\psi_{s1,d} - \psi_{s1,s}) + \frac{kT}{q} (Q_{is} - Q_{id}) \right]. \tag{13.17}$$

The surface potential and inversion charge in (13.17) are calculated using the approximation described in the previous section. $\psi_{s1,s}$ and Q_{is} are calculated at $V_{ch} = 0$; $\psi_{s1,d}$ and Q_{id} are calculated at $V_{ch} = V_{ds}$.

[4]The core model assumes μ is constant. High field effects are accounted for later through the incorporation of real device effects.

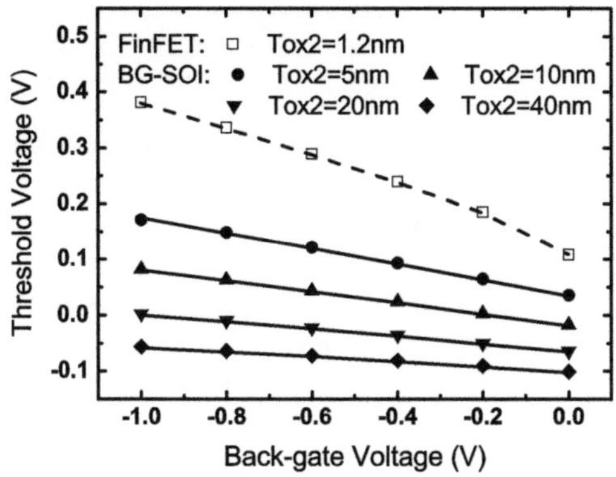

Fig. 13.7 Threshold voltage versus front-gate bias with varying back-oxide thickness. *Solid lines* and *closed symbols*: asymmetric structure ($T_{si} = 15$ nm, $T_{ox1} = 1.2$ nm, $V_{ch} = 0$, $\Phi_{g1} = 4.17$ V, $\Phi_{g2} = 5.29$ V); *dashed lines* and *open symbols*: symmetric structure ($T_{ox1} = T_{ox2} = 1.2$ nm, $T_{si} = 15$ nm, $\Phi_{g1} = 4.4$ V, $\Phi_{g2} = 4.4$ V; *Symbols*: TCAD; *Lines*: Model). The threshold voltage is extracted using a constant current definition ($I_{ds} = 100$ nA $\cdot W/L$)

Fig. 13.8 Drain current versus front-gate voltage for different front dielectric thicknesses ($T_{si} = 15$ nm, $T_{ox2} = 20$ nm, $V_{bg} = 0$ V, $V_{ds} = 50$ mV, $\Phi_{g1} = 4.17$ V, $\Phi_{g2} = 5.29$ V; *Symbols*: TCAD; *Lines*: Model)

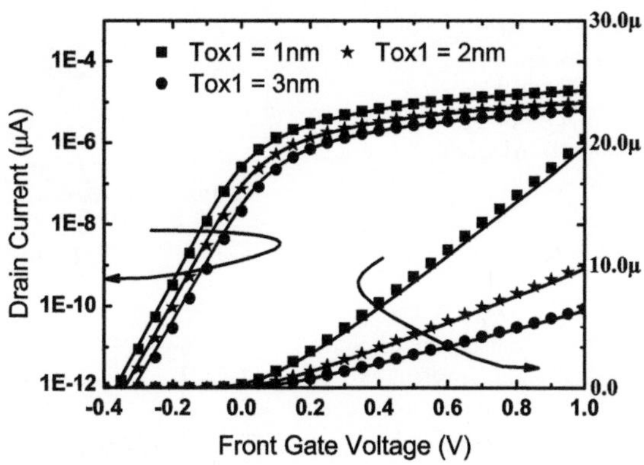

Figures 13.7–13.11 demonstrates the accuracy of the drain current model by comparing it to TCAD [36] without using any fitting parameters. In Fig. 13.7, V_{th} is plotted versus V_{bg} for independent double-gate devices with both symmetric and asymmetric structures. The model agrees well with TCAD for both cases. The larger slope for thinner back-oxide devices is due to the stronger coupling from the back side.

Figure 13.8 shows I_{ds} versus V_{fg} for different T_{ox1}, demonstrating the scalability of the model to different dielectric thicknesses. Figure 13.9 shows I_{ds} versus V_{ds} with different V_{fg}. Figure 13.10 shows transconductance (g_m) versus V_{fg} at both low and high V_{ds}.

In Fig. 13.11, the transconductance efficiency, g_m/I_{ds} is plotted versus V_{fg}. g_m/I_{ds} is an important design metric that characterize the maximum transconductance that the device can provided at a given bias current. At low V_{fg} it should

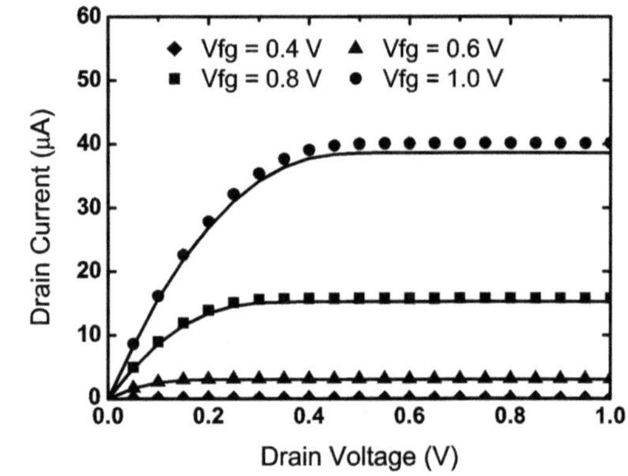

Fig. 13.9 Drain current versus drain voltage for different front-gate bias ($T_{si} = 15$ nm, $T_{ox1} = T_{ox2} = 1.2$ nm, $V_{bg} = -1$ V, $\Phi_{g1} = 4.4$ V, $\Phi_{g2} = 4.4$ V; *Symbols*: TCAD; *Lines*: Model)

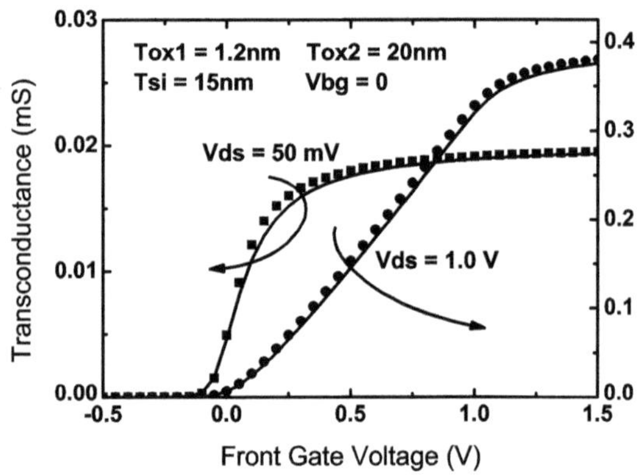

Fig. 13.10 Transconductance versus front-gate voltage at $V_{ds} = 50$ mV and $V_{ds} = 1.0$ V. The transconductance in TCAD is extracted from small-signal simulations ($T_{si} = 15$ nm, $T_{ox1} = 1.2$ nm, $T_{ox2} = 20$ nm, $V_{bg} = 0$ V, $\Phi_{g1} = 4.17$ V, $\Phi_{g2} = 5.29$ V; *Symbols*: TCAD; *Lines*: Model)

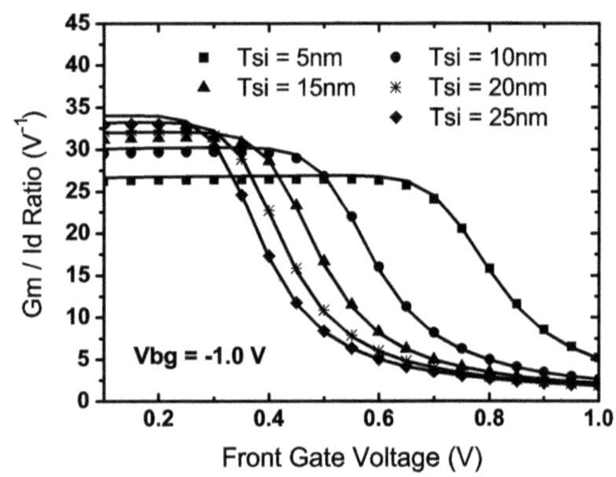

Fig. 13.11 Transconductance efficiency (g_m/I_{ds}) versus front-gate voltage for different silicon body thicknesses ($T_{ox1} = T_{ox2} = 1.2$ nm, $V_{bg} = -1$ V, $\Phi_{g1} = 4.4$ V, $\Phi_{g2} = 4.4$ V; *Symbols*: TCAD; *Lines*: Model)

saturate to a constant, whose value is given by

$$\frac{g_m}{I_{ds}} \approx \frac{\frac{d}{dV_{fg}} \exp(\frac{qV_{fg}}{nkT})}{\exp(\frac{qV_{fg}}{nkT})} = \frac{q}{nkT} \tag{13.18}$$

where n is the inverse subthreshold slope. $\frac{q}{nkT}$ increases with increasing T_{si} because n decreases with decreasing T_{si} without considering short channel effects. This is exactly what we see in Fig. 13.11.

13.2.4 Capacitance Model

We model the C-V using a charge-based approach [37] to ensure charge conservation. The charge associated with each terminal is modeled. The capacitive current flowing into each terminal is expressed as the time derivative of charge.

$$I_x = \frac{dQ_x}{dt} = \sum_y C_{xy} \frac{dV_y}{dt} \tag{13.19}$$

where $x, y = d, fg, bg, s$; each transcapacitance is defined as[5]

$$C_{xy} = \frac{\partial Q_x}{\partial V_y}. \tag{13.20}$$

The charge associated with the front gate can be calculated as:

$$Q_{fg} = W \int_0^L C_{ox1} \left[V_{fg} - \Delta\Phi_1 - \psi_{s1}(y) \right] dy. \tag{13.21}$$

To perform this integration the relation between front surface potential $\psi_{s1}(y)$ and position y is needed. This can be obtained by applying current continuity to (13.17).

$$I_{ds} \cdot y = \mu W \left\{ \frac{Q_{is} + Q_i(y)}{2} \left[\psi_{s1}(y) - \psi_{s1,s} \right] + \frac{kT}{q} \left[Q_{is} - Q_i(y) \right] \right\}. \tag{13.22}$$

Since $Q_i(y)$ is unknown, the capacitor divider approximation is used to relate the front surface potential ψ_{s1} and charge Q_i

$$Q_i(y) = C_{ox1} \left[V_{fg} - \Delta\Phi_1 - \psi_{s1}(y) \right] + \frac{C_{ox2} C_{si}}{C_{ox2} + C_{si}} \left[V_{bg} - \Delta\Phi_2 - \psi_{s1}(y) \right]. \tag{13.23}$$

Combining (13.22) and (13.23) and noting that $Q_{is} = Q_i(\psi_{s1}(y) = \psi_{s1,s})$, we obtain the position dependence of surface potential:

[5] An alternative definition is $C_{xy} = -\frac{\partial Q_x}{\partial V_y}$. See, e.g., [38].

$$y = \mu C_{ox1} \frac{W}{I_{ds}} \left[\psi_{s1}(y) - \psi_{s1,s}\right] \left\{ V_{fg} - \Delta\Phi_1 - \frac{\psi_{s1,s} + \psi_{s1}(y)}{2} + \gamma_c \cdot \left[V_{bg} - \Delta\Phi_2 - \frac{\psi_{s1,s} + \psi_{s1}(y)}{2} \right] + \frac{kT}{q}(1+\gamma_c) \right\} \quad (13.24)$$

where $\gamma_c = (C_{ox2} \parallel C_{si})/C_{ox1}$.

Substituting (13.24) in (13.21) and performing integration, we obtain the expression for the front-gate charge:

$$Q_{fg} = C_{ox1} W L \left\{ V_{fg} - \Delta\Phi_1 - \frac{\psi_{s1,s} + \psi_{s1,d}}{2} + \frac{B(\psi_{s1,d} - \psi_{s1,s})^2}{6[A - B(\psi_{s1,d} + \psi_{s1,s})]} \right\} \quad (13.25)$$

where

$$A = V_{fg} - \Delta\Phi_1 + \gamma_c \cdot (V_{bg} - \Delta\Phi_2) + \frac{kT}{q}(1+\gamma_c), \quad (13.26)$$

$$B = \frac{1+\gamma_c}{2}. \quad (13.27)$$

The charge associated with the back gate can be simply calculated by replacing $\psi_{s1,s(d)}$ with $\psi_{s2,s(d)}$, swapping $(V_{fg} - \Delta\Phi_1)$ and $(V_{bg} - \Delta\Phi_2)$, and swapping C_{ox1} and C_{ox2} in (13.25), following an argument of symmetry.

The front- and back-gate charges are further partitioned into a source component and a drain component according to the Ward-Dutton charge partition method [37]. The drain charge associated with the front gate is given by

$$Q_{d1} = -W \int_0^L C_{ox1} \left[V_{fg} - \Delta\Phi_1 - \psi_{s1}(y) \right] \frac{y}{L} dy. \quad (13.28)$$

After using (13.24) and integrating we obtain

$$Q_{d1} = -\frac{C_{ox1} W L}{2} \left\{ V_{fg} - \Delta\Phi_1 - \frac{\psi_{s1,s} + \psi_{s1,d}}{2} + \frac{B(\psi_{s1,d} - \psi_{s1,s})^2}{30[A - B(\psi_{s1,s} + \psi_{s1,d})]} - \frac{(5A - 4B\psi_{s1,d} - 6B\psi_{s1,s})(A - 2B\psi_{s1,d})(\psi_{s1,d} - \psi_{s1,s})}{30[A - B(\psi_{s1,s} + \psi_{s1,d})]^2} \right\}. \quad (13.29)$$

Similarly, the drain charge associated with the back-gate, Q_{d2} is obtained by replacing $\psi_{s1,s(d)}$ with $\psi_{s2,s(d)}$, swapping $(V_{fg} - \Delta\Phi_1)$ and $(V_{bg} - \Delta\Phi_2)$, and swapping C_{ox1} and C_{ox2} in (13.29)

The total drain charge is the sum of Q_{d1} and Q_{d2}. Since Q_s, Q_d, Q_{fg} and Q_{bg} must sum up to 0, the source charge can be calculated as

$$Q_s = -Q_{fg} - Q_{bg} - Q_d. \quad (13.30)$$

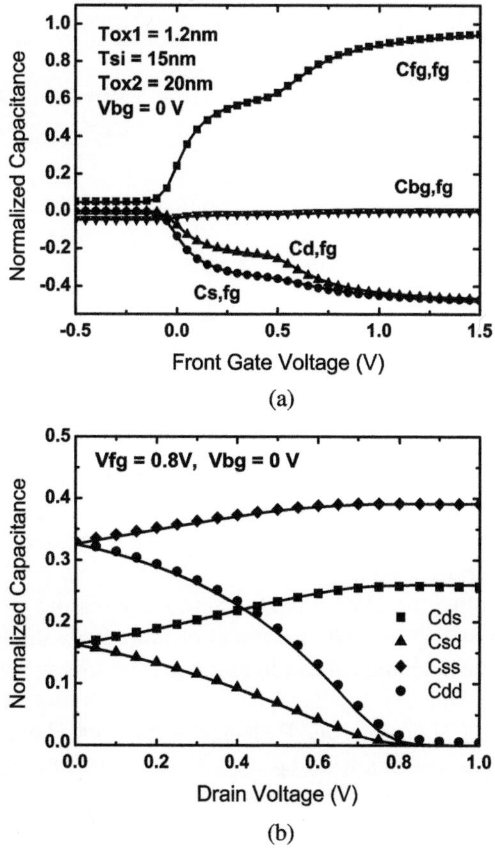

Fig. 13.12 Transcapacitances (**a**) $C_{fg,fg}$, $C_{d,fg}$, $C_{s,fg}$ and $C_{bg,fg}$ versus V_{fg} at $V_{ds} = 0.5$ V and $V_{bg} = 0$ V. (**b**) C_{ds}, C_{sd}, C_{ss} and C_{dd} versus V_{ds} at $V_{fg} = 0.8$ and $V_{bg} = 0$ V. All capacitances are normalized to $C_{ox1}WL$ ($T_{si} = 15$ nm, $T_{ox1} = 1.2$ nm, $T_{ox2} = 20$ nm, $\Phi_{g1} = 4.17$ V, $\Phi_{g2} = 5.29$ V; *Symbols*: TCAD; *Lines*: Model)

Figure 13.12 verifies the C-V model by comparing transcapacitances with small-signal TCAD simulations without using fitting parameters. In Fig. 13.12(a), transcapacitances $C_{fg,fg}$, $C_{s,fg}$, $C_{d,fg}$ and $C_{bg,fg}$ are normalized to $C_{ox1}WL$ and plotted versus V_{fg}. We see a smooth transition from sub-threshold to saturation and to linear. In Fig. 13.12(b), transcapacitances C_{sd}, C_{ds}, C_{ss} and C_{dd} are plotted versus V_{ds}. At $V_{ds} = 0$, C_{ss} and C_{dd} are equal; C_{sd} and C_{ds} are equal, reflecting the source-drain symmetry of the C-V model. Model symmetry is important for predicting correct distortion metrics for circuits operating at $V_{ds} = 0$ especially for analog and RF applications.

13.3 Core Model for the Common Multi-gate MOSFET

In this section we present the core model of BSIM-CMG [20]. The common multi-gate (CMG) is a special case of a multi-gate device with all gates electrically connected and the gate stacks having identical dielectric thicknesses and gate work functions.

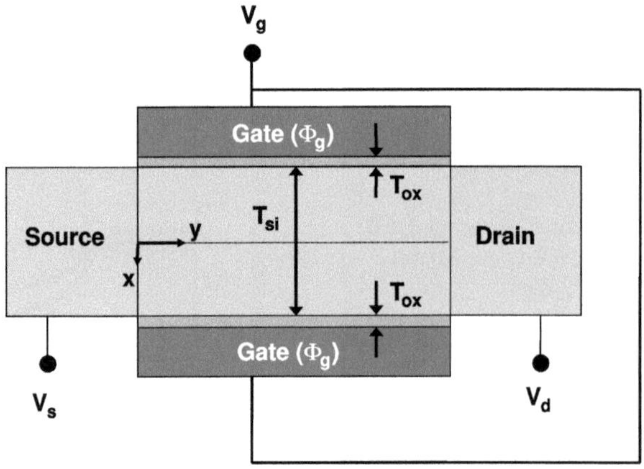

Fig. 13.13 Basic framework for modeling common double-gate MOSFETs illustrated in 2D

The derivation of the CMG core model is similar to that of IMG. The surface potential is obtained by solving the Poisson's equation in one dimension at both the source and the drain ends. I-V and C-V are derived and expressed as function of the surface potentials.

Unlike the IMG model, where body doping is modeled as a bias-independent shift in V_{th}, in CMG the effect of finite body doping is taken into account during surface potential calculation for better accuracy. State-of-the-art CMOS technology offers multiple V_{th} flavors, e.g. high V_{th} N-channel and P-channel transistors and low V_{th} N-channel and P-channel transistors. Having different body doping concentrations, devices with different V_{th} co-exist in the same die. The high V_{th} devices have slower speed but consume less leakage or standby power. Circuit designers use the high V_{th} devices in circuits where speed is not critical and use the low V_{th} devices where it is. This scheme will be also used in common multi-gate CMOS technologies. The low V_{th} CMG devices do not have heavy body doping as explained in Sect. 13.1. The high V_{th} CMG devices must be heavily doped.

13.3.1 Basic Modeling Framework

Figure 13.13 illustrates the basic modeling framework for the common multi-gate MOSFET. This view corresponds to a horizontal cross section of the double-gate FinFET (Fig. 13.1(c)). The front- and back-gate stacks have identical dielectric thicknesses (T_{ox}), dielectric constants (ε_{ox}) and gate work functions (Φ_g). The two gates are connected together and always biased at the same voltage.

Besides the common double-gate device, BSIM-CMG also considers triple-gate, quadruple-gate and all-around gate devices. The triple-gate FinFET (Fig. 13.1(a), (b)) shares the same core model with the double-gate FinFET. The effect of the top gate is taken into account separately, as we will discuss later. The all-around gate MOSFET (Fig. 13.1(d)) is modeled using a separate core model [39]. The Quadruple-gate MOSFET shares the same core model with the all-around gate since

Table 13.1 Geometry options of BSIM-CMG (selected using parameter *GEOMOD*)

Multi-gate type	Cross-sectional shape	GEOMOD
Double-gate	Rectangle with gates on two sides	0
Triple-gate	Rectangle with gates on the top and two sides	1
Quadruple-gate	Rectangle with gates on the top, the bottom and two sides	2
All-around-gate	Circular with the gate wrapped around	3

their cross sectional geometries are similar. The users may specify the geometry by setting the model parameter *GEOMOD* (Table 13.1).

13.3.2 Surface Potential Calculation

The Poisson's equation based on the potential definition in (13.1) is:

$$\frac{d^2\psi(x)}{dx^2} = \frac{qN_c}{\varepsilon_{si}} \exp\left[\frac{q(\psi(x) - V_{ch})}{kT}\right] + qN_A \tag{13.31}$$

where N_A is the net doping concentration in the p-type body.

The electric field at the silicon/oxide interface is derived by integrating (13.31):

$$\frac{d\psi(x)}{dx} = \sqrt{\frac{2qN_c}{\varepsilon_{si}}\frac{kT}{q}\left\{\exp\left[\frac{q(\psi(x) - V_{ch})}{kT}\right] - \exp\left[\frac{q(\psi_0 - V_{ch})}{kT}\right]\right\} + \frac{2qN_A}{\varepsilon_{si}}[\psi(x) - \psi_0]}. \tag{13.32}$$

We have used the fact that $\frac{d\psi(x)}{dx} = 0$ at the center of the body ($x = 0$).

If (13.32) is integrated once more, we will obtain a closed-form expression of $\psi(x)$ as function of position x. Unfortunately this integration is difficult to perform. To further simplify the problem, we write the total surface potential as

$$\psi(x) = \psi_1(x) + \psi_{pert}(x) \tag{13.33}$$

where $\psi_1(x)$ is the solution at $N_A = 0$. $\psi_{pert}(x)$ is a perturbation (correction) due to doping [40, 41]. The exact analytical solution $\psi_1(x)$ is known [42]:

$$\psi_1(x) = \psi_0 - \frac{2kT}{q} \ln\left\{\cos\left[\sqrt{\frac{q^2}{2\varepsilon_{si}kT}N_c \exp\left[\frac{q(\psi_0 - V_{ch})}{kT}\right]} \cdot x\right]\right\} \tag{13.34}$$

where ψ_0 is a constant to be determined from the boundary condition.

Consider a fully-depleted double-gate device with negligible inversion carriers. The potential drop due to the constant depletion charge is

$$\psi_{bulk} = \frac{qN_A}{2\varepsilon_{si}}\left(\frac{T_{si}}{2}\right)^2. \tag{13.35}$$

On the other hand, if the device is partially-depleted, the voltage drop ψ_c is smaller than ψ_{bulk} and may be approximately determined by solving

$$V_g - \Delta\Phi - \phi_B - \frac{E_g}{2} - \psi_c = \frac{\sqrt{2\varepsilon_{si} q N_A \psi_c}}{C_{ox}} \triangleq \gamma\sqrt{\psi_c}. \tag{13.36}$$

The solution is,

$$\psi_c = \left[\sqrt{\frac{\gamma^2}{4} + V_g - \Delta\Phi - \phi_B - \frac{E_g}{2}} - \frac{\gamma}{2}\right]^2 \tag{13.37}$$

where $\Delta\Phi$ is the work function of the gate relative to that of the source and $\phi_B = \frac{kT}{q}\ln\frac{N_A}{n_i}$.

$\psi_{pert}(T_{si}/2)$ is taken as the minimum of the two:

$$\psi_{pert}(T_{si}/2) = \min(\psi_{bulk}, \psi_c). \tag{13.38}$$

The boundary condition at the silicon/oxide interface is needed to determine ψ_0. It is,

$$\left.\frac{d\psi(x)}{dx}\right|_{x=T_{si}/2} = \frac{C_{ox}}{\varepsilon_{si}}\left[V_g - \Delta\Phi - \psi(T_{si}/2)\right]. \tag{13.39}$$

We also define

$$\beta = \sqrt{\frac{q^2 N_c}{2\varepsilon_{si} kT}\exp\left[\frac{q(\psi_0 - V_{ch})}{kT}\right]}\cdot\frac{T_{si}}{2}. \tag{13.40}$$

Equating (13.32) and (13.39) at $x = T_{si}/2$ and substituting (13.33), (13.34) and (13.40) into the final expression, we have

$$\begin{aligned}
&\frac{V_g - \Delta\Phi - V_{ch} - \psi_{pert}}{2\frac{kT}{q}} \\
&= \ln\beta - \ln\cos\beta - \ln\left(\sqrt{\frac{q^2 N_c}{2\varepsilon_{si} kT}}\cdot\frac{T_{si}}{2}\right) \\
&= \frac{2\varepsilon_{si} T_{ox}}{\varepsilon_{ox} T_{si}}\sqrt{\frac{\exp(\frac{q\psi_{pert}}{kT}) - 1}{\cos^2\beta} + \frac{\psi_{bulk}}{(\frac{kT}{q})^2}\left(\psi_{pert} - 2\frac{kT}{q}\ln\cos\beta\right)}
\end{aligned} \tag{13.41}$$

where ψ_{pert} is a short-hand form of $\psi_{pert}(T_{si}/2)$.

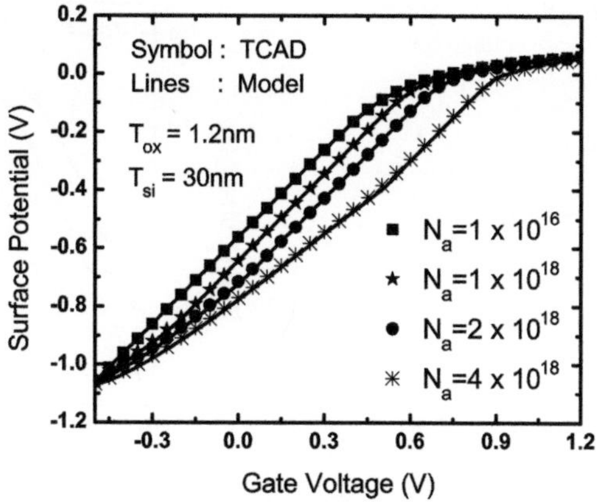

Fig. 13.14 Surface potential versus gate voltage for different body doping concentrations ($T_{si} = 30$ nm, $T_{ox} = 1.2$ nm, $\Phi_g = 4.61$ eV)

Equation (13.41) is solved using two steps of Householder's cubic iteration method [43].

The surface potential solution is verified with TCAD simulations without using fitting parameters (Fig. 13.14). The slope change in the sub-threshold region is due to the transition from a partially-depleted body to a fully-depleted body. As doping concentration becomes higher, partial depletion spans wider ranges of V_g.

13.3.3 Drain Current Model

The core I-V model in BSIM-CMG is based on the drift-diffusion formulation without using the charge-sheet approximation

$$I_d = \mu \cdot W_{eff} \cdot Q_{inv}(y) \frac{dV_{ch}}{dy}. \tag{13.42}$$

Analytical derivation is carried out. A closed form expression of the drain current is obtained as follows [33]:

$$I_d = \mu \cdot \frac{W_{eff}}{L} \cdot \left[f(\psi_{s,s}) - f(\psi_{s,d}) \right] \tag{13.43}$$

where $f(\psi_{s,s(d)})$ is given by:

$$f(\psi_{s,s(d)}) = \frac{Q_{inv,s(d)}^2}{2C_{ox}} + 2\frac{kT}{q} Q_{inv,s(d)} - \frac{kT}{q}\left(5C_{si}\frac{kT}{q} + Q_{bulk}\right) \cdot \ln\left(5C_{si}\frac{kT}{q} + Q_{bulk} + Q_{inv,s(d)}\right) \tag{13.44}$$

where

$$Q_{inv,s(d)} = C_{ox} \cdot (V_g - V_{fb} - \psi_{s,s(d)}) - Q_{bulk}, \tag{13.45}$$

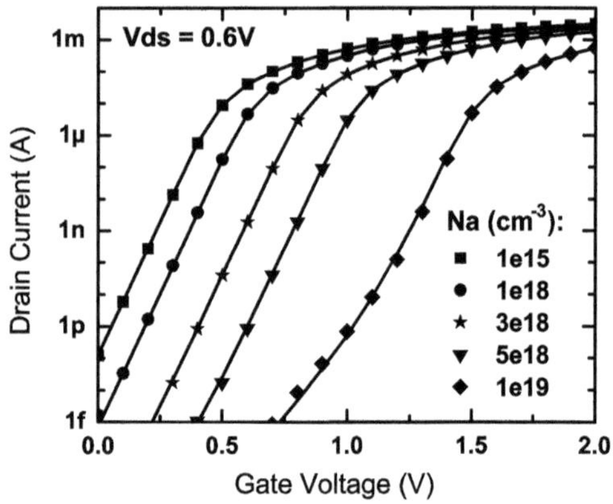

Fig. 13.15 Drain current versus gate voltage for different body doping concentrations. Both partially-depleted and fully-depleted devices are modeled well

$$Q_{bulk} = \sqrt{2qN_A\varepsilon_{si}\psi_{pert}}. \tag{13.46}$$

As in BSIM-IMG, the final drain current I_d is a function of surface potentials at the source ($\psi_{s,s}$) and drain ($\psi_{s,d}$) ends. $\psi_{s,s}$ and $\psi_{s,d}$ are determined by solving (13.41) once at the source end ($V_{ch} = 0$) and once at the drain end ($V_{ch} = V_{ds}$).

The I-V model is verified against TCAD simulations without using any fitting parameters. Figure 13.15 tests the accuracy of $I_d - V_{gs}$ over a wide range of body doping. As shown in the figures BSIM-CMG can predict drain current accurately in both full-depletion and partial-depletion regions.

13.3.4 Capacitance Model

Like BSIM-IMG, the C-V model derivation is also based on Ward-Dutton charge partition and current continuity. Without going through a detailed derivation again, the final charge expressions are given as follows:

$$Q_g = W_{eff}LC_{ox}\left[V_{gs} - \Delta\Phi - \frac{\psi_{s,s} + \psi_{s,d}}{2} + \frac{(\psi_{s,d} - \psi_{s,s})^2}{6(B - \psi_{s,d} - \psi_{s,s})}\right], \tag{13.47}$$

$$Q_d = W_{eff}LC_{ox}\left(\frac{V_{gs} - \Delta\Phi - \frac{Q_{bulk}}{C_{ox}}}{2} - \frac{\psi_{s,s} + \psi_{s,d}}{4} + \frac{(\psi_{s,d} - \psi_{s,s})^2}{60(B - \psi_{s,d} - \psi_{s,s})}\right.$$
$$\left. + \frac{(5B - 4\psi_{s,d} - 6\psi_{s,s})(B - 2\psi_{s,d})(\psi_{s,s} - \psi_{s,d})}{60(B - \psi_{s,d} - \psi_{s,s})^2}\right), \tag{13.48}$$

$$Q_s = -(Q_g + Q_{bulk} + Q_d). \tag{13.49}$$

Interested readers may refer to [33] for details.

13.4 Real Device Effects

In previous sections, the basic I-V and C-V of the ideal long channel multi-gate MOSFET are presented. The physically-derived core model provides BSIM-MG with continuous and smooth current and charge functions in terms of gate, drain and source voltages and the first-order dependencies on physical parameter values such as T_{ox}, T_{si} and body doping. However, to capture the electrical characteristics of state-of-the-art nanoscale devices, real device effects (such as field-dependent mobility, short channel effects, quantum mechanical effects, and many others) must also be taken into account through important modifications of the core I-V and C-V expressions. These modifications should satisfy several requirements:

1. The bias and geometry dependences of the physical effects in question are accurately modeled.
2. The I-V, C-V and their high order derivatives remains continuous in all bias ranges and parameter values.
3. Source/drain symmetry is maintained after the modification.
4. The added computation does not introduce significant speed penalty. In other words, the model must remain computationally efficient.

In this section we will briefly discuss the incorporation of real device effects into BSIM-MG.

BSIM-CMG and BSIM-IMG are designed to share as many real device effect equations as possible. This simplifies the work of model users when switching from one to another. This also makes it easier to fairly compare the performance of IMG and CMG circuit performances using BSIM-MG.

13.4.1 Quantum Mechanical Effects

Quantum mechanical effects has significant impact on state-of-the-art MOSFETs, long channel and short channel, due to the high normal electric field at the body-insulator interface and the small gate dielectric thickness. This is manifested in the I-V and C-V in two ways:

1. With the quantization of energy levels, a larger gate bias is required to reach the same inversion carrier density. This effectively increases the device V_{th} (and makes V_{th} increase with V_{gs}).
2. The carrier confined in the potential well created by the electric field no longer has its peak concentration at the surface. Instead, the charge centroid is at a distance away from the interface. This effectively increases the oxide thickness. The amount of the increase, i.e., the inversion charge centroid, decreases with increasing V_{gs}.

For multi-gate devices, the quantization effect is further complicated due to the extra structural confinement by the thin body.

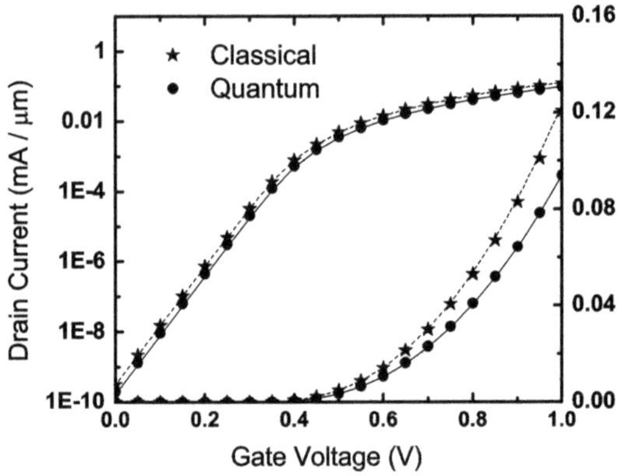

Fig. 13.16 Model and 2D TCAD simulation results for drain current versus gate voltage for the classical and quantum cases. In TCAD quantum effects can be intentionally switched off to observe the classical solution ($T_{si} = 10$ nm, $T_{ox} = 1.1$ nm, $N_a = 10^{16}$ cm^{-3}, $L = 10$ µm; *Symbols*: TCAD, *Lines*: Model; *EOT* = 1.42 nm to account for quantum effects)

Figure 13.16 shows TCAD simulation results of I_d-V_{gs} for a common double-gate device. Curves with quantum effects model turned on/off are both shown to illustrate the effect of quantum confinement. In the sub-threshold region a shift in V_{th} is observed. Additionally, in the strong inversion region the drive current is reduced due to the effective degradation in oxide capacitance and the increase of V_{th}.

To model the V_{th} shift due to energy states quantization, we focus on the sub-threshold region.

For common double-gate devices, the external potential is approximated using a 1-dimensional infinite potential well with a constant potential inside the body. The quantized electron energy levels are:

$$E_n = \frac{n^2\hbar^2\pi^2}{2m_x T_{si}^2} \tag{13.50}$$

where m_x is the electron effective mass in the confinement direction and $n = 1, 2, 3, \ldots$.

For independent double-gate devices, on the other hand, electric field is large even in sub-threshold. The exact analytical solution to the infinite potential well with a linearly-varying potential at the bottom of the well is not known to the authors' knowledge. However, the electron wave functions may be approximated by modifying the ground state wave function of the CMG case with an exponential term [44]:

$$\Psi(x) \propto \sin\left(\frac{n\pi x}{T_{si}}\right) \exp\left(\frac{b_n x}{2}\right). \tag{13.51}$$

Using the variational method in quantum mechanics, upper bounds of energy levels are calculated by minimizing the variational parameters b_n.

The energy levels are approximated as [44]:

$$E_n = \frac{n^2\hbar^2\pi^2}{2m_x T_{si}} \left\{ 1 + \left(\frac{b_n T_{si}}{n\pi}\right)^2 \left[3 - \frac{4}{3} \frac{1}{1 + (\frac{b_n T_{si}}{n\pi})^2} \right] \right\} \tag{13.52}$$

where b_n is given by

$$b_n \approx \left[\left(n - \frac{1}{4}\right) \cdot \frac{2m_x q\mathcal{E}}{\hbar^2}\right]^{1/3} \tag{13.53}$$

and $\mathcal{E}$ is the electric field in the body.

The surface potential correction $\Delta\psi$ is calculated by integrating the product of 2D density of states and the Boltzmann distribution. For common double-gate FETs,

$$\Delta\psi = \frac{E_1}{q} - \frac{kT}{q}\ln\left[\frac{gm_d kT}{\pi\hbar^2 N_c T_{si}} \cdot \left(1 + \frac{g'm'_d}{gm_d} e^{\frac{E_1 - E'_1}{kT}} + e^{\frac{E_1 - E_2}{kT}} + \frac{g'm'_d}{gm_d} e^{\frac{E_1 - E'_2}{kT}}\right)\right]. \tag{13.54}$$

For independent double-gate FETs [44],

$$\Delta\psi = \frac{E_1}{q} - \frac{kT}{q}\ln\left[\frac{gm_d}{\pi\hbar^2 N_c} \cdot \frac{q\mathcal{E}}{1 - \exp(-\frac{q\mathcal{E}T_{si}}{kT})} \times \left(1 + \frac{g'm'_d}{gm_d} e^{\frac{E_1 - E'_1}{kT}} + e^{\frac{E_1 - E_2}{kT}} + \frac{g'm'_d}{gm_d} e^{\frac{E_1 - E'_2}{kT}}\right)\right]. \tag{13.55}$$

Here both the primed valley and the unprimed valley up to the second subband are considered. g and g' are the degeneracy of the umprimed and primed valley, respectively. m_d and m'_d are the density of states effective mass of the two valleys. E_n and E'_n are the subband energy levels.

Quantum confinement also has an impact on the equivalent oxide thickness (*EOT*). In BSIM-MG this is considered by an effective change in oxide thickness, ΔT_{ox}. We denote the equivalent oxide thickness as $EOT = T_{ox} + \Delta T_{ox}$. A constant ΔT_{ox} is usually adequate, as illustrated in Fig. 13.16.

Bias dependence of ΔT_{ox} may be incorporated for better accuracy. We have developed a method to capture the bias effects of ΔT_{ox} [33, 45]. However, it requires the surface potential to be solved using Newton Raphson iteration. BSIM-MG does not employ internal Newton Raphson iteration for model efficiency reasons. A more computationally efficient bias-dependent formulation for ΔT_{ox} is being developed.

13.4.2 Short-Channel Effects

Perhaps the most significant physical effect that influences the dependence of V_{th} on geometry (such as T_{si} or L) in short-L devices is the short channel effect (SCE). This includes V_{th} roll-off at smaller gate lengths, drain induced barrier lowering (DIBL) and sub-threshold slope (SS) degradation. These phenomena are well-known in planar MOSFETs models and are widely-used for circuit simulation [46–48]:

$$\Delta V_{th}(SCE) = -\frac{0.5 \cdot DVT0}{\cosh(DVT1 \cdot \frac{L}{\lambda}) - 1} \cdot (V_{bi} - \phi_B - 0.45), \tag{13.56}$$

Table 13.2 Scale length for different cross-sectional geometry

Model	GEOMOD	Type	Scale Length (λ) formula
BSIM-CMG	0	Double-gate [49]	$\lambda = \sqrt{\frac{\gamma}{2}(1+\frac{T_{si}}{4\gamma T_{ox}})T_{si}\cdot T_{ox}}$
–	1	Triple-gate [50]	$\lambda = \frac{1}{\sqrt{\frac{1}{\frac{\gamma}{2}(1+\frac{T_{si}}{4\gamma T_{ox}})T_{si}\cdot T_{ox}}+\frac{1}{4H_{eff}^2}}}$
–	2	Quadruple-gate [50]	$\lambda = \frac{0.5}{\sqrt{\frac{1}{\frac{\gamma}{2}(1+\frac{T_{si}}{4\gamma T_{ox}})T_{si}\cdot T_{ox}}+\frac{1}{4H_{eff}^2}}}$
–	3	All-around-gate	$\lambda = \sqrt{\frac{\gamma}{2}(1+\frac{R}{2\gamma T_{ox}})R\cdot T_{ox}}$
BSIM-IMG	0/1	Independent double-gate	$\lambda = \sqrt{\frac{\gamma}{2}T_{si}T_{ox1}[1+\frac{\gamma\cdot(T_{ox1}-T_{ox2})}{\gamma(T_{ox1}+T_{ox2})+T_{si}}]}$

Note: $H_{eff} = \sqrt{\frac{H_{fin}}{8}\cdot(H_{fin}+2\gamma T_{ox})}$. R is the radius of the cylindrical body

$$\Delta V_{th}(DIBL) = -\frac{0.5\cdot ETA0}{\cosh(DSUB\cdot\frac{L}{\lambda})-1}\cdot V_{ds}, \tag{13.57}$$

$$C_{dsc} = \frac{0.5(CDSC+CDSCD)\cdot V_{ds}}{\cosh(DVT1\cdot\frac{L}{\lambda})-1}, \tag{13.58}$$

$$n = 1+\frac{CIT+C_{dsc}}{(2C_{si})\parallel C_{ox}}. \tag{13.59}$$

The capitalized quantities on the right hand sides of these expressions are adjustable parameters that allow the user to obtain a better fit to measured data.

It was found that for L being a given multiple of λ, the degree of short channel effect is more or less the same. As technology advances, λ is reduced and SCE starts at a shorter L.

For bulk MOSFETs, λ is given as (assuming the junction depth X_j is large) [46]

$$\lambda = \sqrt{\gamma T_{ox} X_{dep}} \tag{13.60}$$

where $\gamma = \varepsilon_{si}/\varepsilon_{ox}$, T_{ox} is the oxide thickness and X_{dep} is the width of the depletion region.

The scale length (λ) for multi-gate FETs is different from bulk MOSFETs and (13.60) must be modified. The values of λ for various *GEOMOD*'s in BSIM-CMG and BSIM-IMG are summarized in Table 13.2.

In BSIM-IMG, λ is derived with the assumption that the inversion carriers are located near at the front surface. On the other hand, BSIM-CMG assumes the inversion carriers concentrate around the center of the body (fin). Therefore λ for BSIM-IMG and BSIM-CMG are different even at $T_{ox1} = T_{ox2}$.

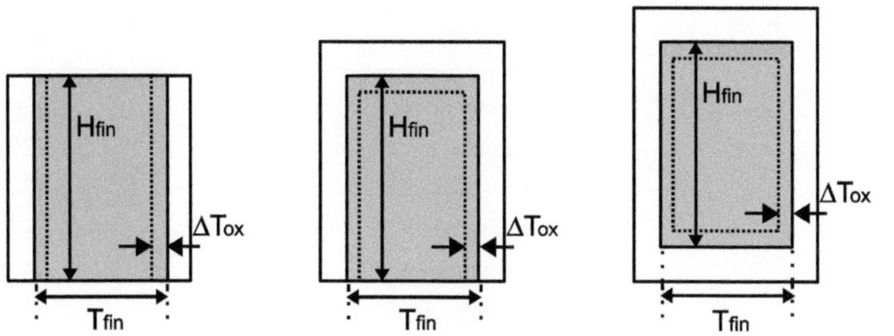

(a) Double-gate (b) Triple-gate (c) Quadruple-gate

Fig. 13.17 Illustration of the effective width model for different geometry types. The *dashed line* represents the charge centroid when the device is biased in strong inversion

Table 13.3 Effective width (W_{eff}) for different cross-sectional geometry

Model	GEOMOD	Type	Effective width (W_{eff}) formula
BSIM-CMG	0	Double-gate	$W_{eff} = 2H_{fin}$
	1	Triple-gate	$W_{eff} = 2H_{fin} + T_{fin} - 4\Delta T_{ox}$
	2	Quadruple-gate	$W_{eff} = 2H_{fin} + T_{fin} - 8\Delta T_{ox}$
	3	All-around-gate	$W_{eff} = 2\pi(R - \Delta T_{ox})$
BSIM-IMG	0	SOI double-gate	$W_{eff} = W$
	1	Independent-gate FinFET	$W_{eff} = H_{fin}$

13.4.3 Effective Width Model

For the common double-gate MOSFET (Fig. 13.17(a)), the effective width of the inversion layer (W_{eff}) is simply twice the height of the fin. Therefore $W_{eff} = 2H_{fin}$ is used for I-V calculation in (13.43) and C-V calculations in (13.47) and (13.48).

For the triple-gate MOSFET (Fig. 13.17(b)) and the quadruple-gate MOSFET (Fig. 13.17(c)), W_{eff} is smaller than the total surface peripheral length because of the finite charge thickness due to quantum mechanical effects. The expressions for W_{eff} for different geometry are summarized in Table 13.3.

13.4.4 Bulk and SOI Substrate Models

In BSIM-MG the user may specify the substrate type through a model parameter, *BULKMOD*. If *BULKMOD* = 1, a bulk multi-gate device is modeled. If *BULKMOD* = 0, a SOI multi-gate device is modeled.

As in conventional MOSFETs, holes created in the body by impact ionization and gate induced drain leakage will flow into the substrate terminal for bulk devices

Table 13.4 Real device effect modules in BSIM-MG

Real device effect module	References
Short channel effect	[20, 33]
Quantum mechanical effect	[20, 33]
Polysilicon depletion	[20]
Mobility degradation	[47]
Velocity Saturation	[47]
Channel length modulation	[47]
Self heating	[48, 51]
Junction leakage current	[47]
Channel thermal noise	[47]
Thermal noise due to parasitics	[47]
Flicker noise	[47, 52]
Impact ionization (body current)	[47]
Gate current	[5, 47, 53]
GIDL/GISL current	[47, 54]
Parasitic gate resistance	[47]
Parasitic Source/drain resistance	[47]
Fringing capacitance	[47]
Junction Capacitance	[47]

shown in Fig. 13.1(a). For SOI devices, however, these components will add to the regular drain to source current.

The substrate usually has little influence on the electrostatics for fully-depleted multi-gate devices. Therefore, the basic I-V and C-V are unchanged with substrate biasing. However, parasitic capacitances between the substrate and the source/drain must be taken into consideration. In addition, junction diodes that exist between the substrate and the source/drain (its importance depends on the multi-gate type) may affect both the current and the charge associated with the substrate terminal. BSIM-MG incorporates a junction diode model for $BULKMOD = 1$ and parasitic source/drain to substrate capacitance models for both $BULKMOD = 0$ and $BULKMOD = 1$.

13.4.5 Other Real Device Effect Models

We have presented short channel effects, quantum effects and effective width as examples of real device effect models. Other real device effects are also important for representing the output characteristics of multi-gate FETs. Table 13.4 lists real device effects considered in BSIM-MG. Useful references are given in the right column for each real device effect.

Since multi-gate device share similar features with planar bulk MOSFETs, many real device effect modules developed and calibrated for planar bulk MOSFETs can be re-used.

13.5 Experimental Verification

BSIM-CMG has been verified with measurements of both SOI and bulk FinFET technologies.

The SOI FinFETs were fabricated on a lightly doped 60 nm thick film with 2 nm SiO_2 dielectric and a strained TiSiN gate [20]. Measured devices had 20 parallel fins and each fin is 22 nm thick. A global parameter extraction methodology is developed to fit devices with gate lengths ranging from 75 nm to 1 μm [55]. Binning [56] is not used to fit the data. Figures 13.18(a)–(d) show I_d-V_{gs} for both n-type and p-type FinFETs in linear and saturation modes at different L's. Figures 13.18(e), (f) show threshold voltage roll-off with L and sub-threshold swing degradation. Good agreement between the model and data is achieved.

Figure 13.19 show I_d-V_{gs} for an SOI FinFET measured at temperatures ranging from −50°C to 200°C. Model agrees with measurements well across the entire temperature range. Gate-induced drain leakage (GIDL) and its temperature dependence are well-modeled, as shown in Fig. 13.19(b).

The bulk FinFETs with moderate doping were fabricated with a TiN gate [20]. Drain current and substrate current of single-fin devices with fin height 27.5 nm and fin width 25 nm are measured. Figure 13.20(a) shows the fitting of I_d-V_{ds}. Good agreement is achieved.

Figure 13.20(b) shows the fitting of substrate current versus V_{gs}. In SOI FinFETs the impact ionization current can not be observed easily. However, in the bulk FinFET the impact ionization current can be easily separated by measuring the substrate current.

13.6 Computational Efficiency

One of the most important requirements of a production worthy compact model like BSIM-MG is its computational efficiency.

From a model development perspective, we may reduce the model evaluation time by using simpler equations. Transcendental functions such as exp(), ln(), sin(), etc. must be avoided whenever possible. In addition, bias independent quantities are calculated only once during the setup phase and stored for later use. BSIM-MG is developed and implemented with these considerations in mind.

Another way to reduce runtime is by simplifying the surface potential calculation. BSIM-CMG offers a more efficient surface potential model that the user may select by setting $COREMOD = 1$. By computing the surface potentials without including the effect of body doping, the overall model runtime is reduced by about 10% compared with $COREMOD = 0$.

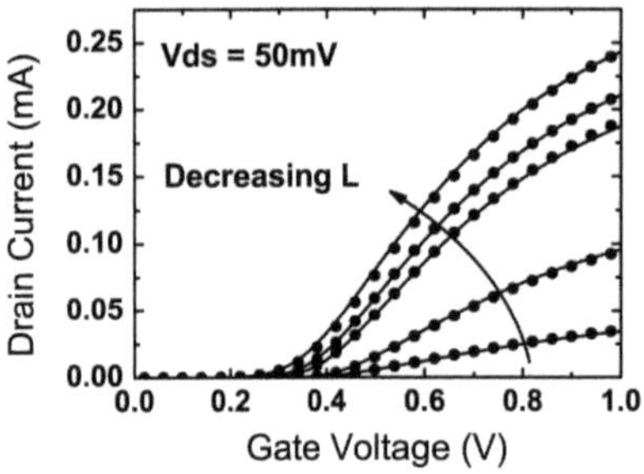

(a) $I_d - V_{gs}$ for n-type FinFET ($V_{ds} = 50$mV)

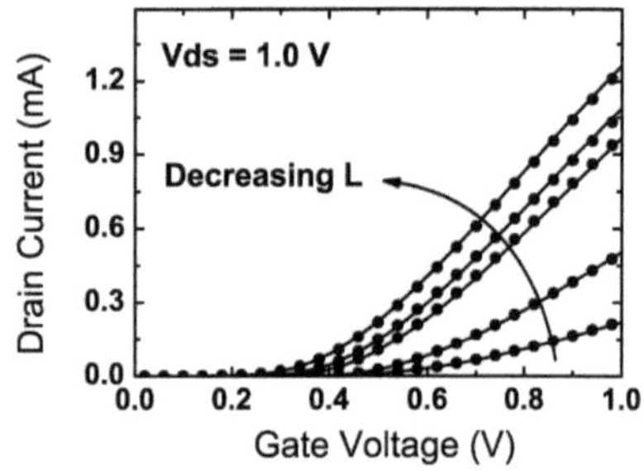

(b) $I_d - V_{gs}$ for n-type FinFET ($V_{ds} = 1.0$V)

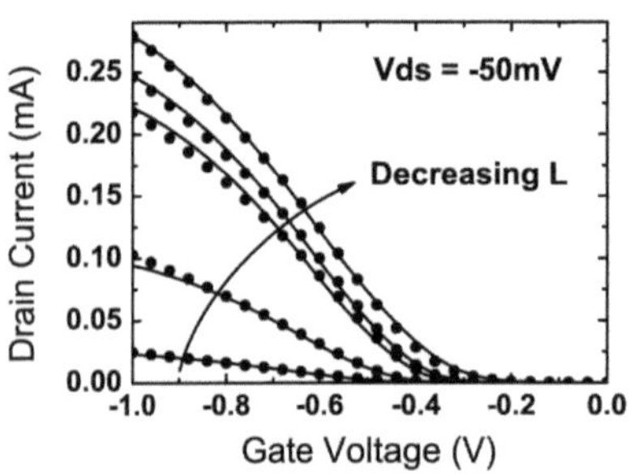

(c) $I_d - V_{gs}$ for p-type FinFET ($V_{ds} = -50$mV)

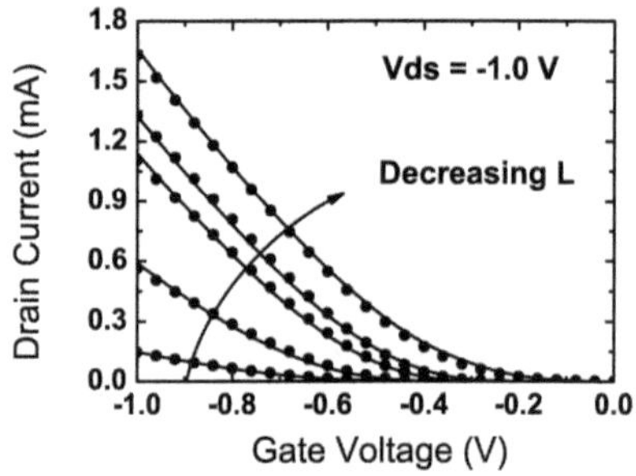

(d) $I_d - V_{gs}$ for p-type FinFET ($V_{ds} = -1.0$V)

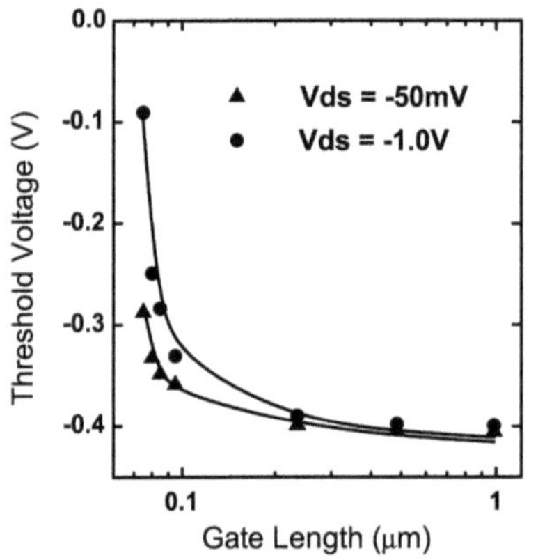

(e) $V_{th} - L$ for p-type FinFET

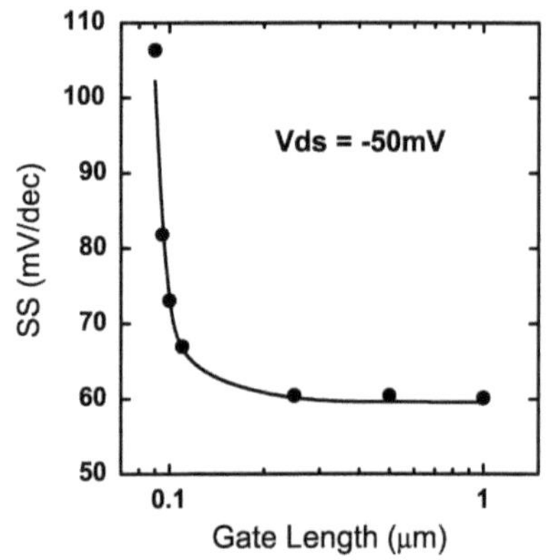

(f) pFET sub-threshold swing degradation

Fig. 13.18 Global parameter extraction results for n-type and p-type SOI FinFETs with 20 parallel fins ($H_{fin} = 60$ nm, $T_{fin} = 22$ nm, $EOT = 2$ nm, length $L = 75$ nm, 85 nm, 90 nm, 235 nm and 1 μm; the body is lightly-doped (2×10^{15} cm^{-3}))

For $COREMOD = 1$, doping is accounted for by shifting the threshold voltage by a bias-independent quantity:

$$\Delta V_{th} = \frac{q N_A T_{FIN}}{2 C_{ox}}. \qquad (13.61)$$

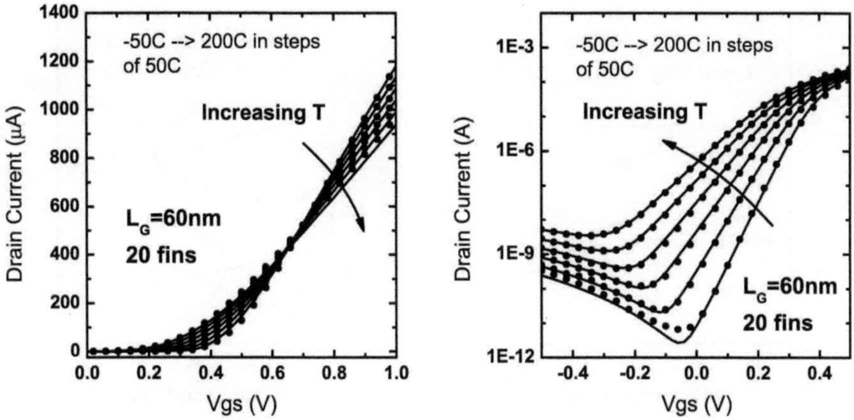

(a) Linear scale (b) Log scale

Fig. 13.19 Temperature dependence verification using SOI FinFETs ($V_{ds} = 1.0$ V, $H_{fin} = 60$ nm, $T_{fin} = 22$ nm, $EOT = 2$ nm, $L = 60$ nm, $N_A = 2 \times 10^{15}$ cm^{-3})

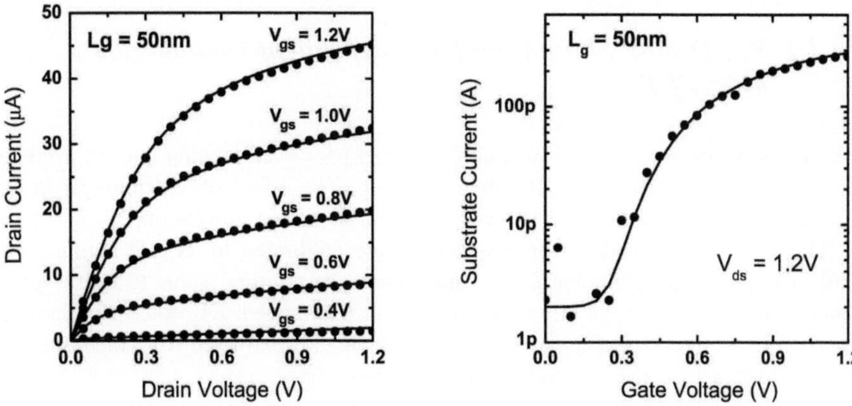

(a) $I_d - V_{ds}$ for n-type FinFET (b) Substrate current for n-type FinFET

Fig. 13.20 Fitting to bulk FinFETs ($H_{fin} = 27.5$ nm, $T_{fin} = 25$ nm, $EOT = 2.4$ nm, $L = 50$ nm, $N_A = 3 \times 10^{17}$ cm^{-3})

In BSIM-IMG, the surface potential calculation does not include doping either. Therefore, it incorporates a similar ΔV_{th} model to consider doping [57]:

$$\Delta V_{th} = \frac{q N_A T_{si}}{C_{ox1}} \cdot \left[1 - \frac{T_{si}}{2(T_{si} + \frac{\varepsilon_{si}}{\varepsilon_{ox}} T_{ox2})} \right]. \tag{13.62}$$

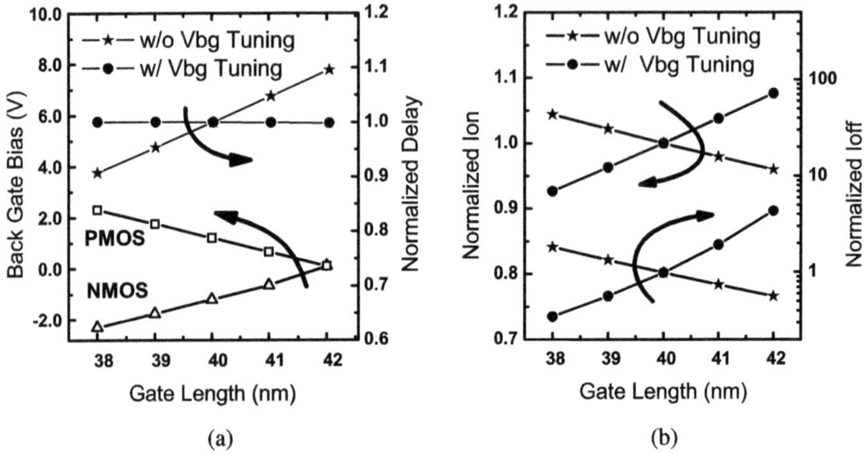

Fig. 13.21 Device variation tuning using back-gated SOI MOSFET. (**a**) The gate delay variation due to gate length variation can be tuned out with V_{bg}. (**b**) The same V_{bg} reverses the trend of both I_{on} and I_{off}

13.7 Simulation Examples

13.7.1 V_{th} Tuning Simulation for Independent Double-gate MOSFETs

As discussed in Sect. 13.1, V_{th} tuning through back-gate biasing can be used to compensate for variability in IC manufacturing. The effect of V_{th} tuning can be simulated with BSIM-IMG, as we will demonstrate in this subsection.

The nominal parameters for BSIM-IMG are calibrated to TCAD simulations [36]. The physical parameter L_g is varied about its nominal value to mimic gate length variation. The remaining parameters are unchanged. DC spice simulations are performed to extract I_{on} and I_{off}. Delay per stage is extracted from the transient waveform of a 17-stage ring oscillator simulation.

The results are shown in Fig. 13.21. Without V_{bg} tuning, gate delay increases with L_g because the gate capacitance increases with increasing L_g and the conduction current decreases with increasing L_g. By biasing the back gates of both NMOS and PMOS devices, the gate-delay variation is tuned out (Fig. 13.21(a)). However, the trends of both I_{on} and I_{off} are reversed (Fig. 13.21(b)) mainly due to the excess V_{bg} needed to compensate capacitance variation.

13.7.2 FinFET SRAM Technology and Simulation Examples

FinFET provides several advantages over the planar MOSFET structure, as we discussed earlier. The difficulties of implementing it into mass production are the risks of treading new grounds and the need to get design tools and circuits developed at

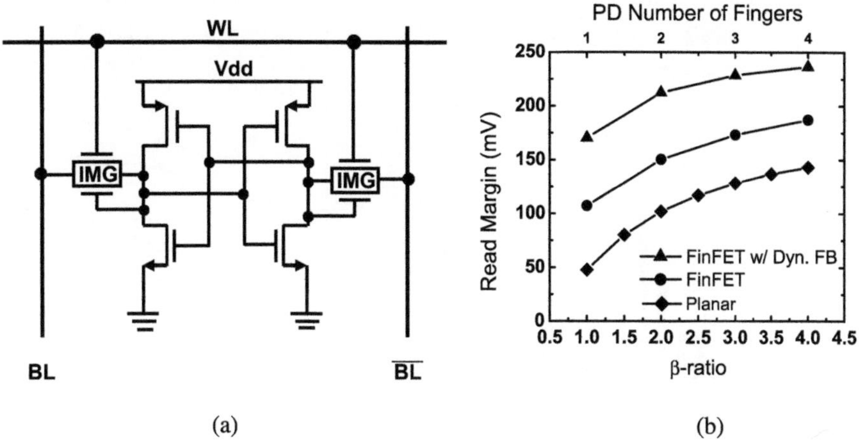

Fig. 13.22 (**a**) Circuit schematic of a 6T FinFET SRAM cell with dynamic feedback [58]. (**b**) Read SNM comparison of the conventional SRAM with planar MOSFETs, the regular FinFET SRAM cell and the dynamic feedback FinFET SRAM cell

the same time as the manufacturing technology. Embedded SRAM cells provide an ideal vehicle for the initial deployment of FinFET because it can benefit more from FinFET than logic circuits initially and does not require extensive design tool and circuit development.

FinFET SRAM has larger static noise margin (SNM) than conventional SRAM due to the better short channel control of FinFETs and therefore larger output impedance. SNM can be further enhanced by employing dynamic feedback [58]. Figure 13.22(a) shows the circuit schematic of a 6T FinFET SRAM with dynamic feedback. By replacing the two access transistors with independent-gate devices and connecting their back-gates to the access nodes, read SNM is improved.

Here we demonstrate the simulation of FinFET SRAM cells using BSIM-IMG and BSIM-CMG. The regular FinFET SRAM is simulated using 6 instances of BSIM-CMG. The dynamic feedback FinFET SRAM is simulated using 4 instances of BSIM-CMG (pull-down and pull-up devices) and 2 instances of BSIM-IMG (access device). Conventional SRAM cells are simulated using BSIM4 [17] as a control group.

Figure 13.22(b) shows the simulation result. The read margin of regular FinFET SRAM is improved compared to conventional SRAM for a given β-ratio. Read margin for dynamic feedback FinFET SRAM is further enhanced. Also, since FinFET only allows discrete number of fins, β-ratio for FinFET SRAM is discretized unlike the conventional SRAM.

13.7.3 Statistical Simulation of FinFET SRAM Cells

As device dimensions shrink from one technology generation to another, variability from device to device, die to die, and wafer to wafer becomes more and more signif-

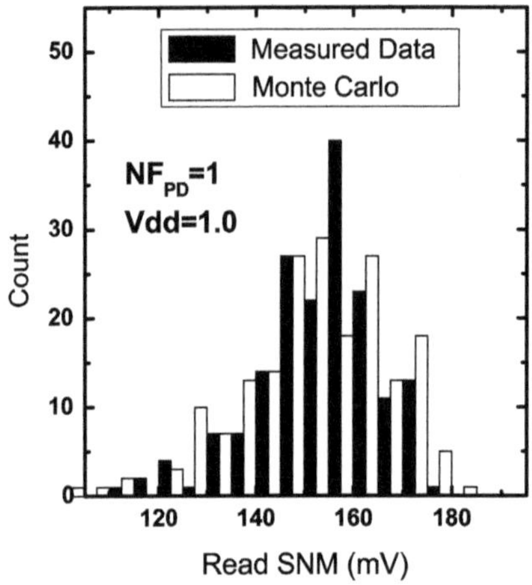

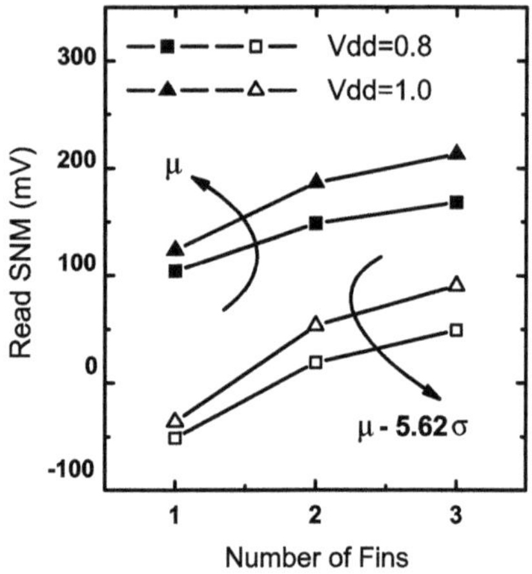

Fig. 13.23 (**a**) Calibration of variation to read static noise margin distribution. (**b**) Read SNM as function of number of pull-down nFET fingers and V_{dd}

icant. Although variation due to RDF is suppressed by using a lightly-doped body in FinFET technologies, other variation sources such as line edge roughness (LER) [59] will still come into play.

BSIM-CMG can be used to model statistical variation in FinFETs, as demonstrated in [60]. Both global and local variation [61] are considered. Global variation in T_{fin}, L, H_{fin} and T_{ox} are taken into account. σ_{TFIN} and σ_L due to local variation are expected to be significant due to LER. They are calibrated by comparing Monte Carlo simulation results with measurements (Fig. 13.23(a)). Once the variation is calibrated we use the statistical model to study the dependence of cell margins and their variation on various design parameters. As an example, Fig. 13.23(b) shows the effect of changing the number of fins of the pull-down nFET (NF_{PD}) and the supply voltage (V_{dd})).

Acknowledgments We would like to express our sincere appreciation to Dr. Mohan Dunga for his pioneering development of BSIM-IMG and BSIM-CMG. We would also like to thank Dr. Weize Xiong and Dr. Rinn Cleavelin at Texas Instrument, Dr. Paul Patruno at SOITEC, Dr. Jiunn-Ren Hwang and Dr. Fu-Liang Yang at Taiwan Semiconductor Manufacturing Corporation for generously sharing their measured FinFET data. The work presented in this chapter would not have been possible without the funding support by Semiconductor Research Corporation (Task ID: 1451.001) and IMPACT, UC Discovery, and its industrial sponsors.

References

1. Frank, D.J., Dennard, R.H., Nowak, E., Solomon, P.M., Taur, Y., Wong, H.-S.P.: Device scaling limits of Si MOSFETs and their application dependencies. Proc. IEEE **89**, 259–288 (2001)

2. Assaderaghi, F., Sinitsky, D., Bokor, J., Ko, P.K., Gaw, H., Hu, C.: High-field transport of inversion-layer electrons and holes including velocity overshoot. IEEE Trans. Electron Devices **44**, 664–671 (1997)
3. Lundstrom, M.: Elementary scattering theory of the Si MOSFET. IEEE Electron Device Lett. **18**, 361–363 (1997)
4. Ghani, T., et al.: A 90 nm high volume manufacturing logic technology featuring novel 45 nm gate length strained silicon CMOS transistors. In: Technical Digest, IEEE International Electron Devices Meeting, pp. 407–410 (2003)
5. Lee, W.-C., Hu, C.: Modeling CMOS tunneling currents through ultrathin gate oxide due to conduction- and valence-band electron and hole tunneling. IEEE Trans. Electron Devices **48**, 1366–1373 (2001)
6. Mistry, K., et al.: 45 nm logic technology with high-k+metal gate transistors, strained silicon, 9 Cu interconnect layers, 193 nm dry patterning, and 100% Pb-free packaging. In: Technical Digest, IEEE International Electron Devices Meeting, pp. 247–250 (2007)
7. Colinge, J.-P.: FinFETs and Other Multi-gate Transistors. Springer, Berlin (2008)
8. Asenov, A.: Random dopant induced threshold voltage lowering and fluctuations in sub-0.1 m MOSFET's: a 3-D "atomistic" simulation study. IEEE Trans. Electron Devices **45**, 2505–2513 (1998)
9. Kim, H.-S., Lee, S.-B., Choi, D.-U., Shim, J.-H., Lee, K.-H., Lee, K.-P., Kim, K.-N., Park, J.-W.: A high-performance 16M DRAM on a thin film SOI. In: Digest of Technical Papers, Symposium on VLSI Technology, pp. 143–144 (1995)
10. Huang, X., Lee, W.-C., Kuo, C., Hisamoto, D., Chang, L., Kedzierski, J., Anderson, E., Takeuchi, H., Choi, Y.-K., Asano, K., Subramanian, V., King, T.-J., Bokor, J., Hu, C.: Sub 50-nm FinFET: PMOS. In: Technical Digest, IEEE International Electron Devices Meeting, pp. 67–70 (1999)
11. Hu, C.: MOSFETs in ICs–scaling, leakage, and other topics. In: Modern Semiconductor Devices for Integrated Circuits. Prentice Hall, New York (2009)
12. Yang, F.-L., Chen, H.-Y., Chen, F.-C., Chan, Y.-L., Yang, K.-N., Chen, C.-J., Tao, H.-J., Choi, Y.-K., Liang, M.-S., Hu, C.: 35 nm CMOS FinFETs. In: Digest of Technical Papers, Symposium on VLSI Technology, pp. 104–105 (2002)
13. von Arnim, K., et al.: A low-power multi-gate FET CMOS technology with 13.9 ps inverter delay, large-scale integrated high performance digital circuits and SRAM. In: Digest of Technical Papers, Symposium on VLSI Technology, pp. 106–107 (2007)
14. Kedzierski, J., et al.: Metal-gate FinFET and fully-depleted SOI devices using total gate silicidation. In: Technical Digest, IEEE International Electron Devices Meeting, pp. 247–250 (2002)
15. Kavalieros, J., Doyle, B., Datta, S., Dewey, G., Doczy, M., Jin, B., Lionberger, D., Metz, M., Rachmady, W., Radosavljevic, M., Shah, U., Zelick, N., Chau, R.: Tri-gate transistor architecture with high-k gate dielectrics, metal gates and strain engineering. In: Digest of Technical Papers, Symposium on VLSI Technology, pp. 50–51 (2006)
16. Sheu, B., Scharfetter, D.L., Ko, P.-K., Jeng, M.-C.: BSIM: Berkeley short-channel IGFET model for MOS transistors. IEEE J. Solid-State Circuits **22**, 558–566 (1987)
17. BSIM (Berkeley Short-channel IGFET Model). http://www-device.eecs.berkeley.edu/~bsim/
18. Su, P., Fung, S.K.H., Tang, S., Assaderaghi, F., Hu, C.: BSIMPD: a partial-depletion SOI MOSFET model for deep-submicron CMOS designs. In: Proc. of the IEEE Custom Integrated Circuits Conference, pp. 197–200 (2000)
19. Chan, M., Su, P., Wan, H., Lin, C.-H., Fung, S.K.-H., Niknejad, A.M., Hu, C., Ko, P.K.: Modeling the floating-body effects of fully depleted, partially depleted, and body-grounded SOI MOSFETs. Solid-State Electron. **48**, 969–978 (2004)
20. Dunga, M.V., Lin, C.-H., Lu, D.D., Xiong, W., Cleavelin, C.R., Patruno, P., Hwang, J.-R., Yang, F.-L., Niknejad, A.M., Hu, C.: BSIM-MG: A versatile multi-gate FET model for mixed-signal design. In: Digest of Technical Papers, Symposium on VLSI Technology, pp. 60–61 (2007)

21. Lu, D.D., Dunga, M.V., Lin, C.-H., Niknejad, A.M., Hu, C.: A multi-gate MOSFET compact model featuring independent-gate operation. In: Technical Digest, IEEE International Electron Devices Meeting, pp. 565–568 (2007)
22. Fossum, J.G., Ge, L., Chiang, M.-H., Trivedi, V.P., Chowdhury, M.M., Matthew, L., Workman, G.O., Nguyen, B.-Y.: A process/physics-based compact model for nonclassical CMOS device and circuit design. Solid-State Electron. **48**, 919–926 (2004)
23. Yu, B., Song, J., Yuan, Y., Lu, W.-Y., Taur, Y.: A unified analytic Drain *V* Current model for multiple-gate MOSFETs. IEEE Trans. Electron Devices **55**, 2157–2163 (2008)
24. Dessai, G., Dey, A., Gildenblat, G., Smit, G.D.J.: Symmetric linearization method for double-gate and surrounding-gate MOSFET models. Solid-State Electron. **53**, 548–556 (2009)
25. Sallese, J.-M., Krummenacher, F., Pregaldiny, F., Lallement, C., Roy, A., Enz, C.: A design oriented charge-based current model for symmetric DG MOSFET and its correlation with the EKV formalism. Solid-State Electron. **49**, 485–489 (2005)
26. Pei, G., Ni, W., Kammula, A.V., Minch, B.A., Kan, E.C.-C.: Physical compact model of DG MOSFET for mixed-signal circuit applications—Part I: model description. IEEE Trans. Electron Devices **50**, 2135–2143 (2003)
27. Ishimura, K., Sadachika, N., Kusu, S., Miura-Mattausch, M.: Compact model HiSIM-DG both for symmetrical and asymmetrical DG-MOSFET structures. In: Proc. Workshop on Compact Modeling (2009)
28. International Technology Roadmap for Semiconductors. http://www.itrs.net/
29. Takayanagi, K., Kondo, Y., Ohnishi, H.: Suspended gold nanowires: ballistic transport of electrons. J. Jpn. Soc. Appl. Phys. Int. (JSAPI) **3**, 3–8 (2001)
30. Fried, D., Duster, J.S., Kornegay, K.T.: High-performance p-type independent-gate FinFETs. IEEE Electron Device Lett. **25**, 199–201 (2004)
31. Yang, I.Y., Vieri, C., Chandrakasan, A., Antoniadis, D.A.: Back-gated CMOS on SOIAS for dynamic threshold voltage control. IEEE Trans. Electron Devices **44**, 822–831 (1997)
32. Liu, H., Taur, Y.: An analytic potential model for symmetric and asymmetric DG MOSFETs. IEEE Trans. Electron Devices, 1161–1168 (2006)
33. Dunga, M.V., Lin, C.-H., Niknejad, A.M., Hu, C.: BSIM-CMG: a compact model for multi-gate transistors. In: FinFETs and Other Multi-gate Transistors, pp. 113–153 (2008)
34. Lu, D.D.: Efficient surface potential calculation for the asymmetric independent double-gate MOSFET. UC Berkeley Master's Report (2007)
35. Brews, J.R.: A charge-sheet model of the MOSFET. Solid-Sate Electron. **21**, 345–355 (1978)
36. Synopsys Inc.: Taurus Process and Device User Manual (2003)
37. Oh, S.-Y., Ward, D.E., Dutton, R.W.: Transient analysis of MOS transistors. IEEE Trans. Electron Devices **27**, 1571–1578 (1980)
38. Tsividis, Y.: Operation and Modeling of the MOS Transistor, 2nd edn. Oxford (1999)
39. Venugopolan, S.: A Compact Model for Cylindrical Gate MOSFET for circuit simulations. UC Berkeley Master's Report (2010)
40. Dunga, M.V.: Nanoscale CMOS modeling. Ph.D. Thesis, UC Berkeley (2007)
41. Dunga, M.V., Lin, C.-H., Xi, X., Lu, D.D., Niknejad, A.M., Hu, C.: Modeling advanced FET technology in a compact model (invited). IEEE Trans. Electron Devices **53**, 1971–1978 (2006)
42. Taur, Y.: An analytical solution to a double-gate MOSFET with undoped body. IEEE Electron Device Lett. **21**, 245–247 (2000)
43. Sebah, P., Gourdon, X.: Newton's method and high order iterations (2001). http://numbers.computation.free.fr/Constants/Algorithms/newton.ps
44. Trivedi, V.P., Fossum, J.G.: Quantum-mechanical effects on the threshold voltage of undoped double-gate MOSFETs. IEEE Electron Device Lett. **26**, 579–582 (2005)
45. Lin, C.-H.: Compact modeling of nanoscale CMOS. Ph.D. Thesis, UC Berkeley (2007)
46. Liu, Z.-H., Hu, C., Huang, J.-H., Chan, T.-Y., Jeng, M.-C., Ko, P.K., Cheng, Y.C.: Threshold voltage model for deep-submicrometer MOSFET's. IEEE Trans. Electron Devices **40**, 86–95 (1993)
47. BSIM4 User's Manual. http://www-device.eecs.berkeley.edu/~bsim3/bsim4_get.html

48. BSIMSOI User's Manual. http://www-device.eecs.berkeley.edu/~bsimsoi/get.html
49. Suzuki, K., Sugii, T.: Analytical models for n^+-p^+ double-gate SOI MOSFET's. IEEE Trans. Electron Devices **42**, 1940–1948 (1995)
50. Pei, G., Kedzierski, J., Oldiges, P., Ieong, M., Kan, E.C.-C.: FinFET design considerations based on 3-D simulation and analytical modeling. IEEE Trans. Electron Devices **49**, 1411–1419 (2002)
51. Jin, W., Fung, S.K.H., Liu, W., Chan, P.C.H., Hu, C.: Self-heating characterization for SOI MOSFET based on AC output conductance. In: Technical Digest, IEEE International Electron Devices Meeting, pp. 175–178 (1999)
52. Hung, K.K., Ko, P.K., Hu, C., Cheng, Y.C.: A unified model for the flicker noise in metal-oxide-semiconductor field-effect transistors. IEEE Trans. Electron Devices **37**, 654–665 (1990)
53. Cao, K.M., Lee, W.-C., Liu, W., Jin, X., Su, P., Fung, S.K.H., An, J.X., Yu, B., Hu, C.: BSIM4 gate leakage model including source-drain partition. In: Technical Digest, IEEE International Electron Devices Meeting, pp. 815–818 (2000)
54. Wann, H.-J., Ko, P.K., Hu, C.: Gate-induced band-to-band tunneling leakage current in LDD MOSFETs. In: Technical Digest, IEEE International Electron Devices Meeting, pp. 147–150 (1992)
55. Yao, S., Morshed, T.H., Lu, D.D., Venugopalan, S., Niknejad, A.M., Hu, C.: A global parameter extraction procedure for multi-gate MOSFETs. To Be Presented in the 23rd International Conference on Microelectronic Test Structures (2010)
56. Cheng, Y., Hu, C.: MOSFET Modeling and BSIM3 User's Guide. Springer, Berlin (1999)
57. Banna, S.R., Chan, P.C.H., Ko, P.K., Nguyen, C.T., Chan, M.: Threshold voltage model for deep-submicrometer fully depleted SOI MOSFET's. IEEE Trans. Electron Devices **42**, 1949–1955 (1995)
58. Guo, Z., Balasubramanian, S., Zlatanovici, R., King, T.-J., Nikolić, B.: FinFET-based SRAM design. In: International Symposium on Low Power Electronics and Design (ISLPED), pp. 2–7 (2005)
59. Oldiges, P., Lin, Q., Petrillo, K., Sanchez, M., Ieong, M., Hargrove, M.: Modeling line edge roughness effects in sub 100 nanometer gate length devices. In: Proc. Int. Conference on Simulation of Semiconductor Devices and Processes (SISPAD), pp. 131–134 (2000)
60. Lu, D.D., Lin, C.-H., Yao, S., Xiong, W., Bauer, F., Cleavelin, C.R., Niknejad, A.M., Hu, C.: Design of FinFET SRAM cells using a statistical compact model. In: Proc. Int. Conference on Simulation of Semiconductor Devices and Processes (SISPAD), pp. 127–130 (2009)
61. Lin, C.-H., Dunga, M.V., Lu, D.D., Niknejad, A.M., Hu, C.: Performance-aware corner model for design for manufacturing. IEEE Trans. Electron Devices **56**, 595–600 (2009)

Chapter 14
Compact Modeling of Double-Gate and Nanowire MOSFETs

Yuan Taur

Abstract This chapter reviews recent developments on compact modeling of double-gate and nanowire MOSFETs. It starts with the core, long-channel drain current models of double-gate and nanowire MOSFETs, derived from the analytic solutions of 1-D Poisson and current continuity equations in Cartesian and cylindrical coordinates, respectively. Explicit and continuous solutions to the implicit parameters in both models have been developed. The short-channel models based on the scale length approach to the boundary value problems of 2-D Poisson's equation in subthreshold are then described, followed by charge and capacitance models for both double-gate and nanowire MOSFETs. A popular, surface-potential based current expression in the literature is examined before concluding the chapter.

14.1 Introduction

Continued scaling of CMOS technology to 10 nm gate length and below calls for a major change in the MOSFET structure. Traditional bulk MOSFETs scale by thinning the gate oxide and raising the body doping. At 20 nm gate length, the gate oxide thickness has reached the tunnelling limit and the body doping level has exceeded 10^{19} cm^{-3}. Instead of doping concentration, double-gate MOSFETs rely on the physical thickness of the silicon film to scale the channel length. For a given gate oxide or high-κ dielectric thickness, a double-gate MOSFET with very thin silicon can ultimately extend scaling beyond bulk CMOS.

In this chapter, recent developments on compact modeling of double-gate and nanowire MOSFETs are reviewed. Nanowire MOSFET, also called gate-all-around or surrounding-gate MOSFET, is a variation of double-gate MOSFET in which a silicon wire with a circular cross-section is surrounded by a cylindrical gate electrode.

Y. Taur (✉)
University of California, San Diego, USA
e-mail: taur@ece.ucsd.edu

G. Gildenblat (ed.), *Compact Modeling*,
DOI 10.1007/978-90-481-8614-3_14, © Springer Science+Business Media B.V. 2010

Section 14.2 starts with the core, long-channel drain current models of double-gate and nanowire MOSFETs. They are derived from the rigorous solutions of 1-D Poisson and current continuity equations in Cartesian and cylindrical coordinates, respectively. Explicit and continuous solutions to the implicit parameters in both models are described. Section 14.3 discusses the short-channel models which are based on the scale length approach to the boundary value problems of 2-D Poisson's equation in both double-gate and nanowire MOSFETs in subthreshold. Subthreshold current of short-channel MOSFETs can be calculated from the lowest order term in the scale length expansion of the 2-D potential function. Section 14.4 covers the charge and capacitance models of double-gate and nanowire MOSFETs. For nanowire MOSFETs, closed form charge expressions for gate, source, and drain can be derived. Section 14.5 addresses a surface-potential based current expression—a popular approximation in the literature that led to a simplified compact model. The chapter is concluded in Sect. 14.6.

14.2 Analytic Potential Models for Double-Gate and Nanowire MOSFETs

This section describes the long-channel drain current models of double-gate [27] and nanowire MOSFETs based on the analytic potential solutions to 1-D Poisson and current continuity equations. They form the core of the more elaborate compact models for these devices.

14.2.1 Analytic Solutions to Double-Gate MOSFETs

For bulk MOSFETs, the 1-D Poisson's equation contains a depletion charge term which renders no analytic solution. As a result, Pao-Sah's double integral [1] for the drain current can only be evaluated numerically given a specific set of device parameters. This had necessitated the charge sheet approximation [2] which formed the basis of all bulk compact models [3]. For double-gate MOSFETs, the silicon body can be undoped because they rely on the silicon thickness to scale down the gate length, not on the depletion layer depth like bulk MOSFETs. In fact, it is preferred to have an undoped or lightly doped silicon film to eliminate impurity scattering and dopant number fluctuation problems in a double-gate device. The absence of the depletion charge term allows the 1-D Poisson's equation of a double-gate MOSFET analytically integrable to yield a closed-form solution to the potential everywhere in the silicon film [4]. By extending this approach, a continuous, analytic I-V model for double-gate MOSFETs has been derived directly from Pao-Sah's integral without the charge sheet approximation [5]. This analytic solution covers all regions of MOSFET operation, thus maintaining strong continuity while retaining the essential device physics.

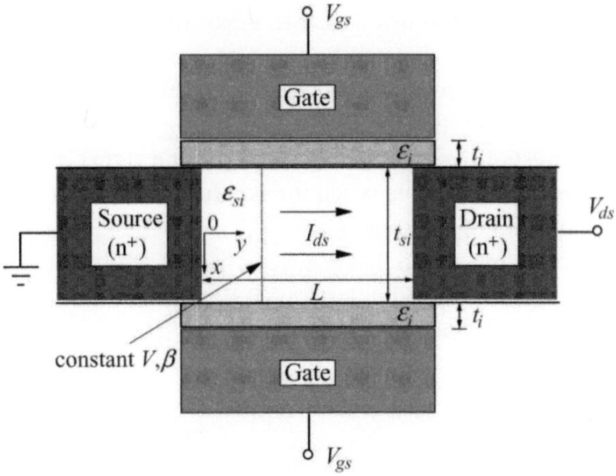

Fig. 14.1 Schematic diagram of a double-gate MOSFET. $V(y)$ is the quasi-Fermi potential at a point in the channel. $V(0) = 0$ at the source and $V(L) = V_{ds}$ at the drain. β is a function of V

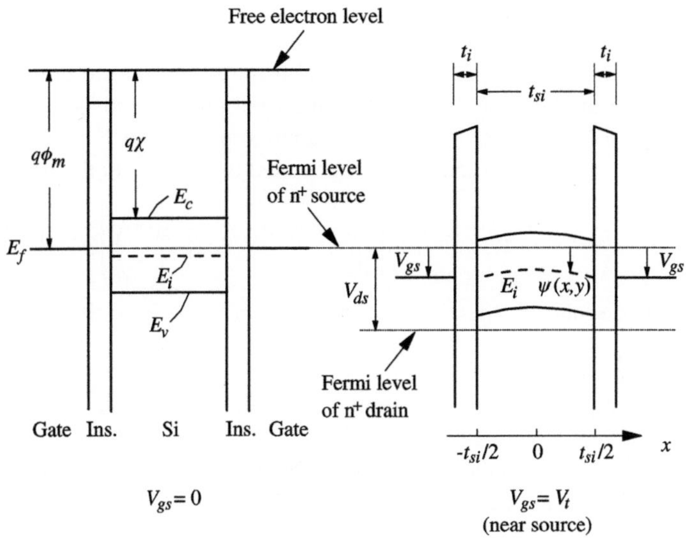

Fig. 14.2 Band diagram along a vertical cut in Fig. 10.10 for two gate voltages. The gate work function in this example is slightly lower than that of intrinsic silicon. For $V_{gs} = V_t$ on the right, the cut is near the source where $V = 0$ (after [6])

Consider an undoped (or lightly doped), symmetric double-gate MOSFET shown schematically in Fig. 14.1. The same voltage is applied to the two gates having the same work function. The band diagrams along a cut perpendicular to the silicon film and the two gates are shown in Fig. 14.2 for two gate voltages [6]. At zero gate voltage below threshold, the bands are essentially flat throughout the silicon film as well as in the gate insulators because both the depletion charge and the inversion charge are negligible. Since there is no contact to the silicon body, the position of Fermi level in the band gap is not determined by the light doping of the body and the usual charge neutrality relationship. Instead, the position of the silicon bands is dictated by the gate work function as shown. The energy levels are referenced to the electron quasi-Fermi level or the conduction band of the n^+ source, represented by the long dotted line in Fig. 14.2. As the gate voltage increases toward the threshold voltage in Fig. 14.2 (on the right), mobile charge or electron density becomes appreciable when the conduction band of the silicon body moves near the Fermi level of the source.

Following Pao-Sah's gradual channel approach [1], Poisson's equation along a vertical cut perpendicular to the silicon film (Fig. 14.1) takes the following form with only the mobile charge (electrons) term:

$$\frac{d^2\psi}{dx^2} = \frac{q}{\varepsilon_{si}} n_i e^{q(\psi - V)/kT}, \tag{14.1}$$

where $\psi(x, y)$ is the electrostatic potential, defined as the intrinsic potential at (x, y) with respect to the Fermi potential at the source, as shown in Fig. 14.2. $V(y)$ is the electron quasi-Fermi potential at y with respect to that of the source. Here we consider an nMOSFET with $q\psi/kT \gg 1$ so that the hole density is negligible.

Since the current flows predominantly from the source to the drain along the y-direction, the gradient of the electron quasi-Fermi potential is also in the y-direction. This justifies the gradual channel approximation that V is constant in the x-direction. Equation (14.1) can then be integrated twice to yield the solution [4]

$$\psi(x) = V - \frac{2kT}{q} \ln\left[\frac{t_{si}}{2\beta}\sqrt{\frac{q^2 n_i}{2\varepsilon_{si} kT}} \cos\left(\frac{2\beta x}{t_{si}}\right)\right], \tag{14.2}$$

where β is a constant (independent of x) to be determined from the boundary condition

$$\varepsilon_i \frac{V_{gs} - \Delta\phi - \psi(x = \pm t_{si}/2)}{t_i} = \pm\varepsilon_{si} \left.\frac{d\psi}{dx}\right|_{x=\pm t_{si}/2}. \tag{14.3}$$

Here V_{gs} is the voltage applied to both gates, and $\Delta\phi$ is the work function of both the top and bottom gate electrodes with respect to that of the intrinsic silicon. In other words, $\Delta\phi = \phi_m - (\chi + E_g/2q)$, where $q\phi_m$ is the gate work function and $q\chi$ is the electron affinity. Substituting (14.2) into (14.3) leads to

$$\frac{q(V_{gs} - \Delta\phi - V)}{2kT} - \ln\left(\frac{2}{t_{si}}\sqrt{\frac{2\varepsilon_{si} kT}{q^2 n_i}}\right) = \ln\beta - \ln(\cos\beta) + \frac{2\varepsilon_{si} t_i}{\varepsilon_i t_{si}} \beta \tan\beta. \tag{14.4}$$

For a given V_{gs}, β can be solved from the implicit equation (14.4) as a function of V. Along the channel direction (y), V varies from the source to the drain. So does β. The functional dependence of $V(y)$ and $\beta(y)$ is determined by the current continuity condition which requires the current $I_{ds} = \mu W Q_i dV/dy =$ constant, independent of V or y. Here μ is the effective mobility, W is the device width, and Q_i is the total mobile charge per unit area including both channels. Integrating $I_{ds}dy$ from the source to the drain and expressing dV/dy as $(dV/d\beta)(d\beta/dy)$, Pao-Sah's integral can be written as

$$I_{ds} = \mu \frac{W}{L} \int_0^{V_{ds}} Q_i(V) dV = \mu \frac{W}{L} \int_{\beta_s}^{\beta_d} Q_i(\beta) \frac{dV}{d\beta} d\beta, \tag{14.5}$$

where β_s, β_d are solutions to (14.4) corresponding to $V = 0$ and $V = V_{ds}$ respectively. From Gauss's law, $Q_i = 2\varepsilon_{si}(d\psi/dx)_{x=tsi/2}$, which equals $2\varepsilon_{si}(2kT/q) \times (2\beta/t_{si}) \tan\beta$ using (14.2). $dV/d\beta$ can also be expressed as a function of β by differentiating (14.4). Substituting these factors in (14.5) and carrying out the integration analytically yield the drain current [5]:

$$\begin{aligned} I_{ds} &= \mu \frac{W}{L} \frac{4\varepsilon_{si}}{t_{si}} \left(\frac{2kT}{q}\right)^2 \int_{\beta_d}^{\beta_s} \left[\tan\beta + \beta\tan^2\beta + \frac{2\varepsilon_{si} t_i}{\varepsilon_i t_{si}} \beta \tan\beta \frac{d}{d\beta}(\beta\tan\beta)\right] d\beta \\ &= \mu \frac{W}{L} \frac{4\varepsilon_{si}}{t_{si}} \left(\frac{2kT}{q}\right)^2 \cdot \left[\beta\tan\beta - \frac{\beta^2}{2} + \frac{\varepsilon_{si} t_i}{\varepsilon_i t_{si}} \beta^2 \tan^2\beta\right]\Bigg|_{\beta_d}^{\beta_s}. \end{aligned} \tag{14.6}$$

The range of β_s, β_d is $(0, \pi/2)$. MOSFET characteristics for all regions: linear, saturation, and subthreshold, can be generated from this continuous, analytic solution [5]. For example, in the linear region above threshold, the LHS of (14.4) $\gg 1$ for both $V = 0$ and V_{ds}, so $\beta_s, \beta_d \sim \pi/2$. The last terms on the RHS of (14.4) and (14.6) dominate, therefore

$$\begin{aligned} I_{ds} &= \mu \frac{\varepsilon_i}{t_i} \frac{W}{L} [(V_{gs} - V_t)^2 - (V_{gs} - V_t - V_{ds})^2] \\ &= 2\mu \frac{\varepsilon_i}{t_i} \frac{W}{L} (V_{gs} - V_t - V_{ds}/2) V_{ds}, \end{aligned} \tag{14.7}$$

where

$$V_t = \Delta\phi + \frac{2kT}{q} \ln\left[\frac{2}{t_{si}} \sqrt{\frac{2\varepsilon_{si} kT}{q^2 n_i}}\right] + \frac{2kT}{q} \ln\left[\frac{q\varepsilon_i t_{si}(V_{gs} - \Delta\phi)}{2\pi\varepsilon_{si} t_i kT}\right]. \tag{14.8}$$

The last term is a second-order term coming from the $\ln[\cos\beta]$ term in (14.4). It is kept here to show that the t_{si} factor cancels the $1/t_{si}$ factor in the previous term so V_t is independent of t_{si}.

In the saturation region, where $\beta_s \sim \pi/2$, but $\beta_d \ll 1$, one obtains

$$I_{ds} = \mu \frac{\varepsilon_i}{t_i} \frac{W}{L} \left[(V_{gs} - V_t)^2 - \frac{8\varepsilon_{si} t_i k^2 T^2}{\varepsilon_i t_{si} q^2} e^{q(V_{gs} - V_t - V_{ds})/kT}\right]. \tag{14.9}$$

Note that in this continuous model, the current approaches the saturation value with a difference term exponentially decreasing with V_{ds}, in contrast to common piecewise models in which the current is made to be constant in saturation.

In the subthreshold region, both $\beta_s, \beta_d \ll 1$, so the $\ln\beta$ term dominates on the RHS of (14.4), and

$$I_{ds} = \mu \frac{W}{L} kT n_i t_{si} e^{q(V_{gs} - \Delta\phi)/kT} \left(1 - e^{-qV_{ds}/kT}\right), \tag{14.10}$$

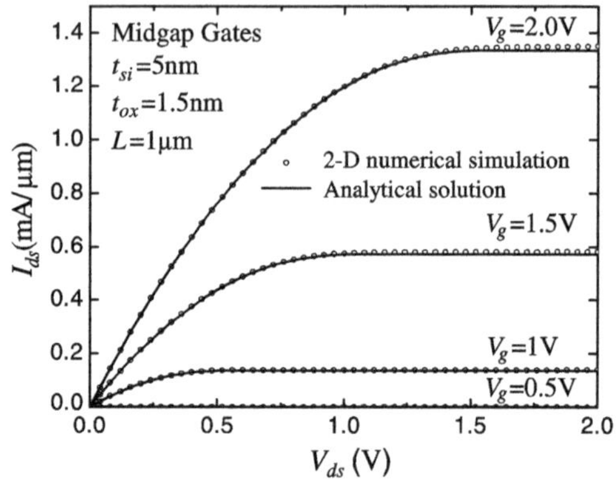

Fig. 14.3 I_{ds}-V_{ds} curves calculated from the analytic model (*solid curves*), compared with the 2-D numerical simulation results (*open circles*)

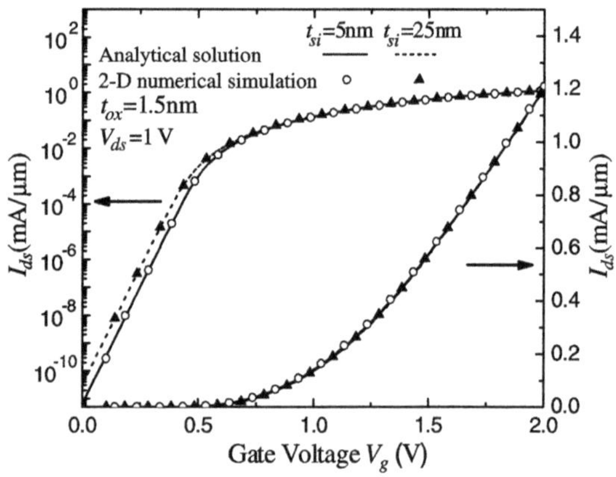

Fig. 14.4 I_{ds}-V_g characteristics obtained from the analytic model for two different values of t_{si} (*solid* and *dashed curves*), compared with the 2-D numerical simulation results (*symbols*). The same currents are plotted on both logarithmic (*left*) and linear (*right*) scales

as would be expected from the basic diffusion current, $J_{diff} = qDdn/dx$. Note that the subthreshold current is proportional to the silicon thickness, but independent of ε_i/t_i—a manifestation of "volume inversion" that the potential in silicon is near constant and directly follows the gate modulation when the mobile carrier density is low. In contrast, the currents above threshold, (14.7) and (14.9), are proportional to ε_i/t_i, but independent of silicon thickness. This is the same as in a bulk MOSFET in that once it becomes energetically favorable for a large electron population in the conduction band, essentially all the gate-induced mobile charge appears at the surface, which electrostatically screens the interior of silicon from the gate.

Figures 14.3 and 14.4 show that the analytic I_{ds}-V_{ds} and I_{ds}-V_g curves computed from this model are in complete agreement with 2-D numerical simulations of a long-channel double-gate MOSFET. No fitting parameters are needed. "Volume inversion," in which the subthreshold current is proportional to t_{si}, is self evident in Fig. 14.4 [5].

While doping in the silicon film is neglected in the above analytic potential model, its first order effect can be incorporated simply as a threshold voltage shift of magnitude $qN_a t_{si}/2C_{ox}$ due to the depletion charge in silicon. P-type dopants shift the threshold positively and n-type negatively. It has been shown that too high a doping level has a negative impact on device characteristics [7].

14.2.2 Analytic Solutions to Nanowire MOSFETs

The compact model for nanowire MOSFETs has been obtained in a similar way as DG MOSFETs. With cylindrical symmetry, Poisson's equation takes the form

$$\frac{1}{\rho}\frac{d}{d\rho}\left(\rho\frac{d\psi}{d\rho}\right)=\frac{q}{\varepsilon_{si}}n_i e^{q(\psi-V)/kT}, \tag{14.11}$$

where ρ is the radial coordinate in the cross sectional plane of the nanowire. Current flows between the source and drain along the y direction. Solving the equation yields the potential at any radius of the nanowire [8],

$$\psi(\rho)=V-\frac{2kT}{q}\ln\left[\frac{r_{si}}{2\sqrt{1-\alpha}}\sqrt{\frac{q^2 n_i}{2\varepsilon_{si}kT}}\left(1-\frac{(1-\alpha)\rho^2}{r_{si}^2}\right)\right] \tag{14.12}$$

where r_{si} is the radius of the nanowire and the parameter α $(0<\alpha<1)$ is determined by the boundary condition at $\rho=r_{si}$ below. In the insulator region, $d(\rho d\psi/d\rho)d\rho=0$, so $\psi\sim\ln\rho$. The radial field $d\psi/d\rho$ at $\rho=r_{si}$ on the insulator side is then given by the gate voltage and surface potential as $[V_{gs}-\Delta\phi-\psi(\rho=r_{si})]/[r_{si}\ln(1+t_i/r_{si})]$, and the boundary condition is

$$\varepsilon_i\frac{V_{gs}-\Delta\phi-\psi(\rho=r_{si})}{r_{si}\ln(1+t_i/r_{si})}=\varepsilon_{si}\left.\frac{d\psi}{d\rho}\right|_{\rho=r_{si}}. \tag{14.13}$$

Substituting (14.12) into the above yields an implicit equation for α,

$$\begin{aligned}&\frac{q(V_{gs}-\Delta\phi-V)}{2kT}+\ln\left(\frac{r_{si}}{2}\sqrt{\frac{q^2n_i}{2\varepsilon_{si}kT}}\right)\\&=\ln\sqrt{1-\alpha}-\ln\alpha+2\frac{1-\alpha}{\alpha}\frac{\varepsilon_{si}}{\varepsilon_i}\ln\left(1+\frac{t_i}{r_{si}}\right).\end{aligned} \tag{14.14}$$

V varies along the channel length direction (y), so does α.

Using (14.12), the inversion charge density per gate area can be expressed as $Q_i=\varepsilon_{si}(d\psi/d\rho)|_{\rho=r_{si}}=4\varepsilon_{si}kT(1-\alpha)/(qr_{si}\alpha)$. The drain current of a nanowire MOSFET can then be evaluated from the current continuity condition and Pao-Sah's integral,

$$I_{ds}=\mu\frac{2\pi r_{si}}{L}\int_0^{V_{ds}}Q_i(V)dV=\mu\frac{2\pi r_{si}}{L}\int_{\alpha_s}^{\alpha_d}Q_i(\alpha)\frac{dV}{d\alpha}d\alpha, \tag{14.15}$$

following a parallel approach to the planar double-gate MOSFET [8]:

$$I_{ds} = \mu \frac{4\pi \varepsilon_{si}}{L} \left(\frac{2kT}{q} \right)^2 \left[\frac{\varepsilon_{si}}{\varepsilon_i} \left(\frac{1-2\alpha}{\alpha^2} \right) \ln\left(1 + \frac{t_i}{r_{si}} \right) + \frac{1}{\alpha} + \frac{\ln \alpha}{2} \right] \Bigg|_{\alpha_d}^{\alpha_s}. \quad (14.16)$$

α_s and α_d are solutions to (14.14) for $V = 0$ and V_{ds}, respectively.

Equations (14.16) and (14.14) cover all regions of nanowire MOSFET operation. An asymptotic expression similar to (14.10) can be obtained in the subthreshold limit where $\alpha_s, \alpha_d \approx 1$. Above threshold behavior like (14.7) and (14.9) occurs when $\alpha_s \approx 0$ [8].

14.2.3 Explicit, Continuous Solutions to the Implicit Equations

The analytic potential models for double-gate and nanowire MOSFETs involve two implicit algebraic equations, (14.4) and (14.14), that can be solved iteratively for β_s, β_d and α_s, α_d. In a compact model, explicit equations are preferred in order to avoid possible convergence problems with iterative methods. In this section, explicit, continuous solutions to (14.4) and (14.14) are described [9]. They follow a similar approach as that employed in the surface potential (PSP) compact model [10].

The first step starts with an initial estimate in the form of an explicit continuous function of gate voltage, structural parameters, etc. Piecewise functions are not acceptable, because they cause discontinuities of derivatives. The second step is to modify the initial guess with higher-order correction terms. We assume that the implicit function to be solved is $f(x; a, b, c) = 0$, where x is the unknown to be solved and a, b, c are the bias voltage and device structural parameters. If the initial estimate $g(a, b, c)$ of x is reasonably accurate, f can be expanded into a Taylor series near g as

$$\sum_{n=0}^{\infty} \frac{1}{n!} \frac{\partial^n f(x; a, b, c)}{\partial x^n} \Bigg|_{x=g(a,b,c)} [x - g(a, b, c)]^n. \quad (14.17)$$

Keeping the expansion to the third order, and letting $h = x - g(a, b, c)$, the equation $f(x; a, b, c) = 0$ becomes

$$f_{g0} + f_{g1} h + \frac{f_{g2}}{2} h^2 + \frac{f_{g3}}{6} h^3 + O(h^4) = 0, \quad (14.18)$$

where $f_{gn} = \frac{\partial^n f(x;a,b,c)}{\partial x^n}|_{x=g(a,b,c)}$.

When h is small, the solution to the cubic equation (14.18) can be approximated as

$$h = -\frac{f_{g0}}{f_{g1}} \left[1 + \frac{f_{g0} f_{g2}}{2 f_{g1}^2} + \frac{f_{g0}^2 (3 f_{g2}^2 - f_{g1} f_{g3})}{6 f_{g1}^4} \right], \quad (14.19)$$

and the solution to $f(x; a, b, c) = 0$ is refined to $x = g(a, b, c) + h(a, b, c)$. If even higher accuracy than the second-step correction is desired, a third-step correction term may be added by repeating the same procedure as the second step:

$$x = g(a, b, c) + h(a, b, c) + w(a, b, c), \tag{14.20}$$

where $w(a, b, c)$ is the new correction term in the third step.

14.2.3.1 Explicit, Continuous Solutions to Double-Gate MOSFETs

To simplify the initial estimate of β in the implicit (14.4), we let $z \equiv \tan\beta$ and $r \equiv \varepsilon_{si} t_i / (\varepsilon_i t_{si})$. The equation to be solved becomes

$$f(z; r, F) = \ln\left[\sqrt{1+z^2}\arctan z\right] + 2rz\arctan z - F = 0, \tag{14.21}$$

where $F \equiv q(V_{gs} - \Delta\phi - V)/(2kT) - \ln((2/t_{si})\sqrt{2\varepsilon_{si}kT/(q^2 n_i)})$. The asymptotic solutions for z are $z = e^F$ as $z \to 0$ and $z = F/(\pi r)$ as $z \to \infty$, respectively. An appropriate initial guess is obtained by connecting the asymptotic behavior as follows [9]:

$$z_1 = \sqrt{\left(\frac{2}{\pi^2 r^2}\right)^2 + \left(\frac{2}{\pi r}\right)^2 [\ln(1+e^{F/2})]^2} - \frac{2}{\pi^2 r^2}. \tag{14.22}$$

Using the initial guess z_1, a second-step correction term is calculated by applying (14.17)–(14.19). For even higher accuracy, a third-step correction is made by repeating the procedure to obtain z_3. The final error of the explicit solution can be assessed by evaluating $f(z_3(r, F); r, F)$ over a wide range of parameters. Examples show that the worst case error is $|f| \approx 1.1 \times 10^{-11}$, indicating extremely high accuracy [9].

14.2.3.2 Explicit, Continuous Solutions to Nanowire MOSFETs

For nanowire MOSFETs, we introduce $z \equiv (1-\alpha)/\alpha$ and $s \equiv 2(\varepsilon_{si}/\varepsilon_i)\ln[1 + (t_i/r_{si})]$ so that the implicit equation (14.14) becomes

$$g(z; s, G) = \ln(\sqrt{z+z^2}) + sz - G = 0, \tag{14.23}$$

where $G \equiv q(V_{gs} - \Delta\phi - V)/(2kT) - \ln((2/r_{si})\sqrt{2\varepsilon_{si}kT/(q^2 n_i)})$. In this case, the asymptotic solutions are $z = e^{2G}$ as $z \to 0$ and $z = G/s$ as $z \to \infty$. An appropriate initial guess is then

$$z_1 = \sqrt{\left(\frac{1}{2s^2}\right)^2 + \left(\frac{1}{s}\right)^2 [\ln(1+e^{G})]^2} - \frac{1}{2s^2}. \tag{14.24}$$

Again, the final solution z_3 is calculated by adding two correction terms to z_1. The worst case error is $|g| \approx 3.2 \times 10^{-13}$, which is extremely small [9].

14.3 Short-Channel Models

This section deals with the 2-D potential models for short-channel double-gate and nanowire MOSFETs. Deriving the potential in a short-channel device involves solving a 2-D Poisson's equation with its associated boundary value problem. A general, analytic solution of this kind is difficult to find. However, in the subthreshold region where mobile charge is negligible, such 2-D potential functions can be solved analytically, in Cartesian coordinates for double-gate [11] and in cylindrical coordinates for nanowire MOSFETs [12]. The subthreshold current, and therefore the threshold voltage of a short-channel device can then be expressed as a function of channel length and drain voltage (DIBL).

14.3.1 Short-Channel Model for Double-Gate MOSFETs

For a lightly-doped DG MOSFET in the subthreshold region, both the fixed and mobile charge densities are negligible. In both the insulator and silicon regions, 2-D Poisson's equation becomes

$$\frac{\partial}{\partial x}\left(\varepsilon(x)\frac{\partial\psi}{\partial x}\right)+\frac{\partial}{\partial y}\left(\varepsilon(x)\frac{\partial\psi}{\partial y}\right)=0, \tag{14.25}$$

where $\varepsilon(x)$ is a step function that equals ε_{si} in the silicon region $(-t_{si}/2 < x < t_{si}/2)$ and ε_i in the insulator regions $(t_{si}/2 < x < t_{si}/2 + t_i$ and $-t_{si}/2 - t_i < x < -t_{si}/2)$. Assuming that the source and drain junctions are abrupt, the boundary conditions are:

$$\begin{aligned}&\text{Top gate:}\quad \psi(-t_{si}/2-t_i, y)=V_{gs}-\Delta\phi \quad 0<y<L,\\ &\text{Bottom gate:}\quad \psi(t_{si}/2+t_i, y)=V_{gs}-\Delta\phi \quad 0<y<L,\\ &\text{Source:}\quad \psi(x,0)=E_g/2q \quad -t_{si}/2<x<t_{si}/2,\\ &\text{Drain:}\quad \psi(x,L)=V_{ds}+E_g/2q \quad -t_{si}/2<x<t_{si}/2.\end{aligned} \tag{14.26}$$

Using superposition, the electrostatic potential in the DG MOSFET can be written as

$$\psi(x,y)=\upsilon(x)+u_L(x,y)+u_R(x,y) \tag{14.27}$$

where $\upsilon(x)$ is the solution to the 1-D equation,

$$\frac{d}{dx}\left(\varepsilon(x)\frac{d\upsilon(x)}{dx}\right)=0, \tag{14.28}$$

and satisfies the top and bottom boundary conditions. u_L and u_R are solutions to (14.25) and satisfy the source side and drain side boundary conditions, respectively. For example, u_L is zero on the top, bottom, and the right (drain side) boundaries, but $\upsilon + u_L$ satisfies the left (source side) boundary condition. Similarly, u_R is zero on the top, bottom, and the left boundaries, but $\upsilon + u_R$ satisfies the right (drain side) boundary condition [6].

For a symmetric DG, the 1-D potential in subthreshold is simply $v(x) = \text{constant} = V_{gs} - \Delta\phi$. u_L and u_R can be written as follows [13]:

$$u_L(x,y) = \begin{cases} \sum_{n=1}^{\infty} b_{Tn} \frac{\sinh[\pi(L-y)/\lambda_n]}{\sinh(\pi L/\lambda_n)} \sin[\pi(x+t_i+t_{si}/2)/\lambda_n] \\ \quad -t_i - \frac{t_{si}}{2} \le x \le -\frac{t_{si}}{2}, \\ \sum_{n=1}^{\infty} b_n \frac{\sinh[\pi(L-y)/\lambda_n]}{\sinh(\pi L/\lambda_n)} \sin[\pi(x+A)/\lambda_n] \quad -\frac{t_{si}}{2} < x \le \frac{t_{si}}{2}, \\ \sum_{n=1}^{\infty} b_{Bn} \frac{\sinh[\pi(L-y)/\lambda_n]}{\sinh(\pi L/\lambda_n)} \sin[\pi(x-t_i-t_{si}/2)/\lambda_n] \\ \quad \frac{t_{si}}{2} \le x \le t_i + \frac{t_{si}}{2} \end{cases} \quad (14.29)$$

such that $u_L(\text{top}) = u_L(\text{bottom}) = u_L(\text{drain}) = 0$, and

$$u_R(x,y) = \begin{cases} \sum_{n=1}^{\infty} c_{Tn} \frac{\sinh(\pi y/\lambda_n)}{\sinh(\pi L/\lambda_n)} \sin[\pi(x+t_i+t_{si}/2)/\lambda_n] \\ \quad -t_i - \frac{t_{si}}{2} \le x \le -\frac{t_{si}}{2}, \\ \sum_{n=1}^{\infty} c_n \frac{\sinh(\pi y/\lambda_n)}{\sinh(\pi L/\lambda_n)} \sin[\pi(x+A)/\lambda_n] \quad -\frac{t_{si}}{2} < x \le \frac{t_{si}}{2}, \\ \sum_{n=1}^{\infty} c_{Bn} \frac{\sinh(\pi y/\lambda_n)}{\sinh(\pi L/\lambda_n)} \sin[\pi(x-t_i-t_{si}/2)/\lambda_n] \quad \frac{t_{si}}{2} \le x \le t_i + \frac{t_{si}}{2} \end{cases} \quad (14.30)$$

such that $u_R(\text{top}) = u_R(\text{bottom}) = u_R(\text{source}) = 0$. To be determined later are the eigenvalues λ_n and constant A.

Poisson's equation (14.25) requires $\psi(x,y)$ and $\varepsilon(\partial\psi/\partial x)$ be continuous in the x direction. At the dielectric boundary between the silicon and the insulator, $x = \pm t_{si}/2$, u_L and $\varepsilon(\partial u_L/\partial x)$ are continuous. Therefore,

$$\begin{aligned} b_{Tn} \sin(\pi t_i/\lambda_n) &= b_n \sin[\pi(A - t_{si}/2)/\lambda_n], \\ -b_{Bn} \sin(\pi t_i/\lambda_n) &= b_n \sin[\pi(A + t_{si}/2)/\lambda_n], \\ \varepsilon_i b_{Tn} \cos(\pi t_i/\lambda_n) &= \varepsilon_{si} b_n \cos[\pi(A - t_{si}/2)/\lambda_n], \\ \varepsilon_i b_{Bn} \cos(\pi t_i/\lambda_n) &= \varepsilon_{si} b_n \cos[\pi(A + t_{si}/2)/\lambda_n]. \end{aligned} \quad (14.31)$$

The above equations yield $A = n\lambda_n/2$ and an eigenvalue equation for the scale length λ_n [11]:

$$\varepsilon_{si} \tan(\pi t_i/\lambda_n) = \varepsilon_i \tan(n\pi/2 - \pi t_{si}/2\lambda_n). \quad (14.32)$$

Note that for a symmetric DG device, only the odd order terms are nonzero. λ_n's form an irregular sequence: $\lambda_1 > \lambda_2 > \lambda_3 > \cdots$ with $\lambda_n/\lambda_1 \sim \frac{1}{n}$. Continuity of u_R and $\varepsilon(\partial u_R/\partial x)$ yields the same eigenvalue equation.

The numerical solutions for the longest (lowest order) eigenvalue λ_1 are plotted in Fig. 14.5 in normalized units with $\varepsilon_i/\varepsilon_{si}$ as a parameter [6]. In the straightforward case where $\varepsilon_i = \varepsilon_{si}$, the scale length is simply $\lambda_1 = t_{si} + 2t_i$, i.e., the physical height of the region between the two gates in Fig. 14.1. For $\varepsilon_i/\varepsilon_{si} < 1$, $\lambda_1 > t_{si} + 2t_i$, and for $\varepsilon_i/\varepsilon_{si} > 1$, $\lambda_1 < t_{si} + 2t_i$. But in no case can λ_1 be smaller than t_{si} or $2t_i$, whichever is larger. This means that even extremely high-κ insulators would have to be physically thin to be useful. In the very high-κ limit, the scale length approaches $2t_i$, insensitive to the silicon thickness as long as $t_{si} < 2t_i$. In the lower right corner

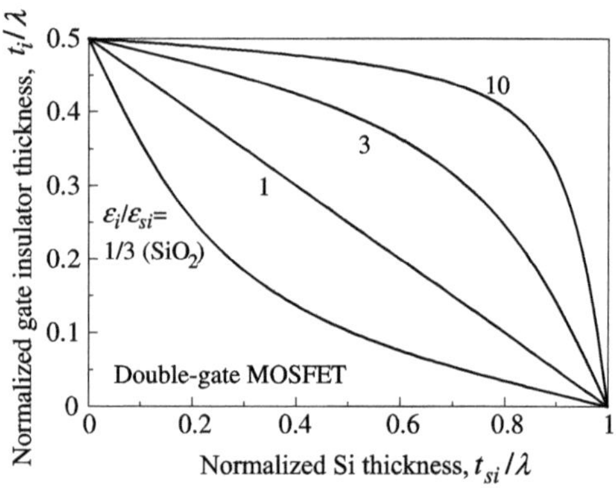

Fig. 14.5 Numerical solutions to (14.32) for different values of $\varepsilon_i/\varepsilon_{si}$

of the curves where silicon is thick and insulator is thin, (14.32) can be approximated to $\lambda_1 = t_{si} + 2(\varepsilon_{si}/\varepsilon_i)t_i$.

From (14.31), b_{Tn} and b_{Bn} can be expressed in terms of b_n so that u_L can be rewritten as: $u_L(x, y) = \sum_{n=1}^{\infty} b_n u_{Ln}(x, y)$. Using the orthogonality relationship between distinct eigenfunctions $u_{Ln}(x, 0)$,

$$\int_{-t_i - t_{si}/2}^{t_i + t_{si}/2} \varepsilon(x) u_{Ln}(x, 0) u_{Lm}(x, 0) dx = 0 \quad \text{except } n = m, \tag{14.33}$$

b_n can be evaluated from the left boundary condition in (14.26). For the dominant λ_1 term,

$$b_1 = \frac{\int_{-\frac{t_{si}}{2} - t_i}^{\frac{t_{si}}{2} + t_i} [\psi(x, 0) - \upsilon(x)] g_1(x) dx}{\int_{-\frac{t_{si}}{2} - t_i}^{\frac{t_{si}}{2} + t_i} u_{L1}(x, 0) g_1(x) dx}, \tag{14.34}$$

where

$$g_1(x) = \begin{cases} \frac{\sin(\pi t_{si}/2\lambda_1)}{\cos(\pi t_i/\lambda_1)} \sin[\pi(x + t_i + t_{si}/2)/\lambda_1] & -t_i - \frac{t_{si}}{2} \le x \le -\frac{t_{si}}{2}, \\ \cos(\pi x/\lambda_1) & -\frac{t_{si}}{2} \le x \le \frac{t_{si}}{2}, \\ -\frac{\sin(\pi t_{si}/2\lambda_1)}{\cos(\pi t_i/\lambda_1)} \sin[\pi(x - t_i - t_{si}/2)/\lambda_1] & \frac{t_{si}}{2} \le x \le t_i + \frac{t_{si}}{2}. \end{cases} \tag{14.35}$$

The integrals in (14.34) can be evaluated to give

$$b_1 = \frac{2\lambda_1^2 \tan(\pi t_i/\lambda_1) \sin(\pi t_{si}/2\lambda_1)}{\pi^2 t_i [\frac{t_{si}}{2} + \frac{\sin(\pi t_{si}/\lambda_1)}{\sin(2\pi t_i/\lambda_1)} t_i]} \left(\frac{E_g}{2q} + \Delta\phi - V_{gs} \right). \tag{14.36}$$

Similarly,

$$c_1 = \frac{2\lambda_1^2 \tan(\pi t_i/\lambda_1) \sin(\pi t_{si}/2\lambda_1)}{\pi^2 t_i [\frac{t_{si}}{2} + \frac{\sin(\pi t_{si}/\lambda_1)}{\sin(2\pi t_i/\lambda_1)} t_i]} \left(\frac{E_g}{2q} + V_{ds} + \Delta\phi - V_{gs} \right). \tag{14.37}$$

Equations (14.29) and (14.30) show that if it is not too close to the source or drain, $u_{Ln} \sim e^{-\pi y/\lambda_n}$, $u_{Rn} \sim e^{-\pi(L-y)/\lambda_n}$, and $(u_{Ln} + u_{Rn})_{\min} \sim e^{-\pi L/2\lambda_n}$. This

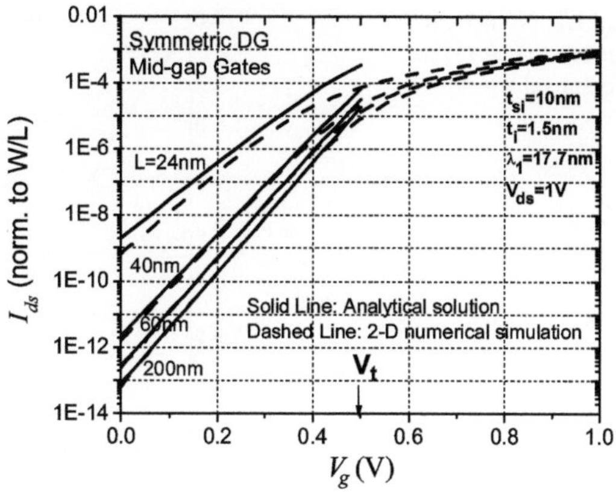

Fig. 14.6 Comparison of short-channel I_{ds}-V_{gs} characteristics calculated from the analytic model to 2-D simulations. Note that the analytic model (*solid lines*) is only valid in the subthreshold

means that for $L > 1.5\lambda_1$, the u series decay rapidly since $\lambda_1/\lambda_n \sim n$. Therefore only the lowest order terms in the u series with the largest eigenvalue λ_1 need to be kept. For extremely short MOSFETs, $L < 1.5\lambda_1$, the higher order terms cannot be neglected. But the short channel effect is too severe for such devices to be useful in practice.

The full expression for the potential in Si in the subthreshold region is then

$$\psi(x,y) = V_{gs} - \Delta\phi + \frac{b_1 \sinh[\pi(L-y)/\lambda_1] + c_1 \sinh(\pi y/\lambda_1)}{\sinh(\pi L/\lambda_1)} \cos(\pi x/\lambda_1). \quad (14.38)$$

Once the 2D potential is solved, the subthreshold current can be derived from the current continuity equation, $I_{ds} = \mu W Q_i dV/dy = \text{constant}$. The inversion charge density per gate area is

$$Q_i = \int_{-t_{si}/2}^{t_{si}/2} n(x,y)dx = \int_{-t_{si}/2}^{t_{si}/2} n_i e^{q[\psi(x,y)-V]/kT} dx. \quad (14.39)$$

Separating the variables V and y in the current continuity equation and integrating over (0, V_{ds}) and (0, L) yield

$$I_{ds} = \frac{\mu W \int_0^{V_{ds}} e^{-qV/kT} dV}{\int_0^L \frac{dy}{\int_{-t_{si}/2}^{t_{si}/2} n_i e^{q\psi(x,y)/kT} dx}} = \frac{\mu W (kT/q)[1 - e^{-qV_{ds}/kT}]}{\int_0^L \frac{dy}{\int_{-t_{si}/2}^{t_{si}/2} n_i e^{q\psi(x,y)/kT} dx}}. \quad (14.40)$$

The subthreshold current can then be calculated analytically as a function of V_{gs} and V_{ds} and L. Figure 14.6 shows the agreement of short-channel I_{ds}-V_{gs} characteristics between the analytic model and 2-D numerical simulations.

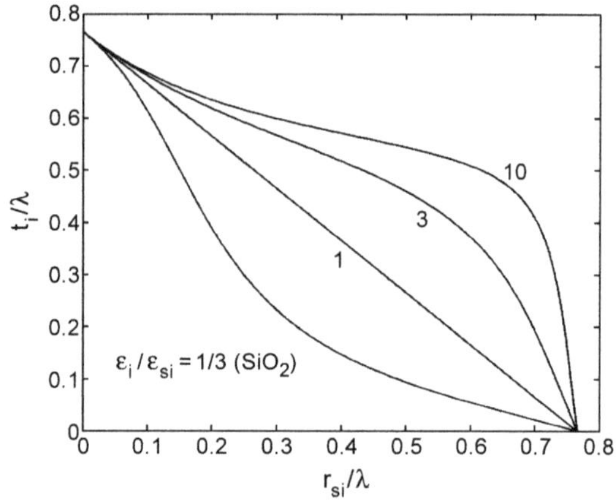

Fig. 14.7 Numerical solutions to (14.43) for different values of $\varepsilon_i/\varepsilon_{si}$

14.3.2 Short-Channel Model for Nanowire MOSFETs

The scale length equation for nanowire MOSFETs [12] has been derived from the 2-D Poisson's equation in cylindrical coordinates,

$$\frac{1}{\rho}\frac{\partial}{\partial\rho}\left[\rho\varepsilon(\rho)\frac{\partial\psi}{\partial\rho}\right]+\frac{\partial}{\partial y}\left[\varepsilon(\rho)\frac{\partial\psi}{\partial y}\right]=0, \tag{14.41}$$

applicable to both the insulator and silicon regions under subthreshold. Here $\varepsilon(\rho) = \varepsilon_{si}$ for $\rho < r_{si}$ and $\varepsilon(\rho) = \varepsilon_i$ for $r_{si} < \rho < r_{si} + t_i$. The boundary conditions are:

$$\begin{aligned}
&\text{Gate:} \quad \psi(r_{si}+t_i, y) = V_{gs} - \Delta\phi \quad 0 < y < L,\\
&\text{Source:} \quad \psi(\rho, 0) = E_g/(2q) \quad 0 < \rho < r_{si},\\
&\text{Drain:} \quad \psi(\rho, L) = V_{ds} + E_g/(2q) \quad 0 < \rho < r_{si}.
\end{aligned} \tag{14.42}$$

The gap regions, $r_{si} < \rho < r_{si} + t_i$, at the source and drain are filled by logarithmic interpolation between the end points.

The general solutions to the homogeneous 2-D (14.41) can be expressed as a series of eigenfunction products with $\sinh[\pi(L-y)/\lambda_n]$ and $\sinh[\pi y/\lambda_n]$ functions for the y dependence and zeroth-order Bessel functions, $J_0(\pi\rho/\lambda_n)$, for the ρ dependence [12]. Following a similar approach as the planar double-gate device in the previous section, the boundary condition that $\varepsilon(\rho)\partial\psi/\partial\rho$ be continuous at $\rho = r_{si}$ yields an eigenvalue equation for the scale length λ:

$$\frac{Y_0'(\pi r_{si}/\lambda)}{J_0'(\pi r_{si}/\lambda)} = \frac{\varepsilon_{si}}{\varepsilon_i}\frac{Y_0(\pi r_{si}/\lambda)}{J_0(\pi r_{si}/\lambda)} + \left(1-\frac{\varepsilon_{si}}{\varepsilon_i}\right)\frac{Y_0[\pi(r_{si}+t_i)/\lambda]}{J_0[\pi(r_{si}+t_i)/\lambda]}, \tag{14.43}$$

where J_0 and Y_0 are Bessel and Neumann functions of the zeroth order and J_0', Y_0' their derivatives. Numerical solutions to (14.43) are plotted in Fig. 14.7 in normalized units for different $\varepsilon_i/\varepsilon_{si}$ values [14].

Using the orthogonality relationship among eigenfunctions of distinct orders as in the double-gate case, the coefficients of the first-order terms in the eigenfunction expansion can be evaluated and the 2-D potential function determined. From which the subthreshold current, DIBL, V_t roll-off, and subthreshold slope can all be obtained as functions of channel length and gate and drain bias voltages [15]. The results of the analytic solutions have been validated by 2-D simulations.

14.4 Charge and Capacitance Models

Unlike bulk MOSFETs, symmetric double-gate and nanowire MOSFETs have only three terminals since the two gates are connected together and there is no body contact. The three terminal charges (intrinsic) Q_g, Q_d, Q_s, associated with the gate, drain and source can be expressed as integrals along the channel direction following Ward and Dutton's linear charge partition method [16].

$$
\begin{aligned}
Q_g &= -W\int_0^L Q_i(y)dy,\\
Q_d &= W\int_0^L \frac{y}{L}Q_i(y)dy, \qquad (14.44)\\
Q_s &= W\int_0^L \left(1-\frac{y}{L}\right)Q_i(y)dy.
\end{aligned}
$$

Using the current continuity condition, $I_{ds} = \mu W Q_i dV/dy = \text{constant}$, and the differential of (14.4), the variable in above integrals can be changed from dy to dV to $d\beta$. Also, $Q_i = 2\varepsilon_{si}(2kT/q)(2\beta/t_{si})\tan\beta$, and y can be expressed as a function of β by replacing L with y and β_d with β in (14.6). The resulting integrals in β cannot be carried out in closed forms for planar double-gate MOSFETs [17].

For nanowire MOSFETs, the corresponding charge integrals in terms of the intermediary variable α (defined in Sect. 14.2.2) can be evaluated explicitly [18]. For example,

$$
\begin{aligned}
Q_g &= -2\pi r_{si}\int_0^L Q_i(y)dy = \frac{\mu}{I_{ds}}(2\pi r_{si})^2\int_{\alpha_s}^{\alpha_d} Q_i^2(\alpha)\frac{dV}{d\alpha}d\alpha\\
&= \frac{\mu}{I_{ds}}(8\pi\varepsilon_{si})^2\left(\frac{kT}{q}\right)^3\left\{\frac{3}{\alpha}-\frac{1}{\alpha^2}+\ln\alpha+4\frac{\varepsilon_{si}}{\varepsilon_i}\ln\left(1+\frac{t_i}{r_{si}}\right)\right.\\
&\quad\times\left.\left[\frac{1}{\alpha^2}-\frac{1}{\alpha}-\frac{1}{3\alpha^3}\right]\right\}\Bigg|_{\alpha_s}^{\alpha_d}, \qquad (14.45)
\end{aligned}
$$

where I_{ds} is given by (14.16). Similarly, Q_d and Q_s can also be expressed as explicit functions of α_s and α_d which are solutions to (14.14) for $V = 0$ and V_{ds}, respectively.

Once all the charges are obtained, the capacitance and transcapacitance terms can be evaluated. For example, $C_{gg} = \partial Q_g/\partial V_{gs}$, $C_{dg} = -\partial Q_d/\partial V_{gs}$, $C_{sd} = -\partial Q_s/\partial V_{ds}$, etc. Figure 14.8 shows the agreement between the analytic capacitance model and 2-D simulation results [18]. Each capacitance component is described by one continuous, analytic function throughout all regions of nanowire MOSFET operation: linear, saturation, and subthreshold.

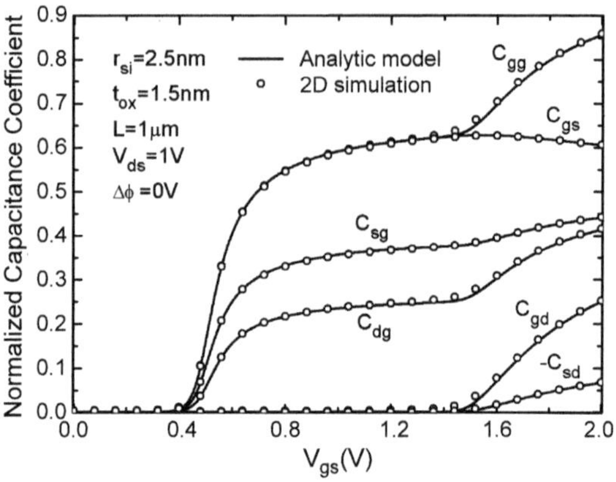

Fig. 14.8 Agreement between analytic capacitance model for nanowire MOSFETs and 2-D numerical simulations. The threshold voltage is 0.4 V. For 0.4 V < V_{gs} < 1.4 V, the device is in saturation. For V_{gs} > 1.4 V, the device is in the linear region. All capacitances are normalized to $2\pi\varepsilon_i L/[\ln(1+t_i/r_{si})]$

14.5 Discussion of Surface-Potential Based Current Expression

Note that the drain current expression used in the analytic potential model for double-gate MOSFETs,

$$I_{ds} = \mu W Q_i \frac{dV}{dy}, \tag{14.46}$$

is based on the drift and diffusion components of the electron current density [6],

$$J_y = -qn\mu\frac{\partial\psi}{\partial y} + \mu kT\frac{\partial n}{\partial y} = -qn\mu\frac{\partial V}{\partial y}, \tag{14.47}$$

where $n(x, y)$ is the electron volume density at (x, y). The second equality in (14.47) follows from the definition of quasi-Fermi potential, $V \equiv \psi - (kT/q) \times \ln(n/n_i)$, i.e., $n = n_i e^{q(\psi-V)/kT}$. The key assumption leading (14.47) to (14.46) is the *gradual-channel approximation* which assumes that V is a function of y only, independent of x [6]. This is justified by the fact that the net electron current flows predominantly in the source-to-drain or y direction. Integration of (14.47) then yields (14.46) with $Q_i(y) \equiv q\int n(x, y)dx$. Note that there is a sign reversal as J_y is defined to be positive in the $+y$ direction while $I_{ds} > 0$ in the $-y$ direction (drain to source).

In the recent literature [19–23], a surface-potential based drift and diffusion current expression has been used in other compact models of double-gate MOSFETs, namely,

$$I_{ds}(s.f.) = \mu W\left(Q_i\frac{d\psi_s}{dy} - \frac{kT}{q}\frac{dQ_i}{dy}\right). \tag{14.48}$$

Here $\psi_s(y) = \psi(x = \pm t_{si}/2, y)$ is the surface potential. This equation originated from the second approximation in Brews's charge sheet model for bulk MOSFETs [2]. ψ_s is related to Q_i through the boundary condition (14.3),

$$Q_i = 2(\varepsilon_i/t_i)(V_{gs} - \Delta\phi - \psi_s). \tag{14.49}$$

This equation establishes a relationship between $d\psi_s$ and dQ_i which allows (14.48) to be integrated:

$$I_{ds}(s.f.) = \mu\frac{W}{L}\left[\frac{Q_i^2}{4(\varepsilon_i/t_i)} + \frac{kT}{q}Q_i\right]\Bigg|_{Q_{id}}^{Q_{is}}, \tag{14.50}$$

where Q_{is} and Q_{id} are the inversion charge sheet densities at the source and drain ends of the channel.

It has been shown that the approximation of $\ln\beta - \ln(\cos\beta) \approx (1/2)\ln(\beta\tan\beta)$ on the right hand side of (14.4) led to the same result as (14.50) [24]. Since $Q_i \propto \beta\tan\beta$, this approximation allows dV to be expressed as a combination of dQ_i/Q_i and dQ_i. All the charge integrals in (14.44) then become easily integrable into closed form functions of Q_{is} and Q_{id}. It forms the basis of a unified charge and capacitance model for double-gate and nanowire MOSFETs [24].

It should be pointed out that (14.48) does not follow from (14.47) because in the drift current term, $\partial\psi/\partial y$ is a function of x and cannot be treated as a constant $(d\psi_s/dy)$ when $n(x, y)$ is integrated to obtain Q_i. Using the analytic potential model in Sect. 14.2.1, it is easy to see the difference between the surface-potential based current, (14.50), and the rigorous solution of (14.46). Substituting $Q_i = 2\varepsilon_{si}(2kT/q)(2\beta/t_{si})\tan\beta$ in (14.50) yields

$$I_{ds}(s.f.) = \mu\frac{W}{L}\frac{4\varepsilon_{si}}{t_{si}}\left(\frac{2kT}{q}\right)^2\left[\frac{\beta\tan\beta}{2} + \frac{\varepsilon_{si}t_i}{\varepsilon_i t_{si}}\beta^2\tan^2\beta\right]\Bigg|_{\beta_d}^{\beta_s}. \tag{14.51}$$

The deviation of $I_{ds}(s.f.)$ from the rigorous solution (14.6) is the approximation of $\beta^2/2$ to $(\beta\tan\beta)/2$ [25].

Asymptotically, both $\beta^2/2$ and $(\beta\tan\beta)/2$ approach the same limit when $\beta \to 0$ (subthreshold). At high overdrives, $\beta \to \pi/2$, however, $\beta^2/2$ remains finite while $(\beta\tan\beta)/2$ diverges. Fortunately, in that case, the current is dominated by the last, $(\varepsilon_{si}t_i/\varepsilon_i t_{si})\beta^2\tan^2\beta$ term so the difference between $\beta^2/2$ and $(\beta\tan\beta)/2$ does not show up prominently. In the intermediate region, however, there is a significant error of (14.51) from (14.6), as shown in Fig. 14.9 [25]. As expected, the worst-case error increases at smaller $\varepsilon_{si}t_i/\varepsilon_i t_{si}$ values. A symmetric linearization method has been applied to reduce the error of the surface-potential based current model for double-gate MOSFETs [26].

14.6 Conclusion

This chapter has reviewed recent developments on compact modeling of double-gate and nanowire MOSFETs. Owing to their highly symmetric structures, analytic solutions have been developed for both the long-channel device over all regions of operation and the short-channel device in subthreshold. The resulting compact models are physics based with no adjustable fitting parameters—hence predictive in nature. They are generic in the sense that they do not depend on any specific process.

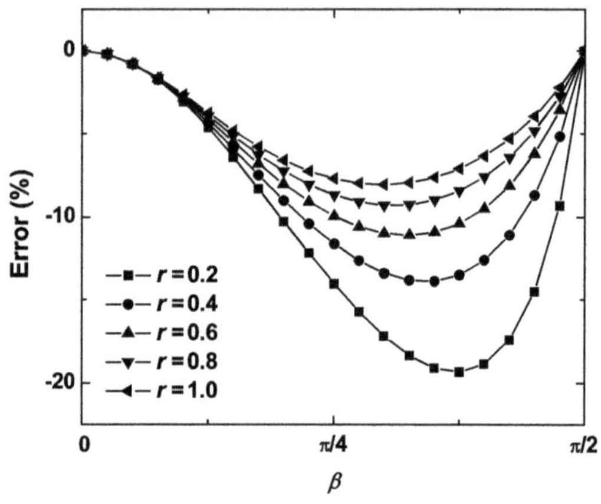

Fig. 14.9 Percentage error of the surface-potential based current, (14.51) as a function of β. The parameter $r \equiv \varepsilon_{si} t_i / \varepsilon_i t_{si}$

With the addition of quantum and transport modules, the full compact model for double-gate MOSFETs has been released and calibrated with experimental FinFET hardware [28]. In the short-channel implementation, the V_t roll-off and DIBL extracted from the subthreshold model are also applied to the bias regions above threshold.

For other variations of the multiple-gate device structures, closed-form solutions do not exist for lack of symmetry. An approximation has been proposed to build the core models for triple-gate, quadruple-gate, pi-gate and omega-gate MOSFETs from some combination of the double-gate and nanowire models [29]. In that perspective, the double-gate and nanowire models discussed in this chapter have laid the foundation for modeling of the more complex multiple-gate MOSFETs in the real world.

References

1. Pao, H.C., Sah, C.T.: Effects of diffusion current on characteristics of metal-oxide (insulator)-semiconductor transistors. Solid-State Electron. **9**, 927–937 (1966)
2. Brews, J.R.: A charge sheet model of the MOSFET. Solid-State Electron. **21**(2), 345–355 (1978)
3. Gildenblat, G., Li, X., Wu, W., Wang, H., Jha, A., van Langevelde, R., Smit, G.D.J., Scholten, A.J., Klaassen, D.B.M.: PSP: An advanced surface-potential-based MOSFET model for circuit simulation. IEEE Trans. Electron Devices **ED-53**, 1979–1993 (2006)
4. Taur, Y.: An analytical solution to a double-gate MOSFET with undoped body. IEEE Electron Device Lett. **21**(5), 245–247 (2000)
5. Taur, Y., Liang, X., Wang, W., Lu, H.: A continuous, analytic drain-current model for DG MOSFETs. IEEE Electron Device Lett. **25**(2), 107–109 (2004)
6. Taur, Y., Ning, T.H.: Fundamentals of Modern VLSI Devices, 2nd edn. Cambridge Univ. Press, Cambridge (2009)
7. Lu, X., Lu, W.-Y., Taur, Y.: Effect of body doping on double-gate MOSFET characteristics. Semicond. Sci. Technol. **22**, 252835 (2008) (6pp)
8. Jimenez, D., Iniguez, B., Sune, J., Marsal, L.F., Pallares, J., Roig, J., Flores, D.: Continuous analytic I-V model for surrounding-gate MOSFETs. IEEE Electron Device Lett. **25**(8), 571–573 (2004)

9. Yu, B., Lu, H., Liu, M., Taur, Y.: Explicit continuous models for double-gate and surrounding-gate MOSFETs. IEEE Trans. Electron Devices **54**(10), 2715–2722 (2007)
10. Chen, T.L., Gildenblat, G.: Analytical approximation for the MOSFET surface potential. Solid State Electron. **45**(2), 335–339 (2001)
11. Liang, X., Taur, Y.: A 2-D analytical solution for SCEs in DG MOSFETs. IEEE Trans. Electron Devices **51**(9), 1385–1391 (2004)
12. Oh, S.-H., Monroe, D., Hergenrother, J.M.: Analytic description of short-channel effects in fully-depleted double-gate and cylindrical, surrounding-gate MOSFETs. IEEE Electron Device Lett. **9**, 445–447 (2000)
13. Frank, D.J., Taur, Y., Wong, H.-S.P.: Generalized scale length for two-dimensional effects in MOSFETs. IEEE Electron Device Lett. **19**, 385–387 (1998)
14. Yu, B., Wang, L., Yuan, Y., Asbeck, P.M., Taur, Y.: Scaling of nanowire transistors. IEEE Trans. Electron Devices **55**, 2846–2858 (2008)
15. Yu, B., Yuan, Y., Song, J., Taur, Y.: A two-dimensional analytical solution for short-channel effects in nanowire MOSFETs. IEEE Trans. Electron Devices **56**, 2357–2362 (2009)
16. Ward, D., Dutton, R.: A charge-oriented model for MOS transistor capacitances. IEEE J. Solid-State Circuits **SC-13**(5), 703–708 (1978)
17. Lu, H., Taur, Y.: An analytic potential model for symmetric and asymmetric DG MOSFETs. IEEE Trans. Electron Devices **53**(5), 1161–1168 (2006)
18. Yu, B., Lu, W.-Y., Lu, H., Taur, Y.: Analytic charge model for surrounding-gate MOSFETs. IEEE Trans. Electron Devices **54**(3), 492–496 (2007)
19. Taur, Y., Richards, P.L.: Parametric amplification and oscillation at 36 GHz using a point-contact Josephson junction. J. Appl. Phys. **48**(3), 1321–1326 (1977)
20. Taur, Y., Kerr, A.R.: Low-noise Josephson mixers at 115 GHz using recyclable point contacts. Appl. Phys. Lett. **32**(11), 775–777 (1978)
21. Dunga, M.V., Lin, C.-H., Xi, X., Lu, D.D., Niknejad, A.M., Hu, C.: Modeling advanced FET technology in a compact model. IEEE Trans. Electron Devices **53**(9), 1971–1978 (2006)
22. Smit, G.D.J., Scholten, A.J., Curatola, G., van Langevelde, R., Gildenblat, G., Klaassen, D.B.M.: PSP-based scalable compact FinFET model. In: NTSI-Nanotech 2007, vol. 3, pp. 520–525 (2007)
23. Sallese, J.-M., Krummenacher, F., Pregaldiny, F., Lallement, C., Roy, A., Enz, C.: A design oriented charge-based current model for symmetric DG MOSFET and its correlation with the EKV formalism. Solid-State Electron. **49**(3), 485–489 (2005)
24. Lu, H., Yu, B., Taur, Y.: A unified charge model for symmetric double-gate and surrounding-gate MOSFETs. Solid-State Electron. **52**(1), 67–72 (2008)
25. Song, J., Yu, B., Yuan, Y., Taur, Y.: A review on compact modeling of multiple-Gate MOSFETs. IEEE Trans. Circuits Syst. I **56**(8), 1858–1869 (2009)
26. Dessai, G., Dey, A., Gildenblat, G., Smit, G.D.J.: Symmetric linearization method for double-gate and surrounding-gate MOSFET model. Solid-State Electron. **53**(5), 548–556 (2009)
27. Ortiz-Conde, A., Garcia-Sanchez, F.J., Muci, J., Malobabic, S., Liou, J.J.: A review of core compact models for undoped double-gate SOI MOSFETs. IEEE Trans. Electron Devices **54**(1), 131–140 (2007)
28. Song, J., Yu, B., Xiong, W., Hsu, C.H., Cleavelin, C.R., Ma, M., Patruno, P., Taur, Y.: Experimental hardware calibrated compact models for 50 nm n-channel FinFETs. In: Conf. SOI, 2007 IEEE, pp. 131–132 (2007)
29. Yu, B., Song, J., Yuan, Y., Taur, Y.: A unified analytic drain current model for multiple-gate MOSFETs. IEEE Trans. Electron Devices **55**(8), 2157–2163 (2008)

Part V
Statistical Modeling

Chapter 15
Modeling of MOS Matching

**Marcel Pelgrom, Hans Tuinhout,
and Maarten Vertregt**

Abstract Circuit operation greatly depends on the ability to control and reproduce transistor and process parameters, such as oxide thickness, dielectric constants, doping levels, width and length. Variation in processing was in the past countered by defining process corners: boundaries in parameter variation that accounted for remaining process tolerances. With the improved control over processing, this batch-to-batch variation is largely under control.

However now a new class of phenomena has appeared: statistical variations. In conventional ICs, analog circuits with a differential operation (e.g. analog-to-digital converters) were already affected by this random parameter spread. The remaining variation between otherwise identical components is generally described by "mismatch" parameters. Next to these random phenomena also systematic errors called "offsets" play an increasingly important role Understanding and mitigating these effects requires statistical means and models.

The chapter will focus on the modeling of systematic and random effects that originate from physical, electrical, thermal and lithographical effects in devices causing intra-die variations.

15.1 Introduction

Circuit design relies heavily on the ability to control and reproduce transistor and process parameters. Variation in processing was in the past taken into account by defining process corners: boundaries in parameter variation that accounted for process tolerances. With the improved control of process technology, this global variation is much better under control.

In analog circuits and memories, Fig. 15.1, the implicit approach to design a robust circuit relies on the mutual equality of basic components like resistors, capacitors, transistors, or units of time. The quality of the design, measured in terms

M. Pelgrom (✉) · H. Tuinhout · M. Vertregt
NXP Semiconductors Research, Prof. Holstlaan 4, 5656 AE Eindhoven, The Netherlands
e-mail: marcel.pelgrom@nxp.com

G. Gildenblat (ed.), *Compact Modeling*,
DOI 10.1007/978-90-481-8614-3_15,

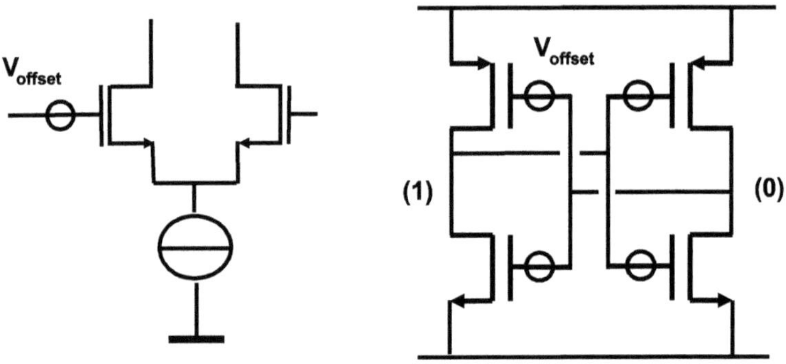

Fig. 15.1 In a differential pair threshold-voltage inequality between the two transistors behaves similar to an offset voltage. Also a static memory cell is sensitive to mutual inequalities of its components leading to a preferential state

of performance or yield, depends strongly on the degree of equality of these components. A different class of phenomena challenges this equality paradigm: statistical variations. Some analog integrated circuits, based on differential operation, were already affected by this random parameter spread. With decreasing device geometries, lower power supply and signal swings the effect of statistical variations begins to dominate the process-corner variations and thereby also forms the major source of concern in high-performance digital circuits. While the global behavior of devices is understood, well-modeled [1–4] and part of the design-flow, the local statistical effects are in their infancy state of adaptation by the designers.

This chapter will review some of the known effects and discuss mitigation of deterministic effects and modeling of statistical phenomena.

15.2 Variability: An Overview

Variability is generally interpreted as a collection of phenomena characterized by uncontrolled parameter variation between individual transistors or components in a circuit. This collection is populated with a large number of effects ranging from offset mechanisms to reliability aspects.

Variability effects can be subdivided along three main axes:

- Time independent versus time variant effects.
- Global variations versus local variations.
- Deterministic versus stochastic (statistical) effects.

Every axis has specific properties that must be considered before defining a successful model and choosing the appropriate design solutions. The first axis that is considered in the control of variability effects is the behavior in the time-domain. Table 15.1 subdivides the variability effects in static, slow, medium and fast changing events. This subdivision is linked to the methodologies a designer has available to mitigate performance loss or yield loss.

Table 15.1 Time dependencies of some variability effects. The lower row lists the methodologies the designer at his disposal has to mitigate these effects

Static	Seconds	Micro-seconds	Pico-seconds
Process corners	Supply voltage	Wiring IR drop	Dyn. IR drop
	Temperature	Temp gradient	Jitter
Lithography	Hot carrier		Substrate noise
Line edge roughness	NBTI, Drift		
Dopant fluctuation	1/f noise	1/f noise	kT noise
WPE, STI Stress	Soft breakdown		
Design margins, calibration	Compensation circuitry	Stretched design margins	Stretched design margins

Table 15.2 From global to local variability

	IC	Sub circuit	Transistor	Atomic
Electrical	Supply voltage	Substrate noise	Wiring IR drop, jitter, cross-talk	
Thermal	Temperature	Temperature gradient	Local heating	Thermal noise
Technology	Process corner	CMP density	Well-proximity	Mechanical stress
Lithography	Line width	Layer density	Lithographic proximity effects	Line-edge roughness
Physics		Wafer gradient	NBTI, soft-breakdown, hot-carrier, stress	Random dopant, mobility variations

Static effects allow a one-time correction of the associated devices in the circuit and can be performed during final test of the product. These corrections can be implemented via non-volatile memories, laser trim or polysilicon fuses. Slow time-dependent mechanisms must be corrected during operation and can be addressed via circuit compensation tricks, such as auto-zeroing and on-chip calibration. However, variations with a time span comparable to the maximum speed of the process cannot be handled in this way. Stretched margins or more power are required.

In the scope of this chapter only phenomena that can be considered static within the observation period are discussed.

Table 15.2 lists variability effects along the physical dimension axis from global to local variations. This list is not complete and new process generations add new phenomena. The global-to-local axis is subdivided into four levels of granularity. Variations on the level of an integrated circuit are normally considered as standard design space variables. Designers are used to incorporate these variations in their Process-Voltage-Temperature "PVT" analysis. Extensive models of components and well-characterized parameter sets cover these aspects.

Table 15.3 Deterministic and stochastic effects

Deterministic	Stochastic
Proximity effects	Line edge roughness
Electrical offsets	Dopant fluctuation
WPE, Stress, STI	Mobility fluctuation

The variations that affect sub-circuits are less commonly incorporated in models and design software. Often the only mechanism available to a designer is a set of design rule checks (DRC) or electrical rules checks (ERC).

Most effects at transistor level are well-modeled as the transistor is the focus point of circuit modeling. Various forms of reliability and hot-carrier effects can be predicted based on the physics involved. Electrical effects on this level of granularity are well understood, but the problem is to judge their relevance within the complexity of the entire circuit. The current variation in a transistor due to substrate noise is trivial if the magnitude of the substrate noise is known. However, establishing that magnitude for a multi-million devices circuit is practically impossible.

An increasing number of transistor-related and atomic-scale effects become relevant in nanometer devices. Most of these effects are well-described and modeled in physics on a micrometer level, but the associated statistical effects are not an integral part of the design flow.

From a statistics point of view these (time-independent) effects can be subdivided in two classes: deterministic and stochastic. In designer's terms: offsets and random matching. As simple as this division seems, there is a complication: a number of phenomena is from a physics point of view deterministic, but due to circuit complexity, a statistical approach is used to serve as a (temporary) fix. An example is wiring stress, where the complexity of concise modeling is too cumbersome.

From a philosophical perspective, the listed random effects are not truly stochastic effects. Here a practical point of view will be used: all effects that are reproducible from die-to-die will be categorized as deterministic. If the variation source changes the behavior of every individual device with respect to the average, the effect will be described with stochastic means, see Table 15.3.

15.3 Deterministic Offsets

Systematic Offsets between pairs of resistors, CMOS transistors or capacitors are due to electrical biasing differences, mechanical stress or lithographic and technological effects in the fabrication process [5].

15.3.1 Offset Caused by Electrical Differences

It may seem trivial, but good matching requires in the first place that matched structures are built of devices made from the same material and are of equal size. This

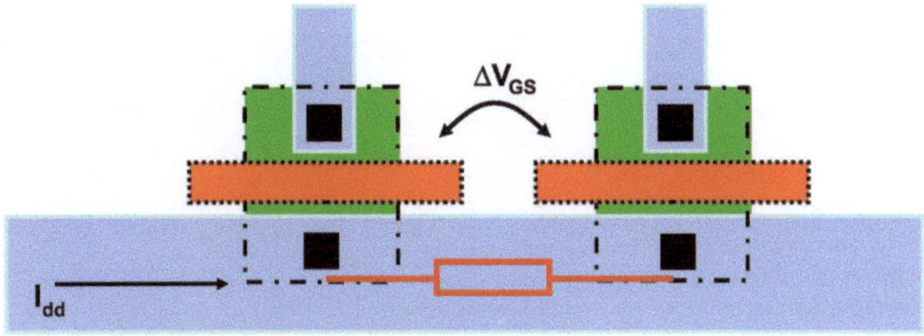

Fig. 15.2 A small resistance in the source connection wire will result in offset in the gate voltage. A common connection point is better from an offset point of view

implies that a 2:1 ratio is constructed with 3 identical elements, connected in a manner that will result in the desired ratio. The required 2:1 ratio applies for every aspect of the combination: area, perimeter, etc. A pitfall can occur when an existing lay-out is scaled and the resulting dimensions rounded off to fit the new lay-out grid. The rounding operation can result in unexpected size deviations in originally perfectly matched devices.

For electrical matching, the voltages on all the elements must be identical. Node voltages are affected by voltage drops in power lines, leakage currents in diodes, substrate coupling, parasitic components, etc. Electrically derived effects must also be considered, e.g. heat gradients due to power dissipation and temporary charging effects due to phenomena as Negative Bias Temperature Instability [6].

Figure 15.2 represents a surprisingly often encountered example of electrical mismatch: a pair of transistors with a wire connecting the sources. The current I_{dd} in this wire will cause a voltage drop between the two source terminals of the transistors. This voltage drop shows up as an offset ΔV_{GS} in the gate-drive voltage. Such problems increase when additional currents are routed via this wire. A star-like connection is well-suited to avoid this problem. A rigorous inspection of the lay-out in which the main current paths have been identified, is always needed when offset problems are suspected.

15.3.2 Offset Caused by Lithography

During the lithography process the structures drawn in the lay-out are transferred to a mask and in the resist. In the next step physical structures are etched into a wafer. During this process there are many crucial details that will affect the quality of the patterning. Figure 15.3 shows the frequencies of free-running ring oscillators of circuits that are measured on a 300-mm wafer. Ring oscillators are most sensitive to gate-length variation, so the frequency map indirectly represents a gate-length map. Most prominent is the middle area, where a 20% higher frequency is measured with respect to the rest. A potential cause is the heat non-uniformity during the high-temperature dopant activation step. A second phenomena is visible as a pattern of

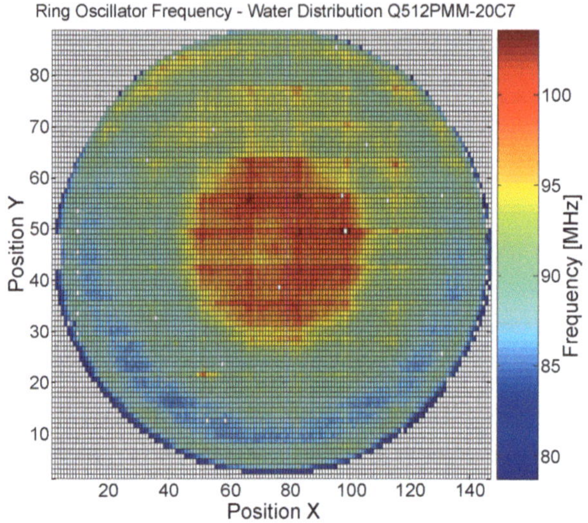

Fig. 15.3 A wafer from a 90-nm CMOS production lot on which a free-running oscillator frequency is measured. The frequency deviation (80–100 MHz) is largely an indication for the variation in gate length due to mask-plate dimensional errors, lithography and the processing steps that determine the electrical gate length (courtesy: B. Ljevar, . NXP)

rectangular shapes. The wafer stepper uses a reticle containing 16 by 7 devices. This ≈400 mm^2 reticle is repeated to pattern the entire wafer. Random effects are averaged over the nearly one hundred stages of the oscillator and are hardly visible.

These large distance effects do not have a direct impact on equality of transistor pairs. Most of this effect is a global variation and becomes part of the tolerance budget in the definition of the parameter corners of a process. In some digital design environment a specific guard band is used to make sure that the potential loss of current drive capability becomes visible to the designer. Yet the example makes clear what the significance is of lithography variations on the overall performance.

15.3.3 Proximity Effects

Figure 15.4 visualizes a proximity effect on a group of lines. The proximity effect is caused by the diffused light from neighboring fields. The line width in the open field will become narrower. Large neighboring structures cause lines to expand. In a precision lay-out dummy structures are placed at distances up to 20–40 μm. Proximity effects can be caused by lithography, but also by depletion of etch liquids or gases.

In 65-nm technologies it is practically impossible to define minimum width lines with acceptable tolerance at random positions. In order to create minimum width patterns pre-distortion is applied to the mask in the form of optical proximity correction (OPC). An example of dimensional deformations of an advanced lithographic tool is visible in the lithography simulation of Fig. 15.5.

Patterns in one layer may sometimes affect the patterning in other layers. During the spinning of the resist fluid, resist may accumulate against altitude differences of

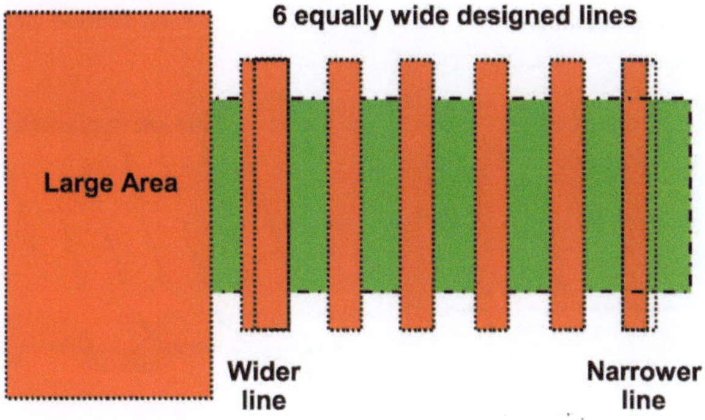

Fig. 15.4 The proximity effect: lines with large neighboring structures grow in size, while lines next to open space shrink

Fig. 15.5 The tips of the structures are enhanced by the OPC tool in this lithography simulation

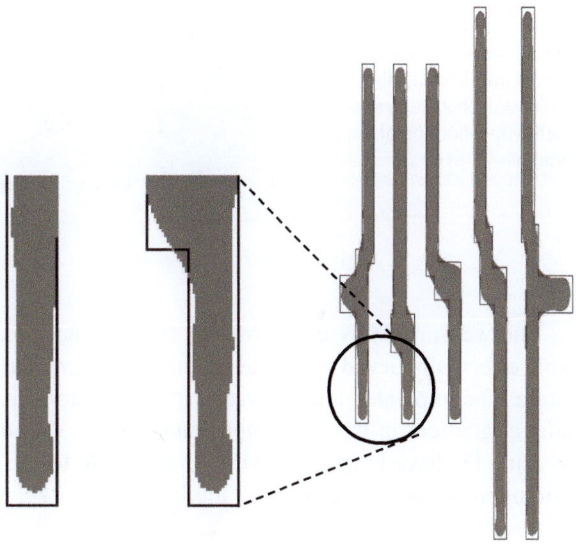

previous layers on a partly processed wafer. This results in circular gradients and is therefore often not easily recognized as a systematic offset.

An ion beam from an implanter that is perfectly aligned with the crystallographic orientation of the substrate will result in ions to penetrate deeply into the lattice. This effect is called "channeling". During the ion-implantation steps in older processes the implantation beam is tilted by some 5–8°. As a result of this non-perpendicular implantation, channeling is avoided but source and drain diffusions will be asymmetrical. The diffusion on one side may extend further underneath the gate than on the other side, see Fig. 15.6. In order to prevent inequalities in currents or overlap capacitors, the directions in which the MOS currents flow must be chosen to run parallel, and not rotated or anti-parallel. In integrated circuit manufacturing there are more processing steps that can cause similar asymmetries.

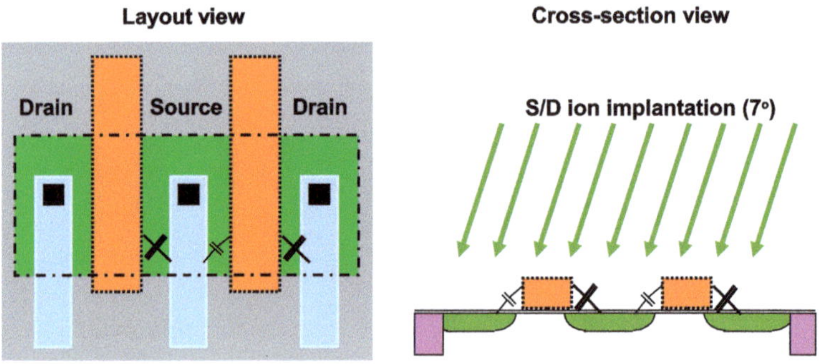

Fig. 15.6 The source and drain diffusions are implanted under an angle. This causes asymmetry for drain and sources

Fig. 15.7 The well-proximity effect occurs during the well implant. The drawn transistor is fabricated afterwards, but resides in a well with a horizontal doping gradient [7, 8]

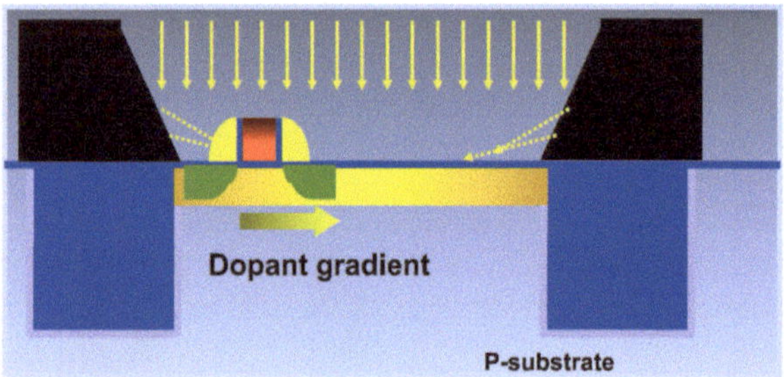

The well-proximity effect in Fig. 15.7 has no direct relation with lithography. This effect is believed to be caused during the implantation of the well in the substrate. The implanted ions interact with the photo-resist boundary and cause a horizontal gradient in the well implantation dose. Variations ranging from 1 μm [7] to 2 μm [8] have been reported. It is advisable to use an overlap well beyond the minimum lay-out design rule for implantation masks, where possible.

15.3.4 Temperature Gradients

Gradients can exist in doping, resistivity, and layer thickness. Although structures tend to decrease in dimensions, situations may occur in which equality is required in a distance in the order of 1 mm. In older processes CMOS thresholds were observed to deviate up to 5 mV over this distance. Resistivity gradients can reach a relative error of several percents over this distance. In advanced processes (<0.18 μm) process control is much better and technology gradients are hardly present.

The temperature distribution across a circuit in operation can be a reason for parameter gradients, see Fig. 15.8. In a System-on-Chip the different blocks show a great variety of power dissipations. On-chip memories shows a relatively low power density. Output drivers, transmitters, high-speed processors, power regulators and

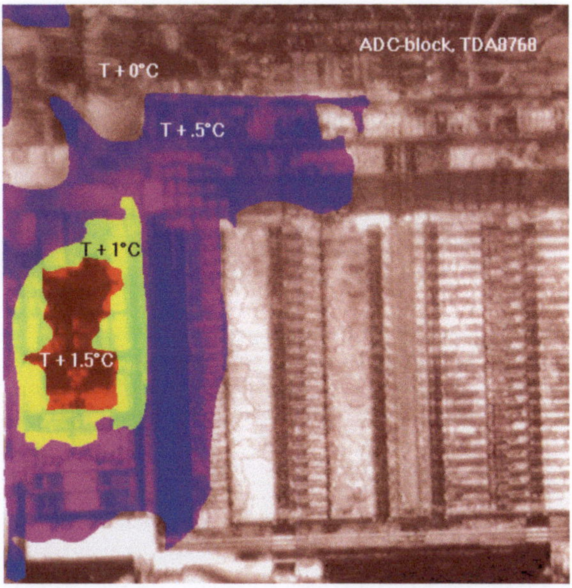

Fig. 15.8 Temperature difference can be visualized by means of a liquid crystal technique. Here the track-and-hold circuit generates the heat, which spreads out via the wiring into the ladders of this analog-to-digital converter

input stages (LNA) may consume much more. In larger chips (50–100 mm^2) with 2–5 Watt power dissipation temperature differences up to 20°C can occur. Local temperature gradients of 2–5°C/mm are possible. With threshold-voltage and diode temperature coefficients of -2 mV/°C, an offset in the order of several milliVolts is well possible. It is therefore important to consider the power distribution when temperature sensitive circuits and heat sources are placed on the same die. The circuit designer can position the critical devices on equal-temperature lines or use cross-coupling.

15.3.5 Offset Caused by Stress

During the fabrication of a circuit, layers are deposited on the substrate. These layers are built of different materials with different thermal expansion coefficients. After the devices have been cooled to room temperature the differences in thermal expansion coefficients lead to mechanical stress. This stress can result in a positive or negative change of the local parameter value. A secondary effect is that the global stress pattern is locally affected by neighboring lay-out features, causing stress modulation in the surrounding components. In high-precision analog design this will lead to undesired systematic offsets.

In resistors and transistors stress predominantly impacts the mobility of carriers. Tensile stress increases the electron mobility and reduces the hole mobility. Compressive stress works opposite and is used to enhance mobility in PMOS transistors. An effect on threshold voltage does occur as well, see Fig. 15.11. Some major causes for stress are:

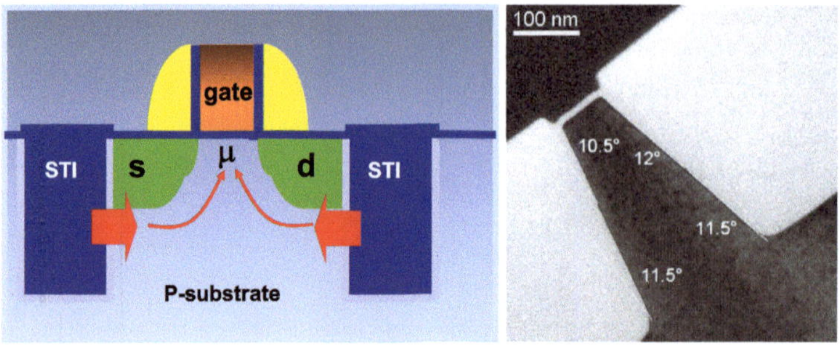

Fig. 15.9 Stress in integrated circuits is caused by thermal expansion of various materials. The blocks of shallow trench isolation (STI) create a considerable stress in the active area of the transistor, causing current variations up to 10%. The photo shows a cross-section, where the slopes of the STI are indicated (courtesy: C. Detcheverry, NXP)

- The presence of the die boundary close to sensitive devices. Typically a distance of several hundreds of microns is safe to avoid die-edge related stress effects.
- Plastic packages are molded around the die. After cooling these packages create rather severe mechanical stress. Special gels or polyimide coatings on top of the die relieve this problem. A simple way to detect package related stress is to heat the package with a hot air flow.
- In modern isolation techniques a trench is etched in the substrate and silicon dioxide is deposited and planarized: Shallow Trench Isolation (STI), Fig. 15.9. The different thermal coefficients of silicon dioxide and the substrate cause mechanical stress. A transistor in the substrate with its diffusion areas is surrounded by STI and experiences this stress. This effect is called "STI-stress" or "LOD-stress" (Length Of Diffusion).
- In a densely packed circuit, there will be many active-to-STI edges. Unrelated edges that are close to a device will influence the stress pattern, this effect is known as "OD-to-OD" stress or "OD-spacing" effect. The Oxide-Definition (OD) mask defines the inverse of the active area.
- In advanced processes an etch-stop layer can be used to create stress that increases the current in a high-performance MOS transistor. The proximity of other structures will influence this effect, called "PS-to-PS" stress or "poly-space" effect.
- Aluminum has a very different thermal expansion coefficient from the dielectric that surrounds it. Asymmetries in wiring will result in stress related offset.

Stress related to Shallow Trench Isolation has been subject of various studies [8–11]. At temperatures of over 1000°C areas of silicon dioxide are formed in the substrate. At that temperature the structure is free of stress or relaxed. After cooling the difference in thermal expansion coefficient causes a compressive mechanical stress that peaks at the border of the active area and the STI formation [9]. The mechanical stress deforms the lattice and causes the mobility in the transistors to vary.

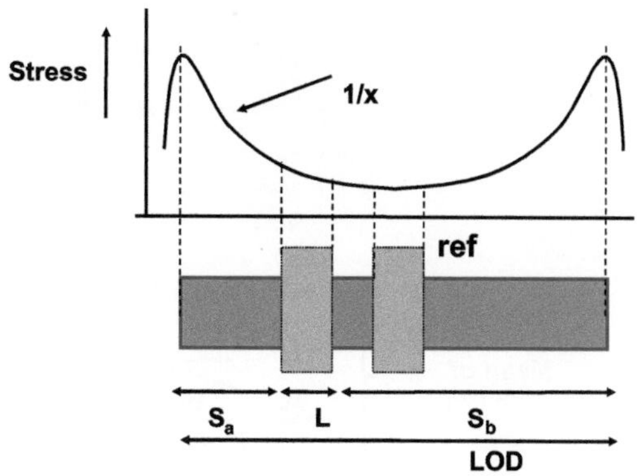

Fig. 15.10 A transistor is placed asymmetrically with respect to the reference device. The edges between active and STI inflict stress in the channel region, however due to the asymmetry the effect of the stress is different for both devices

Figure 15.10 shows two transistors that are placed asymmetrically with respect to the edges of the active area and the STI isolation. The mobility of the device μ_{eff} is usually modeled with an inverse distance model [9]:

$$S = \frac{1}{S_a + L/2} + \frac{1}{S_b + L/2}, \tag{15.1}$$

$$\mu_{eff} = \frac{1 + K_{s,\mu} S}{1 + K_{s,\mu} S_{ref}} \mu_{ref}, \tag{15.2}$$

$$V_T = V_{T,ref} + K_{s,VT}(S - S_{ref}) \tag{15.3}$$

where the suffix "*ref*" indicates the reference device and K_s is a process parameter. The parameter S reflects the distance relation. The stress from the STI edges affects also the doping profile under the transistor. This phenomenon is not fully understood, however the idea that a lattice deformation changes the diffusion of the doping atoms, seems plausible.

In [11] an experiment is reported where the STI-to-active edge of the source and drain is varied. Figure 15.11 shows current factor deviations up to 12% and threshold voltage variations of 10 mV. These observations are technology specific but similar effects are reported in [8, 9].

Next to the effect of the STI-to-active edge of the transistor itself, also neighboring edges will modulate the mechanical stress pattern. This may be less relevant in a digital circuit, however, in precision analog design these effects must be taken into account. The generalized model uses:

$$S = \sum_{i=1}^{n} \frac{\pm W_i}{2W} \frac{1}{S_i + L/2} \tag{15.4}$$

where the first term relates the width of the edge to the gate-width and the $\pm$ sign to an STI-active edge or an active-STI edge.

The poly-space effect [12] in Fig. 15.12 is caused by the stress related to the use of the etch-stop layer. This layer applies compressive or tensile stress to the transistor in order to increase the current drive capabilities. Also stress coming from neighboring devices will influence the stress pattern under the critical transistors. Again an inverse distance model applies.

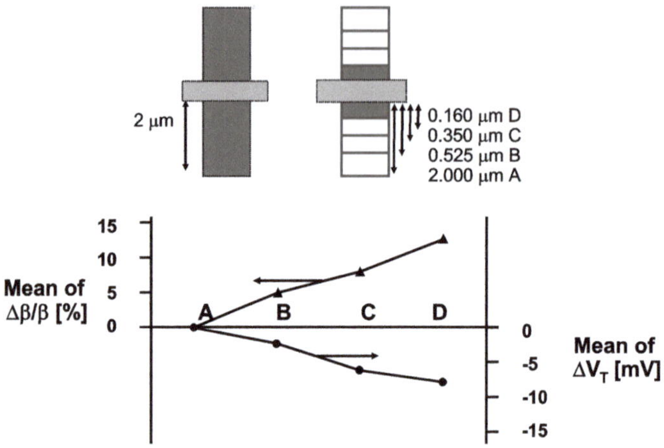

Fig. 15.11 An experiment shows the influence of the STI edge on the drain current and threshold voltage [11]. Top: a 65-nm 2/0.5 μm NMOS reference transistor is designed with the STI edge of the source and drain at 2.0 μm. A second device has a similar STI distance (A) or at 0.525 (B), 0.35 (C), and 0.16 μm (D)

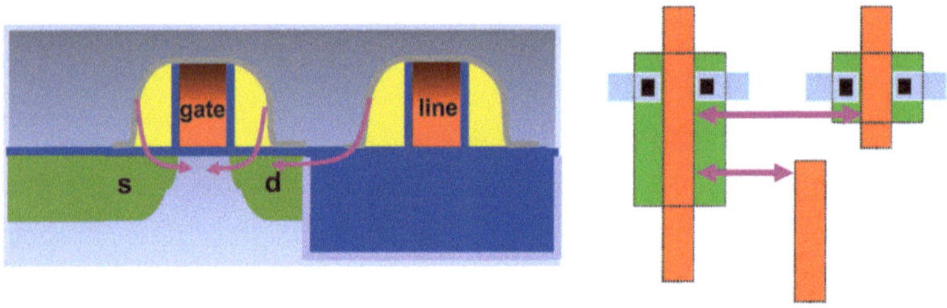

Fig. 15.12 The poly-space effect is caused by the etch-stop layer

The wiring pattern causes transistors to show offset. The coverage of transistors with metal layers can lead to mobility reduction due to incomplete annealing of interface states [13] and to variations in the stress pattern. When a wiring pattern is placed at different spacings on one side of a bipolar pair [14], the resulting current variation is in the order of 1%. Figure 15.13 suggests that the impact of the wiring pattern halves for every 10 μm distance up to 40 μm. This example again shows that a regular, symmetrical and consistent lay-out is required for analog circuits that should yield offsets below 1%.

15.3.6 Offset Mitigation

Systematic deviations are mostly identified during the initial design trials of a process. Sometimes optimizations in an established process flow still lead to unexpected deviations. Measures for overcoming them are found in an extensive study of

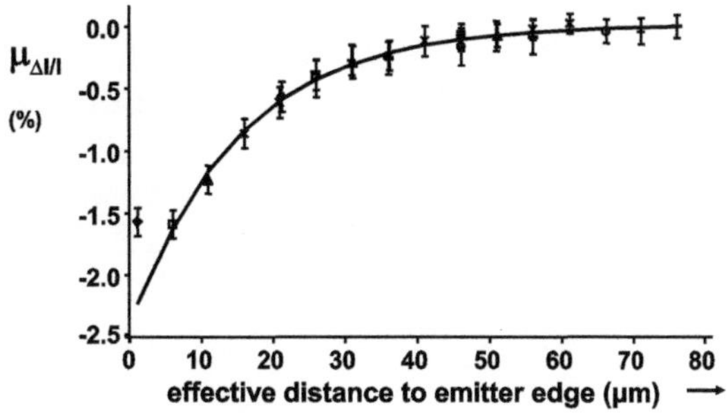

Fig. 15.13 An aluminum wire placed at a certain distance of the emitter causes current deviations, that can be measured up to 40 μm

Table 15.4 Guide lines for the design of equal components

- Equal components are of the same material, have the same form and dimensions
- The potentials, temperatures, pressures and other environmental factors are identical
- Currents in components run in parallel, not anti-parallel or perpendicular
- Only use cross-coupled structures if there is a clear reason for that (e.g. temperature gradient)
- Keep wiring away from the components
- Use star-connected wiring for power, clock and signal
- Apply dummy symmetrical structures up to 20 μm away from sensitive structures
- Keep supply and ground wiring together and take care that no other circuits dump their return current in a ground line
- Check on voltage drops in power lines
- Stay 200 μm away from the die edges to reduce stress from packaging

the fabrication process. As dimensions shrink, some offset effects are incorporated in the device model descriptions that quantify the impact [4]. Despite the complex nature of some variations, a number of guidelines can be formulated to minimize the effect of these offset causes, see Table 15.4 [15].

Common centroid structures are used to reduce the gradient effects [16]. Applying a common centroid geometry is not trivial, an asymmetry in the wiring scheme can easily cause more problems than are solved, see Fig. 15.14. On the left side is a standard cross-coupled differential pair with common source. The right side shows in-line common centroid structures. The lower structure is exactly common centroid with the disadvantage that the outer devices need dummy structures to compensate for their lack of neighbors. The upper-right structure needs no dummy structures, at the cost of a small spacing between the common centroid points.

By following these design guidelines the effects of systematic errors can be significantly reduced. The obtainable limits in a production environment differ per component and can be summarized as:

- Resistors: The absolute value suffers from process variations and temperature. Yet, the relative accuracy of matched resistors is in the order of 10^{-3}–10^{-4} depending on type (diffused is better than polysilicon), size, and environment. Sensitive to substrate coupling.

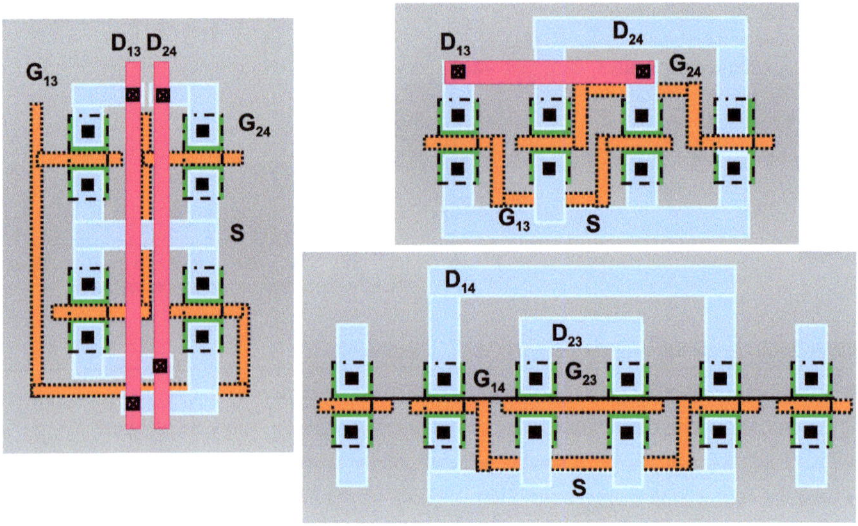

Fig. 15.14 Some common centroid arrangements

- Capacitors: The absolute value is usually well-defined in a double poly-silicon or MIM process. Also horizontally arranged capacitors such as fringe capacitors reach excellent performance. The relative accuracy of capacitors is in the order of 10^{-4} for >1 pF sizes. Minimum usable sizes in design are limited by parasitic elements, relative accuracy or the kT/C noise floor. In the application the net effect of the capacitor is sensitive to different parasitic couplings, which can be mitigated with stray-capacitor insensitive circuit topologies. Often capacitors are seen as a low-power solution, but handling charges requires large peak currents during transfers, so the power of the surrounding circuits limits the low-power ambition of capacitor based circuit solutions.
- Transistors: The current is sensitive to temperature, process spread and variability effects. The relative accuracy in current is in the order of 10^{-3}. Back-gate modulation by substrate noise and $1/f$ noise must be considered.
- Time: with a more or less fixed timing variation or jitter (1–5 ps_{rms}), the best accuracy is achieved for low signal bandwidths.

15.4 Random Matching

15.4.1 Random Fluctuations in Devices

Parameters that reflect the behavior of devices, are the result of the combination of a large number of microscopic processes. The conductivity of resistors and transistors and the capacitance of capacitors is built up of a large number of single events: e.g.

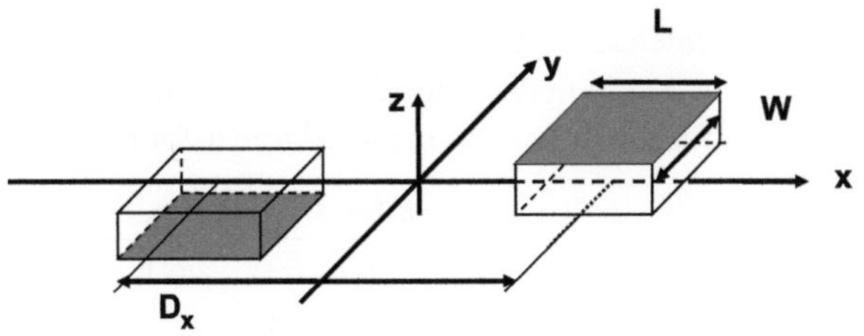

Fig. 15.15 Definition of the area function $h(x, y)$

the presence of ions in the conduction path, the local distances between the polysilicon grains that form capacitor plates, etc. Already in 1961 W. Shockley recognized that these atomic processes can lead to random fluctuations of device parameters. Various authors have investigated random effects in specific structures: capacitors [17–20], resistors [21, 22], MOS transistors [23–26] and bipolar devices [27].

In a general parameter fluctuation model [28] a parameter P describes some physical property of a device. P is composed of a deterministic and random varying function resulting in varying values of P at different coordinate pairs (x, y) on the wafer. The average value of the parameter over any area is given by the weighted integral of $P(x, y)$ over this area. The actual difference between two parameters P of two identically sized areas at coordinates (x_1, y_1) and (x_2, y_2) is:

$$\begin{aligned}\Delta P(x_{12}, y_{12}) &= P(x_1, y_1) - P(x_2, y_2)\\ &= \frac{1}{\text{area}}\left[\int\int_{\text{area}(x1,y1)} P(x', y')dx'dy'\right.\\ &\quad \left. - \int\int_{\text{area}(x2,y2)} P(x', y')dx'dy'\right]. \end{aligned} \tag{15.5}$$

This integral can be interpreted as the convolution of double box functions formed by the integral boundaries (or the device dimensions) with the "mismatch source" function $P(x, y)$. In the Fourier domain the convolution transforms in a multiplication and allows separating the geometry-dependent part from the mismatch source:

$$\Delta\mathcal{P}(\omega_x, \omega_y) = \mathcal{G}(\omega_x, \omega_y)\mathcal{P}(\omega_x, \omega_y). \tag{15.6}$$

Now the mismatch generating process $\mathcal{P}(\omega_x, \omega_y)$ can be regarded as a source that generates spatial frequencies that are spatially filtered by the device geometry dependence function $\mathcal{G}(\omega_x, \omega_y)$. These two components are analyzed separately.

The geometry function as shown in Fig. 15.15 for a pair of rectangular devices with area WL is defined as:

$$h(x, y) = \begin{cases} \frac{1}{WL}, & (D_x/2 - L/2) < x < (D_x/2 + L/2), -W/2 < y < W/2,\\ \frac{-1}{WL}, & (-D_x/2 - L/2) < x < (-D_x/2 + L/2), -W/2 < y < W/2,\\ 0, & \text{elsewhere}. \end{cases} \tag{15.7}$$

For convenience it has been assumed that both areas are at a distance D_x along the x-axis. Some mathematical manipulation (see Appendix) results in a geometry function for the difference in paired transistor parameters (Fig. 15.15):

$$\mathcal{G}(\omega_x, \omega_y) = \frac{\sin(\omega_x L/2)}{\omega_x L/2} \frac{\sin(\omega_y W/2)}{\omega_y W/2} [2 \sin(\omega_x D_x/2)]. \tag{15.8}$$

This geometry function has a zero value for $\omega_x = 0$, thereby eliminating the global value of the parameter from the calculations. The geometry functions for other geometries are found in the same way, e.g. a cross-coupled group of four transistors, Fig. 15.14 (left), has a geometry function where the last term in brackets in (15.8) is replaced by $[\cos(\omega_x D_x/2) - \cos(\omega_y D_y/2)]$.

After this analysis of the geometry dependence the specification of the random contribution to $P(x, y)$ or $\mathcal{P}(\omega_x, \omega_y)$ has to be formulated.

Two classes of distinct physical mismatch causes are considered here as examples of local and global variations. Every mismatch-generating physical process that fulfills the mathematical properties of these classes, results in a similar behavior at the level of mismatching transistor parameters.

The first class is a random process on a parameter P characterized by:

- The total mismatch of parameter P is composed of mutually independent events of the mismatch generating process.
- The effects on the parameter are so small that the contributions to the parameter are linear.
- The correlation distance between the events is small compared to the size of the device (basically saying that boundary effects can be ignored).

Statistically the values of parameter ΔP are described by a Poisson process that converges for a large number of events to a Gaussian distribution with zero mean. In the frequency domain this type of spatial random processes is modeled as spatial "white noise". A process with these properties is described in the Fourier domain as a constant value for all spatial frequencies.

The combination of this mismatch generating process and the paired-transistor geometry function results in a description of the power or the variance of the difference ΔP in parameter P between the two instances (see Appendix):

$$\sigma^2_{\Delta P} = \frac{A_P^2}{WL} \tag{15.9}$$

A_P is the area proportionality constant for parameter ΔP. The proportionality constant can be measured and used to predict the mismatch variance of a circuit.

Many known processes that cause mismatching parameters fulfill in first order the above-mentioned mathematical constraints: distribution of ion-implanted, diffused or substrate ions (Random Dopant Fluctuations), local mobility fluctuations, polysilicon and oxide granularity, oxide charges, etc.

Equation (15.9) describes the statistical properties of area *averaged* or *relative* values of parameter P. The *absolute* number of events (like the charge in a MOS channel) is proportional to the area of the device WL. Therefore differences in the sums of atomic effects obey a Gaussian distribution with zero mean and

$$\sigma_{\Delta P,abs} = A_P\sqrt{WL}. \tag{15.10}$$

In analyzing statistical effects it is important to consider whether the parameter is an absolute quantity (e.g. total amount of ions) or is relative (averaged) to the device area (e.g. threshold voltage).

Apart from theoretical derivations and measurements, 3-D device simulations are applied to analyze the impact of random dopants, line-edge roughness and polysilicon granularity in advanced processes [29].

The assumption of a short correlation distance in the above process implies that no relation exists between matching and the distance D_x between two transistors. Wafer maps show, however, all sorts of parameter-value distributions that originate from wafer fabrication. The second class of mismatch is a deterministic process but, as the original placement of dies on a wafer is unknown after packaging, the effect of this parameter value distribution is modeled as an *additional* stochastic process with a long correlation distance. In the Fourier domain this effect is described as a fixed low-frequency contribution with a spatial frequency inversely proportional to the wafer diameter. The representation of parameter fluctuations in the Fourier domain allows easy determination of the power contents, which in turn can be interpreted as the variance (σ^2) of the stochastic parameter, see Appendix.

$$\sigma^2_{\Delta P} = \frac{A_P^2}{WL} + S_P^2 D_x^2. \tag{15.11}$$

S_P describes the variation of parameter P with the spacing. Note that the probability density function of e.g. a circular parameter pattern is not Gaussian and only randomized because the link to the position on the wafer will be lost after packaging.[1]

For a group of four cross-coupled transistors is found in a similar way:

$$\sigma^2_{\Delta P} \approx \frac{A_P^2}{2WL} + S_P^2 D_x^2 \frac{D_y^2}{\text{wafer diameter}^2}. \tag{15.12}$$

The effect of the doubled gate area and the reduction of the linear components in the gradient is obvious.

15.4.2 MOS Threshold Mismatch

The threshold voltage is given by:

$$V_T - V_{FB} = \frac{Q_B}{C_{ox}} = \frac{qN_x z_d}{C_{ox}} = \frac{\sqrt{2q\epsilon N_x \phi_b}}{C_{ox}} \tag{15.13}$$

[1]The importance of this model extension is more in signaling a potential problem in the process than in accurately modeling a phenomenon.

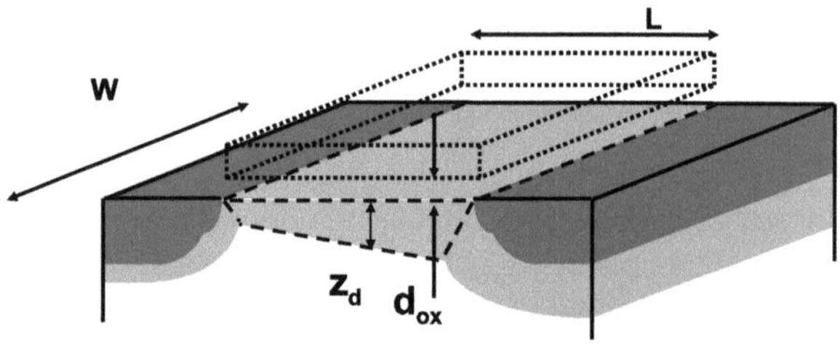

Fig. 15.16 A cross-section through a MOS transistor indicating the depletion region

where ϵ is the permittivity, N_x the dope concentration and ϕ_b the Fermi potential. If the depletion region of a transistor (see Fig. 15.16) is defined by its width W, length L and a depletion region depth $z_d = \sqrt{2\epsilon\phi_b/qN_x}$, then the volume of the depletion region is (in first order): WLz_d. Different impurities are active in this region, with concentrations around 10^{16}–10^{18} cm^{-3}. N_x contains acceptor and donor ions from the intrinsic substrate dope, the well, threshold adjust, punch-through implantation.[2] In the variance analysis it is important to note that the total number of charged ions and other charge contributions must be considered, not the net resulting charge.

The variance of the number of atoms is now approximated by Poisson statistics:

$$\sigma_c^2 = \mu_c \quad \mu_c = WLz_dN_x \quad \Rightarrow \quad \sigma_c = \sqrt{WLz_dN_x}. \tag{15.14}$$

The threshold variance can now be derived from (15.13) by considering that the variance of a threshold voltage equals the variance of the charge in the depletion region multiplied by the partial derivative of the threshold versus the charge

$$\sigma_{\text{single VT}} = \sigma_c \frac{\partial(V_T)}{\partial(WLz_dN_x)}. \tag{15.15}$$

As matching usually occurs between pairs of transistors, the variance of the *difference between two transistors* is [28, 30]:

$$\sigma_{\Delta VT} = \sqrt{2}\sigma_{\text{single VT}} = \frac{qd_{ox}\sqrt{2N_xz_d}}{\epsilon_{ox}\sqrt{WL}} = \frac{A_{VT}}{\sqrt{WL}} \propto \frac{d_{ox}\sqrt[4]{N_x}}{\sqrt{WL}}. \tag{15.16}$$

This function is commonly depicted as a linear relation between $\sigma_{\Delta VT}$ and $1/\sqrt{\text{area}}$. Figure 15.17 shows an example of the measured dependence for $\sigma_{\Delta VT}$ versus $1/\sqrt{\text{area}}$. The slope of the line equals the parameter A_{VT}. In this plot the lay-out dimensions were used. For the smallest sizes the effective gate area is smaller due to underdiffusion and channel encroachment. In order to test the hypothesis that depletion charge is the dominant factor in threshold matching, Table 15.5 compares the A_{VT} coefficients as measured and as calculated using the above formula. The quantity N_xz_d was derived from process simulation which was tuned with accurate C/V measurements [31]. In these relatively straight-forward 0.6 μm and 0.8 μm CMOS processes the fit is good for three out of four A_{VT} coefficients. The deviation of the 0.6 μm NMOST shows that other factors except random dopant fluctuation

[2]For ease of understanding only a uniformly distributed dopant is assumed, more complicated distributions must be numerically evaluated.

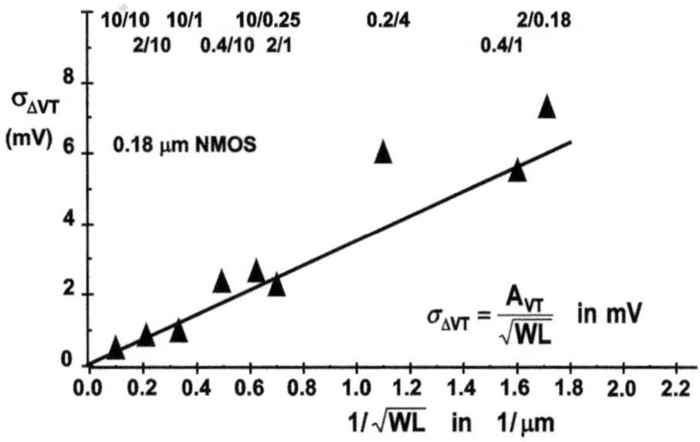

Fig. 15.17 The standard deviation of the NMOS threshold and the relative current factor versus the inverse square root of the area, for a 0.18 μm CMOS process

Table 15.5 Comparison of measured and calculated threshold mismatch coefficients

	A_{VT} measured	A_{VT} calculated
0.8 μm data:		
NMOST	10.7 mV μm	10.6 mV μm
PMOST	18.0 mV μm	18.6 mV μm
0.6 μm data:		
NMOST	11.0 mV μm	7.4 mV μm
PMOST	8.5 mV μm	8.6 mV μm

play a role. Such increases can be attributed to insufficient annealing of interface states [13].

The basic threshold-voltage mismatch model has been extended by various authors. More geometry dependence factors can be included to address deep-submicron effects [32]. A fundamental limit for dopant fluctuation related mismatch was derived in [33]. Reference [34] shows that the threshold-voltage mismatch coefficients are not affected by temperature. Work reported in [11] indicates that there is no relation between deterministic variations and random dopant fluctuations.

In deep sub-micron processes the short channel effects in the channel are controlled by means of "halo" or "pocket" implants. These implants are self-aligned with the gate stack and can introduce some significant variations in the local doping profiles. Next to their own variation, the self-aligned feature prints any line-edge roughness in the doping profile. The pocket implants defy the uniform dopant hypothesis for the calculation of the threshold mismatch of (15.9). An additional term can be included for the variation due to the pocket implant:

$$\sigma^2_{\Delta VT} = \frac{A^2_{VT}}{WL} + \frac{B^2_{VT}}{f(W,L)} \tag{15.17}$$

where the function $f(W,L)$ of the width and length still needs to be established.

15.4.3 Current Factor Mismatch

The matching properties of the current factor are derived by examining the mutually independent components W, L, μ and C_{ox}:

$$\frac{\sigma_{\Delta\beta}^2}{\beta^2} = \frac{\sigma_W^2}{W^2} + \frac{\sigma_L^2}{L^2} + \frac{\sigma_{C_{ox}}^2}{C_{ox}^2} + \frac{\sigma_{\mu_n}^2}{\mu_n^2}. \tag{15.18}$$

The mismatch-generating processes for the gate oxide and the mobility are treated in accordance with (15.9). The variations in W and L originate from line-edge roughness. The analysis of edge-roughness is a one-dimensional variant of the analysis in the previous section and leads to $\sigma^2(L) \propto 1/W$ and $\sigma^2(W) \propto 1/L$.

$$\frac{\sigma_{\Delta\beta}^2}{\beta^2} = \frac{A_W^2}{W^2L} + \frac{A_L^2}{WL^2} + \frac{A_\mu^2}{WL} + \frac{A_{C_{ox}}^2}{WL} \approx \frac{A_\beta^2}{WL} \tag{15.19}$$

where A_W, A_L, A_μ and $A_{C_{ox}}$ are process-related constants.

A significant contribution of gate-oxide thickness variation would lead to a negative correlation between the threshold voltage mismatch and the current factor mismatch. Such a correlation is generally not observed. If W and L are large enough the respective contributions will also disappear. At gate-lengths below 65-nm, simulations [29] indicate some role for edge roughness. This role is in measurements hard to identify in the presence of large threshold mismatch. In [28] it was assumed that the matching of the current factor is determined by local variations of the mobility. Many experiments show that mobility affecting measures (e.g. placing the devices under an angle) indeed lead to a strong increase in current factor mismatch. The relative mismatch in the current factor can be approximated by the inverse-area description as seen in the last part of (15.19).

In contrast to the threshold random fluctuation the absolute current factor variation is a function of temperature. However, the relative current factor mismatch as formulated in (15.19) is according to [34] much less sensitive to temperature.

15.4.4 Current Mismatch in Strong and Weak Inversion

Considering only the threshold and current factor variations,[3] the variance of the difference in drain currents ΔI between two equally sized MOS devices can be calculated. Using the generalized statistical method described in [35]:

$$\left(\frac{\sigma_{\Delta I}}{I}\right)^2 = \left(\frac{dI}{dV_T}\right)^2 \sigma_{\Delta VT}^2 + \left(\frac{dI}{d\beta}\right)^2 \sigma_{\Delta\beta}^2.$$

[3] The contribution of mobility reduction factor θ is next in line.

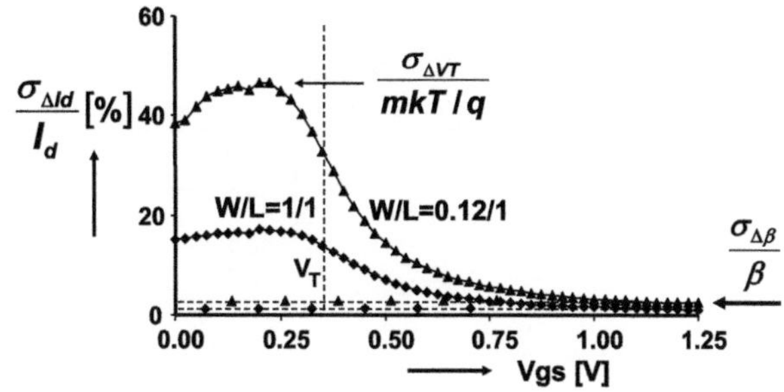

Fig. 15.18 The relative current mismatch for two 65-nm technology transistor geometries swept over the full voltage range. Measurements by N. Wils

For strong inversion this equation is combined with the simple square-law current model:

$$\left(\frac{\sigma_{\Delta I}}{I}\right)^2 = \left(\frac{2\sigma_{\Delta VT}}{V_{GS}-V_T}\right)^2 + \left(\frac{\sigma_{\Delta\beta}}{\beta}\right)^2. \tag{15.20}$$

Equation (15.20) suggests an infinite current mismatch if the gate voltage equals the threshold voltage, however, in that mode of operation the weak inversion model is applicable and levels off the maximum current mismatch. In weak inversion the current is modeled as an exponential function. Due to the low current level, the current factor mismatch is of less importance:

$$\left(\frac{\sigma_{\Delta I}}{I}\right)^2 = \left(\frac{q\sigma_{\Delta VT}}{mkT}\right)^2. \tag{15.21}$$

Although the threshold mismatch itself shows no temperature dependence, the current mismatch certainly will vary due to the denominator term. Figure 15.18 shows an example of the current mismatch relative to the drain current. At high gate-source voltages the current factor mismatch in the strong inversion equation (15.20) dominates. At lower gate-source voltages the threshold-related term in this equation gains importance.

In a 65-nm process $A_{VT} = 3.5$ mV µm. So $\sigma_{\Delta VT} = 10$ mV for a 0.12/1 device. Equation (15.21) predicts a relative current mismatch of 40%, which is confirmed by the measurement in Fig. 15.18. Except for extremely low current densities where the depletion layer width is shrinking, the observation can be made that the same value for $\sigma_{\Delta VT}$ applies for both the strong and the weak inversion regime. This example shows that the operating regime where the mismatch parameters of the transistor are extracted, has a marginal effect on the accuracy[4] of the prediction in other regimes, as confirmed in e.g. [25, 26, 36].

The standard deviation of the current difference between the 0.12/1 µm transistors reaches ≈40% of the current in the sub-threshold regime. Obviously this would imply a reverse drain current below -2.5σ of a Gaussian distribution. At these levels of mismatch the assumption that a small Gaussian distributed threshold mismatch voltage will turn into a Gaussian approximation of the mismatch current is

[4] Accuracy means that the standard deviation of a circuit parameter is within 10% of the prediction, see Sect. 15.4.7.

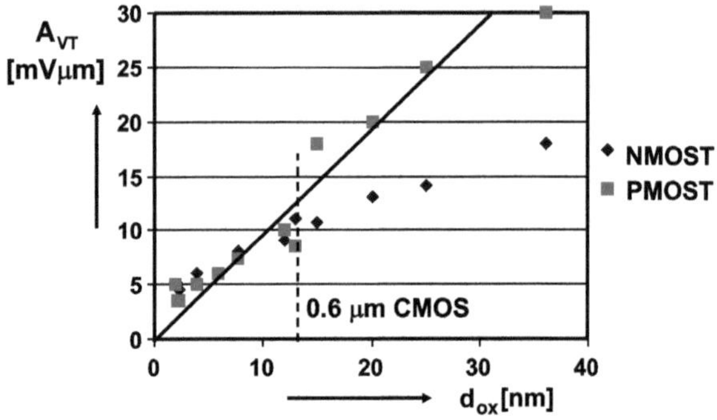

Fig. 15.19 Development of the threshold mismatch factor A_{VT} for NMOS and PMOS transistors as a function of the nominal oxide thickness of the process. The processes span 65-nm up to 1.6 μm CMOS

not valid anymore. The probability density function of the current mismatch needs here a log-normal distribution.

15.4.5 Mismatch for Various Processes

In Fig. 15.19 the threshold mismatch coefficient A_{VT} is plotted as a function of the nominal oxide thickness. As predicted by (15.16) the mismatch coefficient becomes lower for thinner gate-oxide thickness. Of course many more changes in the device architecture took place, still the oxide-thickness seems to be the dominant parameter. The large PMOS transistor coefficients for >0.6 μm CMOS generations is caused by the compensating implants: the N- and PMOS transistor threshold adjust and n-well implants. The quantity $N_x = (N_a + N_d)$ is relevant for matching, while the net value $(N_a - N_d)$ determines the threshold in the PMOS transistor. Beyond the 0.6 μm node a twin well construction with a dedicated well implant for the PMOS transistor is used that avoids compensating charges.

In Fig. 15.19 the diagonal line indicates an A_{VT} factor increase of 1 mV μm for every nm of gate insulator thickness. This line is a first order estimate of what a well-engineered process should bring.

Over the same process range the current mismatch factor A_β varies between 1.2% μm and 2% μm.

Figure 15.20 shows a simulation of the threshold voltage of 200 NMOS and PMOS 90-nm devices in their process corners. Although the corner structure is still visible, it is clear that for small transistors in advanced processes the mismatch is of the same magnitude as the process corner variation. The plot is somewhat misleading as it combines global process corner variation and local mismatch. A simple "root-mean-square" addition of these two variation sources ignores the fundamental difference between the two.

The analysis of matching in various processes allows to compare in Fig. 15.21 the MOS mismatch to the development of power supply voltage. A transistor with a size of $1.5L_{min}^2$ was chosen. During the process development from the 2.5 μm process to the 0.35 μm process both the mismatch and minimum gate-length have reduced. The power supply remained fixed at 5 volt for the micron range process

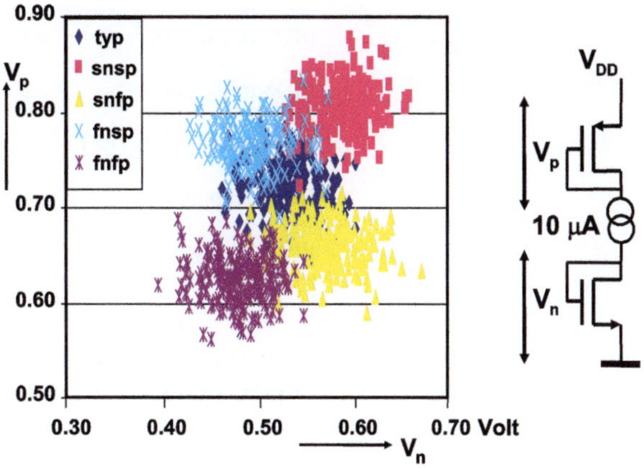

Fig. 15.20 Simulation of 200 0.2/0.1 P- and NMOS transistors in their 90-nm process corners. The notation "snfp" refers to slow NMOS and fast PMOS transistors. The variation due to mismatch is of equal importance as the process variation

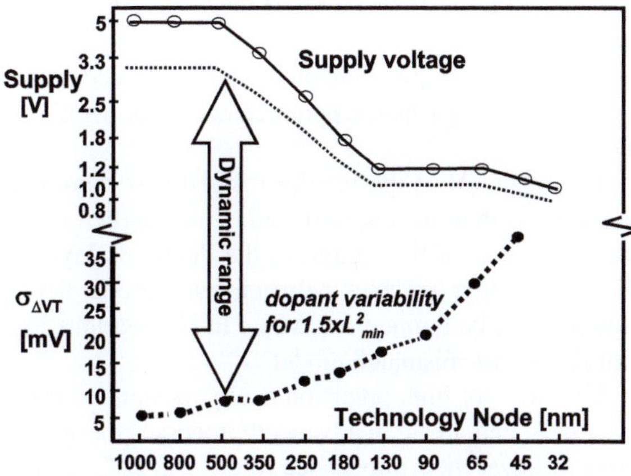

Fig. 15.21 Development of power supply voltage and the measured NMOS threshold matching of a transistor 1.5 times the minimum size through various process generations

generations with a signal swing of 2.5–3 volts. Circuits that relied on analog CMOS performance such as analog-to-digital converters could improve their performance in these process generations by not fully following the line-width reduction.

At the 0.35 μm CMOS node the maximum electrical fields in intrinsic transistors were reached for both the vertical gate-oxide field and the lateral field controlling the charge transport. For this reason and in order to reduce power consumption, the power supply voltage was lowered. On the other hand, the need to tailor the internal fields in the transistor, has led to less uniform and higher implantation channel dope. As can be expected from the theoretical background, the slower scaling of the gate-oxide thickness made that the threshold matching factor A_{VT} stopped decreasing. This became especially pronounced in 65–32 nm technologies, where pocket implants create an additional mismatch source. Shrinking the area of analog blocks in sub-micron processes is clearly an important economical issue, but in combination with a rising mismatch coefficient this will lead to lower performance. The reduction in the signal-to-matching coefficient ratio in sub-micron CMOS will necessitate changes in the system, design or technology. In order to maintain high quality signal

Table 15.6 An overview of matching models and value ranges

MOS transistors	$\sigma_{\Delta VT} = \frac{A_{VT}}{\sqrt{WL}}$	$A_{VT} = 1$ mV µm/nm
MOS transistors	$\frac{\sigma_{\Delta\beta}}{\beta} = \frac{A_\beta}{\sqrt{WL}}$	$A_\beta = 1$–2% µm
Bipolar transistors (BJT)	$\sigma_{\Delta Vbe} = \frac{A_{Vbe}}{\sqrt{WL}}$	$A_{be} = 0.3$ mV µm
Diffused/poly resistors	$\frac{\sigma_{\Delta R}}{R} = \frac{A_R}{\sqrt{WL}}$	$A_R = 0.5/5\%$ µm
Plate, fringe capacitors	$\frac{\sigma_{\Delta C}}{C} = \frac{A_R}{\sqrt{C}}$	$A_C = 0.3\%\sqrt{\text{fF}}$

processing, some enhancements to the standard processes are needed, such as the use of high-voltage devices or precision resistors and capacitors.

15.4.6 Application to Other Components

In Table 15.6 matching parameters of various components are listed. In the previous paragraphs the mismatch parameters for the MOS transistor have been extensively discussed.

The behavior of the bipolar transistor is dominated by the number of dopants in the base that are not depleted. The fluctuation of this number, comparable to the fluctuation of the charge in the depletion layer of the MOS transistor, causes the base-emitter voltages between two bipolar devices to mismatch. Therefore a variance can be defined for ΔV_{be}. In [27] various experiments have confirmed the validity of this mismatch model.

Resistors for high precision analog design are formed by polysilicon or diffused n- or p- doped areas. In advanced processes these layers are covered with a silicide layer to lower their impedance to the 2–5 $\Omega/\Box$ level. A special mask is applied to prevent the deposition of silicide in order to obtain sheet resistances of 50–500 $\Omega/\Box$. Resistors suffer from area related mismatch and from edge roughness. The general description for the relative mismatch is therefore:

$$\frac{\sigma^2_{\Delta R}}{R^2} = \frac{A^2_W}{W^2L} + \frac{A^2_L}{WL^2} + \frac{A^2_\mu}{WL} \approx \frac{A^2_R}{WL}. \tag{15.22}$$

The mobility variation mechanism includes impurity/doping scatter and in the case of polysilicon resistors also includes grain boundary disturbance. The last mechanism is important as the mismatch coefficient A_R increases to 5% µm, while the diffused resistors allow some 0.5% µm. An additional factor in resistor design are the head-end connections [21, 22]. These connections introduce edge roughness, but more importantly also impose stress on the entire structure. A careful design or an additional margin is needed.

In [18] mismatch of capacitors was attributed to edge effects and to area non-uniformities. The first error would result in a line variation $\sigma_{\Delta C} \propto \sqrt{L}, \sqrt{W}$. The area non-uniformities comply to the mathematical conditions of the general model resulting in $\sigma_{\Delta C} \propto \sqrt{WL}$. At small dimensions the edge effects will dominate, but at reasonable sized capacitors >0.1 pF, the area effects become dominant. The description of capacitor mismatch in Table 15.6 is different from the resistor model. For resistors the W/L ratio determines the value allowing to choose the device area independent of the resistor value. The capacitor value is always proportional to the area WL.

15.4.7 Modeling Remarks

The present model describes random variations of devices by means of two statistical variables on the device-parameter level. This intuitive description allows easy communication between foundries, designers and architects to balance power, area and performance of mismatch-sensitive circuits. The standard deviations of Monte-Carlo simulated circuit parameters and from measured circuits agree within approximately 10%, see e.g. Fig. 15.26. This number is based on quantities of around hundred measured samples of a circuit fabricated in a stable process, without foundry liability margins. A significantly better accuracy requires quadratically more samples and e.g. corrections for the effective channel width and length.

In [28] the mismatch contributions of the back-bias factor and the mobility reduction coefficient have been added. Also other authors [24, 32] have proposed methods to get a closer prediction of mismatch in the various operating modes. One problem is that extra mismatch factors are not easy separable from others. E.g. the mismatch in the mobility reduction coefficient θ and the source series resistance R_s are to a certain extent interchangeable: $\theta \Leftrightarrow \beta R_s$.

From a mathematical perspective, the threshold mismatch coefficient in an I_D-V_{GS} plot corresponds to the variation in the back-extrapolated zero crossing of the drain current, while the current factor mismatch coefficient describes the slope of the curve. The zero-order and first-order mismatch components are absorbed in threshold voltage and current factor mismatch coefficients. Identifying other mismatch effects requires a large statistical effort to realize sufficient significance.

Alternative model approaches concentrate on the physical parameters e.g. mobility in stead of current factor. Reference [24] analyzes the individually extracted (BSIM) model parameters of test devices by means of a Principal Component Analysis (PCA). This technique assigns the overall mismatch to the most relevant model parameter and than re-iterates to assign the remaining variation to other model parameters. Depending on the data points, this approach can lead to unexpected cross-correlations with little physical significance. In [37] the measured I-V plots are directly used as the input for a PCA analysis. A large numerical effort seems imperative. The work indicates that just a few principal components sufficiently describe the statistical behavior [37, Fig. 5]. This observation is in line with e.g. [38].

15.5 Measuring Offset and Mismatch

Parametric mismatch characterization is a form of the standard parametric measurements that are applied widely for the development and control of IC-technologies.

However, mismatch characterization does require additional attention in three areas, namely test structure layout consistency, measurement precision (measurement systems and measurement algorithms), and proper use of statistics. Moreover, these three aspects are strongly related [39]. The mismatch mechanism of interest and the required measurement precision (1%, or 0.001% parametric differences) largely determine the test structure and suitable measurement approach. The available test structure determines to a large extend which measurement method and measurement system can be used (and vice-versa). Reliable statistical methods mean cautious evaluation of results, but also imply long overall measurement times and substantial silicon test area consumption. In this section, each of these three areas will be addressed in more depth.

15.5.1 Matched Pair Test Structures

Two main categories of matched pair test structures are distinguishable. In the first place there are the perfectly matched pairs for studying intrinsic mismatch fluctuations (random matching). This category consists of perfectly symmetrical laid out matched devices, with both components placed at a distance small enough to mitigate effects of parametric gradients across a die. Typically, for a well engineered production worthy process, distances up to 100 μm should have no noticeable effect on mismatch if the devices have identical (symmetrical) connections and are placed in an identical layout environment. For this category it is essential to have different pairs available with different active areas and different area versus perimeter ratios. Preferably, a substantial area range of at least two to three decades should be made available, for instance ranging from below 0.1 μm^2 to over 100 μm^2. The smallest ones being representative for the relatively large random device fluctuations as observed in digital library blocks and memories, while the larger devices serve the high precision analog and mixed-signal modeling requirements. Figure 15.22 gives an example of a perfectly symmetrical probe-pad defined matched pair test structure for DC parametric mismatch characterization. The double rail connection to the common source pad assures identical access resistance (star connection) to the two transistors of the pair. And the standardized frame allows placing transistor pairs with different dimensions in the same environment, assuring equality of parasitic resistances in the test structure for all pairs (consistency!). From a mechanical stress related standpoint it is beneficial to extend some of the metal connection lines to provide a more symmetrical layout environment around the transistors.

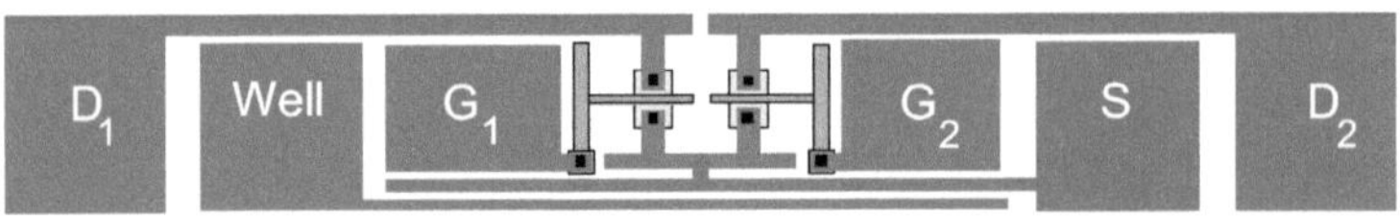

Fig. 15.22 Example of a perfectly symmetrical pad-based single matched MOS transistor pair common Source test structure approach

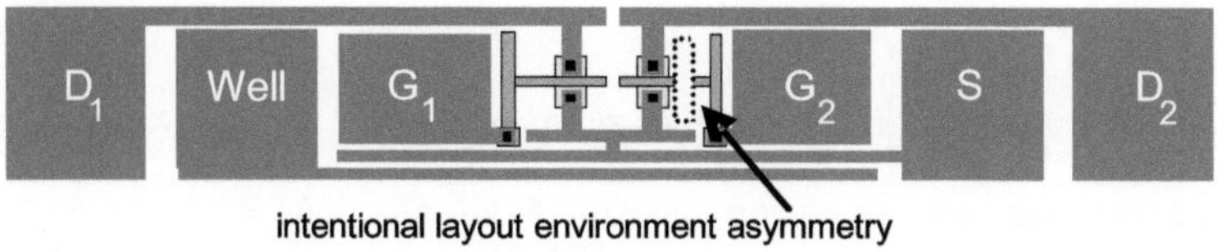

Fig. 15.23 Test structure approach for investigating layout environment effects

This symmetrical metalization pattern is also used to provide sufficient layer density to accommodate CMP (Chemical Mechanical Polishing) related design rules. This avoids unexpected surprises due to uncontrolled dummy tiles that are placed by chip-finishing tools [40].

Secondly there is the category of test structures aimed at characterizing layout options and layout environment rules. These are primarily intended for verifying "what if" rules, i.e. to check what would happen if the transistors are *not* placed in an exactly identical layout environment, if metal routing *must* be placed over transistors, or if devices ratios can *not* be made through parallel or series combinations of identical components. This category can be used to study most of the technology related effects discussed in the previous sections.

Figure 15.23 is an example of a matched pair test structure for studying effects of transistor layout modifications and layout environment effects. In the shown example, one of the transistors of a pair is disturbed, in this case by placing a layout asymmetry (a metal plate, a high topography stack, a well) close to the transistor. Other variability effects can be studied by applying a different layout option (active region shape [11], multiple fingers, other metalization, etc.) to one transistor of the pair. When the mismatch measurements on the resulting population of (asymmetrical) matched pairs yield a statistically significant systematic mismatch, the layout environment effect is relevant. If an asymmetrically disturbed pair has the same mismatch standard deviation as the ideal (symmetrical) pair of the same dimensions, this means that the disturbance effect is constant over the population.

15.5.2 Mismatch Measurement Precision Considerations

Parametric mismatch measurements often require higher measurement precisions than needed for standard process control measurements. For the latter, accuracies and repeatabilities of 0.25% to 1% are generally acceptable. Mismatch standard deviations of larger transistor pairs on the other hand, can easily be as low as 0.01% to 0.1%. It is helpful that for characterizing mismatch, one is less interested in the absolute accuracy of a measurement than in the difference of the measurement on the two devices of the pair. This means that the key performance indicator for mismatch measurements is the Short Term Repeatability (STR) of a particular observable. The STR is generally limited by measurement system noise and/or system and/or slow temperature drifts (of devices or equipment). A useful rule of thumb for parametric mismatch characterization is that if the standard deviation of the measurement

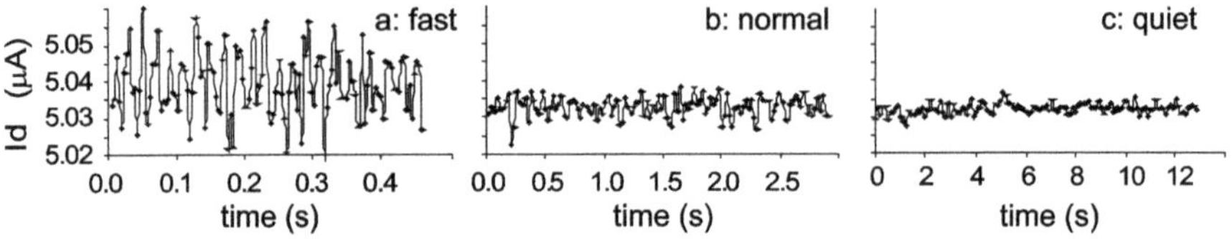

Fig. 15.24 Example of drain current measurement for three different measurement system settings (fast, normal, quiet). Measurement times per data point are 3.7, 23.6 (1 PLC integration) and 103.9 ms respectively. These correspond to relative current fluctuation standard deviations of 1700, 550 and 290 ppm respectively

system noise is less than one third of the parametric mismatch standard deviation under investigation, the system contribution will be negligible. Apart from spending more money on state-of-the-art high precision measurement equipment, there are a couple of measures to improve the measurement repeatability. Usually, the first available measure is to set the integration time of the meter to a longer time. At the cost of additional measurement time (Fig. 15.24), it generally proves possible to reduce the measurement noise contribution to an observable substantially.

A second measure to improve mismatch measurements even further is by using multiple observations for each current or voltage reading. Obviously this is merely an extension of the measure above. When the STR is limited by white measurement system noise, the STR will generally improve roughly with the square-root of the number of observations used.

Finally, for extremely accurate measurements or measurements that are done really at the limits of the measurement system's sensitivity, one can rely on the help of statistics to separate measurement noise and mismatch. This is for instance done in the so-called DUT-1-2-1-2 approach [41]. It proves possible to obtain reliable statistical mismatch estimators even when the measurement noise is much higher than the 'one third' rule of thumb given above.

15.5.3 Statistics for Mismatch Characterizations

Since mismatch characterization deals with random as well as systematic effects, statistics plays an extremely important role for proper modeling of mismatch. Obviously, whole textbooks are filled about the wonderful art of statistical data analysis. In the context of this chapter, here some of the most important issues concerning mismatch statistics are recapitulated.

- For calculating a standard deviation from a set of observations (a population) that is expected to be normally distributed (Gaussian), always verify the shape of the population. Making a normal scaled cumulative probability plot (Fig. 15.25 with the horizontal scale based on an "inverse-normal" scaling of the quantile is a better way of verifying this. A cumulative probability plot is constructed by plotting on the vertical axis the sorted n measured data points and on the

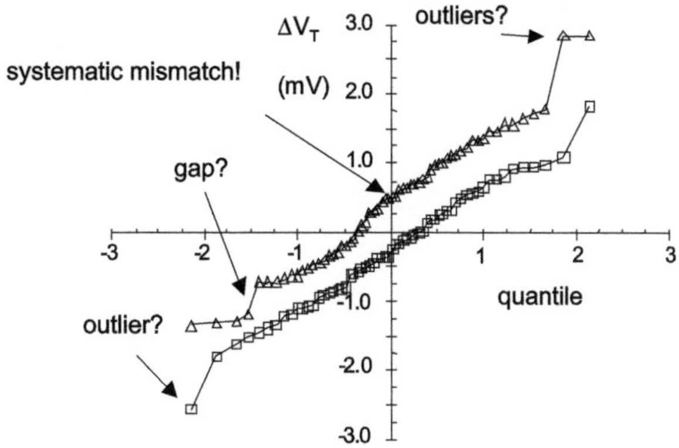

Fig. 15.25 Two examples of normal scaled cumulative probability plots of VT mismatch measurements of two populations with approximately equal standard deviations but significantly differently medians. The slope of the curve represents the standard deviation

horizontal axis n data points from an ideal normal distribution. The intersection of the measured data points with the ideal zero gives the median estimating the μ, while the slope represents the standard deviation estimating the σ. The linearity of the plot is a measure for the fit of the data to the Gaussian distribution. This statistical method avoids overestimation due to outliers of conventional standard deviation procedures. A comprehensive treatise of the general aspects of random distributions and specifics of their estimation are found in [35, 42].

- Outliers can affect the average and standard deviation estimators quite severely. They can be caused by defects, measurement system hick-ups, mobile phone or WLAN transmitter signals, and many more disturbances one is not really interested in when characterizing statistical device mismatch. Medians (middle values) are less sensitive to outliers. Several approaches can be used to filter-out outliers such as 3-sigma filtering, blunt axing or Grubbs' test for outliers.
- A classical statistics mistake that is still encountered quite often in the literature on matching and variability is combining data from un-equal populations. Calculated variances will increase significantly when medians of the populations that are combined (e.g. from different test chips, different wafers, different lots) are not identical. Randomness of the mismatch versus wafer position should also be verified.
- For properly designed and laid out (symmetrical) test structures, the median of a population of mismatch observations should be negligible. A first order rule of thumb is when the median is smaller than 1-sigma, the systematic mismatch is generally not significant.
- Determination of significance or statistical uncertainty forms a crucial, albeit often forgotten, element of proper statistical analysis. Basically one should always realize that a measured population can only be a relatively small sample of the infinite population one is trying to describe. A rule of thumb is that the statistical uncertainty of a standard deviation of a normally distributed population theoretically is $\sigma/\sqrt{2N}$ in which N is the population size. This implies that in order to justify a three-digit precision, a population of at least a million matched pairs is required. An elegant way to take into account the real uncertainties based on the

actually measured population (including suspected outliers and gaps) is using of the so-called "bootstrap technique".

15.6 Consequences for Design

A designer has several options to deal with global variations in his circuit. Next to rigorous simulation, the circuit itself can be designed in a way that effects of global variations are minimized. These techniques include differential design and replica-biasing. Unfortunately these methods are ineffective for battling local variations. Here the designer has to rely on statistical simulations.

The threshold and current factor based model that was defined in the previous sections describes the statistical properties of random mismatch in devices. In the design phase of a circuit these abstract notions of statistical variations on parameters must be translated into design decisions. Application to electronic design includes techniques as Monte-Carlo simulation, hand calculation and various other techniques.

15.6.1 Analog Design

Figure 15.26 shows an example of a bandgap circuit that is used to generate a reference voltage for analog-to-digital conversion. This voltage is generated by amplifying the voltage difference between two diodes with largely deviating current densities and adding it to the absolute value of a forward biased diode. In a correctly designed circuit the amplification has a value where the negative temperature coefficient of the diode voltage is compensated by the positive temperature coefficient of the voltage difference. The caveat in the circuit is in the random offset of the operational amplifier input. This offset is also amplified and will create significant differences in the output voltage of the circuit. In Fig. 15.26 the measured output voltages of a few hundred devices is compared to the Monte-Carlo simulation of a similar number of samples. The MOS model in the simulation is equipped with a mismatch model as in (15.16).

A second example is shown in Fig. 15.27. The differential non-linearity curve (DNL) is a measure for the error that an analog-to-digital converter makes at every code transition. If the DNL reaches the ± 1 limits non-monotonicity in the analog-to-digital transfer curve can occur. In high speed converters MOS threshold mismatch of the input pair of the comparators is the dominant contributor to differential non-linearity. It is imperative that this error must be made small. The measurement of the first prototype(left) shows significant deviations with excursions down to -1. After analyzing the design with a Monte-Carlo simulation, the major contributors were located and correctly dimensioned, resulting in the measured curve on the right. Today the Monte-Carlo analysis is a standard tool for high-performance circuits in analog design.

Fig. 15.26 The probability distribution of the output voltage of a bandgap circuit in CMOS. The measured circuit is compared to the Monte-Carlo simulation. Without any tuning the mean and standard deviation are well in line

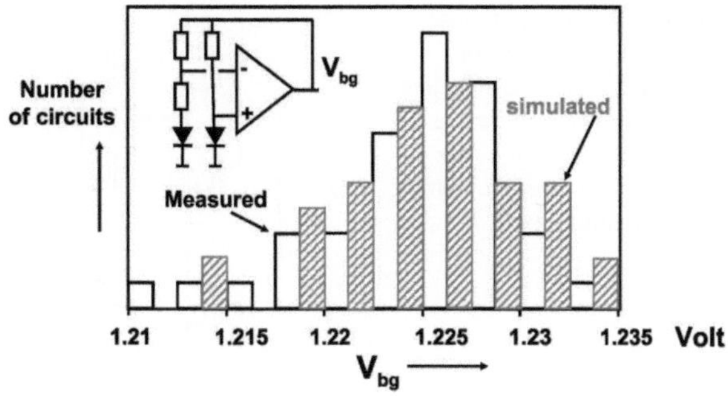

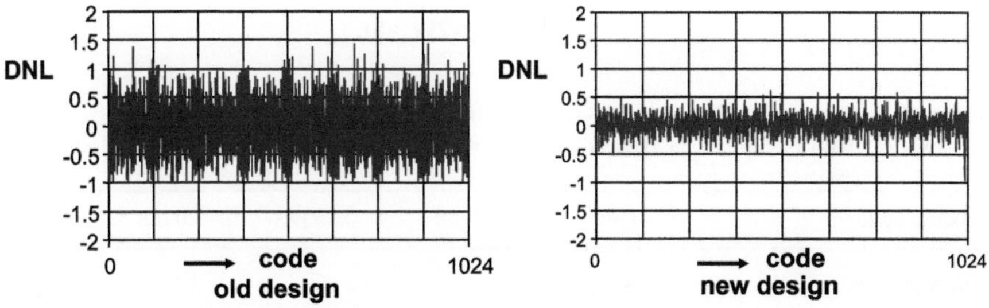

Fig. 15.27 The differential non-linearity is a quality parameter of this 10-bit analog-to-digital converter

Fig. 15.28 The yield on monotonicity of 7 to 10-bit analog-to-digital converters versus the standard deviation at the input of the comparators in a full-flash architecture

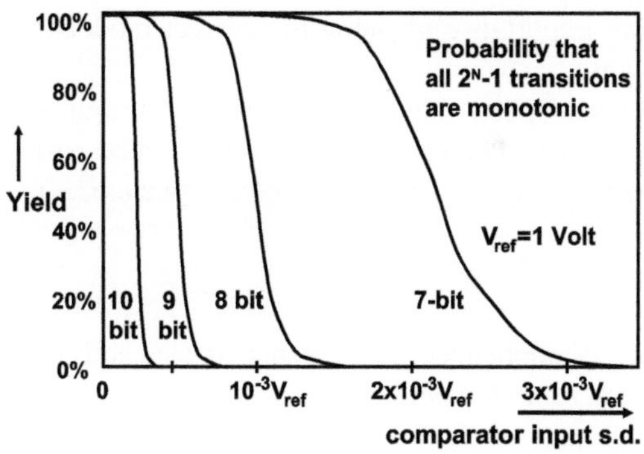

Higher resolution analog-to-digital converters pose more severe demands on the DNL and consequently on the matching of the devices. Figure 15.28 shows the standard deviation at the input of a comparator versus the obtainable yield in "full-flash" analog-to-digital conversion [43]. These converters use comparators to quantize the value of the analog signal. Random parameter fluctuations of the transistors inside a comparator can be combined into an input-referred mismatch source. This mismatch source can cause erroneous switching of adjacent comparators, resulting in yield loss. The shape of the curve is characteristic for yield plots. The transition from excellent yield numbers to commercially uninteresting levels is very sharp. It

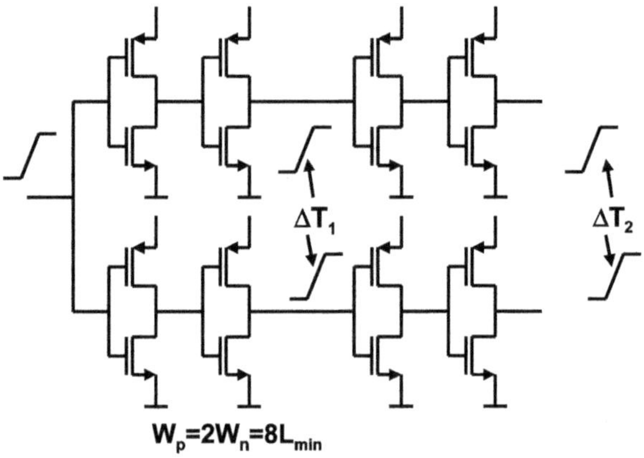

Fig. 15.29 An input pulse is applied to two chains of inverters. Due to mismatches in the transistors there is a time difference at the two outputs. This example mimics a critical timing path

Table 15.7 The simulated standard deviation of the difference in arrival time of the two pulses in the inverter chain of Fig. 15.29. The bottom row shows the standard deviation for scaled feature size

Process node	0.25 μm	0.18 μm	0.13 μm	90 nm	65 nm
Clock period	10 ns	5 ns	2 ns	1 ns	0.5 ns
$\sigma_{\Delta T2}$ $C_{load} = 50$ fF	16 ps	21 ps	38 ps	68 ps	88 ps
$\sigma_{\Delta T2}$ $C_{load} = 50..15$ fF	16 ps	16 ps	22 ps	33 ps	32 ps

shows importance of using the correct models and parameter values in precision designs.

Power, speed and accuracy span the design space e.g. [44]. The idea that accuracy must be balanced against power can be easily understood by considering that the voltage uncertainty on the gate capacitance can be described as an energy term [45, 46]:

$$E_{\sigma VT} = C_{gate}\sigma^2_{\Delta VT} = C_{ox}A^2_{VT} = 4.5 \times 10^{-19} \text{ Joule} \tag{15.23}$$

which is independent of the transistor size and corresponds to about $E_{\sigma VT} = 100$ kT Joule at room temperature. This energy can be seen as the energy required to toggle a latch pair of transistors in meta-stable condition into a desired position with a one-σ certainty.

15.6.2 Digital Design

Also digital designers experience that for small devices the random component can exceed the process corner variation. An example is shown in Fig. 15.29 and Table 15.7 [47]. A pulse is applied to two sets of inverters and it is expected that the outputs will change state simultaneously. Due to mismatch between the transistors of both rows of inverters, a random skew will exist in the difference of arrival time between various samples of this circuit. In Table 15.7 the standard deviation of this random skew is compared to the clock period in five technologies. From an effect in the 0.1% range the random skew can take up to 10% of the clock period in a 65 nm process.

In digital timing chains, an increasing chain length will linearly increase the deterministic skew, while the random component will go up via a square-root relation. The relative impact of random variations reduces. It should also be noted that the "root-mean-square" addition mechanism favors large contributors. The random timing skew in a long chain can be dominated by just one small-sized inverter.

Also in memory structures statistical effects play a major role. On the level of a SRAM cell in Fig. 15.1 these effects can be evaluated as in any other analog circuit. The additional problem in memory structures is the large amount of cells. This requires simulating statistical distributions up to 7σ. This is not practical in Monte-Carlo simulations. Special statistical acceleration mechanisms (importance sampling, Latin Hypercube sampling) allow to sample tails of statistical distributions [48]. Memory designs are affected by mismatch in several ways. Threshold variations influence the margins for the read and write operations. Moreover low-threshold devices create (exponentially) large leakage currents. The choice for the size of the transistors in an SRAM cell and in the sense amplifier critically depends on an accurate prediction of the mismatch coefficients.

15.7 Conclusion

Equally designed components can show differences due to a large number of factors. Despite the manifold of possible causes there is some classification of these variability effects possible: in time, in dimensions and in statistical nature. Important deterministic effects causing "offset" in a circuit have been briefly discussed.

Within the scope of variability, statistical matching of components is one of the most critical subjects. With the help of statistical theory and device physics, a general model for the description of a class of random phenomena is presented. This model is applied to the MOS threshold voltage and current factor. The comparison to measurements in various CMOS technologies indicates its applicability from 2 µm down to sub-micron technologies with a reasonable accuracy. The accurate description of offset and random mismatch paves the path for high-performance mixed-signal design.

Acknowledgments The authors are grateful for being able to use the insights and results of their colleagues at NXP Research. Without the useful discussions and critical comments this chapter would not exist.

Appendix: Derivation of Spatial Behavior

The Fourier transform is used to analyze the behavior of spatially distributed functions. The spatial Fourier transform and its reverse form in two dimensions are defined as:

$$H(\omega_x, \omega_y) = \int_{x=-\infty}^{\infty} \int_{y=-\infty}^{\infty} h(x, y) e^{-j\omega_x x} e^{-j\omega_y y} dy dx$$
$$= \int_{x=-\infty}^{\infty} \int_{y=-\infty}^{\infty} h(x, y) e^{-j2\pi f_x x} e^{-j2\pi f_y y} dy dx,$$
$$h(x, y) = \frac{1}{4\pi^2} \int_{\omega_x=-\infty}^{\infty} \int_{\omega_y=-\infty}^{\infty} H(\omega_x, \omega_y) e^{j\omega_x x} e^{j\omega_y y} d\omega_y d\omega_x$$
$$= \int_{f_x=-\infty}^{\infty} \int_{f_y=-\infty}^{\infty} H(f_x, f_y) e^{j2\pi f_x x} e^{j2\pi f_y y} df_y df_x.$$

Inserting the geometry function as defined in Fig. 15.15:

$$H(\omega_x, \omega_y) = \frac{1}{WL} \int_{y=-W/2}^{W/2} \left(\int_{x=-L/2+D/2}^{L/2+D/2} e^{-j\omega_x x} e^{-j\omega_y y} dx \right.$$
$$\left. - \int_{x=-L/2-D/2}^{L/2-D/2} e^{-j\omega_x x} e^{-j\omega_y y} dx \right) dy.$$

As the x and y components are independent, the y component can be solved separately:

$$H(\omega_x, \omega_y)$$
$$= \frac{e^{-j\omega_y y}}{-j\omega_y WL} \Bigg|_{y=W/2}^{y=-W/2} \left(\int_{x=-L/2+D/2}^{L/2+D/2} e^{-j\omega_x x} dx - \int_{x=-L/2-D/2}^{L/2-D/2} e^{-j\omega_x x} dx \right)$$
$$= \frac{2 \sin \omega_y W/2}{\omega_y WL} \left(\frac{e^{-j\omega_x x}}{-\omega_x} \Bigg|_{x=-L/2+D/2}^{x=+L/2+D/2} - \frac{e^{-j\omega_x x}}{-j\omega_x} \Bigg|_{x=-L/2-D/2}^{x=+L/2-D/2} \right)$$
$$= \left(\frac{2 \sin \omega_y W/2}{-\omega_y \omega_x WL} \right) \Big\{ [\cos(\omega_x (L/2 + D/2)) - j \sin(\omega_x (L/2 + D/2))]$$
$$- [\cos(\omega_x (-L/2 + D/2)) - j \sin(\omega_x (-L/2 + D/2))]$$
$$- [\cos(\omega_x (L/2 - D/2)) - j \sin(\omega_x (L/2 - D/2))]$$
$$+ [\cos(\omega_x (-L/2 - D/2)) - j \sin(\omega_x (-L/2 - D/2))] \Big\}$$
$$= \left(\frac{2 \sin \omega_y W/2}{-\omega_y \omega_x WL} \right) [\cos(\omega_x (L/2 + D/2)) - \cos(\omega_x (-L/2 + D/2))$$
$$- \cos(\omega_x (L/2 - D/2)) + \cos(\omega_x (-L/2 - D/2))]$$
$$= \left(\frac{4 \sin \omega_y W/2}{-\omega_y \omega_x WL} \right) \big[\cos(\omega_x (L/2 + D/2)) - \cos(\omega_x (-L/2 + D/2)) \big]$$
$$= \left(\frac{4 \sin \omega_y W/2}{-\omega_y \omega_x WL} \right) \{ [\cos(\omega_x L/2) \cos(\omega_x D/2) - \sin(\omega_x L/2) \sin(\omega_x D/2)]$$

$$- [\cos(\omega_x(-L/2))\cos(\omega_x D/2) - \sin(\omega_x(-L/2))\sin(\omega_x D/2)]\}$$

$$= \frac{8\sin(\omega_y W/2)\sin(\omega_x(L/2))\sin(\omega_x(D/2))}{\omega_y \omega_x W L}.$$

Using the trigonometric identity: $\cos(a+b) = \cos(a)\cos(b) - \sin(a)\sin(b)$. After some re-arrangement:

$$\mathcal{G}(\omega_x, \omega_y) = \frac{\sin(\omega_x L/2)}{\omega_x L/2}\frac{\sin(\omega_y W/2)}{\omega_y W/2}[2\sin(\omega_x D_x/2)],$$
$$\sigma^2_{\Delta P} = \frac{1}{4\pi^2}\int_{\omega_y=-\infty}^{\omega_y=\infty}\int_{\omega_x=-\infty}^{\omega_x=\infty} \left|\mathcal{G}(\omega_x,\omega_y)\right|^2 \left|\mathcal{P}(\omega_x,\omega_y)\right|^2 d\omega_x d\omega_y. \quad (15.24)$$

The variance of parameter ΔP between two rectangular devices is then found by substitution of (15.8) and the above-described models for the long and short correlation distance variations in (15.24). Mathematically the white noise model is described in the Fourier domain as a constant: $\mathcal{P}(\omega_x, \omega_y) = \mathcal{N}$

$$\sigma^2_{\Delta P} = \frac{1}{4\pi^2}\int_{\omega_y=-\infty}^{\omega_y=\infty}\int_{\omega_x=-\infty}^{\omega_x=\infty} \mathcal{N}^2 \times \left[\frac{\sin(\omega_x L/2)}{\omega_x L/2}\frac{\sin(\omega_y W/2)}{\omega_y W/2} 2\sin(\omega_x D_x/2)\right]^2 d\omega_x d\omega_y.$$

Considering that the ω_x and ω_y dimensions can be separated:

$$\sigma^2_{\Delta P} = \frac{\mathcal{N}^2}{4\pi^2}\int_{\omega_x=-\infty}^{\omega_x=\infty}\left[\frac{\sin(\omega_x L/2)}{\omega_x L/2} 2\sin(\omega_x D_x/2)\right]^2 d\omega_x \times \int_{\omega_y=-\infty}^{\omega_y=\infty}\left[\frac{\sin(\omega_y W/2)}{\omega_y W/2}\right]^2 d\omega_y.$$

Using the standard integrals (CRC handbook 1984, p. 289 form 628 and 630):

$$\int_{x=0}^{\infty}\frac{\sin(ax)\sin(bx)}{x^2}dx = \frac{a\pi}{2} \quad (a \le b), \quad (15.25)$$

$$\int_{x=0}^{\infty}\left[\frac{\sin(ax)}{x}\right]^2 dx = \frac{a\pi}{2} \quad (15.26)$$

the second integral is easy to solve, the first integral can be re-written into a series of squared sine waves, resulting in $a\pi/2$ if $(a = L/2)$ represents the smaller of the two sine coefficients.

$$\sigma^2_{\Delta P} = \frac{\mathcal{N}^2}{4\pi^2} \times \frac{\pi/2 \times L/2}{(L/2)^2} 2^2 \times \frac{\pi W/2}{(W/2)^2} = \frac{2\mathcal{N}^2}{WL}.$$

This is the first part of the equation. For solving the second part the gradient on the wafer must be modeled. A polar description would describe accurately the circular gradient. As this example only sensitivity parallel to the x axis is considered $\omega_x = 1/2W_D$ where W_D is the wafer diameter. Now $\mathcal{P}(\omega_x) = \delta(1/2W_D)$

$$\sigma_{\Delta P}^2 = \frac{1}{4\pi^2} \int_{\omega_x=-\infty}^{\omega_x=\infty} \delta(1/2W_D) \left[2\sin(\omega_x D_x/2) \frac{\sin(\omega_x L/2)}{\omega_x L/2} \right]^2 d\omega_x.$$

It will be clear that the delta function appears at a very low spatial frequency. For that low frequency, the $(\sin x/x)$ term can be considered to approach "1". leaving:

$$\sigma_{\Delta P}^2 = \frac{1}{4\pi^2} \int_{\omega_x=-\infty}^{\omega_x=\infty} \delta(1/2W_D) \left[2\sin(\omega_x D_x/2) \right]^2 d\omega_x \approx \left(\frac{D_x}{2W_D} \right)^2.$$

Combining both results:

$$\sigma_{\Delta P}^2 = \frac{A_P^2}{WL} + S_P^2 D_x^2.$$

References

1. Sze, S.M.: Physics of Semiconductor Devices, 2nd edn. Wiley, New York (1981). (3rd edn. 2006, ISBN: 978-0-471-14323-9)
2. Brews, J.R.: MOSFET hand analysis using BSIM. IEEE Circuits Devices Mag. 28–36 (2006)
3. Enz, C.C., Krummenacher, F., Vittoz, E.A.: An analytical MOS transistor model valid in all regions of operations and dedicated to low-voltage and low-current applications. Analog Integr. Circuits Signal Process. J. 83–114 (1995)
4. Gildenblat, G., Xin, Li, Wu, W., Hailing, Wang, Jha, A., van Langevelde, R., Smit, G.D.J., Scholten, A.J., Klaassen, D.B.M.: PSP: An advanced surface-potential-based MOSFET model for circuit simulation. IEEE Trans. Electron Devices 1979–1993 (2006)
5. Gregor, R.W.: On the relationship between topography and transistor matching in an analog CMOS technology. IEEE Trans. Electron Devices 275–282 (1992)
6. Stathis, J.H., Zafar, S.: The negative bias temperature instability in MOS devices: A review. Microelectron. Reliab. 270–286 (2006)
7. Hook, T.B., Brown, J., Cottrell, P., Adler, E., Hoyniak, D., Johnson, J., Mann, R.: Lateral ion implant straggle and mask proximity effect. IEEE Trans. Electron Devices 1946–1951 (2003)
8. Drennan, P.G., Kniffin, M.L., Locascio, D.R.: Implications of proximity effects for analog design. In: IEEE Custom Integrated Circuits Conference 2006, pp. 169–176 (2006)
9. Bianchi, R.A., Bouche, G., Roux-dit-Buisson, O.: Accurate modeling of trench isolation induced mechanical stress effects on MOSFET electrical performance. In: Digest. International Electron Devices Meeting, pp. 117–120 (2002)
10. Su, K.-W., et al.: A scaleable model for STI mechanical stress effect on layout dependence of MOS electrical characteristics. In: Proceedings of the IEEE Custom Integrated Circuits Conference, pp. 245–248 (2003)
11. Wils, N., Tuinhout, H.P., Meijer, M.: Characterization of STI edge effects on CMOS variability. IEEE Trans. Semicond. Manuf. 59–65 (2009)
12. Ge, L., Adams, V., Loiko, K., Tekleab, D., Bo, X.-Z., Foisy, M., Kolagunta, V., Veeraraghavan, S.: Modeling and simulation of poly-space effects in uniaxially-strained etch stop layer stressors. In: IEEE International SOI Conference, pp. 25–26 (2007)
13. Tuinhout, H.P., Pelgrom, M.J.M., Penning de Vries, R., Vertregt, M.: Effects of metal coverage on MOSFET matching. In: Technical Digest International Electron Devices Meting, pp. 735–739 (1996)

14. Tuinhout, H.P., Bretveld, A., Peters, W.C.M.: Measuring the span of stress asymmetries on high-precision matched devices. In: International Conference on Microelectronic Test Structures, pp. 117–122 (2004)
15. Pelgrom, M.J.M., Vertregt, M., Tuinhout, H.P.: Matching of MOS Transistors. MEAD Course Material (1998–2009)
16. Mao-Feng et al.: Current mirror layout strategies for enhancing matching performance. Analog Integr. Circuits Signal Process. **28**, 9–26 (2001)
17. McCreary, J.L.: Matching properties, and voltage and temperature dependence of MOS capacitors. IEEE J. Solid-State Circuits 608–616 (1981)
18. Shyu, J.-B., Temes, G.C., Yao, K.: Random errors in MOS capacitors. IEEE J. Solid-State Circuits 1070–1076 (1982)
19. Tuinhout, H.P., Elzinga, H., Brugman, J.T., Postma, F.: Accurate capacitor matching measurements using floating-gate test structures. In: IEEE International Conference on Microelectronic Test Structures, pp. 133–137 (1994)
20. Aparicio, R., Hajimiri, A.: Capacity limits and matching properties of integrated capacitors. IEEE J. Solid-State Circuits 384–393 (2002)
21. Drennan, P.G.: Diffused resistor mismatch modeling and characterization. In: Bipolar/BiCMOS Circuits and Technology Meeting, pp. 27–30 (1999)
22. Tuinhout, H.P., Hoogzaad, G., Vertregt, M., Roovers, R.L.J., Erdmann, C.: Design and characterisation of a high precision resistor ladder test structure. In: IEEE International Conference on Microelectronic Test Structures, pp. 223–228 (2002)
23. Lakshmikumar, K.R., Hadaway, R.A., Copeland, M.A.: Characterization and modeling of mismatch in MOS transistors for precision analog design. IEEE J. Solid-State Circuits 1057–1066 (1986)
24. Michael, C., Ismail, M.: Statistical modeling of device mismatch for analog MOS integrated circuits. IEEE J. Solid-State Circuits 154–166 (1992)
25. Forti, F., Wright, M.E.: Measurement of MOS current mismatch in the weak inversion region. IEEE J. Solid-State Circuits 138–142 (1994)
26. Croon, J.A., Sansen, W., Maes, H.E.: Matching Properties of Deep Sub-Micron MOS Transistors. Springer, Dordrecht (2005). ISBN 0-387-24314-3
27. Tuinhout, H.P.: Improving BiCMOS technologies using BJT parametric mismatch characterisation. In: Bipolar/BiCMOS Circuits and Technology Meeting, pp. 163–170 (2003)
28. Pelgrom, M.J.M., Duinmaijer, A.C.J., Welbers, A.P.G.: Matching properties of MOS transistors. IEEE J. Solid-State Circuits 1433–1440 (1989)
29. Brown, A.R., Roy, G., Asenov, A.: Poly-Si-gate-related variability in decananometer MOSFETs with conventional architecture. IEEE Trans. Electron Devices 3056–3063 (2007)
30. Mizuno, T., Okamura, J., Toriumi, A.: Experimental study of threshold voltage fluctuation due to statistical variation of channel dopant number in MOSFETs. IEEE Trans. Electron Devices 2216–2221 (1994)
31. Pelgrom, M.J.M., Vertregt, M.: CMOS technology for mixed signal ICs. Solid-State Electron. 967–974 (1997)
32. Bastos, J., Steyaert, M., Roovers, R., Kinget, P., Sansen, W., Graindourze, B., Pergoot, A., Janssens, E.: Mismatch characterization of small size MOS transistors. In: Proc. IEEE Int. Conf. on Microelectronic Test Structures, pp. 271–276 (1995)
33. Stolk, P.A., Widdershoven, F.P., Klaassen, D.B.M.: Modeling statistical dopant fluctuations in MOS transistors. IEEE Trans. Electron Devices 1960–1971 (1998)
34. Andricciola, P., Tuinhout, H.P.: The temperature dependence of mismatch in deep-submicrometer bulk MOSFETs. IEEE Electron Device Lett. 690–692 (2009)
35. Papoulis, A.: Probability, Random Variables, and Stochastic Processes, student edn. McGraw Hill, New York (1965). McGraw-Hill, 4th edn. (2001), ISBN 0-07-366011-6
36. Croon, J.A., Tuinhout, H.P., Difrenza, R., Knol, J., Moonen, A.J., Decoutere, S., Maes, H.E., Sansen, W.: A comparison of extraction techniques for threshold voltage mismatch. In: Proc. IEEE Int. Conf. on Microelectronic Test Structures, pp. 235–240 (2002)
37. Takeuchi, K., Hane, M.: Statistical compact model parameter extraction by direct fitting to variations. IEEE Trans. Electron Devices 1487–1493 (2008)

38. Cheng, B., Roy, S., Asenov, A.: Statistical compact model parameter extraction strategy for Intrinsic parameter fluctuation. In: Grasser, T., Selberherr, S. (eds.) Simulation on Semiconductor Processes and Devices, pp. 301–304. Springer, New York (2007)
39. Tuinhout, H.P.: Electrical characterisation of matched pairs for evaluation of integrated circuit technologies. Ph.D. Thesis, Delft University of Technology (2005). http://repository.tudelft.nl/file/82893/025295
40. Tuinhout, H.P., Vertregt, M.: Characterization of systematic MOSFET current factor mismatch caused by metal CMP dummy structures. IEEE Trans. Semicond. Manuf. 302–310 (2001)
41. Tuinhout, H.P., van Rossem, F., Wils, N.: High-precision on-wafer backend capacitor mismatch measurements using a benchtop semiconductor characterization system. In: IEEE International Conference on Microelectronic Test Structures, pp. 3–8 (2009)
42. Rey, W.J.J.: Introduction to Robust and Quasi-robust Statistical Methods. Springer, Berlin (1983). ISBN 0-387-12866-2
43. Pelgrom, M.J.M., v. Rens, A.C., Vertregt, M., Dijkstra, M.B.: A 25-Ms/s 8-bit CMOS A/D converter for embedded application. IEEE J. Solid-State Circuits 879–886 (1994)
44. Vertregt, M., Scholtens, P.C.S.: Assessment of the merits of CMOS technology scaling for analog circuit design. In: Proceedings ESSCIRC, pp. 57–64 (2004)
45. Pelgrom, M.J.M.: Low-power high-speed A/D conversion. In: ESSCIRC94, Low-power Workshop (1994)
46. Kinget, P., Steyaert, M.: Impact of transistor mismatch on the speed accuracy power trade-off. Custom Integrated Circuits Conference (1996)
47. Pelgrom, M.J.M., Tuinhout, H.P., Vertregt, M.: Transistor matching in analog CMOS applications. Invited Paper on International Electron Devices Meeting, pp. 915–918 (1998)
48. Doorn, T.S., ter Maten, E.J.W., Croon, J.A., Di Bucchianico, A., Wittich, O.: Importance sampling Monte Carlo simulations for accurate estimation of SRAM yield. In: European Solid-State Circuits Conference, pp. 230–233 (2008)

Chapter 16
Statistical Modeling Using Backward Propagation of Variance (BPV)

Colin C. McAndrew

Abstract This chapter reviews statistical modeling for circuit simulation, with an emphasis on the backward propagation of variance (BPV) technique. Sources of variability are reviewed, and a formulation based on uncorrelated process parameters is presented. A general procedure for modeling variances of electrical performances is detailed, including handling of nonlinearities and correlations, and numerical properties of the procedure are discussed. Approaches to generating corner models are reviewed, and the importance of properly modeling correlations is demonstrated.

16.1 Introduction

Semiconductor processes are inherently stochastic; there are variations in the electrical performances of the devices (resistors, capacitors, transistors, etc.) that are manufactured in a particular process, and therefore in the performances of integrated circuits (ICs) that are constructed from these devices. This variability in manufacturing can lead to reduced performance, or even failure to function, of products, which reduces the revenue generated from essentially fixed wafer-cost manufacturing facilities. The economic viability of the semiconductor industry is therefore intimately tied to being able to manufacture properly functioning, and profitable, parts in processes that unavoidably exhibit statistical fluctuations. This means that the design of products needs to take this variability into account [15, 25], and as IC design is based extensively on models and simulation, the models used for IC design need to include statistical variability. This chapter presents general principles and approaches used for variability modeling of semiconductor devices.

There are many sources of variability in the manufacture of semiconductor devices; a taxonomy of these is presented. Variability is manifest at several levels of

C.C. McAndrew (✉)
Freescale Semiconductor, Tempe, AZ 85284, USA
e-mail: Colin.McAndrew@freescale.com

G. Gildenblat (ed.), *Compact Modeling*,
DOI 10.1007/978-90-481-8614-3_16, © Springer Science+Business Media B.V. 2010

abstraction; this is reviewed, and we show that modeling variability is best done at the level of "process parameters," which are the essentially uncorrelated physical parameters that account for the vast majority of the observed fluctuations in the electrical behavior of devices and circuits. Examples of process parameters are lateral geometry variations, vertical geometry variations (including in oxide thickness), and material properties such as sheet resistance or doping density, flatband voltage, and recombination and generation lifetimes.

Different approaches to statistical modeling are detailed, including principle component analysis (PCA), forward propagation of variation (FPV), and backward propagation of variation (BPV). The need for different statistical models for different types of statistical analyses (corner simulation, distributional simulation, mismatch simulation) is reviewed. We show that the BPV paradigm can be used for self-consistent global and local statistical model characterization, and can account for nonlinearities and parameter correlations while being based on independent, normally distributed parameters.

One of the difficult aspects of variability modeling is determining reasonable values for parameter distributions; simply measuring a large volume of wafers and computing means and standard deviations is not sufficient. We discuss common problems with statistical data, and show that defining specifications for variation, derived from engineering judgment and experience as well as measured data, is preferable to blindly basing statistical models on measured data.

The effect of local statistical variation on non-statistical modeling, especially for modeling trends such as threshold voltage as a function of geometry, is also described.

16.2 Sources of Statistical Variability

There are many sources of statistical variation in IC manufacturing processes. The results of performing an ion implantation, anneal, oxidation growth, or indeed any of the other myriad IC manufacturing steps can vary between different but supposedly identical manufacturing tools, and even between ostensibly identical steps run in the same tool at different times. This leads to differences between wafers from different fabs, and between wafers from different lots run in the same fab.

Steps like anneals and oxidations are done in furnaces, where temperature distributions and reactant concentrations can be nonuniform, which leads to differences between individual wafers within a lot, and between different die within a wafer; there can even be variations across a single die.

At a microscopic level several further sources of variability arise. These days it is difficult to print fine features accurately, so techniques like optical proximity correction (OPC) and phase-shift masking (PSM) are used, to try to make the final on-wafer patterns appear as close to the physical design intent as possible. However, the processes are not perfect and the outcomes depend on adjacent patterns, which leads to differences in the final shapes, and hence electrical behaviors, of individual devices within a die that are supposed to be identical. The rates of chemical reactions such as etches depends on the local concentration of the reactant(s) involved, which varies with the local density of the material it is reacting with, a phenomenon called "loading" (higher local density depletes the reactant concentration faster). The densities of the patterns and materials involved in such steps vary over a die, hence so does the outcome of the step. Imperfections in the lenses in optical projection systems and microscopic variability in the masks used in patterning also lead to differences between ostensibly identical devices within a die. Note that all of these within-die variations are in some sense repeatable: they will be the same for one device within a reticle (at least if processed through the same tool; lens imperfections will differ between different projection systems). However, during design the exact position and local environment of a device is often not known, so the net result of the described effects is stochastic.

There is one final source of within-die (i.e. between device) variability that is not in any sense repeatable. At a microscopic level there are uncontrollable atomistic variations between devices. The edges of vertical structures such as the gates of MOSFETs cannot be perfectly smooth, so there is line-edge-roughness (LER) that causes variability in the final shape, and therefore in the effective electrical dimensions (which control electrical behavior) of devices. Ion implantation involves discrete particles, hence there is variability in number and location of the implanted ions; this is called random dopant fluctuation. Atomistic variations can also affect the oxide thickness in MOSFETs. The results of these atomistic, within die, fluctuations are called local variation or mismatch, as opposed to the global (between die) variation from the mechanisms listed above. A key characteristic of local variations is that they average over the area of a device (for random dopant fluctuations and oxide thickness) or the length or width of a device (for LER in the width and length dimensions, respectively), and they therefore have a reciprocal dependence on geometry [1, 10, 26]. Although local variation used to be primarily of concern for high precision analog ICs, where the design techniques used critically depend on matching between components, since about the 130 nm node it has become of significant importance for digital IC design as well; indeed, mismatch in interconnect (not in the transistors) was a significant barrier to pushing high-speed microprocessors over the 1 GHz clock speed barrier [23].

There are also variations from die edge and stress related effects, from packaging, from aging and burn-in, and from other sources; here we consider only the global and local variations outlined above, and the total variation in performance of a device has contributions from both.

Figure 16.1 shows measured zero-bias threshold voltage V_{t0} from adjacent, identical MOSFETs, for small area and large area devices, from 4 separate wafer lots (represented by the different symbols). Several important characteristics are observable. First, the amount of variability is larger for the small devices than for the large devices. This is expected because not only are the small devices more sensitive to global variations in effective width W and length L, but the local variation is, as discussed above, larger for smaller devices. Second, the degree of correlation between two adjacent devices is clearly much higher for the large devices than for

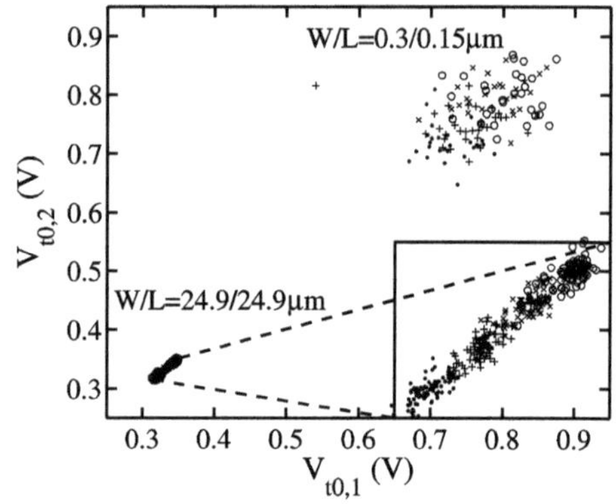

Fig. 16.1 Global and local variability. Different symbols are data from different wafer lots. *Inset* is an expanded view of data from the wide/long devices

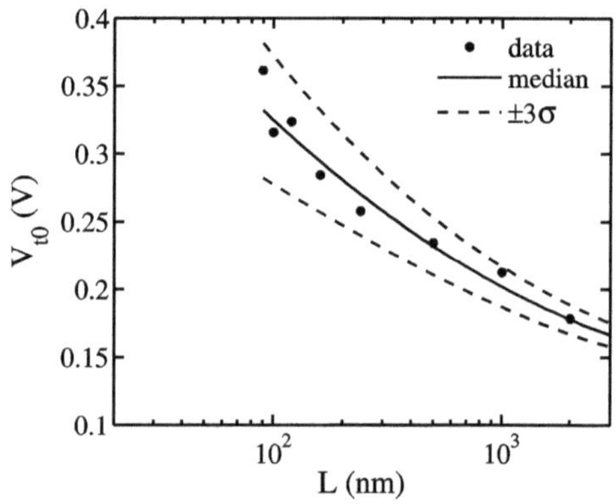

Fig. 16.2 Threshold voltage vs. channel length

the small devices, as evidenced by the data transforming from essentially a line of points to more of a cloud of points. Third, and related, a blow-up of the data from the large devices is shown in the inset at the bottom right; note that the distinction between the different wafer lots is much clearer for the big area than for the small area device. These last two points verify that the overall variation in the devices has transitioned from being primarily from (correlated) global variation for the large devices to being primarily from (uncorrelated) local variation for the small devices.

Before discussing statistical modeling, it is important to note that statistical models need to be built on top of reasonable SPICE models, which (as we will see) need to be sufficiently accurate to model sensitivities properly. There are two aspects in which statistical variability can interfere with the generation of "reasonable" SPICE models, and to remove this interference requires modification of some historic practices in SPICE model parameter extraction.

Figure 16.2 shows threshold voltage as a function of channel length. A standard step in MOSFET parameter extraction is to fit such a curve, under the expectation that halo doping and charge sharing effects, which determine the $V_{t0}(L)$ behavior, should vary smoothly with geometry. Because of local variation, selecting a set of devices of different lengths from one "golden" site on a wafer does not necessarily provide data that embody the expected smooth behavior, and extracting the parameters that control $V_{t0}(L)$ from such "noisy" data would not give appropriate parameter values. Rather, data from multiple sites need to measured, and the median of the $V_{t0}(L)$ characteristics over those sites should be used for the purpose of parameter extraction, rather than the raw data from one site.

A similar situation holds for parameter extraction from MOSFET dc $I(V)$ curves in weak inversion. These characteristics vary approximately exponentially with fluctuations in threshold voltage, which are unavoidable for small devices and lead to log-normal distributions for the drain current. Simply averaging data over multiple devices to arrive at an effective "representative" device does not work, as the mean and median differ for a log-normal distribution. Corrections to measured data can be applied [31], for data from multiple devices connected in parallel (which minimizes test time). However, these corrections are based on up to sixth order derivatives, so it can be preferable to collect data from multiple individual devices and use the median value for parameter extraction.

16.3 Statistical Modeling Basis

Pedantically, there is not really such a thing as a "statistical model." Rather, there are different types of statistical simulation techniques, such as corner, Monte Carlo (MC), or propagation of variance analysis, and each of these requires different types of statistical models. Corner (or "skew" or "worst-case") statistical models are an enumerated set of parameter values that attempt to, when simulated, bracket the expected manufacturing variation of circuit performances. MC simulation models are distributional models that, through statistical sampling, enable the statistical distributions of circuit performances, and hence circuit yield, to be simulated. Propagation of variance enables analytic prediction of the statistics of circuit performances, and is useful for mismatch analysis. Fortuitously, a single formulation for statistical models can be used for all of these types of analysis.

The goal of statistical modeling, for all types of statistical circuit analysis, is to enable accurate simulation of the statistical variations of all important figures of merit ("performances") of ICs. The potential number of different figures of merit for all classes of circuit types is large, and is increasing over time as more types of circuits and additional figure of merit are developed. Therefore, it can be difficult to base general statistical models directly on measures of circuit performance (although the use of circuit performances for statistical modeling can be useful, especially for restricted classes of circuits such as digital CMOS circuits). Consequently, statistical models are more generally developed based on device, rather than circuit, performances, which should be chosen to ensure they are the primary quantities that control circuit performances. We will denote the vector $\mathbf{e} = (e_1, e_2, \ldots, e_{N_e})$ as the N_e electrical performances to be modeled statistically, and although mostly below the individual performances e_i will be measures of device electrical performance it should be understood that the formalism is general and some or all e_i can be measures of circuit electrical performance.

Physically, the manufacturing parameters that control the variations in device (and circuit) performance are the settings on the manufacturing tools and the times, temperatures, and material properties of the manufacturing steps. Examples include the energies and doses of ion implantation steps, the times and temperatures of anneal and oxide growth steps, and the times, temperatures, and chemical concentrations of etch steps. There can be thousands of such manufacturing parameters, and in practice it is not feasible to directly simulate the effect of each such parameter on electrical performance. So the "true" process settings are not a suitable basis for statistical modeling for circuit design.

In simulation, the parameters that control the predicted behavior for a specific device and model are the SPICE model parameters for that device. For the purposes of both statistical modeling and statistical simulation and analysis it is by far simpler to use independent and normally distributed statistical parameters (we will show below that even for correlated parameters and non-normal distributions it is still possible to formulate statistical models in terms of uncorrelated parameters with normal distributions). Unfortunately, not all SPICE model parameters are uncorrelated.

Some MOSFET models, e.g. PSP [8], are indeed formulated based on physical and independent model parameters, such as oxide thickness t_{ox}, flatband voltage V_{fb}, effective substrate doping N_s, low-field mobility μ_0, and channel length variation Δ_L (the difference between the design channel length L_m and the effective electrical channel length L), and these model parameters may be directly used as the basis for statistical modeling. However, for most SPICE models some model parameters are correlated and so cannot be directly used as a basis for statistical modeling.

For example, in threshold voltage based MOSFET models V_{t0} is commonly used as a model parameter. This can be written as a function of independent parameters [30]

$$V_{t0} = V_{fb} + \phi_0 + \gamma\sqrt{\phi_0} \tag{16.1}$$

where $\phi_0 = 2\phi_F + \Delta\phi$, $\phi_F = \phi_t \ln(N_s/n_i)$ is the Fermi potential, $\gamma = \sqrt{2q\epsilon_s N_s}/C'_{ox}$ is the body effect coefficient, $\phi_t = kT/q$ is the thermal voltage, $\Delta\phi$ is several ϕ_t, n_i is the intrinsic carrier concentration, q is the magnitude of the electronic charge, $C'_{ox} = \epsilon_{ox}/t_{ox}$ is the oxide capacitance per unit area, k is Boltzmann's constant, and ϵ_s and ϵ_{ox} are the permittivities of silicon and silicon dioxide, respectively.

More importantly, for bipolar transistor (BJT) models the model parameters, such as the saturation current I_s, forward Early voltage V_{af}, and low bias base transit time τ_b, are highly correlated through physical dependencies on underlying parameters such as the base thickness t_b, emitter size variation Δ_e, and base and collector doping levels N_b and N_c, respectively. Physical analysis [3] shows that, to first order,

$$I_s \propto \frac{(L_e + \Delta_e)(W_e + \Delta_e)}{N_b t_b}, \tag{16.2}$$

where L_e and W_e are the emitter length and width, respectively,

$$V_{af} \propto \frac{N_b t_b}{\sqrt{N_c}}, \tag{16.3}$$

and

$$\tau_b \propto t_b^2. \tag{16.4}$$

Where SPICE model parameters are not independent, but are functions of more physical parameter as just described, the latter can be introduced into SPICE model parameter files as variables, and then the SPICE model parameters can be written as functions of these physical parameters (the capability to define variables and functional mappings, such as those outlined above, is available in all modern circuit simulators). In this way correlated SPICE model parameters can be mapped back to uncorrelated physical parameters, which are then used as the basis for statistical modeling [3].

The N_p independent parameters that control the **e** will be called "process parameters" and will be denoted as the vector $\mathbf{p} = (p_1, p_2, \ldots, p_{N_p})$.

For numerical stability during statistical simulation it is important that the **e** and **p** be scaled; on a typical computer double precision arithmetic has about 16 digits of precision, so variations in t_{ox}, which are of the order of 10^{-10} m, become zero when compared to variations in doping levels, which are roughly of order 10^{22} to 10^{24} m^{-3}. For many parameters and electrical performances it is therefore best to consider their variations to be relative to a nominal value; this has the added benefit of being able to be easily mentally checked for reasonableness (it is much easier to think of 3% as being a reasonable variation than trying to quickly work out whether a sheet resistance variation of 126 $\Omega/\square$ is reasonable compared to a nominal value of 4200 $\Omega/\square$). For such parameters, for the purpose of simulation it is convenient to separate the variation into global and local components, and write the value of the parameter p_i, including both types of statistical variation, as

$$p_i = p_{i,nom}\left(1 + 0.01 N_{S,p_i}\sigma_{p_i} + 0.01 N_{mmS,p_i}\sigma_{mmp_i}(\mathbf{g})\right). \tag{16.5}$$

Here $p_{i,nom}$ is the nominal value of p_i, σ_{p_i} is the standard deviation of the global variation in p_i in terms of percentage, and σ_{mmp_i} is the standard deviation of the local (i.e. mismatch) variation in p_i, again in terms of percentage. The latter is a function of geometric layout attributes **g** such as length, width, multiplicity factor, etc. For the purpose of statistical simulation the statistical parameters are N_{S,p_i}, the number of standard deviations of perturbation in the global component of variation, and N_{mmS,p_i}, the number of standard deviations of perturbation in the local component of variation. For MC statistical simulation these generally are varied within the range $[-3, 3]$ and so are naturally scaled to order unity, which ensures numerical stability. In general, lateral geometry variations and MOSFET V_{fb} or V_{t0} should be treated as having absolute, rather than relative, variations (but they should still be scaled to order unity), all other parameters and performances should be handled in terms of relative variations. Below we will present most equations in terms of absolute variations as this significantly simplifies the notation; it should be understood that all but the quantities just mentioned are normalized to their nominal values.

There are several convenient aspects to formulating the statistical modeling basis as shown in (16.5). First, as just noted, it has good numerical scaling properties and for statistical simulation only N_{S,p_i} and N_{mmS,p_i} (for each individual device included for mismatch analysis) need to be perturbed. Second, if the process mean shifts, then it is only necessary to adjust $p_{i,nom}$ to re-center the models. Third, if the process variability changes then it is only necessary to adjust σ_{p_i} and $\sigma_{mmp_i}(\mathbf{g})$ to recharacterize the statistical spread in the models. Fourth, it enables analysis of variation due to mismatch alone, without the need to include global variation, which is often useful for analog circuits. Finally, and perhaps most importantly, it naturally embodies the geometry dependence of the variability: the local variation in a process parameter increases as the geometry decreases, and the total variance (the square of the standard deviation) in p_i is

$$\sigma^2_{p_i,total} = \sigma^2_{p_i} + \sigma^2_{mmp_i}(\mathbf{g}). \tag{16.6}$$

For technologies with minimum feature sizes greater than about 0.5 μm $\sigma_{mmp_i} \ll \sigma_{p_i}$ so it is quite reasonable to ignore the variation in $\sigma_{p_i,total}$ with geometry. For technologies with minimum feature sizes less than about 0.1 μm the local variation in a parameter can exceed the global variation, and for accurate modeling and simulation it becomes critical to take into account the geometry dependence of the total variation.

16.4 Statistical Modeling Requires Engineering Judgment

Although we will present a solid and comprehensive theoretical framework for statistical modeling, in practice statistical modeling is not an exact science but critically relies on engineering judgment and experience. By definition reliable, accurate, and long-term statistical data are not available for a specific manufacturing process until it has been running in all expected fabs for some period of time, where possible from multiple products (commonly fab engineers will "tweak" a process for individual parts to improve performance or yield, so possible product-specific design deficiencies may skew data from one or a small number of parts). But at such a mature point in the life cycle of a process designs are transitioning to the next generation process in development, and fewer designs are being done in the stable, well characterized, mature process. So in a form of Catch-22, statistical models are most important early in the life cycle of a process, when many designs in that process are being done, but at this time there is uncertainty in the knowledge of the statistical variability of the process.

In a nut shell, statistical data (i.e. the standard deviation of fab data) in practice are often not known to a high degree of accuracy, so investing large amounts of effort to generate statistical models that are more precise than the accuracy of the data on which they are based is not a useful exercise.

The definition of statistical process variation therefore needs to be done through specifications, and not through detailed numerical analysis of limited data from a still-evolving process. This requires engineering judgment. Factors that contribute to development of the specifications of course include measured data, but should include extrapolations from previous similar technologies, understanding of overall industry and tool capability trends, to factor in expected improvements in manufacturing control, TCAD simulations, and technology requirements (there is a "creative" tension between design and manufacturing, with the former wanting allowable parametric variability to be as small as possible, and the latter wanting it to be as large as possible; clearly there is an appropriate and realistic middle ground that does not make either design or manufacturing uncompetitive).

Below we will work in terms of the standard deviation σ or variance σ^2; however, these need to be understood to be not simply measured values, they are derived from specifications that are developed based on engineering judgment.

16.5 Modeling Parameter Correlations Using Uncorrelated Parameters

As we have stated, it is much easier both for statistical modeling and for statistical simulation to work in terms of uncorrelated parameters. Unfortunately, many SPICE model parameters are correlated, and this is especially problematic for BJT models. For BJTs it has been demonstrated that modeling can be done in terms of uncorrelated parameters by abstracting the SPICE model parameters to a set of underlying and independent process parameters, and then introducing physically based mappings from these to the SPICE model parameters [3]. The SPICE model parameter correlations then follow from the mappings that are defined. Given the imprecisions we have noted in the knowledge of statistical variations, the mappings do not have to be perfect, reasonable first order physical mappings have proven to be completely acceptable.

For MOSFETs there is an additional requirement: to capture important correlations between PMOS and NMOS transistors, as many circuits depend on both types of devices. In practice, there are partial correlations between the electrical performances of the different transistor types, so it is not at first obvious how to avoid setting up simulation models without introducing partial correlations between the parameters of the models of the different transistor types. It has been shown [20] that this can be done by introducing additional process parameters, some of which are common to, and therefore perfectly correlated between, PMOS and NMOS devices, and some of which are independent, and therefore uncorrelated, between PMOS and NMOS devices. This results in a set of independent process parameters that allow modeling of partial correlations. The introduction of covariances into the statistical characterization procedure, to be presented in the next section, has been shown to lead to a general procedure that allows characterization of the underlying parameters [22, 29]. Specific analytic solution has been presented for MOSFET channel length variation in [20, 29]; it is instructive to understand how such analyses work, and we will now demonstrate this for MOSFET gain factor variation [20] as it leads to a rather interesting capability: to be able to characterize the statistical variation in t_{ox} without the need to make capacitance measurements.

The gain factor for wide/long MOSFETs, $\beta_r = \mu_0 C'_{ox} W/L$, the nonsaturation (low V_{DS}) transconductance divided by V_{DS}, is a key parameter for analog IC design (here W is the effective electrical channel width, the design width W_m less the variation Δ_W). To model partial correlations between β_r for PMOS and NMOS devices we do not need to introduce additional parameters (as is done in Sect. 16.8 below for correlation modeling of Δ_L between PMOS and NMOS), as t_{ox} and μ_0 are already available as model parameters for most MOSFET models. We consider the former to be completely correlated between PMOS and NMOS, and the latter to be separate and uncorrelated. The question then is: how can these parameters be characterized statistically? (We consider wide/long devices so β_r is not significantly affected by Δ_L and Δ_W variations.)

If we have the gain factors β_{rP} and β_{rN} of both PMOS and NMOS devices, respectively (added "P" and "N" subscripts will denote PMOS and NMOS), we can form the ratio of these, and for the same geometry devices as t_{ox} is common we have

$$\beta_{rR} = \frac{\beta_{rP}}{\beta_{rN}} = \frac{\mu_{0P}}{\mu_{0N}}. \tag{16.7}$$

Now, it may not seem that there is any more information in knowledge of β_{rR} than in knowing the values of the two measurements from which it is formed, and this is true of any single measurement case. However, statistically we have an additional piece of information that is implicit in forming β_{rR}, the correlation between β_{rP} and β_{rN}, and we can leverage that to unravel the statistical variations of the three underlying process parameters from what seems like only two measurements.

Considering both global and mismatch components of variation, and including also measurement errors $\epsilon_{\beta_{rP}}$ and $\epsilon_{\beta_{rN}}$, the total variance in the gain factors and their ratio are

$$\begin{aligned}
\sigma^2_{\beta_{rP}} &= \sigma^2_{t_{ox}} + \sigma^2_{mmt_{oxP}} + \sigma^2_{\mu_{0P}} + \sigma^2_{mm\mu_{0P}} + \sigma^2_{\epsilon_{\beta_{rP}}},\\
\sigma^2_{\beta_{rN}} &= \sigma^2_{t_{ox}} + \sigma^2_{mmt_{oxN}} + \sigma^2_{\mu_{0N}} + \sigma^2_{mm\mu_{0N}} + \sigma^2_{\epsilon_{\beta_{rN}}},\\
\sigma^2_{\beta_{rR}} &= \sigma^2_{mmt_{oxP}} + \sigma^2_{mmt_{oxN}} + \sigma^2_{\mu_{0P}} + \sigma^2_{mm\mu_{0P}} + \sigma^2_{\mu_{0N}} + \sigma^2_{mm\mu_{0N}} + \sigma^2_{\epsilon_{\beta_{rP}}} + \sigma^2_{\epsilon_{\beta_{rN}}}
\end{aligned} \tag{16.8}$$

where we have explicitly recognized that although global variations in t_{ox} are the same for PMOS and NMOS, and so cancel in the computation of $\sigma^2_{\beta_{rR}}$, the mismatch contributions from the two devices are separate and do not cancel (mismatch variations are by definition local to a device and so are uncorrelated between devices, and they are also independent of the global variation).

The above equations can be solved for the global process parameter variances

$$\begin{aligned}
\sigma^2_{t_{ox}} &= \tfrac{1}{2}\left(\sigma^2_{\beta_{rP}} + \sigma^2_{\beta_{rN}} - \sigma^2_{\beta_{rR}}\right),\\
\sigma^2_{\mu_{0P}} &= \tfrac{1}{2}\left(\sigma^2_{\beta_{rP}} - \sigma^2_{\beta_{rN}} + \sigma^2_{\beta_{rR}}\right) - \sigma^2_{mm\mu_{0P}} - \sigma^2_{mmt_{oxP}} - \sigma^2_{\epsilon_{\beta_{rP}}},\\
\sigma^2_{\mu_{0N}} &= \tfrac{1}{2}\left(\sigma^2_{\beta_{rN}} - \sigma^2_{\beta_{rP}} + \sigma^2_{\beta_{rR}}\right) - \sigma^2_{mm\mu_{0N}} - \sigma^2_{mmt_{oxN}} - \sigma^2_{\epsilon_{\beta_{rN}}}.
\end{aligned} \tag{16.9}$$

If the device geometries are large, so the mismatch variances (which depend reciprocally on area) are small, and the measurements are reliable, so $\sigma_{\epsilon_{\beta_{rP}}}$ and $\sigma_{\epsilon_{\beta_{rN}}}$

Table 16.1 NMOS and PMOS β_r correlation analysis results. Models results are from a 10,000 sample MC simulation

Parameter	Value	Fab Value	Performance	Fab Data	Model
$\rho(\beta_{rP}, \beta_{rN})$	0.775	–	$\rho(\beta_{rP}, \beta_{rN})$	0.775	0.777
$\sigma_{\mu_{0P}}$	0.69%	–	$\sigma_{\beta_{rP}}$	1.32%	1.29%
$\sigma_{\mu_{0N}}$	0.51%	–	$\sigma_{\beta_{rN}}$	1.23%	1.22%
$\sigma_{t_{ox}}$	1.12%	–	$\sigma_{\beta_{rR}}$	0.86%	0.84%
$\sigma_{t_{oxP}}$	–	1.25%		–	–
$\sigma_{t_{oxN}}$	–	1.22%		–	–

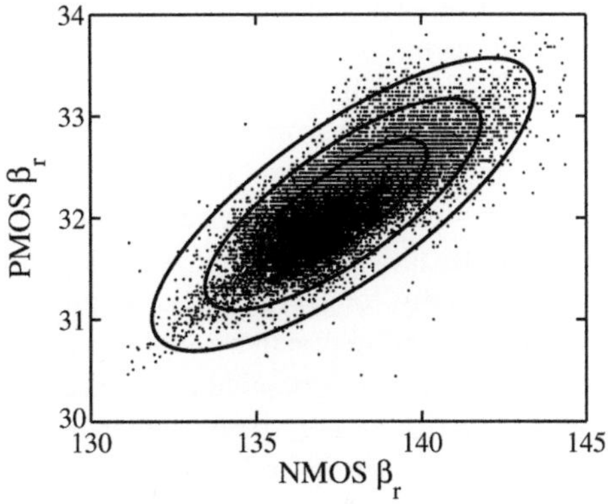

Fig. 16.3 Measured NMOS and PMOS β_r with 1, 2, and 3σ ellipses from a 10,000 MC simulation based on the model parameters of Table 16.1

are small, (16.9) gives three equations that can be directly solved for the global variances of t_{ox}, μ_{0P}, and μ_{0N}.

Note that this procedure directly characterizes the variability in t_{ox} strictly from dc measurements. The oxide thickness also affects the threshold voltage, but so do V_{fb} and N_s, it affects the body effect, but again so does N_s, and it affects β_r, but so does μ_0, so there are confounding effects besides t_{ox} in β_r, V_{t0}, and the body effect: It is not possible to unravel the variability in t_{ox} directly from measurements of these quantities; it is forming the ratio β_{rR} and applying the above analysis that enables this statistically. Also, note that all measurement errors, and any mismatch variances, are assigned to the variances of the mobilities. The variance of the global component of t_{ox} is thus "error-free;" of course, this is only within the context of the above simplified model.

The important point here is that by appropriate selection of the electrical performances to be analyzed, which must include correlation (which is implicit in β_{rR}), the variances of "hidden" underlying process parameters can be determined. Table 16.1 gives the results of the above analysis on approximately 20,000 manufacturing samples, and these are shown, along with 1, 2, and 3σ ellipses from a 10,000 sample MC simulation based on the extracted variances of t_{ox}, μ_{0P}, and μ_{0N}, in Fig. 16.3. Clearly the correlation structure is accurately captured in the model. Direct measurements of t_{ox} were also available, separately from large area PMOS and NMOS structures, and the standard deviations of these are also included in Table 16.1. The $\sigma_{t_{ox}}$ value extracted from the correlation analysis is very close to the directly measured values, and as could be expected from the discussion above the extracted value is slightly less than the measured values, as any contributions from measurement error and mismatch are not included in the extracted t_{ox} variance.

16.6 Theoretical Formulation of BPV

In this section we present the complete theoretical formulation of the BPV procedure. The original formulation [17–19, 21] (we will denote this as linear BPV) was based on a first-order sensitivity analysis and did not take into account correlations between components of **e** or correlations between components of **p**. As detailed in [20] it is possible, with some thought, to handle correlations between parameters by introducing mappings from uncorrelated p_i to intermediate parameters formed as combinations of these p_i, and in this way partial correlations between parameters can be implemented through uncorrelated parameters. So the last of the approximations that underlie the original BPV formulation is not an issue. However, for large variations, especially for BJTs, the assumption that the $\mathbf{e}(\mathbf{p})$ relationship is linear can be inaccurate, and in practice it can be important to model correlations between the e_m properly. We therefore include the recent extensions to the BPV formalism to handle nonlinearities in the $\mathbf{e}(\mathbf{p})$ dependencies [22, 28] and to include correlations between the e_m as quantities to be modeled explicitly [22, 29].

Consider a process where variations about the mean of the fundamental process parameters cause variations in the (device or circuit) electrical performances. We have, using a second order Taylor approximation, for one component e_m of **e**

$$e_m(\mathbf{p}) = e_m(\bar{\mathbf{p}}) + \sum_{i=1}^{N_p} s_{m,i}\delta p_i + \sum_{i,j=1}^{N_p} s_{m,ij}\delta p_i \delta p_j \tag{16.10}$$

where $\delta p_i = p_i - \bar{p}_i$, the first and second order sensitivities are

$$s_{m,i} = \left(\frac{\partial e_m}{\partial p_i}\right)_{\mathbf{p}=\bar{\mathbf{p}}} \tag{16.11}$$

and

$$s_{m,ij} = \frac{1}{2}\left(\frac{\partial^2 e_m}{\partial p_i \partial p_j}\right)_{\mathbf{p}=\bar{\mathbf{p}}} \tag{16.12}$$

respectively, and

$$\bar{\mathbf{p}} = (\bar{p}_1, \bar{p}_2, \ldots, \bar{p}_{N_p}) \tag{16.13}$$

represents the mean values of the process parameters. Note that because the p_i are assumed to be independent and normally distributed $\bar{\mathbf{p}}$ is also the median of **p**. (The mean is the expected value $\langle p_i \rangle$, the median is the 50th percentile; these are the same for any symmetric distribution).

The linear BPV procedure determines the variances σ_i^2 of the p_i (i.e. the second order central moments $\langle (p_i - \langle p_i \rangle)^2 \rangle$) such that using a linear $\mathbf{e}(\mathbf{p})$ mapping (i.e. taking $s_{m,ij} = 0$ in (16.10)) the variances of the e_m are fitted. Note that for simplicity of notation the subscript for the variance of p_i is just i and not p_i; the variance of e_m is denoted as $\sigma_{e_m}^2$. If the mappings are now allowed to be nonlinear, as in (16.10), this implies that the e_m are no longer normally distributed. This is manifest in their distributions being skewed, i.e. they are no longer symmetric. Therefore, to characterize this asymmetry we need to include, as a fitting target for statistical modeling, the skewness of (one or more) e_m. Several measures of skewness of a statistical distribution have been proposed; we use that from reference [32]

$$\gamma_{e_m} = \frac{\langle (e_m - \langle e_m \rangle)^3 \rangle}{\sigma_{e_m}^3}. \tag{16.14}$$

The covariance between e_m and e_n is

$$\sigma_{e_m,e_n} = \langle (e_m - \langle e_m \rangle)(e_n - \langle e_n \rangle) \rangle. \tag{16.15}$$

It is possible to consider the correlation $\rho_{e_m,e_n} = \sigma_{e_m,e_n}/(\sigma_{e_m}\sigma_{e_n})$ instead, but it is more convenient for statistical model characterization to use the covariance. (For reporting and analyzing results, the correlation may be used as it is often the more convenient quantity for that purpose).

Assuming that the p_i are independent, evaluation of the above moments based on (16.10) gives, for the mean of e_m

$$\mu_{e_m} = e_m(\bar{\mathbf{p}}) + \sum_{i=1}^{N_p} s_{m,ii}\sigma_i^2, \tag{16.16}$$

for the variance of e_m (this is a propagation of variance expression)

$$\sigma_{e_m}^2 = \sum_{i=1}^{N_p} \left(s_{m,i}^2 + 2\sum_{j=1}^{N_p} s_{m,ij}^2 \sigma_j^2 \right) \sigma_i^2, \tag{16.17}$$

for the skewness of e_m

$$\gamma_{e_m} = \frac{1}{\sigma_{e_m}^3} \sum_{i,j=1}^{N_p} \left(6 s_{m,i} s_{m,j} s_{m,ij} + 8 \sum_{k=1}^{N_p} s_{m,ij} s_{m,jk} s_{m,ki} \sigma_k^2 \right) \sigma_i^2 \sigma_j^2, \tag{16.18}$$

and for the covariance between e_m and e_n

$$\sigma_{e_m,e_n} = \sum_{i=1}^{N_p} \left(s_{m,i} s_{n,i} + 2\sum_{j=1}^{N_p} s_{m,ij} s_{n,ij} \sigma_j^2 \right) \sigma_i^2. \tag{16.19}$$

Derivation of the above relationships is somewhat tedious, but follows directly from (16.10) and the definition of each quantity.

It is not necessary for skewness for every e_m to be specified as a fitting target for the statistical characterization, and likewise not every possible covariance (there are $N_p(N_p - 1)/2$ of them) needs to be considered. If there are $N_{e,s}$ skewness targets and $N_{e,c}$ covariance targets, then the above gives $2N_e + N_{e,s} + N_{e,c}$ equations in $2N_p$ unknowns, the means $\bar{p}_i$ and variances σ_i^2 of the N_p process parameters.

Note that in the absence of the quadratic terms in (16.10) all $s_{m,ij}$ are zero, so (16.16) and (16.17) reduce to

$$\mu_{e_m} = e_m(\bar{\mathbf{p}}) \tag{16.20}$$

and

$$\sigma_{e_m}^2 = \sum_{i=1}^{N_p} s_{m,i}^2 \sigma_i^2, \tag{16.21}$$

respectively. The first of these is N_e equations in the N_p unknowns, the $\bar{p}_i$. The $\mathbf{e}$ are determined by simulation, based on values for $\mathbf{p}$. Nonlinear least squares optimization can be used to determine appropriate values for $\bar{\mathbf{p}}$; if $\mathbf{e}(\mathbf{p})$ were *exactly* linear it is possible to use a linear least squares process [7]. However, in general $\mathbf{e}(\mathbf{p})$ are not perfectly linear, and it is convenient to have a single program that works both for linear and nonlinear cases, so nonlinear least squares optimization is used in practice (the `dn2fb` routine [4] of the Port subroutine library has proven robust for this task for more than two decades). An obvious condition to be able to uniquely solve (16.20) is that $N_e \geq N_p$. More precisely, if the sensitivity matrix S is

$$S = \left(\frac{\partial \mathbf{e}}{\partial \mathbf{p}} \right) \tag{16.22}$$

then the system of equations (16.20) has a unique least squares solution provided the number of non-zero singular values of (i.e. positive square roots of the eigenvalues of) the matrix $S^T S$ (where the superscript T indicates the matrix transpose operation) is greater than or equal to N_p, i.e. S is of full rank [7]. In more engineering terms, this means that the $\mathbf{p}$ are mathematically observable in the measured $\mathbf{e}$. The second of the above expressions (16.21) is also N_e equations in N_p unknowns; however, in this case the unknowns are the σ_i^2 and the relations are exactly linear, so these linear equations can be directly solved if $N_e = N_p$, or can be solved in a least squares sense using the Moore-Penrose pseudo-inverse [7] if $N_e > N_p$.

Sometimes there are certain p_i, such as t_{ox} for a MOSFET for example, that are measured directly. Their means and variances are therefore known *a priori* and do not need to be solved for as part of the statistical characterization process. These are called forward propagation of variance (FPV) parameters. Nevertheless, they do affect the central moments and covariances of the e_m. If n_p of the $\mathbf{p}$ fall in this category (without loss of generality we assume the first n_p), then we can denote $\mathcal{F} = 1, 2, \ldots, n_p$ as the set of subscript indices for the FPV components of $\mathbf{p}$ and $\mathcal{B} = n_p + 1, n_p + 2, \ldots, N_p$ as the set of subscript indices for the BPV components of $\mathbf{p}$. Accounting for FPV parameters in (16.16) through (16.19) gives

$$\mu_{e_m} - \sum_{i \in \mathcal{F}} s_{m,ii}\sigma_i^2 = e_m(\bar{\mathbf{p}}) + \sum_{i \in \mathcal{B}} s_{m,ii}\sigma_i^2, \tag{16.23}$$

$$\sigma_{e_m}^2 - \sum_{i\in\mathcal{F}} \left(s_{m,i}^2 + 2\sum_{j\in\mathcal{F}} s_{m,ij}^2 \sigma_j^2 \right) \sigma_i^2$$

$$= \sum_{i\in\mathcal{B}} \left(s_{m,i}^2 + 4\sum_{j\in\mathcal{F}} s_{m,ij}^2 \sigma_j^2 + 2\sum_{j\in\mathcal{B}} s_{m,ij}^2 \sigma_j^2 \right) \sigma_i^2, \tag{16.24}$$

$$\sigma_{e_m}^3 \gamma_{e_m} - \sum_{i,j\in\mathcal{F}} \left(6 s_{m,i} s_{m,j} + 8 \sum_{k\in\mathcal{F}} s_{m,jk} s_{m,ki} \sigma_k^2 \right) s_{m,ij} \sigma_i^2 \sigma_j^2$$

$$= \sum_{i,j\in\mathcal{B}} \left(6 s_{m,i} s_{m,j} + 8 \sum_{k\in\mathcal{B}} s_{m,jk} s_{m,ki} \sigma_k^2 + 24 \sum_{k\in\mathcal{F}} s_{m,jk} s_{m,ki} \sigma_k^2 \right) s_{m,ij} \sigma_i^2 \sigma_j^2$$

$$+ \sum_{i\in\mathcal{B}} \sum_{j\in\mathcal{F}} \left(12 s_{m,i} s_{m,j} + 24 \sum_{k\in\mathcal{F}} s_{m,jk} s_{m,ki} \sigma_k^2 \right) s_{m,ij} \sigma_i^2 \sigma_j^2, \tag{16.25}$$

$$\sigma_{e_m,e_n} - \sum_{i\in\mathcal{F}} \left(s_{m,i} s_{n,i} + 2\sum_{j\in\mathcal{F}} s_{m,ij} s_{n,ij} \sigma_j^2 \right) \sigma_i^2$$

$$= \sum_{i\in\mathcal{B}} \left(s_{m,i} s_{n,i} + 4\sum_{j\in\mathcal{F}} s_{m,ij} s_{n,ij} \sigma_j^2 + 2\sum_{j\in\mathcal{B}} s_{m,ij} s_{n,ij} \sigma_j^2 \right) \sigma_i^2. \tag{16.26}$$

These forms for the BPV equations are defined so that, apart from the sensitivities, the quantities on the left of the equals signs involve only the known **e** and FPV components of **p** and the quantities on the right of the equals signs contain the unknown BPV components. For the former, there are $2N_e + N_{e,s} + N_{e,c}$ knowns, and for the latter there are $2(N_p - n_p)$ unknowns. These can be solved for numerically provided that there are at least as many unknown as known quantities (there can be more), and the means and variances of the **p** are "observable," in the sense that the matrix of coefficients relating the unknowns to the knowns is of full rank.

16.7 BPV Requirements

The BPV procedure has some basic requirements on its inputs, and will not work unless these are met. First, the underlying SPICE models must be reasonable, so that the sensitivities, which are computed from these models, are reasonable. If an underlying SPICE model has unphysical behavior, such as V_{t0} becoming unphysically small or large for short device lengths say, then this will give unphysical values for some sensitivities which will cause the procedure to fail. In practice this happens occasionally; it indicates that the underlying SPICE model needs to be improved.

Second, we have noted that the matrix of sensitivities must be of full rank. In fact it should be mathematically well-conditioned [7]; if the condition number of the matrix of sensitivities is too high then the procedure will fail. This happens extremely rarely; generally it is fairly apparent what **e** are suitable to make the **p** observable. Remember also to follow the normalization procedures discussed in Sect. 16.3 (use relative variations for **p** and **e**, and in the few cases where this should not be done scale quantities to order of unity).

Third, the variances specified for **e** and any FPV **p** must be physically self-consistent. This is often the most difficult requirement to meet, particularly as there is imprecision in the knowledge of these variances. Inconsistencies in specifications for variances generally surface as a solution variance that is computed to be negative. This is physically impossible; when this does occur (and it does happen on occasion) the specified variances and the computed SPICE model sensitivities need to be investigated to determine the root cause of the problem, and the specified variances adjusted to be physically consistent.

A simple example illustrates a common problem. Consider the variation in the transconductance g_m of a wide/long MOSFET. It is directly proportional to μ_0 and inversely proportional to t_{ox}, so we have

$$\sigma^2_{\delta g_m/g_m} \approx \sigma^2_{\delta \mu_0/\mu_0} + \sigma^2_{\delta t_{ox}/t_{ox}}. \tag{16.27}$$

Often t_{ox} is considered to be an FPV parameter, and a typical value cited by manufacturing is 10% for 3σ variation. When pressed for accurate data on g_m variation for a wide/long device, a 3σ variation of 5–6% will be admitted to be reasonable. However, from (16.27) this would imply that $\sigma^2_{\delta \mu_0/\mu_0}$ is negative. This is physically impossible, variances must be positive, and the problem is caused by improper specification of the variances of t_{ox} and g_m (the variance of the latter must be greater than the variance of the former).

Assuring physically reasonable and internally self-consistent specifications for variations can be difficult. Using FPV predictions to guide selection of appropriate values, and to help detect possible issues with underlying SPICE models, is useful in practice.

16.8 BPV Application Examples

The analysis of Sect. 16.6 has provided the theory that underlies the BPV procedure for determining the variances and mean values of **p**. We now provide specific application examples.

The first example is NMOS and PMOS transistors in a 0.5 μm BiCMOS process [22, 29]. It was desired to model the correlation between the NMOS and PMOS devices properly, as this is important for many circuit applications. The electrical performances targeted were the threshold voltages of wide/long and wide/short devices, V_{t0r} and V_{t0s}, respectively (the "r" and "s" in the subscripts denote a "reference" (wide/long) device and a "short" (wide/short) device, respectively), the gain factor β_r, and the saturated drain current I_{sats}, which is the current of the short device measured with both gate and drain at the rated supply voltage. In addition, the covariances of β_r between PMOS and NMOS devices and of I_{sats} between PMOS and NMOS were specified as BPV fitting targets.

Table 16.2 Results of correlated PMOS and NMOS BPV characterization

Parameter	Value	Performance	Fab Data	Model
$\sigma_{t_{ox}}$ (FPV)	1.43%	$\rho(\beta_{rP}, \beta_{rN})$	0.596	0.594
σ_{C_D}	0.017	$\rho(I_{satsP}, I_{satsN})$	0.718	0.719
$\sigma_{O_{DP}}$	0.011	$\sigma_{I_{satsP}}$	3.9%	4.1%
$\sigma_{O_{DN}}$	0.008	$\sigma_{I_{satsN}}$	2.4%	2.4%
$\sigma_{V_{fbP}}$	0.015	$\sigma_{V_{t0rP}}$	0.015	0.015
$\sigma_{V_{fbN}}$	0.009	$\sigma_{V_{t0rN}}$	0.009	0.009
$\sigma_{\mu_{0P}}$	1.10%	$\sigma_{\beta_{rP}}$	1.8%	1.8%
$\sigma_{\mu_{0N}}$	1.04%	$\sigma_{\beta_{rN}}$	1.7%	1.7%
$\sigma_{dV_{tP}}$	0.686	$\sigma_{V_{t0sP}}$	0.017	0.017
$\sigma_{dV_{tN}}$	0.001	$\sigma_{V_{t0sN}}$	0.010	0.010

The process parameters used for modeling were t_{ox}, taken to be the same for PMOS and NMOS devices, and the flatband voltages, low-field mobilities, and length dependence of threshold voltage (d_{V_t}) of both PMOS and NMOS devices. For modeling effective channel length variation, following [20] separate parameters were introduced for the critical dimension variation C_D of the poly gates, taken to be the same for PMOS and NMOS devices, and the out-diffusion lengths O_{DP} and O_{DN} of the PMOS and NMOS devices, respectively, giving

$$\Delta_{LP} = C_D + O_{DP}, \tag{16.28}$$

$$\Delta_{LN} = C_D + O_{DN} \tag{16.29}$$

and these parameters were calculated in the model parameter files and used to set the appropriate model parameter values.

Numerical results of the statistical characterization are summarized in Table 16.2, and graphical 1, 2, and 3σ yield ellipses are compared to measured data in Figs. 16.4 and 16.5. The modeled results are from a 10,000 sample MC simulation based on the underlying SPICE model and the statistical model parameters of Table 16.2. The accuracy of the BPV statistical model is clear, in particular the correlation structure in the data is captured accurately.

In general, variations in semiconductor IC manufacturing technologies should be reasonably small otherwise it can be difficult or impossible to design economical, high-yield circuits in them. If variations are small then nonlinearities are small, so the need to take nonlinearities into account is often less important than the need to account for correlations, as in the previous example. However, BJTs often exhibit larger variability than MOSFETs, especially in the base current I_b and the current gain $\beta = I_c/I_b$, where I_c is the collector current, and especially at low bias when the non-ideal component of base current I_{ben} dominates the ideal component of base current I_{bei}. In fact, the variability in I_{ben} can be so large that it is problematic to model it via (16.5) as this can lead to unphysical negative values being generated during MC simulation. A parameter with such a wide variation is generally

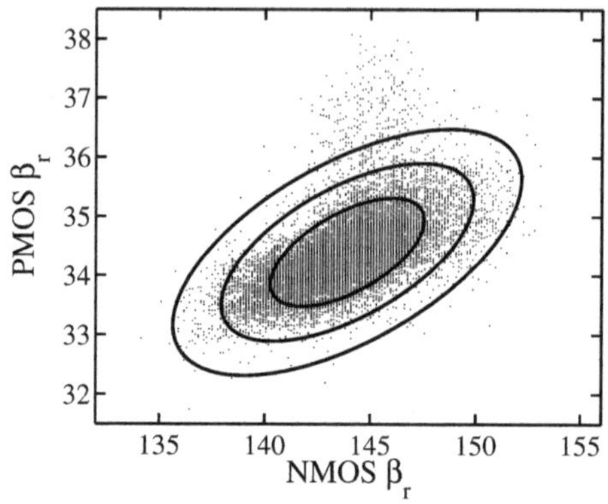

Fig. 16.4 Correlation between β_{rP} and β_{rN}. Points are fab data, with 1, 2, and 3σ ellipses from a 10,000 sample MC simulation

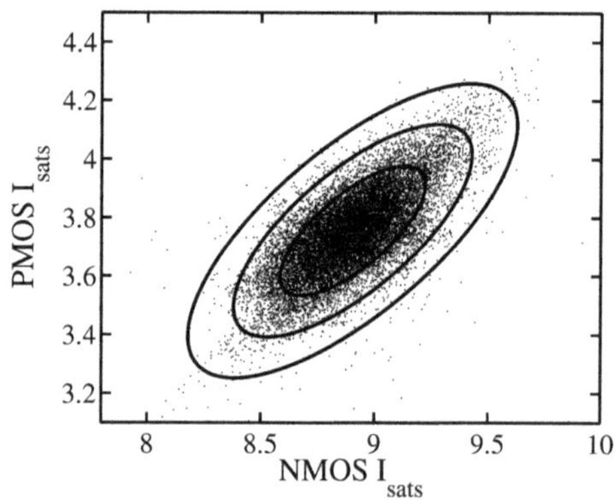

Fig. 16.5 Correlation between I_{satsP} and I_{satsN}. Points are fab data, with 1, 2, and 3σ ellipses from a 10,000 sample MC simulation

observed to have a log-normal distribution, and this can be accounted for by still basing the statistical modeling on an independent normally distributed parameter, but mapping it to the associated process parameter via an exponential mapping, so (16.5) is modified to be

$$p_i = p_{i,nom} \exp\left(0.01 N_{S,p_i} \sigma_{p_i} + 0.01 N_{mmS,p_i} \sigma_{mmp_i}(\mathbf{g})\right). \qquad (16.30)$$

For BJT modeling [28] the process parameters used were the density of the ideal component of base current J_{bei}, the emitter size variation Δ_e, and the pinched base sheet resistance ρ_{psbe} [3]. The last is proportional to the product $N_b t_b$ in (16.2) and (16.3), and these parameters are only separable through data such as the intrinsic transition frequency (which depends primarily on the base thickness and not the base doping) which was not included in the data available from manufacturing; so the composite product was used. The electrical performances fitted were β, the base-emitter voltage V_{bes} at a fixed emitter current, and the Early voltage V_{af}.

From experience working with MOS transistors it may seem that it is not possible to determine a geometric size variation based on only a single geometry device; for MOSFETs at least two different channel length devices are needed to determine Δ_L, and similarly multiple device widths are needed to characterize Δ_W. In contrast, from (16.2) and (16.3) it is apparent that for a BJT the Early voltage depends on

Table 16.3 Results of nonlinear BJT BPV characterization

Performance	Fab μ	Model μ	Fab σ	Model σ
β	80.0	79.3	30.7%	28.4%
V_{af}	46.4 V	45.6 V	39.2%	37.8%
V_{bes}	0.85 V	0.85 V	10.0 mV	9.26 mV

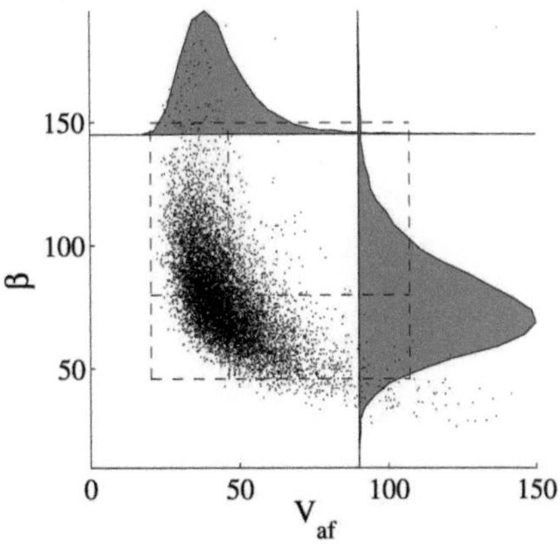

Fig. 16.6 BJT β and V_{af} statistical model showing nonlinear behavior. *Points* are results from a 10,000 sample MC simulation, *shaded curves* are marginal distributions from this simulation. *Dashed lines* are (asymmetric) fab specifications

ρ_{psbe} but, to first order, does *not* depend on the emitter size, whereas the collector current, which is proportional to I_s, depends on both Δ_e and ρ_{psbe}. Therefore, by fitting V_{bes} (or I_c at a fixed base-emitter voltage, to which it has essentially a one-to-one correspondence) and V_{af} from a single device the variance of Δ_e can indeed be unraveled by the BPV procedure.

Table 16.3 and Fig. 16.6 give the numerical and graphical BJT results for this case, where the dashed lines in Fig. 16.6 are the fab specifications for the mean (the middle line in each direction) and 99.7% yield boundaries (this would be $\pm 3\sigma$ for a normal distribution). The expected inverse relationship between β and V_{af} is captured by the underlying mappings from process parameters to SPICE model parameters, and from the degree of variation, evidenced by the non-normality of the modeled distributions, it is apparent that the nonlinearity in the mapping has been taken into account properly. This is the reason the fab specifications are asymmetric, and the mean falls at a point higher than the peak in the marginal distributions because they have positive skewness. This behavior is captured well by the BPV model, as it should be: the skewnesses of the performances were included as fitting targets. Assuming a linear relationship for this example, and not including skewness as a fitting target, gives errors in modeling the standard deviations of β and V_{af} of about 30% [28]. Note that for this example the means of the process parameters are not calculated in an initial step, as in the linear BPV case [18, 19], but are self-consistently solved as part of (16.23) or, for cases like the present where there are no FPV parameters, (16.16).

There is nothing in the description of the BPV formalism that specifies exactly what the **e** should be, only that they should make the **p** mathematically observable.

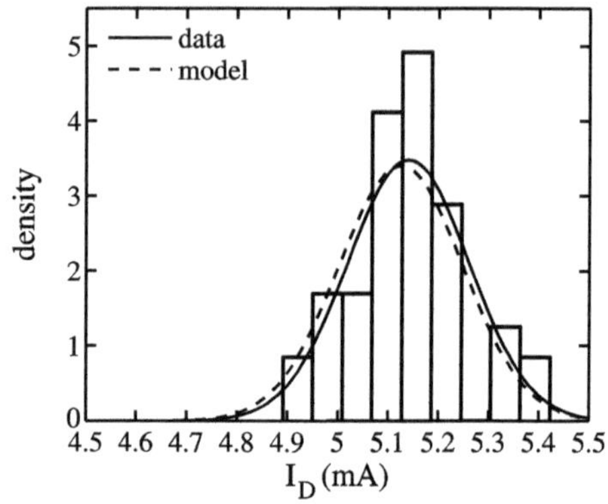

Fig. 16.7 NMOS I_{sat} BPV modeling results. Histogram is from measured data; distributions are fits to data and to a 5000 sample MC simulation based on the BPV model

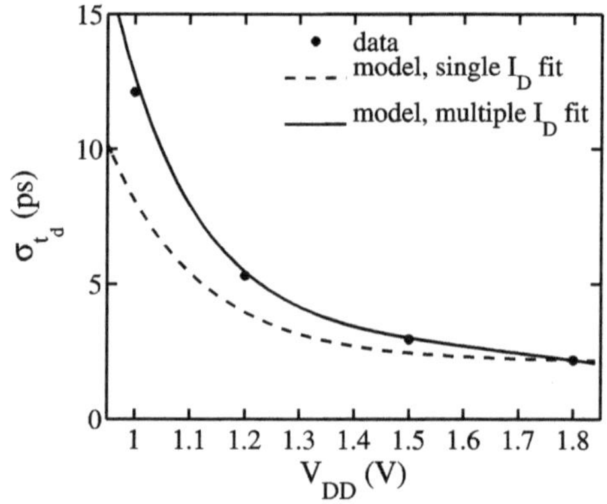

Fig. 16.8 Ring oscillator delay variability vs. supply voltage

Consequently, although it seems logical that circuit performances should be able to be used, instead of or in addition to device performances, this has only recently been experimentally verified [11]. The procedure was tested using the PSP model [8] on a 0.18 μm BiCMOS process.

Individual PMOS and NMOS device characteristics were measured at multiple sites across multiple wafers, and ring oscillator gate delays were measured at the same sites, as a function of supply voltage. The device modeling targets were the variances of V_{t0} for wide/long, wide/short, narrow/long, and narrow/short devices, and I_{sat} for wide/long and narrow/short devices. The circuit modeling target was the variance of ring oscillator delays at the maximum rated supply voltage. The statistical distributions of the device performances were modeled well, an example is shown in Fig. 16.7, and the ring oscillator delay variability at the high supply bias is modeled accurately, see Fig. 16.8. However, the variation in delay at lower supply voltages was not modeled as well (dashed curve in Fig. 16.8).

To try to improve the fitting, additional currents at lower gate voltages were included in the BPV process (but *not* additional ring oscillator delays at these biases). The revised model was then used to simulate the ring oscillator delay variations; this is the solid curve in Fig. 16.8. Clearly, over-specifying the problem by includ-

Table 16.4 Parameters and scaling for BJT mismatch characterization. A_E and P_E are emitter area and perimeter, respectively

Parameter	Description	σ_{mm}^2 Scaling
ρ_{psbe}	pinched base sheet resistance	$1/A_E$
J_{bei}	ideal base current density	$1/A_E$
J_{bei}	non-ideal base current density	$1/P_E$
ρ_{xb}	extrinsic base sheet resistance	$1/A_E$
Δ_e	emitter size variation	$1/P_E$

ing more **e**'s than **p**'s has improved the overall modeling accuracy, with essentially no loss of accuracy modeling delay variability at the highest supply voltage.

Besides including circuit performances for BPV statistical modeling, the formalism does not specify what type of statistical fluctuations, global or local or both, are being fitted. BPV has also been successfully applied to mismatch modeling, first for BJTs [6, 16] and later for MOSFETs [5]. The key thing that has emerged from this is that, being based on physical SPICE models, BPV is able to model behavior over geometry and bias that cannot even be qualitatively captured by mismatch models based on directly measured mismatch in quantities like V_{t0} and gain factor.

The targets used for BPV mismatch modeling are generally measured mismatch variances in currents, rather than V_{t0} etc., over a range of biases and geometries. This gives typically several tens or hundreds of electrical performances. The parameters are the variances of the underlying process parameters, with an appropriate physical geometric scaling. (It is possible to implement empirical geometric scaling rules to try to better fit measured data, but this can be dangerous as mismatch characterization requires a significant amount of measurement time, and so is generally done based on a limited number of geometries. If empirical scaling relations are included care needs to be exercised so that outside of the range characterized the models asymptotically have the expected physical behavior.)

For a vertical *npn* BJT the parameters in Table 16.4, with the geometric variance scaling specified there, were determined using a least squares fit to measured I_c and I_b mismatch. Figures 16.9 and 16.10 show the model results compared to some of the measured data, along with results (dashed lines) from mismatch models based on I_c and β model parameters rather than on the detailed physical scalable model of [3].

It is apparent from the figures that the simple I_c and β approach to mismatch modeling does not capture some of the qualitative behavior seen in the data. Two features specifically stand out. First, at the lowest current levels, there is an increase in β mismatch. This comes from the non-ideal component of base current, which depends on recombination in the base-emitter space charge region, and hence is predominantly a perimeter effect (which explains the scaling for J_{ben} in Table 16.4—in the space charge region the greatest number of traps, and hence most of the recombination, is at the surface, and the extent of the space charge region at the surface scales with perimeter and not area; such surface effects also often exhibit large variability). So this increase is expected, and is captured by the physical model (not

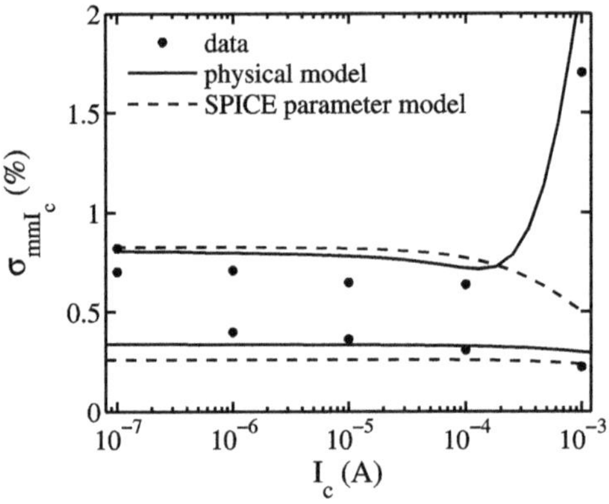

Fig. 16.9 BJT I_c mismatch over bias and geometry (smaller device at top)

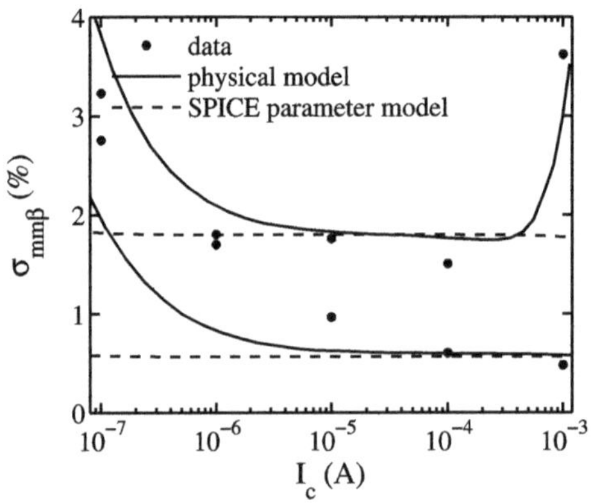

Fig. 16.10 BJT β mismatch over bias and geometry (smaller device at top)

perfectly, but better than by the simple β mismatch model also shown for comparison).

Second, there is a noticeable increase in I_c mismatch at high current for the smallest device. This arises because as the bias level increases BJT behavior becomes more affected by parasitic series resistance than at lower bias. Initially, a higher ρ_{psbe} will give a higher collector current (which varies inversely with base doping and thickness, see (16.2)). As the bias level increases the increased base resistance from the higher ρ_{psbe} will cause a larger debiasing in the higher current device, therefore lowering its current and lowering the mismatch. This is the slow initial drop in I_c mismatch as I_c increases. At the highest current level mismatch in the extrinsic base resistance becomes dominant, causing the mismatch to increase significantly (note that the collector current levels used for biasing are independent of device size, so the current density, and hence the effect of series resistance, is larger for the smaller emitter area device). The simple I_c and β mismatch approach does not capture this. It does appear to capture the initial decrease, but this is serendipitous and for the wrong physical reason: this is caused by the base current in the

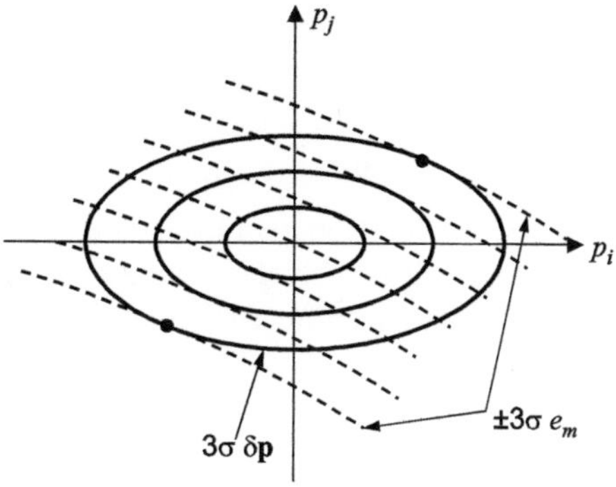

Fig. 16.11 Equi-probability contours for **p** (*solid lines*) and constant value contours for e_m (*dashed lines*). *Dots* are extreme cases

model used (the SPICE Gummel-Poon model) being artificially linked to the collector current (it actually arises from a completely different physical mechanism). This is why there is not a concomitant increase in β, as is seen in the data and is captured by the physically based model.

16.9 Corner Models

One of the foundations of statistical modeling and simulation over the past decades has been the use of "corner" statistical models. In a nutshell, these are an enumerated set of model parameter values, i.e. a discrete selection of points in the possible manufacturing space for **p**, that when simulated are intended to bracket the expected manufacturing variations in circuit performances. Pedantically, it is impossible to guarantee that simulation with a small number of predefined statistical corner models will represent accurately the manufacturing variation of all measures of performance for all types of circuits. We will elaborate on this statement below. Practically, corner models can be very effective, as long as their limitations are understood.

Consider an electrical performance e_m. What constitute "worst-case" points, in the manufacturing space **p**, for e_m? There are two ways of looking at this. The first, formulated from a statistical design point of view, posits that these are the points on the 3σ (or some other specified) contours for e_m that are the most likely to be encountered during manufacturing [24]. That is, they are the points that maximize the probability of **p** occurring subject to e_m being at $\pm 3\sigma$ (the $\pm$ being used to indicate one is a "worst-case" extreme and one is a "best-case" extreme). These are where the dashed equi-e_m contours in Fig. 16.11, for the specified corner values of e_m, are orthogonal to the equi-probability contours for **p**. The second, formulated from a statistical modeling point of view, posits that these are the points at a specified level, generally 3σ, of probability of being encountered in manufacturing space that gives the extreme values for e_m.

Under the assumption that **e**(**p**) are well approximated by a linear relationship, i.e. that $s_{m,ij}$ are zero in (16.10), the two viewpoints for what constitutes an extreme case file are equivalent. Formally then, assuming such a linear relationship, we want to find the solution to the constrained optimization problem [18, 24]

$$\max_{\delta \mathbf{p}} \delta e_m \quad \text{subject to} \quad \delta \mathbf{p}^T C^{-1} \delta \mathbf{p} = z^2 \tag{16.31}$$

where z is the specified number of standard deviations of variation of $\mathbf{p}$ in manufacturing space and C is the variance-covariance matrix of $\mathbf{p}$.

To solve this optimization problem, form the Lagrangian function

$$\mathcal{L} = \mathbf{s}^T \delta \mathbf{p} + \lambda \left(\delta \mathbf{p}^T C^{-1} \delta \mathbf{p} - z^2 \right) \tag{16.32}$$

where $\mathbf{s} = \partial e_m / \partial \mathbf{p}$. At the solution the derivatives of $\mathcal{L}$ with respect to all the parameter variations and the Lagrange multiplier λ must be zero, hence we have

$$\frac{\partial \mathcal{L}}{\partial \delta \mathbf{p}} = \mathbf{s} + 2\lambda C^{-1} \delta \mathbf{p} = \mathbf{0}, \tag{16.33}$$

$$\frac{\partial \mathcal{L}}{\partial \lambda} = \delta \mathbf{p}^T C^{-1} \delta \mathbf{p} - z^2 = 0. \tag{16.34}$$

Solving these gives two explicit solutions

$$\delta \mathbf{p} = \pm z \frac{C\mathbf{s}}{\sqrt{\mathbf{s}^T C \mathbf{s}}} \tag{16.35}$$

which, for the case of uncorrelated process parameters, as we have, simplify to

$$\delta p_i = \pm z \frac{\sigma_i^2 \frac{\partial e_m}{\partial p_i}}{\sqrt{\sum_{j=1}^{N_p} \sigma_j^2 (\frac{\partial e_m}{\partial p_j})^2}}. \tag{16.36}$$

Here the two solutions, one for $+z$ and one for $-z$, give the highest and lowest values for e_m, but the signs of the sensitivities need to be examined to know which corner is high and which is low. These are performance-specific corners, in that they are specifically targeted for e_m and different, separate corner models are needed for different e_n.

Intuitively, (16.36) makes sense. If e_m depends on only one p_i then all $\partial e_m / \partial p_j$ are zero except for $i = j$ and the solution is $\delta p_i = \pm z \sigma_i$ with all other parameter perturbations being zero. If e_m depends on the p_i equally, so that all the sensitivities are the same, then the solution is along the "diagonals" in $\mathbf{p}$-space, i.e. with the same amount of perturbation applied to each p_i, scaled by the variances.

In reality, the assumption that $\mathbf{e}(\mathbf{p})$ is linear is not always reasonable, and iterative solution of a sequence of problems (16.31) may be required, where the linearization at each step is done around the present candidate solution for the worst-case corner and not about the nominal value of $\mathbf{p}$ [14]. Alternatively, knowing the variances of the process parameters, a brute-force Monte Carlo simulation can be run for the device or circuit performances of interest, the results rank-ordered, and the nearest Monte Carlo simulation to the desired $z\sigma$ points selected as the appropriate corner models [12].

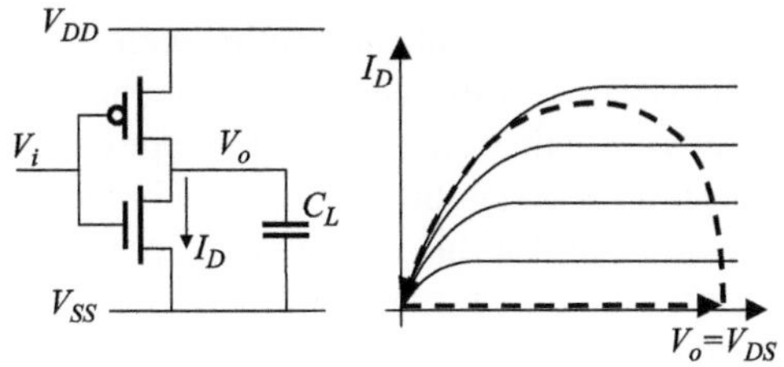

Fig. 16.12 CMOS inverter and trajectory of NMOS transistor operation during switching (*dashed arrows*)

Note that the procedures specified above do not distinguish between whether the e_m are device or circuit electrical performances, so they can be applied during design, to generate specific corner models for individual circuits and circuit performances [14, 24], or they can be applied to predefined device performances, based on an engineering knowledge of which of these are most correlated to circuit performances important for IC design.

Usually corner models are desired that are somewhat generic, so that simulation of the (hopefully small number of) corners for any circuit will give a reasonable estimate of the variability of all important circuit performances. From our analysis it is clear that to do this accurately you need to take into account the sensitivities $\partial e_m/\partial p_j$, otherwise there is no guarantee that specific perturbations in the p_i will lead to quantitatively controlled variations in e_m. Nevertheless, generic corner models that do not quantitatively take these sensitivities into account are extremely useful and have been the most widely used approach to statistical IC design for many years. The main reason for this is that they have been very successful for the design of digital CMOS ICs. Figure 16.12 shows a simple CMOS inverter, along with an approximate $I_D(V_{DS})$ trajectory followed by the pull-down NMOS device during switching. As $I = CdV/dt$, to predict the gate switching delay we need to accurately model the capacitance (of the devices themselves plus of any load) and the current, and the latter should be the drain current at reasonably high V_{GS} and V_{DS} (i.e. the I_{sat} of the transistor) as that is approximately where the device is biased during a large portion of a switching event (in practice, as shown in Fig. 16.12, slightly lower gate and drain biases are perhaps more relevant, and indeed there has been a transition to concentrate more on such biases than with the gate and drain at the supply voltage V_{DD}, which is where I_{sat} is measured). Now, other digital gates besides inverters operate in a similar manner, are generally constructed from minimum channel length devices (and therefore should have similar sensitivities to variations in p_i), and other important measures of circuit performance, such as power dissipation and leakage, are highly correlated with the current drive strength. So in practice targeting corner models for digital CMOS based primarily on current drive strength and capacitance has proven to be accurate and useful. The current drive strength also correlates well with some measures of analog circuit performance, such as the slew rate of an amplifier, so "digital" corner models can also be useful for (some) analog circuit performances. However, as clearly shown in [9], there are e_m for analog circuits whose statistical variabilities are most definitely not well predicted by the digital corners; some, such as unity gain frequency, can have less variability, but others, like phase margin and gain margin, can have their variability significantly under-predicted by digital corners. In summary: corner models can be very useful, but need to be used with a good understanding of what they can represent, and more importantly what they are *not* able to represent accurately: They guarantee nothing about the level of variability they predict for an arbitrary measure of performance for any circuit. Note that this in contrast to performance-specific corner models computed from (16.36), which do guarantee the amount of variability, but only for one e_m at a time.

The simplest way of generating "generic" corner models is by introducing defined variations, generally $\pm z\sigma$ where z is often 3, in each p_i in such a way that they cause the same sign variation in e_m. This approach is forward propagation of variation (FPV). Sometimes this is done directly in SPICE model parameters, but as discussed in Sect. 16.3 there can be underlying correlations between these, especially for BJTS, so this should be done in the (uncorrelated) process parameters $\mathbf{p}$. There are several practical problems that arise when applying this technique. First, it is not always easy to directly measure every p_i, for example collector doping in a vertical *npn* BJT or the base width of a lateral *pnp* BJT. Second, procedures to determine some p_i can involve special test structures or special bias configurations that are designed to make a parameter measurable; it is preferable to determine variation based on normal devices under normal operating conditions, rather than from special structures or extreme bias conditions that are not representative of how a device will be used in circuits (e.g. the open-collector method for determining emitter resistance in BJTs). Third, generally multiple methods have been proposed to measure a parameter, and each method will give a different value for p_i and its variation, which when blindly used as perturbations for corner models will give different predictions of device and circuit variability, yet these should not depend on how you measure a parameter. Fourth, using the same variability for p_i, no matter how it is measured, in different models such as BSIM4 [13] or PSP [8] will give different predictions of device and circuit variability, yet these should not depend on which model you use (at least conceptually; different models are of course not identical). Fifth, every time an additional parameter is included in the FPV process, the corner models become wider and less likely to be encountered during manufacturing. In fact if a $\pm z\sigma$ variation is introduced into each of our N_p parameters the effective probability of encountering such a corner is $\pm z\sqrt{N_p}\sigma$. This can be compensated for by appropriately reducing z as N_p increases. Sixth, because the FPV procedure does not take into account the sensitivities of the e_m to the p_i, as in (16.10), there is no guarantee of what are the actual variability levels of the e_m, in terms of their standard deviations, in the resultant models. That is, even if somewhat reasonable corner models are produced by the FPV procedure, as is usually (but not always) the case, there is no quantitative control of where the corners are, in $\mathbf{e}$-space.

The resolution to the problems just listed is to reverse the way corner models are generated: define them in $\mathbf{e}$-space rather than $\mathbf{p}$-space, and use nonlinear least-squares optimization to determine the values of $\mathbf{p}$ such that $\mathbf{e}$ are at specified $\pm z\sigma$ levels. This is the BPV approach to corner model generation. The primary caution that must be exercised is that appropriate physical correlations between the com-

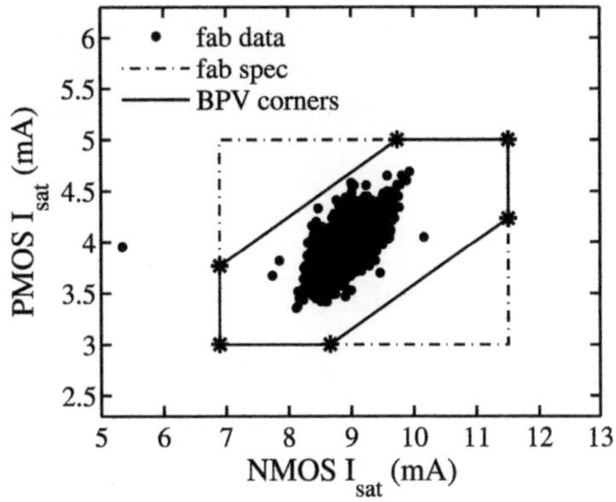

Fig. 16.13 Corner models generated by BPV

ponents of **e** should not be violated, e.g. forcing a MOSFET I_{off} (the drain current with $V_{GS} = 0$ and $V_{DS} = V_{DD}$) to be low and I_{sat} to be high at the same time. Figure 16.13 shows corner models generated by this procedure (the marked corners of the hexagon) compared to defined fab limits, for PMOS and NMOS I_{sat}. The corner models *exactly* match the specifications because the nonlinear least squares optimization used to generate the corner models guarantees this (threshold voltage and peak g_m were also fitted, and in this example $N_e = N_p$ so the fitting for every e_m is exact; this would not necessarily be the case if $N_e > N_p$). The specifications for the fab limits were not based on the data sample shown in Fig. 16.13, but on a larger set of data.

Principal Component Analysis (PCA) is also used for statistical modeling [25, 27]. In this procedure, rather than using physical analysis and understanding to arrive at a set of uncorrelated process parameters (which are recognized, as we stress, as being much more convenient to deal with than correlated variables), PCA takes as input sets of (correlated) model parameters extracted from a statistical sample of devices, and then determines a smaller set of uncorrelated principal components, selected from the eigenvectors of the variance-covariance matrix C, and a linear mapping from these into **p**. The model parameters can be determined from a fairly complete parameter extraction procedure [27], but this can be rather time consuming and can be affected by imprecision (i.e. "noise") in the parameter extraction procedure (for models with hundreds of parameters it is easy to get many different parameter sets that fit the same data to essentially the same degree of accuracy). Simplified and more efficient approaches have emerged, that use manufacturing test data, rather than full device $I(V)$ and $C(V)$ data, and determine a small number of key model parameters to fit these data [2, 25]. This efficiently gives reliable parameter values that are suitable for the purpose of statistical modeling. Combinations, at defined $\pm z\sigma$ levels, are then used to simulate device or circuit performances, a composite, weighted measure of overall performance is formed, and the parameter sets that give the maximum and minimum values for this composite performance measure are selected as the corner files [27]. As with all corner model generation procedures, what is reasonably encompassed by such PCA models depends on the device and circuit performances, and weightings, used to form the composite measure of performance, and gives no guarantee of accuracy for modeling circuits or measures of circuit performance excluded from the composite performance measure computation.

16.10 Why Modeling Correlations is Important

Consider modeling a MOSFET where V_{t0} and the gain factor $k' = \mu C'_{ox}$ are taken to be the controlling statistical parameters, and they are assumed to be independent. This is a common assumption for statistical modeling, especially for mismatch modeling, but it is incorrect because V_{t0} and k' are correlated through t_{ox}. Under the assumption that they *are* uncorrelated we have for the variation in e_m, assuming $s_{m,ij} = 0$,

$$\delta e_m = \frac{\partial e_m}{\partial V_{t0}} \delta V_{t0} + \frac{\partial e_m}{\partial k'} \delta k' \tag{16.37}$$

therefore propagation of variance gives

$$\sigma_{e_m}^2 = \left(\frac{\partial e_m}{\partial V_{t0}} \right)^2 \sigma_{V_{t0}}^2 + \left(\frac{\partial e_m}{\partial k'} \right)^2 \sigma_{k'}^2. \tag{16.38}$$

If instead we were to assume that t_{ox} is the only controlling statistical variable, then

$$\delta e_m = \left(\frac{\partial e_m}{\partial V_{t0}} \frac{\partial V_{t0}}{\partial t_{ox}} + \frac{\partial e_m}{\partial k'} \frac{\partial k'}{\partial t_{ox}} \right) \delta t_{ox} \tag{16.39}$$

and propagation of variance gives

$$\sigma_{e_m}^2 = \left(\frac{\partial e_m}{\partial V_{t0}} \frac{\partial V_{t0}}{\partial t_{ox}} + \frac{\partial e_m}{\partial k'} \frac{\partial k'}{\partial t_{ox}} \right)^2 \sigma_{t_{ox}}^2 \tag{16.40}$$

and expanding and substituting appropriate terms gives

$$\sigma_{e_m}^2 = \left(\frac{\partial e_m}{\partial V_{t0}} \right)^2 \sigma_{V_{t0}}^2 + \left(\frac{\partial e_m}{\partial k'} \right)^2 \sigma_{k'}^2 + 2 \frac{\partial e_m}{\partial V_{t0}} \frac{\partial V_{t0}}{\partial t_{ox}} \frac{\partial e_m}{\partial k'} \frac{\partial k'}{\partial t_{ox}} \sigma_{t_{ox}}^2. \tag{16.41}$$

This is clearly different from (16.38) in that it includes an additional term, which arises from the correlation between V_{t0} and k' through t_{ox}.

There are additional factors besides t_{ox} that contribute to variability in V_{t0} and k', so the situation in reality is not as simple as depicted above. But the qualitative conclusion is the same: ignoring correlations between parameters leads to inaccurate prediction of the variability of electrical performances. Depending on the signs of the sensitivities in (16.41) the variability may either be under-predicted (if the effects "cancel;" this is bad as it means a design will have *more* variation than expected and so may have excessive yield loss) or over-predicted (if the effects "add;" this is bad as it means that a circuit may be over-designed and so use excessive area or current). Accurate modeling and simulation of variability therefore requires proper handling of correlations; this is why it is best to use V_{fb}, N_s, t_{ox}, and μ_0 as process parameters for statistical MOSFET modeling, rather than V_{t0} and k'.

16.11 Conclusions

In this chapter we have reviewed statistical modeling for circuit simulation, with an emphasis on the BPV technique. We summarized the recently developed "full" BPV expressions that include explicit handling of nonlinearities and correlations between electrical performances, and noted that circuit performances, as well as device performances, can be used as a basis for statistical modeling within the BPV framework. Of these extensions, inclusion of handling of correlations is more important, and we showed how ignoring correlations can lead to inaccurate modeling of variability. The theoretical framework may look complex, but the principal of the procedure is fairly straightforward, in fact it is almost cheating: compute the variances and corner values for **p** that give the desired results in terms **e**. The overall procedure runs in several minutes on a typical engineering workstation.

References

1. Asenov, A., Kaya, S., Brown, A.R.: Intrinsic parameter fluctuations in decananometer MOSFETs introduced by gate line edge roughness. IEEE Trans. Electron Devices **50**(5), 1254–1260 (2003)
2. Chen, J.C., Hu, C., Wan, C.P., Bendix, P., Kapoor, A.: E-T based statistical modeling and compact statistical circuit simulation methodologies. In: IEDM Tech. Digest, pp. 635–638 (1996)
3. Davis, W.F., Ida, R.T.: Statistical IC simulation based on independent wafer extracted process parameters and experimental designs. In: IEEE Bipolar Circuits and Technology Meeting (BCTM), pp. 262–265 (1989)
4. Dennis, J.E., Gay, D.M., Welsch, R.E.: An adaptive nonlinear least-squares algorithm. Assoc. Comput. Mach. Trans. Math. Softw. (TOMS) **7**(3), 348–368 (1981). http://www.netlib.org/port/index.html
5. Drennan, P., McAndrew, C.C.: Understanding MOSFET mismatch for analog design. IEEE J. Solid-State Circuits **38**(3), 450–456 (2003)
6. Drennan, P., McAndrew, C.C., Bates, J.: A comprehensive vertical BJT mismatch model. In: IEEE Bipolar Circuits and Technology Meeting (BCTM), pp. 83–86 (1998)
7. Forsythe, G.E., Malcolm, M.A., Moler, C.B.: Computer Methods for Mathematical Computations. Prentice-Hall, New York (1977)
8. Gildenblat, G., Li, X., Wu, W., Wang, H., Jha, A., van Langevelde, R., Smit, G.D.J., Scholten, A.J., Klaassen, D.B.M.: PSP: an advanced surface-potential-based MOSFET model for circuit simulation. IEEE Trans. Electron Devices **53**(9), 1979–1993 (2006)
9. Krick, J.: Statistical transistor SPICE modeling in advanced CMOS technologies. In: IEEE/ACM International Conf. on Computer Aided Design, Compact Modeling of Variability Workshop (2008)
10. Lakshmikumar, K.R., Hadaway, R.A., Copeland, M.A.: Characterization and modeling of mismatch in MOS transistors for precision analog design. IEEE J. Solid-State Circuits **21**(6), 1057–1066 (1986)
11. Li, X., McAndrew, C.C., Wu, W., Chaudhry, S., Victory, J., Gildenblat, G.: Statistical modeling with the PSP MOSFET model. IEEE Trans. Comput.-Aided Des. Integr. Circuits Syst. **29**(4), 599–606 (2010)
12. Lin, C.H., Dunga, M.V., Lu, D.D., Niknejad, A.M., Hu, C.: Performance-aware corner model for design and manufacturing. IEEE Trans. Electron Devices **56**(4), 595–600 (2009)
13. Liu, W., Hu, C.: BSIM4: Theory and Engineering of MOSFET Modeling for IC Simulation. World Scientific, Singapore (2008)

14. Lokanathan, A.N., Brockman, J.B.: Efficient worst case analysis of integrated circuits. In: Proc. IEEE Custom Integrated Circuits Conf., pp. 237–240 (1995)
15. Maly, W., Strowjas, A.J.: Statistical simulation of the IC manufacturing process. IEEE Trans. Comput.-Aided Design Integr. Circuits Syst. **CAD-1**(3), 120–131 (1982)
16. McAndrew, C.C.: Compact device modeling for circuit simulation. In: IEEE Bipolar Circuits and Technology Meeting (BCTM), pp. 28–31 (1997)
17. McAndrew, C.C.: Statistical circuit modeling. In: Proc. International Conf. on Simulation of Semiconductor Processes and Devices, pp. 288–295 (1998)
18. McAndrew, C.C.: Efficient statistical modeling for circuit simulation. In: Reis, R., Jess, J. (eds.) Design of Systems on a Chip: Devices and Components, pp. 97–122. Kluwer Academic, Dordrecht (2004)
19. McAndrew, C.C.: Statistical modeling for circuit simulation. In: IEEE International Symp. on Quality Electronic Design (ISQED), pp. 357–362, March 2003
20. McAndrew, C.C., Drennan, P.: Device correlation: modeling using uncorrelated parameters, characterization using ratios and differences. In: Tech. Proc. Nanotechnology Conf., pp. 698–702 (2006)
21. McAndrew, C.C., Bates, J., Ida, R.T., Drennan, P.: Efficient statistical BJT modeling, why *beta* is more that I_c/I_b. In: Proc. IEEE Custom Integrated Circuits Conf., pp. 151–158 (1997)
22. McAndrew, C.C., Li, X., Stevanovic, I., Gildenblat, G.: Extensions to backward propagation of variance (BPV) for statistical modeling. IEEE Des. Test Comput. **27**(2) (2010)
23. Mehotra, V., Nassif, S.R., Boning, D., Chung, J.: Modeling the effects of manufacturing variation on high-speed microprocessor interconnect performance. In: IEDM Tech. Digest, pp. 767–770 (1998)
24. Nassif, S.R.: Statistical worst-case analysis for integrated circuits. In: Director, S.W., Maly, W., Strowjas, A. (eds.) Statistical Approach to VLSI, pp. 233–253 (1994)
25. Orshansky, M., Nassif, S.R., Boning, D.: Design for Manufacturability and Statistical Design, A Constructive Approach. Springer, Berlin (2008)
26. Pelgrom, M.J.M., Duinmaijer, A.C.J., Welbers, A.P.G.: Matching properties of MOS transistors. IEEE J. Solid-State Circuits **24**(5), 1433–1440 (1989)
27. Power, J.A., Donellan, B., Mathewson, A., Lane, W.A.: Relating statistical MOSFET model parameter variabilities to IC manufacturing process fluctuations enabling realist worst-case design. IEEE Trans. Semicond. Manuf. **7**(3), 306–318 (1994)
28. Stevanovic, I., McAndrew, C.C.: Quadratic backward propagation of variance (QPBV) for nonlinear statistical circuit modeling. IEEE Trans. Comput.-Aided Des. Integr. Circuits Syst. **28**(9), 1428–1432 (2009). Erratum: IEEE Trans. Comput.-Aided Des. Integr. Circuits Syst. **28**(12), 1896 (2009)
29. Stevanovic, I., Li, X., McAndrew, C.C., Green, K.R., Gildenblat, G.: Statistical modeling of inter-device correlations with BPV. Solid-State Electron. **54**(8), 796–800 (2010).
30. Tsividis, Y., McAndrew, C.: Operation and Modeling of the MOS Transistor, 3rd edn. Oxford University Press, London (2010)
31. Watts, J., Pino, W., Trombley, H.: Modeling small MOSFETs using ensemble devices. In: Tech. Proc. Workshop on Compact Modeling, pp. 703–707 (2006)
32. Weiss, E.W.: Skewness. Wolfram MathWorld. http://mathworld.wolfram.com/Skewness.html

Index

DOI 10.1007/978-90-481-8614-3, © Springer Science+Business Media B.V. 2010

P

Q

R

S

MIX
Papier aus verantwortungsvollen Quellen
Paper from responsible sources
FSC® C105338

If you have any concerns about our products,
you can contact us on
ProductSafety@springernature.com

In case Publisher is established outside the EU,
the EU authorized representative is:
Springer Nature Customer Service Center GmbH
Europaplatz 3, 69115 Heidelberg, Germany

Printed by Libri Plureos GmbH
in Hamburg, Germany